EMOTION IN POSTTRAUMATIC STRESS DISORDER

EMOTION IN POSTTRAUMATIC STRESS DISORDER

Etiology, Assessment, Neurobiology, and Treatment

Edited by

MATTHEW T. TULL
Professor, Department of Psychology, University of Toledo, Toledo, OH, United States

NATHAN A. KIMBREL
Associate Professor, Department of Psychiatry and Behavioral Sciences, Duke University School of Medicine, Durham, NC, United States

Academic Press is an imprint of Elsevier
125 London Wall, London EC2Y 5AS, United Kingdom
525 B Street, Suite 1650, San Diego, CA 92101, United States
50 Hampshire Street, 5th Floor, Cambridge, MA 02139, United States
The Boulevard, Langford Lane, Kidlington, Oxford OX5 1GB, United Kingdom

© 2020 Elsevier Inc. All rights reserved.

No part of this publication may be reproduced or transmitted in any form or by any means, electronic or mechanical, including photocopying, recording, or any information storage and retrieval system, without permission in writing from the publisher. Details on how to seek permission, further information about the Publisher's permissions policies and our arrangements with organizations such as the Copyright Clearance Center and the Copyright Licensing Agency, can be found at our website: www.elsevier.com/permissions.

This book and the individual contributions contained in it are protected under copyright by the Publisher (other than as may be noted herein).

Notices
Knowledge and best practice in this field are constantly changing. As new research and experience broaden our understanding, changes in research methods, professional practices, or medical treatment may become necessary.

Practitioners and researchers must always rely on their own experience and knowledge in evaluating and using any information, methods, compounds, or experiments described herein. In using such information or methods they should be mindful of their own safety and the safety of others, including parties for whom they have a professional responsibility.

To the fullest extent of the law, neither the Publisher nor the authors, contributors, or editors, assume any liability for any injury and/or damage to persons or property as a matter of products liability, negligence or otherwise, or from any use or operation of any methods, products, instructions, or ideas contained in the material herein.

Library of Congress Cataloging-in-Publication Data
A catalog record for this book is available from the Library of Congress

British Library Cataloguing-in-Publication Data
A catalogue record for this book is available from the British Library

ISBN 978-0-12-816022-0

For information on all Academic Press publications
visit our website at https://www.elsevier.com/books-and-journals

Publisher: Nikki Levy
Acquisition Editor: Nikki Levy
Editorial Project Manager: Barbara Makinster
Production Project Manager: Bharatwaj Varatharajan
Cover Designer: Christian J. Bilbow

Typeset by SPi Global, India

Working together
to grow libraries in
developing countries

www.elsevier.com • www.bookaid.org

Contents

Contributors   *xiii*
Introduction: Understanding the role of emotion in the etiology, assessment,
neurobiology, and treatment of posttraumatic stress disorder   *xix*
   Matthew T. Tull and Nathan A. Kimbrel

Section 1 Emotions

1. Assessment of emotion and emotion-related processes in PTSD   3
Meghan E. McDevitt-Murphy, Rebecca J. Zakarian, and Cecilia C. Olin

Assessment of emotion and emotion-related processes in PTSD	3
Assessment of emotions in PTSD	4
Emotion-related processes	17
Conclusion	30
References	31

2. Anxiety and fear in PTSD   43
Lori A. Zoellner, Heidi J. Ojalehto, Peter Rosencrans, Rosemary W. Walker, Natalia M. Garcia, Ifrah S. Sheikh, and Michele A. Bedard-Gilligan

Anxiety and fear	43
References	57

3. Anger in PTSD   65
Kirsten H. Dillon, Elizabeth E. Van Voorhees, and Jean C. Beckham

Introduction	65
Defining anger	66
Associations between PTSD and anger	66
Associations over time: Is anger a cause or consequence of PTSD?	68
Theoretical models to explain the association between anger and PTSD	70
Treatment outcomes and implications	73
Conclusions and future directions	80
References	81

4. Sadness and depression in PTSD   89
Blair E. Wisco, Cameron P. Pugach, and Faith O. Nomamiukor

Overview	89
Comorbidity of PTSD and depression	89

Sadness in PTSD	90
Anhedonia/emotional numbing in PTSD	94
Implications for PTSD presentation	99
Rumination	102
Social functioning deficits	103
Suicide	105
Treatment implications	106
Conclusions	108
References	109

5. Disgust in PTSD — 117
Alyssa C. Jones, C. Alex Brake, and Christal L. Badour

Disgust	117
Relevance of disgust to trauma and PTSD	120
Models of disgust and PTSD	122
Disgust proneness	125
Peritraumatic disgust	127
Posttraumatic disgust	128
Clinical assessment and treatment implications	131
Cultural considerations	135
Conclusion	136
References	137

6. Shame and guilt in PTSD — 145
Katherine C. Cunningham

Introduction	145
Shame and guilt as self-evaluative social emotions	146
Differentiating shame and guilt	147
Theoretical models of trauma-related guilt and shame	148
Challenges in measurement	149
Guilt and PTSD	150
Shame and PTSD	153
Sex differences	155
Shame, guilt, and suicide risk	156
Shame and guilt in PTSD treatment	157
Complimentary treatment approaches for shame and guilt	159
Conclusions and future directions	162
References	164

Section 2 Biological bases of emotional responding and dysfunction

7. Neurobiology and neuromodulation of emotion in PTSD — 175
Lysianne Beynel, Lawrence G. Appelbaum, and Nathan A. Kimbrel

Brain regions and networks implicated in PTSD — 175
Noninvasive and nonconvulsive brain stimulation techniques — 183
Noninvasive and convulsive brain stimulation techniques — 195
Invasive brain stimulation techniques — 197
Future directions — 201
References — 204

8. Genetic influences on PTSD — 211
Kaitlin E. Bountress, Leslie A. Brick, Shannon Cusack, Christina M. Sheerin, Nicole R. Nugent, and Ananda B. Amstadter

Behavioral genetics of PTSD — 211
Molecular genetic studies — 212
Novel statistical genetic procedures using GWAS data — 215
Other genomic platforms — 230
DNA methylation — 231
Gene expression — 233
Unique considerations for PTSD genomics — 236
Conclusions and future directions — 237
References — 239

9. Psychophysiology of emotional responding in PTSD — 251
Brittney P. Innocente, Leah T. Weingast, Renie George, and Seth Davin Norrholm

Introduction — 251
Psychophysiological indices revealing emotion dysregulation in PTSD — 252
Translational psychophysiological research: Fear conditioning — 255
Translational fear conditioning models — 256
Fear extinction (extinction learning and extinction recall) — 261
New directions: Individual differences in extinction learning — 263
Psychobiological factors that moderate fear processing — 276
Conclusion — 282
References — 283

Section 3 Difficulties in responding and relating to emotion

10. **Emotion regulation difficulties in PTSD** 295
    Matthew T. Tull, Ariana G. Vidaña, and James E. Betts

    Definitions of emotion regulation 296
    Integrating models of emotion regulation 298
    Evaluating emotion regulation strategy and abilities models in
    the context of PTSD 299
    Emotion regulation strategies and PTSD 300
    Emotion regulation abilities in PTSD 302
    Regulation of positive emotion in PTSD 304
    Implications of emotion regulation research for the treatment of PTSD 305
    Conclusion 306
    References 307

11. **The regulatory role of attention in PTSD from an information processing perspective** 311
    Joseph R. Bardeen

    Introduction: Overview of attentional deployment as a form of
    emotion regulation 311
    Information processing in PTSD: The fear network 313
    Attentional bias to threat: Bottom-up, top-down, or both? 315
    Vigilance-avoidance versus attention maintenance 316
    Dual-process models of ABT: Considering the role of attentional control 317
    Measuring ABT in PTSD 320
    Attentional control as a trauma-related regulatory mechanism 326
    Attention bias modification 333
    References 336

12. **Distress tolerance in PTSD** 343
    Anka A. Vujanovic and Maya Zegel

    Introduction 343
    Theoretical framework 344
    Defining and measuring DT 346
    DT and PTSD symptoms 346
    Applications to suicidal ideation and behavior 359
    Applications to substance use 362
    Limitations and future directions 366
    References 369

13. Emotional granularity in PTSD — 377
Michael K. Suvak, Regina M. Musicaro, and Hilary Hodgdon

Introduction: Emotional granularity and emotion differentiation	377
Emotional granularity, related constructs, and PTSD	380
PTSD and emotional granularity: A preliminary road map	383
PTSD and arousal focus	389
PTSD and valence focus	392
Psychological constructionism, PTSD, and emotional granularity	393
Conclusion	396
References	399

14. Experiential avoidance and PTSD — 409
Holly K. Orcutt, Anthony N. Reffi, and Robyn A. Ellis

What is experiential avoidance?	409
Summary	413
Experiential avoidance and posttraumatic stress disorder	413
Summary	421
Possible mechanisms of the EA PTSD relationship	421
Summary	426
Malleability of experiential avoidance	426
Summary	428
Conclusions	428
Limitations and future directions	430
References	431

15. Emotion-driven impulsivity in PTSD — 437
Nicole H. Weiss, Shannon R. Forkus, Svetlana Goncharenko, and Ateka A. Contractor

Posttraumatic stress disorder and emotion-driven impulsivity	437
Conceptualizing emotion-driven impulsivity	437
Emotion-driven impulsivity and PTSD	439
Psychological mechanisms	440
Biological mechanisms	443
Empirical evidence	445
Important avenues for future research	448
Clinical implications	450
Acknowledgment	451
References	451

Section 4 Treatment and cultural considerations

16. Prolonged exposure for PTSD: Impact on emotions   463
Katie A. Ragsdale, Lauren B. McSweeney, and Sheila A.M. Rauch

Prolonged Exposure (PE)	463
Emotional processing theory	466
The influence of PE on specific emotions	469
Case study	474
Impact of culture on PE on emotion	481
Limitations and future research	482
Conclusion	485
Acknowledgments and disclosures	485
References	486

17. Emotion in cognitive processing therapy for PTSD   491
Colleen Martin, Laura Stayton, and Kathleen M. Chard

Emotion in cognitive processing therapy	491
Specific emotional changes in CPT	494
Mechanisms of action in CPT	502
Limitations and future directions	505
References	506

18. Skills Training in Affective and Interpersonal Regulation (STAIR) Narrative Therapy: Making meaning while learning skills   513
Kile M. Ortigo, Ashley Bauer, and Marylene Cloitre

Treatment development and rationale	514
Theoretical frame	516
An overview of STAIR Narrative Therapy	522
Evidence and applications of STAIR Narrative Therapy	535
Future directions and adaptations of STAIR Narrative Therapy	538
Acknowledgments	538
References	539

19. Acceptance-based behavioral therapy for PTSD   545
Elizabeth Coe, Sonja V. Batten, and Eric C. Meyer

Overview of acceptance-based behavioral therapies	546
Benefits of acceptance-based behavioral therapies	555
Summary of the literature, limitations, and areas for future research	558

In closing	560
References	561

20. Self-compassion in PTSD — 567
Christine Braehler, and Kristin Neff

Defining self-compassion	567
Self-compassion and PTSD	571
Self-compassion, shame, and PTSD	572
Self-compassion and the treatment of PTSD	574
Understanding fears and barriers to self-compassion	576
Safely navigating unchartered attachment trauma territory	578
Principles for selecting a starting point in treatment	581
Integrating self-compassion into treatment of complex PTSD	584
Conclusion	591
References	591

21. Culture, PTSD, and emotion regulation: An anthropological perspective — 597
Andrea Chiovenda, Devon E. Hinton and Byron J. Good

Introduction	597
Emotion regulation in context: Part 1	599
Emotion regulation in context: Part 2	603
Emotion regulation in context: Part 3	609
Conclusion	611
References	612

Index *615*

Contributors

Ananda B. Amstadter
Virginia Institute for Psychiatry and Behavioral Genetics, Virginia Commonwealth University, Richmond, VA, United States

Lawrence G. Appelbaum
Department of Psychiatry and Behavioral Sciences, Duke University School of Medicine, Durham, NC, United States

Christal L. Badour
University of Kentucky, Lexington, KY, United States

Joseph R. Bardeen
Department of Psychology, Auburn University, Auburn, AL, United States

Sonja V. Batten
Booz Allen Hamilton, Washington, DC, United States

Ashley Bauer
National Center for Posttraumatic Stress Disorder—Dissemination & Training Division, VA Palo Alto Health Care System, Palo Alto, CA, United States

Jean C. Beckham
Durham Veterans Affairs Medical Center; Department of Psychiatry & Behavioral Sciences, Duke University Medical Center, Durham, NC, United States

Michele A. Bedard-Gilligan
Psychiatry and Behavioral Sciences, University of Washington, Seattle, WA, United States

James E. Betts
Department of Psychology, University of Toledo, Toledo, OH, United States

Lysianne Beynel
Department of Psychiatry and Behavioral Sciences, Duke University School of Medicine, Durham, NC, United States

Kaitlin E. Bountress
Virginia Institute for Psychiatry and Behavioral Genetics, Virginia Commonwealth University, Richmond, VA, United States

Christine Braehler
University of Glasgow, United Kingdom

C. Alex Brake
University of Kentucky, Lexington, KY, United States

Leslie A. Brick
Department of Psychiatry and Human Behavior; Department of Neurology, Warren Alpert Medical School of Brown University, Providence, RI, United States

Kathleen M. Chard
Cincinnati VA Medical Center, Cincinnati, OH, United States

Andrea Chiovenda
Department of Global Health and Social Medicine, Harvard Medical School, Boston, MA, United States

Marylene Cloitre
National Center for Posttraumatic Stress Disorder—Dissemination & Training Division, VA Palo Alto Health Care System, Palo Alto; Stanford University Department of Psychiatry and Behavioral Sciences, Stanford, CA, United States

Elizabeth Coe
Warriors Research Institute, Baylor Scott & White Health, Waco, TX, United States

Ateka A. Contractor
Department of Psychology, University of North Texas, Denton, TX, United States

Katherine C. Cunningham
Salem Veterans Affairs Medical Center, Salem; Fralin Biomedical Research Institute at Virginia Tech Carilion, Roanoke, VA, United States

Shannon Cusack
Virginia Institute for Psychiatry and Behavioral Genetics; Department of Psychology, Virginia Commonwealth University, Richmond, VA, United States

Kirsten H. Dillon
Durham Veterans Affairs Medical Center; Department of Psychiatry & Behavioral Sciences, Duke University Medical Center, Durham, NC, United States

Robyn A. Ellis
Northern Illinois University, DeKalb, IL, United States

Shannon R. Forkus
Department of Psychology, University of Rhode Island, Kingston, RI, United States

Natalia M. Garcia
Psychology, University of Washington, Seattle, WA, United States

Renie George
Mental Health Service Line, Atlanta Veterans Administration Health Care System, Decatur, GA, United States

Svetlana Goncharenko
Department of Psychology, University of Rhode Island, Kingston, RI, United States

Byron J. Good
Department of Global Health and Social Medicine, Harvard Medical School, Boston, MA, United States

Devon E. Hinton
Center for Anxiety and Traumatic Stress Disorders, Massachusetts General Hospital, Harvard Medical School, Boston, MA, United States

Hilary Hodgdon
Psychology Department, Suffolk University, Boston; The Trauma Center at Justice Resource Institute, Brookline, MA, United States

Brittney P. Innocente
Mental Health Service Line, Atlanta Veterans Administration Health Care System, Decatur, GA, United States

Alyssa C. Jones
University of Kentucky, Lexington, KY, United States

Nathan A. Kimbrel
Department of Psychiatry and Behavioral Sciences, Duke University School of Medicine, Durham, NC, United States

Colleen Martin
Cincinnati VA Medical Center, Cincinnati, OH, United States

Meghan E. McDevitt-Murphy
The University of Memphis, Memphis, TN, United States

Lauren B. McSweeney
Emory University School of Medicine, Atlanta, GA, United States

Eric C. Meyer
Warriors Research Institute, Baylor Scott & White Health; U.S. Department of Veterans Affairs, VISN 17 Center of Excellence for Research on Returning War Veterans, Waco; Department of Psychiatry and Behavioral Science, Texas A&M University Health Science Center, College of Medicine, College Station; Department of Psychology and Neuroscience, Baylor University, Waco, TX, United States

Regina M. Musicaro
Psychology Department, Suffolk University, Boston, MA, United States

Kristin Neff
University of Texas at Austin, United States

Faith O. Nomamiukor
University of North Carolina at Greensboro, Greensboro, NC, United States

Seth Davin Norrholm
Mental Health Service Line, Atlanta Veterans Administration Health Care System, Decatur; Department of Psychiatry and Behavioral Sciences, Emory University School of Medicine, Atlanta, GA, United States

Nicole R. Nugent
Department of Psychiatry and Human Behavior; Department of Pediatrics, Warren Alpert Medical School of Brown University, Providence, RI, United States

Heidi J. Ojalehto
Psychiatry and Behavioral Sciences, University of Washington, Seattle, WA, United States

Cecilia C. Olin
The University of Memphis, Memphis, TN, United States

Holly K. Orcutt
Northern Illinois University, DeKalb, IL, United States

Kile M. Ortigo
National Center for Posttraumatic Stress Disorder—Dissemination & Training Division, VA Palo Alto Health Care System; Center for Existential Exploration, Palo Alto, CA, United States

Cameron P. Pugach
University of North Carolina at Greensboro, Greensboro, NC, United States

Katie A. Ragsdale
Emory University School of Medicine, Atlanta, GA, United States

Sheila A.M. Rauch
Emory University School of Medicine, Atlanta; Atlanta VA Healthcare System, Decatur, GA, United States

Anthony N. Reffi
Northern Illinois University, DeKalb, IL, United States

Peter Rosencrans
Psychology, University of Washington, Seattle, WA, United States

Christina M. Sheerin
Virginia Institute for Psychiatry and Behavioral Genetics, Virginia Commonwealth University, Richmond, VA, United States

Ifrah S. Sheikh
Psychology, University of Washington, Seattle, WA, United States

Laura Stayton
Cincinnati VA Medical Center, Cincinnati, OH, United States

Michael K. Suvak
Psychology Department, Suffolk University, Boston, MA, United States

Matthew T. Tull
Department of Psychology, University of Toledo, Toledo, OH, United States

Ariana G. Vidaña
Department of Psychology, University of Toledo, Toledo, OH, United States

Elizabeth E. Van Voorhees
Durham Veterans Affairs Medical Center; Department of Psychiatry & Behavioral Sciences, Duke University Medical Center, Durham, NC, United States

Anka A. Vujanovic
Department of Psychology, University of Houston, Houston, TX, United States

Rosemary W. Walker
Psychology, University of Washington, Seattle, WA, United States

Leah T. Weingast
Mental Health Service Line, Atlanta Veterans Administration Health Care System, Decatur, GA, United States

Nicole H. Weiss
Department of Psychology, University of Rhode Island, Kingston, RI, United States

Blair E. Wisco
University of North Carolina at Greensboro, Greensboro, NC, United States

Rebecca J. Zakarian
The University of Memphis, Memphis, TN, United States

Maya Zegel
Department of Psychology, University of Houston, Houston, TX, United States

Lori A. Zoellner
Psychology, University of Washington, Seattle, WA, United States

Introduction: Understanding the role of emotion in the etiology, assessment, neurobiology, and treatment of posttraumatic stress disorder

Matthew T. Tull[a], Nathan A. Kimbrel[b]
[a]Department of Psychology, University of Toledo, Toledo, OH, United States
[b]Department of Psychiatry and Behavioral Sciences, Duke University School of Medicine, Durham, NC, United States

Since its inclusion in the third edition of the *Diagnostic and Statistical Manual of Mental Disorders* (DSM; American Psychiatric Association, 1980), posttraumatic stress disorder (PTSD) has been recognized as a psychological disorder characterized by disruptions in the experience of and responses to emotion. Yet, despite consensus across models for the relevance of emotional dysfunction in understanding PTSD (Cahill & Foa, 2007; Keane & Barlow, 2002), the specific deficits in emotional experience and processing exhibited by individuals with PTSD and their role in the development and maintenance of PTSD remained poorly understood at the turn of the century (Litz, Orsillo, Kaloupek, & Weathers, 2000). This was further evident in the DSM-IV (American Psychiatric Association, 2000) diagnostic criteria for PTSD, which arguably did not capture the full extent of the diverse emotion dysfunction observed in PTSD. Moreover, despite early recognition of PTSD's broad impact on emotional experience, PTSD has largely been considered and discussed as a disorder of fear and anxiety (e.g., Keane & Barlow, 2002; Shin & Handwerger, 2009). Until the publication of the DSM-5 (where PTSD is currently categorized as a trauma- and stressor-related disorder; American Psychiatric Association, 2013), PTSD was classified as an anxiety disorder, and early theories on the development and maintenance of PTSD largely emphasized fear-related processes in PTSD (Foa & Kozak, 1986).

However, in the past 15 years, there has been an exponential increase in research on emotion and emotion regulation (Barrett, 2013; Gross, 2015; Tull & Aldao, 2015). As research on emotion has progressed, so has research on the role of emotions in PTSD. This rapidly growing body of research has demonstrated the importance of considering emotions besides fear and anger in understanding PTSD pathology, such as shame, guilt, sadness, and disgust (e.g., Beck et al., 2011; Engelhard, Olatunji, & de Jong, 2011; Lee, Scragg, & Turner, 2001). In addition, research has shown that well-established cognitive-behavioral treatments for PTSD are not only effective at reducing excessive fear and anxiety among individuals with PTSD but also effective at reducing excesses in and problems with other emotions, such as shame, guilt, and anger (Cahill, Rauch, Hembree, & Foa, 2003; Resick, Nishith, Weaver, Astin, & Feuer, 2002; Stapleton, Taylor, & Asmundson, 2006). Researchers are also increasingly finding support for the relevance of specific emotional processes (e.g., emotion regulation, distress intolerance, and impulsivity) in the development and maintenance of PTSD (Roley, Contractor, Weiss, Armour, & Elhai, 2017; Seligowski, Lee, Bardeen, & Orcutt, 2015; Vujanovic & Bernstein, 2011), and there is growing understanding of the psychophysiological and neurobiological markers of emotional dysfunction in PTSD (Yehuda & LeDoux, 2007).

These findings have led to modifications of existing theories or the development of new theories on PTSD that take into account all emotions and their role in the development and maintenance of PTSD (Frewen & Lanius, 2006; Rauch & Foa, 2006) and the investigation of treatments for PTSD that specifically target emotion dysregulation (Cloitre et al., 2010; Orsillo & Batten, 2005; Steil, Dyer, Priebe, Kleindienst, & Bohus, 2011). In addition, with the publication of the DSM-5 (American Psychiatric Association, 2013), changes were made to the operationalization of PTSD that deemphasized the specific role of fear and anxiety in PTSD while including symptoms that specifically mention the role of other emotions. For example, to be considered as a traumatic event, an event is no longer required to include the experience of fear-spectrum emotions (fear, helplessness, or horror). In addition, the DSM-5 conceptualization of PTSD now includes a symptom characterized by the experience of persistent negative emotions, such as fear, horror, anger, guilt, or shame. The growth of research on emotion and emotional processes in PTSD and recent changes to the symptom structure of PTSD that highlight the relevance of multiple emotions will no doubt lead to a more nuanced understanding of the development, maintenance, and treatment of PTSD. It is for this reason that we decided there was a need for a text that organizes the large body of theoretical and empirical literature on PTSD and emotion.

This book is organized into four sections, with each composed of multiple chapters written by leading experts in the field that cover specific topics associated with the section content area. The first section of the book reviews the theoretical and empirical literature concerning the relation between specific emotions and PTSD. This section begins with a chapter that describes the current state of the art with respect to the assessment of emotion and emotion-related processes in PTSD. This is then followed by a series of chapters that cover the role of specific emotions in PTSD, including anxiety and fear, anger, sadness and depression, disgust, and shame and guilt.

The second section of the book describes the neurobiological underpinnings of emotion and emotion regulation in PTSD. This section begins with an overview of the current literature on the neurobiology and neuromodulation of emotional dysfunction in PTSD. Notably, in addition to describing the current literature on neurobiology, this chapter provides a thorough summary of a variety of neuromodulation techniques (e.g., repetitive transcranial magnetic stimulation and direct transcranial stimulation) that have only begun to be applied to PTSD in recent years. The next chapter in this section provides a comprehensive summary of the current state of the literature concerning the genetic and epigenetic basis of PTSD. This section then concludes with a chapter exploring the benefits of using objective psychophysiological paradigms to characterize PTSD and how the use of such paradigms can lead to a better understanding of PTSD symptoms and the development of more effective PTSD treatments.

The third section of the book provides a review of specific difficulties in responding and relating to emotional experience in PTSD and their associated consequences. This section begins with a review of the theoretical and empirical literature on emotion regulation difficulties in PTSD. The next chapter reviews the regulatory role of attention in PTSD from an information processing perspective, followed by chapters on distress intolerance and PTSD, emotional granularity in PTSD, experiential avoidance and PTSD, and emotion-driven impulsivity and PTSD. The chapters on emotional granularity and emotion-driven impulsivity deserve particular attention in this section as they cover a relatively new area of research in PTSD, highlighting the relevance of this particular line of investigation and its implications for understanding the symptoms, associated problems (e.g., substance abuse), and treatment of PTSD.

The final section of the book discusses different treatment approaches for PTSD, their influence on emotion and emotion dysfunction, and the role of culture in understanding the relation between PTSD and emotion. This section begins with a chapter describing prolonged exposure (PE) for PTSD and the impact of this treatment on ameliorating the diverse emotional consequences of traumatic exposure. This chapter is followed by a discussion of cognitive processing therapy (CPT) and how this PTSD treatment can also be used to effectively influence emotional responding in PTSD. Next, Skills Training in Affective and Interpersonal Regulation (STAIR) Narrative Therapy is presented, including an overview of the theoretical foundation of this treatment, its specific components, and recent applications of this treatment approach. The next chapter describes the growing body of evidence supporting the potential use of acceptance-based behavioral therapy to treat PTSD, followed by a chapter on self-compassion therapy, a relatively new treatment approach that may hold promise for individuals with PTSD who have not responded to previous therapeutic approaches. This section concludes with a chapter discussing how an anthropological approach can be used to better understand the ways in which culture might impact the expression and regulation of emotion in PTSD.

Although we do believe this book is comprehensive in its coverage of PTSD and emotion, we also recognize that it is in no way exhaustive in its review of the literature. Research on the role of emotion in the development, maintenance, and treatment of PTSD is being produced at a pace much too rapid to cover in its entirety. That said, it is our expectation that these chapters effectively represent the current state of the literature on PTSD and emotion, and by integrating diverse bodies of literature, it is our hope that this text will lay the foundation for future research on PTSD and emotion. We also hope that this text will provide a guide for clinicians interested in understanding the broad difficulties in the experience and regulation of emotion that may be exhibited by patients with PTSD and how existing treatments for PTSD can effectively be used to target these difficulties.

References

American Psychiatric Association. (1980). *Diagnostic and statistical manual of mental disorders* (3rd ed.). Washington, DC: American Psychiatric Association.

American Psychiatric Association. (2000). *Diagnostic and statistical manual of mental disorders, 4th edition, text revision*. Washington, DC: American Psychiatric Association.

American Psychiatric Association. (2013). *Diagnostic and statistical manual of mental disorders* (5th ed.). Washington, DC: American Psychiatric Association.

Barrett, L. F. (2013). Psychological construction: The Darwinian approach to the science of emotion. *Emotion Review, 5*, 379–389.

Beck, J. G., McNiff, J., Clapp, J. D., Olsen, S. A., Avery, M. L., & Hagewood, J. H. (2011). Exploring negative emotion in women experiencing intimate partner violence: Shame, guilt, and PTSD. *Behavior Therapy*, *42*, 740–750.

Cahill, S. P., & Foa, E. B. (2007). Psychological theories of PTSD. In M. J. Friedman, T. M. Keane, & P. A. Resick (Eds.), *Handbook of PTSD: Science and practice* (pp. 55–77). New York: Guilford Press.

Cahill, S. P., Rauch, S. A., Hembree, E. A., & Foa, E. B. (2003). Effect of cognitive behavioral treatments for PTSD on anger. *Journal of Cognitive Psychotherapy*, *17*, 113–131.

Cloitre, M., Stovall-McClough, K. C., Nooner, K., Zorbas, P., Cherry, S., Jackson, C. L., … Petkova, E. (2010). Treatment for PTSD related to childhood abuse: A randomized controlled trial. *American Journal of Psychiatry*, *167*, 915–924.

Engelhard, I. M., Olatunji, B. O., & de Jong, P. J. (2011). Disgust and the development of posttraumatic stress among soldiers deployed to Afghanistan. *Journal of Anxiety Disorders*, *25*, 58–63.

Foa, E. B., & Kozak, M. J. (1986). Emotional processing of fear: Exposure to correct information. *Psychological Bulletin*, *99*, 20–35.

Frewen, P. A., & Lanius, R. A. (2006). Toward a psychobiology of posttraumatic self-dysregulation: Reexperiencing, hyperarousal, dissociation, and emotional numbing. *Annals of the New York Academy of Sciences*, *1071*, 110–124.

Gross, J. J. (2015). Emotion regulation: Current status and future prospects. *Psychological Inquiry*, *26*, 1–26.

Keane, T. M., & Barlow, D. H. (2002). Posttraumatic stress disorder. In D. H. Barlow (Ed.), *Anxiety and its disorders: The nature and treatment of anxiety and panic* (pp. 418–452). New York: Guilford Press.

Lee, D. A., Scragg, P., & Turner, S. (2001). The role of shame and guilt in traumatic events: A clinical model of shame-based and guilt-based PTSD. *British Journal of Medical Psychology*, *74*, 451–466.

Litz, B. T., Orsillo, S. M., Kaloupek, D., & Weathers, F. (2000). Emotional processing in posttraumatic stress disorder. *Journal of Abnormal Psychology*, *109*, 26–39.

Orsillo, S. M., & Batten, S. V. (2005). Acceptance and commitment therapy in the treatment of posttraumatic stress disorder. *Behavior Modification*, *29*, 95–129.

Rauch, S. A. M., & Foa, E. (2006). Emotional processing theory (EPT) and exposure therapy for PTSD. *Journal of Contemporary Psychotherapy*, *36*, 61–65.

Resick, P. A., Nishith, P., Weaver, T. L., Astin, M. C., & Feuer, C. A. (2002). A comparison of cognitive-processing therapy with prolonged exposure and a waiting condition for the treatment of chronic posttraumatic stress disorder in female rape victims. *Journal of Consulting and Clinical Psychology*, *70*, 867–879.

Roley, M. E., Contractor, A. A., Weiss, N. H., Armour, C., & Elhai, J. D. (2017). Impulsivity facets' predictive relations with DSM-5 PTSD symptom clusters. *Psychological Trauma Theory Research Practice and Policy*, *9*, 76–79.

Seligowski, A. V., Lee, D. J., Bardeen, J. R., & Orcutt, H. K. (2015). Emotion regulation and posttraumatic stress symptoms: A meta-analysis. *Cognitive Behaviour Therapy*, *44*, 87–102.

Shin, L. M., & Handwerger, K. (2009). Is posttraumatic stress disorder a stress-induced fear circuitry disorder? *Journal of Traumatic Stress*, *22*, 409–415.

Stapleton, J., Taylor, S., & Asmundson, G. (2006). Effects of three PTSD treatments on anger and guilt: Exposure therapy, eye movement desensitization and reprocessing, and relaxation training. *Journal of Traumatic Stress*, *19*, 369–375.

Steil, R., Dyer, A., Priebe, K., Kleindienst, N., & Bohus, M. (2011). Dialectical behavior therapy for posttraumatic stress disorder related to childhood sexual abuse: A pilot study of an intensive residential treatment program. *Journal of Traumatic Stress*, *24*, 102–106.

Tull, M.T., & Aldao, A. (2015). Editorial overview: New directions in the science of emotion regulation. *Current Opinion in Psychology, 3*, iv–x.
Vujanovic, A. A., & Bernstein, A. (2011). Traumatic stress. In M. J. Zvolensky, A. Bernstein, & A. A. Vujanovic (Eds.), *Distress tolerance: Theory, research, and clinical applications* (pp. 126–148). New York, NY, US: Guilford Press.
Yehuda, R., & LeDoux, J. (2007). Response variation following trauma: A translational neuroscience approach to understanding PTSD. *Neuron, 56*, 19–32.

SECTION 1

Emotions

CHAPTER 1

Assessment of emotion and emotion-related processes in PTSD

Meghan E. McDevitt-Murphy, Rebecca J. Zakarian, Cecilia C. Olin
The University of Memphis, Memphis, TN, United States

Assessment of emotion and emotion-related processes in PTSD

Posttraumatic stress disorder (PTSD) is a complex disorder characterized by intense emotions across several domains, including not only prominent fear and anxiety but also sadness, guilt, and shame. PTSD has also been characterized by emotional numbing and anhedonia (i.e., a loss of interest or pleasure), both of which reflect a diminished experience of positive emotion. Additionally, some research has pointed to emotion-related processes (e.g., emotion regulation and experiential avoidance) as risk factors that may contribute to the onset or maintenance of PTSD following trauma exposure. Key to research and clinical work with these emotions and processes is a thorough assessment of PTSD. There are two psychometrically sound, DSM-correspondent, assessment instruments available from the National Center for PTSD (www.ncptsd.org): the Clinician-Administered PTSD Scale for DSM-5 (CAPS-5; Weathers, Blake, et al., 2013), a structured interview, and the PTSD Checklist (PCL-5; Weathers, Litz, et al., 2013), a self-report questionnaire. For clinicians and researchers who also wish to thoroughly assess the emotions and emotion-related processes that accompany or contribute to PTSD or have implications for psychotherapy, this chapter offers insight into the available, research-supported assessment options. Herein, we briefly review key emotions and emotion-related processes that have relevance for PTSD and recommend assessment instruments for each domain based on the available research. The recommended measures presented in this chapter do not reflect an exhaustive list, but instead represent some of the measures with the most support in the published literature.

Assessment of emotions in PTSD

We begin with the assessment of emotions in the context of PTSD. In this section, we include emotions that are either components of the diagnostic criteria for PTSD or frequently present alongside PTSD. In general the most well-established assessment methods for most constructs are questionnaires. All recommended measures appear in Table 1.

Anger

Individuals with PTSD frequently report problems with anger, ranging from the manifestation of high anger emotional states to behavioral dyscontrol in the form of aggressive, violent, or risky behaviors (American Psychiatric Association, 2000, 2013). Heightened irritability and angry outbursts are frequent features of the disorder and have been implicated among some trauma survivors as the most impairing feature of PTSD's emotional landscape (e.g., Biddle, Elliott, Creamer, Forbes, & Devilly, 2002; Rosen, Adler, & Tiet, 2013). In a metaanalytic review of anger expression in PTSD, Olatunji, Ciesielski, and Tolin (2010) examined over 2000 patients with anxiety disorders across 28 studies and found that anger was uniquely heightened in patients with PTSD relative to other anxiety disorders, where high levels of irritability were also found. Moreover, results from the study indicated that individuals with PTSD distinctively struggle with the expression and control of anger, suggesting that problems with anger may be part of a broader network of emotional dysregulation and behavioral dysfunction.

The eroding effects of anger on interpersonal relationships (Kubany, Bauer, Muraoka, Richard, & Read, 1995) and links between anger and further negative outcomes (e.g., comorbidity (Gonzalez, Novaco, Reger, & Gahm, 2016) and aggressive/violent behavior (Jakupcak et al., 2007)) suggest that the assessment of anger in clinical contexts is necessary for comprehensive trauma care. Researchers studying trauma populations would also benefit from further investigation of this important emotional feature of PTSD emotional profiles. Additionally, certain populations of trauma survivors have been more frequently studied (e.g., combat veterans) than others, suggesting that there are unique gaps in the literature that warrant greater inquiry to understand the contextual role of anger across trauma types and populations.

Like other measures intended to capture emotional expression, measures of anger are frequently conceptualized in terms of state anger (e.g., specific, cued anger responses) and trait anger (e.g., the dispositional tendency to experience anger). Similarly, although anger is often conceptualized as a

Table 1 Assessment of emotion in PTSD.

Measure	Citation	Construct measured	Trauma samples studied	Psychometric evidence	Administration details
Anxiety					
State–Trait Anxiety Inventory (STAI)	Spielberger (1983)	State and trait anxiety	Civilian trauma survivors, veterans	Evidence of reliability and validity; state anxiety and trait anxiety reflect distinct constructs	40 items total. Each of 20 items is rated both with respect to "right now" (state) and "generally" (trait). Items rated on a 4-point scale
Anger					
State–Trait Anger Expression Inventory (STAXI-II)	Spielberger (1999)	State anger, trait anger, anger expression (out/in), anger control (out/in)	9/11 disaster workers, veterans with PTSD, combat veterans, adolescent survivors of sexual violence	Extensive factor structure analysis, international adaptation, and psychometric evaluation; normative data derived from large clinical and nonclinical samples	57 items rated on a 4-point Likert scale. May be administered online or on paper. Gender-normed T-scores available
Buss-Perry Aggression Questionnaire (BPAQ)	Buss and Perry (1992)	Anger, hostility, aggression	Has been used extensively with survivors of a wide range of traumas, including childhood abuse and sexual assault; prisoners of war; and combat veterans	Demonstrated strong convergent validity, discriminant validity, construct validity, test-retest reliability, and internal consistency	Anger subscale: 8 items rated on 5-point Likert scale

Continued

Table 1 Assessment of emotion in PTSD—cont'd

Measure	Citation	Construct measured	Trauma samples studied	Psychometric evidence	Administration details
Fear					
NPU-Threat Test	Schmitz and Grillon (2012)	Fear, anxiety	Small sample of patients with PTSD	Discriminates between state fear and state anxiety	Lab-administered physiological task. Measures startle response to predictable or unpredictable shock
Shame & guilt					
Internalized Shame Scale	Cook (2001)	Internalized (trait) shame, general self-esteem	Survivors of intimate partner violence, combat veterans, treatment-seeking veterans with PTSD, sexual assault survivors	Performed well on indices of internal consistency, temporal stability, convergent validity, discriminant validity	30 items rated on 5-point Likert scale
Tests of Self-Conscious Affect (TOSCA-3)	Tangney, Dearing, Wagner, and Gramzow (2000)	Shame proneness, guilt proneness, blame, unconcern	Has been used extensively with trauma survivors, including military, interpersonal, and childhood traumas	Strong test-retest reliability, internal consistency, and convergent/divergent validity	15 vignettes with item response likelihood rated on a 5-point scale; adolescent and child versions are also available
Trauma-Related Guilt Inventory	Kubany et al. (1996)	Global guilt, guilt distress, guilt cognitions	Vietnam veterans, post-9/11 veterans, survivors of intimate partner violence, refugees	Performed well on indices of internal consistency, temporal stability, and convergent validity	32 items rated on 5-point Likert scale

Trauma-Related Shame Inventory	Øktedalen, Hagtvet, Hoffart, Langkaas, and Smucker (2014)	Internal and external trauma-related shame	Inpatient and outpatient treatment seekers diagnosed with PTSD, trauma-exposed veterans, adult sexual assault survivors	Demonstrated strong construct validity and convergent validity	24 items rated on 4-point Likert scale
Various					
Positive and Negative Affect Scale	Watson, Clark, and Tellegen (1988)	Items assess positive or negative emotion. PANAS-X includes scales assessing fear, hostility, and guilt, in addition to sadness, joviality, self-assurance, attentiveness, shyness, fatigue, serenity, and surprise	Has been used extensively with a range of trauma samples	All versions have shown evidence of discriminant and convergent validity	Three versions exist: the PANAS has 20 items, the International PANAS-Short Form has 10 items, and the PANAS-X has 60 items and 11 scales. Items for all versions are rated on a 5-point scale

prelude to hostility or aggression, the constructs of anger, aggression, and hostility are best understood as distinct, albeit related, phenomena. As such the following provides a selection of well-validated, frequently-used self-report measures of anger in the context of PTSD assessment. For a broader review of measures of anger, see Fernandez, Day, and Boyle (2015).

The State-Trait Anger Expression Inventory (STAXI; Spielberger & Sydeman, 1994; STAXI-II; Spielberger, 1999) is an internationally used, self-report measure of state and trait anger. The original STAXI is composed of 44 items that make up 5 major scales (anger-state, anger-trait, anger in, anger-out, and anger control) with 2 subscales (anger temperament and anger reaction). The STAXI-II comprises 57 items organized into 6 major scales (anger-state, anger-trait, anger expression-out, anger expression-in, anger control-out, and anger control-in) and an expression index. Both versions of the STAXI can be administered via paper-and-pencil or online. Online administrations generate a report that provides raw scores, gender-normed T-scores and percentile conversions, and brief interpretative text. A sample interpretative report is provided at the publisher's website. The STAXI and STAXI-II are intended for use with individuals between 16 and 63 years of age; child and adolescent versions are also available. Both measures have also been adapted for Spanish-speakers and have been translated into multiple languages.

The STAXI has been widely used and evaluated in clinical and nonclinical populations (e.g., Lievaart, Franken, & Hovens, 2016), including a range of trauma survivor samples such as adult survivors of interpersonal assault (e.g., Galovski, Mott, Young-Xu, & Resick, 2011), civilians exposed to war (Thabet, Abu Tawahina, El Sarraj, & Vostanis, 2008), crime victims (Orth & Maercker, 2009), child abuse survivors (Cloitre, Koenen, Cohen, & Han, 2002), and Vietnam veterans (Lasko, Gurvits, Kuhne, Orr, & Pitman, 1994). The STAXI-II has similarly been used among trauma populations, including treatment-seeking adolescent survivors of sexual violence (Kaczkurkin, Asnaani, Zhong, & Foa, 2016), veterans with PTSD (Owens, Chard, & Cox, 2008; Rauch et al., 2009; Roberge, Allen, Taylor, & Bryan, 2016), disaster responders (Palmieri, Weathers, Difede, & King, 2007), and 9/11 survivors (Difede et al., 2014). Factor analyses of the STAXI have supported its content structure (e.g., Forgays, Forgays, & Spielberger, 1997), and normative data for the STAXI and STAXI-II were derived from clinical and nonclinical samples (e.g., Spielberger, 1999).

The Buss-Perry Aggression Questionnaire (BPAQ; Buss & Perry, 1992) is an updated and revised version of the historically widely used Buss-Durkee

Hostility Inventory (BDHI; Buss & Durkee, 1957). The BPAQ measures anger, hostility, and aggression. The anger subscale of the BPAQ comprises 8 items that are scored on a 5-point Likert scale in which respondents are asked to rate how characteristic an item is of them (e.g., "Sometimes I fly off the handle for no good reason"). An internationally used measure of the anger-hostility-aggression triumvirate, the BPAQ evidenced strong convergent validity, divergent validity, and construct validity in initial psychometric evaluations (Buss & Perry, 1992). Subsequent investigations of the scale's psychometric properties demonstrated strong internal consistency, test-retest reliability, convergence with other measures of aggression, and divergence with social desirability measures (Harris, 1997). Additionally, the measure has been adapted internationally including for Dutch (Meesters, Muris, Bosma, Schouten, & Beuving, 1996), Japanese (Nakano, 2001), and Greek (Tsorbatzoudis, 2006) audiences with corresponding psychometric evaluations. The BPAQ has been used frequently among trauma survivors, including veterans with military sexual trauma-related PTSD (David, Simpson, & Cotton, 2006), children of mothers exposed to genocide (Roth, Neuner, & Elbert, 2014), combat veterans and former prisoners of war (Kip et al., 2013; Savic, Knezevic, Damjanovic, Spiric, & Matic, 2012), incarcerated persons endorsing childhood trauma exposure (Cima, Smeets, & Jelicic, 2008), and treatment seekers with comorbid substance use disorders (Barrett, Mills, & Teesson, 2011).

Anxiety

Anxiety and fear are related; anxiety reflects a state of moderately elevated arousal that may persist for an extended period of time, as opposed to briefer flares of acute fear, which may be more intense but shorter-lived. Anxiety is typically characterized by physiological arousal, apprehension, and worry. Historically, anxiety has been measured with self-report, physiological assessment, and behavioral measures. In terms of self-report measures, we recommend the State-Trait Anxiety Inventory (STAI; Spielberger, 1983), which assesses a litany of anxiety-related symptoms with one page asking respondents to consider how they generally feel (i.e., trait anxiety) and the other page asking individuals how they feel "right now" (i.e., state anxiety). Items include statements such as "I feel calm" (reverse scored), "I feel strained," "I feel frightened," and "I feel nervous" each rated on a 4-point scale from "not at all" to "very much so." The STAI has been used extensively in research and clinical settings. STAI scores are generally correlated with measures of PTSD, but these correlations are not as strong as those between different measures of PTSD, suggesting that, while there is some shared variance, the STAI is not redundant with measures of PTSD. In a sample of trauma survivors, the STAI showed stronger test-retest reliability and similar internal consistency reliability relative to other measures of anxiety (Adkins, Weathers, McDevitt-Murphy, & Daniels, 2008). Barnes, Harp, and Jung (2002) reviewed over 800 studies that reported psychometric properties for the STAI and reported satisfactory internal consistency and temporal stability across a range of populations.

Fear

Fear is a prominent emotion in PTSD, particularly fear resulting from trauma-related stimuli. While fear and anxiety are somewhat overlapping, fear refers specifically to acute hyperarousal, generally occurring in the presence of real or perceived threat as opposed to anxious apprehension or worry, which may be of a lower intensity but more persistent across time. Although research has suggested that PTSD shows more similarities to disorders characterized by a high degree of generalized negative affect (e.g., major depressive disorder [MDD] and generalized anxiety disorder), compared with those characterized as fear disorders (e.g., panic or phobia; Watson, 2005), cue-elicited fear remains a salient aspect of PTSD.

A behavioral task called the "NPU-threat test" was recently developed to discriminate between anxiety and fear (Schmitz & Grillon, 2012). This lab paradigm involves using different stimuli to systematically elicit responses consistent with fear or anxiety. A detailed protocol is provided by Schmitz and Grillon (2012). The name of the test is derived from the three conditions of (a) no aversive event (N), (b) predictable aversive conditions (P), and (c) unpredictable aversive events (U). Briefly each condition lasts 120 seconds, and over each testing session, participants experience the P and U conditions twice, separated by the N condition. During the P condition, participants are presented with a visual cue and an aversive stimulus (typically a shock) such that the cue reliably predicts the shock. In the U condition, shocks are not preceded by a cue and are therefore unpredictable. Fear and anxiety responses are assessed using the startle reflex. Specifically, electromyogram is used to measure the eyeblink during each elicited startle response. The responses elicited in the P and U conditions are thought to reflect fear-potentiated and anxiety-potentiated startle responses, respectively. The startle responses during P sessions are generally stronger than those during N sessions, and the startle responses during U sessions are stronger than those during both P and N sessions.

The NPU paradigm is similar to procedures used to model fear and anxiety in animals, making it well suited for translational research. The validity of the NPU-threat test has been supported by research showing that the startle response in the U condition correlated with trait anxiety and anxiety sensitivity (Stegmann, Reicherts, Andreatta, Pauli, & Wieser, 2019). Self-report of anxiety using a Likert scale during the task, however, has not always corresponded to physiological anxiety response (Grillon et al., 2008). Some research suggests the NPU-threat test may discriminate between PTSD and generalized anxiety disorder (Grillon et al., 2009), although it has not been extensively evaluated in samples of individuals with PTSD, and there are few studies that have compared the physiological responses elicited to self-reported emotion ratings.

Because fear generally presents as a state rather than a trait, it is difficult to meaningfully assess the affective experience of fear in real time without employing a trigger. In the context of PTSD, individuals may experience frequent fear cued by the presence of trauma-related stimuli or reminders. Trauma-related triggers may include environmental stimuli or interoceptive experiences (internal experiences such as memories, emotions, or bodily sensations that are associated with the traumatic event in some way). In a setting where the individual feels safe and is not actively being triggered, their self-reported fear rating may be low, but this conveys little about the frequency and intensity of the person's experience of fear when it occurs outside of the ecological boundaries of the laboratory setting.

Trauma script-driven imagery is a paradigm for assessing individuals' responses to a salient trauma-related reminder. Typically, this paradigm involves the researcher recording a second-person narrative of the respondent's index trauma. While listening to the recording, respondents' reactivity is assessed vis a vis various physiological channels. The trauma script paradigm has been used to elicit PTSD symptoms in studies using psychophysiological assessment (McDonagh-Coyle et al., 2001), positron-emission tomography (PET; Barkay et al., 2012; Rauch, van der Kolk, Fisler, & Alpert, 1996), and functional magnetic resonance imagery (fMRI; Lanius et al., 2003). These studies have assessed brain activation and/or physiological arousal during script exposure, which are related to fear, though not exclusively. Studies of trauma script-driven imagery often include self-rated assessments of different emotional states, in which the individual is asked to rate their experience of a set of specific emotions in the moment. In some studies, researchers used a standardized measure like the Positive and Negative Affect Schedule (PANAS; Watson et al., 1988), while in others researchers used a locally derived set of questions to assess adverse emotional states (e.g., McDonagh-Coyle et al., 2001).

Although the script-driven imagery paradigm is a viable method for eliciting fear, it has most often been employed for the opportunity to observe physiological phenomena in real time. Given that few studies have used standardized self-report assessments of responses to script-driven imagery, Hopper, Frewen, Sack, Lanius, and Van der Kolk (2007) developed the Responses to Script-Driven Imagery Scale (RSDI), which assesses reexperiencing, avoidance, and dissociative symptoms. Although the RSDI does not assess the subjective experience of fear, per se, reexperiencing symptoms are frequently accompanied by fear, and both avoidance and dissociation reflect attempts to cope with intense negative affect. Thus the RSDI alone would not provide a measure of the intensity of fear, but could provide additional useful information as a supplement to a self-reported scale of fear and other negative emotions, such as the PANAS.

Guilt and shame

Although fear and anxiety have historically been emphasized as core emotions in PTSD, other negative affective experiences have gained recognition more recently as important emotional components of the disorder. Guilt and shame were both explicitly mentioned for the first time in the diagnostic criteria for PTSD in DSM-5 (APA, 2013; see also Lee, Scragg, & Turner, 2001), although both have been recognized for decades as part of the posttraumatic emotional experience for many survivors (e.g., Herman, 1992; Lewis, 1971). Individuals with PTSD report high levels of guilt and shame across trauma experiences, including combat trauma (e.g., Crocker, Haller, Norman, & Angkaw, 2016), sexual assault (Vidal & Petrak, 2007), intimate partner violence (e.g., Beck et al., 2011), child abuse (e.g., Feiring, Taska, & Lewis, 2002; Wolfe, Sas, & Wekerle, 1994), and even noninterpersonal traumas such as natural disasters (e.g., Carmassi et al., 2017).

Guilt and shame are classified as self-conscious emotions, or emotions that require some degree of developmental self-reflection and self-evaluation (pride and embarrassment are also self-conscious feelings; Tangney, 1999). Both guilt and shame involve a negative self-evaluation informed by deeply held societal constructions of morality, honor, and belonging (Tangney, 1999; Wilson, Drožđek, & Turkovic, 2006), wherein the self is determined to, in some way, fail to meet desired or expected standards. Although they are often considered in tandem, guilt and shame are distinct. Whereas guilt refers to a negative evaluation of an action or behavior (e.g., participating in or failing to act to prevent an event that transgresses one's moral convictions), shame orients the negative appraisal specifically toward the individual (Lewis, 1971; Tangney, 1996). In other words, guilt is about an action (e.g., I *committed* this *act*), and shame is about the individual (e.g., *I* committed this act). Although both posttraumatic guilt and shame are associated with broader PTSD symptomatology (e.g., Andrews, Brewin, Rose, & Kirk, 2000; Crocker et al., 2016; Leskela, Dieperink, & Thuras, 2002), guilt may also be ameliorated through acts of reparation. Shame, on the other hand, tends to invite isolation and withdrawal (Leskela et al., 2002).

Self-report measures of shame and guilt can be organized into two types: measures that assess cued emotional states (e.g., feelings of shame or guilt when cued by specific contexts or events) and measures of the dispositional tendency to experience shame or guilt (e.g., shame or guilt proneness; Tangney, 1996). Clinicians may particularly benefit from measures that capture guilt or shame proneness or attribution styles as the persistent experience of these negative emotional states is also implicated in other forms of psychopathology. Here, we recommend measures of shame and guilt based on the frequency with which they are used in the contemporary trauma literature and their psychometric characteristics. For a broader review of measures of shame and guilt in the context of PTSD, the reader is directed to Robins, Noftle, and Tracy (2007).

The Trauma-Related Guilt Inventory (TRGI; Kubany et al., 1996) is a 32-item questionnaire designed to specifically measure guilt associated with a traumatic event. Using a 5-point Likert scale, respondents rate how true each statement is for them, ranging from "extremely true" to "not at all true." The TRGI assesses three domains of trauma-related guilt, including hindsight bias/responsibility (e.g., "I blame myself for something I did, thought, or felt"), violation of personal standards (e.g., "I did something that went against my values"), the lack of justification (e.g., scoring low on the item "If I knew today only what I knew when the event occurred, I would do exactly the same thing"), and a general distress factor (e.g., "What happened causes me emotional pain"). In initial psychometric evaluations the TRGI showed strong evidence of internal consistency and temporal stability and subsequently demonstrated strong convergent validity among combat veterans and domestic violence survivors (Kubany et al., 1996). The TRGI is a widely used measure of trauma-related guilt cognitions.

The Internalized Shame Scale (ISS; Cook, 1987, 1994, 2001) is a widely used, 30-item self-report inventory of trait shame designed for use in both clinical and research settings. Twenty-four items measure trait shame (or internalized shame), and 6 items measure general self-esteem, which is believed to negatively correlate with shame. The measure has demonstrated strong internal consistency and high temporal stability (del Rosario & White, 2006; Rybak & Brown, 1996). The ISS originated in the field of treatment for alcohol use problems but since has expanded broadly and been used in a wide range of clinical samples including individuals with trauma experiences and PTSD (e.g., Beck et al., 2011; Crocker et al., 2016).

The Trauma-Related Shame Inventory (TRSI; Øktedalen et al., 2014) is another contemporary measure of dispositional shame, specifically organized around the experience of a trauma. The 24-item measure includes two subscales distinguishing between internal shame, wherein the shame derives from the self, and external shame, in which the shame is experienced as coming from another. Respondents indicate how true each statement is for them on a 4-point Likert scale ranging from "not true of me" to "completely true of me." Example items include "I am ashamed of myself because of what happened to me" (internal shame) and "If others knew what happened to me, they would be disgusted with me" (external shame)." Initial validation efforts demonstrated strong construct validity and convergent validity (Øktedalen et al., 2014), suggesting that the TRSI may be a promising measure for further evaluation and for the assessment of trauma-specific shame experiences.

The Tests of Self-Conscious Affect (TOSCA; Tangney, Wagner, & Gramzow, 1989) are a battery of self-report questionnaires for assessing dimensions of self-conscious emotions, including guilt, shame, blame, and pride. Now in the third iteration (TOSCA-3; Tangney et al., 2000), the format is similar to its original conception: respondents read a series of brief scenarios that might occur as part of one's daily life and then rate on a 5-point scale how likely it is that they would react in the manner given, with the aim of determining the frequency or proneness of experiencing the given emotion. The original TOSCA was designed for use with adults, but additional versions have been developed for use with children aged 8–12 (TOSCA-C; Tangney, Wagner, Burggraf, Gramzow, & Fletcher, 1990) and adolescents aged 12–20 (TOSCA-A; Tangney, Wagner, Gavlas, & Gramzow, 1991). Psychometric evaluation of the TOSCA supports the distinction between shame and guilt as separate constructs with clinically relevant implications (e.g., Luyten, Fontaine, & Corveleyn, 2002; Watson, Gomez, & Gullone, 2016).

Sadness

As the cardinal emotion associated with depressive disorders, sadness is an important emotion to consider in the context of PTSD. Epidemiological research suggests that the rate of co-occurring major depressive disorder (MDD) among individuals with PTSD is high, with 42.8% of those diagnosed with PTSD also meeting criteria for comorbid MDD in the National Comorbidity Survey Replication (Rojas, Bujarski, Babson, Dutton, & Feldner, 2014). In a nationally representative sample of adolescents, there was a substantial gender difference, with 47.3% of boys with PTSD and 70.6% of girls with PTSD also meeting criteria for co-occurring MDD (Kilpatrick et al., 2003). Although sadness is a prominent affect in MDD, it is not necessary or sufficient to warrant a diagnosis. Thus an assessment of MDD symptoms may not provide the clinician or researcher with a clear sense of the respondent's experience of sadness itself. Sadness is important to assess in the context of PTSD, as it may be particularly prominent when the traumatic event involved a significant loss (Dalgleish & Power, 2004). Many measures of depression include questions querying sadness (e.g., the Beck Depression Inventory, revised; Beck, Steer, & Brown, 1996), but scales of basic emotions would provide a more efficient way to assess this affective experience if that is the goal.

Sadness is sometimes assessed during trauma script-driven imagery tasks that investigate brain regions that correspond to different emotional experiences (Liberzon & Martis, 2006 provide a review). The PANAS (Watson et al., 1988) has been published in three versions. The original PANAS is a 20-item measure where each item specifies a distinct emotion and respondents are asked to indicate the extent to which they have felt that emotion in the past week. While both positively valenced (e.g., interested, alert, excited, and proud) and negatively valenced (e.g., distressed, upset, guilty, and irritable) emotions are included, there is not an item specifying "sad" on the original version of the PANAS nor on the 10-item PANAS-Short Form (Thompson, 2007). The expanded, 60-item version of the PANAS (PANAS-X; Watson & Clark, 1999), however, includes a scale assessing sadness that is composed of 5 items (i.e., sad, blue, downhearted, alone, and lonely). There are 11 scales on this version of the PANAS that cover 4 negative emotions (fear, hostility, and guilt, in addition to sadness) and 3 positive emotions (joviality, self-assurance, and attentiveness). There are also four other scales assessing shyness, fatigue, serenity, and surprise (Watson & Clark, 1999). While the PANAS and PANAS-SF will provide scores of general negative affect, they will not permit the assessment of sadness in particular; the PANAS-X would be required for this.

Negative and positive affect

The International Affective Picture System (IAPS; Lang, Bradley, & Cuthbert, 1997; Lang, Bradley, & Cuthbert, 2008) offers an alternative way to assess negative affect broadly defined. This assessment method involves presenting participants with emotionally evocative or neutral images and rating their responses on three dimensions: valence, arousal, and dominance (Lang et al., 1997). Participants are asked to view a subset of images for a given amount of time, and their emotional reaction is recorded. The IAPS currently includes 956 images (Lang et al., 2008), each of which is categorized by the emotional valence it is designed to evoke, including positive images (e.g., a woman on a beach; IAPS #4200), neutral images (e.g., a basket; IAPS #7010), and negative images (e.g., a burn victim; IAPS #3100; Lang & Bradley, 2007). Responses to the images have been normed in samples of college students in the United States (Lang et al., 2008), children (aged 7–9, 10–12, and 13–14; Lang et al., 1997), and older adults (Grühn and Scheibe, 2008) and internationally (e.g., Deák, Csenki, & Révész, 2010; Lasaitis, Ribeiro, & Bueno, 2008). There are several methods available for evaluating participants' reactions to the IAPS, the most common being the Self-Assessment Manikin (SAM; Bradley & Lang, 1994), which allows individuals to rate the valence of their emotion, their arousal, and the dominance of their emotion using a series of figures. Alternatively, researchers have also used facial electromyography, skin conductance, and heart rate to evaluate emotionality, arousal, and attention. Still, others have measured reactions using fMRI (e.g., McLaughlin et al., 2014; van Rooij et al., 2015; van Rooij, Kennis, Vink, & Geuze, 2016). Several studies using the IAPS have found that, in response to emotionally evocative stimuli, individuals with PTSD demonstrate greater negative emotionality and blunted positive emotionality when compared with controls (e.g., Amdur, Larsen, & Liberzon, 2000; Wolf, Miller, & McKinney, 2009).

The IAPS uniquely offers a method of experimentally eliciting emotions to assess a wide range of emotion-related constructs. Within the PTSD literature alone, the IAPS has been used to assess a number of emotion-related constructs believed to be central to the conceptualization of PTSD. Specifically, some studies have assessed numbing and heightened negative emotional reactivity, whereas others have used the IAPS to evaluate the neural correlates of positive and negative emotional responses in individuals with PTSD (e.g., McLaughlin et al., 2014; van Rooij et al., 2015, 2016). Still, others have used the images to elicit particular negative emotions to evaluate emotion regulation (Shepherd and Wild, 2014). Some researchers have even used the IAPS as a trauma analogue in healthy controls (e.g., Krans, Reinecke, Jong, Näring, & Becker, 2012; Oulton, Takarangi, & Strange, 2016). Overall the IAPS offers an experimental assessment method that may provide qualitatively different information relative to self-report measures for a variety of constructs.

In addition to the array of negatively valenced affective states we have described and that are often associated with PTSD, a disorder that loads heavily on measures of negative affect, it may also be important in some contexts to assess the experience of positive affect in individuals with PTSD. Research has suggested that positive affect and negative affect do not necessarily reflect opposite ends of the same spectrum. Rather, they seem to be orthogonal bipolar dimensions on which meaningful variability may be observed (Tellegen, Watson, & Clark, 1999). Research suggests the capacity for positive emotion is hindered in major depressive disorder (MDD) to a greater extent than in PTSD. Individuals with co-occurring PTSD and MDD show similar levels of negative affect to those with PTSD alone but lower levels of positive affect, suggesting that negative affect may be a shared factor between the two diagnoses, while suppressed positive affect may be more unique to MDD (Post, Zoellner, Youngstrom, & Feeny, 2011). The capacity for positive affect seems to relate to reward function, or one's ability to derive enjoyment from activities or experiences, the flip side of anhedonia (Nawijn et al., 2015), and anhedonia may be one of the most deleterious aspects of PTSD. In the absence of rewarding activities to motivate behavior, a person with PTSD may have little incentive to overcome avoidance and isolation. Thus, in addition to ratings of negative affective states and of anhedonia specifically, there may be value in measuring one's capacity for positive emotion. The various forms of the PANAS each offer a broad scale reflecting positive affect. Depending on the level of granularity desired, the short form offers a quick method to assess a few aspects of positive emotion, whereas the PANAS-X includes three separate component scales loading on the broad positive emotion factor (Watson & Clark, 1999).

Emotion-related processes

In addition to the assessment of key affective dimensions that are related to PTSD, researchers and clinicians have found benefit in assessing several emotion-related processes that also have relevance to PTSD. These largely reflect constructs that are thought to function as trait-like individual difference variables that may influence vulnerability to PTSD following exposure to a traumatic event. In each case, we provide a brief description of the construct and recommend assessment instruments that have been supported by the literature. A summary of measures assessing these emotion-related processes appears in Table 2.

Alexithymia

Alexithymia is an emotional deficit made up of four primary features: difficulty identifying, naming, describing, and/or expressing emotions and feelings; difficulty distinguishing between emotions or feelings and bodily or physical sensations; difficulty with symbolism that manifests as an inability to experience, describe, or express fantasies and dreams; and a marked focus on external events as opposed to internal events or an inclination toward thinking externally as opposed to internally (Taylor, 1984). These deficits are known to interfere with treatment. The loss of introspection can result in reduced engagement for the client and difficulty for the clinician, while the confusion of feelings for physical symptoms can result in medical help seeking instead of psychological (Taylor, 1984). In the context of PTSD, "posttraumatic alexithymia" has specifically been defined as difficulty identifying, expressing, or regulating one's responses to traumatic events, resulting in either not knowing what one is feeling or not feeling anything at all (Frewen, Pain, Dozois, & Lanius, 2006).

Most of the literature regarding alexithymia has evaluated this construct using the 20-item Toronto Alexithymia Scale (TAS-20; Bagby, Parker, & Taylor, 1994; Bagby, Taylor, & Parker, 1994). This self-report scale includes items rated on a 5-point scale (i.e., 1 = *strongly disagree* and 5 = *strongly agree*) that can be factored into three subscales: difficulty identifying feelings, difficulty describing feelings, and externally oriented thinking, all key elements of alexithymia. Psychometric evaluations of the TAS-20 have demonstrated adequate internal consistency and test-retest reliability for the overall measure and subscales (Bagby, Parker, & Taylor, 1994). With regard to construct validity, there is evidence of convergent, discriminant, and concurrent validity (Bagby, Taylor, & Parker, 1994). As evidence of convergent validity, the TAS-20 has been found to be negatively related to openness to experience (in particular, openness to feelings; $r=-0.49$) and positively related to negative affect ($r=0.27$). With regard to discriminant validity, the TAS-20 has been found to be unrelated to personality traits such as agreeableness ($r=-0.09$), conscientiousness ($r=0.21$), or sensation-seeking ($r=0.07$). Evidence of concurrent validity was evaluated by comparing scores with clinician ratings based on two subscales of the 17-item interviewer-rated

Table 2 Measures of emotion-related processes.

Measure	Citation	Constructs measured	Trauma samples used	Psychometric evidence	Administration details
Alexithymia					
Toronto Alexithymia Scale (TAS)	Bagby, Parker, and Taylor (1994)	Alexithymia; three subscales: difficulty identifying feelings, difficulty describing feelings, and externally oriented thinking	Survivors of childhood trauma and sexual assault, police officers, holocaust survivors, general psychiatric cases	Evidence of internal consistency, test-retest reliability, and convergent and discriminant validity	20 items rated on 5-point scale
Anhedonia					
Snaith-Hamilton Pleasure Scale	Snaith et al. (1995)	Hedonic reward	Nontreatment-seeking smokers with mixed trauma	Evidence of convergent and discriminant validity, internal consistency, and test-retest reliability	14 items rated dichotomously
Hedonic Deficit and Interference Scale	Frewen, Dean, and Lanius (2012)	Hedonic deficit, negative affective interference	Young adult survivors of child abuse	Strong internal consistency, evidence of convergent validity with measures of hedonic deficit, expected correlations with measures of positive and negative affect	21 items rated on 11-point scale

Continued

Table 2 Measures of emotion-related processes—cont'd

Measure	Citation	Constructs measured	Trauma samples used	Psychometric evidence	Administration details
Anxiety sensitivity					
Anxiety Sensitivity Index-3	Taylor et al. (2007)	Anxiety sensitivity; three subscales: physical concerns, cognitive concerns, and social concerns	Veteran and civilian trauma, women with comorbid PTSD and substance use disorders	Evidence of reliability and validity has been demonstrated for ASI-3 and predecessors	18 items rated on 5-point scale
Distress tolerance					
Distress Tolerance Scale	Simons and Gaher (2005)	Perceived emotional distress tolerance; four subscales: tolerance, appraisal, absorption, and regulation	Trauma-exposed community members, veterans from multiple eras, survivors of sexual trauma, natural disasters, and intimate partner violence	Evidence of test-retest reliability, convergent, and discriminant validity	15 items rated on 5-point Likert scale
Mirror Tracing Task	Quinn, Brandon, and Copeland (1996)	Actual psychological distress tolerance	Trauma-exposed community members, veterans from multiple eras, smokers with comorbid PTSD	Evidence that the task successfully induces stress (subjective and physiological). Correlates strongly with other behavioral measures of distress tolerance but minimally with self-report measures	Lab-administered behavioral task: multiple trials of tracing images on a computer as though looking at them through a mirror. A buzzer sounds when the participant makes a mistake. They can stop the task at any time. Longer endurance of the task indicates greater distress tolerance

Paced Auditory Serial Addition Task–Computer-based	Gronwall (1977); Lejuez, Kahler, and Brown (2003)	Actual psychological distress tolerance	Trauma-exposed community members, undergraduates and inpatients, survivors of sexual assault	Evidence that the task successfully induces stress (subjective and physiological); convergent with other behavioral measures of distress tolerance, but not with self-report measures; found to be more reliable than paper-administered versions of the PASAT	Lab-administered behavioral task: multiple trials varying in difficulty in which participants must add numbers as they appear on a screen. They can stop the task at any time. Longer endurance of the task indicates greater distress tolerance
Emotion regulation					
Difficulties in Emotion Regulation Scale	Gratz and Roemer (2004)	Emotion regulation; six subscales: nonacceptance, goals, impulse, awareness, strategies, and clarity	Has been used extensively with trauma survivors, including military, interpersonal, childhood traumas	Evidence of convergent, discriminant, and predictive validity and internal consistency and test-retest reliability	36 items rated on a 5-point scale of frequency
Experiential avoidance					
Acceptance and Action Questionnaire, Revised (AAQ-II)	Bond et al. (2011)	Psychological flexibility/experiential avoidance	Combat veterans, civilian trauma survivors	Evidence of reliability and convergent validity. Discriminant validity has been criticized	7 items rated on 7-point Likert scale. Higher scores suggest greater experiential avoidance

Continued

Table 2 Measures of emotion-related processes—cont'd

Measure	Citation	Constructs measured	Trauma samples used	Psychometric evidence	Administration details
Multidimensional Experiential Avoidance Questionnaire (MEAQ)	Gámez et al. (2011)	Psychological flexibility/experiential avoidance; 6 subscales: behavioral avoidance, distress aversion, procrastination, distraction and suppression, repression and denial, distress endurance	Limited use with trauma samples. Has been used in civilian trauma survivors	Evidence of reliability and construct validity, including strong convergent and discriminant validity evidence	62 items rated on a 6-point scale. Higher scores suggest greater experiential avoidance. A brief version was published (Gámez et al., 2014), the BEAQ has 15 items

Beth Israel Hospital Psychosomatic Questionnaire (BIQ; Sifneos, 1973): affect awareness ($r=0.53$) and operatory thinking ($r=0.48$). Total and subscale scores on the TAS-20 were significantly correlated to clinician ratings on both BIQ subscales. Note, however, that Marchesi, Ossola, Tonna, and De Panfilis (2014) found evidence that scores on the TAS-20 are highly sensitive to an individual's current level of distress, which may indicate that it is a measure of negative affect rather than alexithymia. Given the dearth of alternative measures of alexithymia, however, the TAS-20 is recommended for use with caution and awareness of this limitation.

Anhedonia/numbing

In the era of DSM-IV, PTSD diagnoses included an "avoidance and numbing" symptom cluster, which included items that specifically targeted both effortful avoidance and emotional numbing (conceptualized as a dampening of emotional responsiveness). In DSM-5, these criteria were restructured such that effortful avoidance appears as an independent criterion, and three items regarding emotional numbing (i.e., "markedly diminished interest or participation in significant activities," "feelings of detachment or estrangement from others," and "persistent inability to experience positive emotions") are now included as part of the "negative alterations in cognition and mood" symptom cluster. This cluster also includes items referring to a broad range of emotions, including shame, guilt, fear, anger, and horror.

Of the emotional numbing symptoms, anhedonia is particularly important to assess. It reflects a loss of interest in activities that one previously enjoyed and/or a loss of pleasure derived from those activities. Although this is not unique to PTSD (anhedonia plays a large role in depressive disorders as well), it appears to be a particularly deleterious aspect of PTSD and has been linked to the social isolation that goes along with PTSD (Nawijn et al., 2015). The converse of anhedonia has been described as "reward function," a construct that has only received attention in the context of PTSD in recent years.

A recent article by Nawijn et al. (2015) provided an exhaustive review of measures of reward function or anhedonia that have been used in the context of PTSD, including both self-report and lab-based measures. Additionally, a review by Rizvi, Pizzagalli, Sproule, and Kennedy (2016) evaluated measures of anhedonia in the context of depression. Both of these reviews recommended the Snaith-Hamilton Pleasure Scale (SHAPS; Snaith et al., 1995) as the gold standard for assessing anhedonia. The SHAPS is a

14-item questionnaire with item responses ranging from "strongly agree" to "strongly disagree." The original scoring rubric entailed assigning scores of 1 for item responses of "disagree" or "strongly disagree" and values of 0 for "agree" and "strongly agree" (Snaith et al., 1995), with total scores ranging from 0 to 14. However, in some research, authors report scoring items on a 4-point scale from 1 to 4 with "definitely agree" rated 1 and "definitely disagree" rated 4 (e.g., Franken, Rassin, & Muris, 2007; Langvik & Borgen Austad, 2019). In both scoring scenarios, higher scores are suggestive of a greater degree of anhedonia. Psychometric analyses suggest that the SHAPS is a reliable and valid measure of anhedonia, showing a pattern of correlations suggestive of convergence with measures of anhedonia and depression and discrimination from measures of positive affect and life satisfaction (Franken et al., 2007). Interestingly, however, the magnitude of all of these correlations was fairly modest. In a sample of smokers, the SHAPS showed statistically significant positive correlations with each facet of PTSD, but somewhat surprisingly the correlation with emotional numbing ($r=0.25$; $P<.01$) was in the same range as the correlations between the SHAPS and other symptom clusters (Mathew, Cook, Japuntich, & Leventhal, 2015). In a sample of combat-exposed veterans, those who met criteria for PTSD had a significantly higher mean score on the SHAPS relative to combat-exposed controls (Yuan et al., 2018). The psychometric properties of the SHAPS have not been explored in samples of individuals with PTSD, but have been found to be strong in samples of individuals with depression and nonclinical samples (Franken et al., 2007; Langvik & Borgen Austad, 2019).

Frewen et al. (2012) speculated that anhedonia in PTSD may reflect a more complex emotional process than has typically been observed and conceptualized. They identified "negative affective interference" as a construct that occurs in individuals with PTSD, wherein not only the experience of positive emotion is blunted (i.e., a hedonic deficit) but also negative emotional states are elicited by ostensibly positive stimuli and the negative emotion preempts the positive affective state. These authors developed a measure that attempts to capture this complex combination of emotional states. The Hedonic Deficit and Interference Scale (HDIS; Frewen et al., 2012) includes 21 items rated on an 11-point scale. Five of the items assess positive emotionality, 5 assess hedonic deficits (difficulty experiencing positive emotions), and 11 items assess interference of negative affect in the context of positive events. The HDIS subscales showed good evidence of internal consistency in the PTSD group (α ranged from 0.85 to 0.93 for the three subscales). Interestingly, in the non-PTSD group, the internal consistency for negative affect interference was weak ($\alpha = 0.56$) although the other two subscales appeared to show adequate internal consistency ($\alpha = 0.84$ and 0.85). Regarding validity, the HDIS showed a pattern of correlations demonstrating convergence with other self-report measures of anhedonia. In fMRI analyses the three subscales of the HDIS showed differential correlations with neural activity, suggesting that negative affective interference may reflect a different neurological substrate than hedonic deficits (Frewen et al., 2012).

Anxiety sensitivity

Anxiety sensitivity has received considerable attention in the literature on anxiety disorders and trauma-related disorders. Anxiety sensitivity is conceptualized as a fear of anxiety-related symptoms due to catastrophic expectations and is thought to contribute to the development and maintenance of anxiety disorders (Reiss, 1991) and PTSD (Marshall, Miles, & Stewart, 2010). Specifically, numerous studies have demonstrated the relevance of anxiety sensitivity to PTSD across a variety of populations (see Olatunji & Wolitzky-Taylor, 2009) Anxiety sensitivity is conceptualized as a dimensional variable with three components: fear of somatic concerns, cognitive dyscontrol, and fear of socially observable symptoms (Wheaton, Deacon, McGrath, Berman, & Abramowitz, 2012). In a sample of trauma-exposed patients with substance use disorders, anxiety sensitivity was correlated with PTSD severity. This study used the trauma script-driven paradigm and found that anxiety sensitivity was correlated with postscript negative affect ratings but not with postscript ratings of cravings for substances (McHugh, Gratz, & Tull, 2017), suggesting that anxiety sensitivity may play a role in the amplitude of fear and anxiety responses exhibited by individuals with PTSD in response to trauma-related cues.

Studies have shown that anxiety sensitivity may be amenable to treatment, with trials of cognitive behavioral therapy demonstrating reductions in anxiety sensitivity over the course of treatment for various anxiety disorders (Gutner, Nillni, Suvak, Wiltsey-Stirman, & Resick, 2013; Smits, Berry, Tart, & Powers, 2008). The Anxiety Sensitivity Index (ASI; Peterson & Reiss, 1992) was developed to assess this construct. The scale has been revised twice, with the current version, the ASI-3 (Taylor et al., 2007) including 18 items that reflect three factors: physical concerns, cognitive concerns, and social concerns. Items are rated on a 5-point scale from 0 (very little) to 4 (very much). Total scores range from 0 to 72. Sample items include "It is important for me not to appear nervous," "When my throat feels tight, I worry that I could choke to death," and "It scares me when I am unable to keep my mind on a task." The ASI-3 is available by request from Dr. Steven Taylor at taylor@unixg.ubc.ca.

Distress tolerance

Distress tolerance refers to one's ability to endure adverse or threatening states and experiences (Brown, Lejuez, Kahler, & Strong, 2002). Individuals with low distress tolerance appraise situations as more distressing and thus engage in avoidance behaviors to reduce their exposure to the aversive experiences. When avoidance is impossible the distress can become overwhelming and can interfere with functioning (Simons & Gaher, 2005). Distress tolerance is typically classified into emotional/psychological distress tolerance and physical distress tolerance. Assessment methods are organized around either evaluating perceived (i.e., the extent to which one believes that they can withstand distress) or actual distress tolerance (i.e., one's demonstrable ability to withstand distress; Marshall-Berenz, Vujanovic, Bonn-Miller, Bernstein, & Zvolensky, 2010). Perceived distress tolerance is measured using self-report measures, whereas actual distress tolerance is measured using behavioral tasks.

Low distress tolerance appears to be a risk factor for PTSD. Theoretically, it fits well into current conceptualizations of PTSD as a disorder of emotional and experiential avoidance (Foa & Kozak, 1986). Deficits in distress tolerance increase the likelihood of avoidance behavior and as such are likely to place an individual at higher risk for PTSD symptoms following a traumatic event (Marshall-Berenz et al., 2010). In the literature, PTSD diagnosis has been associated with lower levels of physical distress tolerance when compared with other anxiety-related diagnoses (Vujanovic, Marshall, Gibson, & Zvolensky, 2010). PTSD symptom severity has also been found to be negatively related to perceived emotional distress tolerance (Marshall-Berenz et al., 2010; Vujanovic et al., 2013). In terms of PTSD symptom clusters (based on DSM-IV), perceived emotional distress tolerance has been found to be negatively related to global symptom severity and also to reexperiencing, avoidance, and hyperarousal symptoms (Vujanovic, Bonn-Miller, Potter, Marshall, & Zvolensky, 2011). Distress tolerance has also been evaluated as a mechanism linking several comorbidities with PTSD, including alcohol use (Marshall-Berenz, Vujanovic, & MacPherson, 2011; Vujanovic, Marshall-Berenz, & Zvolensky, 2011), marijuana use (Potter, Vujanovic, Marshall-Berenz, Bernstein, & Bonn-Miller, 2011), and suicidal behavior (Anestis, Tull, Bagge, & Gratz, 2012).

The most widely used measure of distress tolerance is the Distress Tolerance Scale (DTS; Simons & Gaher, 2005), which evaluates perceived emotional distress tolerance. It includes 15 items designed to evaluate four subconstructs: *tolerance*, one's perceived ability to tolerate emotional distress ("I can't handle feeling distressed or upset"); *appraisal*, one's subjective appraisal of the acceptability of the distress ("My feelings of distress or being upset are not acceptable"); *absorption*, the amount of attention absorbed by the negative emotions ("When I feel distressed or upset I cannot help but concentrate on how bad the distress actually feels"); and *regulation*, one's efforts to relieve the distress ("When I feel distressed or upset, I must do something about it immediately"). Participants rate these items on a 5-point Likert-type scale (1 = *strongly agree* and 5 = *strongly disagree*). Regarding validity, the DTS showed evidence of convergence through positive correlations with expectancies regarding emotion regulation ($r=0.54$) and mood acceptance ($r=0.47$) and a negative relationship with emotion dysregulation ($r=-0.51$). Discriminant validity was evidenced by a smaller relationship with mood typicality ($r=0.17$). Regarding reliability, test-retest over a 6-month period was good ($r=0.61$).

Behavioral measures of actual distress tolerance are also widely available. Two of the most commonly used are the Mirror Tracing Task (Quinn et al., 1996) and the Computerized Paced Auditory Serial Addition Task (PASAT; Lejuez et al., 2003). The Mirror Tracing Task is simple to administer and shows evidence of being a valid method for frustration and stress (i.e., increases subjective stress, heart rate, and blood pressure; Quinn et al., 1996). Participants are asked to simply trace objects on a screen using their mouse, but the screen presents a reverse image so it appears as though the participant is looking at it through a mirror, thus increasing the difficulty of the task. Individuals are given practice trials, which have relatively simple objects, and experimental trials, in which the objects are more complex. When the participant makes a mistake, a buzzer sounds. Participants are asked to engage with the task until they have completed it, but they are given the option to give up on a given trial at any time so that they can move forward. Shorter duration of time spent on trials is believed to be indicative of greater frustration and lower capacity for distress tolerance.

The PASAT was originally designed to assess information processing capacity (Gronwall, 1977), but it quickly became apparent that the task induces stress (Bornovalova et al., 2008; Gratz, Rosenthal, Tull, Lejuez, & Gunderson, 2010; Holdwick Jr. & Wingenfeld, 1999). As a result, with some modification, it has become a common measure of one's ability to endure stress (Brown et al., 2002). In the computer-based version of the task (PASAT-C; Lejuez et al., 2003), participants are shown a series of numbers very briefly on a screen. During the time between each number, participants are asked to type in the sum of the most recently shown number and the number prior. They then must forget that sum as soon as the next number appears, because once again they must sum the current number and the number previous (e.g., 3, 4 [correct answer=7], 2 [6], 5 [7]). The task has three trials or "levels," each with increasing difficulty (i.e., latency between number presentation reduces as the task progresses) and length. During the third trial the participant has the option of pressing an escape button to end the task. As with the Mirror Tracing Task, participants who spend less time on trials either by moving forward to the next trial or ending the task are exhibiting low thresholds for distress tolerance. Psychometric evaluations of this task have found it to be more reliable than previous paper-administered versions, which varied in procedure (Lejuez et al., 2003). Evaluations of construct validity have found that it successfully induces stress in a laboratory setting and offers comprehensive insight into behavioral, cognitive, and physiological responses to stress (Lejuez et al., 2003).

Emotion regulation

PTSD has frequently been associated with deficits in emotion regulation. These deficits seem to confer risk for worse functional impairment beyond the effect of PTSD symptoms (Cloitre, Miranda, Stovall-McClough, & Han, 2005). Gross (1998) defined emotion regulation as modulation of one's emotional experience by "shaping" *which* emotion one experiences, *when* one experiences it, and *how* one experiences and/or expresses it. Thus there are three core features of emotion regulation in this conceptualization: goals, strategies, and outcomes (Gross, 2014). In the context of PTSD specifically, symptoms are related to negative attitudes about emotional expression and a tendency to suppress or withhold negative emotions (Moore, Zoellner, & Mollenholt, 2008; Nightingale & Williams, 2000; Roemer, Litz, Orsillo, & Wagner, 2001).

Several measures of emotion regulation have been developed, but the most commonly used is the Difficulties in Emotion Regulation Scale (DERS; Gratz & Roemer, 2004), which was designed to evaluate four features of emotion regulation: awareness and understanding of one's emotions, the ability to accept said emotions, the ability to attenuate impulsive behavior and engage in goal-directed behavior while experiencing negative emotions, and the ability to flexibly use emotion regulation strategies to modulate (though not eliminate) emotional responses as necessary (Gratz & Roemer, 2004). It was also developed to allow for *flexible* measurement of effective strategy use, as an alternative to measures that denote universally (rather than situationally) adaptive or maladaptive strategies. The resulting measure includes 36 items that can be broken down into 6 factors: *nonacceptance* (i.e., nonacceptance of emotional responses), *goals* (i.e., difficulties engaging in goal-directed behavior), *impulse* (i.e., impulse control difficulties), *awareness* (i.e., the lack of emotional awareness), *strategies* (i.e., limited access to emotion regulation strategies), and *clarity* (i.e., the lack of emotional clarity). Evaluations of reliability indicated that internal consistency is generally strong for the full scale and adequate for the subscales, and test-retest reliability has been found to be good (Gratz & Roemer, 2004). Indices of construct and predictive validity were also found to be adequate (Gratz & Roemer, 2004). Both the overall measure ($r=-0.69$) and the subscales (r's $=-0.34$ to -0.69) evidenced significant convergence with other measures of emotion regulation (i.e., the Negative Mood Regulation scale; r's $=-0.34$ to -0.69). Further convergent validity was evidenced through significant correlations between emotional avoidance and the DERS overall ($r=0.60$) and all subscales (r's $=0.32-0.56$). Predictive validity was demonstrated through significant correlations with behavioral indices of dysregulation (i.e., self-harm among men [$r=0.26$] and women [$r=0.20$] and intimate partner violence among men [$r=0.34$]).

Alternative measures tend to focus on fewer features of emotion regulation (e.g., on specific emotion regulation strategies) but may be more appropriate depending on the context for assessment. If a structured or semistructured interview is more appropriate for one's assessment goals, there are two options available: the Emotion Regulation Interview (Werner, Goldin, Ball, Heimberg, & Gross, 2011) and the Semistructured Emotion Regulation Interview (Lee, Weathers, Sloan, Davis, & Domino, 2017). Note, however, that neither of these have been evaluated or used in samples of individuals with PTSD.

Experiential avoidance

Experiential avoidance refers to a general tendency to avoid undesirable internal states. Experiential avoidance includes behavioral avoidance, the use of cognitive strategies to avoid thinking about distressing cues, and dissociation from one's emotional state. Although it has been considered to be an aspect of emotion regulation (e.g., Seligowski, Lee, Bardeen, & Orcutt, 2015), experiential avoidance has often been examined outside of the framework of emotion regulation and is not included as a facet of emotion regulation on the DERS. Results from a metaanalysis suggest that there was a large effect size across studies of shared variance between PTSD symptoms and experiential avoidance. Experiential avoidance is thought to be the inverse of emotional acceptance and is also closely related to psychological flexibility in the ACT framework. The Acceptance and Action Questionnaire, revised (AAQ-II; Bond et al., 2011), is the most current version of the gold-standard measure of experiential avoidance and psychological inflexibility. The psychometrics of the AAQ-II have been investigated in samples of adults with anxiety and depression and reflect an improvement over its predecessor (Rochefort, Baldwin, & Chmielewski, 2018). The AAQ-II is conceptualized as a measure of psychological flexibility, a construct that overlaps with experiential avoidance; these terms are used interchangeably (along with emotional acceptance) in this literature (Rochefort et al., 2018).

A newer measure, the Multidimensional Experiential Avoidance Questionnaire (MEAQ; Gámez et al., 2011), was introduced to tap into a broad range of aspects of emotional avoidance. The MEAQ includes 62 measures and 6 subscales (behavioral avoidance, distress aversion, procrastination, distraction and suppression, repression and denial, and distress endurance). The MEAQ has shown evidence of strong internal consistency for the full scale and subscales (Bond et al., 2011; Rochefort et al., 2018). In an extensive examination of the psychometric properties of both the MEAQ and the AAQ-II, the MEAQ showed a stronger profile of convergent and discriminant validity evidenced by correlations with measures of experiential avoidance, mindfulness, neuroticism, and negative affect. The MEAQ and AAQ-II only correlated moderately with each other in both of the samples studied ($r=0.50$ in college students and $r=0.59$ in an online sample of adults). The AAQ-II showed stronger correlations with measures of neuroticism and negative affect than with other measures of experiential avoidance (Rochefort et al., 2018). Concerns have been raised about the discriminant validity of the AAQ-II (Wolgast, 2014).

Conclusion

The current conceptualization of PTSD as a traumatic stress reaction distinct from other anxiety disorders or disorders of negative affect is reflective of the emergence of research implicating emotion and emotional processes as increasingly central to the presentation and course of PTSD. PTSD is associated with a broad and often severe range of negative emotions, most of which are not assessed in detail by diagnostic measures of PTSD. Additionally, PTSD has shown associations with many emotion-related processes, and it is unclear whether these processes are risk factors, contributing factors, or consequences of PTSD. The relations between these processes and the development of PTSD have generally not been extensively studied, and evidence that would bear on the temporal sequencing of these processes and the development of PTSD is lacking. It is clear that thorough assessment of emotion and emotion-related processes in the context of trauma and PTSD is important both for understanding the complex clinical presentations of individuals with PTSD and for moving the field forward in terms of understanding how PTSD symptoms, emotional experiences, and emotional processes relate to each other. Researchers and clinicians alike would benefit from the inclusion of comprehensive assessments of emotion and emotion-related processes in their work to understand, conceptualize, and treat PTSD.

References

Adkins, J. W., Weathers, F. W., McDevitt-Murphy, M., & Daniels, J. B. (2008). Psychometric properties of seven self-report measures of posttraumatic stress disorder in college students with mixed civilian trauma exposure. *Journal of Anxiety Disorders, 22*(8), 1393–1402. https://doi.org/10.1016/j.janxdis.2008.02.002.

Amdur, R. L., Larsen, R., & Liberzon, I. (2000). Emotional processing in combat-related posttraumatic stress disorder: A comparison with traumatized and normal controls. *Journal of Anxiety Disorders, 14*(3), 219–238. https://doi.org/10.1016/S0887-6185(99)00035-3.

American Psychiatric Association. (2000). *Diagnostic and statistical manual of mental disorders: DSM-IV-TR*. Washington, DC: American Psychiatric Association.

American Psychiatric Association. (2013). *Diagnostic and statistical manual of mental disorders* (5th ed.). Washington, DC: American Psychiatric Association.

Andrews, B., Brewin, C. R., Rose, S., & Kirk, M. (2000). Predicting PTSD symptoms in victims of violent crime: The role of shame, anger, and childhood abuse. *Journal of Abnormal Psychology, 10*, 69. https://doi.org/10.1037/0021-843X.109.1.69.

Anestis, M. D., Tull, M. T., Bagge, C. L., & Gratz, K. L. (2012). The moderating role of distress tolerance in the relationship between posttraumatic stress disorder symptom clusters and suicidal behavior among trauma-exposed substance users in residential treatment. *Archives of Suicide Research, 16*, 198–211. https://doi.org/10.1080/13811118.2012.695269.

Bagby, R. M., Parker, J. D., & Taylor, G. J. (1994). The twenty-item Toronto Alexithymia Scale—I. Item selection and cross-validation of the factor structure. *Journal of Psychosomatic Research, 38*, 23–32. https://doi.org/10.1016/0022-3999(94)90005-1.

Bagby, R. M., Taylor, G. J., & Parker, J. D. (1994). The twenty-item Toronto Alexithymia Scale—II. Convergent, discriminant, and concurrent validity. *Journal of Psychosomatic Research, 38*, 33–40. https://doi.org/10.1016/0022-3999(94)90006-X.

Barkay, G., Freedman, N., Lester, H., Louzoun, Y., Sapoznikov, D., Luckenbaugh, D., … Bonne, O. (2012). Brain activation and heart rate during script-driven traumatic imagery in PTSD: Preliminary findings. *Psychiatry Research: Neuroimaging, 204*(2–3), 155–160. https://doi.org/10.1016/j.pscychresns.2012.08.007.

Barnes, L. L. B., Harp, D., & Jung, W. S. (2002). Reliability generalization of scores on the Spielberger State-Trait Anxiety Inventory. *Educational and Psychological Measurement*, 62(4), 603–618. https://doi.org/10.1177/0013164402062004005.

Barrett, E. L., Mills, K. L., & Teesson, M. (2011). Hurt people who hurt people: Violence amongst individuals with comorbid substance use disorder and post traumatic stress disorder. *Addictive Behaviors*, 36(7), 721–728. https://doi.org/10.1016/j.addbeh.2011.02.005.

Beck, A. T., Steer, R. A., & Brown, G. (1996). *Beck Depression Inventory-II manual*. San Antonio, TX: Psychological Corporation.

Beck, J. G., McNiff, J., Clapp, J. D., Olsen, S. A., Avery, M. L., & Hagewood, J. H. (2011). Exploring negative emotion in women experiencing intimate partner violence: Shame, guilt, and PTSD. *Behavior Therapy*, 42(4), 740–750. https://doi.org/10.1016/j.beth.2011.04.001.

Biddle, D., Elliott, P., Creamer, M., Forbes, D., & Devilly, G. J. (2002). Self-reported problems: A comparison between PTSD-diagnosed veterans, their spouses, and clinicians. *Behaviour Research and Therapy*, 40(7), 853–865. https://doi.org/10.1016/S0005-7967(01)00084-5.

Bond, F. W., Hayes, S. C., Baer, R. A., Carpenter, K. M., Guenole, N., Orcutt, H. K., ... Zettle, R. D. (2011). Preliminary psychometric properties of the Acceptance and Action Questionnaire–II: A revised measure of psychological inflexibility and experiential avoidance. *Behavior Therapy*, 42(4), 676–688. https://doi.org/10.1016/j.beth.2011.03.007.

Bornovalova, M. A., Gratz, K. L., Daughters, S. B., Nick, B., Delany-Brumsey, A., Lynch, T. R., ... Lejuez, C. W. (2008). A multimodal assessment of the relationship between emotion dysregulation and borderline personality disorder among inner-city substance users in residential treatment. *Journal of Psychiatric Research*, 42(9), 717–726. https://doi.org/10.1016/j.jpsychires.2007.07.014.

Bradley, M. M., & Lang, P. J. (1994). Measuring emotion: The Self-Assessment Manikin and the semantic differential. *Journal of Behavior Therapy and Experimental Psychiatry*, 25(1), 49–59. https://doi.org/10.1016/0005-7916(94)90063-9.

Brown, R. A., Lejuez, C. W., Kahler, C. W., & Strong, D. R. (2002). Distress tolerance and duration of past smoking cessation attempts. *Journal of Abnormal Psychology*, 111(1), 180–185. https://doi.org/10.1037/0021-843X.111.1.180.

Buss, A. H., & Durkee, A. (1957). An inventory for assessing different kinds of hostility. *Journal of Consulting Psychology*, 21(4), 343–349. https://doi.org/10.1037/h0046900.

Buss, A. H., & Perry, M. (1992). The aggression questionnaire. *Journal of Personality and Social Psychology*, 63(3), 452–459. https://doi.org/10.1037/0022-3514.63.3.452.

Carmassi, C., Bertelloni, C. A., Gesi, C., Conversano, C., Stratta, P., Massimetti, G., ... Dell'Osso, L. (2017). New DSM-5 PTSD guilt and shame symptoms among Italian earthquake survivors: Impact on maladaptive behaviors. *Psychiatry Research*, 251, 142–147. https://doi.org/10.1016/j.psychres.2016.11.026.

Cima, M., Smeets, T., & Jelicic, M. (2008). Self-reported trauma, cortisol levels, and aggression in psychopathic and non-psychopathic prison inmates. *Biological Psychology*, 78(1), 75–86. https://doi.org/10.1016/j.biopsycho.2007.12.011.

Cloitre, M., Koenen, K. C., Cohen, L. R., & Han, H. (2002). Skills training in affective and interpersonal regulation followed by exposure: A phase-based treatment for PTSD related to childhood abuse. *Journal of Consulting and Clinical Psychology*, 70(5), 1067. https://doi.org/10.1037//0022-006X.70.5.1067.

Cloitre, M., Miranda, R., Stovall-McClough, K. C., & Han, H. (2005). Beyond PTSD: Emotion regulation and interpersonal problems as predictors of functional impairment in survivors of childhood abuse. *Behavior Therapy*, 36(2), 119–124. https://doi.org/10.1016/S0005-7894(05)80060-7.

Cook, D. R. (1987). Measuring shame: The Internalized Shame Scale. *Alcoholism Treatment Quarterly*, 4(2), 197–215. https://doi.org/10.1300/J020v04n02_12.

Cook, D. R. (1994). *Internalized shame scale: Technical manual*. North Tonawanda, NY: Multi-Health Systems.

Cook, D. R. (2001). *Internalized shame scale: Technical manual*. Toronto: Multi-Health Systems.

Crocker, L. D., Haller, M., Norman, S. B., & Angkaw, A. C. (2016). Shame versus trauma-related guilt as mediators of the relationship between PTSD symptoms and aggression among returning veterans. *Psychological Trauma Theory Research Practice and Policy, 8*(4), 520–527. https://doi.org/10.1037/tra0000151.

Dalgleish, T., & Power, M. J. (2004). Emotion-specific and emotion-non-specific components of posttraumatic stress disorder (PTSD): Implications for a taxonomy of related psychopathology. *Behaviour Research and Therapy, 42*(9), 1069–1088. https://doi.org/10.1016/j.brat.2004.05.001.

David, W. S., Simpson, T. L., & Cotton, A. J. (2006). Taking charge: A pilot curriculum of self-defense and personal safety training for female veterans with PTSD because of military sexual trauma. *Journal of Interpersonal Violence, 21*(4), 555–565. https://doi.org/10.1177/0886260505285723.

Deák, A., Csenki, L., & Révész, G. (2010). Hungarian ratings for the International Affective Picture System (IAPS): A cross-cultural comparison. *Empirical Text and Culture Research, 4*, 90–101.

del Rosario, P. M., & White, R. M. (2006). The Internalized Shame Scale: Temporal stability, internal consistency, and principal components analysis. *Personality and Individual Differences, 41*(1), 95–103. https://doi.org/10.1016/j.paid.2005.10.026.

Difede, J., Cukor, J., Wyka, K., Olden, M., Hoffman, H., Lee, F. S., & Altemus, M. (2014). D-cycloserine augmentation of exposure therapy for post-traumatic stress disorder: A pilot randomized clinical trial. *Neuropsychopharmacology, 39*(5), 1052. https://doi.org/10.1038/npp.2013.317.

Feiring, C., Taska, L., & Lewis, M. (2002). Adjustment following sexual abuse discovery: The role of shame and attributional style. *Developmental Psychology, 38*(1), 79–92. https://doi.org/10.1037/0012-1649.38.1.79.

Fernandez, E., Day, A., & Boyle, G. J. (2015). Measures of anger and hostility in adults. In G. Boyle, D. H. Saklofske, & G. Matthews (Eds.), *Measures of personality and social psychological constructs*. (pp. 74–100). San Diego, CA: Elsevier Academic Press. https://doi.org/10.1016/B978-0-12-386915-9.00004-8.

Foa, E. B., & Kozak, M. J. (1986). Emotional processing of fear: Exposure to corrective information. *Psychological Bulletin, 99*(1), 20–35. https://doi.org/10.1037/0033-2909.99.1.20.

Forgays, D. G., Forgays, D. K., & Spielberger, C. D. (1997). Factor structure of the state-trait anger expression inventory. *Journal of Personality Assessment, 69*(3), 497. https://doi.org/10.1207/s15327752jpa6903_5.

Franken, I. H. A., Rassin, E., & Muris, P. (2007). The assessment of anhedonia in clinical and non-clinical populations: Further validation of the Snaith-Hamilton Pleasure Scale (SHAPS). *Journal of Affective Disorders, 99*(1–3), 83–89. https://doi.org/10.1016/j.jad.2006.08.020.

Frewen, P. A., Dean, J. A., & Lanius, R. A. (2012). Assessment of anhedonia in psychological trauma: Development of the Hedonic Deficit and Interference Scale. *European Journal of Psychotraumatology, 3*, https://doi.org/10.3402/ejpt.v3i0.8585.

Frewen, P. A., Pain, C., Dozois, D. J. A., & Lanius, R. A. (2006). Alexithymia in PTSD. *Annals of the New York Academy of Sciences, 1071*(1), 397–400. https://doi.org/10.1196/annals.1364.029.

Galovski, T. E., Mott, J., Young-Xu, Y., & Resick, P. A. (2011). Gender differences in the clinical presentation of PTSD and its concomitants in survivors of interpersonal assault. *Journal of Interpersonal Violence, 26*(4), 789–806. https://doi.org/10.1177/0886260510365865.

Gámez, W., Chmielewski, M., Kotov, R., Ruggero, C., & Watson, D. (2011). Development of a measure of experiential avoidance: The multidimensional experiential avoidance questionnaire. *Psychological Assessment, 23*(3), 692–713. https://doi.org/10.1037/a0023242.supp. [Supplemental].

Gámez, W., Chmielewski, M., Kotov, R., Ruggero, C., Suzuki, N., & Watson, D. (2014). The brief experiential avoidance questionnaire: Development and initial validation. *Psychological Assessment, 26*(1). http://dx.doi.org/10.1037/a0034473.

Gonzalez, O. I., Novaco, R. W., Reger, M. A., & Gahm, G. A. (2016). Anger intensification with combat-related PTSD and depression comorbidity. *Psychological Trauma Theory Research Practice and Policy, 8*(1), 9–16. https://doi.org/10.1037/tra0000042.

Gratz, K. L., & Roemer, L. (2004). Multidimensional assessment of emotion regulation and dysregulation: Development, factor structure, and initial validation of the Difficulties in Emotion Regulation Scale. *Journal of Psychopathology and Behavioral Assessment, 26*(1), 41–54. https://doi.org/10.1023/B:JOBA.0000007455.08539.94.

Gratz, K. L., Rosenthal, M. Z., Tull, M. T., Lejuez, C. W., & Gunderson, J. G. (2010). An experimental investigation of emotional reactivity and delayed emotional recovery in borderline personality disorder: The role of shame. *Comprehensive Psychiatry, 51*(3), 275–285. https://doi.org/10.1016/j.comppsych.2009.08.005.

Grillon, C., Lissek, S., Rabin, S., McDowell, D., Dvir, S., & Pine, D. S. (2008). Increased anxiety during anticipation of unpredictable but not predictable aversive stimuli as a psychophysiologic marker of panic disorder. *The American Journal of Psychiatry, 165*, 898–904. https://doi.org/10.1176/appi.ajp.2007.07101581.

Grillon, C., Pine, D. S., Lissek, S., Rabin, S., Bonne, O., & Vythilingam, M. (2009). Increased anxiety during anticipation of unpredictable aversive stimuli in posttraumatic stress disorder but not in generalized anxiety disorder. *Biological Psychiatry, 66*(1), 47–53. https://doi.org/10.1016/j.biopsych.2008.12.028.

Gronwall, D. M. (1977). Paced auditory serial-addition task: A measure of recovery from concussion. *Perceptual and Motor Skills, 44*(2), 367–373. https://doi.org/10.2466/pms.1977.44.2.367.

Gross, J. J. (1998). The emerging field of emotion regulation: An integrative review. *Review of General Psychology, 2*, 271–299. https://doi.org/10.1037/1089-2680.2.3.271.

Gross, J. J. (2014). Emotion regulation: Conceptual and empirical foundations. In J. J. Gross (Ed.), *Handbook of emotion regulation.* (2nd ed.)(pp. 3–20). New York, NY: Guilford Press.

Grühn, D., & Scheibe, S. (2008). Age-related differences in valence and arousal ratings of pictures from the International Affective Picture System (IAPS): Do ratings become more extreme with age? *Behavior Research Methods, 40*(2), 512–521. https://doi.org/10.3758/BRM.40.2.512.

Gutner, C. A., Nillni, Y. I., Suvak, M., Wiltsey-Stirman, S., & Resick, P. A. (2013). Longitudinal course of anxiety sensitivity and PTSD symptoms in cognitive-behavioral therapies for PTSD. *Journal of Anxiety Disorders, 27*, 728–734. https://doi.org/10.1016/j.janxdis.2013.09.010.

Harris, J. A. (1997). A further evaluation of the aggression questionnaire: Issues of validity and reliability. *Behaviour Research and Therapy, 35*, 1047–1053. https://doi.org/10.1016/S0005-7967(97)00064-8.

Herman, J. L. (1992). Complex PTSD: A syndrome in survivors of prolonged and repeated trauma. *Journal of Traumatic Stress, 5*(3), 377–391. https://doi.org/10.1002/jts.2490050305.

Holdwick, D. J., Jr., & Wingenfeld, S. A. (1999). The subjective experience of PASAT testing: Does the PASAT induce negative mood? *Archives of Clinical Neuropsychology, 14*(3), 273–284. https://doi.org/10.1093/arclin/14.3.273.

Hopper, J. W., Frewen, P. A., Sack, M., Lanius, R. A., & Van der Kolk, B. A. (2007). The Responses to Script-Driven Imagery Scale (RSDI): Assessment of state posttraumatic symptoms for psychobiological and treatment research. *Journal of Psychopathology and Behavioral Assessment, 29*, 249–268.

Jakupcak, M., Conybeare, D., Phelps, L., Hunt, S., Holmes, H. A., Felker, B., … McFall, M. E. (2007). Anger, hostility, and aggression among Iraq and Afghanistan war veterans reporting PTSD and subthreshold PTSD. *Journal of Traumatic Stress, 20*(6), 945–954. https://doi.org/10.1002/jts.20258.

Kaczkurkin, A. N., Asnaani, A., Zhong, J., & Foa, E. B. (2016). The moderating effect of state anger on treatment outcome in female adolescents with PTSD. *Journal of Traumatic Stress, 29*(4), 325–331. https://doi.org/10.1002/jts.22116.

Kilpatrick, D. G., Ruggiero, K. J., Acierno, R., Saunders, B. E., Resnick, H. S., & Best, C. L. (2003). Violence and risk of PTSD, major depression, substance abuse/dependence, and comorbidity: Results from the National Survey of Adolescents. *Journal of Consulting and Clinical Psychology, 71*(4), 692–700. https://doi.org/10.1037/0022-006X.71.4.692.

Kip, K. E., Rosenzweig, L., Hernandez, D. F., Shuman, A., Sullivan, K. L., Long, C. J., … Sahebzamani, F. M. (2013). Randomized controlled trial of accelerated resolution therapy (ART) for symptoms of combat-related post-traumatic stress disorder (PTSD). *Military Medicine, 178*(12), 1298–1309. https://doi.org/10.7205/MILMED-D-13-00298.

Krans, J., Reinecke, A., Jong, P. J. d., Näring, G., & Becker, E. S. (2012). Analogue trauma results in enhanced encoding of threat information at the expense of neutral information. *Journal of Anxiety Disorders, 26*(6), 656–664. https://doi.org/10.1016/j.janxdis.2012.05.003.

Kubany, E. S., Bauer, G. B., Muraoka, M. Y., Richard, D. C., & Read, P. (1995). Impact of labeled anger and blame in intimate relationships. *Journal of Social and Clinical Psychology, 14*(1), 53–60. https://doi.org/10.1521/jscp.1995.14.1.53.

Kubany, E. S., Haynes, S. N., Abueg, F. R., Manke, F. P., Brennan, J. M., & Stahura, C. (1996). Development and validation of the Trauma-Related Guilt Inventory (TRGI). *Psychological Assessment, 8*(4), 428–444. https://doi.org/10.1037/1040-3590.8.4.428.

Lang, P., & Bradley, M. M. (2007). The international affective picture system (IAPS) in the study of emotion and attention. In J. A. Coan & J. J. B. Allen (Eds.), *Handbook of emotion elicitation and assessment* (pp. 29–46). New York, NY: Oxford University Press.

Lang, P. J., Bradley, M. M., & Cuthbert, B. N. (1997). International affective picture system (IAPS): Technical manual and affective ratings. *NIMH Center for the Study of Emotion and Attention, 1*, 39–58.

Lang, P. J., Bradley, M. M., & Cuthbert, B. N. (2008). *International affective picture system (IAPS): Affective ratings of pictures and instruction manual.* Technical Report A-8 Gainesville, FL: University of Florida.

Langvik, E., & Borgen Austad, S. (2019). Psychometric properties of the Snaith–Hamilton Pleasure Scale and a facet-level analysis of the relationship between anhedonia and extraversion in a nonclinical sample. *Psychological Reports, 122*(1), 360–375. https://doi.org/10.1177/0033294118756336.

Lanius, R. A., Williamson, P. C., Hopper, J., Densmore, M., Boksman, K., Gupta, M. A., … Menon, R. S. (2003). Recall of emotional states in posttraumatic stress disorder: An fMRI investigation. *Biological Psychiatry, 53*(3), 204–210. https://doi.org/10.1016/S0006-3223(02)01466-X.

Lasaitis, C., Ribeiro, R. L., & Bueno, O. F. A. (2008). Normas Brasileiras para o International Affective Picture System (IAPS)—Estudo comparativo dos novos estímulos para avaliações afetivas entre sujeitos brasileiros e norte-americanos = Brazilian norms for the International Affective Picture System (IAPS)—Comparison of the affective ratings for new stimuli between Brazilian and North-American subjects. *Jornal Brasileiro de Psiquiatria, 57*(4), 270–275. https://doi.org/10.1590/S0047-20852008000400008.

Lasko, N. B., Gurvits, T. V., Kuhne, A. A., Orr, S. P., & Pitman, R. K. (1994). Aggression and its correlates in Vietnam veterans with and without chronic posttraumatic stress disorder. *Comprehensive Psychiatry, 35*(5), 373–381.

Lee, D. A., Scragg, P., & Turner, S. (2001). The role of shame and guilt in traumatic events: A clinical model of shame-based and guilt-based PTSD. *British Journal of Medical Psychology, 74*(4), 451–466. https://doi.org/10.1348/000711201161109.

Lee, D. J., Weathers, F. W., Sloan, D. M., Davis, M. T., & Domino, J. L. (2017). Development and initial psychometric evaluation of the Semi-Structured Emotion Regulation Interview. *Journal of Personality Assessment, 99*(1), 56–66. https://doi.org/10.1080/00223891.2016.1215992.

Lejuez, C. W., Kahler, C. W., & Brown, R. A. (2003). A modified computer version of the Paced Auditory Serial Addition Task (PASAT) as a laboratory-based stressor. *The Behavior Therapist, 26*(4), 290–293. Retrieved from http://ezproxy.memphis.edu/login?url=http://search.ebscohost.com/login.aspx?direct=true&db=psyh&AN=2003-06608-006&site=ehost-live.

Leskela, J., Dieperink, M., & Thuras, P. (2002). Shame and posttraumatic stress disorder. *Journal of Traumatic Stress, 15*(3), 223–226. https://doi.org/10.1023/A:1015255311837.

Lewis, H. B. (1971). *Shame and guilt in neurosis*. New York: International Universities Press.

Liberzon, I., & Martis, B. (2006). Neuroimaging studies of emotional responses in PTSD. In R. Yehuda (Ed.), Vol. 1071. *Psychobiology of posttraumatic stress disorders: A decade of progress* (pp. 87–109). Malden: Blackwell Publishing.

Lievaart, M., Franken, I. H., & Hovens, J. E. (2016). Anger assessment in clinical and non-clinical populations: Further validation of the State–Trait Anger Expression Inventory-2. *Journal of Clinical Psychology, 72*(3), 263–278. https://doi.org/10.1002/jclp.22253.

Luyten, P., Fontaine, J. R. J., & Corveleyn, J. (2002). Does the Test of Self-Conscious Affect (TOSCA) measure maladaptive aspects of guilt and adaptive aspects of shame? An empirical investigation. *Personality and Individual Differences, 33*(8), 1373–1387. https://doi.org/10.1016/S0191-8869(02)00197-6.

Marchesi, C., Ossola, P., Tonna, M., & De Panfilis, C. (2014). The TAS-20 more likely measures negative affects rather than alexithymia itself in patients with major depression, panic disorder, eating disorders and substance use disorders. *Comprehensive Psychiatry, 55*(4), 972–978. https://doi.org/10.1016/j.comppsych.2013.12.008.

Marshall, G. N., Miles, J. N. V., & Stewart, S. H. (2010). Anxiety sensitivity and PTSD symptom severity are reciprocally related: Evidence from a longitudinal study of physical trauma survivors. *Journal of Abnormal Psychology, 119*(1), 143–150. https://doi.org/10.1037/a0018009.

Marshall-Berenz, E. C., Vujanovic, A. A., Bonn-Miller, M. O., Bernstein, A., & Zvolensky, M. J. (2010). Multimethod study of distress tolerance and PTSD symptom severity in a trauma-exposed community sample. *Journal of Traumatic Stress, 23*(5), 623–630. https://doi.org/10.1002/jts.20568.

Marshall-Berenz, E. C., Vujanovic, A. A., & MacPherson, L. (2011). Impulsivity and alcohol use coping motives in a trauma-exposed sample: The mediating role of distress tolerance. *Personality and Individual Differences, 50*(5), 588–592. https://doi.org/10.1016/j.paid.2010.11.033.

Mathew, A. R., Cook, J. W., Japuntich, S. J., & Leventhal, A. M. (2015). Post-traumatic stress disorder symptoms, underlying affective vulnerabilities, and smoking for affect regulation. *The American Journal on Addictions, 24*(1), 39–46. https://doi.org/10.1111/ajad.12170.

McDonagh-Coyle, A., McHugo, G. J., Friedman, M. J., Schnurr, P. P., Zayfert, C., & Descamps, M. (2001). Psychophysiological reactivity in female sexual abuse survivors. *Journal of Traumatic Stress, 14*(4), 667–683. https://doi.org/10.1023/A:1013081803429.

McHugh, R. K., Gratz, K. L., & Tull, M. T. (2017). The role of anxiety sensitivity in reactivity to trauma cues in treatment-seeking adults with substance use disorders. *Comprehensive Psychiatry, 78*, 107–114. https://doi.org/10.1016/j.comppsych.2017.07.011.

McLaughlin, K. A., Busso, D. S., Duys, A., Green, J. G., Alves, S., Way, M., & Sheridan, M. A. (2014). Amygdala response to negative stimuli predicts PTSD symptom onset following a terrorist attack. *Depression and Anxiety, 31*(10), 834–842. https://doi.org/10.1002/da.22284.

Meesters, C., Muris, P., Bosma, H., Schouten, E., & Beuving, S. (1996). Psychometric evaluation of the Dutch version of the Aggression Questionnaire. *Behaviour Research and Therapy, 34*(10), 839–843.

Moore, S. A., Zoellner, L. A., & Mollenholt, N. (2008). Are expressive suppression and cognitive reappraisal associated with stress-related symptoms? *Behaviour Research and Therapy, 46*(9), 993–1000. https://doi.org/10.1016/j.brat.2008.05.001.

Nakano, K. (2001). Psychometric evaluation on the Japanese adaptation of the Aggression Questionnaire. *Behaviour Research and Therapy, 39*(7), 853–858.

Nawijn, L., van Zuiden, M., Frijling, J. L., Koch, S. B. J., Veltman, D. J., & Olff, M. (2015). Reward functioning in PTSD: A systematic review exploring the mechanisms underlying anhedonia. *Neuroscience & Biobehavioral Reviews, 51*, 189–204. https://doi.org/10.1016/j.neubiorev.2015.01.019.

Nightingale, J., & Williams, R. M. (2000). Attitudes to emotional expression and personality in predicting post-traumatic stress disorder. *British Journal of Clinical Psychology, 39*(3), 243–254. https://doi.org/10.1348/014466500163266.

Øktedalen, T., Hagtvet, K. A., Hoffart, A., Langkaas, T. F., & Smucker, M. (2014). The Trauma Related Shame Inventory: Measuring trauma-related shame among patients with PTSD. *Journal of Psychopathology and Behavioral Assessment, 36*(4), 600–615. https://doi.org/10.1007/s10862-014-9422-5.

Olatunji, B. O., Ciesielski, B. G., & Tolin, D. F. (2010). Fear and loathing: A meta-analytic review of the specificity of anger in PTSD. *Behavior Therapy, 41*(1), 93–105. https://doi.org/10.1016/j.beth.2009.01.004.

Olatunji, B. O., & Wolitzky-Taylor, K. B. (2009). Anxiety sensitivity and the anxiety disorders: A meta-analytic review and synthesis. *Psychological Bulletin, 135*, 974–999.

Orth, U., & Maercker, A. (2009). Posttraumatic anger in crime victims: Directed at the perpetrator and at the self. *Journal of Traumatic Stress, 22*(2), 158–161. https://doi.org/10.1002/jts.20392.

Oulton, J. M., Takarangi, M. K. T., & Strange, D. (2016). Memory amplification for trauma: Investigating the role of analogue PTSD symptoms in the laboratory. *Journal of Anxiety Disorders, 42*, 60–70. https://doi.org/10.1016/j.janxdis.2016.06.001.

Owens, G. P., Chard, K. M., & Cox, T. A. (2008). The relationship between maladaptive cognitions, anger expression, and posttraumatic stress disorder among veterans in residential treatment. *Journal of Aggression, Maltreatment & Trauma, 17*(4), 439–452. https://doi.org/10.1080/10926770802473908.

Palmieri, P. A., Weathers, F. W., Difede, J., & King, D. W. (2007). Confirmatory factor analysis of the PTSD Checklist and the Clinician-Administered PTSD Scale in disaster workers exposed to the World Trade Center Ground Zero. *Journal of Abnormal Psychology, 116*(2), 329. https://doi.org/10.1037/0021-843X.116.2.329.

Peterson, R. A., & Reiss, S. (1992). *Anxiety Sensitivity Index revised manual.* Worthington, OH: IDS Publishing.

Post, L. M., Zoellner, L. A., Youngstrom, E., & Feeny, N. C. (2011). Understanding the relationship between co-occurring PTSD and MDD: Symptom severity and affect. *Journal of Anxiety Disorders, 25*(8), 1123–1130. https://doi.org/10.1016/j.janxdis.2011.08.003.

Potter, C. M., Vujanovic, A. A., Marshall-Berenz, E. C., Bernstein, A., & Bonn-Miller, M. O. (2011). Posttraumatic stress and marijuana use coping motives: The mediating role of distress tolerance. *Journal of Anxiety Disorders, 25*(3), 437–443. https://doi.org/10.1016/j.janxdis.2010.11.007.

Quinn, E. P., Brandon, T. H., & Copeland, A. L. (1996). Is task persistence related to smoking and substance abuse? The application of learned industriousness theory to addictive behaviors. *Experimental and Clinical Psychopharmacology, 4*(2), 186–190. https://doi.org/10.1037/1064-1297.4.2.186.

Rauch, S. A., Defever, E., Favorite, T., Duroe, A., Garrity, C., Martis, B., & Liberzon, I. (2009). Prolonged exposure for PTSD in a Veterans Health Administration PTSD clinic. *Journal of Traumatic Stress, 22*(1), 60–64. https://doi.org/10.1002/jts.20380.

Rauch, S. L., van der Kolk, B. A., Fisler, R. E., & Alpert, N. M. (1996). A symptom provocation study of posttraumatic stress disorder using positron emission tomography and script-driven imagery. *Archives of General Psychiatry, 53*(5), 380–387. https://doi.org/10.1001/archpsyc.1996.01830050014003.

Reiss, S. (1991). Expectancy model of fear, anxiety, and panic. *Clinical Psychology Review*, *11*(2), 141–153.
Rizvi, S. J., Pizzagalli, D. A., Sproule, B. A., & Kennedy, S. H. (2016). Assessing anhedonia in depression: Potentials and pitfalls. *Neuroscience & Biobehavioral Reviews*, *65*, 21–35. https://doi.org/10.1016/j.neubiorev.2016.03.004.
Roberge, E. M., Allen, N. J., Taylor, J. W., & Bryan, C. J. (2016). Relationship functioning in Vietnam veteran couples: The roles of PTSD and anger. *Journal of Clinical Psychology*, *72*(9), 966–974. https://doi.org/10.1002/jclp.22301.
Robins, R. W., Noftle, E. E., & Tracy, J. L. (2007). Assessing self-conscious emotions: A review of self-report and nonverbal measures. In J. L. Tracy, R. W. Robins, & J. P. Tangney (Eds.), *The self-conscious emotions: Theory and research* (pp. 443–467). New York, NY: Guilford Press.
Rochefort, C., Baldwin, A. S., & Chmielewski, M. (2018). Experiential avoidance: An examination of the construct validity of the AAQ-II and MEAQ. *Behavior Therapy*, *49*(3), 435–449. https://doi.org/10.1016/j.beth.2017.08.008.
Roemer, L., Litz, B. T., Orsillo, S. M., & Wagner, A. W. (2001). A preliminary investigation of the role of strategic withholding of emotions in PTSD. *Journal of Traumatic Stress*, *14*(1), 149–156. https://doi.org/10.1023/A:1007895817502.
Rojas, S. M., Bujarski, S., Babson, K. A., Dutton, C. E., & Feldner, M. T. (2014). Understanding PTSD comorbidity and suicidal behavior: Associations among histories of alcohol dependence, major depressive disorder, and suicidal ideation and attempts. *Journal of Anxiety Disorders*, *28*(3), 318–325. https://doi.org/10.1016/j.janxdis.2014.02.004.
Rosen, C., Adler, E., & Tiet, Q. (2013). Presenting concerns of veterans entering treatment for posttraumatic stress disorder. *Journal of Traumatic Stress*, *26*(5), 640–643. https://doi.org/10.1002/jts.21841.
Roth, M., Neuner, F., & Elbert, T. (2014). Transgenerational consequences of PTSD: Risk factors for the mental health of children whose mothers have been exposed to the Rwandan genocide. *International Journal of Mental Health Systems*, *8*(1), 12. https://doi.org/10.1186/1752-4458-8-12.
Rybak, C. J., & Brown, B. M. (1996). Assessment of internalized shame: Validity and reliability of the Internalized Shame Scale. *Alcoholism Treatment Quarterly*, *14*(1), 71–83. https://doi.org/10.1300/J020V14N01_07.
Savic, D., Knezevic, G., Damjanovic, S., Spiric, Z., & Matic, G. (2012). The role of personality and traumatic events in cortisol levels—Where does PTSD fit in? *Psychoneuroendocrinology*, (7), 937. https://doi.org/10.1016/j.psyneuen.2011.11.001.
Schmitz, A., & Grillon, C. (2012). Assessing fear and anxiety in humans using the threat of predictable and unpredictable aversive events (the NPU-threat test). *Nature Protocols*, *7*(3), 527–532. https://doi.org/10.1038/nprot.2012.001.
Seligowski, A. V., Lee, D. J., Bardeen, J. R., & Orcutt, H. K. (2015). Emotion regulation and posttraumatic stress symptoms: A meta-analysis. *Cognitive Behaviour Therapy*, *44*, 87–102.
Shepherd, L., & Wild, J. (2014). Emotion regulation, physiological arousal and PTSD symptoms in trauma-exposed individuals. *Journal of Behavior Therapy and Experimental Psychiatry*, *45*(3), 360–367. https://doi.org/10.1016/j.jbtep.2014.03.002.
Sifneos, P. E. (1973). The prevalence of 'alexithymic' characteristics in psychosomatic patients. *Psychotherapy and Psychosomatics*, *22*(2–6), 255–262. https://doi.org/10.1159/000286529.
Simons, J. S., & Gaher, R. M. (2005). The Distress Tolerance Scale: Development and validation of a self-report measure. *Motivation and Emotion*, *29*(2), 83–102. https://doi.org/10.1007/s11031-005-7955-3.
Smits, J. A. J., Berry, A. C., Tart, C. D., & Powers, M. B. (2008). The efficacy of cognitive-behavioral interventions for reducing anxiety sensitivity: A meta analytic review. *Behaviour Research and Therapy*, *46*(9), 1047–1054. https://doi.org/10.1016/j.brat.2008.06.010.

Snaith, R. P., Hamilton, M., Morley, S., Humayan, A., Hargreaves, D., & Trigwell, P. (1995). A scale for the assessment of hedonic tone the Snaith–Hamilton Pleasure Scale. *British Journal of Psychiatry, 167*(1), 99–103. https://doi.org/10.1192/bjp.167.1.99.

Spielberger, C. D. (1983). *Manual of the State-Trait Anxiety Inventory*. Palo Alto, CA: Consulting Psychologists Press.

Spielberger, C. D. (1999). *STAXI-2: State-Trait Anger Expression Inventory-2: Professional manual*. Odessa, FL: Psychological Assessment Resources.

Spielberger, C. D., & Sydeman, S. J. (1994). State-Trait Anxiety Inventory and State-Trait Anger Expression Inventory. In M. E. Maruish (Ed.), *The use of psychological testing for treatment planning and outcome assessment* (pp. 292–321). Hillsdale, NJ: Lawrence Erlbaum Associates, Inc..

Stegmann, Y., Reicherts, P., Andreatta, M., Pauli, P., & Wieser, M. J. (2019). *The effect of trait anxiety on attentional mechanisms in combined context and cue conditioning and extinction learning*. https://doi.org/10.31234/osf.io/h4vn.

Tangney, J. P. (1996). Conceptual and methodological issues in the assessment of shame and guilt. *Behaviour Research and Therapy, 34*(9), 741–754. https://doi.org/10.1016/0005-7967(96)00034-4.

Tangney, J. P. (1999). The self-conscious emotions: Shame, guilt, embarrassment and pride. In T. Dalgleish & M. J. Power (Eds.), *Handbook of cognition and emotion*. (pp. 541–568). Chichester: Wiley. https://doi.org/10.1002/0470013494.ch26.

Tangney, J. P., Dearing, R. L., Wagner, P. E., & Gramzow, R. (2000). *The test of self-conscious affect-3 (TOSCA-3)*. Fairfax, VA: George Mason University.

Tangney, J. P., Wagner, P. E., Burggraf, S. A., Gramzow, R., & Fletcher, C. (1990). *The test of self-conscious affect for children (TOSCA-C)*. Fairfax, VA: George Mason University.

Tangney, J. P., Wagner, P. E., Gavlas, J., & Gramzow, R. (1991). *The test of self-conscious affect for adolescents (TOSCA-A)*. Fairfax, VA: George Mason University.

Tangney, J. P., Wagner, P. E., & Gramzow, R. (1989). *The test of self-conscious affect (TOSCA)*. Fairfax, VA: George Mason University.

Taylor, G. J. (1984). Alexithymia: Concept, measurement, and implications for treatment. *The American Journal of Psychiatry, 141*(6), 725–732. https://doi.org/10.1176/ajp.141.6.725.

Taylor, S., Zvolensky, M. J., Cox, B. J., Deacon, B., Heimberg, R. G., Ledley, D. R., … Cardenas, S. J. (2007). Robust dimensions of anxiety sensitivity: Development and initial validation of the Anxiety Sensitivity Index-3. *Psychological Assessment, 19*, 176–188. https://doi.org/10.1037/1040-3590.19.2.176.supp. [Supplemental].

Tellegen, A., Watson, D., & Clark, L. A. (1999). On the dimensional and hierarchical structure of affect. *Psychological Science, 10*, 297–303. https://doi.org/10.1111/1467-9280.00157.

Thabet, A. A., Abu Tawahina, A., El Sarraj, E., & Vostanis, P. (2008). The relationship between siege of Gaza Strip, anger, and psychological symptoms. *Arabpsynet E Journal, 20*, 174–184.

Thompson, E. R. (2007). Development and validation of an internationally reliable short-form of the positive and negative Affect Schedule (PANAS). *Journal of Cross-Cultural Psychology, 38*(2), 227–242. https://doi.org/10.1177/0022022106297301.

Tsorbatzoudis, H. (2006). Psychometric evaluation of the Greek version of the Aggression Questionnaire. *Perceptual and Motor Skills, 102*(3), 703–718. https://doi.org/10.2466/pms.102.3.703-718.

van Rooij, S. J. H., Kennis, M., Vink, M., & Geuze, E. (2016). Predicting treatment outcome in PTSD: A longitudinal functional MRI study on trauma-unrelated emotional processing. *Neuropsychopharmacology, 41*(4), 1156–1165. https://doi.org/10.1038/npp.2015.257.

van Rooij, S. J. H., Rademaker, A. R., Kennis, M., Vink, M., Kahn, R. S., & Geuze, E. (2015). Neural correlates of trauma-unrelated emotional processing in war veterans with PTSD. *Psychological Medicine, 45*(3), 575–587. https://doi.org/10.1017/S0033291714001706.

Vidal, M. E., & Petrak, J. (2007). Shame and adult sexual assault: A study with a group of female survivors recruited from an East London population. *Sexual and Relationship Therapy*, *22*(2), 159–171. https://doi.org/10.1080/14681990600784143.

Vujanovic, A. A., Bonn-Miller, M. O., Potter, C. M., Marshall, E. C., & Zvolensky, M. J. (2011). An evaluation of the relation between distress tolerance and posttraumatic stress within a trauma-exposed sample. *Journal of Psychopathology and Behavioral Assessment*, *33*(1), 129–135. https://doi.org/10.1007/s10862-010-9209-2.

Vujanovic, A. A., Hart, A. S., Potter, C. M., Berenz, E. C., Niles, B., & Bernstein, A. (2013). Main and interactive effects of distress tolerance and negative affect intensity in relation to PTSD symptoms among trauma-exposed adults. *Journal of Psychopathology and Behavioral Assessment*, *35*(2), 235–243. https://doi.org/10.1007/s10862-012-9325-2.

Vujanovic, A. A., Marshall, E. C., Gibson, L. E., & Zvolensky, M. J. (2010). Cognitive–affective characteristics of smokers with and without posttraumatic stress disorder and panic psychopathology. *Addictive Behaviors*, *35*(5), 419–425. https://doi.org/10.1016/j.addbeh.2009.12.005.

Vujanovic, A. A., Marshall-Berenz, E. C., & Zvolensky, M. J. (2011). Posttraumatic stress and alcohol use motives: A test of the incremental and mediating role of distress tolerance. *Journal of Cognitive Psychotherapy*, *25*(2), 130–141. https://doi.org/10.1891/0889-8391.25.2.130.

Watson, D. (2005). Rethinking the mood and anxiety disorders: A quantitative hierarchical model for DSM-V. *Journal of Abnormal Psychology*, *114*(4), 522–536. https://doi.org/10.1037/0021-843X.114.4.522.

Watson, D., & Clark, L. A. (1999). *The PANAS-X: Manual for the positive and negative affect schedule-expanded form* Retrieved from https://ir.uiowa.edu/cgi/viewcontent.cgi?article=1011&context=psychology_pubs.

Watson, D., Clark, L. A., & Tellegen, A. (1988). Development and validation of brief measures of positive and negative affect: The PANAS scales. *Journal of Personality and Social Psychology*, *54*(6), 1063–1070. https://doi.org/10.1037/0022-3514.54.6.1063.

Watson, S. D., Gomez, R., & Gullone, E. (2016). The shame and guilt scales of the Test of Self-Conscious Affect-Adolescent (TOSCA-A): Psychometric properties for responses from children, and measurement invariance across children and adolescents. *Frontiers in Psychology*, *7*, 635.

Weathers, F. W., Blake, D. D., Schnurr, P. P., Kaloupek, D. G., Marx, B. P., & Keane, T. M. (2013). *The Clinician-Administered PTSD Scale for DSM-5 (CAPS-5)* Interview available from the National Center for PTSD at www.ptsd.va.gov.

Weathers, F. W., Litz, B. T., Keane, T. M., Palmieri, P. A., Marx, B. P., & Schnurr, P. P. (2013). *The PTSD Checklist for DSM-5 (PCL-5)* Scale available from the National Center for PTSD at www.ptsd.va.gov.

Werner, K. H., Goldin, P. R., Ball, T. M., Heimberg, R. G., & Gross, J. J. (2011). Assessing emotion regulation in social anxiety disorder: The emotion regulation interview. *Journal of Psychopathology and Behavioral Assessment*, *33*(3), 346–354. https://doi.org/10.1007/s10862-011-9225-x.

Wheaton, M. G., Deacon, B. J., McGrath, P. B., Berman, N. C., & Abramowitz, J. S. (2012). Dimensions of anxiety sensitivity in the anxiety disorders: Evaluation of the ASI-3. *Journal of Anxiety Disorders*, *26*, 401–408.

Wilson, J. P., Droždek, B., & Turkovic, S. (2006). Posttraumatic shame and guilt. *Trauma, Violence & Abuse*, *7*(2), 122–141. https://doi.org/10.1177/1524838005285914.

Wolf, E. J., Miller, M. W., & McKinney, A. E. (2009). Emotional processing in PTSD: Heightened negative emotionality to unpleasant photographic stimuli. *The Journal of Nervous and Mental Disease*, *197*(6), 419–426. https://doi.org/10.1097/NMD.0b013e3181a61c68.

Wolfe, D. A., Sas, L., & Wekerle, C. (1994). Factors associated with the development of posttraumatic stress disorder among child victims of sexual abuse. *Child Abuse & Neglect*, *18*(1), 37–50. https://doi.org/10.1016/0145-2134(94)90094-9.

Wolgast, M. (2014). What does the acceptance and action questionnaire (AAQ-II) really measure? *Behavior Therapy*, *45*, 831–839.

Yuan, H., Phillips, R., Wong, C. K., Zotev, V., Misaki, M., Wurfel, B., … Bodurka, J. (2018). Tracking resting state connectivity dynamics in veterans with PTSD. *NeuroImage: Clinical*, *19*, 260–270. https://doi.org/10.1016/j.nicl.2018.04.014.

CHAPTER 2

Anxiety and fear in PTSD[☆]

Lori A. Zoellner[a], Heidi J. Ojalehto[b], Peter Rosencrans[a], Rosemary W. Walker[a], Natalia M. Garcia[a], Ifrah S. Sheikh[a], Michele A. Bedard-Gilligan[b]

[a]Psychology, University of Washington, Seattle, WA, United States
[b]Psychiatry and Behavioral Sciences, University of Washington, Seattle, WA, United States

Anxiety and fear

One of the best conceptualizations on the relationship among anxiety, fear, and trauma exposure is predatory imminence theory (Fanselow & Lester, 1988; Perusini & Fanselow, 2015), where anxiety, fear, and panic are on a continuum of a coordinated brain and bodily defensive response to perceived or actual proximity to threat. Anxiety can be viewed as a future-oriented state associated with preparation for a potential threatening event, whereas fear can be viewed as an alarm response to present or imminent danger (e.g., Craske et al., 2009). Notably the learning and neural mechanisms that underlie fear are relatively well conserved across species, with extensive animal and human research on fear acquisition, expression, generalization, avoidance, and extinction (e.g., Mahan & Ressler, 2012; Milad & Quirk, 2012).

Exposure to a traumatic event, involving actual or threatened death, serious injury, sexual violation, or harm to physical integrity, is considered the proximal cause for trauma- and stressor-related disorders. Classical or Pavlovian conditioning provides a strong associative model for basic learning processes that can lead to the persistence of pathological anxiety, fear, and trauma-related symptoms. In fear learning models the experience of perceived or actual danger via an unconditioned stimulus (US) leads previously neutral stimuli to signal danger (conditioned stimuli, CS+). For example, a previously neutral dog paired with an aversive experience (e.g., a painful bite at the park) can come to elicit conditioned anxiety and fear even in nondangerous situations. Fear generalization further explains the process by which the same individual comes to fear not only that specific dog but also related stimuli that may signal danger. For instance the individual may learn

---

[☆]This chapter was funded in part by R34DA040034 (Bedard-Gilligan, PI).

to fear all dogs, all four-legged animals, or all parks. This overgeneralization of conditioned fear and avoidance of potential future threats may prevent the individual from leaving their home, thereby resulting in functional impairment and diminished quality of life.

A criticism of a classical conditioning model of post-traumatic stress disorder (PTSD) is that, if the traumatic event is the etiologic agent, then event exposure should predictably result in disorder. Yet, even after the most severe traumatic events, only a minority of individuals develop persistent reactions. To account for the range of responding following trauma exposure, the impaired fear extinction hypothesis (e.g., Milad et al., 2008; Milad & Quirk, 2012) suggests that individuals who develop chronic PTSD have deficiencies in learning new inhibitory associations to trauma-related reminders (CSs). Impaired extinction describes a process by which repeated presentations of the CS in the absence of the US continues to evoke a conditioned reaction of anxiety, fear, or panic. Thus reminders continue to signal potential danger to the trauma survivor with PTSD, even when no danger, is present. This hypothesis is consistent with the observed pattern of reactions after trauma exposure, where following severe traumatic events, the vast majority of individuals have initial reactions that dissipate over days or weeks and only a minority have reactions that persist and result in symptoms of PTSD (e.g., Rosellini et al., 2018).

In this chapter, we will review basic neurobiological understandings of fear, particularly as evidenced in PTSD, and examine sensitivity to threat, fear overgeneralization, fear-related avoidance, and impaired extinction of fear. Although anxiety and fear are not the only emotions that occur following trauma exposure, they are common, innate reactions to severe trauma and result in a cascade of basic learned associations and defensive responding that underlie hallmark symptoms of chronic PTSD. An integrated associative learning model of PTSD can be seen in Fig. 1.

Neurobiology of anxiety and fear in PTSD

Hallmark symptoms of PTSD include the expression of fear or anxiety that occurs despite the absence of actual danger in the environment or occurs out of proportion to the actual danger posed by the environment. Although a range of emotions are associated with PTSD symptomology, associative learning as a result of fear is understood to be at the core of PTSD development, maintenance, and amelioration. For this reason, understanding how maladaptive fear-based responses occur on a neurobiological level is critical. The amygdala functions as the core neural structure in stressful

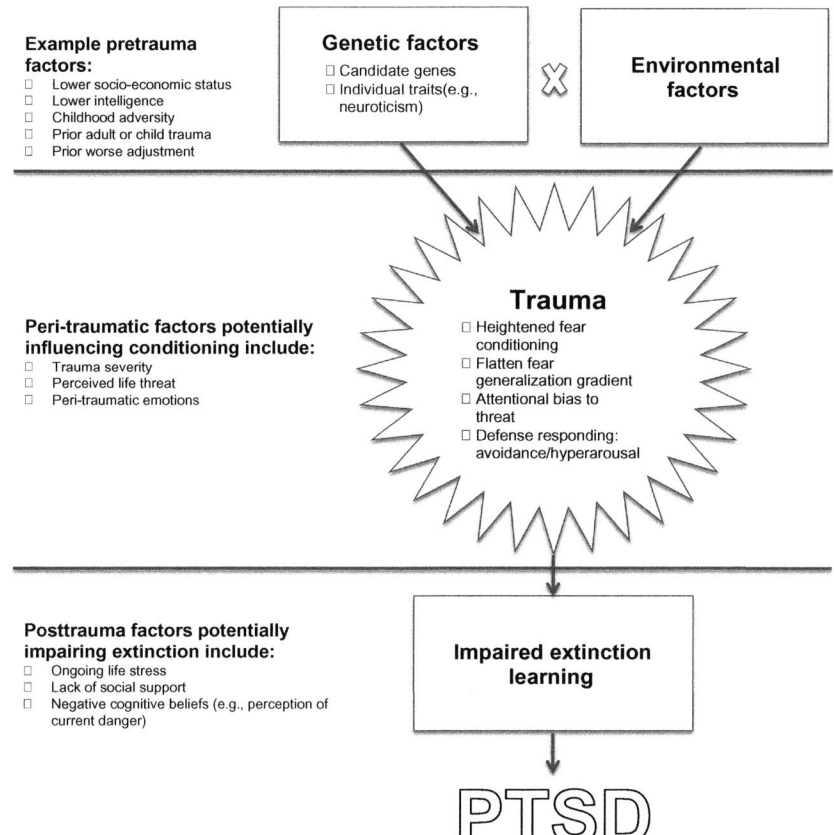

Fig. 1 Conditioning and extinction model of PTSD. A conditioning and extinction model of PTSD argues for genetic and environmental factors contributing to an increased likelihood of traumatic conditioning, and the effects of conditioning contribute to resistance to post-trauma extinction learning. *(Model adapted and used with permission from Zoellner, L. A., Graham, B., & Bedard-Gilligan, M. A. (2016). Trauma- and stressor-related disorders. In J. E. Maddux, B. A. Winstead, J. E. Maddux, B. A. Winstead (Eds.), Psychopathology: Foundations for a contemporary understanding (4th ed.) (pp. 162–181). New York, NY, USA: Routledge/Taylor & Francis Group).*

and traumatic situations. When faced with any type of potentially threatening stimuli, the thalamus initially processes it and directs projections to the amygdala lead to fear excitation promoting a quick, fast response to potential threat (e.g., Lissek et al., 2014). With this excitation a coordinated brain and behavioral fear response occurs, including activation of the anterior insula and the anterior cingulate cortex/supplementary motor area, which are associated with startle responses, initial defensive behaviors, and response from the sympathetic and parasympathetic nervous systems as well as the hypothalamic-pituitary-adrenal (HPA) axis (e.g., Yehuda, 2001). Projections from the thalamus also go to the visual cortex and then the hippocampus for schematic matching to detect threat or absence of threat. Following these projections an alternative, somewhat slower process, occurs that results in either fear inhibition via the ventromedial prefrontal cortex or fear excitation as per the preceding text (e.g., Lissek, Bradford, et al., 2014). Ultimately, this quick and coordinated response to threat has served as an adaptive tool for humans to survive; however, such responses become maladaptive when the brain excessively detects threat from harmless or ambiguous stimuli.

Brain regions associated with fear conditioning and extinction are consistently implicated in PTSD, namely, the sensory cortex, dorsal thalamus, and lateral and central nucleus of the amygdala, hippocampus, and structures in the medial prefrontal cortex (mPFC), including the anterior cingulate cortex. In animal and human Pavlovian fear conditioning studies, the amygdala has been found to regulate fear learning and receive projections from the prefrontal cortex and hippocampus (Mahan & Ressler, 2012). Based on LeDeoux's proposed fear system, the amygdala functions not only as a fear-based response hub but also as a "defensive survival circuit" that establishes a "defensive motivation state" in the brain to ensure survival (LeDoux, 2015). In comparison with trauma-exposed individuals without PTSD, individuals with PTSD consistently show decreased medial prefrontal cortex (mPFC) activation and increased amygdala activation (e.g., Patel, Spreng, Shin, & Girard, 2012). Often their relationship is conceptualized as one of a car "accelerator" and "brake" model of fear activation or expression (VanElzakker, Dahlgren, Davis, Dubois, & Shin, 2013). Essentially, the prelimbic cortex acts as a fear response accelerator during conditioning, and the ventral infralimbic cortex acts as a brake during extinction. Either an overly reactive accelerator and/or impaired braking can result in hyperresponsiveness to stimuli. Consistent with this model of reduced top-down control of the amygdala, there are gray matter reductions in the anterior cingulate cortex and ventromedial prefrontal cortex, the left temporal pole/middle temporal gyrus, and the left hippocampus in patients with PTSD compared with trauma-exposed individuals without PTSD (Kühn & Gallinat, 2013).

The hippocampus, associated with explicit memories of the traumatic event, learned responses to contextual cues, and pattern schema matching, has projections to both the ventral medial prefrontal cortex and amygdala. In particular, in PTSD, there is increased hippocampal activation to trauma and emotional stimuli (e.g., Gilbertson et al., 2002) and slightly lower hippocampal volume compared with trauma-exposed individuals without PTSD (Logue et al., 2018). Lower hippocampal volume also may be a vulnerability factor for developing PTSD (Gilbertson et al., 2002). Finally the HPA axis, including the hypothalamus, pituitary gland, and adrenal gland, is also involved with stress and the regulation of emotions. Yet, the literature on the role of the stress hormone cortisol in PTSD is equivocal at best, with a large meta-analysis showing no differences in basal cortisol between individuals who were exposed to trauma in adulthood and non-trauma-exposed controls (Klaassens, Giltay, Cuijpers, van Veen, & Zitman, 2012) and another meta-analysis failing to show cortisol measured early after trauma exposure predicting the development of PTSD (Morris, Hellman, Abelson, & Rao, 2016). However, some studies suggest that adult trauma exposure, but not PTSD per se, is associated with a stronger HPA-axis feedback response (Klaassens et al., 2012). Taken together, brain regions associated with fear excitation and inhibition processes in promoting behavioral response to perceived or imminent threat are consistently implicated in individuals with PTSD.

Heightened sensitivity to threat

Sensitivity to threat may be both a vulnerability factor and a consequence of trauma exposure or PTSD. Broadly, there is a small to moderate effect for higher acquired fear responses to conditioned safety cues compared with those without PTSD or other anxiety disorders (Duits et al., 2015), and, relative to trauma-exposed controls, individuals with PTSD show elevated physiological responding to personally relevant trauma scripts (Pole, 2007). Two factors that are implicated in this sensitivity to threat are anxiety sensitivity and attentional bias to threat.

Anxiety sensitivity

Anxiety sensitivity refers to a cognitive-affective individual difference factor of the fear of bodily sensations, fearing that these sensations have harmful consequences, sometimes labeled as "fear of fear" (Reiss, Peterson, Gursky, & McNally, 1986). Specifically, this cognitive-affective factor is considered conceptually distinct from anxiety, reflecting the fear of anxiety and anxious arousal symptoms. In a large meta-analysis of anxiety sensitivity, there was a moderate to large effect size between anxiety sensitivity and PTSD symptoms, comparable with those seen in panic disorder and generalized anxiety disorder (Naragon-Gainey, 2010). Less is known about the prospective role of anxiety sensitivity in the development of PTSD, though recent prospective research showed that in military personnel, predeployment anxiety sensitivity predicted postdeployment stress-related symptoms (Cobb, Lancaster, Meyer, Lee, & Telch, 2017). More broadly, anxiety sensitivity may be a transdiagnostic mechanism for both development and recovery of a variety of disorders, including PTSD (e.g., Gallagher, 2017). Related to this, anxiety sensitivity is often implicated as a factor underlying the high co-occurrence of PTSD and substance use disorders; yet, this relationship may be complex and instead reflect an indirect relationship in that anxiety sensitivity may moderate or mediate the association between PTSD and substance use (Vujanovic et al., 2018). Similarly, anxiety sensitivity is a potential mediator underlying the relationship between suicidality and PTSD (e.g., Boffa et al., 2018). Finally, preliminary evidence suggests that interventions targeting anxiety sensitivity modestly reduced post-traumatic stress symptoms (Short et al., 2017). Conceptually, whether a vulnerability factor, a consequence of trauma exposure, or a consequence of PTSD itself (e.g., Cobb et al., 2017; Thormar et al., 2010), anxiety sensitivity may make individuals more vulnerable to experiencing intrusive memories, exaggerated anxiety, or fear responses to cues and reminders of trauma exposure (e.g., Olatunji & Fan, 2015). Indeed, anxiety sensitivity may increase risk for intrusive memories via heightened physiological arousal in response to both the trauma and trauma reminders (Olatunji & Fan, 2015). In turn, this may motivate the use of avoidance strategies to reduce distress and prevent opportunities for new corrective learning to occur, preventing symptom reduction. Anxiety sensitivity may also interfere with treatment for PTSD, as it has been found to predict treatment dropout in individuals with co-occurring PTSD and substance use disorders (e.g., Belleau et al., 2017).

Attentional bias to threat

Individuals with PTSD consistently show attentional biases toward threat, preferentially allocating attentional resources toward threat-related cues relative to neutral cues (e.g., Constans, 2005; McNally, 1998). In the presence of threat, these biases are adaptive, enabling the detection of danger in the environment and facilitating a defensive response. However, in the absence of actual threat, these biases are maladaptive, as they interfere with the processing of corrective information and disrupt downstream cognitive processes, factors that may otherwise inhibit a fear response (e.g., Fani et al., 2012; White, Suway, Pine, Bar-Haim, & Fox, 2011).

Information processing biases have been assigned a prominent role in the etiology and maintenance of anxiety disorders, including PTSD

(Ehlers & Clark, 2000). Specifically, cognitive models of PTSD posit that information processing biases toward threat-related information may lead individuals to perceive a state of current threat, accompanied by increased arousal and fear, which, in turn, may maintain or trigger other PTSD symptoms such as intrusive memories and avoidance (e.g., Ehlers & Clark, 2000). Similarly, Foa, Steketee, and Rothbaum (1989) hypothesized that individuals with PTSD exhibit attentional biases toward trauma-related stimuli because this material is easily activated in fear structures formed after trauma. The more elaborated or overgeneralized these fear structures are, the more likely an individual is to perceive benign trauma-related cues as threatening and to have an exaggerated fear response to the cue.

Generally, consistent across the literature, a number of studies have found evidence for attentional bias toward threat, with individuals with PTSD exhibiting slower reaction times for trauma-related words compared with neutral words in the emotional Stroop task (e.g., Kaspi, McNally, & Amir, 1995) and faster reaction times to target probes replacing trauma-related stimuli compared with probes replacing neutral stimuli in the dot probe task (e.g., Fani et al., 2012). However, several studies have failed to replicate these findings (e.g., Elsesser, Sartory, & Tackenberg, 2005; Kimble, Frueh, & Marks, 2009). The vigilance-avoidance model postulates that anxious individuals initially orient attention toward threatening stimuli and then subsequently engage in avoidance of these cues to ameliorate anxiety and fear (Mogg, Mathews, & Weinman, 1987). In contrast the attention-maintenance model proposes a delayed disengagement from threat-related stimuli once attended to (e.g., Weierich, Treat, & Hollingworth, 2008). This sustained attention toward threat may increase distress and interrupt other cognitive processes that might inhibit or minimize a fear response.

Consistent with this vigilance-avoidance model, several studies have found evidence for facilitated attention to threat in PTSD, evidenced by a tendency to initially fixate on trauma or threat-related stimuli over neutral stimuli during eye tracking paradigms (e.g., Felmingham, Rennie, Manor, & Bryant, 2011; Thomas, Gonsalvez, & Johnstone, 2013). Further, one of these studies also found later increased avoidance of threat in individuals with PTSD, evidenced by decreased subsequent fixation on trauma-related images (Thomas et al., 2013). However, another eye tracking study found sustained attention toward threat stimuli consistent with the attention-maintenance model (Lee & Lee, 2012). Additional evidence for delayed disengagement in PTSD was also found using a dot probe task modified to capture temporal changes in attention (Bardeen & Orcutt, 2011) and using reaction time to identifying discrepant stimuli embedded in either trauma or neutral stimuli (e.g., Pineles, Shipherd, Mostoufi, Abramovitz, & Yovel, 2009). Although there is growing evidence for delayed disengagement as a robust component of attentional bias in PTSD, facilitated attention to threat remains a commonly observed phenomenon in individuals with PTSD.

Attentional bias, whether it is a facilitated attention toward threat or delayed disengagement from threat, has important clinical implications in PTSD. Attentional bias toward threat can interfere with the processing of other relevant information in the environment, preventing the individual from attending to neutral or safe signals and leading to exaggerated fear responding (Fani et al., 2012). Difficulty disengaging from trauma-related stimuli may maintain PTSD, as an inability to disengage from trauma reminders may contribute to increased and prolonged distress and fear (Foa, Feske, Murdock, Kozak, & McCarthy, 1991; Pineles et al., 2009). In turn, this may lead to exaggerated fear responding and promote the use of avoidance behaviors to decrease distress (Foa et al., 1991). Furthermore, attentional bias has been linked to interpretation bias (White et al., 2011), leading individuals with PTSD to interpret even innocuous stimuli in the environment as threatening.

Fear overgeneralization

Fear is a remarkably adaptive response to danger, allowing avoidance or escape of threatening situations. However, fear can also be maladaptive when it persists in the absence of danger. Similarly the capacity to broadly generalize fear learning is adaptive in that it helps organisms survive by learning to anticipate potentially dangerous situations and possibly avoid or escape them altogether (Mineka, 1992). Conversely, fear generalization that is overly broad may lead to excessive threat detection, even in the face of innocuous cues. Thus the construct of fear overgeneralization, defined as the spread of conditioned fear to stimuli that are not inherently dangerous but in some way resemble dangerous cues, is a key construct for understanding why individuals can struggle to feel safe in objectively nonthreatening situations (Dunsmoor, Mitroff, & LaBar, 2009).

Theoretical explanations for pathways leading to fear overgeneralization include poor discrimination between conditioned danger cues and safety cues (e.g., Lissek et al., 2005), poor transfer from safety cues to ambiguous stimuli (e.g., Jovanovic, Kazama, Bachevalier, & Davis, 2012), or deficits in using contextual information to modulate fear responding (e.g., Waters & Craske, 2016). Regardless of the pathway the spread of fear from a CS+ to nondangerous stimuli results in a proliferation of fear eliciting cues. This proliferation of cues causes individuals to react fearfully to nonthreatening stimuli and to struggle with identifying and utilizing safety signals in the environment (e.g., Jovanovic et al., 2010; Lissek et al., 2005).

Although the spread of fear has long been documented within the animal literature (e.g., McLaren & Mackintosh, 2002; Pavlov, 1927; Pearce, 1987), the study of fear generalization in humans has grown substantially in recent years (see Dunsmoor & Paz, 2015; Dymond, Dunsmoor, Vervliet, Roche, & Hermans, 2015). Lissek et al. (2008) has used a well-established fear generalization paradigm to demonstrate how fear spreads across a generalization gradient, wherein conditioned fear spreads to increasingly dissimilar stimuli. Using this paradigm, studies have demonstrated fear overgeneralization in clinical samples with anxiety- and stressor-based disorders, such as panic disorder (Lissek et al., 2010), generalized anxiety disorder (GAD; Lissek et al., 2014), and social anxiety disorder (Ahrens et al., 2016). Consistent with this, fear overgeneralization has also been demonstrated in individuals with high self-reported state and trait anxiety (Haddad, Pritchett, Lissek, & Lau, 2012) and trait neuroticism (Garcia & Zoellner, 2017; Lommen, Engelhard, & Van den Hout, 2010). Across the board, there is ample evidence for fear generalization as a critical conditioning abnormality across individuals with elevated anxiety and related traits and disorders.

A diverse array of experimental paradigms has established fear generalization as a robust phenomenon with strong effects observed across different stimuli and paradigms. For instance, fear generalization has been shown to spread both perceptually and conceptually, with studies showing generalization across categories (e.g., types of animals, Dunsmoor, Martin, & LaBar, 2012) and semantic meaning (e.g., "broth" to "soup"; Boyle, Roche, Dymond, & Hermans, 2016). Fear generalization has also been tested with a variety of stimuli such as faces (e.g., Dunsmoor et al., 2009; Garcia & Zoellner, 2017), shapes (e.g., Lissek et al., 2008), sounds (Norrholm et al., 2014), and colors (Dunsmoor & LaBar, 2013). Neuroimaging studies lend further convergence by showing neural activation patterns that mirror fear generalization gradients (Dunsmoor, White, & LaBar, 2011; Lissek, Bradford, et al., 2014) and evidence that fear generalization may engage similar brain networks that underlie fear learning and inhibition processes (Dymond et al., 2015). This breadth of research across simple paradigms and complex fear learning models highlights fear generalization as a critical construct for understanding the spread of fear.

Excessive avoidance

Avoidance is an evolutionarily adaptive, innate defensive response to realistic threat or danger (Bolles, 1970, 1971). Early theories proposed that avoidance was motivated by fear and maintained operantly, reinforced by fear reduction (Hull, 1943; Mowrer, 1951). However, the relationship between fear and avoidance is more complex than originally proposed. For one, avoidance responses have been found to continue even after fear extinction, which presumably reduces fear (Solomon, Kamin, & Wynne, 1953). The desynchrony between fear and avoidance has been explained via the role of informational factors, such as learned safety signals, outcome expectancy (Lovibond, 2006; Seligman & Johnston, 1973), and negative occasion setters (i.e., stimuli or contexts that disambiguate associative relationships; De Houwer, Crombez, & Baeyens, 2005), which are all proposed to inhibit a fear response. Thus although fear and avoidance often go hand in hand, avoidance sometimes occurs in the absence of fear.

Although avoidance serves to protect from danger, avoidance becomes maladaptive when it is excessive, provoked by relatively harmless stimuli, and interferes with daily functioning (Arnaudova, Kindt, Faneslow, & Beckers, 2017). Indeed, avoidance is a transdiagnostic feature of anxiety- and stressor-related disorders (Barlow, 2004). PTSD is characterized by efforts to avoid thoughts and feelings related to the trauma and to avoid situations, places, or people that serve as reminders of the trauma. Individuals with PTSD use behavioral and cognitive avoidance strategies, such as distraction and thought suppression, to avoid thinking about their trauma (e.g., Ehlers & Clark, 2000). Although behavioral avoidance patterns in individuals with PTSD tend to be idiosyncratic and specific to the trauma experienced, high rates of avoidance are reported for visual reminders of the trauma (e.g., images, pictures, and television content), sensory reminders of the trauma (e.g., tastes, smells, and sensations), crowded places, being at home in the dark, and certain social interactions such as smiling at strangers, intimate relationships, and making new friends (van Minnen & Hagenaars, 2010).

Behavioral and cognitive avoidance have been consistently associated with the development, severity, and persistence of PTSD. Indeed, prospective studies show that higher cognitive avoidance (Dunmore, Clark, & Ehlers, 2001; Ehlers, Mayou, & Bryant, 1998) and higher avoidance coping (Benotosch et al., 2000; Schuster, Park, & Frisman, 2011) predict more severe PTSD symptoms. The reverse relationship has also been found, with higher initial PTSD symptoms predicting more avoidance coping 10 months later (Tiet et al., 2006). Further, Schuster and colleagues showed that higher

PTSD symptoms were associated with more avoidant coping, which, in turn, mediated the relationship between previous trauma and subsequent symptom severity (Schuster et al., 2011).

In addition to self-report and interview measures of avoidance, as seen in other anxiety disorders, (e.g., Heuer, Rinck, & Becker, 2007), individuals with PTSD also have automatic avoidance tendencies. In one such study, compared with healthy controls, women with PTSD after sexual trauma were found to show more avoidance of high-threat sexual pictures using the approach-avoid task (AAT), which involves pushing (avoiding) or pulling (approaching) a joystick in response to images on the screen, with higher self-report arousal associated with more avoidance of trauma-related images (Fleurkens, Rinck, & Van Minnen, 2014). In a unique longitudinal study, using participants exposed to rocket attacks, longitudinal analyses showed that attentional bias away from threat (i.e., avoidance of the stimuli) predicted higher PTSD severity 1 year later (Wald et al., 2011).

Although avoidance has primarily been conceptualized as an innate defensive response, major consequences of avoidance are the loss of potential positive and rewarding experiences. Indeed, in recent approach-avoid conflict studies, in which conflict is elicited between approaching potential positive consequences and avoiding potential negative consequences, individuals with higher anxiety make greater disadvantageous decisions to avoid fear-related stimuli (e.g., Pittig, Schulz, Craske, & Alpers, 2014). Given the often costly effects of avoidance, there is growing consensus that approach motivation may also play an important role in avoidance. Indeed, newer theories of PTSD propose that PTSD is characterized by an imbalanced approach-avoid system, representing disruption to adaptive processing of fear and reward (Stein & Paulus, 2009). Consistent with this theory, reward deficits are found in PTSD (Nawijn et al., 2015), and anhedonia is estimated to be present in 63%–75% of patients with PTSD (Carmassi et al., 2014). Further, higher avoidance in individuals with PTSD has been associated with higher depression severity (van Minnen & Hagenaars, 2010). Given that incentives are the main reason people approach their fears, a hyposensitivity or lack of interest in rewards may be an important determining factor related to avoidance behavior in those with PTSD.

Impaired fear extinction

Extensions of conditioning models posit that the persistence of fear-related PTSD symptoms is linked to impairments in fear inhibition (e.g., Christianson et al., 2012; Milad et al., 2009). Fear inhibition can occur via two distinct processes, extinction learning and safety signal learning (Christianson et al., 2012). During extinction learning the CS comes to signal the absence of the aversive event, resulting in a new ambiguous meaning that outcompetes the original danger association and inhibits fear responses (Bouton, 2004). In contrast, safety signal learning entails associating stimuli other than the danger CS with the absence of an aversive event. Thus extinction learning involves learning competing signals (danger vs safety) for the same CS, whereas safety signal learning involves distinct CSs with competing signals (Christianson et al., 2012). Impairment in both inhibitory processes is thought to be implicated in the development and persistence of PTSD symptoms. An alternative to inhibitory models, the role of memory reconsolidation, reactivating and altering previously consolidated fear memories, has also been put forth as a means to erase rather than modify learned associations developed during conditioning (e.g., Nader, Schafe, & Le Doux, 2000), though the promise of extending this work to human clinical applications has not yet been realized (Drexler & Wolf, 2018). Regardless, learning how not to be afraid, that is, inhibit previously learned associations, may be a critical process facilitating natural and therapeutic recovery after trauma exposure.

Contemporary approaches to studying safety signal learning utilize conditional discrimination, or AX+/BX− paradigms, in which a physiological measure such as fear-potentiated startle or skin conductance reactivity is used as an index of fear response. Using this paradigm, Vietnam veterans with higher PTSD symptom severity showed an impaired ability to discriminate danger and safety cues and an inability to transfer inhibition compared with veterans with low severity and healthy controls, suggesting an inability to apply learned safety signals across contexts in those with PTSD (Jovanovic et al., 2009). These effects further appear localized to PTSD and not depression. Again, using this paradigm, participants with PTSD or co-occurring PTSD and major depressive disorder (MDD), but not participants with MDD alone and healthy controls, failed to show inhibition to the safety signal and less transfer of inhibition (Jovanovic et al., 2010), with elevated responding on inhibition transfer trials being associated with more severe hyperarousal symptoms in those with PTSD. Based on this and other convergent research, some have argued that impaired fear inhibition may be a biomarker of PTSD (Jovanovic et al., 2012).

Extinction learning paradigms have been used to demonstrate impaired fear inhibition in PTSD. These paradigms typically differentiate between extinction learning, that is, the acquisition of a fear inhibition response within a single training session, and extinction recall, that is, the transfer of extinction learning across training sessions or contexts. Emerging evidence suggests broader impairment in extinction recall rather than extinction learning in PTSD (e.g., Norrholm & Jovanovic, 2011). Yet, both phenomena are clinically relevant as impaired extinction learning may interfere with in the moment attempts to learn more adaptive responding to conditioned trauma cues (e.g., petting a dog in a therapist office), while impaired extinction recall may interfere with one's ability to transfer extinction learning to broader contexts (e.g., petting a dog at a local park). In studies of civilian- and combat-related PTSD, there has been consistent evidence for elevated startle responses to the CS+ during extinction trials and delayed, or protracted, extinction learning trials, with these elevated responses being associated with higher severity reexperiencing symptoms (Norrholm et al., 2006, 2011). Furthermore, elevated fear responses in individuals with PTSD at the end of extinction training persisted when extinction recall was tested 24 h later (Norrholm et al., 2006, 2011). Similarly, Milad et al. (2008) studied extinction learning and recall in trauma-exposed, PTSD-discordant twins and found no differences between PTSD and non-PTSD participants' learning of extinction responses but instead found that the presence of PTSD was associated with elevated fear responses during extinction recall 24 h later. These findings were replicated in a study that included PTSD patients and trauma-exposed controls with a wide range of trauma histories (Milad et al., 2009), arguing for impaired extinction recall in PTSD.

In summary, impaired fear inhibition in PTSD is related to both the ability to discriminate between safety and danger signals across contexts and the ability to learn when a stimulus that previously predicted danger no longer does so. Impaired fear inhibition models of PTSD and of exposure therapy posit that repeated exposure to the CSs without the occurrence of the traumatic event makes the meaning of the CS ambiguous and allows for contextual cues to disambiguate the meaning, even arguing that exposure interventions are cognitive interventions that specifically change the expectancy or appraisal of harm (Hofmann, 2008). Indeed, there is a strong evidence from animal and analogue samples that fear inhibition learning mediates the reduction of conditioned fear responses in paradigms that mirror exposure processes (e.g., Norrholm et al., 2011; Norrholm & Jovanovic, 2011) and strong evidence for the lasting effects of exposure-based therapy in PTSD (e.g., Kline, Cooper, Rytwinksi, & Feeny, 2017), and meaning changes differentially underlie symptom changes in exposure therapy compared with pharmacotherapy for PTSD (e.g., Cooper, Zoellner, Roy-Byrne, Mavissakalian, & Feeny, 2017). Some researchers (e.g., de Kleine, Smits, Hendriks, Becker, & van Minnen, 2015; Rothbaum et al., 2014; Zoellner et al., 2017) have used exposure augmentation studies involving cognitive enhancers such as d-cycloserine (DCS) and methylene blue to examine whether these enhancers facilitate extinction learning, which in turn may mediate greater symptom reductions, though evidence to date is limited in the clinical utility to enhance long-term outcomes.

Implications of anxiety and fear in PTSD

Associative learning processes at the time of trauma exposure and its aftermath often have a long-standing impact on the lives of trauma survivors. This initial learning is easily acquired so as to be an adaptive part of a defensive responding system aimed at avoiding future danger, alerting to potential danger, and quickly responding in the face of acute threat. However, the persistence of these learned reactions when danger is no longer present is not adaptive. This learning not only does produce overt PTSD symptoms such as reexperiencing of the trauma memory and cue reactivity to trauma reminders but also has downstream cascade effects of defensive responding, including hyperarousal symptoms and avoidance of trauma thoughts, feelings, and reminders.

Associative fear learning models also postulate changes in the meaning or appraisals of cues pointing to safety or danger, including one's own competency, the trustworthiness of others, and the overall dangerousness of the world. Sometimes the downstream effects of this defensive responding are more difficult to ascertain, as over years overt avoidance shifts to habits and trauma–reminder cues are no longer even encountered in daily life. The erosion of social support also can easily be understood within associative learning models, where some characteristics of individuals may be overt cues, people viewed as untrustworthy or dangerous, and limited social activities due to the fact that avoidance no longer provides opportunities for new corrective learning. Further, there is likely a bidirectional relationship between anxiety and depression, where depression and related appraisals of hopelessness and guilt often emerge after the presence of fear and anxiety and vice versa (Jacobson & Newman, 2017). Ultimately, many of the symptoms of PTSD can be understood within a framework of associative learning, where initial fear conditioning persists and there is a failure of extinction learning or extinction recall.

Finally, conceptualizing PTSD from an associative learning framework directly informs efforts to prevent the development of PTSD following trauma exposure and therapeutic principles for persistent trauma-related symptoms. This fear learning framework capitalizes on strong animal research to generate translational hypotheses to develop new treatment approaches and improve existing interventions. For clinicians, it highlights a targeted mechanism approach focusing on new, inhibitory learning after trauma exposure. Clinical targets should promote new, corrective learning about the meaning of the event and trauma-related cues, including more adaptive and balanced beliefs about one's self and the dangerousness of others or the world. These new associations can be learned through a wide variety of techniques such as psychoeducation, approaching nondangerous trauma reminders (e.g., returning to nondangerous pretrauma routines and behaviors), recounting the trauma memory (e.g., disclosing to others, writing, and revisiting the memory with a therapist), and carefully examining existing beliefs and generating alternative beliefs. Helping individuals process and learn new meaning associated with the trauma ultimately helps decrease avoidance, increases the ability to inhibit fear reactions to ambiguous or objectively safe cues, prevents further fear overgeneralization, and ultimately promotes remission of PTSD symptoms.

References

Ahrens, L. M., Pauli, P., Reif, A., Muhlberger, A., Langs, G., Aalderink, T., & Wieser, M. J. (2016). Fear conditioning and stimulus generalization in patients with social anxiety disorder. *Journal of Anxiety Disorders*, *44*, 36–46. https://doi.org/10.1016/j.janxdis.2016.10.003.

Arnaudova, I., Kindt, M., Faneslow, M., & Beckers, T. (2017). Pathways towards the proliferation of avoidance in anxiety and implications for treatment. *Behavior Research and Therapy*, *96*, 3–13. https://doi.org/10.1016/j.brat.2017.04.004.

Bardeen, J. R., & Orcutt, H. K. (2011). Attentional control as a moderator of the relationship between posttraumatic stress symptoms and attentional threat bias. *Journal of Anxiety Disorders*, *25*(8), 1008–1018. https://doi.org/10.1016/j.janxdis.2011.06.009.

Barlow, D. H. (2004). *Anxiety and its disorders: The nature and treatment of anxiety and panic*. New York: Guilford Press.

Belleau, E. L., Chin, E. G., Wanklyn, S. G., Zambrano-Vazquez, L., Schumacher, J. A., & Coffey, S. F. (2017). Pre-treatment predictors of dropout from prolonged exposure therapy in patients with chronic posttraumatic stress disorder and comorbid substance use disorders. *Behaviour Research and Therapy*, *91*, 43–50. https://doi.org/10.1016/j.brat.2017.01.011.

Benotosch, E. G., Brailey, K., Vasterling, J. J., Uddo, M., Constans, J. I., & Sutker, P. B. (2000). War zone stress, personal, and environmental resources, and PTSD symptoms in Gulf War Veterans: A longitudinal perspective. *Journal of Abnormal Psychology*, *109*(2), 205–213. https://doi.org/10.1037//0021-843X.109.2.205.

Boffa, J. W., Stanley, I. H., Smith, L. J., Mathes, B. M., Tran, J. K., Buser, S. J., … Vujanovic, A. A. (2018). Posttraumatic stress disorder symptoms and suicide risk in male firefighters: The mediating role of anxiety sensitivity. *Journal of Nervous and Mental Disease*, *206*(3), 179–186. https://doi.org/10.1097/NMD.0000000000000779.

Bolles, R. C. (1970). Species-specific defense reactions and avoidance learning. *Psychological Review*, *77*, 32–48. https://doi.org/10.1037/h0028589.

Bolles, R. C. (1971). Species-specific defense reactions. In F. R. Brush (Ed.), *Aversive conditioning and learning* (pp. 183–233). New York, NY: Academic Press.

Bouton, M. E. (2004). Context and behavioral processes in extinction. *Learning & Memory, 11*(5), 485–494. https://doi.org/10.1101/lm.78804.

Boyle, S., Roche, B., Dymond, S., & Hermans, D. (2016). Generalisation of fear and avoidance along a semantic continuum. *Cognition and Emotion, 30*(2), 340–352. https://doi.org/10.1080/02699931.2014.1000831.

Carmassi, C., Akiskal, H. S., Bessonov, D., Massimetti, G., Calderani, E., Stratta, P., … Dell, L. (2014). Gender differences in DSM-5 versus DSM-IV-TR PTSD prevalence and criteria comparison among 512 survivors to the L'Aquila earthquake. *Journal of Affective Disorders, 160*, 55–61. https://doi.org/10.1016/j.jad.2014.02.028.

Christianson, J. P., Fernando, A. B. P., Kazama, A. M., Jovanovic, T., Ostroff, L. E., & Sangha, S. (2012). Inhibition of fear by learned safety signals: A mini-symposium review. *Journal of Neuroscience, 32*(41), 14118–14124. https://doi.org/10.1523/JNEUROSCI.3340-12.2012.

Cobb, A. R., Lancaster, C. L., Meyer, E. C., Lee, H.-J., & Telch, M. J. (2017). Pre-deployment trait anxiety, anxiety sensitivity and experiential avoidance predict war-zone stress-evoked psychopathology. *Journal of Contextual Behavioral Science, 6*(3), 276–287. https://doi.org/10.1016/j.jcbs.2017.05.002.

Constans, J. I. (2005). Information-processing biases in PTSD. In J. J. Vasterling & C. R. Brewin (Eds.), *Neuropsychology of PTSD: Biological, cognitive, and clinical perspectives* (pp. 105–130). New York: Guilford Press.

Cooper, A. A., Zoellner, L. A., Roy-Byrne, P., Mavissakalian, M. R., & Feeny, N. C. (2017). Do changes in trauma-related beliefs predict PTSD symptom improvement in prolonged exposure and sertraline? *Journal of Consulting and Clinical Psychology, 85*(9), 873. https://doi.org/10.1037/ccp0000220.

Craske, M. G., Rauch, S. L., Ursano, R., Prenoveau, J., Pine, D. S., & Zinbarg, R. E. (2009). What is an anxiety disorder? *Depression and Anxiety, 26*(12), 1066–1085. https://doi.org/10.1002/da.20633.

De Houwer, J., Crombez, G., & Baeyens, F. (2005). Avoidance behavior can function as a negative occasion setter. *Journal of Experimental Psychology: Animal Behavior Processes, 31*(1), 101–106. https://doi.org/10.1037/0097-7403.31.1.101.

de Kleine, R. A., Smits, J. A. J., Hendriks, G.-J., Becker, E. S., & van Minnen, A. (2015). Extinction learning as a moderator of D-cycloserine efficacy for enhancing exposure therapy in posttraumatic stress disorder. *Journal of Anxiety Disorders, 34*, 63–67. https://doi.org/10.1016/j.janxdis.2015.06.005.

Drexler, M. S., & Wolf, O. T. (2018). Behavioral disruption of memory reconsolidation: From bench to bedside and back again. *Behavioral Neuroscience, 132*(1), 13.

Duits, P., Cath, D. C., Lissek, S., Hox, J. J., Hamm, A. O., Engelhard, I. M., … Baas, J. M. (2015). Updated meta-analysis of classical fear conditioning in the anxiety disorders. *Depression and Anxiety, 32*(4), 239–253. https://doi.org/10.1002/da.22353.

Dunmore, E., Clark, D. M., & Ehlers, A. (2001). A prospective investigation of the role of cognitive factors in persistent posttraumatic stress disorder (PTSD) after physical or sexual assault. *Behaviour Research and Therapy, 39*(9), 1063–1084. https://doi.org/10.1016/S0005-7967(00)00088-7.

Dunsmoor, J. E., & LaBar, K. S. (2013). Effects of discrimination training on fear generalization gradients and perceptual classification in humans. *Behavioral Neuroscience, 127*(3), 350. https://doi.org/10.1037/a0031933.

Dunsmoor, J. E., Martin, A., & LaBar, K. S. (2012). Role of conceptual knowledge in learning and retention of conditioned fear. *Biological Psychology, 89*(2), 300–305. https://doi.org/10.1016/j.biopsycho.2011.11.002.

Dunsmoor, J. E., Mitroff, S. R., & LaBar, K. S. (2009). Generalization of conditioned fear along a dimension of increasing fear intensity. *Learning & Memory, 16*(7), 460–469. https://doi.org/10.1101/lm.1431609.

Dunsmoor, J. E., & Paz, R. (2015). Fear generalization and anxiety: Behavioral and neural mechanisms. *Biological Psychiatry*, *78*(5), 336–343. https://doi.org/10.1016/j.biopsych.2015.04.010.

Dunsmoor, J. E., White, A. J., & LaBar, K. S. (2011). Conceptual similarity promotes generalization of higher order fear learning. *Learning & Memory*, *18*(3), 156–160. https://doi.org/10.1101/lm.2016411.

Dymond, S., Dunsmoor, J. E., Vervliet, B., Roche, B., & Hermans, D. (2015). Fear generalization in humans: Systematic review and implications for anxiety disorder research. *Behavior Therapy*, *46*(5), 561–582. https://doi.org/10.1016/j.beth.2014.10.001.

Ehlers, A., & Clark, D. M. (2000). A cognitive model of posttraumatic stress disorder. *Behaviour Research and Therapy*, *38*(4), 319–345. https://doi.org/10.1016/S0005-7967(99)00123-0.

Ehlers, A., Mayou, R. A., & Bryant, B. (1998). Psychological predictors of chronic posttraumatic stress disorder after motor vehicle accidents. *Journal of Abnormal Psychology*, *107*(3), 508. https://doi.org/10.1037/0021-843X.107.3.508.

Elsesser, K., Sartory, G., & Tackenberg, A. (2005). Initial symptoms and reactions to trauma-related stimuli and the development of posttraumatic stress disorder. *Depression and Anxiety*, *21*(2), 61–70. https://doi.org/10.1002/da.20047.

Fani, N., Tone, E. B., Phifer, J., Norrholm, S. D., Bradley, B., Ressler, K. J., … Jovanovic, T. (2012). Attention bias toward threat is associated with exaggerated fear expression and impaired extinction in PTSD. *Psychological Medicine*, *42*(3), 533–543. https://doi.org/10.1017/S0033291711001565.

Fanselow, M. S., & Lester, L. S. (1988). A functional behavioristic approach to aversively motivated behavior: Predatory imminence as a determinant of the topography of defensive behavior. In R. C. Bolles & M. D. Beecher (Eds.), *Evolution and learning* (pp. 185–212). Hillsdale, NJ, USA: Lawrence Erlbaum Associates, Inc.

Felmingham, K. L., Rennie, C., Manor, B., & Bryant, R. A. (2011). Eye tracking and physiological reactivity to threatening stimuli in posttraumatic stress disorder. *Journal of Anxiety Disorders*, *25*(5), 668–673. https://doi.org/10.1016/j.janxdis.2011.02.010.

Fleurkens, P., Rinck, M., & Van Minnen, A. (2014). Implicit and explicit avoidance in sexual trauma victims suffering from posttraumatic stress disorder: A pilot study. *European Journal of Psychotraumatology*, *5*(1), 21359. https://doi.org/10.3402/ejpt.v5.21359.

Foa, E. B., Feske, U., Murdock, T. B., Kozak, M. J., & McCarthy, P. R. (1991). Processing of threat-related information in rape victims. *Journal of Abnormal Psychology*, *100*(2), 156.

Foa, E. B., Steketee, G., & Rothbaum, B. O. (1989). Behavioral/cognitive conceptualizations of post-traumatic stress disorder. *Behavior Therapy*, *20*(2), 155–176. https://doi.org/10.1016/S0005-7894(89)80067-X.

Gallagher, M. W. (2017). Transdiagnostic mechanisms of change and cognitive-behavioral treatments for PTSD. *Current Opinion in Psychology*, *14*, 90–95. https://doi.org/10.1016/j.copsyc.2016.12.002.

Garcia, N. M., & Zoellner, L. A. (2017). Ignorance is bliss: Increasing predictability increases fear generalization in individuals with higher neuroticism. *Cognition and Emotion*, *10*, 1–16.

Gilbertson, M. W., Shenton, M. E., Ciszewski, A., Kasai, K., Lasko, N. B., Orr, S. P., et al. (2002). Smaller hippocampal volume predicts pathologic vulnerability to psychological trauma. *Nature Neuroscience*, *5*, 1242–1247. https://doi.org/10.1038/nn958.

Haddad, A. D. M., Pritchett, D., Lissek, S., & Lau, J. Y. F. (2012). Trait anxiety and fear responses to safety cues: Stimulus generalization or sensitization? *Journal of Psychopathology and Behavioral Assessment*, *34*(3), 232–331. https://doi.org/10.1007/s10862-012-9284-7.

Heuer, K., Rinck, M., & Becker, E. S. (2007). Avoidance of emotional facial expressions in social anxiety: The approach–avoidance task. *Behaviour Research and Therapy*, *45*(12), 2990–3001. https://doi.org/10.1016/j.brat.2007.08.010.

Hofmann, S. G. (2008). Cognitive processes during fear acquisition and extinction in animals and humans: Implications for exposure therapy of anxiety disorders. *Clinical Psychology Review*, *28*(2), 199–210. https://doi.org/10.1016/j.cpr.2007.04.009.

Hull, C. L. (1943). *Principles of behavior*. New York, NY: Appleton-Century-Crofts.
Jacobson, N. C., & Newman, M. G. (2017). Anxiety and depression as bidirectional risk factors for one another: A meta-analysis of longitudinal studies. *Psychological Bulletin*, *143*(11), 1155.
Jovanovic, T., Kazama, A., Bachevalier, J., & Davis, M. (2012). Impaired safety signal learning may be a biomarker of PTSD. *Neuropharmacology*, *62*(2), 695–704. https://doi.org/10.1016/j.neuropharm.2011.02.023.
Jovanovic, T., Norrholm, S. D., Blanding, N. Q., Davis, M., Duncan, E., Bradley, B., & Ressler, K. J. (2010). Impaired fear inhibition is a biomarker of PTSD but not depression. *Depression and Anxiety*, *27*(3), 244–251. https://doi.org/10.1002/da.20663.
Jovanovic, T., Norrholm, S. D., Fennell, J. E., Keyes, M., Fiallos, A. M., Myers, K. M., … Duncan, E. J. (2009). Posttraumatic stress disorder may be associated with impaired fear inhibition: Relation to symptom severity. *Psychiatry Research*, *167*(1–2), 151–160. https://doi.org/10.1016/j.psychres.2007.12.014.
Kaspi, S. P., McNally, R. J., & Amir, N. (1995). Cognitive processing of emotional information in posttraumatic stress disorder. *Cognitive Therapy and Research*, *19*(4), 433–444. https://doi.org/10.1007/BF02230410.
Kimble, M. O., Frueh, B. C., & Marks, L. (2009). Does the modified Stroop effect exist in PTSD? Evidence from dissertation abstracts and the peer reviewed literature. *Journal of Anxiety Disorders*, *23*(5), 650–655.
Klaassens, E. R., Giltay, E. J., Cuijpers, P., van Veen, T., & Zitman, F. G. (2012). Adulthood trauma and HPA-axis functioning in healthy subjects and PTSD patients: A meta-analysis. *Psychoneuroendocrinology*, *37*(3), 317–331. https://doi.org/10.1016/j.psyneuen.2011.07.003.
Kline, A. C., Cooper, A. A., Rytwinksi, N. K., & Feeny, N. C. (2017). Long-term efficacy of psychotherapy for posttraumatic stress disorder: A meta-analysis of randomized controlled trials. *Clinical Psychology Review*.
Kühn, S., & Gallinat, J. (2013). Gray matter correlates of posttraumatic stress disorder: A quantitative meta-analysis. *Biological Psychiatry*, *73*, 70–74. https://doi.org/10.1016/j.biopsych.2012.06.029.
LeDoux, J. E. (2015). *Anxious: Using the brain to understand and treat fear and anxiety*. Penguin.
Lee, J. H., & Lee, J. H. (2012). Attentional bias to violent images in survivors of dating violence. *Cognition & Emotion*, *26*(6), 1124–1133. https://doi.org/10.1080/02699931.2011.638906.
Lissek, S., Biggs, A. L., Rabin, S. J., Cornwell, B. R., Alvarez, R. P., Pine, D. S., & Grillon, C. (2008). Generalization of conditioned fear-potentiated startle in humans: Experimental validation and clinical relevance. *Behaviour Research and Therapy*, *46*(5), 678–687. https://doi.org/10.1016/j.brat.2008.02.005.
Lissek, S., Bradford, D. E., Alvarez, R. P., Burton, P., Espensen-Sturges, T., Reynolds, R. C., & Grillon, C. (2014). Neural substrates of classically conditioned fear-generalization in humans: A parametric fMRI study. *Social Cognitive and Affective Neuroscience*, *9*(8), 1134–1142. https://doi.org/10.1093/scan/nst096.
Lissek, S., Kaczkurkin, A. N., Rabin, S., Geraci, M., Pine, D. D., & Grillon, C. (2014). Generalized anxiety disorder is associated with overgeneralization of classically conditioned fear. *Biological Psychiatry*, *75*(11), 909–915. https://doi.org/10.1016/j.biopsych.2013.07.025.
Lissek, S., Powers, A. S., McClure, E. B., Phelps, E. A., Woldehawariat, G., Grillon, C., & Pine, D. S. (2005). Classical fear conditioning in the anxiety disorders: A meta-analysis. *Behaviour Research and Therapy*, *43*(11), 1391–1424. https://doi.org/10.1016/j.brat.2004.10.007.
Lissek, S., Rabin, S., Heller, R. E., Lukenbaugh, M. A., Geraci, M., Pine, D. S., & Grillon, C. (2010). Overgeneralization of conditioned fear as a pathogenic marker of panic disorder. *American Journal of Psychiatry*, *167*(1), 47–55. https://doi.org/10.1176/appi.ajp.2009.09030410.

Logue, M. W., van Rooij, S. H., Dennis, E. L., Davis, S. L., Hayes, J. P., Stevens, J. S., ... Morey, R. A. (2018). Smaller hippocampal volume in posttraumatic stress disorder: A multisite ENIGMA-PGC study: Subcortical volumetry results from posttraumatic stress disorder consortia. *Biological Psychiatry, 83*(3), 244–253. https://doi.org/10.1016/j.biopsych.2017.09.006.

Lommen, M. J. J., Engelhard, I. M., & Van den Hout, M. A. (2010). Neuroticism and threat avoidance: Better safe than sorry? *Personality and Individual Differences, 49*(8), 1001–1006. https://doi.org/10.1016/j.paid.2010.08.012.

Lovibond, P. (2006). Fear and avoidance: An integrated expectancy model. In M. G. Craske, D. Hermans, & D. Vansteenwegen (Eds.), *Fear and learning: From basic processes to clinical implications*, (pp. 117–132). Washington, DC, US: American Psychological Association. https://doi.org/10.1037/11474-006.

Mahan, A. L., & Ressler, K. J. (2012). Fear conditioning, synaptic plasticity and the amygdala: Implications for posttraumatic stress disorder. *Trends in Neurosciences, 35*(1), 24–35. https://doi.org/10.1016/j.tins.2011.06.007.

McLaren, I. P. L., & Mackintosh, N. J. (2002). Associative learning and elemental representation: II. Generalization and discrimination. *Animal Learning & Behavior, 30*, 177–200. https://doi.org/10.3758/BF03192828.

McNally, R. J. (1998). Experimental approaches to cognitive abnormality in posttraumatic stress disorder. *Clinical Psychology Review, 18*(8), 971–982. https://doi.org/10.1016/S0272-7358(98)00036-1.

Milad, M. R., Orr, S. P., Lasko, N. B., Chang, Y., Rauch, S. L., & Pitman, R. K. (2008). Presence and acquired origin of reduced recall for fear extinction in PTSD: Results of a twin study. *Journal of Psychiatric Research, 42*(7), 515–520. https://doi.org/10.1016/j.jpsychires.2008.01.017.

Milad, M. R., Pitman, R. K., Ellis, C. B., Gold, A. L., Shin, L. M., Lasko, N. B., ... Rauch, S. L. (2009). Neurobiological basis of failure to recall extinction memory in posttraumatic stress disorder. *Biological Psychiatry, 66*(12), 1075–1082. https://doi.org/10.1016/j.biopsych.2009.06.026.

Milad, M. R., & Quirk, G. J. (2012). Fear extinction as a model for translational neuroscience: Ten years of progress. *Annual Review of Psychology, 63*, 129–151. https://doi.org/10.1146/annurev.psych.121208.131631.

Mineka, S. (1992). Evolutionary memories, emotional processing and the emotional disorders. *The Psychology of Learning and Motivation, 28*, 161–206. https://doi.org/10.1016/S0079-7421(08)60490-9.

Mogg, K., Mathews, A., & Weinman, J. (1987). Memory bias in clinical anxiety. *Journal of Abnormal Psychology, 96*(2), 94. https://doi.org/10.1037/0021-843X.96.2.94.

Morris, M. C., Hellman, N., Abelson, J. L., & Rao, U. (2016). Cortisol, heart rate, and blood pressure as early markers of PTSD risk: A systematic review and meta-analysis. *Clinical Psychology Review, 49*, 79–91. https://doi.org/10.1016/j.cpr.2016.09.001.

Mowrer, O. H. (1951). Two-factor learning theory: Summary and comment. *Psychological Review, 58*(5), 350. https://doi.org/10.1037/h0058956.

Nader, K., Schafe, G. E., & Le Doux, J. E. (2000). Fear memories require protein synthesis in the amygdala for reconsolidation after retrieval. *Nature, 406*(6797), 722.

Naragon-Gainey, K. (2010). Meta-analysis of the relations of anxiety sensitivity to the depressive and anxiety disorders. *Psychological Bulletin, 136*(1), 128.

Nawijn, L., van Zuiden, M., Frijling, J. L., Koch, S. B. J., Veltman, D. J., & Olff, M. (2015). Reward functioning in PTSD: A systematic review exploring the mechanisms underlying anhedonia. *Neuroscience & Biobehavioral Reviews, 51*, 189–204. https://doi.org/10.1016/j.neubiorev.2015.01.019.

Norrholm, S. D., & Jovanovic, T. (2011). Translational fear inhibition models as indices of trauma-related psychopathology. *Current Psychiatry Reviews, 7*(3), 194–204. https://doi.org/10.2174/157340011797183193.

Norrholm, S. D., Jovanovic, T., Briscione, M. A., Anderson, K. M., Kwon, C. K., Warren, V. T., ... Bradley, B. (2014). Generalization of fear-potentiated startle in the presence of auditory cues: A parametric analysis. *Frontiers in Behavioral Neuroscience, 8*, 361.

Norrholm, S. D., Jovanovic, T., Olin, I. W., Sands, L. A., Karapanou, I., Bradley, B., & Ressler, K. J. (2011). Fear extinction in traumatized civilians with posttraumatic stress disorder: Relation to symptom severity. *Biological Psychiatry, 69*(6), 556–563. https://doi.org/10.1016/j.biopsych.2010.09.013.

Norrholm, S. D., Jovanovic, T., Vervliet, B., Myers, K. M., Davis, M., Rothbaum, B. O., & Duncan, E. J. (2006). Conditioned fear extinction and reinstatement in a human fear-potentiated startle paradigm. *Learning & Memory, 13*(6), 681–685. https://doi.org/10.1101/lm.393906.

Olatunji, B. O., & Fan, Q. (2015). Anxiety sensitivity and post-traumatic stress reactions: Evidence for intrusions and physiological arousal as mediating and moderating mechanisms. *Journal of Anxiety Disorders, 34*, 76–85. https://doi.org/10.1016/j.janxdis.2015.06.002.

Patel, R., Spreng, R. N., Shin, L. M., & Girard, T. A. (2012). Neurocircuitry models of posttraumatic stress disorder and beyond: A meta-analysis of functional neuroimaging studies. *Neuroscience and Biobehavioral Reviews, 36*(9), 2130–2142. https://doi.org/10.1016/j.neubiorev.2012.06.003.

Pavlov, I. P. (1927). *Conditioned reflexes: An investigation of the physiological activity of the cerebral cortex*. London: Oxford University Press.

Pearce, J. M. (1987). A model for stimulus generalization in Pavlovian conditioning. *Psychological Review, 94*, 61–73. https://doi.org/10.1037/0033-295X.94.1.61.

Perusini, J. N., & Fanselow, M. S. (2015). Neurobehavioral perspectives on the distinction between fear and anxiety. *Learning & Memory, 22*(9), 417–425. https://doi.org/10.1101/lm.039180.115.

Pineles, S. L., Shipherd, J. C., Mostoufi, S. M., Abramovitz, S. M., & Yovel, I. (2009). Attentional biases in PTSD: More evidence for interference. *Behaviour Research and Therapy, 47*(12), 1050–1057. https://doi.org/10.1016/j.brat.2009.08.001.

Pittig, A., Schulz, A. R., Craske, M. G., & Alpers, G. W. (2014). Acquisition of behavioral avoidance: Task-irrelevant conditioned stimuli trigger costly decisions. *Journal of Abnormal Psychology, 123*(2), 314–329. https://doi.org/10.1037/a0036136.

Pole, N. (2007). The psychophysiology of posttraumatic stress disorder: A meta-analysis. *Psychological Bulletin, 133*, 725–746. https://doi.org/10.1037/0033-2909.133.5.725.

Reiss, S., Peterson, R. A., Gursky, D. M., & McNally, R. J. (1986). Anxiety sensitivity, anxiety frequency and the prediction of fearfulness. *Behaviour Research and Therapy, 24*(1), 1–8.

Rosellini, A. J., Liu, H., Petukhova, M. V., Sampson, N. A., Aguilar-Gaxiola, S., Alonso, J., ... Kessler, R. C. (2018). Recovery from DSM-IV post-traumatic stress disorder in the WHO World Mental Health surveys. *Psychological Medicine, 48*(3), 437–450. https://doi.org/10.1017/S0033291717001817.

Rothbaum, B. O., Price, M., Jovanovic, T., Norrholm, S. D., Gerardi, M., Dunlop, B., ... Ressler, K. J. (2014). A randomized, double-blind evaluation of D-cycloserine or alprazolam combined with virtual reality exposure therapy for posttraumatic stress disorder in Iraq and Afghanistan war veterans. *American Journal of Psychiatry, 171*(6), 640–648. https://doi.org/10.1176/appi.ajp.2014.13121625.

Schuster, J., Park, C. L., & Frisman, L. K. (2011). Trauma exposure and PTSD symptoms among homeless mothers: Predicting coping and mental health outcomes. *Journal of Social and Clinical Psychology, 30*(8), 887–904. https://doi.org/10.1521/jscp.2011.30.8.887.

Seligman, M. E., & Johnston, J. C. (1973). A cognitive theory of avoidance learning. In F. J. McGuigan & D. B. Lumsden (Eds.), *Contemporary approaches to conditioning and learning*. V. H. Winston & Sons: Oxford, England.

Short, N. A., Boffa, J. W., Norr, A. M., Albanese, B. J., Allan, N. P., & Schmidt, N. B. (2017). Randomized clinical trial investigating the effects of an anxiety sensitivity intervention on posttraumatic stress symptoms: A replication and extension. *Journal of Traumatic Stress*, *30*(3), 296–303. https://doi.org/10.1002/jts.22194.

Solomon, R. L., Kamin, L. J., & Wynne, L. C. (1953). Traumatic avoidance learning: The outcomes of several extinction procedures with dogs. *The Journal of Abnormal and Social Psychology*, *48*(2), 291. https://doi.org/10.1037/h0058943.

Stein, M. B., & Paulus, M. P. (2009). Imbalance of approach and avoidance: The yin and yang of anxiety disorders. *Biological Psychiatry*, *66*(12), 1072–1074. https://doi.org/10.1016/j.biopsych.2009.09.023.

Thomas, S. J., Gonsalvez, C. J., & Johnstone, S. J. (2013). Neural time course of threat-related attentional bias and interference in panic and obsessive–compulsive disorders. *Biological Psychology*, *94*(1), 116–129. https://doi.org/10.1016/j.biopsycho.2013.05.012.

Thormar, S. B., Gersons, B. P. R., Juen, B., Marschang, A., Djakababa, M. N., & Olff, M. (2010). The mental health impact of volunteering in a disaster setting: A review. *Journal of Nervous and Mental Disease*, *198*(8), 529–538. https://doi.org/10.1097/NMD.0b013e3181ea1fa9.

Tiet, Q. Q., Rosen, C., Cavella, S., Moos, R. H., Finney, J. W., & Yesavage, J. (2006). Coping, symptoms, and functioning outcomes of patients with posttraumatic stress disorder. *Journal of Traumatic Stress*, *19*(6), 799–811. https://doi.org/10.1002/jts.20185.

van Minnen, A., & Hagenaars, M. A. (2010). Avoidance behaviour of patients with posttraumatic stress disorder. Initial development of a questionnaire, psychometric properties and treatment sensitivity. *Journal of Behavior Therapy and Experimental Psychiatry*, *41*(3), 191–198. https://doi.org/10.1016/j.jbtep.2010.01.002.

VanElzakker, M., Dahlgren, M. K., Davis, C. F., Dubois, S., & Shin, L. M. (2013). From Pavlov to PTSD: The extinction of conditioned fear in rodents, humans, and anxiety disorders. *Neurobiology of Learning and Memory*, *113*, 3–18. https://doi.org/10.1016/j.nlm.2013.11.014.

Vujanovic, A. A., Farris, S. G., Bartlett, B. A., Lyons, R. C., Haller, M., Colvonen, P. J., & Norman, S. B. (2018). Anxiety sensitivity in the association between posttraumatic stress and substance use disorders: A systematic review. *Clinical Psychology Review*, *62*, 37–55. https://doi.org/10.1016/j.cpr.2018.05.003.

Wald, I., Shechner, T., Bitton, S., Holoshitz, Y., Charney, D. S., Muller, D., … Bar-Haim, Y. (2011). Attention bias away from threat during life threatening danger predicts PTSD symptoms at one-year follow-up. *Depression and Anxiety*, *28*(5), 406–411. https://doi.org/10.1002/da.20808.

Waters, A. M., & Craske, M. G. (2016). Towards a cognitive-learning formulation of youth anxiety: A narrative review of theory and evidence and implications for treatment. *Clinical Psychology Review*, *50*, 50–66. https://doi.org/10.1016/j.cpr.2016.09.008.

Weierich, M. R., Treat, T. A., & Hollingworth, A. (2008). Theories and measurement of visual attentional processing in anxiety. *Cognition and Emotion*, *22*(6), 985–1018. https://doi.org/10.1080/02699930701597601.

White, L. K., Suway, J. G., Pine, D. S., Bar-Haim, Y., & Fox, N. A. (2011). Cascading effects: The influence of attention bias to threat on the interpretation of ambiguous information. *Behaviour Research and Therapy*, *49*(4), 244–251. https://doi.org/10.1016/j.brat.2011.01.004.

Yehuda, R. (2001). Biology of posttraumatic stress disorder. *The Journal of Clinical Psychiatry*, *62*(Suppl. 17), 41–46.

Zoellner, L. A., Telch, M., Foa, E. B., Farach, F. J., McLean, C. P., Gallop, R., … Gonzalez-Lima, F. (2017). Enhancing extinction learning in posttraumatic stress disorder with brief daily imaginal exposure and methylene blue: A randomized controlled trial. *The Journal of Clinical Psychiatry*, *78*(7), e782–e789.

CHAPTER 3

Anger in PTSD

Kirsten H. Dillon[a,b], Elizabeth E. Van Voorhees[a,b], Jean C. Beckham[a,b]
[a]Durham Veterans Affairs Medical Center, Durham, NC, United States
[b]Department of Psychiatry & Behavioral Sciences, Duke University Medical Center, Durham, NC, United States

Introduction

Although posttraumatic stress disorder (PTSD) has long been considered a primarily fear-based disorder, there is considerable evidence to indicate that anger is also central to the experience of PTSD (Beckham, Moore, & Reynolds, 2000; Olatunji, Ciesielski, & Tolin, 2010; Orth & Wieland, 2006). For example, in a large national sample of adults with a history of interpersonal trauma, problems with anger were found to be more predictive of PTSD than persistent fear (Badour, Resnick, & Kilpatrick, 2015). Anger has also been found to contribute to interpersonal difficulties among individual with PTSD, including aggressive behavior. A large study of US Army soldiers found that risk of aggression was elevated among soldiers with PTSD and high levels of anger, but not among those with low anger (Wilk, Quartana, Clarke-Walper, Kok, & Riviere, 2015). Similarly, PTSD combined with high levels of hostility has been found to predict aggression in Iraq/Afghanistan-era veterans, whereas PTSD combined with low levels of hostility did not (Van Voorhees et al., 2016). Consistent with these studies, in a metaanalysis on associations between PTSD and intimate relationship difficulties, Taft, Watkins, Stafford, Street, and Monson (2011) found that the relationship between PTSD and interpersonal aggression was mediated by trait anger.

Anger has also been associated with diminished functioning among individuals with PTSD across a range of domains and settings. In one study of Vietnam veteran couples, PTSD was associated with higher rates of anger, which were, in turn, predictive of poorer relationship functioning (Roberge, Allen, Taylor, & Bryan, 2016). PTSD severity was not predictive of relationship functioning, indicating that anger had a more detrimental effect on relationships than PTSD symptoms. With respect to occupational functioning, unemployed veterans with PTSD have been found to report higher levels of anger than those who were employed when covarying for PTSD severity (Frueh, Henning, Pellegrin, & Chobot, 1997). Anger can also impede successful outcomes from PTSD treatment. Indeed, pretreatment levels of anger have been found to account for both higher levels of dropout (Rizvi, Vogt, & Resick, 2009; Stevenson & Chemtob, 2000) and less PTSD symptom reduction over the course of psychotherapy (Foa, Riggs, Massie, & Yarczower, 1995; Forbes, Creamer, Hawthorne, Allen, & McHugh, 2003; Speckens, Ehlers, Hackmann, & Clark, 2006; Taylor et al., 2001).

Defining anger

Anger can be defined as a negatively valenced emotion that is experienced in response to provocation or threat that includes cognitive, somatic, and behavioral components (DiGiuseppe & Tafrate, 2003). Anger is similar to fear but is associated with approach- rather than retreat-related behaviors. When discussing anger in the context of PTSD, it is important to acknowledge that anger is a multidimensional construct that has been divided into several domains (Spielberger, 1999). Spielberger (1999) differentiates between state anger, trait anger, anger control, and anger expression. State anger is the degree to which an individual is experiencing anger at the present moment, whereas trait anger is a more general assessment of an individual's predisposition to react to situations with anger. Anger control refers to the extent to which an individual monitors and controls the expression of anger. Anger expression refers to the tendency to either suppress anger (anger-in) or to express anger outwardly in an aggressive manner (anger-out). Hostility is a construct closely related to anger. Whereas anger is an emotion, hostility refers to a general predisposition or tendency to mistrust and dislike others. Measures of anger and hostility are highly correlated (Eckhardt, Norlander, & Deffenbacher, 2004).

Associations between PTSD and anger

In their 2006 meta-analysis, Orth and Wieland examined the relationship between anger, hostility, and PTSD. Across 39 studies, they found a large effect for the relationship between anger/hostility and PTSD (mean effect size of $r=0.48$). When examining specific anger expression and suppression variables, they found medium to large effects for the relationship of PTSD with anger-in ($r=0.53$) and anger control ($r=-0.44$) and medium effects for PTSD with anger-out ($r=0.29$). They also examined several potential moderators of the relationship between anger/hostility and PTSD, including time since trauma and trauma type. Results indicated that the amount of time that had passed since the trauma had a logarithmic effect on the strength of the association. Specifically, immediately after the traumatic event, the relationship between anger/hostility and PTSD was low; however, the association increased strongly during the first few months following the trauma, before eventually converging at a limiting value.

Orth and Wieland (2006) also found the strongest associations between anger/hostility and PTSD among military samples (mean effect size of $r=0.56$), suggesting that anger/hostility may be particularly problematic among military veterans with PTSD. Based on these findings, however, it is not possible to determine whether military-related traumatic events are more likely to lead to anger or whether this stronger association is due to selection variables (e.g., are individuals who join the military more likely to have experienced trauma prior to military service, or are they more prone to anger generally?). Interestingly, research in a representative sample of Reserve and National Guard personnel found that, while anger was associated with both nondeployment- and deployment-related traumas (Worthen et al., 2014), these associations differed by gender (Worthen et al., 2015): in men the association between anger was strongest for deployment-related traumas, whereas for women the association was strongest for civilian traumas.

Olatunji et al. (2010) conducted a meta-analytic review to examine the extent to which anger was specific to PTSD versus an artifact of other anxiety symptoms. They examined rates of anger across the anxiety disorders and found that the relationship between PTSD and anger had a larger effect size than that of the other anxiety disorders. Similar to the findings of Orth and Wieland (2006), they found that anger-in (anger suppression), anger-out (outward expression of anger), and anger control (ability to cope with anger) differentiated between PTSD and non-PTSD whereas state and trait anger did not.

Considering that irritability and outbursts of anger are included in the diagnostic criteria for PTSD, several studies have examined whether the relationship between anger and PTSD is a product of measurement overlap. In a sample of Vietnam combat veterans, Novaco and Chemtob (2002) found that anger accounted for more than 40% of the variance in PTSD (when not including anger), above and beyond age, education, and combat exposure. Other studies have also confirmed that the association between PTSD and anger remains when the item assessing irritability or anger outbursts is removed from the PTSD assessment, among both samples of military personnel (Lommen, Engelhard, Schoot, & Hout, 2014) and samples of female assault victims (Orth, Cahill, Foa, & Maercker, 2008).

Associations over time: Is anger a cause or consequence of PTSD?

While associations between anger and PTSD have been consistently documented, the directionality of this relationship is far less clear. Several longitudinal studies have been conducted to examine whether anger is a cause or consequence of PTSD. One of the first studies to investigate this association assessed female crime victims. This study found that higher state anger and anger-in 2 weeks after the assault was associated with the development of PTSD 1 month later (Riggs, Dancu, Gershuny, Greenburg, & Foa, 1992). Similarly, Feeny, Zoellner, and Foa (2000) assessed female assault victims and found that anger expression 1 month post assault was associated with PTSD severity 2 months later. In another more long-term follow-up study of disaster relief workers from the attack on the World Trade Center on September 11, 2001, state anger levels 1 year post trauma were associated with PTSD severity 1 year later, suggesting that anger plays a role in the maintenance of PTSD (Jayasinghe, Giosan, Evans, Spielman, & Difede, 2008).

Whereas the studies discussed earlier assessed anger using Spielberger's scales (Spielberger, 1983, 1999), other research has examined whether other aspects of anger are predictive of PTSD. Andrews, Brewin, Rose, and Kirk (2000) examined the role of self-directed and other directed anger among victims of violent crime and found that anger with others, but not with self, was associated with PTSD 1 month post trauma. When predicting PTSD 6 months after trauma (and covarying for 2-month PTSD symptoms), anger at others was no longer a significant predictor. The authors argued that it is possible that anger at others is related to later PTSD through its association with early symptoms, suggesting that anger at others plays a role in the development of PTSD. In a 3-year follow-up study of motor vehicle accident victims, anger cognitions (i.e., "Others have harmed me") at all time points (3 months, 1 year, and 3 years post trauma) were found to predict PTSD symptoms at 3 years post trauma, suggesting that anger cognitions may serve as a maintaining factor for PTSD (Mayou, Ehlers, & Bryant, 2002).

The aforementioned studies examined whether anger predicted subsequent PTSD, but did not explore the reverse relationship. Orth et al. (2008) sought to further examine the temporal relationship between anger and

PTSD symptoms, testing whether state anger predicted subsequent PTSD or vice versa in a sample of crime victims. Contrary to the findings described earlier, they found that PTSD symptoms predicted subsequent anger, but anger did not predict subsequent PTSD symptoms. They further found that the relationship between PTSD and anger was mediated by trauma-related rumination. More recently, we have largely replicated these results over a shorter time period using ecological momentary assessment methods (Van Voorhees et al., 2018). Specifically, using cross-lagged analyses, our team found that individuals with PTSD endorsed significantly more instances of hostile and irritable affect over the week-long assessment period. Moreover, whereas PTSD symptoms predicted hostile/irritable affect over a period of several hours, the reverse relationship was not observed. This suggests that day-to-day exposure to trauma cues may lead to chronically elevated levels of anger-related affect and arousal, which, in turn, contributes to increased risk for verbal and physical aggression.

While there is some evidence that anger *after trauma* predicts subsequent PTSD, less is known about the association between anger *prior to trauma exposure* and subsequent PTSD. A few studies have been able to examine this in samples of first responders (e.g., police and firefighters) or soldiers prior to training. For example, Heinrichs et al. (2005) conducted a 2-year prospective study of firefighters and found that high hostility and low self-efficacy prior to training predicted PTSD symptoms 24 months later (accounting for 42% of variance). van Zuiden et al. (2011) also found that soldiers with high hostility and low self-directedness prior to deployment to Afghanistan experienced higher levels of PTSD symptoms 6 months after deployment. Finally, Meffert et al. (2008) followed police recruits and found not only that trait anger during training predicted PTSD symptoms 12 months later but also that PTSD symptoms at 1 year predicted increased state anger, suggesting that anger is both a vulnerability factor and a consequence of PTSD.

Lommen et al. (2014) assessed Dutch soldiers 2 month prior to their deployment to Afghanistan and found that predeployment trait anger predicted PTSD symptoms at 2 months post deployment and indirectly at 9 months (through PTSD symptoms from the prior assessment). This relationship remained when accounting for trauma severity experienced in Afghanistan, content overlap (the anger item was excluded from PTSD assessments), and predeployment PTSD symptoms; however, the finding did not remain statistically significant when neuroticism was included in the model. These findings support a diathesis-stress model for anger and PTSD and suggest that neuroticism and trait anger are vulnerability factors for PTSD, with neuroticism being a stronger predictor.

It is important to note, however, that anger is a facet of neuroticism (Costa & McCrae, 1992), so this finding may have been impacted by measurement overlap. Interestingly, Lommen et al. (2014) did not find that postdeployment trait anger predicted subsequent PTSD symptoms or vice versa. These findings differ from the longitudinal studies presented earlier in this chapter. It is possible that these discrepant findings may be related to the fact that this was a military sample whereas the other studies used civilian samples. Additionally, it is unclear if this null finding is a product of the anger domain assessed, since the other significant findings were not specific to trait anger.

More recently, Lin et al. (2017) conducted a longitudinal study assessing anger-related neurobehavioral markers for PTSD in healthy Israeli soldiers. During their baseline assessment (prior to military training), they viewed an anger-inducing film with simultaneous brain imaging (functional magnetic resonance imaging and electroencephalography) and self-reported state anger. They found that both state anger during the film and amygdala response were each uniquely associated with greater PTSD symptoms 1 year later, suggesting that anger response and amygdala reactivity may be vulnerability factors for PTSD. Taken together, these studies indicate that high hostility/anger prior to trauma exposure may be a vulnerability factor for developing PTSD. However, a notable limitation of these studies is that all of the studies in this area to date have been conducted in high-risk samples (e.g., firefighters and soldiers) and the generalizability of these findings beyond such high-risk samples is unknown.

Theoretical models to explain the association between anger and PTSD

Given the strong associations between anger and PTSD symptoms over time, several theoretical models have been proposed. In the succeeding texts, we outline several of the predominant theories:

Fear avoidance theory

One prominent theory concerning the role of anger in PTSD is the fear avoidance theory (Foa et al., 1995; Riggs et al., 1992). This theory proposes that anger serves as a form of avoidance from unwanted memories, thoughts, and feelings associated with the trauma (e.g., fear). Whereas fear is a vulnerable emotion, anger is a mobilizing emotion that may be experienced as more positive and acceptable. Furthermore, anger and fear have been conceptualized as incompatible reactions to threat (e.g., Foa et al., 1995).

According to this theory, high anger may impede natural emotional processing of the traumatic event by inhibiting fear activation. In this way, high anger may serve to maintain PTSD symptoms and prevent recovery from PTSD. In support of this theory, a study of 12 female assault victims with PTSD found that those who reported more anger at baseline displayed less fear expression during exposure therapy and benefited less from treatment compared with those with lower baseline anger (Foa et al., 1995). Subsequent treatment studies have found inconsistent support for this theory, however, with some finding evidence that baseline anger reduces the effectiveness of PTSD treatment (Forbes et al., 2003, 2005), whereas others have not observed this relationship (Cahill, Rauch, Hembree, & Foa, 2003; Taylor, 2003; Van Minnen, Arntz, & Keijsers, 2002).

"Survival mode" theory

Another prominent theoretical model of the relationship between PTSD and anger is the "survival mode" theory proposed by Chemtob, Novaco, Hamada, Gross, and Smith (1997) in the context of military- and combat-related PTSD. This model theorizes that PTSD-related anger is triggered when individuals adopt a "survival mode" of functioning wherein perceived threats trigger increased arousal, hostile appraisals, and hostile behaviors (Chemtob, Novaco, Hamada, Gross, & Smith, 1997; Novaco, Swanson, Gonzalez, Gahm, & Reger, 2012). Responding to threat with anger can be protective and adaptive in high-risk environments like combat because it can be energizing and promote a sense of mastery. It could also potentially protect against vulnerable emotions such as fear, grief, and guilt (Gerlock, 1994); however, continuing to inflexibly enter survival mode in response to threat becomes maladaptive in civilian situations where threat is no longer as pervasive. Furthermore a cycle emerges wherein people with PTSD are vulnerable to perceive threat in benign or ambiguous situations. Such a bias can lead to more anger and can result in fearful or angry responses from others, which then confirms the original bias and leads to greater readiness to perceive threat in the future (Chemtob, Novaco, Hamada, Gross, & Smith, 1997; Novaco & Chemtob, 2002). Findings suggesting that the association between anger and PTSD is strongest among military samples offer support for this "survival mode" theory (Orth & Wieland, 2006). However, Worthen et al. (2015) found evidence that the elevated association of combat-related traumas and anger may only apply to men. In women the relationship was reversed, such that nondeployment-related traumas were most strongly associated with anger.

Social information processing theories

Related to the "survival mode" theory, social information processing models of PTSD and anger propose that individuals with PTSD exhibit cognitive biases that underlie their difficulties with anger and aggression (e.g., Constans, 2005; Taft et al., 2015). One specific type of bias that may link PTSD and anger is the *hostile attribution bias* (Taft et al., 2015) or *hostile interpretation bias* (Dillon, Allan, Cougle, & Fincham, 2015), which refers to the tendency to interpret ambiguous interpersonal situations as hostile (Wilkowski & Robinson, 2008, 2010). For example, in a case where someone gets bumped into in a crowd, the person with hostile interpretation bias may be more likely to interpret being bumped as an aggressive action rather than as a mistake (Wilkowski & Robinson, 2010). There is a strong literature basis supporting the link between hostile interpretation biases and increased anger, starting with early work by Dodge (1980), who identified this bias in aggressive children. Since then, researchers have demonstrated that hostile interpretation bias is linked to trait anger and aggression in adults as well (Bond, Verheyden, Wingrove, & Curran, 2004; Epps & Kendall, 1995; Hazebroek, Howells, & Day, 2001; Wenzel & Lystad, 2005). While little empirical work has been done to directly link hostile interpretation bias in PTSD with PTSD-related anger, at least one study in veterans has found that hostile cognitions mediated the relationship between PTSD symptoms at baseline and aggression 6 months later (Van Voorhees et al., 2016).

Cognitive appraisal theories

Cognitive models of PTSD emphasize the importance of trauma-related appraisals in the development and maintenance of PTSD and negative posttraumatic beliefs about the self, others, and the world (Ehlers & Clark, 2000; Foa, Ehlers, Clark, Tolin, & Orsillo, 1999; Resick, Monson, & Chard, 2014). Research suggests that the same maladaptive posttraumatic appraisals that underlie PTSD are also associated with increased posttraumatic anger. For example, in one study of civilians with PTSD, negative posttraumatic beliefs about the self and the world as measured by the Posttraumatic Cognitions Inventory (PTCI; Foa et al., 1999) were associated with anger expression (Whiting & Bryant, 2007). Examples of items from the negative beliefs about the self and world subscales of the PTCI include the following: "I am a weak person," and "You can never know who will harm you" (Foa et al., 1999). In another study of male combat veterans in Australia, posttraumatic beliefs (again, using the PTCI) were found to partially mediate the relationship between PTSD and anger variables (Germain, Kangas, Taylor, & Forbes, 2015). Specifically, negative beliefs about the self partially mediated the relationship between PTSD and anger suppression and anger control, and negative beliefs about the world partially mediated the relationship between PTSD and anger expression.

Treatment outcomes and implications
Impact of anger on PTSD treatment efficacy

There is mounting evidence that high levels of anger interfere with the effectiveness of therapy for PTSD (Foa et al., 1995; Forbes et al., 2003, 2005; Jaycox & Foa, 1996). Foa et al. (1995) first reported this phenomenon in a sample of 12 female assault victims with PTSD treated with prolonged exposure therapy (PE), noting that participants with higher anger at baseline displayed less fear during exposure and less recovery over the course of therapy. These findings are consistent with fear avoidance theory in that PE is based upon the premise that recovery from PTSD requires activation of the underlying fear structure associated with a traumatic event, accompanied by simultaneous exposure to corrective, contextualizing information that allows the patient to recognize that the trauma memory does not pose a current threat (Foa, Huppert, & Cahill, 2006; Foa & Kozak, 1986). If anger in response to threat inhibits the experience of fear during exposure, then the fear activation necessary for critical emotional processing cannot take place, and symptom relief does not occur (Foa et al., 2006; Foa & Kozak, 1986). Some subsequent studies are consistent with this interpretation in combat veterans (Forbes et al., 2003) and peacekeepers (Forbes et al., 2005), where it was observed that high comorbid anger diminished the effectiveness of PTSD treatment. However, notably, this relationship has not been consistently observed (Cahill et al., 2003; Taylor, 2003; Van Minnen et al., 2002).

The fear inhibition theory of anger in PTSD is also consistent with the survival mode model of anger in combat-related PTSD: anger and aggression are adaptive in the context of combat precisely because they protect against the retreat-related emotions of fear, guilt, and grief (Chemtob, Novaco, Hamada, Gross, & Smith, 1997; Gerlock, 1994). From this perspective, it would not be surprising that anger would be activated over fear in therapeutic contexts when trauma-related material is confronted. More recent research, however, suggests that the relationship between anger and PTSD treatment outcomes may be more complex. In a study investigating mechanisms of anger and treatment outcome in Vietnam combat veterans with PTSD, Forbes et al. (2008) found that alcohol misuse and fear of anger together explained the variance in the association between high levels of pretreatment anger and poorer PTSD treatment outcome. That is, rather than the anger itself preventing recovery, it was the fear of losing control of anger that seemed to block full engagement with the therapeutic process. In explaining their findings, Forbes and colleagues noted that, in military samples, anger in response to threat is associated with the use of extreme and potentially lethal aggression. Thus they reasoned that accessing traumatic memories in therapy may feel threatening to combat veterans because of the perceived high cost of losing control of anger and that this sense of threat was likely to be particularly salient in the context of the disinhibiting effects of alcohol (Forbes et al., 2008). Forbes and colleagues' observations in Vietnam combat veterans were partially replicated in a more recent study in Iraq and Afghanistan veterans, where fear of losing emotional control was associated with dropout from cognitive processing therapy (CPT). Interestingly, however, among those veterans who completed treatment, fear of anger at baseline was negatively associated with PTSD severity at post-treatment (Miles, Smith, Maieritsch, & Ahearn, 2015).

Impact of PTSD treatment on anger

Given the attenuating impact of anger on psychotherapy treatment outcomes for PTSD, one might expect the effect of treatment for PTSD on anger outcomes to be similarly limited. Whether this is actually the case is unclear. Cahill et al. (2003) compared treatment with PE alone, stress inoculation therapy alone (SIT; a cognitive behavioral intervention for anger), combined PE/SIT, and wait-list control on outcomes on the State-Trait Anger Expression Inventory (STAXI; Spielberger, 1983). While these authors reported that all treatment groups improved on state anger with the SIT alone group showing the greatest benefit, findings were difficult to interpret because outcomes on trait anger or anger expression were not presented. Stapleton, Taylor, and Asmundson (2006) compared PE, eye movement desensitization and reprocessing therapy (EMDR), and relaxation training in 60 participants with PTSD and found small improvements in both trauma-related anger and trait anger in all three treatment groups, but no differences between groups on anger outcomes. They also observed worsening of anger in a subset of participants across groups. Finally, Ford, Grasso, Greene, Slivinsky, and DeViva (2018) compared PE with a present-centered affect regulation therapy (TARGET) in 31 male combat veterans with PTSD and found small reductions in trait anger in both groups.

Studies of the effects of nonexposure-based PTSD treatments on anger outcomes may be somewhat more promising. Resick et al. (2008) conducted a dismantling study of CPT in which they randomized 150 women with PTSD to full CPT, CPT without the trauma narrative, or written trauma narrative only and observed small but significant reductions on anger-in scores but no changes on anger-out scores in all three groups. Forbes et al. (2012) examined the effects of CPT on anger in 59 veterans and found significant improvements in anger on the Dimensions of Anger Reaction Scale over the effects of treatment as usual, with moderate to large effect sizes in the CPT group. Finally a CPT treatment study in a sample of 139 female survivors of interpersonal violence found significant improvements in state anger, trait anger, anger-in, and anger control over the course of therapy in those who completed a course of therapy, though these gains were not observed among those who terminated therapy early (Galovski, Elwood, Blain, & Resick, 2014). Interestingly, findings also suggested that recovery from PTSD in response to CPT may be related to how effective the therapy is at addressing some dimensions of anger as well. That is, while changes in state anger and anger-in did not differentiate between those who did and did not benefit from treatment, CPT treatment "responders" showed significantly greater improvement in trait anger and anger control than did treatment "nonresponders."

Taken together, these findings suggest that, while anger may interfere with the ability to engage with and benefit from therapy for PTSD, for those who do engage, improvements in anger are linked to recovery from PTSD as well. Further, it appears that therapies that focus on cognitive restructuring may be more effective at addressing comorbid anger than therapies that emphasize exposure. This is consistent with the underlying theoretical orientations of each approach: while anger would be expected to interfere with activation of the fear structure and thus impede recovery in exposure-based therapies, cognitive therapies that target trauma-related cognitions would be expected to challenge maladaptive fear- and anger-related cognitions with similar efficacy. However, it is important to keep in mind that few studies have directly addressed the question of anger outcomes in PTSD treatment, and therefore any conclusions about the relative effectiveness of one approach to treating PTSD over another in impacting anger outcomes should be considered tentative.

Treatments for anger in PTSD

The past decade has seen an increase in research on therapies directly targeting anger and aggression in PTSD. While meta-analyses and reviews of the literature suggest that therapy can be effective in treating anger problems and reducing aggression generally, a consistent theme in these reviews has been that there is considerable variance in outcomes depending upon the patient population sampled (Del Vecchio & O'Leary, 2004; DiGiuseppe & Tafrate, 2003). Given the evidence that PTSD-related anger may be unique in terms of both information processing and physiological arousal regulation (Chemtob, Novaco, Hamada, Gross, & Smith, 1997; Taft et al., 2007, 2015; Teten et al., 2010), it is reasonable to assume that anger management treatment may need to be tailored to address the needs of individuals with PTSD.

The majority of the work that has been done to evaluate treatments for anger in individuals with PTSD has been in veteran samples, with cognitive behavioral therapy (CBT) being the most commonly employed approach. The first trial examining the effects of therapy for anger in veterans with PTSD was conducted by Chemtob and colleagues (Chemtob, Novaco, Hamada, & Gross, 1997) in a small sample of male Vietnam veterans with PTSD and severe anger problems. These researchers developed a cognitive behavioral intervention based upon the survival mode model and compared veterans treated with 12 individual sessions of this intervention ($n=8$) to veterans in a routine clinical care control condition ($n=7$) on a range of anger outcome measures. Despite a high (46%) dropout rate, greater improvements were found in anger control in the active treatment condition at posttreatment and 18-month follow-up in both completers and dropouts. However, no differences between conditions were observed for any of the subscales of the Novaco Anger Scale; on anger-in, anger-out, or trait anger subscales of the STAXI-2; or on reactions to anger provocation scenes. Over a decade later, Shea, Lambert, and Reddy (2013) adapted Chemtob and colleagues' cognitive behavioral protocol and compared it with a supportive intervention in 23 Iraq and Afghanistan war veterans with anger problems and at least 1 other PTSD hyperarousal symptom. Among those who completed treatment, greater improvement was observed in the cognitive behavioral versus supportive intervention from pre- to posttreatment on the Anger Expression Index of the STAXI-2 and on each of the four subscales of this index (expression-in, expression-out, anger control-in, anger and control-out).

While several other published studies have examined the effects of cognitive behavioral approaches to anger management therapy for veterans with PTSD, none have included a control group. Each of these uncontrolled trials have reported significant reductions in anger outcomes from pre- to posttreatment, including state anger (Gerlock, 1994; Marshall et al., 2010; Morland et al., 2010), trait anger (Gerlock, 1994; Marshall et al., 2010), anger expression (Morland et al., 2010), and physical aggression (Marshall et al., 2010). Two more uncontrolled studies examined the effects of therapy programs for PTSD that included anger management groups as part of the treatment, and these also reported significant decreases in anger expression (Turner, Beidel, & Frueh, 2005) and self-reported aggression (Bolton et al., 2004).

Taken together, these findings suggest that cognitive behavioral therapies targeting anger in individuals with PTSD may hold promise for reducing a range of anger outcomes, at least in veteran samples. However, the question of mechanisms of action contributing to outcomes has been addressed in only two studies to date. One of these studies found that improvement in physiological arousal calming, but not development of cognitive coping and behavioral control skills, was associated with reduction in anger symptoms over the course of therapy (Mackintosh, Morland, Frueh, Greene, & Rosen, 2014). These authors also found that therapeutic alliance had an impact on anger outcomes, though this effect was largely mediated by its relationship with arousal calming skills. Interestingly, another study found that antisocial personality characteristics reduced the impact of anger management on anger outcomes. The authors attributed this effect to the diminished ability of individuals with antisocial characteristics to effectively form a therapeutic working alliance (Marshall et al., 2010).

PTSD and intimate partner violence: The strength at home program

Intimate partner violence (IPV) has been found to be difficult to treat, with at least two reviews suggesting that treatment has limited effects at preventing recidivism over the effects of being arrested (Babcock, Green, & Robie, 2004; Stover, Meadows, & Kaufman, 2009). Few IPV treatment programs, however, have accounted for the role of trauma in perpetuating the cycle of violence (Barrett, Mills, & Teesson, 2011; Beckham, Feldman, Kirby, Hertzberg, & Moore, 1997; Gillikin et al., 2016; Teten et al., 2010). A recent exception is the Strength at Home Program (SAH), a group intervention designed to reduce partner aggression in male veterans with PTSD. This 12-week cognitive behavioral group program incorporates elements addressing information processing deficits that impact anger and aggression in male veterans with PTSD, including hostile attribution bias and survival mode of functioning. Taft and colleagues compared SAH with enhanced treatment as usual (ETAU) in a sample of 135 male veterans with a recent history of physical IPV or ongoing court involvement. Medium to large effect size reductions in physical and psychological aggression were observed in the SAH group, and small effect size reductions on these outcomes were observed in the ETAU group, with small to medium between group effect sizes observed favoring SAH (Taft, Macdonald, Creech, Monson, & Murphy, 2016). While these results await replication, considering the role trauma and PTSD play in IPV perpetration among males who have used domestic violence, this approach appears to hold promise for addressing this serious and difficult-to-treat problem.

The role of gender in treatment for anger in PTSD

None of the earlier cited studies of treatments developed to address anger in PTSD have included women in the samples. The absence of women in studies of anger treatments does not reflect a lack of need: A study of anger problems and PTSD in Reserve and National Guard service members found similarly high rates of self-reported anger problems in men and women (Worthen et al., 2014), and women with PTSD have been found to perpetrate general aggression and interpersonal violence at higher rates than women without PTSD (Kirby et al., 2008). In addition, female Iraq/Afghanistan-era veterans are more likely to endorse aggression in a family context than their male counterparts (Sullivan & Elbogen, 2013). It also appears untenable to assume that interventions developed to target anger in men will work equally well for women. While many of the stereotypes about women's experience of anger do not hold up to empirical scrutiny (Averill, 1983; Kring, 2000), there is evidence that women express anger differently than men: Women have been found to have more difficulty than men expressing anger at the object of their distress, they may cry more often in response to anger, and they have been found to be more likely to feel embarrassed or ashamed after anger displays than men (Kring, 2000). Finally, women's anger may be more intimately related with powerlessness and economic and social inequality than men's (Carmony & DiGiuseppe, 2003; Ross & Van Willigen, 1996; Strachan & Dutton, 1992; Thomas, 1989; Thomas & Gonzalez-Prendes, 2009).

Critically, there is evidence that the association between PTSD and anger may differ between men and women veterans. The study by Worthen et al. (2015) cited earlier found that, while the survival mode model was supported for male veterans, it did not account for PTSD-related anger in female veterans. Specifically, among women, PTSD associated with trauma that occurred in a nondeployment context was more strongly associated with anger than was PTSD associated with deployment-related trauma

(Worthen et al., 2015). Given that the survival mode model forms the foundation of the majority of anger interventions for individuals with PTSD, this finding points to a critical need to better understand anger in women with PTSD before implementing anger treatments that may not address women's needs.

Van Voorhees and colleagues recently completed the first trial of anger management therapy for veterans with PTSD to include women in the sample, and the findings support the assertion that overlooking gender differences in PTSD-related anger may have critical implications for treatment. Our small pilot study compared 12 sessions of group cognitive behavioral therapy for anger in veterans with PTSD (CBT; $n=19$) with 12 sessions of group present-centered therapy (PCT; $n=17$). As with prior trials the CBT intervention was theoretically grounded in the survival mode model of anger and aggression in PTSD. Results in male veterans replicated previous findings, such that in both treatment arms men showed decreased scores on the Dimensions of Anger Reaction scores with large effect sizes (Van Voorhees, Dillon, Wilson, Dennis, et al., 2019). Women veterans, however, expressed dissatisfaction with the cognitive behavioral treatment approach, and all of the women assigned to the CBT arm dropped out by the ninth session. Women also started the treatment with higher anger scores than men and showed significantly less improvement in anger scores over time. Qualitative analyses of posttreatment interviews with women veterans suggested that, while they found the opportunity to connect with other women veterans and to be listened to as healing, some of them experienced the CBT material as invalidating and inappropriate to their experience (Van Voorhees, Dillon, Wilson, Medenblik, et al., 2019).

Novel approaches to treating anger in PTSD

Given the limitations of the current approaches to treating anger in individuals with PTSD, it will be important to continue to explore novel treatment approaches that target specific deficits common to PTSD and anger. We are currently developing and pilot testing a mobile intervention designed to reduce hostile interpretation biases among veterans with PTSD and problematic anger. This intervention utilizes interpretation bias modification techniques to guide participants to make benign, rather than hostile, interpretations of ambiguous situations (Cougle et al., 2017; Hawkins & Cougle, 2013; Smith, Dillon, & Cougle, 2018). This treatment is derived from a larger research literature on the use of cognitive bias modification (CBM) techniques to reduce attention and interpretation biases that contribute to psychopathology (Woud & Becker, 2014). Several CBM interventions have been employed to target other biases central to PTSD (see Woud, Verwoerd, and Krans (2017) for a systematic review). We theorize that this mobile intervention, entitled Mobile Intervention for Reducing Anger (MIRA), will be particularly beneficial to individuals with PTSD and high anger since it has been developed to efficiently reduce a specific processing bias central to information processing theories of anger in PTSD.

Conclusions and future directions

Anger has been consistently associated with PTSD across populations (Olatunji et al., 2010; Orth & Wieland, 2006), with the strongest associations found in military samples (Orth & Wieland, 2006). Research suggests that the relationship between anger and PTSD is bidirectional, with directionality in part determined by the time points at which anger and PTSD were measured. Several theoretical models have been developed to explain the association between anger and PTSD, including the fear avoidance, survival mode, social information processing, and cognitive appraisal theories.

Anger in PTSD has been associated with poorer relationship functioning (Roberge et al., 2016), occupational functioning (Frueh et al., 1997), and aggressive behavior (Taft et al., 2011; Van Voorhees et al., 2016; Wilk et al., 2015). There is also evidence that anger may diminish the effectiveness of treatment for PTSD, particularly exposure-based treatments (Foa et al., 1995; Forbes et al., 2003, 2005; Jaycox & Foa, 1996), and that it may also interfere with therapeutic alliance (Mackintosh et al., 2014; Marshall et al., 2010). While some studies have suggested that cognitive treatments for PTSD may be more effective than exposure-based treatments at reducing anger, more research is necessary to evaluate the reliability and size of potential effects. Cognitive behavioral therapy targeting PTSD-related anger may be effective in men, at least in veterans. However, these effects may not generalize to women with anger and PTSD. Additional research is needed to better understand anger in PTSD in women.

Taken together, these findings suggest that treatments for anger in PTSD, while leading to statistically significant reductions on some anger outcomes, need further development to maximize clinical outcomes. It will be important in future research to develop evidence based treatments grounded in theory to further our understanding and treatment of anger in PTSD across multiple trauma populations. Additionally, assessment of a range of functional outcomes will be helpful to include in future research studies. For example, assessments of relationship, social and occupational functioning, and quality of life may be useful to include when evaluating treatment effects.

Research in the area of anger and PTSD has established that anger is an important target for assessment and treatment, and an extensive base of theoretical and clinical information in this field of inquiry has been created. Additional research aimed at better addressing anger among individuals with PTSD, and maximizing improvements in symptoms and functioning is needed.

References

Andrews, B., Brewin, C. R., Rose, S., & Kirk, M. (2000). Predicting PTSD symptoms in victims of violent crime: The role of shame, anger, and childhood abuse. *Journal of Abnormal Psychology*, *109*(1), 69–73.

Averill, J. R. (1983). Studies on anger and aggression. Implications for theories of emotion. *American Psychologist*, *38*(11), 1145–1160.

Babcock, J. C., Green, C. E., & Robie, C. (2004). Does batterers' treatment work? A meta-analytic review of domestic violence treatment. *Clinical Psychology Review*, *23*(8), 1023–1053. https://doi.org/10.1016/j.cpr.2002.07.001.

Badour, C. L., Resnick, H. S., & Kilpatrick, D. G. (2015). Associations between specific negative emotions and DSM-5 PTSD among a National Sample of interpersonal trauma survivors. *Journal of Interpersonal Violence*, *32*(11), 1620–1641. https://doi.org/10.1177/0886260515589930.

Barrett, E. L., Mills, K. L., & Teesson, M. (2011). Hurt people who hurt people: Violence amongst individuals with comorbid substance use disorder and post traumatic stress disorder. *Addictive Behaviors*, *36*(7), 721–728. [http://dx.doi.org/S0306-4603(11)00076-1].

Beckham, J. C., Feldman, M. E., Kirby, A. C., Hertzberg, M. A., & Moore, S. D. (1997). Interpersonal violence and its correlates in Vietnam veterans with chronic posttraumatic stress disorder. *Journal of Clinical Psychology*, *53*(8), 859–869. https://doi.org/10.1002/(SICI)1097-4679(199712)53:8.

Beckham, J. C., Moore, S. D., & Reynolds, V. (2000). Interpersonal hostility and violence in Vietnam combat veterans with chronic posttraumatic stress disorder: A review of theoretical models and empirical evidence. *Aggression and Violent Behavior*, *5*(5), 451–466.

Bolton, E. E., Lambert, J. F., Wolf, E. J., Raja, S., Varra, A. A., & Fisher, L. M. (2004). Evaluating a cognitive-behavioral group treatment program for veterans with posttraumatic stress disorder. *Psychological Services*, *1*(2), 140–146. https://doi.org/10.1037/1541-1559.1.2.140.

Bond, A. J., Verheyden, S. L., Wingrove, J., & Curran, H. V. (2004). Angry cognitive bias, trait aggression and impulsivity in substance users. *Psychopharmacology*, *171*(3), 331–339.

Cahill, S. P., Rauch, S. L., Hembree, E. A., & Foa, E. B. (2003). Effects of cognitive-behavioral treatments for PTSD on anger. *Journal of Cognitive Psychotherapy: An International Quarterly*, *17*(2), 113–131.

Carmony, T. M., & DiGiuseppe, R. (2003). Cognitive induction of anger and depression: The role of power, attribution, and gender. *Journal of Rational-Emotive and Cognitive-Behavior Therapy*, *21*(2), 105–118. https://doi.org/10.1023/A:1025099315118.

Chemtob, C. M., Novaco, R. W., Hamada, R. S., & Gross, D. M. (1997). Cognitive-behavioral treatment for severe anger in posttraumatic stress disorder. *Journal of Consulting and Clinical Psychology*, *65*(1), 184–189.

Chemtob, C. M., Novaco, R. W., Hamada, R. S., Gross, D. M., & Smith, G. (1997). Anger regulation deficits in combat-related posttraumatic stress disorder. *Journal of Traumatic Stress, 10*(1), 17–36.

Constans, J. I. (2005). Information-processing biases in PTSD. In J. J. Vasterling & C. R. Brewin (Eds.), *Neuropsychology of PTSD: Biological, cognitive, and clinical perspectives* (pp. 105–130). New York: The Guilford Press.

Costa, P. T., & McCrae, R. R. (1992). Normal personality assessment in clinical practice: The NEO personality inventory. *Psychological Assessment, 4*, 5.

Cougle, J. R., Summers, B. J., Allan, N. P., Dillon, K. H., Smith, H. L., Okey, S. A., & Harvey, A. M. (2017). Hostile interpretation training for individuals with alcohol use disorder and elevated trait anger: A controlled trial of a web-based intervention. *Behaviour Research and Therapy, 99*, 57–66. https://doi.org/10.1016/j.brat.2017.09.004.

Del Vecchio, T., & O'Leary, K. D. (2004). Effectiveness of anger treatments for specific anger problems: A meta-analytic review. *Clinical Psychology Review, 24*(1), 15–34. https://doi.org/10.1016/j.cpr.2003.09.006.

DiGiuseppe, R., & Tafrate, R. (2003). Anger treatments for adults: A meta-analytic review. *Clinical Psychology: Science and Practice, 10*(1), 70–84.

Dillon, K. H., Allan, N. P., Cougle, J. R., & Fincham, F. D. (2015). Measuring hostile interpretation bias: The WSAP-hostility scale. *Assessment, 23*(6), 707–719. https://doi.org/10.1177/1073191115599052.

Dodge, K. A. (1980). Social cognition and children's aggressive behavior. *Child Development, 51*(1), 162–170. https://doi.org/10.2307/1129603.

Eckhardt, C., Norlander, B., & Deffenbacher, J. (2004). The assessment of anger and hostility: A critical review. *Aggression and Violent Behavior, 9*(1), 17–43. https://doi.org/10.1016/S1359-1789(02)00116-7.

Ehlers, A., & Clark, D. M. (2000). A cognitive model of posttraumatic stress disorder. *Behaviour Research and Therapy, 38*(4), 319–345. https://doi.org/10.1016/S0005-7967(99)00123-0.

Epps, J., & Kendall, P. C. (1995). Hostile attributional bias in adults. *Cognitive Therapy and Research, 19*(2), 159–178.

Feeny, N. C., Zoellner, L. A., & Foa, E. B. (2000). Anger, dissociation, and posttraumatic stress disorder among female assault victims. *Journal of Traumatic Stress, 13*(1), 89–100. https://doi.org/10.1023/A:1007725015225.

Foa, E. B., Ehlers, A., Clark, D. M., Tolin, D. F., & Orsillo, S. M. (1999). The posttraumatic cognitions inventory (PTCI): Development and validation. *Psychological Assessment, 11*(3), 303–314. https://doi.org/10.1037/1040-3590.11.3.303.

Foa, E. B., Huppert, J. D., & Cahill, L. P. (2006). Emotional processing theory: An update. In B. O. Rothbaum (Ed.), *Pathological anxiety: Emotional processing in etiology and treatment* (pp. 3–22). New York: Guilford Press.

Foa, E. B., & Kozak, M. J. (1986). Emotional processing of fear: Exposure to corrective information. *Psychological Bulletin, 99*(1), 20–35.

Foa, E. B., Riggs, D. S., Massie, E. D., & Yarczower, M. (1995). The impact of fear activation and anger on the efficacy of exposure treatment for posttraumatic stress disorder. *Behavior Therapy, 26*, 487–499.

Forbes, D., Bennett, N., Biddle, D., Crompton, D., McHugh, T., Elliott, P., & Creamer, M. (2005). Clinical presentations and treatment outcomes of peacekeeper veterans with PTSD: Preliminary findings. *American Journal of Psychiatry, 162*(11), 2188–2190. https://doi.org/10.1176/appi.ajp.162.11.2188.

Forbes, D., Creamer, M., Hawthorne, G., Allen, N., & McHugh, T. (2003). Comorbidity as a predictor of symptom change after treatment in combat-related posttraumatic stress disorder. *Journal of Nervous and Mental Disease, 191*(2), 93–99. https://doi.org/10.1097/01.NMD.0000051903.60517.98.

Forbes, D., Lloyd, D., Nixon, R. D. V., Elliott, P., Varker, T., Perry, D., ... Creamer, M. (2012). A multisite randomized controlled effectiveness trial of cognitive processing therapy for military-related posttraumatic stress disorder. *Journal of Anxiety Disorders, 26*(3), 442–452. https://doi.org/10.1016/j.janxdis.2012.01.006.

Forbes, D., Parslow, R., Creamer, M., Allen, N., McHugh, T., & Hopwood, M. (2008). Mechanisms of anger and treatment outcome in combat veterans with posttraumatic stress disorder. *Journal of Traumatic Stress, 21*(2), 142–149. https://doi.org/10.1002/jts.20315.

Ford, J. D., Grasso, D. J., Greene, C. A., Slivinsky, M., & DeViva, J. C. (2018). Randomized clinical trial pilot study of prolonged exposure versus present centred affect regulation therapy for PTSD and anger problems with male military combat veterans. *Clinical Psychology & Psychotherapy*, https://doi.org/10.1002/cpp.2194.

Frueh, B. C., Henning, K. R., Pellegrin, K. L., & Chobot, K. (1997). Relationship between scores on anger measures and PTSD symptomatology, employment, and compensation-seeking status in combat veterans. *Journal of Clinical Psychology, 53*(8), 871–878. https://doi.org/10.1002/(SICI)1097-4679(199712)53:8.

Galovski, T. E., Elwood, L. S., Blain, L. M., & Resick, P. A. (2014). Changes in anger in relationship to responsivity to PTSD treatment. *Psychological Trauma: Theory, Research, Practice and Policy, 6*(1), 56–64. https://doi.org/10.1037/a0031364.

Gerlock, A. A. (1994). Veterans' responses to anger management intervention. *Issues in Mental Health Nursing, 15*(4), 393–408.

Germain, C. L., Kangas, M., Taylor, A., & Forbes, D. (2015). The role of trauma-related cognitive processes in the relationship between combat-PTSD symptom severity and anger expression and control. *Australian Journal of Psychology, 68*(2), 73–81. https://doi.org/10.1111/ajpy.12097.

Gillikin, C., Habib, L., Evces, M., Bradley, B., Ressler, K. J., & Sanders, J. (2016). Trauma exposure and PTSD symptoms associate with violence in inner city civilians. *Journal of Psychiatric Research, 83*, 1–7. https://doi.org/10.1016/j.jpsychires.2016.07.027.

Hawkins, K. A., & Cougle, J. R. (2013). Effects of interpretation training on hostile attribution bias and reactivity to interpersonal insult. *Behavior Therapy, 44*(3), 479–488. https://doi.org/10.1016/j.beth.2013.04.005.

Hazebroek, J. F., Howells, K., & Day, A. (2001). Cognitive appraisals associated with high trait anger. *Personality and Individual Differences, 30*(1), 31–45.

Heinrichs, M., Wagner, D., Schoch, W., Soravia, L. M., Hellhammer, D. H., & Ehlert, U. (2005). Predicting posttraumatic stress symptoms from Pretraumatic risk factors: A 2-year prospective follow-up study in firefighters. *American Journal of Psychiatry, 162*(12), 2276–2286. https://doi.org/10.1176/appi.ajp.162.12.2276.

Jayasinghe, N., Giosan, C., Evans, S., Spielman, L., & Difede, J. (2008). Anger and posttraumatic stress disorder in disaster relief workers exposed to the 9/11/01 World Trade Center disaster: One-year follow-up study. *The Journal of Nervous and Mental Disease, 196*(11), 844–846. https://doi.org/10.1097/NMD.0b013e31818b492c.

Jaycox, L., & Foa, E. (1996). Obstacle in implementing exposure therapy for PTSD: Case discussions and practical solutions. *Clinical Psychology and Psychotherapy, 3*(3), 176–184.

Kirby, A. C., Hertzberg, B. P., Collie, C. F., Yeatts, B., Dennis, M. F., McDonald, S. D., ... Beckham, J. C. (2008). Smoking in help-seeking veterans with PTSD returning from Afghanistan and Iraq. *Addictive Behaviors, 33*(11), 1448–1453. https://doi.org/10.1016/j.addbeh.2008.05.007.

Kring, A. M. (2000). Gender and anger. In A. H. Fischer, ... (Eds.), *Gender and emotion: Social psychological perspectives* (pp. 211). Cambridge: Cambridge University Press.

Lin, T., Gilam, G., Raz, G., Or-Borichev, A., Bar-Haim, Y., Fruchter, E., & Hendler, T. (2017). Accessible neurobehavioral anger-related markers for vulnerability to post-traumatic stress symptoms in a population of male soldiers. *Frontiers in Behavioral Neuroscience, 11*, 38. https://doi.org/10.3389/fnbeh.2017.00038.

Lommen, M. J. J., Engelhard, I. M., Schoot, R., & Hout, M. A. (2014). Anger: Cause or consequence of posttraumatic stress? A prospective study of Dutch soldiers. *Journal of Traumatic Stress*, *27*(2), 200–207. https://doi.org/10.1002/jts.21904.

Mackintosh, M. A., Morland, L. A., Frueh, B. C., Greene, C. J., & Rosen, C. S. (2014). Peeking into the black box: Mechanisms of action for anger management treatment. *Journal of Anxiety Disorders*, *28*(7), 687–695. https://doi.org/10.1016/j.janxdis.2014.07.001.

Marshall, A. D., Martin, E. K., Warfield, G. A., Doron-Lamarca, S., Niles, B. L., & Taft, C. T. (2010). The impact of antisocial personality characteristics on anger management treatment for veterans with PTSD. *Psychological Trauma: Theory, Research, Practice, and Policy*, *2*(3), 224–231. https://doi.org/10.1037/a0019890.

Mayou, R. A., Ehlers, A., & Bryant, B. (2002). Posttraumatic stress disorder after motor vehicle accidents: 3-year follow-up of a prospective longitudinal study. *Behaviour Research and Therapy*, *40*(6), 665–675. https://doi.org/10.1016/S0005-7967(01)00069-9.

Meffert, S. M., Metzler, T. J., Henn-Haase, C., McCaslin, S., Inslicht, S., Chemtob, C., … Marmar, C. R. (2008). A prospective study of trait anger and PTSD symptoms in police. *Journal of Traumatic Stress*, *21*(4), 410–416. https://doi.org/10.1002/jts.20350.

Miles, S. R., Smith, T. L., Maieritsch, K. P., & Ahearn, E. P. (2015). Fear of losing emotional control is associated with cognitive processing therapy outcomes in U.S. veterans of Afghanistan and Iraq. *Journal of Traumatic Stress*, *28*(5), 475–479. https://doi.org/10.1002/jts.22036.

Morland, L., Greene, C., Rosen, C., Foy, D., Reilly, P., Shore, J., … Frueh, B. (2010). Telemedicine for anger management therapy in a rural population of combat veterans with posttraumatic stress disorder: A randomized noninferiority trial. *Journal of Clinical Psychiatry*, *71*(7), 855–863. https://doi.org/10.4088/JCP.09m05604blu.

Novaco, R. W., & Chemtob, C. M. (2002). Anger and combat-related posttraumatic stress disorder. *Journal of Traumatic Stress*, *15*(2), 123–132. https://doi.org/10.1023/A:1014855924072.

Novaco, R. W., Swanson, R. D., Gonzalez, O. I., Gahm, G. A., & Reger, M. D. (2012). Anger and postcombat mental health: Validation of a brief anger measure with US soldiers post-deployed from Iraq and Afghanistan. *Psychological Assessment*, *24*(3), 661–675.

Olatunji, B. O., Ciesielski, B. G., & Tolin, D. F. (2010). Fear and loathing: A meta-analytic review of the specificity of anger in PTSD. *Behavior Therapy*, *41*(1), 93–105. [http://dx.doi.org/S0005-7894(09)00056-2].

Orth, U., Cahill, S. P., Foa, E. B., & Maercker, A. (2008). Anger and posttraumatic stress disorder symptoms in crime victims: A longitudinal analysis. *Journal of Consulting and Clinical Psychology*, *76*(2), 208–218. [http://dx.doi.org/2008-03290-004].

Orth, U., & Wieland, E. (2006). Anger, hostility, and posttraumatic stress disorder in trauma-exposed adults: A meta-analysis. *Journal of Consulting and Clinical Psychology*, *74*(4), 698–706. [http://dx.doi.org/2006-09621-007].

Resick, P. A., Galovski, T. E., O'Brien Uhlmansiek, M., Scher, C. D., Clum, G. A., & Young-Xu, Y. (2008). A randomized clinical trial to dismantle components of cognitive processing therapy for posttraumatic stress disorder in female victims of interpersonal violence. *Journal of Consulting and Clinical Psychology*, *76*(2), 243–258 [http://dx.doi.org/2008-03290-007].

Resick, P. A., Monson, C. M., & Chard, K. M. (2014). *Cognitive processing therapy: Veteran/military version*. Washington, DC: Department of Veterans' Affairs.

Riggs, D. S., Dancu, C. V., Gershuny, B. S., Greenburg, D., & Foa, E. B. (1992). Anger and posttraumatic stress disorder in female crime victims. *Journal of Traumatic Stress*, *5*(4), 613–625.

Rizvi, S. L., Vogt, D. S., & Resick, P. A. (2009). Cognitive and affective predictors of treatment outcome in cognitive processing therapy and prolonged exposure for posttraumatic stress disorder. *Behaviour Research and Therapy*, *47*(9), 737–743.

Roberge, E. M., Allen, N. J., Taylor, J. W., & Bryan, C. J. (2016). Relationship functioning in Vietnam veteran couples: The roles of PTSD and anger. *Journal of Clinical Psychology*, https://doi.org/10.1002/jclp.22301.

Ross, C. E., & Van Willigen, M. (1996). Gender, parenthood, and anger. *Journal of Marriage and Family*, 58(3), 572–584. https://doi.org/10.2307/353718.

Shea, M. T., Lambert, J., & Reddy, M. K. (2013). A randomized pilot study of anger treatment for Iraq and Afghanistan veterans. *Behaviour Research and Therapy*, 51(10), 607–613. https://doi.org/10.1016/j.brat.2013.05.013.

Smith, H. L., Dillon, K. H., & Cougle, J. R. (2018). Modification of hostile interpretation bias in depression: A randomized controlled trial. *Behavior Therapy*, 49(2), 198–211. https://doi.org/10.1016/j.beth.2017.08.001.

Speckens, A. E., Ehlers, A., Hackmann, A., & Clark, D. M. (2006). Changes in intrusive memories associated with imaginal reliving in posttraumatic stress disorder. *Journal of Anxiety Disorders*, 20(3), 328–341.

Spielberger, C. (1983). *Manual for the state trait anxiety inventory (STAI)*. Palo Alto: Consulting Psychologists.

Spielberger, C. D. (1999). *State-Trait Anger Expression Inventory-2: STAXI-2*. Lutz, Florida: PAR, Psychological Assessment Resources.

Stapleton, J. A., Taylor, S., & Asmundson, G. J. (2006). Effects of three PTSD treatments on anger and guilt: Exposure therapy, eye movement desensitization and reprocessing, and relaxation training. *Journal of Traumatic Stress*, 19(1), 19–28. https://doi.org/10.1002/jts.20095.

Stevenson, V. E., & Chemtob, C. M. (2000). Premature treatment termination by angry patients with combat-related post-traumatic stress disorder. *Military Medicine*, 165(4), 422–424.

Stover, C. S., Meadows, A. L., & Kaufman, J. (2009). Interventions for intimate partner violence: Review and implications for evidence-based practice. *Professional Psychology: Research and Practice*, 40(3), 223–233. https://doi.org/10.1037/a0012718.

Strachan, C. E., & Dutton, D. G. (1992). The role of power and gender in anger responses to sexual jealousy. *Journal of Applied Social Psychology*, 22(22), 1721–1740. https://doi.org/10.1111/j.1559-1816.1992.tb00973.x.

Sullivan, C. P., & Elbogen, E. B. (2013). PTSD symptoms and family versus stranger violence in Iraq and Afghanistan veterans. *Law and Human Behavior*, https://doi.org/10.1037/lhb0000035.

Taft, C., Kaloupek, D., Schumm, J., Marshall, A., Panuzio, J., King, D., & Keane, T. (2007). Posttraumatic stress disorder symptoms, physiological reactivity, alcohol problems, and aggression among military veterans. *Journal of Abnormal Psychology*, 116(3), 498–507. [http://dx.doi.org/2007-11737-007].

Taft, C. T., Macdonald, A., Creech, S. K., Monson, C. M., & Murphy, C. M. (2016). A randomized controlled clinical trial of the strength at home men's program for partner violence in military veterans. *Journal of Clinical Psychiatry*, 77(9), 1168–1175. https://doi.org/10.4088/JCP.15m10020.

Taft, C., Watkins, L. E., Stafford, J., Street, A. E., & Monson, C. M. (2011). Posttraumatic stress disorder and intimate relationship problems: A meta-analysis. *Journal of Consulting and Clinical Psychology*, 79(1), 22–33. [http://dx.doi.org/2011-01388-003].

Taft, C. T., Weatherill, R. P., Scott, J. P., Thomas, S. A., Kang, H. K., & Eckhardt, C. I. (2015). Social information processing in anger expression and partner violence in returning U.S. veterans. *Journal of Traumatic Stress*, 28(4), 314–321. https://doi.org/10.1002/jts.22017.

Taylor, S. (2003). Outcome predictors for three PTSD treatments: Exposure therapy, EMDR, and relaxation training. *Journal of Cognitive Psychotherapy*, 17, 149–162.

Taylor, S., Fedoroff, I. C., Koch, W. J., Thordarson, D. S., Fecteau, G., & Nicki, R. M. (2001). Posttraumatic stress disorder arising after road traffic collisions: Patterns of response to cognitive–behavior therapy. *Journal of Consulting and Clinical Psychology, 69*(3), 541–551.

Teten, A. L., Miller, L. A., Stanford, M. S., Petersen, N. J., Bailey, S. D., Collins, R. L., ... Kent, T. A. (2010). Characterizing aggression and its association to anger and hostility among male veterans with post-traumatic stress disorder. *Military Medicine, 175*(6), 405–410.

Teten, A. L., Schumacher, J. A., Taft, C. T., Stanley, M. A., Kent, T. A., Bailey, S. D., ... White, D. L. (2010). Intimate partner aggression perpetrated and sustained by male Afghanistan, Iraq, and Vietnam veterans with and without posttraumatic stress disorder. *Journal of Interpersonal Violence, 25*(9), 1612–1630. https://doi.org/10.1177/0886260509354583.

Thomas, S. P. (1989). Gender differences in anger expression: Health implications. *Research in Nursing and Health, 12*(6), 389–398. https://doi.org/10.1002/nur.4770120609.

Thomas, S. A., & Gonzalez-Prendes, A. A. (2009). Powerlessness, anger, and stress in African American women: Implications for physical and emotional health. *Health Care for Women International, 30*(1–2), 93–113. https://doi.org/10.1080/07399330802523709.

Turner, S. M., Beidel, D. C., & Frueh, B. C. (2005). Multicomponent behavioral treatment for chronic combat-related posttraumatic stress disorder: Trauma management therapy. *Behavior Modification, 29*(1), 39–69. https://doi.org/10.1177/0145445504270872.

Van Minnen, A., Arntz, A., & Keijsers, G. P. J. (2002). Prolonged exposure in patients with chronic PTSD: Predictors of treatment outcome and dropout. *Behaviour Research and Therapy, 40*, 439–457.

Van Voorhees, E. E., Dennis, P. A., Elbogen, E. B., Fuemmeler, B., Neal, L. C., Calhoun, P. S., & Beckham, J. C. (2018). Characterizing anger-related affect in individuals with post-traumatic stress disorder using ecological momentary assessment. *Psychiatry Research, 261*, 274–280. https://doi.org/10.1016/j.psychres.2017.12.080.

Van Voorhees, E. E., Dennis, P. A., Neal, L. C., Hicks, T. A., Calhoun, P. S., Beckham, J. C., & Elbogen, E. B. (2016). Posttraumatic stress disorder, hostile cognitions, and aggression in Iraq/Afghanistan era veterans. *Psychiatry, 79*(1), 70–84. https://doi.org/10.1080/00332747.2015.1123593.

Van Voorhees, E. E., Dillon, K. H., Wilson, S., Dennis, P. A., Neal, L. C., Mendenblik, A., & Beckham, J. C. (2019). A comparison of group anger management treatments for combat veterans with PTSD: Results from a quasi-experimental trial. *Journal of Interpersonal Violence*, https://doi.org/10.1177/0886260519873335. [advance online publication].

Van Voorhees, E. E., Dillon, K. H., Wilson, S. M., Medenblik, A. M., Dennis, P. A., Neal, L. C., ... Beckham, J. C. (2019). Gender differences in outcomes of group therapy for anger in combat veterans with posttraumatic stress disorder (submitted for publication).

van Zuiden, M., Kavelaars, A., Rademaker, A. R., Vermetten, E., Heijnen, C. J., & Geuze, E. (2011). A prospective study on personality and the cortisol awakening response to predict posttraumatic stress symptoms in response to military deployment. *Journal of Psychiatric Research, 45*(6), 713–719. https://doi.org/10.1016/j.jpsychires.2010.11.013.

Wenzel, A., & Lystad, C. (2005). Interpretation biases in angry and anxious individuals. *Behaviour Research and Therapy, 43*(8), 1045–1054.

Whiting, D., & Bryant, R. A. (2007). Role of appraisals in expressed anger after trauma. *Clinical Psychologist, 11*(1), 33–36. https://doi.org/10.1080/13284200601178136

Wilk, J. E., Quartana, P. J., Clarke-Walper, K., Kok, B. C., & Riviere, L. A. (2015). Aggression in US soldiers post-deployment: Associations with combat exposure and PTSD and the moderating role of trait anger. *Aggressive Behavior, 41*(6), 556–565. https://doi.org/10.1002/ab.21595.

Wilkowski, B. M., & Robinson, M. D. (2008). The cognitive basis of trait anger and reactive aggression: An integrative analysis. *Personality and Social Psychology Review, 12*(1), 3–21.

Wilkowski, B. M., & Robinson, M. D. (2010). The anatomy of anger: An integrative cognitive model of trait anger and reactive aggression. *Journal of Personality, 78*(1), 9–38.

Worthen, M., Rathod, S. D., Cohen, G., Sampson, L., Ursano, R., Gifford, R., … Ahern, J. (2014). Anger problems and posttraumatic stress disorder in male and female National Guard and Reserve Service members. *Journal of Psychiatric Research, 55*, 52–58. https://doi.org/10.1016/j.jpsychires.2014.04.004.

Worthen, M., Rathod, S. D., Cohen, G., Sampson, L., Ursano, R., Gifford, R., … Ahern, J. (2015). Anger and posttraumatic stress disorder symptom severity in a trauma-exposed military population: Differences by trauma context and gender. *Journal of Traumatic Stress, 28*(6), 539–546. https://doi.org/10.1002/jts.22050.

Woud, M. L., & Becker, E. S. (2014). Editorial for the special issue on cognitive bias modification techniques: An introduction to a time Traveller's tale. *Cognitive Therapy and Research, 38*(2), 83–88. https://doi.org/10.1007/s10608-014-9605-0.

Woud, M. L., Verwoerd, J., & Krans, J. (2017). Modification of cognitive biases related to posttraumatic stress: A systematic review and research agenda. *Clinical Psychology Review, 54*, 81–95. https://doi.org/10.1016/j.cpr.2017.04.003.

CHAPTER 4

Sadness and depression in PTSD

Blair E. Wisco, Cameron P. Pugach, Faith O. Nomamiukor
University of North Carolina at Greensboro, Greensboro, NC, United States

Overview

Sadness is an emotional reaction to loss or failure characterized by the subjective feeling of unhappiness, negative thoughts, and a behavioral tendency toward inaction or withdrawal. Sadness is considered core to our experience as human beings; to the extent that some emotions can be considered "basic" emotions, sadness is among their ranks (Ekman, 1992; Izard, 1992). In relation to psychopathology, sadness is most readily associated with depression, given that prolonged sad mood is one of the two cardinal symptoms of major depressive disorder (MDD) (American Psychiatric Association, 2013). Sadness may also play an important role in PTSD, but sadness has received relatively less attention in the context of PTSD compared with depression.

In this chapter, we summarize the empirical evidence for the role of sadness and sadness-related difficulties in the phenomenology of PTSD. We start by summarizing epidemiological evidence that depressive disorders are highly comorbid with PTSD, motivating a focus on sadness in PTSD. We then discuss the available research examining the specific emotion of sadness in the development and maintenance of PTSD, which is surprisingly limited given the widespread recognition of depression-PTSD comorbidity. Next, we describe the relatively more developed literature on anhedonia, or the absence of pleasure, in PTSD. We then discuss implications of sadness and anhedonia for PTSD presentation. Specifically, we summarize the evidence that sadness and/or depression can lead to dissociation, rumination, social difficulties, and even suicide among individuals with PTSD. Finally, we discuss the implications of these findings for PTSD treatment.

Comorbidity of PTSD and depression

Epidemiological data indicate that PTSD and depression are highly comorbid, with approximately half of individuals diagnosed with PTSD also meeting criteria for comorbid depression (Kessler, Sonnega, Bromet, Hughes, &

Nelson, 1995; Rytwinski, Scur, Feeny, & Youngstrom, 2013). Estimates of comorbidity have varied widely and have differed depending on a number of characteristics including the type of population assessed and how PTSD and depression are measured. In nationally representative samples of the US population, rates of depression among individuals diagnosed with PTSD have ranged from 35% (Pietrzak, Goldstein, Southwick, & Grant, 2011) to 48% (Kessler et al., 1995). In a comprehensive meta-analysis of *DSM-IV-TR*-referenced diagnoses assessed using clinical interviews, Rytwinski et al. (2013) found that, on average, 52% of individuals with current *DSM-IV-TR* PTSD also met criteria for current *DSM-IV-TR* MDD (95% CI = 48%–56%; Rytwinski et al., 2013). They also found that depression rates were higher in military samples (compared with civilian samples) and for PTSD resulting from interpersonal trauma (compared with natural disasters).

Although comorbidity data using *DSM-5* diagnostic criteria are in the early stages, recent studies suggest that PTSD-depression comorbidity rates are similar to prior studies using *DSM-IV* referenced measures. In a nationally representative sample of US military veterans, 42% of veterans with lifetime *DSM-5* PTSD reported lifetime history of MDD, with even higher comorbidity for past-month diagnoses (57%; Wisco et al., 2016). Similar findings were reported from the veteran subsample of National Epidemiologic Survey on Alcohol and Related Conditions (NESARC)-III, another nationally representative sample, with 61.72% of veterans with *DSM-5* PTSD also meeting criteria for a comorbid mood disorder (including MDD, persistent depression, and/or bipolar I or II disorder; Smith, Goldstein, & Grant, 2016). Thus, early evidence suggests that depression–PTSD comorbidity is fairly consistent across recent DSM versions, although more research is needed, particularly with *DSM-5* diagnoses in nonveteran samples.

Sadness in PTSD

Compared with the literature on depression and PTSD, the role of sadness in PTSD has received relatively little attention, perhaps due to dominant theories of PTSD as a fear-based disorder (e.g., Foa & Kozak, 1986). For example, at the time we are writing this chapter, a keyword search in PsycInfo for "depression" and "PTSD or posttraumatic stress disorder or post-traumatic stress or posttraumatic stress disorder" yields 2275 results. After replacing "depression" with "sadness," the same search yields just seven results. However, this keyword search is perhaps a bit misleading, because many PTSD studies have included an assessment of sadness even if sadness was not the primary focus of the study. Some studies have examined sadness in the context of peritraumatic emotions (emotions experienced at the time of the traumatic event) and their relation to the development of PTSD/PTSD symptoms, and others have examined trait-like sad mood or momentary experiences of sadness in daily life among those diagnosed with PTSD versus other disorders. The available literature indicates that sadness is a prominent, if not unique, emotion associated with PTSD.

In the literature on peritraumatic emotions, sadness is a common immediate reaction to traumatic events and one significantly associated with PTSD symptoms. In one study of police officers and a convenience comparison sample of community participants, 75% of police officers and 88% of community participants endorsed peritraumatic sadness at time of their index trauma (Brunet et al., 2001). These rates of endorsement were similar to those observed for peritraumatic anger and helplessness (70%–90%) and higher than other peritraumatic emotions, such as fear (52%; Brunet et al., 2001). A study of veterans of the wars in Iraq and Afghanistan found a similar but slightly different pattern—in that study, anger (86%) was the most commonly endorsed peritraumatic emotion, followed by fear and sadness (70%–79%) and then by disgust, helplessness, numbness, and horror (60%–68%; Engel-Rebitzer et al., 2017). Other studies have asked participants to select which of many possible peritraumatic emotions was "dominant" at the time of their worst trauma. In two studies of trauma-exposed undergraduates, sadness was the most frequently reported "dominant" peritraumatic emotion, and participants who endorsed sadness reported similar levels of PTSD symptoms as participants who endorsed other dominant peritraumatic emotions, such as fear, anger, or disgust (Berntsen & Rubin, 2007; Hathaway, Boals, & Banks, 2010). Thus sadness is not only frequently endorsed as one of many peritraumatic emotions but also most commonly selected as the *main* emotion that participants associate with their trauma, when forced to choose just one emotion.

Providing further support for a role of peritraumatic sadness in PTSD, Lancaster, Melka, and Rodriguez (2011) found that peritraumatic sadness was significantly associated with PTSD symptoms in a sample of trauma-exposed undergraduates, even after controlling for other peritraumatic emotions. In that study, anger, guilt, and disgust also significantly predicted PTSD, indicating that sadness was significantly but not uniquely related to PTSD symptoms. However, in one of the few longitudinal studies on this topic, peritraumatic sadness did not significantly predict later PTSD among veterans of the wars in Iraq and Afghanistan, after controlling for baseline PTSD, depression, and other peritraumatic emotions (Engel-Rebitzer et al., 2017). In that study, numbness was the only peritraumatic emotion associated with later PTSD, although we should note that the sample was drawn from a PTSD registry with very high rates of PTSD at baseline, complicating interpretation of these results. Thus future longitudinal research is needed to clarify whether peritraumatic sadness prospectively predicts the development of PTSD.

The literature on peritraumatic sadness in PTSD is marked by some significant limitations—peritraumatic emotions are, by necessity, examined retrospectively, and many of the studies cited earlier relied upon self-report measures of both traumatic events and PTSD severity. Clinician-administered measures may be better able to distinguish PTSD-specific symptoms from general distress or comorbid conditions such as depression. Despite these limitations the available evidence suggests that sadness is a frequent peritraumatic emotional response and is associated, at least retrospectively, with similar levels of current PTSD symptoms as peritraumatic fear and other negative emotions, highlighting the importance of focusing on sadness in future research on PTSD.

Other studies examining sadness in PTSD have measured emotional responses to trauma reminders, trait-like mood, or day-to-day experiences of sadness. In a study of naturally occurring trauma reminders, Reynolds and Brewin (1999) examined a range of emotional responses to intrusive memories among individuals diagnosed with PTSD and/or depression using a semistructured interview. They found that both groups reported a number of negative emotions in response to their intrusive memories, with anger being the emotion most frequently reported, followed by sadness, fear, helplessness, and guilt (Reynolds & Brewin, 1999). Similarly, symptom provocation studies using script-driven imagery, in which scripts of the index trauma are written, audio-recorded, and played back to the participant while subjective and physiological responses are recorded, have found that trauma scripts typically elicit a range of negative emotions among individuals diagnosed with PTSD (e.g., Pineles et al., 2013). Emotional responses to trauma scripts include not only sadness but also fear, anger, and disgust, and these emotions are stronger for individuals diagnosed with PTSD than trauma-exposed individuals who did not develop PTSD (Pineles et al., 2013).

In one of the few studies of trait-like sadness in PTSD, Finucane, Dima, Ferreira, and Halvorsen (2012) compared four groups (PTSD, depression, chronic pain, and healthy controls) on the basic emotion scale (Dalgleish & Power, 2004), a self-report measure of the trait experience of several emotions. They found that the healthy control group reported happiness as their most frequently endorsed emotion, followed by fear, then anger, then sadness, and disgust (which did not differ from each other). The PTSD group reported that fear was the most frequently endorsed emotion, followed by anger, sadness, disgust, and happiness. Surprisingly the depressed group reported a similar pattern (fear, anger, sadness, disgust, and then happiness). When directly compared with the healthy control group, both the depressed and PTSD groups reported significantly more sadness and less happiness. The PTSD group also reported significantly more fear, anger, and disgust than the healthy control group. Although this pattern could reflect greater severity in the PTSD group (the groups were not matched on severity or demographic characteristics, introducing potential confounds), these results are broadly consistent with the idea that sadness is a common but not unique negative emotion associated with PTSD.

Several experience-sampling studies have also been conducted to examine the day-to-day emotional experiences of individuals with PTSD (see Chun, 2016 for a review). Experience sampling designs offer many advantages over retrospective report of momentary emotions, in terms of enhanced accuracy and possibly less biased memory. Although several experience-sampling studies have been conducted in PTSD populations, most have assessed either specific PTSD symptoms or general mood states (e.g., negative affect), and have not included assessment of specific emotions such as sadness (Chun, 2016). To our knowledge, only one experience-sampling study has assessed the momentary experience of sadness in PTSD. Golier, Yehuda, Schmeidler, and Siever (2001) assessed male combat veterans diagnosed with PTSD and two comparison groups of men diagnosed with major depressive disorder (MDD) and men with no history of mental illness. They conducted experience sampling over a 24-h period while blood samples were collected. They found that men with PTSD and MDD reported significantly more sadness than the healthy control group, and did not significantly differ from each other. In contrast, the PTSD group reported higher levels of anxiety than the MDD group, who reported higher anxiety than the healthy controls.

Taken together, the available literature suggests that individuals diagnosed with PTSD frequently experience sadness at the time their index traumatic event occurs, when they are later reminded of that event, and generally in their day-to-day lives. That being said, sadness is only one of a number of negative emotions frequently experienced by individuals with PTSD, with fear, anger, and disgust also emerging as important emotions for consideration. Although sadness is not the only negative emotion experienced by individuals with PTSD, the dearth of research focusing on sadness specifically is surprising given its frequent endorsement in this population.

Anhedonia/emotional numbing in PTSD

As mentioned previously, prolonged sad mood is one of two cardinal symptoms of major depressive disorder (MDD) according to the *DSM-5* (American Psychiatric Association, 2013). The other cardinal symptom of MDD is anhedonia, or the loss of interest or pleasure in activities that one previously enjoyed. Whereas sadness indicates the presence of a negative emotional state, anhedonia indicates the absence of positive emotions. In the context of PTSD, anhedonia has received considerably more attention than sadness, although anhedonia has typically not been studied on its own, but rather as one of a cluster of "emotional numbing" symptoms of PTSD. Emotional numbing in PTSD has been defined in different ways but typically consists of the following symptoms: a diminished interest or participation in activities, detachment or estrangement from others, and a restricted range of affect (now called persistent inability to experience positive emotions). The current section will begin with a discussion of emotional numbing definitions across *DSM* PTSD nosology, followed by contemporary theories of emotional numbing, and will conclude with an examination of the impact of emotional numbing on psychosocial functioning among those with PTSD.

The nosology of PTSD, as defined in *DSM-IV* (American Psychiatric Association, 1994), spurred a great deal of discussion pertaining to the categorization and composition of PTSD (McHugh & Treisman, 2007; McNally, 2009; North, Suris, Davis, & Smith, 2009; for a review, see Pai, Suris, & North, 2017; Spitzer, First, & Wakefield, 2007). Among other criteria, *DSM-IV* stipulated that an individual must meet a requisite number of 17 PTSD symptoms comprising three distinct symptom clusters: reexperiencing, avoidance/numbing, and arousal. Briefly the reexperiencing cluster (Criterion B) included five symptoms all related to persistent trauma-related recollections, the avoidance/numbing cluster (Criterion C) was composed of efforts to avoid trauma-related thoughts or situations, an inability to recall important aspects of the trauma, diminished interest in activities, feelings of detachment or estrangement from others, a restricted range of affect, and a sense of a foreshortened future, and the arousal cluster (Criterion D) consisted of impaired sleep, irritability or outbursts of anger, impaired concentration, hypervigilance, and an exaggerated startle response (American Psychiatric Association, 1994).

Many researchers critiqued the *DSM-IV*'s organization of PTSD symptoms into these three symptom clusters, particularly the avoidance/numbing and arousal clusters. Dissent from the three-factor model was largely derived from the results of a line of factor analytic studies, which yielded two four-factor models that yielded better fit than the *DSM-IV*'s three cluster approach (King, Leskin, King, & Weathers, 1998; Simms, Watson, & Doebbeling, 2002). The emotional numbing model (King et al., 1998), which produced structurally similar symptom clusters to the presently defined *DSM-5* PTSD criteria, consisted of the reexperiencing and arousal clusters originally put forth by *DSM-IV*, but separated Criterion C into two separate clusters of avoidance and emotional numbing, the latter of which was composed of five symptoms: an inability to recall important aspects of the trauma, diminished interest in activities, feelings of detachment and/or estrangement from others, a restricted range of affect, and a sense of foreshortened future. Indeed, evidence has indicated that the avoidance and emotional numbing clusters differentially predict treatment response and treatment outcome correlates, psychopathology, and physiological indices of attention, supporting their distinctiveness (Asmundson, Stapleton, & Taylor, 2004). The second model, termed the dysphoria model (Simms et al., 2002), found evidence for an eight-symptom factor called dysphoria, which groups together the emotional numbing symptoms with three symptoms from the arousal cluster.

While the emotional numbing and dysphoria models have received comparable empirical support (Elhai & Palmieri, 2011), the *DSM-5* elected to create a new criterion termed "negative alterations in cognition in mood" (NACM), around which symptoms of the two models converged. Thus the NACM cluster includes the primary emotional numbing symptoms previously described (i.e., an inability to recall important aspects of the trauma, diminished interest or participation in activities, feelings of detachment or estrangement from others, and a persistent inability to experience positive emotions), as well as commonly observed trauma-related sequelae such as persistent and exaggerated negative beliefs about oneself, others, or the world; persistent distorted cognitions of self-blame for the causes or consequences of the traumatic event; and a persistent negative emotional state (e.g., fear, horror, and anger; American Psychiatric Association, 2013). Although emotional numbing symptoms are subsumed within the larger NACM cluster within *DSM-5*, factor analyses do suggest that the anhedonia loads onto a different factor than the other NACM symptoms, offering further support for the distinction of emotional numbing within PTSD (Armour et al., 2015; Pietrzak et al., 2015; Seligowski & Orcutt, 2016; but see also Rasmussen, Verkuilen, Jayawickreme, Wu, & McCluskey, 2018, for a critique). Thus the literature supporting a role for anhedonia in PTSD is somewhat complex given that no single definition of emotional numbing has predominated over time, but generally supports the fact that individuals with PTSD experience emotional numbing symptoms comprising anhedonia and related difficulties.

Theoretical underpinnings of anhedonia/emotional numbing in PTSD

Emotional numbing is unique when considering PTSD symptoms. The classical reexperiencing, avoidance, and hyperarousal clusters maintained in *DSM-5* facilitate negative subjective, physiological, and behavioral experiences that contribute to heightened levels of distress. By contrast, emotional numbing involves reductions in goal-oriented behavior in response to positive stimuli and incentives (Litz, 1992; Litz, Orsillo, Kaloupek, & Weathers, 2000). These fundamental disparities in emotional experience when considering PTSD symptom clusters are reflected by research showing that, while reexperiencing, avoidance, and hyperarousal symptoms are moderately correlated and statistically coalesce to form a PTSD factor, emotional numbing symptoms (along with some nonspecific arousal symptoms) are distinguishable from this factor and tend to be more strongly associated with depression and anxiety (Gros, Simms, & Acierno, 2010; Kashdan, Elhai, & Frueh, 2006; Simms et al., 2002).

Consequently, emotional numbing has been studied as a mechanism explaining the high comorbidity between PTSD and MDD. Specifically the emotional numbing symptoms of PTSD, typically defined as diminished interest in activities, feelings of detachment or estrangement from others, and a restricted range of affect, have been shown to be conceptually and empirically related to depressive anhedonia over and above any other PTSD symptom cluster (Kashdan et al., 2006). In the context of depression, anhedonia is conceptualized as a failure to generate emotional responses to objectively positive stimuli (Rottenberg, Kasch, Gross, & Gotlib, 2002) and is characterized by low levels of positive affect and disinterest in previous pleasurable activities. Thus both emotional numbing and anhedonia appear to reflect diminished reward functioning (i.e., reduced positive affect, reward sensitivity, and movement toward appetitive goals) common across depression and PTSD (Davidson, 1994; Nawijn et al., 2015).

Past researchers have suggested that broad-based emotion regulation deficits, particularly concerning emotional expression and emotional stimulation, may account for emotional numbing and anhedonia (Kashdan et al., 2006). In fact, Litz and Gray (2001) proposed a model of emotional numbing wherein individuals with PTSD maintain complete access to their pretraumatic emotional repertoires, but emotional numbing symptoms arise due to inhibited emotional expression (rather than an inability to feel positive emotions). The authors further suggest that those with PTSD require higher stimulation to feel positive affect compared with individuals without PTSD (Litz & Gray, 2001). This theory is corroborated by accounts of veterans with PTSD reporting substantially higher rates of actively withholding emotional expression compared with veterans without PTSD (Roemer, Litz, Orsillo, & Wagner, 2001). Additional support for this theory comes from research showing that individuals with PTSD have difficulty identifying and describing feelings, particularly positive emotions (Frewen, Dozois, Neufeld, & Lanius, 2008, 2011). Moreover, individuals with PTSD may have difficulty describing their emotions because of various negative beliefs (e.g., beliefs causing shame or fear) that serve to inhibit the expression and to a lesser extent the identification of emotions (Frewen et al., 2011).

Taken together, results indicate that, in the context of PTSD, emotional numbing may not most aptly be described as numbing of all emotions, but rather numbing of positive emotions specifically. Theory suggests that individuals with PTSD do not suffer from core information processing deficits in the ability to feel or express positive emotions (Litz & Gray, 2001). Rather, hyperarousal and sensitization to negative affect may compete with the expression of positive affect. In essence, individuals with PTSD may require positive stimuli of a heightened intensity threshold to generate positive emotional experiences, and these individuals may also be particularly likely to hold maladaptive beliefs that inhibit the expression and identification of such positive emotions (Frewen et al., 2008, 2011; Roemer et al., 2001). Additionally, emotional numbing appears to be functionally related to depressive anhedonia, with both constructs potentially manifesting as a result of emotion regulation deficits and reflecting diminished reward functioning—and in turn diminished positive affect—seen in PTSD and depression. However, while research has indicated that the symptom overlap may partially explain comorbidity between PTSD and depression, no study has systematically differentiated emotional numbing in PTSD from anhedonia in depression on conceptual, functional, or empirical grounds. Efforts to do so are important as they may yield important insights into the emotional processes underlying PTSD and depression.

Costs of emotional numbing in PTSD

Relative to other PTSD symptoms, one might expect particularly strong associations between emotional numbing and psychosocial impairment, given the diminished positive affect associated with emotional numbing. In comparison with negative emotions, positive emotions are related to the broadening and building of social, physical, and intellectual resources (Fredrickson, 1998), suggesting that positive emotions are particularly important for psychosocial functioning. Consistent with expectations, research has supported associations between the emotional numbing symptoms of PTSD and impairment in psychosocial functioning across a range of trauma-exposed populations including combat veterans (Hassija, Jakupcak, & Gray, 2012), women exposed to intimate partner violence (Birkley, Eckhardt, & Dykstra, 2016), motor vehicle accident survivors (Clapp, Beck, Palyo, & Grant, 2008; Kuhn, Blanchard, & Hickling, 2003), Cambodian refugees (Palmieri, Marshall, & Schell, 2007), and disaster workers and college students indirectly exposed to the September 11th attacks (Baschnagel, O'Conner, Colder, & Hawk, 2005; Hunnicutt-Ferguson et al., 2018).

Emotional numbing symptoms have been linked to various aspects of impaired global adjustment and distress. In a review of studies examining emotional numbing symptoms among Iraq and Afghanistan combat veterans, Hassija et al. (2012) showed that the numbing symptoms of the four-factor emotional numbing model (King et al., 1998) are uniquely associated with family problems including role-related readjustment, poor sexual functioning, and increased risk for suicidality. In studies that utilized the four-factor dysphoria model (Simms et al., 2002), the dysphoria factor was independently and strongly associated with a range of psychosocial variables including greater psychosocial difficulties (i.e., family, peer, work, school, and financial problems), reduced perceptions of unit and postdeployment social support, increased suicidal ideation, and decreased psychological resilience (e.g., Pietrzak, Goldstein, Malley, Rivers, & Southwick, 2010). Results from another study using the dysphoria model showed that the dysphoria factor was the most salient predictor of future intimate relationship and family adjustment problems. In the one study that examined psychosocial functioning using the three-factor model of *DSM-IV* PTSD, results showed that avoidance/numbing symptoms were the most robust predictor of interpersonal, occupational, and social functioning, as well as overall life satisfaction (Hassija et al., 2012). Finally, research with the seven-factor model of *DSM-5* PTSD indicates that the anhedonia factor is most strongly related to quality of life measures.

Importantly, similar results have been reported among diverse trauma-exposed populations. For example, a recent meta-analysis by Birkley et al. (2016) reported results from 23 studies demonstrating that emotional numbing symptoms as assessed by the four-factor emotional numbing model are moderately to largely associated with problems involving relationship functioning and intimacy; intimate partner violence; and parent, child, and family functioning. In an examination of motor vehicle accident survivors, the emotional numbing factor was associated with decreased life satisfaction and role functioning (Clapp et al., 2008). Finally, in a study of disaster recovery workers in the World Trade Center attacks, numbing/avoidance symptoms were prospectively associated with increased subjective distress and deficits in social functioning (Hunnicutt-Ferguson et al., 2018).

Taken together, results suggest that emotional numbing symptoms—assessed by both the emotional numbing and dysphoria models of PTSD—impact the presence and severity of a broad range of psychosocial functioning across a range of trauma-exposed populations. In many cases, emotional numbing symptoms were uniquely related to important aspects of psychosocial functioning above and beyond the effects of other PTSD symptoms. Although this has yet to be empirically tested, we suggest that the unique relations between emotional numbing and psychosocial functioning may be due to the diminished positive affect specific to emotional numbing.

Implications for PTSD presentation

The common comorbidity between depression and PTSD and the frequent experiences of sadness and anhedonia/emotional numbing among those diagnosed with PTSD have important implications for the phenomenology of PTSD. Dissociation has long been recognized as a potential trauma sequela, and dissociation has often been conceptualized as an extreme form of emotional numbing. Thus, it is important to consider to what extent emotional numbing symptoms and/or depression comorbidity may explain the experience of dissociative symptoms in PTSD. Additionally, some symptoms more typically associated with sadness or depression (rumination, social withdrawal, and suicide) are now being increasingly studied in relation to PTSD. Here, we consider the extent to which depression comorbidity may contribute to dissociation, rumination, social withdrawal, and suicide in the lives of individuals with PTSD.

Dissociation

Dissociation is a common psychological response during or following a traumatic event and is characterized by individuals feeling detached from reality and disconnected from their own thoughts, emotions, actions, or identity (American Psychiatric Association, 2013). Common symptoms of dissociation include gaps in one's memory and difficulty in social or occupational functioning (American Psychiatric Association, 2013). Dissociation is related to many trauma-related psychological disturbances such as PTSD, emotional numbing, and depression (Armour, Karstoft, & Richardson, 2014; Feeny, Zoellner, Fitzgibbons, & Foa, 2000; Harvey & Bryant, 1999; McCanlies, Sarkisian, Andrew, Burchfiel, & Violanti, 2017). In fact, an emerging body of literature suggests that dissociation, emotional numbing, depression, and PTSD are all bidirectionally interrelated to one another (Armour et al., 2014; Feeny et al., 2000; Harvey & Bryant, 1999; McCanlies et al., 2017), and past research demonstrates that numbing symptoms are especially related to increased dissociative symptoms (Feeny et al., 2000; Harvey & Bryant, 1999). In addition, research also shows that the comorbidity of major depression and PTSD are also strongly related to elevated dissociative symptoms (Armour et al., 2014; Johnson, Pike, & Chard, 2001).

Research indicates that a subset of individuals with PTSD have a unique symptom presentation with strong dissociative symptoms (Steuwe, Lanius, & Frewen, 2012; Wolf et al., 2012, 2012). For these reasons, the *DSM-5* has now added a dissociative subtype of PTSD (American Psychiatric Association, 2013). Hallmark symptoms of the dissociative subtype of PTSD are depersonalization and derealization. Depersonalization symptoms make individuals feel that they are having "out of body" experiences, whereas derealization symptoms cause individuals to feel that their environment or certain events are not real (American Psychiatric Association, 2013). The dissociative subtype of PTSD was added because strong evidence suggests that about 12%–30% of individuals with PTSD have either depersonalization or derealization symptoms (Steuwe et al., 2012; Wolf, Lunney, et al., 2012; Wolf, Miller, et al., 2012). Additionally, specific symptoms of the dissociative subtype of PTSD were found to respond differently to common PTSD treatments, with research suggesting that cognitive restructuring and exposure based therapies are more useful for this subtype (Cloitre, Petkova, Wang, & Lu Lassell, 2012; Lanius, Brand, Vermetten, Frewen, & Spiegel, 2012; Resick, Suvak, Johnides, Mitchell,

& Iverson, 2012). In addition, individuals with the dissociative subtype of PTSD are more likely to endorse greater exposure to childhood and adult sexual trauma compared with individuals with nondissociative PTSD (Wolf, Miller, et al., 2012).

Understanding the relationship between dissociation and PTSD

When present, dissociative symptoms are thought to play a significant role in the maintenance and development of PTSD (Brewin, 2001; Feeny et al., 2000; van der Hart, Bolt, & van der Kolk, 2005; Wolf, Lunney, et al., 2012; Wolf, Miller, et al., 2012). Researchers theorize that dissociation can play a key role in PTSD for some populations because it causes incomplete cognitive processing of the traumatic experience, disrupting memory storage and retrieval, thereby leading to the development of PTSD (Brewin, 2001; van der Hart et al., 2005).

Past research demonstrates that dissociation has a strong relationship to increased PTSD symptoms (Armour et al., 2014; Brewin, 2001; Feeny et al., 2000; Ozer, Best, Lipsey, & Weiss, 2003; van der Hart et al., 2005), but there is debate as to whether dissociation is an independent predictor of PTSD even after controlling for other symptoms such as depression or emotional numbing. A meta-analysis by Ozer et al. (2003), covering 16 studies concerning dissociation, found that peritraumatic dissociation (PD), or dissociation during or immediately following a trauma, is a strong predicator of PTSD symptoms (Ozer et al., 2003). However, as noted by van der van der Velden and Wittmann (2008), one limitation to this meta-analysis is that they did not examine the independent predictive value of PD or control for possible effects of pretrauma risk factors (van der Velden & Wittmann, 2008). van der Velden and Wittmann's (2008) systematic literature review of 17 studies found that, after accounting for prior psychiatric disturbances, like depression or emotional numbing, PD no longer predicted PTSD symptoms (van der Velden et al., 2006; Wittmann, Moergeli, & Schnyder, 2006). Rather, prior psychiatric disturbances were better predictors of PTSD symptomatology than PD. This finding demonstrates that analyzing dissociation and PTSD symptoms without controlling for other psychological problems, such as depression or numbing, can lead researchers to erroneously overestimate the predictive power of dissociation on PTSD.

Thus dissociation is a well-recognized trauma reaction that has long been studied in the context of PTSD. Dissociation is particularly associated with the emotional numbing symptoms of PTSD, and researchers have recognized the possible overlap between dissociation, emotional numbing, and comorbid depression in PTSD presentation. The available evidence suggests that dissociation, emotional numbing, and depression are separable constructs, but the extent to which they offer independent contributions to the prediction of PTSD symptoms remains somewhat unclear.

Rumination

In contrast to dissociation, which is typically associated with emotional numbing, symptoms such as rumination have been conceptualized primarily as reactions to sadness. Rumination was first described as a risk factor for depression in the response styles theory (Nolen-Hoeksema, 1991), which defined rumination as a cognitive response to distress that involves repetitively and passively thinking about the symptoms of distress, the meaning of distress, and the causes and consequences of distress. More recently, however, rumination has been identified as a transdiagnostic form of repetitive negative thought that involves disorder-specific content and has been shown to predict multiple forms of psychopathology, including PTSD (Aldao, Nolen-Hoeksema, & Schweizer, 2010; Ehlers & Clark, 2000; Ehring & Watkins, 2008; Nolen-Hoeksema, Wisco, & Lyubomirsky, 2008; Szabo, Warnecke, Newton, & Valentine, 2017). Multiple types of rumination have been described, all of which are united by core ruminative processes yet differentiated by the content of their thoughts (Siegle, Moore, & Thase, 2004; Treynor, Gonzalez, & Nolen-Hoeksema, 2003). Most relevant to the experience of depression is brooding, which involves passive, abstract negative self-evaluations. Brooding has been consistently linked to negative outcomes including the onset of depression and has been functionally described as a conscious cognitive process through which individuals believe they will garner insight, meaning, and an ability to solve problems related to their distress (Nolen-Hoeksema et al., 2008). However, brooding is negatively related to problem solving, promotes immobilization via indecision, and is positively related to suppression and avoidance of distressing feelings and thoughts (Hong, 2007; Nolen-Hoeksema et al., 2008; Ward, Lyubomirsky, Sousa, & Nolen-Hoeksema, 2003).

The PTSD literature has also identified trauma-focused rumination, which involves repetitive and recurrent thinking about the trauma and its consequences, as a form of rumination particularly relevant to PTSD (Michael, Halligan, Clark, & Ehlers, 2007). Across an impressive body of cross-sectional, prospective, and experimental research, trauma-focused rumination has been clearly identified as a maintenance factor for PTSD (Michael et al., 2007). However, there is still contention regarding the incremental validity of measuring trauma-focused rumination versus more general measures of perseverative negative thinking, with a recent meta-analysis of rumination in PTSD demonstrating that trait rumination (i.e., brooding) is more robustly associated with PTSD symptoms than trauma-focused rumination (Szabo et al., 2017). These results indicate that, relative to event-specific measures of rumination, trait rumination may be more effective in capturing the maladaptive aspects of rumination involved in the maintenance of PTSD symptoms. Finally, when considering the functionality of rumination in PTSD, theorists have suggested that rumination serves as a maladaptive avoidant response to processing the traumatic event in a manner that produces a sense of serious and current threat (Ehlers & Clark, 2000). This cognitive model of PTSD posits that negative appraisals of the trauma and its sequelae, combined with a disturbance in the elaboration and contextualization of the trauma memory, promotes engagement in dysfunctional coping strategies intended to reduce short-term distress. However, in this context, rumination inhibits cognitive change and instead paradoxically maintains and/or increases PTSD symptoms over time (Ehlers & Clark, 2000).

Collectively, rumination—and brooding in particular—appears to be an important factor involved in the experience of both depression and PTSD. Despite these findings a number of important questions still exist pertaining to ruminative processes in PTSD. For example, there is still debate regarding the most effective means of measuring rumination and the functional role of rumination (i.e., meaning-making versus avoidance functions) among trauma-exposed individuals. Additionally, given the substantial comorbidity between PTSD and depression, taken together with independent findings implicating rumination in both of these disorders, one question that has yet to be addressed is whether rumination serves a more pronounced role among individuals who experience PTSD and depression relative to those who experience only PTSD or depression.

Social functioning deficits

Social functioning deficits are defined as an individual's inability to properly fulfill social roles (e.g., as a parent, employee, or spouse). Both PTSD and depression are associated with social functioning deficits, and individuals with comorbid PTSD and depression may be at particularly heightened risk (Bolton et al., 2004). A cross-sectional study by Wingo et al. (2017) on 264 veterans found that greater severity in PTSD or depression was significantly related to increased impairment in social functioning. The comorbidity of PTSD and depression can work together to further exacerbate social functioning symptoms, because both of these disorders have hallmark symptoms related to social functioning impairments. For example, one major symptom of PTSD is avoidant behaviors, which can cause people to disengage in social environments. Avoidant behaviors can account for why those with PTSD often have increased difficulty connecting on an intimate level in close relationships (Bolton et al., 2004; Hanley, Leifker, Blandon, & Marshall, 2013; Solomon, Dekel, & Zerach, 2008; Taft, Watkins, Stafford, Street, & Monson, 2011). In addition, depression symptoms, such as diminished interest in social or work activities, disturbed sleep, fatigue, and concentration difficulties, also impact an individual's ability to function properly in social interactions (Bolton et al., 2004). Therefore, the combined effects of having comorbid PTSD and depression could work together to exacerbate social dysfunction, accounting for why individuals with this comorbidity have increased difficulty with social functioning. Social functioning impairment has been found to predict relapses in both depression and PTSD (Fontana & Rosenheck, 2010; Judd et al., 2000; Renner, Cuijpers, & Huibers, 2014; Vittengl, Clark, & Jarrett, 2009), suggesting a vicious cycle between symptoms and functional impairment that maintain both over time.

PTSD, depression, and social impairment in intimate relationships

Research indicates that individuals with PTSD and/or depression have considerable social functioning impairment, especially in intimate relationships. In a national study on 2538 married participants (Whisman, 1999), PTSD and depression were found to be more strongly related to marriage dissatisfaction than other forms of psychopathology (i.e., bipolar disorder, dysthymia, panic disorder, social phobia, alcohol abuse, drug abuse, and generalized anxiety disorder). This finding suggests that individuals with PTSD and/or depression struggle to maintain intimate relationships more so than individuals with other psychiatric conditions.

Studies also indicate that those with PTSD, in particular, have greater intimacy problems and increased difficulty providing support to their romantic partners compared with individuals without PTSD (Hanley et al., 2013; Solomon et al., 2008; Taft et al., 2011) In a meta-analysis of over 31 studies, medium-level associations between PTSD and intimate relationship problems were found (Taft et al., 2011). However, increased social support from one's partner can help buffer the negative social impairment effects of PTSD (Wu, Chen, Weng, & Wu, 2009). Social support might help individuals with PTSD connect better with romantic partners because social support helps individuals with PTSD cope with trauma (Wu et al., 2009). We speculate that emotional numbing symptoms of PTSD, specifically, might be particularly implicated in difficulties in intimate relationships. Intimacy requires individuals to be vulnerable and talk about their feelings. Emotional numbing symptoms may inhibit emotional expression, which may in turn interfere with development of close relations. The extent to which emotional numbing, specifically, explains the strong associations between PTSD and social functioning impairment represents an important direction for future research in this area.

Suicide

Suicide is an especially pertinent concern among those with depression and/or PTSD, with research indicating that these populations have an increased risk for suicidality (Bolton & Robinson, 2010; Cougle, Resnick, & Kilpatrick, 2009; Oquendo et al., 2005; Panagioti, Gooding, & Tarrier, 2012; Stevens et al., 2013). Suicide, one of the top 10 leading causes of death in the United States, is a major public health concern (Centers for Disease Control and Prevention, 2016). Men are 3.5 times more likely to die from suicide than women, and white men, in particular, are the most susceptible demographic group for suicidality, accounting for 70% of all suicides in 2016 in America (Centers for Disease Control and Prevention, 2016).

The association between depression and suicide has long been recognized, and suicidal ideation is even included as a symptom of major depressive disorder according to the DSM (American Psychiatric Association, 2013). Research indicates that depression is the greatest risk factor for suicide out of all the other DSM disorders, accounting for about 40% of all suicide attempts (Bolton & Robinson, 2010). However, research also indicates that PTSD is independently predictive of suicidality, even after controlling for depression and other comorbid psychiatric conditions (LeBouthillier, McMillan, Thibodeau, & Asmundson, 2015). A meta-analysis on 59 studies found that PTSD was related to suicidality among various population including war veterans, survivors of interpersonal victimization, PTSD samples with mixed traumas, psychiatric populations, and community samples (Panagioti et al., 2012).

PTSD may be related to suicide at least partially because of emotional numbing (Davis, Witte, & Weathers, 2014; Guerra & Calhoun, 2011). Past research shows that among all the symptoms of PTSD, the emotional numbing symptoms are most strongly related to suicidality (Davis et al., 2014; Guerra & Calhoun, 2011). For example, one study of 434 female undergraduates found that the emotional numbing symptom of detachment or estrangement from others was more predictive of suicidal ideation than any other PTSD symptom (Davis et al., 2014). Thus PSTD is independently associated with suicidality, above and beyond the effects of comorbid depression, and this association may be driven at least in part by the emotional numbing symptoms of PTSD.

Suicidality and comorbid PTSD and depression

PTSD and depression both independently increase an individual's risk for developing suicidality and may further increase risk when these conditions occur comorbidly (Cougle et al., 2009; Oquendo et al., 2005; Panagioti et al., 2012; Stevens et al., 2013). A study of 3805 women with a history of lifetime suicide attempts found that only 16% of them did not have at least one diagnosis (PTSD or depression) and 45.2% had comorbid PTSD and MDD (Cougle et al., 2009). Additionally, a meta-analysis of 13 studies found that the comorbidity of PTSD and depression was related to increased suicidality (Panagioti et al., 2012). Therefore comorbid PTSD and depression is related to greater suicide risk in comparison with groups with either PTSD or depression alone (Cougle et al., 2009; Oquendo et al., 2005; Panagioti et al., 2012; Stevens et al., 2013).

Treatment implications

As reviewed earlier, sadness, emotional numbing/anhedonia, and depression have important implications for the presentation of PTSD, and people diagnosed with comorbid PTSD and depression typically have more severe symptoms and worse impairment than people with either disorder on its own. Fortunately the treatment outcome literature offers hope that empirically supported treatments developed for PTSD also work well for individuals with comorbid PTSD and depression. The American Psychological Association's Clinical Practice Guidelines for PTSD summarize the strength of evidence for a number of different psychotherapies and medications for adults diagnosed with PTSD (American Psychological Association, Guideline Development Panel for the Treatment of PTSD in Adults, 2017). The guidelines "strongly recommend" four treatments for adults with PTSD: cognitive behavioral therapy, cognitive processing therapy, cognitive therapy, and prolonged exposure. These treatments are all considered forms of trauma-focused cognitive behavioral therapy and include similar core components such as psychoeducation, exposure to internal and/or external trauma cues, and cognitive restructuring of dysfunctional beliefs. The expert panel concluded that there was at least moderate evidence that *all four* of the "strongly recommended" therapies also offered medium to large benefits for comorbid depression.

The clinical practice guidelines also "recommend" other PTSD treatments for which the evidence is weaker but still sufficient to recommend over no treatment. These "recommended" treatments include three additional psychotherapies (brief eclectic psychotherapy, eye movement desensitization and reprocessing therapy [EMDR], and narrative exposure therapy) and four medications (fluoxetine, paroxetine, sertraline, and venlafaxine). Of these "recommended" treatments, only two offered at least moderate evidence of medium-to-large effects on comorbid depression: EMDR and fluoxetine. Thus clinicians treating a patient with comorbid PTSD and depression would be well advised to start with one of the "strongly recommended" trauma-focused therapies before moving to other treatment approaches.

Despite these encouraging findings, even our most strongly recommended treatments do not work for every patient with comorbid PTSD and depression. Recent research has focused on enhancing treatment outcomes for this particularly severe patient population by combining different types of psychotherapy. Much of this work has examined the potential efficacy of combining a strongly recommended PTSD treatment (prolonged exposure) with an empirically supported treatment for depression (behavioral activation; Jakupcak et al., 2006; Strachan, Gros, Ruggiero, Lejuez, & Acierno, 2012). Behavioral activation treatment for depression involves increasing positive reinforcement by scheduling and participating in rewarding activities. Although not typically a part of trauma-focused therapy, behavioral activation has been identified as a potentially useful complement to existing PTSD treatment approaches, particularly for individuals with comorbid depression. Early results are promising (Jakupcak et al., 2006; Strachan et al., 2012), although future work is needed to demonstrate whether the combination of behavioral activation and exposure offers additional benefit beyond the already demonstrated effects of prolonged exposure on depression symptoms.

Given the well-established efficacy of antidepressant medication in treating depression, recent research has also examined whether the combination of trauma-focused therapy and antidepressant medication performs better than trauma-focused therapy alone. Unfortunately, there have been few studies examining the combination of therapy and mediation in the treatment of PTSD (Rauch et al., 2019). In a large randomized controlled trial that compared trauma-focused therapy (prolonged exposure), antidepressant medication (sertraline), and their combination, Rauch and colleagues (2018) found that therapy, medication, and their combination were all equally effective in treating PTSD, indicating no benefit of combined treatment (depression outcomes were not reported in this study). Similarly equivocal outcomes for combination treatments have been reported in other randomized controlled trials (Popiel, Zawadzki, Praglowska, & Teichman, 2015; Rothbaum et al., 2006). Thus current evidence is insufficient to recommend for or against combining trauma-focused therapy and medication in the treatment of comorbid PTSD and depression, and there is a clear need for additional research to guide clinical decision-making.

Conclusions

In this chapter, we reviewed the roles of depression, sadness, and emotional numbing/anhedonia in the experience of PTSD. There is consistent evidence that depressive disorders are highly comorbid with PTSD, that individuals with PTSD frequently experience the emotion of sadness (along with a number of other negative emotions), and that individuals with PTSD have difficulty experiencing positive emotions (often studied as part of the broader construct of "emotional numbing"). Many difficulties commonly associated with PTSD, such as dissociation, rumination, social withdrawal, and even suicide, may be driven at least in part by comorbid depression, sadness, and/or emotional numbing. Fortunately, effective treatments for PTSD have been developed, and the evidence is clear that successful PTSD treatment also reduces symptoms of depression, offering hope for individuals who experience these two disorders together.

Although it is widely accepted that sadness is a common reaction to trauma and that depression and PTSD are highly comorbid, no clear theoretical model exists linking the specific emotion of sadness to PTSD. Such a theoretical model would ideally draw upon basic affective science examining the functions of sadness and the literature on how extreme or dysfunctional experiences of sadness relate to different forms of psychopathology. Another limitation is the tendency for PTSD researchers to study the broad construct of "emotional numbing," which lumps anhedonia together with other related symptoms, rather than evaluating the specific role of anhedonia in the development and maintenance of PTSD. The literature on emotional numbing symptoms of PTSD is further complicated by the many varying definitions of emotional numbing that have been used across time, making it difficult to compare results across studies. The literature would benefit from a clearer definition of anhedonia as it relates to PTSD and could profit from the work done in the depression and schizophrenia literatures on conceptual models of anhedonia as it relates to different forms of psychopathology. An overarching conceptual model of the role of sadness and anhedonia following trauma and its implications for later development of PTSD and related conditions, including depression, is sorely needed to guide future research in this area.

References

Aldao, A., Nolen-Hoeksema, S., & Schweizer, S. (2010). Emotion-regulation strategies across psychopathology: A meta-analytic review. *Clinical Psychology Review, 30*, 217–237. https://doi.org/10.1016/j.cpr.2009.11.004.

American Psychiatric Association. (1994). *Diagnostic and statistical manual of mental disorders* (4th ed.). Washington, DC: American Psychiatric Association.

American Psychiatric Association. (2013). *Diagnostic and statistical manual of mental disorders* (5th ed.). Washington, DC: American Psychiatric Association.

American Psychological Association, Guideline Development Panel for the Treatment of PTSD in Adults. (2017). Clinical practice guideline for the treatment of Posttraumatic Stress Disorder (PTSD) in adults. Retrieved from http://www.apa.org/about/offices/directorates/guidelines/ptsd.pdf.

Armour, C., Karstoft, K.-I., & Richardson, J. D. (2014). The co-occurrence of PTSD and dissociation: Differentiating severe PTSD from dissociative-PTSD. *Social Psychiatry and Psychiatric Epidemiology, 49*(8), 1297–1306. https://doi.org/10.1007/s00127-014-0819-y.

Armour, C., Tsai, J., Durham, T. A., Charak, R., Biehn, T. L., Elhai, J. D., & Pietrzak, R. H. (2015). Dimensional structure of DSM-5 posttraumatic stress symptoms: Support for a hybrid anhedonia and externalizing behaviors model. *Journal of Psychiatric Research, 61*, 106–113. https://doi.org/10.1016/j.jpsychires.2014.10.012.

Asmundson, G. J., Stapleton, J. A., & Taylor, S. (2004). Are avoidance and numbing distinct PTSD symptom clusters? *Journal of Traumatic Stress, 17*, 467–475. https://doi.org/10.1007/s10960-004-5795-7.

Baschnagel, J. S., O'Conner, R. M., Colder, C. R., & Hawk, L. W. (2005). Factor structure of posttraumatic stress among Western New York undergraduates following the September 11th terrorist attack on the World Trade Centre. *Journal of Traumatic Stress, 18*, 677–684. https://doi.org/10.1002/jts.20076.

Berntsen, D., & Rubin, D. C. (2007). When a trauma becomes a key to identity: Enhanced integration of trauma memories predicts posttraumatic stress disorder symptoms. *Applied Cognitive Psychology, 21*, 417–431. https://doi.org/10.1002/acp.1290.

Birkley, E. L., Eckhardt, C. I., & Dykstra, R. E. (2016). Posttraumatic stress disorder symptoms, intimate partner violence, and relationship functioning: A meta-analytic review. *Journal of Traumatic Stress, 29*, 397–405. https://doi.org/10.1002/jts.22129.

Bolton, D., Hill, J., O'Ryan, D., Udwin, O., Boyle, S., & Yule, W. (2004). Long-term effects of psychological trauma on psychosocial functioning. *Journal of Child Psychology and Psychiatry, 45*(5), 1007–1014. https://doi.org/10.1111/j.1469-7610.2004.t01-1-00292.x.

Bolton, J. M., & Robinson, J. (2010). Population-attributable fractions of Axis I and Axis II mental disorders for suicide attempts: Findings from a representative sample of the adult, noninstitutionalized US population. *American Journal of Public Health, 100*(12), 2473–2480. https://doi.org/10.2105/AJPH.2010.192252.

Brewin, C. R. (2001). A cognitive neuroscience account of posttraumatic stress disorder and its treatment. *Behaviour Research and Therapy, 39*(4), 373–393. https://doi.org/10.1016/S0005-7967(00)00087-5.

Brunet, A., Weiss, D. S., Metzler, T. J., Best, S. R., Neylan, T. C., Rogers, C., ... Marmar, C. R. (2001). The peritraumatic distress inventory: A proposed measure of PTSD criterion A2. *American Journal of Psychiatry, 158*, 1480–1485. https://doi.org/10.1176/appi.ajp.158.9.1480.

Centers for Disease Control and Prevention. (2016). WISQARS leading causes of death reports, 1981–2017. Retrieved from: https://www.nimh.nih.gov/health/statistics/suicide.shtml.

Chun, C. A. (2016). The expression of posttraumatic stress symptoms in daily life: A review of experience sampling methodology and daily diary studies. *Journal of Psychopathology and Behavioral Assessment, 38*, 406–420. https://doi.org/10.1007/s10862-016-9540-3.

Clapp, J. D., Beck, J. G., Palyo, S. A., & Grant, D. M. (2008). An examination of the synergy of pain and PTSD on quality of life: Additive or multiplicative effects? *Pain, 138*, 301–309. https://doi.org/10.1016/l.pain.2008.01.001.

Cloitre, M., Petkova, E., Wang, J., & Lu Lassell, F. (2012). An examination of the influence of a sequential treatment on the course and impact of dissociation among women with PTSD related to childhood abuse. *Depression and Anxiety, 29*, 709–717. https://doi.org/10.1002/da.21920.

Cougle, J. R., Resnick, H., & Kilpatrick, D. G. (2009). PTSD, depression, and other comorbidity in relation to suicidality: Cross-sectional and prospective analyses of a national probability sample of women. *Depression and Anxiety, 26*(12), 1151–1157. https://doi.org/10.1002/da.20621.

Dalgleish, T., & Power, M. J. (2004). Emotion-specific and emotion-non-specific components of posttraumatic stress disorder (PTSD): Implications for a taxonomy of related psychopathology. *Behaviour Research and Therapy, 42*, 1069–1088.

Davidson, R. J. (1994). Asymmetric brain function, affective style, and psychopathology: The role of early experience and plasticity. *Development and Psychopathology, 6*, 741–758. https://doi.org/10.1017/S0954579400004764.

Davis, M. T., Witte, T. K., & Weathers, F. W. (2014). Posttraumatic stress disorder and suicidal ideation: The role of specific symptoms within the framework of the interpersonal-psychological theory of suicide. *Psychological Trauma Theory Research Practice and Policy, 6*(6), 610–618. https://doi.org/10.1037/a0033941.

Ehlers, A., & Clark, D. M. (2000). A cognitive model of posttraumatic stress disorder. *Behavior Research and Therapy, 38*, 319–345. https://doi.org/10.1016/S0005-7967(99)00123-0.

Ehring, T., & Watkins, E. R. (2008). Repetitive negative thinking as a transdiagnostic process. *International Journal of Cognitive Therapy, 1*, 192–205. https://doi.org/10.1521/ijct.2008.1.3.192.

Ekman, P. (1992). An argument for basic emotions. *Cognition and Emotion, 6*, 169–200. https://doi.org/10.1080/02699939208411068.

Elhai, J. D., & Palmieri, P. A. (2011). The factor structure of posttraumatic stress disorder: A literature update, critique of methodology, and agenda for future research. *Journal of Anxiety Disorders, 25*, 849–854. https://doi.org/10.1016/j.janxdis.2011.04.007.

Engel-Rebitzer, E., Bovin, M. J., Black, S. K., Rosen, R. C., Keane, T. M., & Marx, B. P. (2017). A longitudinal examination of peritraumatic emotional responses and their association with posttraumatic stress disorder and major depressive disorder among veterans. *Journal of Trauma & Dissociation, 18*, 679–692. https://doi.org/10.1080/15299732.2016.1267683.

Feeny, N. C., Zoellner, L. A., Fitzgibbons, L. A., & Foa, E. B. (2000). Exploring the roles of emotional numbing, depression, and dissociation in PTSD. *Journal of Traumatic Stress*, *13*(3), 489–498. https://doi.org/10.1023/A:1007789409330.

Finucane, A. M., Dima, A., Ferreira, N., & Halvorsen, M. (2012). Basic emotion profiles in healthy, chronic pain, depressed and PTSD individuals. *Clinical Psychology & Psychotherapy*, *19*, 14–24. https://doi.org/10.1002/cpp.733.

Foa, E. B., & Kozak, M. J. (1986). Emotional processing of fear: Exposure to corrective information. *Psychological Bulletin*, *99*, 20–35. https://doi.org/10.1037/0033-2909.99.1.20.

Fontana, A., & Rosenheck, R. (2010). War zone veterans returning to treatment: Effects of social functioning and psychopathology. *The Journal of Nervous and Mental Disease*, *198*(10), 699–707. https://doi.org/10.1097/NMD.0b013e3181f4ac88.

Fredrickson, B. L. (1998). What good are positive emotions? *Review of General Psychology*, *2*, 300–319. https://doi.org/10.1037/1089-2680.2.3.300.

Frewen, P. A., Dozois, D. J. A., Neufeld, R. W. J., & Lanius, R. A. (2008). Meta-analysis of alexithymia in posttraumatic stress disorder. *Journal of Traumatic Stress*, *21*, 243–246. https://doi.org/10.1002/jts.20320.

Frewen, P. A., Dozois, D. J. A., Neufeld, R. W. J., & Lanius, R. A. (2011). Disturbances of emotional awareness and expression in posttraumatic stress disorder: Meta-mood, emotion regulation, mindfulness and interference of emotional expressiveness. *Psychological Trauma Theory Research Practice and Policy*, *4*, 152–161. https://doi.org/10.1037/a0023114.

Golier, J. A., Yehuda, R., Schmeidler, J., & Siever, L. J. (2001). Variability and severity of depression and anxiety in posttraumatic stress disorder and major depressive disorder. *Depression and Anxiety*, *13*, 97–100.

Gros, D. F., Simms, L. J., & Acierno, R. (2010). Specificity of posttraumatic stress disorder symptoms: An investigation of comorbidity between posttraumatic stress disorder symptoms and depression in treatment-seeking veterans. *Journal of Nervous and Mental Disease*, *198*, 885–890. https://doi.org/10.1097/NMD.0b013e3181fe7410.

Guerra, V. S., & Calhoun, P. S. (2011). Examining the relation between posttraumatic stress disorder and suicidal ideation in an OEF/OIF veteran sample. *Journal of Anxiety Disorders*, *25*(1), 12–18. https://doi.org/10.1016/j.janxdis.2010.06.025.

Hanley, K. E., Leifker, F. R., Blandon, A. Y., & Marshall, A. D. (2013). Gender differences in the impact of posttraumatic stress disorder symptoms on community couples' intimacy behaviors. *Journal of Family Psychology*, *27*(3), 525–530. https://doi.org/10.1037/a0032890.

Harvey, A. G., & Bryant, R. A. (1999). The relationship between acute stress disorder and posttraumatic stress disorder: A 2-year prospective evaluation. *Journal of Consulting and Clinical Psychology*, *67*(6), 985–988. https://doi.org/10.1037/0022-006X.67.6.985.

Hassija, C. M., Jakupcak, M., & Gray, M. J. (2012). Numbing and dysphoria symptoms of posttraumatic stress disorder among Iraq and Afghanistan war veterans: A review of findings and implications for treatment. *Behavior Modification*, *36*, 834–856. https://doi.org/10.1177/01454455124453735.

Hathaway, L. M., Boals, A., & Banks, J. B. (2010). PTSD symptoms and dominant emotional response to a traumatic event: An examination of DSM-IV criterion A2. *Anxiety, Stress, and Coping*, *23*, 119–126. https://doi.org/10.1080/10615800902818771.

Hong, R. Y. (2007). Worry and rumination: Differential associations with anxious and depressive symptoms and coping behaviors. *Behavior Research and Therapy*, *45*, 277–290. https://doi.org/10.1016/j.brat.2006.03.006.

Hunnicutt-Ferguson, K., Wyka, K. E., Peskin, M., Cukor, J., Olden, M., & Difede, J. (2018). Posttraumatic stress disorder, functional impairment, and subjective distress in world trade center disaster workers. *Journal of Traumatic Stress*, *31*, 234–243. https://doi.org/10.1002/jts.22268.

Izard, C. E. (1992). Basic emotions, relations among emotions, and emotion-cognition relations. *Psychological Review*, *99*, 561–565. https://doi.org/10.1037/0033-295X.99.3.561.

Jakupcak, M., Roberts, L., Martell, C., Mulick, P., Michael, S., Reed, R., & McFall, M. (2006). A pilot study of behavioral activation for veterans with posttraumatic stress disorder. *Journal of Traumatic Stress, 19*, 387–391. https://doi.org/10.1002/jts.20125.

Johnson, D. M., Pike, J. L., & Chard, K. M. (2001). Factors predicting PTSD, depression, and dissociative severity in female treatment-seeking childhood sexual abuse survivors. *Child Abuse and Neglect, 25*, https://doi.org/10.1016/S0145-2134(00)00225-8.

Judd, L. L., Akiskal, H. S., Zeller, P. J., Paulus, M., Leon, A. C., Maser, J. D., ... Keller, M. B. (2000). Psychosocial disability during the long-term course of unipolar major depressive disorder. *Archives of General Psychiatry, 57*(4), 375. https://doi.org/10.1001/archpsyc.57.4.375.

Kashdan, T. B., Elhai, J. D., & Frueh, B. C. (2006). Anhedonia and emotional numbing in combat veterans with PTSD. *Behavior Research and Therapy, 44*, 457–467. https://doi.org/10.1016/j.brat.2005.03.001.

Kessler, R. C., Sonnega, A., Bromet, E., Hughes, M., & Nelson, C. B. (1995). Posttraumatic stress disorder in the national comorbidity survey. *Archives of General Psychiatry, 52*, 1048–1060. https://doi.org/10.1001/archpsyc.1995.03950240066012.

King, D. W., Leskin, G. A., King, L. A., & Weathers, F. W. (1998). Confirmatory factor analysis of the clinician-administered PTSD scale: Evidence for the dimensionality of posttraumatic stress disorder. *Psychological Assessment, 10*(2), 90–96. https://doi.org/10.1037/1040-3590.10.2.90.

Kuhn, E., Blanchard, E. B., & Hickling, E. J. (2003). Posttraumatic stress disorder and psychosocial functioning within two samples of MVA survivors. *Behaviour Research and Therapy, 41*(9), 1105–1112. https://doi.org/10.1016/S0005-7976(03)00071-8.

Lancaster, S. L., Melka, S. E., & Rodriguez, B. F. (2011). Emotional predictors of PTSD symptoms. *Psychological Trauma Theory Research Practice and Policy, 3*, 313–317. https://doi.org/10.1037/a0022751.

Lanius, R. A., Brand, B., Vermetten, E., Frewen, P. A., & Spiegel, D. (2012). The dissociative subtype of posttraumatic stress disorder: Rationale, clinical and neurobiological evidence, and implications. *Depression and Anxiety, 29*, 1–8. https://doi.org/10.1002/da.21889.

LeBouthillier, D. M., McMillan, K. A., Thibodeau, M. A., & Asmundson, G. J. G. (2015). Types and number of traumas associated with suicidal ideation and suicide attempts in PTSD: Findings from a U.S. nationally representative sample. *Journal of Traumatic Stress, 28*(3), 183–190. https://doi.org/10.1002/jts.22010.

Litz, B. T. (1992). Emotional numbing in combat-related post-traumatic stress disorder: A critical review and reformulation. *Clinical Psychology Review, 12*, 417–432. https://doi.org/10.1016/0272-7358(92)90125-R.

Litz, B. T., & Gray, M. J. (2001). Emotional numbing in posttraumatic stress disorder: Current and future research directions. *Australian and New Zealand Journal of Psychiatry, 36*, 198–204. https://doi.org/10.1046/j.1440-1614.2002.01002.x.

Litz, B. T., Orsillo, S. M., Kaloupek, D., & Weathers, F. (2000). Emotional processing in posttraumatic stress disorder. *Journal of Abnormal Psychology, 109*, 26–39. https://doi.org/10.1037/0021-843X.109.1.26.

McCanlies, E. C., Sarkisian, K., Andrew, M. E., Burchfiel, C. M., & Violanti, J. M. (2017). Association of peritraumatic dissociation with symptoms of depression and posttraumatic stress disorder. *Psychological Trauma Theory Research Practice and Policy, 9*(4), 479–484. https://doi.org/10.1037/tra0000215.

McHugh, P. R., & Treisman, G. (2007). PTSD: A problematic diagnostic category. *Journal of Anxiety Disorders, 21*, 211–222. https://doi.org/10.1016/j.janxdis.2006.09.003.

McNally, R. J. (2009). Can we fix PTSD in DSM-V? *Depression and Anxiety, 26*, 597–600. https://doi.org/10.1002/da.20586.

Michael, T., Halligan, S. L., Clark, D. M., & Ehlers, A. (2007). Rumination and posttraumatic stress disorder. *Depression and Anxiety, 24*, 307–317. https://doi.org/10.1002/da.20228.

Nawijn, L., van Zuiden, M., Frijling, J. L., Koch, S. B. J., Veltman, D. J., & Olff, M. (2015). Reward functioning in PTSD: A systematic review exploring the mechanisms underlying anhedonia. *Neuroscience and Biobehavioral Reviews, 51*, 189–204. https://doi.org/10.1016/j.neubiorev.2015.01.019.

Nolen-Hoeksema, S. (1991). Responses to depression and their effects on the duration of depressive episodes. *Journal of Abnormal Psychology, 100*, 569–582. https://doi.org/10.1037/0021-843X.100.4.569.

Nolen-Hoeksema, S., Wisco, B. E., & Lyubomirsky, S. (2008). Rethinking rumination. *Perspectives on Psychological Science, 3*, 400–424. https://doi.org/10.1111/j.1745-6924.2008.00088.x.

North, C. S., Suris, A. M., Davis, M., & Smith, R. P. (2009). Toward validation of the diagnosis of posttraumatic stress disorder. *American Journal of Psychiatry, 166*, 34–41. https://doi.org/10.1176/appi.ajp.2008.08050644.

Oquendo, M., Brent, D. A., Birmaher, B., Greenhill, L., Kolko, D., Stanley, B., … Mann, J. J. (2005). Posttraumatic stress disorder comorbid with major depression: Factors mediating the association with suicidal behavior. *American Journal of Psychiatry, 162*(3), 560–566. https://doi.org/10.1176/appi.ajp.162.3.560.

Ozer, E. J., Best, S. R., Lipsey, T. L., & Weiss, D. S. (2003). Predictors of posttraumatic stress disorder and symptoms in adults: A meta-analysis. *Psychological Bulletin, 129*(1), 52–73. https://doi.org/10.1037/0033-2909.129.1.52.

Pai, A., Suris, A. M., & North, C. S. (2017). Posttraumatic stress disorder in the DSM-5: Controversy, change, and conceptual considerations. *Behavioral Science, 7*(7), 1–7. https://doi.org/10.3390/bs7010007.

Palmieri, P. A., Marshall, G. N., & Schell, T. L. (2007). Confirmatory factor analysis of posttraumatic stress symptoms in Cambodian refugees. *Journal of Traumatic Stress, 20*, 207–216. https://doi.org/10.1002/jts.20196.

Panagioti, M., Gooding, P. A., & Tarrier, N. (2012). A meta-analysis of the association between posttraumatic stress disorder and suicidality: The role of comorbid depression. *Comprehensive Psychiatry, 53*(7), 915–930. https://doi.org/10.1016/j.comppsych.2012.02.009.

Pietrzak, R. H., Goldstein, M. B., Malley, J. C., Rivers, A. J., & Southwick, S. M. (2010). Structure of posttraumatic stress disorder symptoms and psychosocial functioning in veterans of operations enduring freedom and Iraqi freedom. *Psychiatry Research, 178*, 323–329. https://doi.org/10.1016/j.psychres.2010.04.039.

Pietrzak, R. H., Goldstein, R. B., Southwick, S. M., & Grant, B. F. (2011). Prevalence and axis I comorbidity of full and partial posttraumatic stress disorder in the United States: Results from wave 2 of the National Epidemiologic Survey on Alcohol and Related Conditions. *Journal of Anxiety Disorders, 25*, 456–465. https://doi.org/10.1016/j.janxdis.2010.11.010.

Pietrzak, R. H., Tsai, J., Armour, C., Mota, N., Harpaz-Rotem, I., & Southwick, S. M. (2015). Functional significance of a novel 7-factor model of DSM-5 PTSD symptoms: Results from the National Health and Resilience in Veterans Study. *Journal of Affective Disorders, 174*, 522–526. https://doi.org/10.1016/j.jad.2014.12.007.

Pineles, S. L., Suvak, M. K., Liverant, G. I., Gregor, K., Wisco, B. E., Pitman, R. K., & Orr, S. P. (2013). Psychophysiologic reactivity, subjective distress, and their associations with PTSD diagnosis. *Journal of Abnormal Psychology, 122*, 635–644. https://doi.org/10.1037/a0033942.

Popiel, A., Zawadzki, B., Praglowska, E., & Teichman, Y. (2015). Prolonged exposure, paroxetine and the combination in the treatment of PTSD following a motor vehicle accident. A randomized clinical trial—The "TRAKT" study. *Journal of Behavior Therapy and Experimental Psychiatry, 48*, 17–26. https://doi.org/10.1016/j.jbtep.2015.01.002.

Rasmussen, A., Verkuilen, J., Jayawickreme, N., Wu, Z., & McCluskey, S. T. (2018). When did posttraumatic stress disorder get so many factors? Confirmatory factor models since DSM-5. *Clinical Psychological Science, 7*, 234–248.

Rauch, S. A., Kim, H. M., Powell, C., Tuerk, P. W., Simon, N. M., Acierno, R., ... Stein, M. B. (2019). Efficacy of prolonged exposure therapy, sertraline hydrochloride, and their combination among combat veterans with posttraumatic stress disorder: a randomized clinical trial. *JAMA Psychiatry*, *76*, 117–126. https://doi.org/10.1001/jamapsychiatry.2018.3412.

Renner, F., Cuijpers, P., & Huibers, M. J. H. (2014). The effect of psychotherapy for depression on improvements in social functioning: A meta-analysis. *Psychological Medicine*, *44*(14), 2913–2926. https://doi.org/10.1017/S0033291713003152.

Resick, P. A., Suvak, M. K., Johnides, B. D., Mitchell, K. S., & Iverson, K. M. (2012). The impact of dissociation on PTSD treatment with cognitive processing therapy. *Depression and Anxiety*, *29*, 718–730. https://doi.org/10.1002/da.21938.

Reynolds, M., & Brewin, C. R. (1999). Intrusive memories in depression and posttraumatic stress disorder. *Behaviour Research and Therapy*, *37*, 201–215. https://doi.org/10.1016/S0005-7967(98)00132-6.

Roemer, L., Litz, B. T., Orsillo, S. M., & Wagner, A. W. (2001). A preliminary investigation of the role of strategic withholding of emotions in PTSD. *Journal of Traumatic Stress*, *14*, 149–156. https://doi.org/10.1023/A:1007895817502.

Rothbaum, B. O., Cahill, S. P., Foa, E. B., Davidson, J. R. T., Compton, J., Connor, K. M., ... Hahn, C. (2006). Augmentation of sertraline with prolonged exposure in the treatment of posttraumatic stress disorder. *Journal of Traumatic Stress*, *19*, 625–638. https://doi.org/10.1002/jts.20170.

Rottenberg, J., Kasch, K. L., Gross, J. J., & Gotlib, I. H. (2002). Sadness and amusement reactivity differentially predict concurrent and prospective functioning in major depressive disorder. *Emotion*, *2*, 135–146. https://doi.org/10.1037//1528-3542.2.2.135.

Rytwinski, N. K., Scur, M. D., Feeny, N. C., & Youngstrom, E. A. (2013). The co-occurrence of major depressive disorder among individuals with posttraumatic stress disorder: A meta-analysis. *Journal of Traumatic Stress*, *26*, 299–309. https://doi.org/10.1002/jts.21814.

Seligowski, A. V., & Orcutt, H. K. (2016). Support for the 7-factor hybrid model of PTSD in a community sample. *Psychological Trauma Theory Research Practice and Policy*, *8*, 218–221. https://doi.org/10.1037/tra0000104.

Siegle, G. J., Moore, P. M., & Thase, M. E. (2004). Rumination: One construct, many features in health individuals, depressed individuals, and individuals with lupus. *Cognitive Therapy and Research*, *28*, 645–668. https://doi.org/10.1023/B:COTR.0000045570.62733.9f.

Simms, L. J., Watson, D., & Doebbeling, B. N. (2002). Confirmatory factor analyses of posttraumatic stress symptoms in deployed and nondeployed veterans of the Gulf War. *Journal of Abnormal Psychology*, *111*, 637–647. https://doi.org/10.1037/0021-843X.111.4.637.

Smith, S. M., Goldstein, R. B., & Grant, B. F. (2016). The association between post-traumatic stress disorder and lifetime DSM-5 psychiatric disorders among veterans: Data from the National Epidemiologic Survey on Alcohol and Related Conditions-III (NESARC-III). *Journal of Psychiatric Research*, *82*, 16–22. https://doi.org/10.1016/j.jpsychires.2016.06.022.

Solomon, Z., Dekel, R., & Zerach, G. (2008). The relationships between posttraumatic stress symptom clusters and marital intimacy among war veterans. *Journal of Family Psychology*, *22*(5), 659–666. https://doi.org/10.1037/a0013596.

Spitzer, R. L., First, M. B., & Wakefield, J. C. (2007). Saving PTSD from itself in DSM-V. *Journal of Anxiety Disorders*, *21*, 233–241. https://doi.org/10.1016/j.janxdis.2006.09.006.

Steuwe, C., Lanius, R. A., & Frewen, P. A. (2012). The role of dissociation in civilian posttraumatic stress disorder: Evidence for a dissociative subtype by latent class and confirmatory factor analysis. *Depression and Anxiety*, *29*, 689–700. https://doi.org/10.1002/da.21944.

Stevens, D., Wilcox, H. C., MacKinnon, D. F., Mondimore, F. M., Schweizer, B., Jancic, D., ... Potash, J. B. (2013). Posttraumatic stress disorder increases risk for suicide attempt in adults with recurrent major depression. *Depression and Anxiety, 30*(10), 940–946. https://doi.org/10.1002/da.22160.

Strachan, M., Gros, D. F., Ruggiero, K. J., Lejuez, C. W., & Acierno, R. (2012). An integrated approach to delivering exposure-based treatment for symptoms of PTSD and depression in OIF/OEF veterans: Preliminary findings. *Behavior Therapy, 43*, 560–569. https://doi.org/10.1016/j.beth.2011.03.003.

Szabo, Y. Z., Warnecke, A. J., Newton, T. L., & Valentine, J. C. (2017). Rumination and posttraumatic stress symptoms in trauma-exposed adults: A systematic review and meta-analysis. *Anxiety, Stress, and Coping, 30*(4), 396–414. https://doi.org/10.1080/10615806.21027.1313835.

Taft, C. T., Watkins, L. E., Stafford, J., Street, A. E., & Monson, C. M. (2011). Posttraumatic stress disorder and intimate relationship problems: A meta-analysis. *Journal of Consulting and Clinical Psychology, 79*(1), 22–33. https://doi.org/10.1037/a0022196.

Treynor, W., Gonzalez, R., & Nolen-Hoeksema, S. (2003). Rumination reconsidered: A psychometric analysis. *Cognitive Therapy and Research, 27*(3), 247–259. https://doi.org/10.1023/A:1023910315561.

van der Hart, O., Bolt, H., & van der Kolk, B. A. (2005). Memory fragmentation in dissociative identity disorder. *Journal of Trauma & Dissociation, 6*(1), 55–70. https://doi.org/10.1300/J229v06n01_04.

van der Velden, P. G., Kleber, R. J., Christiaanse, B., Gersons, B. P. R., Marcelissen, F. G. H., Drogendijk, A. N., ... Meewisse, M. L. (2006). The independent predictive value of peritraumatic dissociation for postdisaster intrusions, avoidance reactions, and PTSD symptom severity: A 4-year prospective study. *Journal of Traumatic Stress, 19*(4), 493–506. https://doi.org/10.1002/jts.20140.

van der Velden, P. G., & Wittmann, L. (2008). The independent predictive value of peritraumatic dissociation for PTSD symptomatology after type I trauma: A systematic review of prospective studies. *Clinical Psychology Review, 28*(6), 1009–1020. https://doi.org/10.1016/j.cpr.2008.02.006.

Vittengl, J. R., Clark, L. A., & Jarrett, R. B. (2009). Deterioration in psychosocial functioning predicts relapse/recurrence after cognitive therapy for depression. *Journal of Affective Disorders, 112*(1-3), 135–143. https://doi.org/10.1016/j.jad.2008.04.004.

Ward, A., Lyubomirsky, L., Sousa, L., & Nolen-Hoeksema, S. (2003). Can't quite commit: Rumination and uncertainty. *Personality and Social Psychology Bulletin, 29*, 96–107. https://doi.org/10.1177/0146167202238375.

Whisman, M. A. (1999). Marital dissatisfaction and psychiatric disorders: Results from the national comorbidity survey. *Journal of Abnormal Psychology, 108*(4), 701–706. https://doi.org/10.1037/0021-843X.108.4.701.

Wingo, A. P., Briscione, M., Norrholm, S. D., Jovanovic, T., McCullough, S. A., Skelton, K., & Bradley, B. (2017). Psychological resilience is associated with more intact social functioning in veterans with post-traumatic stress disorder and depression. *Psychiatry Research, 249*, 206–211. https://doi.org/10.1016/j.psychres.2017.01.022.

Wisco, B. E., Marx, B. P., Miller, M. W., Wolf, E. J., Mota, N. P., Krystal, J. H., ... Pietrzak, R. H. (2016). Probable posttraumatic stress disorder in the U.S. veteran population according to DSM-5: Results from the National Health and Resilience in Veterans Survey. *Journal of Clinical Psychiatry, 77*, 1503–1510. https://doi.org/10.4088/JCP.15m10188.

Wittmann, L., Moergeli, H., & Schnyder, U. (2006). Low predictive power of peritraumatic dissociation for PTSD symptoms in accident survivors. *Journal of Traumatic Stress, 19*(5), 639–651. https://doi.org/10.1002/jts.20154.

Wolf, E. J., Lunney, C. A., Miller, M. W., Resick, P. A., Friedman, M. J., & Schnurr, P. P. (2012). The dissociative subtype of PTSD: A replication and extension. *Depression and Anxiety*, *29*, 679–688. https://doi.org/10.1002/da.21946.

Wolf, E. J., Miller, M. W., Reardon, A. F., Ryabchenko, K. A., Castillo, D., & Freund, R. (2012). A latent class analysis of dissociation and posttraumatic stress disorder: Evidence for a dissociative subtype. [Research Support, N.I.H., Extramural Research Support, U.S. Gov't, Non-P.H.S.]. *Archives of General Psychiatry*, *69*, 698–705. https://doi.org/10.1001/archgenpsychiatry.2011.1574.

Wu, C.-H., Chen, S.-H., Weng, L.-J., & Wu, Y.-C. (2009). Social relations and PTSD symptoms: A prospective study on earthquake-impacted adolescents in Taiwan. *Journal of Traumatic Stress*, *22*(5), 451–459. https://doi.org/10.1002/jts.20447.

CHAPTER 5

Disgust in PTSD

Alyssa C. Jones, C. Alex Brake, Christal L. Badour
University of Kentucky, Lexington, KY, United States

Disgust

Among the basic human emotions systematically defined by Ekman (1992), disgust is perhaps the most historically overlooked and understudied emotion within psychology. Although literature on disgust can trace its roots as far back as Darwin (2003) and his seminal work, *The Expression of the Emotions in Man and Animals* (1872/2003), the study of disgust and its impact on psychopathology has only recently gained a foothold in the empirical literature. Disgust has long been recognized as a negatively valenced, universally experienced, and evolutionarily adaptive emotion that compels individuals to withdraw from environmental stimuli associated with harmful consumables such as rotting food and bodily products such as feces. This core disgust response can be observed in humans and animals alike, serving a protective function against the ingestion of potentially harmful substances (Rozin & Fallon, 1987). In this function, disgust is also characterized by a number of physiological responses. For example, facial changes during disgust elicitation often include a scrunched nose and downturned mouth—changes theorized to limit inhalation of foul odors and facilitate removal of harmful or contaminated substances via the corners of the mouth. Disgust is also frequently associated with nausea or vomiting—physiological responses that promote ejection of harmful consumables (Rozin, Haidt, & McCauley, 2009).

Initial conceptualization of disgust as an emotion specific to food rejection likely explains why disgust has long been neglected in the psychological literature. Matchett and Davey (1991) first recognized the important role of disgust in specific phobias—particularly those related to certain animals and insects. Their work expanded the food rejection model into a broader disease-avoidance model, suggesting that disgust reactions may emerge not only from consumables, but also additionally from objects associated with uncleanliness or contagion. This disease-avoidance model had at least two important secondary implications that laid the groundwork for disgust as a broader psychopathological phenomenon. First, Matchett and Davey's work demonstrated that disgust is communicable. Unlike other emotions, disgust is typically transmitted by environmental stimuli, and it can be easily conditioned to otherwise neutral stimuli, thereby spreading its influence. This process of learned transmittal may explain why most adults, but not children, display concerns about contagion (Rozin & Nemeroff, 1990). Second, the disease-avoidance model and other research has suggested that elicitation of disgust and its impact on behaviors is not always rational; many individuals will refuse a drink that has contacted a sterilized cockroach or candy that resembles feces, despite knowledge that these stimuli present no risk of contagion (Rozin, Millman, & Nemeroff, 1986).

In their comprehensive model of disgust, Rozin and colleagues (Rozin et al., 2009; Rozin & Fallon, 1987) have argued that disgust's innate ability to easily spread and endure in the absence of rational contamination risk may explain why disgust can also be experienced in situations involving interpersonal, moral, and existential threats as well. Their model defines four distinct disgust subtypes: (1) core disgust, (2) animal-reminder disgust, (3) interpersonal disgust, and (4) moral disgust (Olatunji & Sawchuk, 2005; Rozin et al., 2009). Whereas core disgust encompasses the evolutionarily adaptive response to protect against biological contaminants and associated objects as outlined previously, reactions of animal-reminder, interpersonal, and moral disgust are uniquely human experiences that are extensions of core disgust origins.

Animal-reminder disgust is based on the premise that we, as humans, prefer to think of ourselves as separate and transcendent from other animals. In particular, the fact that humans, like animals, are creatures of flesh and blood and thus susceptible to death is an aversive reality. Consequently, reminders of our animal nature, including certain sexual acts, bodily injury or mutilation, body orifices, and stimuli associated with death and decay (e.g., maggots and cadavers), may elicit particularly strong disgust reactions and thus motivate individuals to avoid conscious awareness of their physical vulnerability and mortality (Haidt, Rozin, McCauley, & Imada, 1997; Rozin et al., 2009). Interpersonal disgust is conceptualized as disgust toward others, particularly individuals of another social group, class, or lifestyle that are considered strange, unfamiliar, undesirable, or immoral (Haidt et al., 1997; Hodson & Costello, 2007; Olatunji & Sawchuk, 2005). Interpersonal disgust is characterized primarily by its repulsion to perceived lifestyle, personality, social, or moral "contaminants" rather than physical contagion, as demonstrated in disgust reactions to wearing an article of clothing from an immoral person (e.g., Hitler, a murderer; Rozin et al., 2009). Interpersonal disgust has also been implicated in forms of prejudice and dehumanization toward dissimilar, socially deviant, or low-status groups (e.g., immigrants and homosexuals; Hodson & Costello, 2007). Third, elicitors of moral disgust are culturally defined and typically involve heinous or extreme acts of moral violation, such as rape, genocide, torture, or cannibalism (Haidt et al., 1997; Rozin et al., 2009). Though moral disgust may overlap with other forms of disgust and other emotions—particularly anger and contempt—Rozin, Lowery, Imada, and Haidt (1999) have distinguished moral disgust as the prominent response to human degradation or defilement of spirituality, purity, or the natural moral order.

Related models of disgust have also emerged in recent years, including a narrower framework presented by Tybur, Lieberman, and Griskevicius (2009) that emphasizes only three domains: pathogen disgust (i.e., aversion to pathogens and contaminants associated with disease), sexual disgust (i.e., aversion to sexual behaviors that put one at risk of disease and/or are unrelated to reproduction), and moral disgust (i.e., aversion to violations of culturally defined morality or ethics). These domains maintain a high degree of overlap with Rozin and colleagues' four subtypes, yet have aimed to differentiate disgust elicitors along slightly different lines. Additional clinical research has explored the separate and interactive impacts of individuals' degree of distress experienced when disgusted—defined as *disgust sensitivity*—and individuals' general frequency or ease of experienced disgust, known as *disgust propensity* (Olatunji, Armstrong, Fan, & Zhao, 2014). Recent work has even advocated for assessment of *disgust proneness*, a higher-order individual difference factor that combines these disgust components (e.g., Olatunji, Tart, Ciesielski, McGrath, & Smits, 2011). Finally, whereas much of disgust research has either focused on individual trait differences (e.g., Olatunji et al., 2014) or disgust directed toward external stimuli (e.g., objects; perpetrators of an assault; Feldner, Frala, Badour, Leen-Feldner, & Olatunji, 2010), studies have begun to examine self-disgust—persistent and maladaptive disgust directed internally at one's characteristics or behaviors—as a unique form of extreme self-criticism similar to shame or guilt but involving more extreme, visceral, and repulsive reactions to oneself (e.g., Badour & Adams, 2015; Brake, Rojas, Badour, Dutton, & Feldner, 2017).

This burgeoning body of literature has led to some notable findings with significant clinical implications in recent decades. Since its early links to specific animal, spider, and insect phobias (e.g., Matchett & Davey,

1991), disgust has subsequently been shown to play an important role in the severity of disorders of blood-injection-injury phobia (Sawchuk, Lohr, Tolin, Lee, & Kleinknecht, 2000), illness anxiety (Davey & Bond, 2006), and contamination-based obsessive compulsive disorder (OCD; Rachman, 2006) above and beyond general negative affect (for a review, see Olatunji, Armstrong, & Elwood, 2017). Although connections between disgust and such disorders fit well within the disease-avoidance model, further research has now identified significant relationships between disgust and a much broader array of clinical presentations, including relevance in general OCD (e.g., Olatunji et al., 2011), sexual dysfunction (e.g., de Jong & Peters, 2009), eating disorders (e.g., Troop & Baker, 2009), depression (e.g., Powell, Simpson, & Overton, 2013), self-harm (e.g., Smith, Steele, Weitzman, Trueba, & Meuret, 2015), and suicide (e.g., Brake et al., 2017). Clearly disgust has found a foothold as a far more complex and psychiatrically relevant emotion than first conceived.

Relevance of disgust to trauma and PTSD

Posttraumatic stress disorder (PTSD) is now included among the growing list of disorders for which disgust is relevant. This development aligns with the recent removal of PTSD from the anxiety disorders section of the 5th edition of the Diagnostic and Statistical Manual of Mental Disorders (DSM-5; American Psychiatric Association, 2013) into a new class of trauma- and stressor-related disorders. This change represents a significant shift in over 30 years of conceptualization of PTSD as a primarily fear- and anxiety-based disorder. Although fear and anxiety still factor heavily into our understanding of the processes involved in the development and maintenance of PTSD, this expansion toward a broader conceptualization of PTSD is an important step toward acknowledging the complex array of emotional experiences that underlie this disorder. Additionally, a new symptom included in the DSM-5 revised Criterion D: Negative Alterations in Cognitions and Mood cluster of PTSD symptoms explicitly recognizes the persistence of various negative mood states (i.e., fear, horror, anger, guilt, and shame) as a characteristic feature of PTSD. Although the role of disgust in PTSD has received less attention in the literature relative to emotions such as anger, shame, and guilt, we believe that the notable absence of disgust from this list is an unfortunate oversight, as evidence continues to mount supporting the importance of disgust in understanding PTSD symptomatology among some individuals. Indeed, evidence suggests that approximately 10% of adults with PTSD will report disgust as their primary experienced negative emotion (Power & Fyvie, 2013), and experienced disgust has been shown to distinguish individuals with PTSD from those with chronic pain, depression, and no disorder or better than happiness, sadness, fear, and anger (Finucane, Dima, Ferreira, & Halvorsen, 2012).

Upon closer examination, the convergence between PTSD and disgust may come as no surprise. Some of the earliest work recognizing the connection between disgust and traumatic experiences traces back to generations old case examples involving war and combat. Rivers (1920) described an officer from the First World War who experienced recurring, intrusive mental images of vomiting after a battlefield explosion had propelled the officer face-first into the decomposing corpse of an enemy soldier; the case also notes the officer's ongoing eating avoidance and disgust-induced physiological symptoms of nausea and vomiting when reexperiencing reminders of the event. In line with the disease-avoidance model of disgust, this case demonstrates how feelings of core disgust at the time of the traumatic event, including its physiological responses, are communicable and associable with even mental reminders of the event. Yet this officer's experience also suggests that elements of animal-reminder disgust (through contact with a human corpse) and moral disgust (by experiencing the visceral atrocities of war firsthand) may also co-occur to strengthen and propagate future disgust-related posttraumatic symptoms.

The experience of disgust in relation to trauma may be more likely for events that involve particular characteristics. These characteristics may include contact with disgust elicitors such as salient peritraumatic (i.e., at the time of trauma) smells, rot/decay, death, and bodily products (e.g., vomit, semen, and blood). Disgust may also be elicited in response to events involving perceived violations of morality and trust, experiences of betrayal, and events involving sexual violation. For instance, individuals may experience intense moral and sexual disgust after a violation such as rape or incest but not following consensual sex (e.g., de Silva & Marks, 1999; Gershuny, Baer, Radomsky, Wilson, & Jenike, 2003; Steil, Jung, & Stangier, 2011). Military personnel may feel disgust toward themselves or their commanding officer after participating in killing or other atrocities of war. Disgust reactions may be particularly salient for moral violations such as killing unarmed civilians compared with an enemy combatant (Litz et al., 2009). These examples have opened the door to the possibility that disgust-based symptoms may come into play in an even wider variety of traumatic event types. For example, individuals witnessing horrific death may feel disgusted when reminded of the killer or the method of death (Gershuny et al., 2003). Those who have experienced repeated physical or emotional abuse may develop heightened self-disgust if they grow to believe they are undesirable or possess repulsive qualities (Rachman, 2006). Encounters with bodily gore or gruesome injury, such as work as a first responder or experiencing a motor vehicle or industrial accident, may also lead to disgust responses (de Silva & Marks, 1999).

Models of disgust and PTSD

To understand how disgust may function within the context of traumatic responding and PTSD, we can draw upon the rich body of accumulated knowledge regarding the role of fear conditioning in PTSD symptom development and maintenance. Similar to fear conditioning models of PTSD, principles from both classical and operant conditioning models explain how disgust is involved in the development and maintenance of PTSD. Strong feelings of disgust during a traumatic event may facilitate the formation of negative associations between previously neutral stimuli present during a traumatic event (conditioned stimulus [CS]) and the aversive stimuli present during a trauma (i.e., unconditioned stimulus [UCS]) via classical conditioning. For example, a victim of sexual assault may experience intense feelings of peritraumatic disgust (unconditioned response [UCR]) at the time of the trauma in response to perceptions of moral and sexual violation and exposure to potentially contaminating bodily products (e.g., saliva, blood, and semen). Previously neutral stimuli (e.g., room where the assault occurred, song playing in the background, and people who resemble the perpetrator) may acquire the ability to elicit disgust (conditioned response [CR]) even after the immediate threat of the traumatic event has passed. Conditioned disgust reactions may persist long after the immediate threat of the traumatic event has passed. Continued disgust may be elicited in response to both external (e.g., people, places, and situations) and internal (e.g., thoughts, memories, emotions, and physical sensations) trauma cues. Operant conditioning models suggest that avoidance of and escape from traumatic reminders provide temporary relief, but in the long term serves to maintain conditioned disgust reactions by preventing extinction learning from taking place (Badour, Feldner, Blumenthal, & Knapp, 2013). In addition to well-studied forms of avoidance and escape in PTSD, such as behavioral avoidance, thought-suppression, and substance-related coping (Foa & Kozak, 1986; Ullman, Relyea, Peter-Hagene, & Vasquez, 2013), persistent disgust reactions may also lead to emotion-specific forms of avoidance and escape such as washing, cleaning, or contamination-relation mental rituals (Badour & Adams, 2015).

Disgust has also been linked to a unique form of conditioning called evaluative conditioning, in which the hedonic value (i.e., pleasantness or unpleasantness) of a disgust-eliciting stimulus (e.g., blood) is transferred to a previously neutral stimulus (Jones, Olson, & Fazio, 2010). In other words, the previously neutral stimulus does not simply "signal" that the disgust-eliciting stimulus may occur again, but the previously neutral stimulus actually becomes disgusting in and of itself. The process of evaluative conditioning has been used to explain why disgust may be "sticky" and more resistant to extinction than conditioned anxiety (Mason & Richardson, 2012; Olatunji, Forsyth, & Cherian, 2007). This may also be why some victims of trauma may go on to feel that they *themselves* are disgusting, persistently contaminated, or dirty, even in the absence of a potential containment. Continuing with the example of sexual assault, a victim may feel unclean even in the absence of trauma reminders, as she has now taken on the disgusting properties of the assault through the morally and sexually violating experience and via contact with potential contaminants from the perpetrator. Her persistent sense of self-disgust may lead to excessive cleaning or washing behaviors—serving as forms of avoidance/escape—to reduce feelings of dirtiness. These and other avoidance behaviors will serve to maintain the distress associated with the trauma memory and continued beliefs that she is dirty and needs to wash to feel better (Badour, Feldner, Babson, Blumenthal, & Dutton, 2013). She may also avoid sexual contact as a result of viewing herself as unwanted by others due to feeling dirty or contaminated.

These resulting feelings of being dirty or contaminated have been termed *mental contamination*. Mental contamination refers to "feelings of dirtiness and urges to wash that arise independently of physical contact with a contaminant" (p. 2805; Herba & Rachman, 2007). Because there is no specific site of contact, mental contamination may lead to an internal, pervasive feeling of dirtiness, leading some individuals to feel tainted or immoral as a person. Contamination-based disgust reactions are conceptualized as a consequence of evaluative conditioning, where the properties of disgust-eliciting stimuli are actually transferred to a previously neutral stimulus (McKay, 2006), which in the case of mental contamination is the self. Mental contamination has been consistently associated with urges to wash or clean; one study found that 70% of women with sexual assault histories experienced urges to wash in response to feelings of internal dirtiness (Fairbrother & Rachman, 2004). Given that washing and cleaning behaviors are forms of avoidance/escape, which ultimately maintain PTSD symptoms by preventing extinction learning, we would expect mental contamination to be associated with increased PTSD symptoms. Indeed, mental contamination has been consistently associated with increased PTSD symptoms, particularly among those with sexual assault histories (Adams, Badour, Cisler, & Feldner, 2014; Badour, Feldner, Babson, Blumenthal, & Dutton, 2013; Badour, Feldner, Blumenthal, & Bujarski, 2013; Fairbrother & Rachman, 2004; Ishikawa, Kobori, & Shimizu, 2015; Olatunji, Elwood, Williams, & Lohr, 2008; Ojserkis, McKay, & Lebeaut, 2018). Despite efforts to wash or clean oneself in response to mental contamination, these internal, pervasive feelings of dirtiness often persist (Coughtrey, Shafran, Knibbs, & Rachman, 2012). Through increased PTSD symptoms, mental contamination has also been associated with an increased risk for engaging in risky behaviors and negative attitudes about help seeking (Brake, Jones, Wakefield, & Badour, 2018). Although mental contamination is not unique to sexual assault, the majority of work on trauma-related mental contamination has been in samples of female sexual assault victims.

Another area where evaluative conditioning may be relevant is for combat veterans and others who engage in or witness events that violate moral beliefs (e.g., killing and torture of enemy combatants), which may be particularly salient given the emphasis on morality and ethics in military culture (Litz et al., 2009). Veterans who experience these types of events often report "violations of conscience" associated with moral emotions that may lead to negative appraisals about themselves as a person (e.g., *I am tainted*, *What I've done cannot be forgiven*; Litz et al., 2009), a process termed *moral injury*. Moral injury is conceptualized as a reaction to engaging in, witnessing, learning about, or failing to prevent acts that violate one's deeply held moral values and beliefs (Litz et al., 2009). Although linked closely to risk for PTSD, moral injury appears to encompass unique difficulties not fully captured by PTSD (Drescher et al., 2011; Litz et al., 2009). As one of the core emotions implicated in moral injury, disgust may be elicited from perceptions of contamination of one's sense of moral purity (Farnsworth, Drescher, Nieuwsma, Walser, & Currier, 2014). Research has consistently demonstrated that moral transgressions elicit feelings of disgust, even in the absence of basic disgust stimuli (e.g., blood and rotten food; see Chapman & Anderson, 2013). A variety of events associated with combat may elicit moral injury, including betrayal (e.g., leadership failures), disproportionate violence (e.g., mistreatment of enemy combatants), civilian incidents (e.g., destruction of civilian property and assault against civilians), and within-rank violence (e.g., friendly fire and military sexual trauma; Drescher et al., 2011).

Cognitive models also offer hypotheses for the role of disgust and disgust-related cognitions in the development and maintenance of PTSD, mental contamination, and moral injury. Specific disgust-related cognitions after a traumatic event may include the following: "I have been contaminated by this event" or "I am a dirty person," which have been documented within both sexual assault (Badour, Feldner, Blumenthal, & Bujarski, 2013) and combat veteran (Litz et al., 2009) samples. These negative cognitions may be based on one's prior views about what it means to be pure, clean, and/or moral (Farnsworth et al., 2014). Negative cognitions may also be associated with beliefs about what will happen if one allows him–/herself to think about the trauma (e.g., *I will fall apart* or *I will lose control*; Ehlers & Clark, 2000). Such cognitions may further perpetuate avoidance and escape, impeding extinction learning, thus maintaining or exacerbating PTSD symptoms (Ehlers & Clark, 2000).

It is important to acknowledge that disgust may not be experienced by all people in response to all types of traumatic events. Fear and anxiety may be more ubiquitously experienced emotions in the context of traumatic events, given that most traumatic events are characterized by threatened or actual death or serious injury to the self or others. However, for some people, in response to certain events, disgust is likely to be very salient.

Disgust proneness

In addition to certain event characteristics being more likely to elicit disgust in the context of trauma, disgust proneness, or individual differences in the general tendency to experience or to be bothered by the emotion of disgust, are likely to influence whether disgust is experienced both during (peritraumatic disgust) and after a traumatic event (posttraumatic disgust). The majority of studies on disgust proneness in trauma and PTSD have focused on disgust propensity and disgust sensitivity. Conceptualized as underlying personality traits, disgust propensity and sensitivity may serve as risk factors for the development of PTSD via a diathesis-stress model (Olatunji et al., 2014). Within this model, one might hypothesize that individuals who are prone to experience disgust (i.e., disgust propensity) may be at a greater risk of experiencing peritraumatic disgust at the time of a trauma, whereas viewing disgust-related experiences as more distressing (i.e., disgust sensitivity) may be responsible for negative appraisals about disgust-related experiences that lead to and maintain PTSD symptoms.

Indeed, disgust propensity has been prospectively associated with increased peritraumatic disgust among deployed veterans, even when controlling for neuroticism (Engelhard, Olatunji, & de Jong, 2011). Preliminary research has also found that disgust propensity was associated with increased intrusions after a distressing film among an undergraduate sample, although replication with a trauma-exposed sample is needed (Bomyea & Amir, 2012). Evidence is mixed with regard to whether disgust propensity actually leads to increased PTSD symptoms, as hypothesized. Among the small group of studies that has examined this, one found increased disgust propensity to be associated with a PTSD diagnosis (Rüsch et al., 2011), whereas others have failed to find such a relation (Badour, Bown, Adams, Bunaciu, & Feldner, 2012; Engelhard et al., 2011). Mixed findings in this area may highlight the fact that it is not simply feeling disgusted that may lead to psychopathology, but rather distress associated with feeling disgusted that may be more important in the development of symptoms.

The literature suggests that disgust sensitivity—the tendency to experience disgust as more distressing—is associated with a range of psychopathological symptoms and has been frequently identified as a more robust predictor of psychopathology than disgust propensity (e.g., Olatunji, Unoka, Beran, David, & Armstrong, 2009; although see Muris et al., 2000). Disgust sensitivity may serve as a risk factor for developing PTSD after trauma by influencing the strength of classically conditioned associations during a traumatic event, leading to stronger peritraumatic disgust conditioning. Individuals higher in disgust sensitivity may also be more likely to engage in avoidance or escape to mollify their disgust experienced in response to trauma cues (Badour & Feldner, 2018; Dalgleish & Power, 2004; Olatunji et al., 2014). As expected, disgust sensitivity has been positively associated with PTSD symptoms in several studies (Badour et al., 2012; Badour, Feldner, Blumenthal, & Bujarski, 2013; Engelhard et al., 2011; van Delft, Finkenauer, Tybur, & Lamers Winkelman, 2016). There is some evidence to suggest that disgust experienced during combat may be more strongly linked to PTSD symptomatology among individuals high in disgust sensitivity (Engelhard et al., 2011). Some evidence also suggests that the relation between disgust propensity and mental contamination among trauma-exposed individuals is higher among those who report increased disgust sensitivity (Ojserkis et al., 2018; Travis & Fergus, 2015). Findings from a study with sexual assault survivors suggest that disgust sensitivity may relate to PTSD via negative disgust-related appraisals (e.g., thoughts about being dirty or contaminated from the assault; Badour, Feldner, Blumenthal, & Bujarski, 2013). However, Engelhard et al. (2011) did not find that disgust sensitivity moderated the association between peritraumatic disgust and subsequent negative evaluations about combat experiences. Another study found that trauma-exposed veterans without PTSD reported lower levels of disgust sensitivity (i.e., disgust *in*sensitivity) compared with trauma-exposed veterans with PTSD, providing support of the potential protective nature of disgust insensitivity (Olatunji et al., 2014).

Despite mixed findings, there is enough evidence to warrant additional research on disgust propensity and disgust sensitivity as it relates to PTSD. Few studies have looked disgust propensity and sensitivity together in the context of PTSD. As such, future research may focus on comprehensively assessing all aspects of disgust proneness, ideally using a combination of cross-sectional and prospective methods, to further underline the role these factors may play in the development and maintenance of PTSD.

Peritraumatic disgust

Emotional and physical reactions at the time of a trauma (i.e., peritraumatic reactions) have long been recognized as key predictors in the development of PTSD (Bovin & Marx, 2011; Kilpatrick et al., 1998). Although PTSD research initially emphasized the role of fear, helplessness, and horror, and more recently anger, guilt, and shame, disgust has also been evidenced among peritraumatic reactions. Peritraumatic disgust has been reported by victims of an array of traumatic events, including flooding (Fredman et al., 2010), physical and sexual assault (Badour, Feldner, Babson, Blumenthal, & Dutton, 2013; Feldner et al., 2010), combat (Engelhard et al., 2011), and industrial accidents (Grunert et al., 1992). Peritraumatic disgust has been associated with increased PTSD symptoms even when accounting for fear, horror, guilt, helplessness, and anger (Lancaster, Melka, & Rodriguez, 2011), and individuals who report experiencing disgust-related emotions (i.e., disgust, guilt, shame) at the time of a trauma have similar levels of PTSD symptoms as those with predominant reactions based in fear or sadness (Hathaway, Boals, & Banks, 2010). There is some evidence to suggest that traumatic events involving sexual victimization may be particularly likely to elicit peritraumatic disgust, even when compared with other events involving interpersonal (but not sexual) violation. Indeed, one study found that adolescent victims of sexual assault were six times more likely to report peritraumatic disgust as compared with victims of physical assault (Feldner et al., 2010). Intensity of peritraumatic disgust experienced toward the perpetrator of a sexual assault has also been linked to severity of PTSD symptoms, while self-focused disgust during a sexual assault has been linked to posttraumatic contamination-based obsessive–compulsive symptoms (Badour et al., 2012).

A major limitation of the peritraumatic literature, however, is reliance on retrospective self-report. One notable limitation of assessing peritraumatic experiences and PTSD symptoms simultaneously is the potential confound that individuals with increased PTSD symptoms may be more likely to associate past events with higher distress and be influenced by negative posttraumatic appraisals and emotions (e.g., Engelhard, van den Hout, & McNally, 2008). Prospective studies are most desirable for capturing the effects of peritraumatic experiences on psychopathology. Unfortunately, only one study to date has used such methodology to specifically study peritraumatic disgust and PTSD symptoms. Engelhard et al. (2011) found that retrospective report of peritraumatic disgust intensity predicted simultaneously assessed PTSD symptoms following combat, above and beyond intensity of peritraumatic fear. However, peritraumatic disgust was no longer predictive of PTSD symptoms when they were reassessed 9 months later. Additional prospective research is warranted to answer questions related to the association between peritraumatic disgust in the development of PTSD, including whether peritraumatic disgust is consistently linked to PTSD or whether its influence on trauma-related symptomology is attenuated over time. In models of PTSD, peritraumatic experiences may influence conditioning between neutral and emotionally distressing stimuli and facilitate the development of the disorder, but persistent negative emotions, appraisals, and associated escape and avoidance are expected to be involved in the maintenance and/or exacerbation of symptoms. Therefore, it is not only peritraumatic disgust but also *posttraumatic* disgust that may predict PTSD symptoms. As such, we now turn our attention to persistent feelings of disgust after trauma (i.e., posttraumatic disgust).

Posttraumatic disgust

Although many individuals may experience peritraumatic disgust, a smaller proportion are expected to continue to experience persistent disgust months or years after a trauma. More intense peritraumatic disgust may lead to greater conditioned reactions to previously neutral cues, and such cues may then be capable of eliciting stronger posttraumatic disgust reactions (e.g., trauma memories or reminders). Posttraumatic disgust elicited by conditioned trauma stimuli may then motivate avoidance or escape behaviors (including washing/cleaning) to reduce distress, ultimately preventing extinction learning, thus maintaining (or exacerbating) PTSD symptoms. Therefore, we might expect posttraumatic disgust to mediate the association between peritraumatic disgust and increased PTSD symptoms. Research has indeed found evidence to suggest that peritraumatic disgust may lead to increased PTSD symptoms through continued posttraumatic disgust reactions (Badour, Feldner, Blumenthal, & Knapp, 2013; Bomyea & Allard, 2017).

Researchers have used several methods to study the relation between posttraumatic disgust and PTSD. One method has been to assess whether individuals with PTSD report higher levels of general disgust than trauma-exposed individuals without PTSD. Research utilizing this approach has generally found that individuals with PTSD report more persistent, frequent feelings of disgust compared with trauma-exposed individuals without PTSD (Finucane et al., 2012; Foy, Sipprelle, Rueger, & Carroll, 1984; Power & Fyvie, 2013). General feelings of self-focused disgust have also been implicated as a potential explanatory factor in the association between PTSD symptoms and suicide risk (Brake et al., 2017) and between a probable PTSD diagnosis and hazardous drinking among college students (Sonnier, Brake, Flores, & Badour, 2019). Relatedly, survivors of childhood sexual abuse have been shown to report more frequent general and weekly feelings of disgust compared with emotions of anger, sadness, and fear (Coyle, Karatzias, Summers, & Power, 2014). Persistent feelings of disgust have also been linked to PTSD symptom severity, general distress, subjective well-being, dissociation, self-harm, and risk to self and others among childhood sexual abuse survivors (Bradley, Karatzias, & Coyle, 2018; Coyle et al., 2014). These findings suggest that persistent feelings of disgust not only are relevant to the severity of PTSD symptoms, but also may pose difficulties that negatively impact other aspects of well-being.

Researchers have also employed laboratory methods to study the relation between disgust and PTSD by measuring disgust responses to either standardized (e.g., film clips/pictures depicting trauma-related content or other emotionally evocative stimuli) or idiographic trauma stimuli (e.g., participant generated narratives of a past traumatic event). Combat veterans with PTSD have been shown to respond to combat film clips with increased disgust, fear, anger, sadness, and guilt when compared with male combat veterans without PTSD and men without a history of combat exposure, suggesting that PTSD may be associated with increased nonspecific negative emotional reactivity (Pitman, van der Kolk, Orr, & Greenberg, 1990). Supporting this finding, another study utilizing emotionally evocative pictures (i.e., not trauma specific) also found that male veterans with PTSD reported increased negative reactivity (i.e., increased feelings of disgust, shame, anger, and sadness) to the pictures compared with male combat veterans without PTSD and men without a history of combat exposure (Amdur, Larsen, & Liberzon, 2000).

More commonly, idiographic trauma imagery has been used to elicit and examine trauma-related disgust. Compared to women without PTSD, women with PTSD report increased disgust in response to trauma imagery (i.e., personalized narrative scripts), even when controlling for anxiety reactivity and peritraumatic fear (Olatunji, Babson, Smith, Feldner, & Connolly, 2009) and even in the absence of elevations in other emotions (e.g., fear, anger, sadness, and shame; Shin et al., 1999). Increased PTSD-related disgust reactivity to idiographic trauma imagery has also been documented in studies of male Vietnam veterans (Pitman et al., 1990; Pitman, Orr, Forgue, de Jong, & Claiborn, 1987), mixed-gender samples of Vietnam combat veterans/nurses (Shin et al., 2004), women with a history of interpersonal trauma (Badour, Feldner, Blumenthal, & Bujarski, 2013; Badour, Feldner, Blumenthal, & Knapp, 2013), and individuals with a mixed traumatic event history (Lanius et al., 2007). However, few studies have failed to demonstrate that disgust reactivity (or other negative emotional reactivity) differentiates individuals with PTSD from trauma-exposed individuals without PTSD (Carson et al., 2000; Lanius et al., 2003; Orr, Pitman, Lasko, & Herz, 1993), while another failed to demonstrate links between PTSD symptoms and trait disgust, moral disgust, or state disgust after writing and reading a personalized trauma script among a mixed-trauma sample of undergraduates (Ojserkis et al., 2014).

Consistent with the idea that disgust is likely to be more germane to certain kinds of traumatic experiences than others, Badour et al. (2011) found that a history of interpersonal trauma, compared with noninterpersonal traumatic events such as accidents/natural disasters, was associated with increased disgust and anger reactivity to trauma imagery, even when controlling for PTSD symptoms and negative affect. Similarly, Badour, Feldner, Babson, Blumenthal, and Dutton (2013) found that PTSD symptom severity was positively related to disgust reactivity in response to trauma imagery among women with either a history of sexual or physical assault, but sexual assault victims reported greater overall disgust reactivity. Moreover, PTSD symptoms were positively related to feelings of dirtiness and urges to wash in response to the trauma imagery, but only for women with a history of sexual trauma.

Although this work is in line with operant conditioning models that suggest persistent posttraumatic disgust reactivity in response to trauma cues is linked to PTSD due to negatively reinforced avoidance and escape, research has yet to explicitly examine how different emotion regulation or coping strategies may impact disgust and other forms of emotional reactivity in response to traumatic event reminders in the laboratory. Future research in this area should additionally consider accounting for specific factors that may influence responses to disgust, particularly with regard to gender. Women with PTSD appear to report stronger feelings of disgust in response to trauma cues compared with women without PTSD and men with and without PTSD, although men with PTSD have higher increases in heart rate in response to ideographic trauma scripts than men without PTSD and women with PTSD (Olatunji, Babson, et al., 2009). Additionally, women may have stronger subjective responses to disgust and increased physiological responding (e.g., skin conductance; Rohrmann, Hopp, & Quirin, 2008). Other factors to consider when conducting trauma-focused disgust research include trauma type (e.g., sexual vs nonsexual assault; Badour et al., 2011; Badour, Feldner, Babson, Blumenthal, & Dutton, 2013; Feldner et al., 2010) and disgust proneness (Engelhard et al., 2011).

Clinical assessment and treatment implications

As the sections of this chapter illustrate, disgust is a highly complex emotion that has been increasingly recognized for its role in responses to trauma, including PTSD, mental contamination, and moral injury. Considering the combined effects of disgust's complexity and the diversity of presentations seen in PTSD, the need for reliable assessment and effective treatment to address trauma-related disgust factors in PTSD is evident. Although diagnostic guidelines and accompanying literature have begun broadening the scope of emotions considered when evaluating PTSD cases, adopting new methods for the systematic assessment of nonfear emotions in standard clinical practice is likely to change gradually. And given that disgust itself is not explicitly identified as a potential negative emotion within the PTSD cluster D criteria, it is all the more essential that practicing clinicians recognize the likelihood that treatment-relevant disgust may often go undetected without deliberate intention to assess for it.

One primary challenge in assessing trauma-related disgust is the limited likelihood that individuals affected by PTSD will voluntarily report disgust in treatment. Much like the historical approach to PTSD as a fear- and anxiety-based disorder in clinical academia, public psychoeducation on PTSD has traditionally emphasized these fear- and anxiety-based features, and public perceptions of PTSD based on news accounts and media depictions typically highlight fear and anxiety as the primary emotional experience (and perhaps, to some degree, anger or guilt). Disgust may be tied to aspects of trauma memories that are more visceral, unpleasant, embarrassing, or shameful to discuss such as contact with bodily fluids, sensory details of events involving death or decay, or experiences involving deeply held moral or sexual transgressions or behavior. As such, patients may be inclined to avoid bringing up disgusting aspects of their traumatic experiences and may not volunteer relevant information unless asked.

Useful assessment of trauma-related disgust, however, cannot easily be accomplished without the proper tools. Although a number of trait and state disgust measures currently exist, very few trauma-specific measures have been developed, though empirical evidence demonstrating that the added utility of disgust-focused trauma measures has started to emerge (e.g., Posttraumatic Experience of Mental Contamination Scale [PEMC]; Brake, Adams, Hood, & Badour, 2019). Clinical assessment and treatment may still benefit from the use of general disgust-focused measures and screening questions currently available to either rule out or identify important disgust-based treatment targets that might otherwise go overlooked. Particularly for traumatic events that are more likely to have involved elicitors of disgust (e.g., sexual assault, events involving death, or moral violation), reported peritraumatic disgust reactions (e.g., vomiting and gagging), or current avoidance behaviors that logically align with disgust aversion (e.g., washing/cleaning, aversion to sex, or other domains perceived as disgusting), querying individuals about any feelings of disgust, and who or what they feel disgusted toward, can often reveal critical details of a comprehensive and informed case formulation.

Assessment of trauma related disgust may also yield useful information about other symptoms beyond PTSD. This is important to consider, given that PTSD exhibits notable comorbidity with other forms of psychopathology also susceptible to disgust. In particular, both PTSD and OCD share symptom characteristics such as intrusive cognitions and avoidance behaviors and have been shown to be highly comorbid (19%–41%), and their co-occurrence has been implicated in increased symptom severity, poorer quality of life, and greater risk for other mood and substance use difficulties (Brown, Campbell, Lehman, Grisham, & Mancill, 2001; Fontenelle et al., 2012; Nacasch, Fostick, & Zohar, 2011; Ojserkis et al., 2017). Specific examples in which both disgust-specific OCD and PTSD may be relevant can include individuals who have experienced sexual violence and may develop intrusive memories that elicit disgust toward themselves or sexual activity in general, leading to increased disgust sensitivity, emergent contamination concerns, and compulsive washing behaviors to manage disgusting stimuli. Aversion to disgusting posttraumatic reminders may develop into a general aversion to the feeling of disgust itself, and thus comorbid PTSD/OCD presentations rooted in a traumatic occurrence may not always appear logical. Other complex cases documenting this comorbidity have described trauma exposures (e.g., military combat, physical/emotional abuse, and death) that have precipitated obsessions about one's disgusting appearance, obsessive responsibility for affecting others with contamination or future morally disgusting behaviors, and compulsions ranging from cleaning/washing to ritualized repetition, checking, or counting to avoid or suppress disgusting thoughts (e.g., de Silva & Marks, 1999; Gershuny et al., 2003).

Beyond the identification of the presence of disgust as a potential treatment target, important questions also remain as to whether best practice interventions currently employed for PTSD are effectively able to target and alleviate trauma-related disgust and associated difficulties of mental contamination and moral injury. Research in other disgust-oriented disorders has suggested that disgust may be less amenable to exposure-based interventions that are routinely delivered to address anxiety and experiential avoidance. Individuals with contamination OCD submitted to exposure tasks have shown more rapid habituation to anxiety than to disgust (McKay, 2006), whereas individuals treated for specific phobia have shown successful reductions in fear-based symptoms but enduring disgust responses to the phobic stimuli up to 1 year after treatment (de Jong, Vorage, & van den Hout, 2000; for a review of disgust treatment in anxiety-based disorders, see Mason & Richardson, 2012). However, very little research has explicitly examined disgust in PTSD-specific treatment contexts or associated reductions in disgust elicitation and PTSD severity. Preliminary work incorporating imaginal exposure into dialectical behavior therapy for women with PTSD and borderline personality disorder found that women did report significant decreases in disgust between (but not within) exposure sessions, but the effect size from the beginning to end of exposure treatment was small, compared with a large effect for reduced fear (Harned, Ruork, Liu, &

Tkachuck, 2015). In a single analogue imaginal exposure session with multiple trials, Badour and Feldner (2016) found that women with a history of sexual trauma showed greater disgust versus anxiety reactivity to the initial exposure trial but similar rates of improvement. The authors suggested that disgust may effectively respond to exposure techniques but may require additional exposure trials to result in adequate disgust amelioration. This study also found that participants who reported greater improvements in both anxiety and disgust reactivity exhibited more improvement in state PTSD symptoms across exposure trials compared with those only reporting improvement in anxiety reactivity, suggesting targeting disgust in PTSD treatment may yield additional symptom improvement above and beyond improvements associated with decreased anxiety. Clinicians may potentially be prematurely terminating exposures if they are only assessing reductions in anxiety among individuals who are also struggling with PTSD-related disgust.

Additional work has begun to explore other cognitive-focused interventions to address posttraumatic symptoms linked to disgust and/or contamination. These intriguing approaches may be particularly useful, given that many forms of posttraumatic disgust and related contamination concerns are thought to emerge from cognitive processes such as memories and mental images, making them perhaps more susceptible to techniques that target these sources directly. Adaptive disclosure (AD) was specifically created to address the difficulties associated with moral injury and seeks to address trauma-related appraisals (e.g., *I am unforgiveable*) by incorporating both exposure to the morally injurious event and strategies to facilitate self-forgiveness, reparation, and reconnection (Litz, Lebowitz, Gray, & Nash, 2017). For instance, clients might visualize a conversation with a compassionate moral authority figure in which they verbalize a perceived transgression and then imagine what this person would say to them. Although disgust was not specifically assessed, preliminary evidence found that AD led to reductions in PTSD symptoms, depression, and negative appraisals as well as increased posttraumatic growth (Gray et al., 2012). Another study, seeking to identify effective strategies to reduce disgust (as well as other moral emotions), compared a traditional cognitive challenge task with comprehensive distancing, a strategy focused on observing one's thoughts from a nonjudgmental, distanced perspective without reacting to them (Ojserkis et al., 2014). For example, an individual might use comprehensive distancing to rephrase the thought "I am permanently tainted by what happened to me" to "I am having the thought that I have been permanently tainted."

Results from this preliminary study suggested that comprehensive distancing successfully reduced both state and moral disgust, shame, and guilt among trauma-exposed undergraduates after reading an idiographic trauma script, although it was not more effective than the cognitive challenge control task. Both AD and comprehensive distancing involve novel methods that may facilitate reductions in moral emotions, which is particularly important given potential limitations of traditional cognitive reappraisal-based approaches in the context of complex reactions to moral violations (Litz et al., 2009; Ojserkis et al., 2014).

Jung, Steil, and colleagues have also shown promising effects piloting their combined cognitive restructuring and imagery modification (CRIM) approach to effectively reduce not only contamination concerns but also PTSD symptoms in adult survivors of childhood sexual abuse (Jung & Steil, 2012, 2013; Steil et al., 2011). With potential benefits as a adjunctive trauma-focused treatment module, CRIM consists of only two sessions involving (1) information on human dermal cell regeneration to help challenge maladaptive beliefs about lingering trauma-related contamination and (2) practice with idiosyncratic mental imagery to help patients "decontaminate" disgusting or contaminated thoughts as they occur (e.g., imagining shedding contaminated or disgusting skin for clean skin; mentally "bathing" in light that purifies the contamination). Such novel approaches hold significant potential for further trauma-focused treatment research.

Cultural considerations

Cross-cultural research suggests that disgust is a universal emotion recognized across many different cultures (Ekman, 1992). However, cultural differences may largely impact what factors are considered disgusting and how disgust is experienced (Olatunji & Sawchuk, 2005). As such, it is important to consider whether and to what extent cultural differences may impact disgust-based reactions to trauma. In nontrauma contexts, greater contamination aversion and disgust sensitivity have been documented among African Americans compared with European Americans in both clinical and nonclinical groups (e.g., Williams, Turkheimer, Schmidt, & Oltmanns, 2005), and recent research has implicated increased disgust sensitivity as a mediating factor that may partially explain racial differences in contamination aversion between these groups (George, Pittenger, Kelmendi, Lohr, & Adams, 2018). A smaller number of findings have also identified similar contamination elevations among Hispanic Americans (e.g., Williams et al.,

2005). The occurrence of such ethnic and racial differences is thought to perhaps result from both current and historic negative stereotypes such as racism, segregation, and interpersonal disgust toward minority groups in the United States. Such findings point to the potential for increased susceptibility among minority groups to trauma-specific disgust and related posttraumatic difficulties. Unfortunately, very little research has explored the interactions among culture, disgust/disgust reactivity, and PTSD. One study found that peritraumatic disgust predicted PTSD symptoms among European American participants, but not among African Americans (Lancaster et al., 2011). Additional research in this area is clearly needed.

New research has also started exploring the implications of religiosity for disgust and mental contamination, as religious traditions may often involve rituals and beliefs based on morality and both physical and mental purity. Findings to date on primarily Christian denominations have yielded few notable differences in trait mental contamination levels between denomination groups (e.g., Lorona & Fergus, 2018). However, preliminary results have suggested that individuals with stronger Islamic beliefs may be more susceptible to mental contamination, perhaps because strict Islamic religious practices adhere to a narrower set of customs related to morality, sexual attitudes, and ritualistic behaviors as compared with other major religious doctrines (Bilekli & Inozu, 2018). More so than general religious affiliation, early findings have pointed to the potential impact that scrupulosity—a distressing and functionally impairing form of guilt associated with violation of religious beliefs—may have on mental contamination (Fergus, 2014), suggesting that religious beliefs that frame morally or spiritually disgusting thoughts as sinful may increase risk for general mental contamination difficulties. Although this new research domain has not yet focused on trauma-specific contexts, these findings may hold important implications for susceptibility to disgust-related difficulties among certain religious groups following trauma experiences. Future research in this area may help facilitate our understanding about the presence of disgust among cultural groups, which may be helpful in assessment and treatment planning.

Conclusion

Disgust is a universally experienced emotion conceptualized as a reaction to stimuli associated with potential harm to one's physical, psychological, and moral well-being. In this chapter, we reviewed the ways in which disgust may be relevant to trauma, noteworthy findings regarding both peritraumatic and posttraumatic disgust responses as they relate to PTSD and other forms of psychopathology, and risk factors for experiencing disgust during a traumatic event, including both disgust proneness and certain event characteristics that are most likely to elicit disgust. In reviewing the literature on posttraumatic disgust responding, we concluded that, while individuals with increased PTSD symptoms evidence higher negative emotion reactivity in general (Amdur et al., 2000; Pitman, van der Kolk, et al., 1990), there also exists evidence of disgust-specific responding in response to trauma cues for these individuals, particularly among individuals with histories of sexual victimization (Badour et al., 2011; Olatunji, Babson, et al., 2009; Shin et al., 1999). The implications of this research are that disgust may indeed be a key factor in the maintenance of PTSD symptoms, particularly in cases involving mental contamination or moral injury, which may not be adequately addressed in traditional PTSD treatments. As such, disgust-based PTSD may warrant specific considerations for assessment and treatment. We hope that after reading this chapter, the relevance of trauma-related disgust to PTSD is clear and inspires additional research to further our understanding in this arena. Areas for future research include improving our understanding of risk factors for experiencing disgust, including cross-sectional and prospective studies that comprehensively assess the influence of disgust propensity and sensitivity on both peri- and posttraumatic disgust and PTSD symptoms. Future research may also aid our understanding of other factors that may impact persistent disgust after trauma, including trauma type, gender, and cultural differences in the experience and perception of disgust. The work that has been done in this area has substantially improved our understanding of disgust-based reactions to trauma, which ultimately informs treatments that may more successfully address the complex difficulties (e.g., mental contamination and moral injury) subsequent to traumatic events. Our understanding of the emotional impact of trauma has come a long way in the last 30 years, and we hope to see continued growth as disgust gains more recognition for its relevance within PTSD.

References

Adams, T., Badour, C., Cisler, J., & Feldner, M. (2014). Contamination aversion and posttraumatic stress symptom severity following sexual trauma. *Cognitive Therapy and Research*, *38*(4), 449–457.

Amdur, R., Larsen, R., & Liberzon, I. (2000). Emotional processing in combat-related posttraumatic stress disorder: A comparison with traumatized and normal controls. *Journal of Anxiety Disorders*, *14*(3), 219–238.

American Psychiatric Association. (2013). *Diagnostic and statistical manual of mental disorders* (5th ed.). Washington, DC: American Psychiatric Association.

Badour, C., & Adams, T. (2015). Contaminated by trauma: Understanding links between self-disgust, mental contamination, and post-traumatic stress disorder. In P. A. Powell, P. G. Overton, & J. Simpson (Eds.), *The revolting self: Perspectives on the psychological, social, and clinical implications of self-directed disgust* (pp. 127–149). London, England: Karnac Books.

Badour, C., Bown, S., Adams, T., Bunaciu, L., & Feldner, M. (2012). Specificity of fear and disgust experienced during traumatic interpersonal victimization in predicting posttraumatic stress and contamination-based obsessive–compulsive symptoms. *Journal of Anxiety Disorders, 26*(5), 590–598.

Badour, C., & Feldner, M. (2016). Disgust and imaginal exposure to memories of sexual trauma: Implications for the treatment of posttraumatic stress. *Psychological Trauma Theory Research Practice and Policy, 8*(3), 267–275.

Badour, C., & Feldner, M. (2018). The role of disgust in posttraumatic stress: A critical review of the empirical literature. *Journal of Experimental Psychopathology, 9*(3). pr-032813.

Badour, C., Feldner, M., Babson, K., Blumenthal, H., & Dutton, C. (2013). Disgust, mental contamination, and posttraumatic stress: Unique relations following sexual versus non-sexual assault. *Journal of Anxiety Disorders, 27*(1), 155–162.

Badour, C., Feldner, M., Babson, K., Smith, R., Blumenthal, H., Trainor, C., … Olatunji, B. (2011). Differential emotional responding to ideographic cues of traumatic interpersonal violence compared to non-interpersonal traumatic experiences. *Journal of Experimental Psychopathology, 2*(3), https://doi.org/10.5127/jep-014711.

Badour, C., Feldner, M., Blumenthal, H., & Bujarski, S. (2013). Examination of increased mental contamination as a potential mechanism in the association between disgust sensitivity and sexual assault-related posttraumatic stress. *Cognitive Therapy and Research, 37*(4), 697–703.

Badour, C., Feldner, M., Blumenthal, H., & Knapp, A. (2013). Preliminary evidence for a unique role of disgust-based conditioning in posttraumatic stress. *Journal of Traumatic Stress, 26*(2), 280–287.

Bilekli, I., & Inozu, M. (2018). Mental contamination: The effects of religiosity. *Journal of Behavior Therapy and Experimental Psychiatry, 58*, 43–50.

Bomyea, J., & Allard, C. (2017). Trauma-related disgust in veterans with interpersonal trauma. *Journal of Traumatic Stress, 30*(2), 149–156.

Bomyea, J., & Amir, N. (2012). Disgust propensity as a predictor of intrusive cognitions following a distressing film. *Cognitive Therapy and Research, 36*(3), 190–198.

Bovin, M., & Marx, B. (2011). The importance of the peritraumatic experience in defining traumatic stress. *Psychological Bulletin, 137*(1), 47.

Bradley, A., Karatzias, T., & Coyle, E. (2018). Derealisation and self-harm strategies are used to regulate disgust, fear and sadness in adult survivors of childhood sexual abuse: Self-harm, derealisation and PTSD. *Clinical Psychology & Psychotherapy*.

Brake, C., Adams, T., Hood, C., & Badour, C. (2019). Posttraumatic mental contamination and the interpersonal psychological theory of suicide: Effects via DSM-5 PTSD symptom clusters. *Cognitive Therapy and Research, 43*, 259–271.

Brake, C., Jones, A., Wakefield, J., & Badour, C. (2018). Mental contamination and trauma: Understanding posttraumatic stress, risky behaviors, and help-seeking attitudes. *Journal of Obsessive-Compulsive and Related Disorders, 17*, 31–38.

Brake, C., Rojas, S., Badour, C., Dutton, C., & Feldner, M. (2017). Self-disgust as a potential mechanism underlying the association between PTSD and suicide risk. *Journal of Anxiety Disorders, 47*, 1–9.

Brown, T., Campbell, L., Lehman, C., Grisham, J., & Mancill, R. (2001). Current and lifetime comorbidity of the DSM-IV anxiety and mood disorders in a large clinical sample. *Journal of Abnormal Psychology, 110*(4), 585–599.

Carson, M., Paulus, L., Lasko, N., Metzger, L., Wolfe, J., Orr, S., & Pitman, R. (2000). Psychophysiologic assessment of posttraumatic stress disorder in Vietnam nurse veterans who witnessed injury or death. *Journal of Consulting and Clinical Psychology, 68*, 890–897.

Chapman, H., & Anderson, A. (2013). Things rank and gross in nature: A review and synthesis of moral disgust. *Psychological Bulletin, 139*(2), 300.

Coughtrey, A., Shafran, R., Knibbs, D., & Rachman, S. (2012). Mental contamination in obsessive–compulsive disorder. *Journal of Obsessive-Compulsive and Related Disorders, 1*(4), 244–250.

Coyle, E., Karatzias, T., Summers, A., & Power, M. (2014). Emotions and emotion regulation in survivors of childhood sexual abuse: The importance of "disgust" in traumatic stress and psychopathology. *European Journal of Psychotraumatology, 5*(1), 23306.

Dalgleish, T., & Power, M. (2004). Emotion-specific and emotion-non-specific components of posttraumatic stress disorder (PTSD): Implications for a taxonomy of related psychopathology. *Behaviour Research and Therapy, 42*(9), 1069–1088.

Darwin, C. (2003). *Expression of the emotions in man and animals.* In *The history of psychology: Fundamental questions* (pp. 188–202). New York: Oxford University Press.

Davey, G., & Bond, N. (2006). Using controlled comparisons in disgust psychopathology research: The case of disgust, hypochondriasis and health anxiety. *Journal of Behavior Therapy and Experimental Psychiatry, 37*(1), 4–15.

de Jong, P., & Peters, M. (2009). Sex and the sexual dysfunctions: The role of disgust and contamination sensitivity. In *Disgust and its disorders: Theory, assessment, and treatment implications* (pp. 253–270). Washington, DC: American Psychological Association.

de Jong, P., Vorage, I., & van den Hout, M. (2000). Counterconditioning in the treatment of spider phobia: Effects of disgust, fear, and valence. *Behaviour Research and Therapy, 38*(11), 1055–1069.

de Silva, P., & Marks, M. (1999). The role of traumatic experiences in the genesis of obsessive–compulsive disorder. *Behaviour Research and Therapy, 37*(10), 941–951.

Drescher, K., Foy, D., Kelly, C., Leshner, A., Schutz, K., & Litz, B. (2011). An exploration of the viability and usefulness of the construct of moral injury in war veterans. *Traumatology, 17*(1), 8–13.

Ehlers, A., & Clark, D. (2000). A cognitive model of posttraumatic stress disorder. *Behaviour Research and Therapy, 38*(4), 319–345.

Ekman, P. (1992). An argument for basic emotions. *Cognition and Emotion, 6*(3–4), 169–200.

Engelhard, I., Olatunji, B., & de Jong, P. (2011). Disgust and the development of posttraumatic stress among soldiers deployed to Afghanistan. *Journal of Anxiety Disorders, 25*(1), 58–63.

Engelhard, I., van den Hout, M., & McNally, R. (2008). Memory consistency for traumatic events in Dutch soldiers deployed to Iraq. *Memory, 16*(1), 3–9.

Fairbrother, N., & Rachman, S. (2004). Feelings of mental pollution subsequent to sexual assault. *Behaviour Research and Therapy, 42*, 173–189.

Farnsworth, J., Drescher, K., Nieuwsma, J., Walser, R., & Currier, J. (2014). The role of moral emotions in military trauma: Implications for the study and treatment of moral injury. *Review of General Psychology, 18*(4), 249.

Feldner, M., Frala, J., Badour, C., Leen-Feldner, E., & Olatunji, B. (2010). An empirical test of the association between disgust and sexual assault. *International Journal of Cognitive Therapy, 3*(1), 11–22.

Fergus, T. (2014). Mental contamination and scrupulosity: Evidence of unique associations among Catholics and Protestants. *Journal of Obsessive-Compulsive and Related Disorders, 3*(3), 236–242.

Finucane, A., Dima, A., Ferreira, N., & Halvorsen, M. (2012). Basic emotion profiles in healthy, chronic pain, depressed and PTSD individuals. *Clinical Psychology & Psychotherapy, 19*(1), 14–24.

Foa, E., & Kozak, M. (1986). Emotional processing of fear: Exposure to corrective information. *Psychological Bulletin*, *99*(1), 20.
Fontenelle, L., Cocchi, L., Harrison, B., Shavitt, R., do Rosário, M., Ferrão, Y., ... Torres, A. (2012). Towards a post-traumatic subtype of obsessive–compulsive disorder. *Journal of Anxiety Disorders*, *26*(2), 377–383.
Foy, D., Sipprelle, R., Rueger, D., & Carroll, E. (1984). Etiology of posttraumatic stress disorder in Vietnam veterans: Analysis of premilitary, military, and combat exposure influences. *Journal of Consulting and Clinical Psychology*, *52*(1), 79.
Fredman, S., Monson, C., Schumm, J., Adair, K., Taft, C., & Resick, P. (2010). Associations among disaster exposure, intimate relationship adjustment, and PTSD symptoms: Can disaster exposure enhance a relationship? *Journal of Traumatic Stress*, *23*(4), 446–451.
George, J., Pittenger, C., Kelmendi, B., Lohr, J., & Adams, T. (2018). Disgust sensitivity mediates the effects of race on contamination aversion. *Journal of Obsessive-Compulsive and Related Disorders*, *19*, 72–76.
Gershuny, B., Baer, L., Radomsky, A., Wilson, K., & Jenike, M. (2003). Connections among symptoms of obsessive-compulsive disorder and posttraumatic stress disorder: A case series. *Behaviour Research and Therapy*, *41*(9), 1029–1041.
Gray, M., Schorr, Y., Nash, W., Lebowitz, L., Amidon, A., Lansing, A., ... Litz, B. (2012). Adaptive disclosure: An open trial of a novel exposure-based intervention for service members with combat-related psychological stress injuries. *Behavior Therapy*, *43*(2), 407–415.
Grunert, B., Hargarten, S., Matloub, H., Sanger, J., Hanel, D., & Yousif, N. (1992). Predictive value of psychological screening in acute hand injuries. *The Journal of Hand Surgery*, *17*(2), 196–199.
Haidt, J., Rozin, P., McCauley, C., & Imada, S. (1997). Body, psyche, and culture: The relationship between disgust and morality. *Psychology and Developing Societies*, *9*(1), 107–131.
Harned, M., Ruork, A., Liu, J., & Tkachuck, M. (2015). Emotional activation and habituation during imaginal exposure for PTSD among women with borderline personality disorder. *Journal of Traumatic Stress*, *28*(3), 253–257.
Hathaway, L., Boals, A., & Banks, J. (2010). PTSD symptoms and dominant emotional response to a traumatic event: An examination of DSM-IV Criterion A2. *Anxiety, Stress, and Coping*, *23*(1), 119–126.
Herba, J., & Rachman, S. (2007). Vulnerability to mental contamination. *Behaviour Research and Therapy*, *45*(11), 2804–2812.
Hodson, G., & Costello, K. (2007). Interpersonal disgust, ideological orientations, and dehumanization as predictors of intergroup attitudes. *Psychological Science*, *18*(8), 691–698.
Ishikawa, R., Kobori, O., & Shimizu, E. (2015). Unwanted sexual experiences and cognitive appraisals that evoke mental contamination. *Behavioural and Cognitive Psychotherapy*, *43*(1), 74–88.
Jones, C., Olson, M., & Fazio, R. (2010). *Evaluative conditioning: The "how" question.* In Vol. 43. *Advances in experimental social psychology* (pp. 205–255). Academic Press.
Jung, K., & Steil, R. (2012). The feeling of being contaminated in adult survivors of childhood sexual abuse and its treatment via a two-session program of cognitive restructuring and imagery modification: A case study. *Behavior Modification*, *36*(1), 67–86.
Jung, K., & Steil, R. (2013). A randomized controlled trial on cognitive restructuring and imagery modifcation to reduce the feeling of being contaminated in adult survivors of childhood sexual abuse suffering from posttraumatic stress disorder. *Psychotherapy and Psychosomatics*, *82*(4), 213–220.
Kilpatrick, D., Resnick, H., Freedy, J., Pelcovitz, D., Resick, P., Roth, S., & van der Kolk, B. (1998). *The posttraumatic stress disorder field trial: Evaluation of the PTSD construct: Criteria A through E.* In Vol. 4. *DSM-IV Sourcebook* (pp. 803–844).

Lancaster, S., Melka, S., & Rodriguez, B. (2011). Emotional predictors of PTSD symptoms. *Psychological Trauma Theory Research Practice and Policy, 3*(4), 313.

Lanius, R., Frewen, P., Girotti, M., Neufeld, R., Stevens, T., & Densmore, M. (2007). Neural correlates of trauma script-imagery in posttraumatic stress disorder with and without comorbid major depression: A functional MRI investigation. *Psychiatry Research: Neuroimaging, 155*(1), 45–56.

Lanius, R., Williamson, P., Hopper, J., Densmore, M., Boksman, K., Gupta, M., ... Menon, R. (2003). Recall of emotional states in posttraumatic stress disorder: An fMRI investigation. *Biological Psychiatry, 53*, 204–210.

Litz, B., Lebowitz, L., Gray, M., & Nash, W. (2017). *Adaptive disclosure: A new treatment for military trauma, loss, and moral injury*. Guilford Publications.

Litz, B., Stein, N., Delaney, E., Lebowitz, L., Nash, W., Silva, C., & Maguen, S. (2009). Moral injury and moral repair in war veterans: A preliminary model and intervention strategy. *Clinical Psychology Review, 29*(8), 695–706.

Lorona, R., & Fergus, T. (2018). How dual-faceted disgust relates to state mental contamination in religious individuals. *Mental Health, Religion and Culture, 21*(2), 139–152.

Mason, E., & Richardson, R. (2012). Treating disgust in anxiety disorders. *Clinical Psychology: Science and Practice, 19*(2), 180–194.

Matchett, G., & Davey, G. (1991). A test of a disease-avoidance model of animal phobias. *Behaviour Research and Therapy, 29*(1), 91–94.

McKay, D. (2006). Treating disgust reactions in contamination-based obsessive–compulsive disorder. *Journal of Behavior Therapy and Experimental Psychiatry, 37*(1), 53–59.

Muris, P., Merckelbach, H., Nederkoorn, S., Rassin, E., Candel, I., & Horselenberg, R. (2000). Disgust and psychopathological symptoms in a nonclinical sample. *Personality and Individual Differences, 29*(6), 1163–1167.

Nacasch, N., Fostick, L., & Zohar, J. (2011). High prevalence of obsessive–compulsive disorder among posttraumatic stress disorder patients. *European Neuropsychopharmacology, 21*(12), 876–879.

Ojserkis, R., Boisseau, C., Reddy, M., Mancebo, M., Eisen, J., & Rasmussen, S. (2017). The impact of lifetime PTSD on the seven-year course and clinical characteristics of OCD. *Psychiatry Research, 258*, 78–82.

Ojserkis, R., McKay, D., Badour, C., Feldner, M., Arocho, J., & Dutton, C. (2014). Alleviation of moral disgust, shame, and guilt in posttraumatic stress reactions: An evaluation of comprehensive distancing. *Behavior Modification, 38*(6), 801–836.

Ojserkis, R., McKay, D., & Lebeaut, A. (2018). Associations between mental contamination, disgust, and obsessive-compulsive symptoms in the context of trauma. *Journal of Obsessive-Compulsive and Related Disorders, 17*, 23–30.

Olatunji, B., Armstrong, T., & Elwood, L. (2017). Is disgust proneness associated with anxiety and related disorders? A qualitative review and meta-analysis of group comparison and correlational studies. *Perspectives on Psychological Science, 12*(4), 613–648.

Olatunji, B., Armstrong, T., Fan, Q., & Zhao, M. (2014). Risk and resiliency in posttraumatic stress disorder: Distinct roles of anxiety and disgust sensitivity. *Psychological Trauma Theory Research Practice and Policy, 6*(1), 50.

Olatunji, B., Babson, K., Smith, R., Feldner, M., & Connolly, K. (2009). Gender as a moderator of the relation between PTSD and disgust: A laboratory test employing individualized script-driven imagery. *Journal of Anxiety Disorders, 23*, 1091–1097.

Olatunji, B., Elwood, L., Williams, N., & Lohr, J. (2008). Mental pollution and PTSD symptoms in victims of sexual assault: A preliminary examination of the mediating role of trauma-related cognitions. *Journal of Cognitive Psychotherapy, 22*(1), 37–47.

Olatunji, B., Forsyth, J., & Cherian, A. (2007). Evaluative differential conditioning of disgust: A sticky form of relational learning that is resistant to extinction. *Journal of Anxiety Disorders, 21*(6), 820–834.

Olatunji, B., & Sawchuk, C. (2005). Disgust: Characteristic features, social manifestations, and clinical implications. *Journal of Social and Clinical Psychology, 24*(7), 932–962.

Olatunji, B., Tart, C., Ciesielski, B., McGrath, P., & Smits, J. (2011). Specificity of disgust vulnerability in the distinction and treatment of OCD. *Journal of Psychiatric Research, 45*(9), 1236–1242.

Olatunji, B., Unoka, Z., Beran, E., David, B., & Armstrong, T. (2009). Disgust sensitivity and psychopathological symptoms: Distinctions from harm avoidance. *Journal of Psychopathology and Behavioral Assessment, 31*(2), 137.

Orr, S., Pitman, R., Lasko, N., & Herz, L. (1993). Psychophysiological assessment of posttraumatic stress disorder imagery in World War II and Korean combat veterans. *Journal of Abnormal Psychology, 102*, 152–159.

Pitman, R., Orr, S., Forgue, D., Altman, B., de Jong, J., & Herz, L. (1990). Psychophysiologic responses to combat imagery of Vietnam veterans with posttraumatic stress disorder versus other anxiety disorders. *Journal of Abnormal Psychology, 99*, 49–54.

Pitman, R., Orr, S., Forgue, D., de Jong, J., & Claiborn, J. (1987). Psychophysiologic assessment of posttraumatic stress disorder imagery in Vietnam combat veterans. *Archives of General Psychiatry, 44*, 970–975.

Pitman, R., van der Kolk, B., Orr, S., & Greenberg, M. (1990). Naloxone-reversible analgesic response to combat-related stimuli in posttraumatic stress disorder. *Archives of General Psychiatry, 47*, 541–544.

Powell, P., Simpson, J., & Overton, P. (2013). When disgust leads to dysphoria: A three-wave longitudinal study assessing the temporal relationship between self-disgust and depressive symptoms. *Cognition & Emotion, 27*(5), 900–913.

Power, M., & Fyvie, C. (2013). The role of emotion in PTSD: Two preliminary studies. *Behavioural and Cognitive Psychotherapy, 41*(2), 162–172.

Rachman, S. (2006). *Fear of contamination: Assessment and treatment* (2006). Oxford: Oxford University Press.

Rivers, W. (1920). *Instinct and the unconscious: A contribution to a biological theory of the psycho-neuroses*. Cambridge: Cambridge University Press.

Rohrmann, S., Hopp, H., & Quirin, M. (2008). Gender differences in psychophysiological responses to disgust. *Journal of Psychophysiology, 22*(2), 65–75.

Rozin, P., & Fallon, A. (1987). A perspective on disgust. *Psychological Review, 94*(1), 23–41.

Rozin, P., Haidt, J., & McCauley, C. (2009). Disgust: The body and soul emotion in the 21st century. In *Disgust and its disorders: Theory, assessment, and treatment implications* (pp. 9–29). Washington, DC: American Psychological Association.

Rozin, P., Lowery, L., Imada, S., & Haidt, J. (1999). The CAD triad hypothesis: A mapping between three moral emotions (contempt, anger, disgust) and three moral codes (community, autonomy, divinity). *Journal of Personality and Social Psychology, 76*(4), 574–586.

Rozin, P., Millman, L., & Nemeroff, C. (1986). Operation of the laws of sympathetic magic in disgust and other domains. *Journal of Personality and Social Psychology, 50*(4), 703–712.

Rozin, P., & Nemeroff, C. (1990). The laws of sympathetic magic: A psychological analysis of similarity and contagion. In J. W. Stigler, R. A. Shweder, & G. Herdt (Eds.), *Cultural psychology: Essays on comparative human development* (pp. 205–232). New York: Cambridge University Press.

Rüsch, N., Schulz, D., Valerius, G., Steil, R., Bohus, M., & Schmahl, C. (2011). Disgust and implicit self-concept in women with borderline personality disorder and posttraumatic stress disorder. *European Archives of Psychiatry and Clinical Neuroscience, 261*(5), 369–376.

Sawchuk, C., Lohr, J., Tolin, D., Lee, T., & Kleinknecht, R. (2000). Disgust sensitivity and contamination fears in spider and blood–injection–injury phobias. *Behaviour Research and Therapy, 38*(8), 753–762.

Shin, L., McNally, R., Kosslyn, S., Thompson, W., Rauch, S., Alpert, N., … Pitman, R. (1999). Regional cerebral blood flow during script-driven imagery in childhood sexual abuse-related PTSD: A PET investigation. *American Journal of Psychiatry, 156*, 575–584.

Shin, L., Orr, S., Carson, M., Rauch, S., Macklin, M., Lasko, N., ... Pitman, R. (2004). Regional cerebral blood flow in the amygdala and medial prefrontal cortex during traumatic imagery in male and female Vietnam veterans with PTSD. *Archives of General Psychiatry*, *61*, 168–176.

Smith, N., Steele, A., Weitzman, M., Trueba, A., & Meuret, A. (2015). Investigating the role of self-disgust in nonsuicidal self-Injury. *Archives of Suicide Research*, *19*(1), 60–74.

Sonnier, H., Brake, C., Flores, J., & Badour, C. (2019). Posttraumatic stress and hazardous alcohol use in trauma-exposed young adults: Indirect effects of self-disgust. *Substance Use & Misuse*, *54*, 1051–1059.

Steil, R., Jung, K., & Stangier, U. (2011). Efficacy of a two-session program of cognitive restructuring and imagery modification to reduce the feeling of being contaminated in adult survivors of childhood sexual abuse: A pilot study. *Journal of Behavior Therapy and Experimental Psychiatry*, *42*(3), 325–329.

Travis, R., & Fergus, T. (2015). The potentiating effect of disgust sensitivity on the relationship between disgust propensity and mental contamination. *Journal of Obsessive-Compulsive and Related Disorders*, *6*, 114–119.

Troop, N., & Baker, A. (2009). *Food, body and soul: The role of disgust in eating disorders*. American Psychological Association.

Tybur, J., Lieberman, D., & Griskevicius, V. (2009). Microbes, mating, and morality: Individual differences in three functional domains of disgust. *Journal of Personality and Social Psychology*, *97*(1), 103–122.

Ullman, S., Relyea, M., Peter-Hagene, L., & Vasquez, A. (2013). Trauma histories, substance use coping, PTSD, and problem substance use among sexual assault victims. *Addictive Behaviors*, *38*(6), 2219–2223.

van Delft, I., Finkenauer, C., Tybur, J., & Lamers-Winkelman, F. (2016). Disgusted by sexual abuse: Exploring the association between disgust sensitivity and posttraumatic stress symptoms among mothers of sexually abused children. *Journal of Traumatic Stress*, *29*(3), 237–244.

Williams, M., Turkheimer, E., Schmidt, E., & Oltmanns, T. (2005). Ethnic identification biases responses to the Padua Inventory for obsessive-compulsive disorder. *Assessment*, *12*(2), 174–185.

CHAPTER 6
Shame and guilt in PTSD

Katherine C. Cunningham
Salem Veterans Affairs Medical Center, Salem, VA, United States
Fralin Biomedical Research Institute at Virginia Tech Carilion, Roanoke, VA, United States

Introduction

Guilt and shame repeatedly surfaced as a central feature in observations of war neuroses, shell shock, and combat stress after each major war (e.g., Haley, 1974; Ludwig, 1947; Rivers, 1922) and in descriptions of rape trauma syndrome (e.g., Burgess, 1983; Dahl, 1989). When these stress syndromes were formally labeled as posttraumatic stress disorder (PTSD) following the Vietnam War, the third edition of the *Diagnostic and Statistical Manual of Mental Disorders* (*DSM-III*; American Psychological Association [APA], 1980) included survivor guilt among the diagnostic criteria. When the *DSM-III* was revised in 1987, survivor guilt was moved to associated features due to the observation that it did not manifest in every case. As PTSD became conceptualized as a fear-based anxiety disorder (Foa, Steketee, & Rothbaum, 1989), guilt and shame remained overlooked in the predominant literature; however, arguments continued to be made for the roles of shame and guilt in the maintenance of PTSD (e.g., Lee, Scragg, & Turner, 2001; Stone, 1992). Based on growing research evidence, guilt and shame were once again added to the diagnostic criteria for PTSD in the *DSM-5* (APA, 2013) under negative alterations in cognition and emotion (Cluster D). Shame and guilt have been consistently associated with PTSD symptom severity (e.g., Cunningham, Davis, Wilson, & Resick, 2018), and emerging research suggests these emotions may play a causal role in the maintenance of PTSD (e.g., Øktedalen, Hoffart, & Langkaas, 2015). Within the context of PTSD, shame and guilt have also been associated with myriad additional deleterious outcomes, including suicidal ideation and behavior (e.g., Bryan, Morrow, Etienne, & Ray-Sannerud, 2013; Cunningham et al., 2018; Cunningham et al., 2017). The present chapter explores the extant literature on guilt and shame in relationship to PTSD, reviews the impact of empirically supported PTSD treatments on these emotions, and discusses complimentary treatment approaches and avenues for future research.

Shame and guilt as self-evaluative social emotions

Fundamentally, emotions are context-dependent interpretations of internal physiological experience. Differences in physiological arousal do not define emotions; rather, specific emotional experience is constructed via prior learning and cognition, for example, situation- and self-related interpretations and attributions (Barrett, 2017a, 2017b; Izard, 1977, 2011). Self-conscious and self-conscious evaluative (henceforth called self-evaluative) emotions are complex experiences that incorporate social mores and beliefs about the self in relationship to others (Lewis, 2008; Tracy & Robins, 2004). Like other emotions, they range in valence from positive to negative and include hubris, pride, empathy, envy, embarrassment, humiliation, guilt, and shame (Izard, 2011; Leary, 2007; Lewis, 1997, 2008; Lewis, Sullivan, Stanger, & Weiss, 1989; Tracy & Robins, 2004). The earliest self-conscious emotions to appear in human development emerge from a sense of the self as separate from others. These are empathy, envy, and embarrassment (Lewis, 2008, 2014; Lewis et al., 1989). As cognitive development and social learning progress, more complex emotions such as pride, guilt, and shame begin to emerge (Lewis et al., 1989). These are the so-called self-evaluative emotions, because they incorporate attributions about the self in relationship to others within the context of social and moral standards (Lewis, 2008, 2014).

As individuals internalize social norms and expectations, including cultural standards of morality (Lewis, 1997, 2014), core beliefs are formed (Beck, 1995). These internalized standards become the ideal against which an individual compares their identity and behavior. From this comparison, self-evaluative emotions are generated (Lewis, 1997, 2008, 2014; Tracy & Robins, 2004). They are predicated on whether the self is perceived as congruent or incongruent with the internalized ideal. In a given situation an individual makes causal attributions reflecting perceived locus of control (internal vs external), globality (generalized vs specific; e.g., "I am" vs "I did"), and stability (unchangeable vs changeable; Lewis, 2008; Tangney & Dearing, 2002; Tracy & Robins, 2004). The individual may engage in an active process of retrospective meaning-making, or this attributional process may occur without awareness, guided by prior learning. They then predict that others will reach the same conclusion (Leary, 2007). For example, a survivor who attributes an assault to their own actions is likely to believe that others will also blame them.

The congruence or incongruence of self-appraisal with social/moral/personal standards determines the valence (positive or negative) of the consequent self-evaluative emotion (Tracy & Robins, 2004). If an event is perceived as congruent with social/moral norms and an individual's self-concept, it will be experienced as a positively valenced self-evaluative emotion, such as pride. If the same event or action is incongruent with the individual's internalized ideal, the individual will feel a negatively valenced self-evaluative emotion, such as guilt or shame. For example, consider someone who experiences increased heart rate, respiration, and perspiration while presenting a public speech. If they believe they have met or exceeded the expectations of their audience through hard work, they feel pride; however, if they believe they have not met expectations due to poor preparation or lack of skills, they are likely to feel shame. In this example the situation and autonomic arousal remain identical: it is the person's knowledge of social expectations and their attributions about their own behavior (i.e., hard work vs laziness) that determined their emotional experience. Thus, self-attributions and interpretations of arousal based on relevant schemata determine which self-evaluative emotion is experienced (Tracy & Robins, 2004). In the context of PTSD, trauma survivors must attempt to incorporate an extremely aversive event, and this process is informed by prior learning and existing schemata, which are used to make sense of what happened and why (Ehlers & Clark, 2000).

Differentiating shame and guilt

To understand shame and guilt in relationship to PTSD, we must understand the differences between these two emotions. Shame and guilt are both negative self-evaluative emotions; they are separated by the self-behavior distinction originally proposed by Lewis (1971). Guilt originates from specific behavior-related attributions (i.e., specific, unstable attributions), whereas shame originates from global identity-related attributions (i.e., global, stable attributions; Lewis, 2008, 2014; Tangney & Dearing, 2002). The common expression is that "I did a bad thing" results in guilt, and "I am a bad person" engenders shame. Gilbert (2004) further distinguished shame and guilt based on variant threats in social relationships. Namely, shame is produced by threats to social status via failure to adhere to social and moral standards. Guilt, however, is related to the violation of the specific sociomoral standard requiring us to not cause harm to others (Gilbert, 2004). This distinction particularly relates to guilt as a retrospective moral emotion (Tangney & Dearing, 2002).

Guilt and shame also lead to different action urges. Izard (1977) suggested that, "while intense shame tends to heighten self-consciousness and decrease the logical-intellective operations of consciousness, guilt tends to be associated with increased cognition as the individual ruminates over the feeling that all is not right with others or with God" (p. 149). Because behavior is modifiable and reparations may be made for harm caused, guilt is believed to promote prosocial behavior (Lewis, 2014; Tangney & Dearing, 2002). In contrast, because shame is associated with unchangeable generalized (i.e., global and stable) attributions related to one's overall worthiness as a person (Lewis, 1971, 1997, 2008; Tangney & Dearing, 2002), it motivates an individual to hide the source of their shame (Lewis, 2014; Tangney & Dearing, 2002). This action urge often leads to social isolation (Tangney & Dearing, 2002) or even a desire to die (Lewis, 2014).

Theoretical models of trauma-related guilt and shame
Guilt

The only proposed model of trauma-related guilt defines it as a combination of affective distress and guilt-related cognitions (Kubany et al., 1995; Kubany & Watson, 2003). In other words, guilt is experienced when a person experiences unpleasant emotional arousal at the memory of a traumatic event and retroactively evaluates their own thoughts, feelings, or actions during the trauma as inappropriate (Kubany & Watson, 2003). This multidimensional model asserts that, in addition to diffuse distress, guilt is composed of cognitions about perceived responsibility for negative outcomes, the lack of justification for actions, perceived violation of values (i.e., moral injury), and assumptions about whether the traumatic event was predictable or preventable (i.e., hindsight bias; Kubany & Watson, 2003). Research has supported that these components (i.e., negative affect and specific types of cognitions) are associated with the experience of trauma-related guilt. Initial studies examining this model found that the distress component, in particular, was strongly correlated with PTSD symptoms (Kubany et al., 1995). Further research replicated similar findings, showing that the distress and guilt cognitions components, as measured by the Trauma-Related Guilt Inventory (TRGI; Kubany et al., 1996), were directly related to PTSD symptoms, whereas posttraumatic (global) guilt was not (Browne, Trim, Myers, & Norman, 2015). Instead, guilt cognitions indirectly affected both trauma-related guilt and PTSD symptoms via distress (Browne et al., 2015). This is not a surprising finding given that physiological arousal, negative affect, and negative cognitions are themselves core symptoms of PTSD.

Shame

Stone (1992) defined trauma-related shame as the dissonance between an individual's self-perception, whether real or ideal self, and their interpretation of a traumatic event. Broadening this definition, others have noted that shame not only is generated by dissonance but also is intensified when a traumatic event is congruent with existing shame-relevant beliefs and schemata (Lee et al., 2001). Trauma characteristics (e.g., betrayal) and cognitive interpretations of the event generate shame (Lewis, 2008; Stone, 1992), which is then maintained or intensified by social reactions and PTSD symptoms in a perpetual cycle (Stone, 1992). This process is difficult to interrupt due to the inherent urge to hide the sources of shame: the trauma, the PTSD, and ultimately one's self. Shame maintains PTSD symptoms, because shame avoidance interferes with processing trauma memories and reduces access to social support and new learning (Joseph, Williams, & Yule, 1997; Lee et al., 2001).

These formulations fit well with the predominant cognitive model of PTSD presented by Ehlers and Clark (2000). In this model the combined nature of the traumatic memory (i.e., associated cues), appraisal of the event, and subsequent symptoms result in a sustained sense of current threat. In the case of trauma-related shame, the threat is to identity and social status. The avoidance of threat leads to cognitive and behavioral patterns that maintain PTSD symptoms and associated distress.

Challenges in measurement

Measurement challenges have played a prominent role in the study of guilt and shame among trauma-exposed samples. It is difficult to compare studies, because some instruments measure guilt or shame proneness, whereas others tap into trauma-specific emotions. The issue of measurement is further confounded by the similarity of these emotions, which produces statistical multicollinearity, threats to construct validity, and theoretical debates. Some have asserted that the existing body of research on guilt has, in reality, examined shame (Lewis, 2014; Tangney & Dearing, 2002). Researchers continuously struggle to clearly differentiate these emotions. Many measures include items that, at face value, reflect both guilt and shame (Tangney, 1996). Confounding items within measures raise the question of which emotion is responsible for an observed effect. With current measures, the ability to accurately capture these emotions remains elusive.

Early research on shame and guilt commonly used the Test of Self-Conscious Affect (TOSCA; Tangney, Wagner, & Gramzow, 1989). The

TOSCA is scenario-based self-report measure that uses vignettes of commonplace situations. Response choices for each situation reflect shame, guilt, externalization, detachment/unconcern, and pride. For some time, this was the predominant measure of shame and guilt used in trauma research. Similarly, Cook developed the Internalized Shame Scale (ISS; Cook, 1988, 1996) to capture general shame proneness via self-report using a 5-point Likert-type scale ranging from "never" to "almost always" true of the respondent. This measure also became a widely used measure of shame in PTSD research.

In 1996, Kubany and colleagues responded to the need for trauma-specific measures by developing the Trauma-Related Guilt Inventory (TRGI; Kubany et al., 1996), which became the standard measure still used today. The TRGI is a self-report measure on which respondents rate the degree to which they believe guilt-related statements apply to them in relationship to a traumatic event. It uses a 5-point Likert-type scale ranging from "never/not at all true" to "extremely/always true." It has three scales (global guilt, guilt distress, and guilt cognitions) and three subscales (hindsight bias/responsibility, wrongdoing, and lack of justification).

Targeted measures of trauma-related shame were not introduced until quite recently, with the appearance of the Trauma-Related Shame Inventory (TRSI; Øktedalen, Hagtvet, Hoffart, Langkaas, & Smucker, 2014) and the Shame and Guilt After Trauma Scale (SGATS; Aakvaag et al., 2016). These measures are promising, but not yet widely utilized. Prior to the availability of these measures, researchers interested in trauma-related shame often adapted items from other measures or interviews. The lack of specific measures led to some studies using a visual analogue scale (e.g., Semb, Strömsten, Sundbom, Fransson, & Henningsson, 2011) or a single item to capture the trauma-related shame (e.g., La Bash & Papa, 2013).

Awareness of the challenges of measurement and the advantages and limitations of available measures provides context for interpreting sometimes disparate research findings. Guilt, in particular, has produced variant results depending on the population, type of measure used (e.g., trait vs trauma-related), and whether shame is included in analyses. Shame has shown more consistent results across studies but has not remained unquestioned.

Guilt and PTSD

Guilt has received persistent, although not consistent, attention in PTSD. Commentary on World War II, combat soldiers observed guilt as a central feature of combat neuroses (e.g., Ludwig, 1947). During and after the Vietnam War, the presence of guilt was used to identify "true" versus "pseudo" cases of combat fatigue (Strange, 1968). Following the post 9/11 conflicts in Iraq and Afghanistan, conversation about trauma-related guilt was again highlighted by the recognition and study of moral injury, a posttraumatic syndrome caused by violation of moral values and believed to have guilt and shame at its core (Litz et al., 2009). Despite this recognition and the extant evidence of guilt's relationship with PTSD, research still struggles to clearly define its role. It remains unclear whether guilt is a causal mechanism of PTSD or an independent posttraumatic outcome cooccurring with PTSD (Pugh, Taylor, & Berry, 2015).

The mixed findings on guilt and PTSD have spanned the spectrum from a positive association, to no association, to a negative association (see Pugh et al., 2015 for a review). Of 27 studies reviewed by Pugh et al. (2015), the majority demonstrated positive effect sizes ranging from small to large (Pugh et al., 2015). Five of the studies did not observe a statistically significant relationship between guilt and PTSD. Three of these measured situational guilt proneness using the TOSCA and may not have adequately captured trauma-related guilt, perhaps accounting for the lack of significant findings. The remaining two studies that did not find a relationship between guilt and PTSD did find effects for shame. Indeed, research suggests that, when the effects of shame are controlled, guilt ceases to have a significant positive correlation with PTSD and sometimes even shows a negative correlation suggestive of a protective role of guilt (e.g., Leskela, Dieperink, & Thuras, 2002). Researchers have suggested that shame-free guilt is less associated with psychopathology than is shame and would be expected to result in more prosocial behaviors, such as making reparation and repairing social relationships (e.g., Lewis, 1971, 2014; Tangney & Dearing, 2002). However, when attempts are made to separate guilt and shame, guilt can still show a positive, although diminished, relationship with PTSD (Cunningham et al., 2018). These findings lead to the question of whether shame-free guilt truly exists in a persistent form or whether unresolved guilt transforms into shame (Lewis, 1971, 2008, 2014).

Kubany et al. (1995, 1996) initially demonstrated that trauma-related guilt was associated with PTSD symptoms among male Vietnam combat veterans and female survivors of interpersonal violence. Although significant in both samples, the association of guilt with PTSD symptoms was stronger among combat veterans (Kubany et al., 1996). Indeed, research consistently shows larger effect sizes among military/veteran samples than among traumatized civilians (Pugh et al., 2015). This difference may be attributable to moral injury (Battles et al., 2018; Beckham, Feldman, & Kirby, 1998). Moral injury (i.e., violation of core moral values) during combat or other deployment experiences appears particularly potent for generating guilt (Beckham et al., 1998; Zerach & Levi-Belz, 2018). Among combat Veterans, witnessing or participating in wartime atrocities is associated with increased guilt and greater PTSD severity, specifically cognitions about responsibility and wrongdoing (Beckham et al., 1998; Dennis et al., 2017; Owens, Chard, & Cox, 2008). Combat guilt has also been particularly associated with the reexperiencing and avoidance symptoms of PTSD (Henning & Frueh, 1997).

When examined among military and veteran samples without accounting for shame, trauma-related guilt is consistently positively associated with PTSD. Findings among civilian samples have been less consistent. For example, guilt has been found to be unrelated to posttraumatic symptoms among samples of female survivors of intimate partner violence (Street & Arias, 2001), prisoners of war (Leskela et al., 2002), and victims of violent crimes (i.e., rape, physical assault, and robbery; Semb et al., 2011).

Longitudinal intervention studies have consistently observed simultaneous reductions in trauma-related guilt and PTSD symptoms from pre- to posttreatment (e.g., Resick et al., 2008). These studies, however, have not established a causal relationship across time. In their review, Pugh et al. (2015) note that, without such empirical support, it remains possible that "the relationship between guilt and PTSD is in fact artefactual, reflecting the fact that both are co-occurring products of trauma" (p. 145) with no causal link between the two.

Recently, Øktedalen et al. (2015) tested the bidirectional relationship between trauma-related guilt and PTSD throughout the course of exposure-based PTSD treatment. Using selected face-valid items, they found that changes in trauma-related guilt (without controlling for shame) predicted changes in PTSD symptom severity 3 days later. The reverse effect was not significant: changes in PTSD symptom severity did not predict changes in trauma-related guilt 4 days later (Øktedalen et al., 2015). These findings were the first to support a potential causal relationship between guilt and PTSD severity; however, the authors did not control for the effects of shame in their analysis of guilt. Based on previous cross-sectional findings in which shame has suppressed the effects of guilt (e.g., Leskela et al., 2002), the question remains whether it is guilt or shame causing the change in PTSD severity.

Studies examining both guilt and shame have often not done so simultaneously. When the effects of shame have been controlled, the effects of guilt on PTSD tend to disappear (e.g., Leskela et al., 2002; Semb et al., 2011). One study (Leskela et al., 2002) reported that guilt shifted from a positive to a negative correlation with PTSD after shame was included. These findings bolster the assertion that guilt may contribute to PTSD insomuch as it is psychometrically confounded with or contaminated by shame, that is, that there may not be "shame-free" guilt (Tangney & Dearing, 2002). Seeking to explore this perplexity, Cunningham et al. (2018) used relative weight analysis (Tonidandel & LeBreton, 2011) to account for multicollinearity and examined the unique contributions of guilt and shame to PTSD symptom severity among a sample of military veterans. They found that together guilt and shame explained 46% of the total variance in PTSD symptom severity; however, the unique contribution of guilt was only 34.8% of that variance explained (Cunningham et al., 2018). Despite these findings, the existence and role of shame-free guilt in PTSD remains an empirical question.

Overall, it is apparent that guilt is associated with trauma and PTSD. It remains unclear whether trauma-related guilt plays a causal role in maintaining PTSD by interfering with recovery and whether guilt is simply one symptom of PTSD or an independent, cooccurring trauma sequela. Further research is needed to distinguish shame-free guilt, elucidate its temporal relationship with PTSD, and examine it within the context of trauma-related shame.

Shame and PTSD

Despite evidence that shame may be more powerful than guilt in driving PTSD, it has received comparatively little research attention. As a result the role of shame has remained less well understood. Research has established shame as a common emotional reaction to trauma (e.g., Aakvaag et al., 2016; Aakvaag, Thoresen, Wentzel-Larsen, Røysamb, & Dyb, 2014; Amstadter & Vernon, 2008; Andrews, Brewin, Rose, & Kirk, 2000; Badour, Resnick, & Kilpatrick, 2017; Brewin, Andrews, & Rose, 2000; Hathaway, Boals, & Banks, 2010; La Bash & Papa, 2013; Lee et al., 2001), particularly events involving violence (Amstadter & Vernon, 2008; Herman, 2011; Stone, 1992).

Among university students, interpersonal trauma was associated with higher levels of peritraumatic shame and PTSD symptoms than was noninterpersonal trauma (La Bash & Papa, 2013). Likewise, greater exposure to prior traumas was associated with higher shame and higher PTSD severity.

Interestingly, cross-sectional path analysis suggested that the effects of prior trauma exposure on PTSD is partially accounted for by shame (La Bash & Papa, 2013). The indirect path through fear, however, did not hold significant. Thus shame may be a mechanism by which interpersonal trauma leads to increased risk for PTSD.

In addition to being a common response to trauma, shame appears to play a central role in PTSD symptom severity. Among women survivors of interpersonal violence, shame was associated with greater PTSD severity (Beck et al., 2011; Street & Arias, 2001). Furthermore, shame accounted for the association between psychological abuse and PTSD among these samples. These findings suggest that the shame engendered by psychological abuse in intimate relationships leads to increased PTSD severity.

In a 6-month longitudinal study, Andrews et al. (2000) examined the predictive power of shame and anger for PTSD severity among a sample of adult violent crime victims (75% men). At 1 month posttrauma, both shame and anger toward other people predicted PTSD symptom severity after accounting for demographic variables (e.g., age, sex, and education). After controlling for prior (1 month) PTSD symptom severity, shame emerged as the only additional significant predictor of PTSD severity at 6 months posttrauma. Thus shame emerged as a significant emotional predictor beyond the effects of prior PTSD symptoms, history of child abuse, demographic characteristics, anger at self, and anger at others (Andrews et al., 2000). Similarly, among a sample of physical assault victims, peritraumatic fear, anger with others, and shame were all significant predictors of maintained PTSD 6 months after trauma exposure (Brewin et al., 2000).

As previously stated, the contribution of shame to PTSD appears to be greater than that of guilt. Among a sample of female university students, shame proneness, but not guilt proneness, predicted PTSD severity (Pineles, Street, & Koenen, 2006). Shame was positively associated with the desire and urge to conceal the trauma, whereas guilt negatively predicted concealment. The concealment predicted by shame appeared to further increase PTSD severity (Pineles et al., 2006). There is also evidence that trauma related shame mediates the observed impact of trauma-related guilt on PTSD (Held, Owens, & Anderson, 2015), further supporting the assertion that trauma-related guilt may contribute to PTSD insomuch as it is contaminated by shame.

Strengthening the evidence for shame, one study showed that shame— but not guilt—mediated treatment effects on PTSD (Ginzburg et al., 2009). Among women survivors of childhood sexual abuse participating in trauma-focused or present-focused therapy groups, treatment ameliorated both shame and guilt and resulted in lowered PTSD severity. The change in shame accounted for approximately 33% of the treatment effects on PTSD (Ginzburg et al., 2009). The same finding was not observed for guilt. This study, however, did not establish the temporal precedence of changes in shame. Øktedalen et al. (2015) examined changes in trauma-related shame and guilt over time and subsequent change in PTSD severity among a clinical sample seeking inpatient treatment for PTSD. Subjects completed 10 weekly sessions of either prolonged exposure therapy (PE) or modified PE that included imagery rescripting. They found that, over the course of treatment, shame scores predicted PTSD severity 3 days later; however, changes in PTSD symptom severity did not predict later changes in trauma-related shame (Øktedalen et al., 2015). Although they found similar results examining guilt separately (described earlier), the within- and between-person effect sizes for shame on PTSD were larger than those for guilt. These findings clearly suggest that shame may play a causal role in PTSD.

Taken together, the extant literature shows that shame can play an influential role in PTSD severity and maintenance. It may even account for the observed effects of trauma-related guilt (Held et al., 2015). Nevertheless, continued research is needed to elucidate the role of shame in PTSD and its relationship with guilt.

Sex differences

Few studies have examined sex differences in the experience of trauma-related guilt and shame or in the relationship of these emotions to PTSD (Aakvaag et al., 2014, 2016; Badour et al., 2017). This is an area in need of further research, as the current literature has left many questions unanswered. Studies have consistently shown that women report significantly more shame and guilt on average than do men (Aakvaag et al., 2014, 2016; Badour et al., 2017). The explanation for this observed difference, however, remains elusive. Among a large national sample of Norwegian survivors of interpersonal violence, a significant interaction between sex and level of polytrauma exposure was observed for shame. Although being female was associated with reporting greater shame, women who reported experiencing multiple types of violence (e.g., child sexual abuse + exposure to parental violence + physical assault + intimate partner violence) endorsed the highest levels of shame. Likewise, men with cumulative trauma exposure reported greater shame compared to men who reported experiencing only one type of violence (Aakvaag et al., 2016). Although the association of sex with shame remained, the effects of violence exposure were stronger. The same interaction was not observed for guilt. This is consistent with findings in a national sample in the United States showing that, although women tended to endorse shame and guilt more frequently than did men, this difference appeared to be driven by the respondents who did not have PTSD (Badour et al., 2017). Among veterans reporting exposure to morally injurious events and experiencing moral injury (which has guilt and shame at core; Farnsworth, Drescher, Nieuwsma, Walser, & Currier, 2014; Litz et al., 2009), sex differences were not observed for moral injury, PTSD severity, or other negative mental health outcomes (Kelley, Braitman, White, & Ehlke, 2018). Although symptoms of moral injury did predict PTSD, sex did not moderate this effect (Kelley et al., 2018). These findings tentatively suggest that observed sex differences may be better explained by trauma characteristics and exposure than by inherent sex differences.

Shame, guilt, and suicide risk

Because PTSD has been associated with increased risk of suicidal ideation (SI), attempts, and death (see Panagioti, Gooding, & Tarrier, 2009 and Pompili et al., 2013 for reviews), research has increasingly focused on identifying the factors driving this connection. Emotions, particularly those of shame and guilt, appear to play a central role in this increased risk (Bryan, Morrow, et al., 2013; Bryan, Ray-Sannerud, Morrow, & Etienne, 2013a, 2013b; Bryan, Rudd, et al., 2013; Cunningham et al., 2017, 2018). Fluid vulnerability theory (FVT; Rudd, 2006) and the interpersonal theory of suicide (IPTS; Joiner, 2005) help explain the importance of emotion in contributing to suicide risk. According to FVT (Rudd, 2006), dynamic changes in cognitive, emotional, physical, and behavioral risk and protective factors determine changes in risk level from individual baseline. Shame and guilt possess qualities that fall into multiple risk categories. Relevant cognitive factors include negative self-cognitions (e.g., self-blame or self-hate), and emotional factors include the emotional experience of guilt and shame and potential difficulty regulating them. Physical factors include autonomic arousal and the associated feeling of internal intensity or agitation. Indeed, when Bryan, Rudd, and Wertenberger (2013) examined self-reported reasons why active duty soldiers attempted suicide, they found that the primary reason was to relieve emotional distress. These findings suggest that the unpleasant intensity and inherent internal attributions central to shame and guilt may help explain why these emotions are powerful predictors of suicidal ideation (Bryan, Morrow, et al., 2013; Bryan et al., 2013a, 2013b; Bryan et al., 2015; Cunningham et al., 2017; Cunningham, Grossmann, et al., 2019; Cunningham, LoSavio, et al., 2019) and behavior (Bryan, Rudd, et al., 2013).

Likewise, the IPTS (Joiner, 2005) provides insight into the role of negative self-evaluative emotions. IPT explains that suicidal ideation becomes possible when individuals believe they are a burden on others, lack a sense of social belonging, and feel hopeless (Joiner, 2005). Shame in particular can contribute to disconnectedness through social withdrawal (shame's action urge), perceived burdensomeness via perceived or feared loss of social status, and hopelessness due to the inability to control or change the source of shame. One study conducted with a sample of military veterans (Cunningham et al., 2018) examined the mediating effect of shame using the internalized shame subscale of the ISS (Cook, 1996). The ISS measures both the intensity of the emotion and related cognitions (e.g., "I feel intensely inadequate…"). The results showed that internalized shame fully accounted for the relationship between PTSD and SI among veterans. The reverse, however, was not true. PTSD severity did not affect the relationship between shame and SI (Cunningham et al., 2018). Research on guilt has also supported the importance of cognitions in SI. A model comparison examining components of guilt showed that guilt cognitions were the most important facet of trauma-related guilt mediating the effects of PTSD symptoms on SI among veterans (Cunningham et al., 2017). Interestingly, trauma-specific guilt-related distress was not a significant factor, suggesting that it is negative self-cognitions (e.g., "what I did was unforgiveable") and not the perceived intensity of guilt that impacts the PTSD-SI relationship.

Shame and guilt in PTSD treatment

Although shame and guilt are not primary targets of evidence-based PTSD treatments, existing "gold-standard" treatments do appear to partially ameliorate these emotions (e.g., Harned, Korslund, Foa, & Linehan, 2012; Resick, Nishith, Weaver, Astin, & Feuer, 2002; Rizvi, Vogt, & Resick, 2009). In a systematic review and metaanalysis of six studies examining PTSD treatment among adult sexual assault survivors, Regehr, Alaggia, Dennis, Pitts, and Saini (2013) found that prolonged exposure therapy (PE; Foa, Hembree, & Rothbaum, 2007) and cognitive processing therapy (CPT; Resick, Monson, & Chard, 2017) produced reductions in guilt. And while a handful of small studies have reported reductions in shame (Balcom, Call, & Pearlman, 2000) and guilt (Stapleton, Taylor, & Asmundson, 2006) following eye movement desensitization and reprocessing therapy (EMDR; Shapiro, 2001), the review and metaanalysis (Regehr et al., 2013) did not reflect these findings.

Prolonged exposure (PE; Foa et al., 2007) is an evidence-based treatment for PTSD in which exposure to traumatic memories and trauma-related triggers results in habituation and a lessening of trauma-related emotions. Although developed from a fear-based model of PTSD, PE has been shown to result in reductions of guilt (Clifton, Feeny, & Zoellner, 2017; Øktedalen et al., 2015) and shame (Øktedalen et al., 2015). Although the presence of pretreatment guilt or shame has not been associated with treatment dropout or iatrogenic outcomes in PE, individuals with higher shame at pretreatment exhibit more PTSD symptoms at posttreatment and follow-up (van Minnen, Arntz, & Keijsers, 2002). Higher guilt, however, has not been observed to impair recovery (van Minnen et al., 2002) and has even been associated with better PE outcomes (Clifton et al., 2017; Rizvi et al., 2009). This suggests that shame, but not guilt, may interfere with the effectiveness of PE. As emotional suppression and avoidance maintain PTSD and interfere with recovery (Clifton et al., 2017; Foa, Huppert, & Cahill, 2006), the impairment of treatment effects may be attributable to the innate desire to hide and avoid that is caused by shame (Lewis, 2014). In contrast, guilt does not appear to interfere with emotional engagement during exposure (Clifton et al., 2017).

Trauma management therapy (TMT; Beidel, Frueh, Neer, & Lejuez, 2017) is a less widely used exposure-based treatment package developed to address military combat trauma. TMT uses flooding to induce habituation. Imaginal exposure sessions are often enhanced by virtual reality to increase engagement and reduce emotional avoidance. Tailored imaginal and in vivo exposures in TMT are complimented by psychosocial skills groups aimed at improving associated areas of impairment, including sleep, anger management, and interpersonal skills (Beidel et al., 2017). Behavioral activation is also included to address the effects of comorbid depression and guilt (Beidel et al., 2019). TMT has been shown to effectively reduce PTSD symptoms related to military trauma (Beidel et al., 2017, 2019). Additionally, there is evidence that it also reduces trauma-related guilt (Beidel et al., 2017; Trachik et al., 2018). In line with findings from treatment studies of PE and CPT, individuals with higher pretreatment guilt exhibited faster improvements in PTSD symptoms over the course of TMT (Trachik et al., 2018). At this time, no TMT studies have examined shame.

Cognitive processing therapy (CPT; Resick et al., 2017) is another gold-standard evidence-based treatment for PTSD that works by helping individuals learn to identify thoughts and emotions and to challenge the inaccurate or unbalanced beliefs responsible for generating maladaptive emotional and behavioral responses. CPT is not explicitly fear focused and directly addresses cognitions that could lead to experiences of guilt and shame, including beliefs about responsibility, blame, and self-esteem. CPT has been shown to reduce experiences of both guilt (Larsen, Fleming, & Resick, 2019; Nishith, Nixon, & Resick, 2005; Resick et al., 2002) and shame (Resick et al., 2008). Indeed, patients exhibiting higher pretreatment global guilt as measured by the TRGI (Kubany et al., 1996) showed greater PTSD symptom improvement at the end of treatment (Rizvi et al., 2009). This cognitive approach appears to provide an advantage over exposure-based treatment. Although PE and CPT do not significantly differ in their impact on PTSD symptom severity and recovery, there is evidence that CPT is more effective at modifying guilt (Resick et al., 2002, 2008).

Complimentary treatment approaches for shame and guilt

Despite the strong empirical foundation for PTSD treatments, approximately 40% of trauma survivors who receive treatment for PTSD do not achieve a resolution of symptoms (in part due to treatment dropout; Hoge et al., 2014). Guilt cognitions and self-blame are common residual symptoms after CPT and PE (Larsen et al., 2019). Although treatment outcome research including shame is sparse, it is reasonable to expect that shame will also linger. Because shame is generated by core beliefs that the self is inherently and irrevocably flawed or worthless, these cognitions may be particularly resistant to traditional treatment approaches. In addition to interfering with treatment, shame may also keep trauma survivors from seeking treatment in the first place. Due to these factors, additional approaches are needed to directly address trauma-related guilt and shame in PTSD. Emotion-focused approaches that enhance mindfulness, distress tolerance, emotion regulation, and self-compassion may prove useful in ameliorating trauma-related guilt and shame.

Acceptance and commitment therapy (ACT; Hayes, Strosahl, & Wilson, 2012) is a transdiagnostic behavioral treatment protocol based in mindfulness and acceptance practices. As a transdiagnostic treatment, ACT has received strong empirical support for depression (Hacker, Stone, & MacBeth, 2016), a disorder often cooccurring with PTSD. Initial evidence suggests that ACT ameliorates PTSD symptoms and may become a viable treatment option for PTSD (see Bean, Ong, Lee, & Twohig, 2017 for a review); however, additional empirical support is needed. ACT is particularly promising for reducing trauma-related shame (Luoma & Platt, 2015) and addressing moral injury (Nieuwsma et al., 2015). Initial research has shown that ACT lowers internalized stigma and feelings of shame (e.g., Luoma, Kohlenberg, Hayes, Bunting, & Rye, 2008; Luoma, Kohlenberg, Hayes, & Fletcher, 2012). Among a sample seeking treatment for substance use disorder (SUD), a randomized controlled trial showed that a brief group-based ACT intervention resulted in higher treatment attendance and better shame-related outcomes compared with treatment as usual (Luoma et al., 2012). A recent longitudinal pilot study examined ACT among a sample of military veterans with SUD and PTSD (Meyer et al., 2018). They found that ACT resulted in reduced clinician-assessed and self-reported PTSD symptom severity at posttreatment, and these gains were maintained at 3-month follow-up (Meyer et al., 2018). Other findings included decreased depression and suicidal ideation and increased abstinence and quality of life (Meyer et al., 2018). Interestingly, they also observed reductions in experiential avoidance (Meyer et al., 2018), which is central to maintaining both PTSD and shame.

Initial evidence should be interpreted cautiously, but research is warranted to examine ACT for trauma-related guilt and shame. The mechanisms of cognitive flexibility, mindfulness, acceptance, and self-compassion inherent to ACT show promise for addressing trauma-related self-evaluative emotions in the context of PTSD.

Compassion-focused therapy (CFT; Gilbert, 2009) is a multimodal therapy approach specifically developed to address shame and self-criticism, as well as underlying emotion regulation mechanisms. CFT operationalizes self-compassion in the form of skills (e.g., attention regulation, mindful imagery, and caring behavior), which through practice develop compassionate attributes within an individual (e.g., distress tolerance, empathy, and nonjudgment). Self-compassion is a point of intervention that may prove central to ameliorating the effects of guilt and shame in PTSD.

Greater self-compassion has been associated with lower PTSD symptom severity (Hiraoka et al., 2015; Meyer et al., 2019). Indeed, there is emerging evidence that increasing self-compassion has direct ameliorating effects on PTSD symptoms by shifting self-criticism and resultant feelings of shame (Hoffart, Øktedalen, & Langkaas, 2015; Meyer et al., 2019). Meyer et al. (2019) demonstrated that self-compassion was a unique predictor of decreased PTSD symptom severity. They explain that this may be due, at least in part, to associated reductions in guilt and shame.

Initial evidence suggests that incorporating self-compassion can enhance established PTSD treatments. Among a community sample receiving PE or modified PE in which imaginal exposure was replaced with self-compassionate rescripting, self-compassion was shown to impact PTSD, such that greater self-compassion predicted lesser PTSD severity (Hoffart et al., 2015). The level of PTSD symptoms, however, did not predict level of self-compassion. This study suggests that, for individuals with intense trauma-related shame, high levels of self-critical trauma-related cognitions, and/or low self-compassion, traditional exposure therapy for PTSD may be enhanced by self-compassion skills. The relationships among guilt, shame, and self-compassion in the context of PTSD are important aspects for future research.

Emotion regulation therapy (ERT; Fresco, Mennin, Heimberg, & Ritter, 2013; Mennin, Fresco, O'Toole, & Heimberg, 2018) is an empirically supported cognitive behavioral treatment developed for psychological disorders that are driven by emotional distress (Fresco et al., 2013; Renna et al., 2018). The primary processes targeted by ERT include motivation (i.e., security vs reward), self-regulation (e.g., attention, emotion, and cognition), and contextual learning (Renna et al., 2018). By learning to notice emotions and emotional processes, individuals are able to practice cognitive and emotional skills to regulate distress. Through this process and experiential exposure, new learning occurs, bolstering the individual's flexibility and increasing their potential behavioral response repertoire (Renna et al., 2018).

ERT targets the negative self-referential cognitive processes that are central to generating and maintaining shame (Fresco et al., 2013; Mennin et al., 2018; Renna et al., 2018). Research has shown that ERT is effective at reducing self-criticism, emotional distress, and other symptoms among individuals with generalized anxiety (GAD) and major depressive disorder (MDD; Fresco et al., 2013; Mennin et al., 2018). In a randomized controlled trial among treatment-seeking adults with GAD, ERT resulted in significantly reduced anxious and depressive symptoms and significantly increased emotion regulation, quality of life, and mindfulness (Mennin et al., 2018). Although further research is needed, observed effect sizes were moderate to large for all variables (Mennin et al., 2018), suggesting that ERT may produce outcomes similar to current best-practice treatments for distress-based emotional disorders (Renna et al., 2018). Depression and anxiety share overlapping symptoms with PTSD (e.g., emotion dysregulation, negative cognitions, intrusive thoughts, and avoidance; APA, 2013); thus expanding research to trauma-exposed populations is warranted. As a transdiagnostic approach targeting shared underlying processes, ERT presents an opportunity for primary or complimentary treatment of trauma-related emotional distress, including guilt and shame.

Conclusions and future directions

The associations of shame and guilt with PTSD have been well established (Cunningham et al., 2018; Gaudet, Sowers, Nugent, & Boriskin, 2016; Pugh et al., 2015), particularly among samples exposed to military trauma, including combat, military sexual trauma, or moral injury (e.g., Cunningham et al., 2018; Gaudet et al., 2016), or interpersonal traumas (e.g., Badour et al., 2017). Research shows that guilt and shame play a role in PTSD severity and maintenance (Cunningham et al., 2018; Øktedalen et al., 2015). Causal associations have been supported by longitudinal research showing that changes in shame and guilt predict subsequent changes in PTSD severity. In contrast, PTSD severity has not been shown to predict shifts in shame and guilt (Øktedalen et al., 2015).

Although these emotions are believed to interfere with treatment engagement and to result in poorer outcomes (e.g., Lee et al., 2001), empirically supported PTSD treatments have been shown to reduce guilt (Resick et al., 2002; Rizvi et al., 2009) and shame (Resick et al., 2008). Research suggests that cognitive approaches, such as cognitive processing therapy, are more effective at reducing guilt and shame than are exposure-based approaches (Resick et al., 2002). Nevertheless, underlying negative self-referent cognitions remain refractory following both treatments (Larsen et al., 2019).

Effective mechanisms of change for trauma-related shame and guilt are not yet well understood. This has led clinicians and researchers alike to suggest that emotion-focused approaches, particularly those utilizing mindfulness and self-compassion skills, may prove to be effective for trauma-related guilt and shame (e.g., Hoffart et al., 2015; Orsillo & Batten, 2005). Future research is needed to explore and tailor effective treatment options for guilt and shame among trauma-exposed individuals. A theoretically informed approach focusing on underlying mechanisms (e.g., social information processing, self-referential cognition, attentional control, and emotion regulation) is essential to inform research and treatment in this area.

Suicide risk is a serious associated outcome of PTSD (see Krysinska & Lester, 2010 for a review), and shame and guilt appear to play a powerful role in increasing this risk. Studies suggest that shame and guilt mediate the relationship between PTSD and suicidal ideation (e.g., Bryan, Morrow, et al., 2013; Bryan et al., 2013a, 2013b; Bryan et al., 2015; Bryan, Rudd, et al., 2013; Cunningham et al., 2017, 2018). Understanding the processes by which these emotions contribute to suicidal ideation and behavior after trauma provides an opportunity to reduce suicide risk among this vulnerable population.

Relatedly, research is needed to examine the relationship of trauma-related guilt and shame to nonsuicidal self-injury (NSSI). NSSI is a strong risk factor for suicidal ideation (e.g., Cunningham, Grossmann, et al., 2019; Cunningham, LoSavio, et al., 2019; Holliday, Smith, & Monteith, 2018) and attempts (e.g., Kimbrel, DeBeer, Meyer, Gulliver, & Morissette, 2016) among trauma-exposed populations. Emerging research suggests that self-criticism, self-dissatisfaction, and shame are common experiences among those who engage in NSSI (Hack & Martin, 2018; Mahtani, Hasking, & Melvin, 2018; VanDerhei, Rojahn, Stuewig, & McKnight, 2013). NSSI appears to be more common among trauma-exposed populations than previously realized (Bentley, Cassiello-Robbins, Vittorio, Sauer-Zavala, & Barlow, 2015; Cunningham, Grossmann, et al., 2019; Cunningham, LoSavio, et al., 2019; Dixon-Gordon, Tull, & Gratz, 2014; Ford & Gómez, 2015; Holliday et al., 2018), which is likely due to its emotion regulatory function (Nock, 2009). Given the strong associations of NSSI with suicide risk, shame, and emotion dysregulation, it is imperative that research seeks to better understand these relationships.

Overall, it is clear that guilt and shame are related to PTSD severity and concurrent suicide risk, and emerging research shows that changes in these emotions have a function in the treatment process. Nevertheless, challenges remain. Research struggles to accurately differentiate and capture negative self-evaluative emotions, and better consensus about definitions and advances in measurement are needed. Furthermore, research must clarify the process by which guilt is transformed into shame and whether shame-free guilt is, in fact, a contributor to PTSD. Additionally, shame has been related to anger and aggression (e.g., Tangney, Wagner, Fletcher, & Gramzow, 1992). Because anger is another emotion with strong empirical support for having a primary role in PTSD (see Chapter 4 for a detailed discussion), a better understanding of the relationship between anger and shame is warranted. Continuing research is also needed to understand the functions of negative self-evaluative emotions in PTSD and to expand treatment approaches and improve effectiveness. Recognizing and addressing shame and guilt among trauma-exposed populations have the potential to greatly improve PTSD prevention and recovery.

References

Aakvaag, H. F., Thoresen, S., Wentzel-Larsen, T., Dyb, G., Røysamb, E., & Olff, M. (2016). Broken and guilty since it happened: A population study of trauma-related shame and guilt after violence and sexual abuse. *Journal of Affective Disorders, 204*, 16–23. https://doi.org/10.1016/j.jad.2016.06.004.

Aakvaag, H. F., Thoresen, S., Wentzel-Larsen, T., Røysamb, E., & Dyb, G. (2014). Shame and guilt in the aftermath of terror: The Utøya Island study. *Journal of Traumatic Stress, 27*, 618–621. https://doi.org/10.1002/jts.21957.

American Psychiatric Association. (1980). *Diagnostic and statistical manual of mental disorders-III* (3th ed.). Washington, DC: APA.

American Psychiatric Association. (2013). *Diagnostic and statistical manual of mental disorders-5* (5th ed.). Washington, DC: APA.

Amstadter, A., & Vernon, L. (2008). Emotional reactions during and after trauma: A comparison of trauma types. *Journal of Aggression, Maltreatment, & Trauma, 16*, 391–408. https://doi.org/10.1080/10926770801926492.

Andrews, B., Brewin, C. R., Rose, S., & Kirk, M. (2000). Predicting PTSD symptoms in victims of violent crime: The role of shame, anger, and childhood abuse. *Journal of Abnormal Psychology, 109*, 69–73. https://doi.org/10.1037//0021-843X.109.1.69.

Badour, C. L., Resnick, H. S., & Kilpatrick, D. G. (2017). Associations between specific negative emotions and DSM-5 PTSD among a national sample of interpersonal trauma survivors. *Journal of Interpersonal Violence, 32*, 1620–1641. https://doi.org/10.1177/0886260515589930.

Balcom, D., Call, E., & Pearlman, D. N. (2000). Eye movement desensitization and reprocessing treatment of internalized shame. *Traumatology, 6*, 69–83. https://doi.org/10.1177/153476560000600202.

Barrett, L. F. (2017a). The theory of constructed emotion: An active inference account of interoception and categorization. *Social Cognitive and Affective Neuroscience, 12*, 1–23. https://doi.org/10.1093/scan/nsw154.

Barrett, L. F. (2017b). *How emotions are made*. New York, NY: Houghton Mifflin Harcourt Publishing Company.

Battles, A. R., Bravo, A. J., Kelley, M. L., White, T. D., Braitman, A. L., & Hamrick, H. C. (2018). Moral injury and PTSD as mediators of the associations between morally injurious experiences and mental health and substance use. *Traumatology, 24*, 246–254. https://doi.org/10.1037/trm0000153.

Bean, R. C., Ong, C. W., Lee, J., & Twohig, M. P. (2017). Acceptance and commitment therapy for PTSD and trauma: An empirical review. *The Behavior Therapist, 40*, 145–150.

Beck, J. S. (1995). *Cognitive therapy: Basics and beyond*. New York, NY: Guilford.

Beck, J. G., McNiff, J., Clapp, J. D., Olsen, S. A., Avery, M. L., & Hagewood, J. H. (2011). Exploring negative emotion in women experiencing intimate partner violence: Shame, guilt, and PTSD. *Behavior Therapy, 42*, 740–750. https://doi.org/10.1016/j.beth.2011.04.001.

Beckham, J. C., Feldman, M. E., & Kirby, A. C. (1998). Atrocities exposure in Vietnam combat veterans with chronic posttraumatic stress disorder: Relationship to combat exposure, symptom severity, guilt, and interpersonal violence. *Journal of Traumatic Stress, 11*, 777–785. https://doi.org/10.1023/A:1024453618638.

Beidel, D. C., Frueh, B. C., Neer, S. M., Bowers, C. A., Trachick, B., Uhde, T. W., & Grubaugh, A. (2019). *Journal of Anxiety Disorders, 61*, 64–74. https://doi.org/10.1016/j.janxdis.2017.08.005.

Beidel, D. C., Frueh, B. C., Neer, S. M., & Lejuez, C. W. (2017). The efficacy of trauma management therapy: A controlled pilot investigation of a three-week intensive outpatient program for combat-related PTSD. *Journal of Anxiety Disorders, 50*, 23–32. https://doi.org/10.1016/j.janxdis.2017.05.001.

Bentley, K. H., Cassiello-Robbins, C. F., Vittorio, L., Sauer-Zavala, S., & Barlow, D. H. (2015). The association between nonsuicidal self-injury and the emotional disorders: A meta-analytic review. *Clinical Psychology Review, 37*, 72–88. https://doi.org/10.1016/j.cpr.2015.02.006.

Brewin, C. R., Andrews, B., & Rose, S. (2000). Fear, helplessness, and horror in posttraumatic stress disorder: Investigating DSM-IV Criterion A2 in victims of violent crimes. *Journal of Traumatic Stress, 13*, 499–509. https://doi.org/10.1023/A:1007741526169.

Browne, K. C., Trim, R. S., Myers, U. S., & Norman, S. B. (2015). Trauma-related guilt: Conceptual development and relationship with posttraumatic stress and depressive symptoms. *Journal of Traumatic Stress, 28*, 134–141. https://doi.org/10.1002/jts.21999.

Bryan, C. J., Morrow, C. E., Etienne, N., & Ray-Sannerud, B. (2013). Guilt, shame, and suicidal ideation in a military outpatient clinical sample. *Depression and Anxiety, 30*, 55–60. https://doi.org/10.1002/da/22002.

Bryan, C. J., Ray-Sannerud, B., Morrow, C. E., & Etienne, N. (2013a). Shame, pride, and suicidal ideation in a military clinical sample. *Journal of Affective Disorders, 147*, 212–216. https://doi.org/10.1016/j.jad.2012.11.006.

Bryan, C. J., Ray-Sannerud, B., Morrow, C. E., & Etienne, N. (2013b). Guilt is more strongly associated with suicidal ideation among military personnel with direct combat exposure. *Journal of Affective Disorders, 148*, 37–41. https://doi.org/10.1016/j.jad.2012.11.044.

Bryan, C. J., Roberge, E., Bryan, A. O., Ray-Sannerud, B., Morrow, C. E., & Etienne, N. (2015). Guilt as a mediator of the relationship between depression and posttraumatic stress with suicide ideation in two samples of military personnel and veterans. *International Journal of Cognitive Therapy, 8*, 143–155. https://doi.org/10.1521/ijct.2015.8.2.143.

Bryan, C. J., Rudd, M. D., & Wertenberger, E. (2013). Reasons for suicide attempts in a clinical sample of active duty soldiers. *Journal of Affective Disorders, 144*, 148–152. https://doi.org/10.1016/j.jad.2012.06.030.

Burgess, A. W. (1983). Rape trauma syndrome. *Behavioral Sciences & The Law, 1*, 97–113. https://doi.org/10.1002/bsl.2370010310.

Clifton, E. G., Feeny, N. C., & Zoellner, L. A. (2017). Anger and guilt in treatment for chronic posttraumatic stress disorder. *Journal of Behavior Therapy and Experimental Psychiatry, 54*, 9–16. https://doi.org/10.1016/j.jbtep.2016.05.003.

Cook, D. R. (1988). Measuring shame: The Internalized Shame Scale. *Alcoholism Treatment Quarterly, 4*, 197–215. https://doi.org/10.1300/j020v04n02-12.

Cook, D. R. (1996). Empirical studies of shame and guilt: The Internalized Shame Scale. In D. L. Nathanson (Ed.), *Knowing feeling: Affect, script, and psychotherapy* (pp. 132–165). New York, NY: W. W. Norton & Company, Inc.

Cunningham, K. C., Davis, J. L., Wilson, S. M., & Resick, P. A. (2018). A relative weights comparison of trauma-related shame and guilt as predictors of DSM-5 posttraumatic stress disorder among US veterans and military members. *British Journal of Clinical Psychology, 57*, 163–176. https://doi.org/10.1111/bjc.12163.

Cunningham, K. C., Farmer, C., LoSavio, S. T., Dennis, P. A., Clancy, C. P., Hertzberg, M. A., … Beckham, J. C. (2017). A model comparison approach to trauma-related guilt as a mediator of the relationship between PTSD symptoms and suicidal ideation among veterans. *Journal of Affective Disorders, 221*, 227–231. https://doi.org/10.1016/j.jad.2017.06.046.

Cunningham, K. C., Grossmann, J. L., Seay, K. B., Dennis, P. A., Clancy, C. P., Hertzberg, M. A., … Kimbrel, N. A. (2019). Nonsuicidal self-injury and borderline personality features as risk factors for suicidal ideation among male veterans with posttraumatic stress disorder. *Journal of Traumatic Stress, 32*, 141–147. https://doi.org/10.1002/jts.22369.

Cunningham, K. C., LoSavio, S. T., Dennis, P. A., Farmer, C., Clancy, C. P., Hertzberg, M. A., … Beckham, J. C. (2019). Shame as a mediator between posttraumatic stress disorder symptoms and suicidal ideation among veterans. *Journal of Affective Disorders, 243*, 216–219. https://doi.org/10.1016/j.jad.2018.09.040.

Dahl, S. (1989). Acute response to rape—A PTSD variant. *Acta Psychiatrica Scandinavica, 80,* 56–62. https://doi.org/10.1111/j.1600-0447.1989.tb05254.x.
Dennis, P. A., Dennis, N. M., Van Voorhees, E. E., Calhoun, P. S., Dennis, M. F., & Beckham, J. C. (2017). Moral transgression during the Vietnam war: A path analysis of the psychological impact of veterans' involvement in wartime atrocities. *Anxiety, Stress, & Coping, 30,* 188–201. https://doi.org/10.1080/10615806.2016.1230669.
Dixon-Gordon, K. L., Tull, M. T., & Gratz, K. L. (2014). Self-injurious behaviors in posttraumatic stress disorder: An examination of potential moderators. *Journal of Affective Disorders, 166,* 359–367. https://doi.org/10.1016/j.jad.2014.05.033.
Ehlers, A., & Clark, D. M. (2000). A cognitive model of posttraumatic stress disorder. *Behavior Research and Therapy, 38,* 319–345.
Farnsworth, J. K., Drescher, K. D., Nieuwsma, J. S., Walser, R. B., & Currier, J. M. (2014). The role of moral emotions in military trauma: Implications for the study and treatment of moral injury. *Review of General Psychology, 18,* 249–262. https://doi.org/10.1037/gpr0000018.
Foa, E. B., Hembree, E. A., & Rothbaum, B. O. (2007). *Treatments that work. Prolonged exposure therapy for PTSD: Emotional processing of traumatic experiences: Therapist guide.* New York, NY: Oxford University Press.
Foa, E. B., Huppert, J. D., & Cahill, S. P. (2006). Emotional processing theory: An update. In B. O. Rothbaum (Ed.), *Pathological anxiety: Emotional processing in etiology and treatment* (pp. 3–24). New York, NY: Guilford Press.
Foa, E. B., Steketee, G. S., & Rothbaum, B. O. (1989). Behavioral/cognitive conceptualizations of posttraumatic stress disorder. *Behavior Therapy, 20,* 155–176. https://doi.org/10.1016/S0005-7894(89)80067-X.
Ford, J. D., & Gómez, J. M. (2015). The relationship of psychological trauma and dissociative and posttraumatic stress disorders to nonsuicidal self-injury and suicidality: A review. *Journal of Trauma & Dissociation, 16,* 232–271. https://doi.org/10.1080/15299732.2015.989563.
Fresco, D. M., Mennin, D. S., Heimberg, R. G., & Ritter, M. (2013). Emotion regulation therapy for generalized anxiety disorder. *Cognitive and Behavioral Practice, 20,* 282–300. https://doi.org/10.1016/j.cbpra.2013.02.001.
Gaudet, C. M., Sowers, K. M., Nugent, W. R., & Boriskin, J. A. (2016). A review of PTSD and shame in military veterans. *Journal of Human Behavior in the Social Environment, 26,* 56–68. https://doi.org/10.1080/10911359.2015.1059168.
Gilbert, P. (2004). Evolution, attractiveness, and the emergence of shame and guilt in a self-aware mind: A reflection on Tracy and Robins. *Psychological Inquiry, 15,* 132–135.
Gilbert, P. (2009). Introducing compassion-focused therapy. *Advances in Psychiatric Treatment, 15,* 199–208. https://doi.org/10.1192/apt.bp.107.005264.
Ginzburg, K., Butler, L. D., Giese-Davis, J., Cavanaugh, C. E., Neri, E., Koopman, C., … Spiegel, D. (2009). Shame, guilt, and posttraumatic stress disorder in adult survivors of childhood sexual abuse at risk for human immunodeficiency virus: Outcomes of a randomized clinical trial of group psychotherapy treatment. *Journal of Nervous and Mental Disorders, 197,* 536–542. https://doi.org/10.1097/NMD.0b013e3181ab2ebd.
Hack, J., & Martin, G. (2018). Expressed emotion, shame, and non-suicidal self-injury. *International Journal of Environmental Research and Public Health, 15,* 890–907. https://doi.org/10.3390/ijerph15050890.
Hacker, T., Stone, P., & MacBeth, A. (2016). Acceptance and commitment therapy – Do we know enough? Cumulative and sequential meta-analyses of randomized controlled trials. *Journal of Affective Disorders, 190,* 551–565. https://doi.org/10.1016/j.jad.2015.10.053.
Haley, S. A. (1974). When a patient reports atrocities: Specific treatment considerations of the Vietnam veteran. *Archives of General Psychiatry, 30,* 191–196. https://doi.org/10.1001/archpsych.1974.01760080051008.

Harned, M. S., Korslund, K. E., Foa, E. B., & Linehan, M. M. (2012). Treating PTSD in suicidal and self-injuring women with borderline personality disorder: Development and preliminary evaluation of dialectical behavioral therapy prolonged exposure protocol. *Behavior Research and Therapy, 50*, 381–386. https://doi.org/10.1016/j.brat.2012.02.011.

Hathaway, L. M., Boals, A., & Banks, J. B. (2010). PTSD symptoms and dominant emotional response to a traumatic event: An examination of DSM-IV Criterion A2. *Anxiety, Stress, & Coping, 23*, 119–126.

Hayes, S. C., Strosahl, K. D., & Wilson, K. G. (2012). *Acceptance and commitment therapy: The process and practice of mindful change* (2nd ed.). New York, NY: Guilford Press.

Held, P., Owens, G. P., & Anderson, S. E. (2015). The interrelationships among trauma-related guilt and shame, disengagement coping, and PTSD in a sample of trauma-seeking substance users. *Traumatology, 21*, 285–292. https://doi.org/10.1037/trm0000050.

Henning, K. R., & Frueh, B. C. (1997). Combat guilt and its relationship to PTSD symptoms. *Journal of Clinical Psychology, 53*, 801–808. https://doi.org/10.1002/(SICI)1097-4679(199712)53:8<801::AID-JCLP3>3.0.CO;2-1.

Herman, J. (2011). Posttraumatic stress disorder as a shame disorder. In R. L. Dearing & J. P. Tangney (Eds.), *Shame in the therapy hour*. Washington, DC: American Psychological Association.

Hiraoka, R., Meyer, E. C., Kimbrel, N. A., DeBeer, B. B., Gulliver, S. B., & Morissette, S. B. (2015). Self-compassion as a prostpective predictor of PTSD symptom severity among trauma-exposed US Iraq and Afghanistan war veterans. *Journal of Traumatic Stress, 28*, 127–133. https://doi.org/10.1002/jts.21995.

Hoffart, A., Øktedalen, T., & Langkaas, T. F. (2015). Self-compassion influences PTSD symptoms in the process of change in trauma-focused cognitive-behavioral therapies: A study of within-person processes. *Frontiers in Psychology, 6*, 1–11. https://doi.org/10.3389/fpsyg.2015.01273.

Hoge, C. W., Grossman, S. H., Auchterlonie, J. L., Riviere, L. A., Milliken, C. S., & Wilk, J. E. (2014). PTSD treatment for soldiers after combat deployment: Low utilization of mental health care and reasons for dropout. *Psychiatric Services, 65*, 997–1004. https://doi.org/10.1176/appi.ps.201300307.

Holliday, R., Smith, N. B., & Monteith, L. L. (2018). An initial investigation of nonsuicidal self-injury among male and female survivors of military sexual trauma. *Psychiatry Research, 268*, 335–339. https://doi.org/10.1015/j.psychres.2018.07.033.

Izard, C. E. (1977). *Human emotions*. New York, NY: Plenum.

Izard, C. E. (2011). Forms and functions of emotions: Matters of emotion-cognition interactions. *Emotion Review, 3*, 371–378. https://doi.org/10.1177/1754073911410737.

Joiner, T. (2005). *Why people die by suicide*. Cambridge, MA: Harvard University Press.

Joseph, S., Williams, R., & Yule, W. (1997). *Understanding post-traumatic stress: A psychosocial perspective on PTSD and treatment*. New York, NY: John Wiley & Sons.

Kelley, M. L., Braitman, A. L., White, T. D., & Ehlke, S. J. (2018). Sex differences in mental health symptoms and substance use and their association with moral injury in veterans. *Psychological Trauma: Theory, Research, Practice, and Policy*, 337–344. https://doi.org/10.1037/tra/0000407.

Kimbrel, N. A., DeBeer, B. B., Meyer, E. C., Gulliver, S. B., & Morissette, S. B. (2016). Nonsuicidal self-injury and suicide attempts in Iraq/Afghanistan war veterans. *Psychiatry Research, 243*, 232–237. https://doi.org/10.1016/j.psychres.2016.06.039.

Krysinska, K., & Lester, D. (2010). Posttraumatic stress disorder and suicide risk: A systematic review. *Archives of Suicide Research, 14*, 1–23. https://doi.org/10.1080/13811110903478997.

Kubany, E. S., Abueg, F. R., Owens, J. A., Brennan, J. M., Kaplan, A. S., & Watson, S. B. (1995). Initial examination of a multidimensional model of trauma-related guilt: Applications to combat veterans and battered women. *Journal of Psychopathology and Behavioral Assessment, 17*, 353–376. https://doi.org/10.1007/BF02229056.

Kubany, E. S., Haynes, S. N., Abueg, F. R., Manke, F. P., Brennan, J. M., & Stahura, C. (1996). Development and validation of the Trauma-Related Guilt Inventory (TRGI). *Psychological Assessment, 8*, 428–444.

Kubany, E. S., & Watson, S. B. (2003). Guilt: Elaboration of a multidimensional model. *The Psychological Record, 53*, 51–90.

La Bash, H., & Papa, A. (2013). Shame and PTSD symptoms. *Psychological Trauma: Theory, Research, Practice, and Policy, 6*, 159–166. https://doi.org/10.1037/z0032637.

Larsen, S. E., Fleming, C. J. E., & Resick, P. A. (2019). Residual symptoms following empirically supported treatment for PTSD. *Psychological Trauma: Theory, Research, Practice, and Policy, 11*, 207–215. https://doi.org/10.1037/tra0000384.

Leary, M. R. (2007). Motivational and emotional aspects of the self. *Annual Review of Psychology, 58*, 317–344. https://doi.org/10.1146/annurev.psych.58.110405.085658.

Lee, D. A., Scragg, P., & Turner, S. (2001). The role of shame and guilt in traumatic events: A clinical model of shame-based and guilt-based PTSD. *British Journal of Medical Psychology, 74*, 451–466. https://doi.org/10.1348/000711201161109.

Leskela, J., Dieperink, M., & Thuras, P. (2002). Shame and posttraumatic stress disorder. *Journal of Traumatic Stress, 15*, 223–226.

Lewis, H. B. (1971). *Shame and guilt in neurosis.* New York, NY: International Universities Press.

Lewis, M. (1997). The self in self-conscious emotions. In J. G. Snodgrass & R. L. Thompson (Eds.), *The self across psychology: Self-recognition, self-awareness, and the self concept* (pp. 119–142). New York, NY: New York Academy of Sciences.

Lewis, M. (2008). Self-conscious emotions: Embarrassment, pride, shame, and guilt. In M. Lewis, J. M. Haviland-Jones, & L. F. Barrett (Eds.), *Handbook of emotions* (3rd ed., pp. 742–756). New York, NY: Guilford.

Lewis, M. (2014). *The rise of consciousness and the development of emotional life.* New York, NY: Guilford.

Lewis, M., Sullivan, M. W., Stanger, C., & Weiss, M. (1989). Self development and self-conscious emotions. *Child Development, 60*, 146–156. https://doi.org/10.2307/1131080.

Litz, B. T., Stein, N., Delaney, E., Lebowitz, L., Nash, W. P., Silva, C., & Maguen, S. (2009). Moral injury and moral repair in war veterans: A preliminary model and intervention strategy. *Clinical Psychology Review, 29*, 695–706. https://doi.org/10.1016/j.cpr.2009.07.003.

Ludwig, A. O. (1947). Neuroses occurring in soldiers after prolonged combat exposure. *Bulletin of the Menninger Clinic, 11*, 15–23.

Luoma, J. B., Kohlenberg, B. S., Hayes, S. C., Bunting, K., & Rye, A. K. (2008). Reducing self-stigma in substance abuse through acceptance and commitment therapy: Model, manual development, and pilot outcomes. *Addiction Research & Theory, 16*, 149–165. https://doi.org/10.1080/16066350701850295.

Luoma, J. B., Kohlenberg, B. S., Hayes, S. C., & Fletcher, L. (2012). Slow and steady wins the race: A randomized clinical trial of acceptance and commitment therapy target shame in substance use disorders. *Journal of Consulting and Clinical Psychology, 80*, 43–53. https://doi.org/10.1037/a0026070.

Luoma, J. B., & Platt, M. G. (2015). Shame, self-criticism, self-stigma, and compassion in acceptance and commitment therapy. *Current Opinion in Psychology, 2*, 97–101. https://doi.org/10.1016/j.copsyc.2014.12.016.

Mahtani, S., Hasking, P., & Melvin, G. A. (2018). Shame and non-suicidal self-injury: Conceptualization and preliminary test of a novel developmental model among emerging adults. *Journal of Youth and Adolescence*, https://doi.org/10.1007/s10964-018-0944-0.

Mennin, D. S., Fresco, D. M., O'Toole, M. S., & Heimberg, R. G. (2018). A randomized controlled trial of emotion regulation therapy for generalized anxiety disorder with and without co-occurring depression. *Journal of Consulting and Clinical Psychology, 86*, 268–281. https://doi.org/10.1037/ccp0000289.

Meyer, E. C., Szabo, Y. Z., Frankfurt, S. B., Kimbrel, N. A., DeBeer, B. B., & Morrisette, S. B. (2019). Predictors of recovery from post-deployment posttraumatic stress disorder symptoms in war veterans: The contribution of psychological flexibility, mindfulness, and self-compassion. *Behavior Research and Therapy, 114*, 7–14. https://doi.org/10.1016/j.brat.2019.01.002.

Meyer, E. C., Walser, R., Hermann, B., La Bash, H., DeBeer, B. B., Morissette, S. B., ... Schnurr, P. P. (2018). Acceptance and commitment therapy for co-occurring posttraumatic stress disorder and alcohol use disorders in veterans: Pilot treatment outcomes. *Journal of Traumatic Stress, 31*, 781–789. https://doi.org/10.1002/jts.22322.

Nieuwsma, J. A., Walser, R. D., Farnsworth, J. K., Drescher, K. D., Meador, K. G., & Nash, W. P. (2015). Possibilities with acceptance and commitment therapy for approaching moral injury. *Current Psychiatry Reviews, 11*, 193–206.

Nishith, P., Nixon, R. D.V., & Resick, P. A. (2005). Resolution of trauma-related guilt following treatment of PTSD in female rape victims: A result of cognitive processing therapy targeting comorbid depression? *Journal of Affective Disorders, 86*, 259–265.

Nock, M. K. (2009). Why do people hurt themselves? New insights into the nature and functions of self-injury. *Current Directions in Psychological Science, 18*, 78–83. https://doi.org/10.1111/j.1467-8721.2009.01612.x.

Øktedalen, T., Hagtvet, K. A., Hoffart, A., Langkaas, T. F., & Smucker, M. (2014). The Trauma-Related Shame Inventory: Measuring trauma-related shame among patients with PTSD. *Journal of Psychopathology and Behavioral Assessment*, https://doi.org/10.1007/s10862-014-9422-5.

Øktedalen, T., Hoffart, A., & Langkaas, T. F. (2015). Trauma-related shame and guilt as time-varying predictors of posttraumatic stress disorder symptoms during imagery exposure and imagery rescripting—A randomized controlled trial. *Psychotherapy Research, 25*, 518–532. https://doi.org/10.1080/10503307.2014.917217.

Orsillo, S. M., & Batten, S. V. (2005). Acceptance and commitment therapy in the treatment of posttraumatic stress disorder. *Behavior Modification, 29*, 95–129. https://doi.org/10.1177/0145445504270876.

Owens, G. P., Chard, K. M., & Cox, T. A. (2008). The relationship between maladaptive cognitions, anger expression, and posttraumatic stress disorder among veterans in residential treatment. *Journal of Aggression, Maltreatment & Trauma, 17*, 439–452. https://doi.org/10.1080/10926770802473908.

Panagioti, M., Gooding, P., & Tarrier, N. (2009). Post-traumatic stress disorder and suicidal behavior: A narrative review. *Clinical Psychology Review, 29*, 471–482. https://doi.org/10.1016/j.cpr.2009.05.001.

Pineles, S. L., Street, A. E., & Koenen, K. C. (2006). The differential relationship of shame-proneness and guilt-proneness to psychological and somatization symptoms. *Journal of Social and Clinical Psychology, 25*, 688–704. https://doi.org/10.1521/jscp.2006.256.688.

Pompili, M., Sher, L., Serafini, G., Forte, A., Innamorati, M., Dominici, G., & Giardi, P. (2013). Posttraumatic stress disorder and suicide risk among veterans: A literature review. *Journal of Nervous and Mental Disorders, 201*, 802–812. https://doi.org/10.1097/NMD.0b013e3182a21458.

Pugh, L. R., Taylor, P. J., & Berry, K. (2015). The role of guilt in the development of post-traumatic stress disorder: A systematic review. *Journal of Affective Disorders, 182*, 138–150. https://doi.org/10.1016/j.jad.2015.04.026.

Regehr, C., Alaggia, R., Dennis, J., Pitts, A., & Saini, M. (2013). Interventions to reduce distress in adult victims of rape and sexual violence: A systematic review. *Research on Social Work Practice, 23*, 257–265. https://doi.org/10.1177/1049731512474103.

Renna, M. E., Quintero, J. M., Soffer, A., Pino, M., Ader, L., Fresco, D. M., & Mennin, D. S. (2018). A pilot study of emotion regulation therapy for generalized anxiety and depression: Findings from a diverse sample of young adults. *Behavior Therapy, 49*, 403–418. https://doi.org/10.1016/j.beth.2017.09.001.

Resick, P. A., Galovski, T. E., Uhlmansiek, M. O., Scher, C. D., Clum, G. A., & Young-Xu, Y. (2008). A randomized clinical trial to dismantle components of cognitive processing therapy for posttraumatic stress disorder in female victims of interpersonal violence. *Journal of Consulting and Clinical Psychology*, *76*, 243–258. https://doi.org/10.1037/0022-006X.76.2.243.

Resick, P. A., Monson, C. M., & Chard, K. M. (2017). *Cognitive processing therapy for PTSD: A comprehensive manual.* New York, NY: Guilford.

Resick, P. A., Nishith, P., Weaver, T. L., Astin, M. C., & Feuer, C. A. (2002). A comparison of cognitive-processing therapy with prolonged exposure and a waiting condition for the treatment of chronic posttraumatic stress disorder in female rape victims. *Journal of Consulting and Clinical Psychology*, *70*, 867–879. https://doi.org/10.1037//0022-006X.70.4.867.

Rivers, W. H. R. (1922). *Instinct and the unconscious: A contribution to a biological theory of the psycho-neuroses* (2nd ed.). London, England: Cambridge University Press.

Rizvi, S. L., Vogt, D. S., & Resick, P. A. (2009). Cognitive and affective predictors of treatment outcome in cognitive processing therapy and prolonged exposure for posttraumatic stress disorder. *Behavior Therapy and Research*, *47*, 737–743. https://doi.org/10.1016/j.brat.2009.06.003.

Rudd, M. D. (2006). Fluid vulnerability theory: A cognitive approach to understanding the process of acute and chronic risk. In T. E. Ellis (Ed.), *Cognition and suicide: Theory, research, and therapy* (pp. 355–368). Washington, DC: American Psychological Association.

Semb, O., Strömsten, L. J., Sundbom, E., Fransson, P., & Henningsson, M. (2011). Distress after a single violent crime: How shame-proneness and event-related shame work together as risk factors for post-victimization symptoms. *Psychological Reports*, *109*, 3–23. https://doi.org/10.2466/02.09.15.16.PR0.109.4.3-23.

Shapiro, F. (2001). *Eye movement desensitization and reprocessing (EMDR): Basic principles, protocols, and procedures.* New York, NY: Guilford Press.

Stapleton, J. A., Taylor, S., & Asmundson, G. J. G. (2006). Effects of three PTSD treatments on anger and guilt: Exposure therapy, eye movement desensitization and reprocessing, and relaxation training. *Journal of Traumatic Stress*, *19*, 19–28. https://doi.org/10.1002/jts.20095.

Stone, A. (1992). The role of shame in post-traumatic stress disorder. *American Journal of Orthopsychiatry*, *62*, 131–136. https://doi.org/10.1037/h0079308.

Strange, R. E. (1968). Combat fatigue versus pseudo-combat fatigue in Vietnam. *Military Medicine*, *133*, 823–826. https://doi.org/10.1093/milmed/133.10.823.

Street, A. E., & Arias, I. (2001). Psychological abuse and posttraumatic stress disorder in battered women: Examining the roles of shame and guilt. *Violence and Victims*, *16*, 65–78.

Tangney, J. P. (1996). Conceptual and methodological issues in the assessment of shame and guilt. *Behavior Research and Therapy*, *34*, 741–754. https://doi.org/10.1016/0005-7967(96)00034-4.

Tangney, J. P., & Dearing, R. L. (2002). *Shame and guilt.* New York, NY: Guilford Press.

Tangney, J. P., Wagner, P., Fletcher, C., & Gramzow, R. (1992). Shamed into anger? The relation of shame and guilt to anger and self-reported aggression. *Journal of Personality and Social Psychology*, *62*, 669–675. https://doi.org/10.1037/0022-3514.62.4.669.

Tangney, J. P., Wagner, P., & Gramzow, R. (1989). *The Test of Self-Conscious Affect.* Fairfax, VA: George Mason University.

Tonidandel, S., & LeBreton, J. M. (2011). Relative importance analysis: A useful supplement to regression analyses. *Journal of Business Psychology*, *26*, 1–9. https://doi.org/10.1007/s10869-010-9204-3.

Trachik, B., Bowers, C., Neer, S. M., Nguyen, V., Frueh, B. C., & Beidel, D. C. (2018). Combat-related guilt and the mechanisms of exposure therapy. *Behavior Research and Therapy*, *102*, 68–77. https://doi.org/10.1016/j.brat.2017.11.006.

Tracy, J. L., & Robins, R. W. (2004). Putting the self into self-conscious emotions: A theoretical model. *Psychological Inquiry*, *15*, 103–125. https://doi.org/10.1207/s15327965pli1502_01.

van Minnen, A., Arntz, A., & Keijsers, G. P. J. (2002). Prolonged exposure in patients with chronic PTSD: Predictors of treatment outcome and dropout. *Behavior Research and Therapy*, *40*, 439–457. https://doi.org/10.1016/S00005-7967(01)00024-9.

VanDerhei, S., Rojahn, J., Stuewig, J., & McKnight, P. E. (2013). The effect of shame-proneness, guilt-proneness, and internalizing tendencies on nonsuicidal self-injury. *Suicide and Life-Threatening Behavior*, *44*, 317–330. https://doi.org/10.1111/sltb.12069.

Zerach, G., & Levi-Belz, Y. (2018). Moral injury process and its psychological consequences among Israeli combat veterans. *Journal of Clinical Psychology*, *74*, 1526–1544. https://doi.org/10.1002/jclp.22598.

SECTION 2

Biological bases of emotional responding and dysfunction

CHAPTER 7

Neurobiology and neuromodulation of emotion in PTSD

Lysianne Beynel, Lawrence G. Appelbaum, Nathan A. Kimbrel
Department of Psychiatry and Behavioral Sciences, Duke University School of Medicine, Durham, NC, United States

Brain regions and networks implicated in PTSD

Numerous neuroimaging studies of PTSD have been conducted during the past three decades. In general, these studies have revealed several key regions of the brain that are consistently associated with PTSD, including the hippocampus, the amygdala, and the medial prefrontal cortex (mPFC). The most common model linking these structures is the "fear extinction model" described in the succeeding text; however, more recent studies suggest that, while still valid, the fear extinction model remains insufficient to explain the totality of symptoms associated with PTSD. Instead, newer approaches implicate additional deficits in three brain networks—the salience network, the central executive network, and the default mode network—as described in the succeeding text in "the triple network model" of PTSD.

The hippocampus

The hippocampus, a gray matter structure involved in declarative memory, memory for episodic events (Tulving & Markowitsch, 1998), and stress regulation, is highly impacted by changes in glucocorticoid levels (Tatomir, Micu, & Crivii, 2014). Response to stress is mediated by the complex network of nervous and endocrine systems that form the hypothalamic-pituitary-adrenal axis. The hypothalamic-pituitary-adrenal axis generates glucocorticoids, a stress hormone strongly associated with the modulation of memory. Whereas an intermediate level of glucocorticoids enhances memory consolidation, higher levels of glucocorticoids impair memory. Studies in animal models have shown that stress-induced increases in glucocorticoids level lead to hippocampal damage (Ohl, Michaelis, Vollmann-Honsdorf, Kirschbaum, & Fuchs, 2000). In humans, studies have reported glucocorticoid-related changes in hippocampal volume, but there is still debate about the directionality of these changes. A metaanalysis conducted by Kitayama, Vaccarino, Kutner, Weiss, and Bremner (2005) showed that both left and right hippocampal volumes were smaller in patients with PTSD compared with both healthy controls and trauma-exposed controls. A more recent and larger metaanalysis conducted by Logue et al. (2018) that included 1868 participants (794 PTSD cases) also reported significantly smaller hippocampi among PTSD participants relative to trauma-exposed control participants (Cohen's $d=0.17$, $P=.00054$). Given the role of the hippocampus in memory, this decrease in hippocampal volume has been proposed to explain the associated memory deficits often observed among patients with PTSD.

Abnormalities in hippocampal activation have also been found in patients with PTSD. Using positron emission tomography (PET) to measure regional cerebral blood flow induced by an explicit memory task, patients with PTSD exhibited greater increases in hippocampal activity compared with healthy control subjects (Shin et al., 2004). In particular, patients with PTSD exhibited greater activation in the hippocampus for the low recall condition compared with healthy subjects, who showed increased regional cerebral blood flow from low to high recall. Hippocampal hyperactivation in the easiest condition might explain its reduced efficiency in more complex tasks and therefore explain the memory deficits reported by patients with PTSD. While it is possible that this hyperactivation compensates for the structural damage, it is notable that this study did not find any correlation between these two factors. Hippocampal hyperactivation associated with trauma-related triggers was also noted by Osuch et al. (2001) who measured regional cerebral blood flow while participants were listening to an audio script of their traumatic event. Listening to the scripts generated flashbacks, and the intensity of the reported flashbacks was found to positively correlate with activation in the left hippocampus. Thus the latter study suggests that, beyond its involvement with memory deficits, hippocampal activation may also be linked with reexperiencing symptoms.

The amygdala

The amygdala is a subcortical collection of nuclei in the temporal lobe that project to the brain stem and hypothalamic regions. This region of the brain has been consistently found to be hyperactive in patients with PTSD. In contrast, studies examining amygdala volume in patients with PTSD have found contradictory results (Pavliša, Papa, Pavić, & Pavliša, 2006). To better understand these discrepancies, Woon and Hedges (2009) metaanalyzed nine studies. They reported that the right amygdala was significantly larger than the left in patients with PTSD, patients exposed to a traumatic event but without PTSD, and subjects who have never been exposed to a trauma. This finding suggests an asymmetrical lateralization of the amygdala, which is preserved in adults with PTSD but does not express as volumetric differences between patients and healthy cohorts; however, a more recent and larger metaanalysis conducted by Logue et al. (2018) found significantly smaller amygdalae in PTSD patients compared with trauma-exposed control participants (Cohen's $d=0.11$, $P=.025$), although this finding did not survive correction for testing of multiple subcortical regions ($P=.0063$).

Despite the observation that the volume of the amygdala does not appear to be greatly impacted by PTSD, a number of studies have shown that its functional activation is highly modified in patients with PTSD. Again, however, contradictory results have been observed, with some studies showing hypoactivity in patients compared with healthy controls (Britton, Phan, Taylor, Fig, & Liberzon, 2005), others showing hyperactivity (Shin et al., 2004), and yet others finding no differences (Lanius et al., 2002). To resolve this ambiguity, Etkin and Wager (2007) metaanalyzed 15 PET and functional magnetic resonance imaging (fMRI) studies. They concluded that there was evidence for amygdala hyperactivation among patients with PTSD compared with healthy controls that was specific to the basal and lateral nuclei of the amygdala, also called the basolateral complex. They also noted that, while the basolateral complex is activated, activity in the central nucleus was comparatively decreased. The authors hypothesized that, while the hypoactivation of the central sulcus might be related to autonomic blunting or dissociation, the hyperactivated basolateral complex may be linked with the acquired fear responses in PTSD. A more recent metaanalysis by Patel, Spreng, Shin, and Girard (2012) confirmed amygdala hyperactivation among patients with PTSD; however, this analysis reported that hyperactivation was only observed in the left amygdala and only in comparison with nontrauma-exposed control participants.

While fMRI studies can only provide correlational information linking brain and behavior, lesion studies can provide information about the causal involvement of a brain region involved in specific symptoms. Koenigs et al. (2008) divided veterans into four groups according to the location of their brain lesion: damage to the amygdala, damage to the ventromedial prefrontal cortex (vmPFC), damage to other brain regions not involving amygdala and vmPFC, and no brain damage. The prevalence of PTSD in veterans with no brain damage and those with brain damage in regions not involving the vmPFC or the amygdala was similar (48% and 40%, respectively). In contrast, none of the patients with an amygdala lesion developed PTSD symptoms, which suggests that removing or disabling the amygdala might constitute a protective factor against PTSD. Thus applying deep brain stimulation over the amygdala—particularly the basolateral nuclei of the amygdala—could represent a reversible way to deactivate this region and effectively treat PTSD symptoms.

The ventromedial prefrontal cortex

The ventromedial prefrontal cortex (vmPFC) is a large brain region including the rostral anterior cingulate cortex (ACC), mPFC, subcallosal cortex, subgenual ACC, and the orbitofrontal cortex. This region is highly linked with the amygdala and plays an important regulatory role by exerting top-down inhibitory control on the amygdala. In a voxel-based morphometry study comparing cortical volumes in victims of the 1995 attack on a Tokyo subway, it was found that nine patients who developed PTSD displayed a gray matter volume reduction in the left ACC, relative to 16 victims who did not develop PTSD (Yamasue et al., 2003). Moreover, this reduction was negatively correlated with PTSD symptom severity, as measured by the Clinician-Administered PTSD Scale (CAPS). This finding was replicated by a study using cortical parcellation of magnetic resonance imaging (MRI), an automated volumetric segmentation procedure (Rauch et al., 2003), which also found reduced volume of pregenual ACC and subcallosal cortex, but not in the dorsal ACC in patients with PTSD. This result is consistent with the idea that the subdivision of the ACC relates to different functions, with the dorsal area corresponding to cognitive-motor functions, the rostral area (pregenual) corresponding to affective functions, and the ventral (subcallosal) area corresponding to visceromotor functions.

These structural abnormalities are also associated with functional abnormalities, but the direction of changes remains unclear. Some studies (Lanius et al., 2002; Liberzon et al., 1999; Rauch et al., 2000; Shin et al., 1997) have observed an equal or greater activation of the ACC in PTSD patients, compared with control participants. Other studies (Bremner et al., 1999; Lanius et al., 2003; Shin et al., 2001), however, have obtained opposite results with lower brain activation in the ACC for patients with PTSD when cued with emotional stimuli in comparison with control subjects. In a recent review focusing on neuroimaging abnormalities associated with affective processing (Negreira & Abdallah, 2019), the authors highlighted the key role that hypoactivation of the vmPFC plays both in symptom provocation and trauma-unrelated emotional paradigms. Moreover, this pattern of vmPFC hypoactivation was found to be consistent across different trauma types, highlighting its potential importance in understanding the neurobiology of PTSD. Also of note the study by Koenigs et al. (2008) described earlier found that vmPFC lesions were also likely to be protective against PTSD, as the prevalence of PTSD in patients presenting with vmPFC lesions was only 18%, which was substantially lower than the 48% rate of PTSD observed among patients without brain damage.

The fear extinction model

The hippocampus, amygdala, and mPFC, along with their respective pathophysiology, have been linked together in the fear extinction model to explain PTSD symptoms. In fear conditioning a neutral stimulus is paired with an aversive event, such as an electrical shock (unconditioned stimulus), to elicit a fear response such as freezing or startle (unconditioned response). After several repetitions of the pairing between the two stimuli, the presentation of the neutral event itself will induce a fear response (i.e., a conditioned response). Fear extinction occurs when, after presenting the neutral event alone, the conditioned response decreases until it is eventually eliminated. PTSD has been associated with a disruption in fear extinction (Rauch, Shin, & Phelps, 2006). According to this model, PTSD symptoms can be explained by a lack of top-down control by the vmPFC on amygdala hyperreactivity to the threat-related traumatic event. Hyperactivation of the amygdala has been proposed to mediate hyperarousal symptoms and helps to explain the indelible quality of the emotional memory for the traumatic event (Rauch et al., 2006), whereas abnormal hippocampal functioning is proposed to underlie deficits in identifying safe contexts and, as such, explain reexperiencing of symptoms and explicit memory difficulties (Bremner, Krystal, Southwick, & Charney, 1995).

Altered connections between the anterior cingulate gyrus and the amygdala may help to explain the top-down inhibition deficits observed. In nonhuman primates the amygdala and the prefrontal cortex are highly connected (Carmichael & Price, 1995). The majority of the afferent fibers to the amygdala originate in the orbitofrontal cortex and the mPFC. In return the amygdala, and more specifically its basolateral complex, has substantial efferent fibers to these two structures. In humans, diffusion tensor imaging—a neuroimaging method that measures the movement of the water molecules in the brain—provides information about white matter tracts that connect gray matter in the brain. The cingulum bundle has been identified as the most prominent white matter tract in the limbic system that connects the ACC to and from the amygdala. Analysis of this tract in patients with PTSD, compared with healthy controls, revealed reduced white matter integrity between the rostral, subgenual, and dorsal anterior region of the left anterior cingulum bundle (Kim et al., 2006). Thus the functional abnormalities that have been observed in patients with PTSD may be explained, at least in part, by altered structural connections between frontal and limbic areas. As mentioned earlier, these subdivisions are involved in both cognitive and affective functions, therefore contributing to PTSD symptoms.

The role of prefrontal regions in emotion regulation

While the fear extinction model has many appealing features, it remains insufficient to explain some PTSD symptoms, such as negative mood and cognition. A metaanalysis conducted by Hayes, Hayes, and Mikedis (2012) explored alternatives to the fear extinction model by investigating symptom provocation studies ($n=10$) separately from cognitive-emotional studies ($n=12$) to distinguish neuronal activation induced by these two specific aspects. Hayes et al. (2012) reported that the mid- and dorsal ACC were hyperactivated, while the medial frontal gyrus (vmPFC), right inferior frontal gyrus, and right precuneus were hypoactivated compared with controls. However, the cognitive-emotional studies were associated with hyperactivity in the supplementary motor area, bilateral amygdala, and medial frontal gyrus and hypoactivity in the pregenual ACC (vmPFC) and medial frontal gyrus (dorsomedial PFC). This metaanalysis therefore highlights the role of the ACC, with an opposite pattern of activation in its dorsal and anterior portion. Under this model, hyperactivation has been proposed to heighten appraisals of potential threats in the environment, whereas hypoactivity in vmPFC regions may reflect dysfunction in emotion regulation. Interestingly the amygdala was found to be hyperactivated only in the emotion-cognition studies and not in the symptom provocation ones. This intriguing result suggests that—contrary to what is suggested by the fear extinction model—the amygdala might not be the core region of dysfunction leading to PTSD and that instead its activation might be related to the type of tasks used. However, this idea should be considered cautiously as the amygdala is small in volume and is highly vulnerable to imaging artifacts. Thus the lack of findings in this metaanalysis could also be explained by limitations inherent in neuroimaging techniques. Another intriguing result is the lack of difference in hippocampal activation, despite the assumption that it represents a key structure in PTSD. The conclusion from this metaanalysis is that, while the fear extinction model still largely supports PTSD symptoms, emotion regulation is also an important factor in PTSD, and therefore, more attention should be given to the role of emotion regulation in PTSD treatment. Notably the findings from Hayes et al. (2012) also highlight the fact that many of the regions implicated in PTSD symptomatology actually belong to a larger salience network and that, instead of focusing on separated brain regions, future models should also consider distributed brain networks.

The role of network connectivity: The "triple network model"

The brain is organized in neural networks that bring together individual regions. Three of these networks have been found to be highly involved in psychiatric disorders, constituting the "triple network model" (Menon, 2011). The default mode network—which includes the mPFC, hippocampus, and the posterior parietal cortex—is highly activated at rest and is involved in self-referencing processes, task-independent internally focused thought, and autobiographical memory (Koch et al., 2016). The central executive network is a frontoparietal network highly involved in executive functions and cognitive control of thoughts and emotion. The salience network, which includes the subgenual ACC, the insula, and the amygdala, is involved in detection of personally salient internal and external stimuli that are used to direct behavior (Dosenbach et al., 2007). According to the triple network model, both cognitive and affective disturbances can be explained by dysfunction in these large brain networks. fMRI allows investigation of functional connectivity between brain regions by looking at the temporal correlation between them and is therefore a useful approach for assessing dysfunction at the network level. Moreover, functional connectivity can be assessed while subjects are performing a task or when they are at rest, revealing task-related functional connectivity or resting-state functional connectivity, respectively.

Task-related functional connectivity
A review of task-related functional connectivity studies focusing on the triple network model conducted by Lanius, Frewen, Tursich, Jetly, and McKinnon (2015) demonstrated an alteration in each of these three

networks. For example, patients with PTSD exhibited decreased connectivity in the central executive network during cognitive tasks, such as working memory (Daniels et al., 2010) and emotional processing tasks (Cisler, Steele, Smitherman, Lenow, & Kilts, 2013). Interestingly, whereas patients with PTSD exhibited decreased connectivity in the salience network and the central executive network, they showed higher connectivity within the default mode network compared with healthy controls during the tasks. It is widely accepted that, when performing a task, both the central executive network and salience network are more activated, whereas the default mode network is deactivated. This natural switching, likely mediated by the anterior insula (Menon & Uddin, 2010), seems not to occur in PTSD and could explain the cognitive deficits associated with PTSD. Lanius et al. (2015) further suggested that cognitive remediation could represent an effective way to restore the functional reorganization of the central executive network. The authors also reviewed abnormalities within the central executive network and concluded that there is altered connectivity within this network as well. Interestingly, patients in remission after 12 sessions of prolonged exposure therapy showed increased functional connectivity within the central executive network compared with nonremitters (Simmons, Norman, Spadoni, & Strigo, 2013). Lanius et al. also found evidence of altered connectivity in the default mode network, which they noted was correlated with PTSD symptom severity in several studies (Birn, Patriat, Phillips, Germain, & Herringa, 2014).

Resting-state functional connectivity
While the review by Lanius et al. (2015) elegantly summarizes the results from task-related connectivity studies, the observed differences between patients and controls could be biased by task-induced differences, such as behavioral performance, level of attention allocated to the task, or type of tasks used. Resting-state connectivity studies limit these biases by measuring connectivity in participants at rest as they let their minds wander. To examine this issue, Koch et al. (2016) conducted a systematic review to investigate the abnormalities in resting-state connectivity associated with PTSD symptoms. Compared with healthy controls (traumatized without PTSD and nontraumatized), patients with PTSD showed increased connectivity within the salience network and decreased connectivity within the default mode network. The increased connectivity within the salience network—which is thought to relate to attention—is consistent with the hypervigilance features of PTSD, whereas the hypoconnectivity observed

within the default mode network was proposed to indicate impaired internally focused thought and autobiographical memory processes. Finally the reduced between-network connectivity observed in these patients is consistent with the lack of top-down inhibitory control from the prefrontal cortex to the amygdala observed in functional studies and the abnormalities in fear response extinction.

Noninvasive and nonconvulsive brain stimulation techniques

Noninvasive and nonconvulsive brain stimulation approaches have been widely used to treat patients with mood disorders, such as major depressive disorder or bipolar disorders. While techniques such as repetitive transcranial magnetic stimulation (rTMS) and transcranial electric stimulation have gained a lot of traction as nonpharmaceutical alternative treatments for PTSD, the following section will highlight that only a few studies have actually tested their efficacy; however, prior to presenting the results of these studies, an overview of noninvasive and nonconvulsive brain stimulation mechanisms of action and the crucial stimulation parameters impacting the induced effects is provided.

Repetitive transcranial magnetic stimulation for PTSD

rTMS is a noninvasive, nonconvulsive brain stimulation technique that uses brief, high-intensity magnetic fields to modify neural activity underneath a stimulating coil. In brief, high-intensity current is rapidly turned on and off in a TMS coil placed over subjects' head, producing a time-varying magnetic field (1.5–2.0T) that lasts for 100–300 μs. This magnetic field generates a current flow in the neural tissue that modifies the excitability of the underlying cortex. While rTMS-induced effects on cortical excitability are quite focal, with depth penetration to about 2 cm below the scalp, rTMS leads to downstream effects through neural network connectivity. rTMS-induced effects on cortical excitability, and therefore on subsequent behavioral outcomes, depend on three main factors: (1) the spatial extent of the field relative to the desired cortical target; (2) the stimulation waveform parameters, including frequency, number, and intensity of the pulses; and (3) the number and spacing of rTMS sessions during both the acute treatment and maintenance phases.

Spatial parameters: Cortical target, targeting approach, and coil shape

The selection of a *cortical target* depends on the underlying neurobiology of the condition to be treated. As described earlier, PTSD is primarily associated with hypoactivation of prefrontal structures and hyperactivation of limbic structures, such as the hippocampus and amygdala; however, given limitations in the depth of penetration of rTMS, neither the amygdala nor the hippocampus can be directly modulated by the TMS field. Instead, all prior studies of rTMS of PTSD have focused on stimulating the PFC. In particular, these studies have targeted: (1) the medial prefrontal cortex because of its direct connectivity with the amygdala, (2) the right dorsolateral prefrontal cortex (DLPFC) to activate the hypothalamic-pituitary-adrenal axis and increase control over autonomic responses and suppress amygdala activation, or (3) the left DLPFC because of its involvement in depressive and cognitive symptoms associated with PTSD.

Different *targeting approaches* exist to reach distinct cortical targets and, therefore, constitute a crucial factor regarding subsequent rTMS-induced effects. There is a trade-off, however, between the precision of different targeting approaches and the cost associated with the devices or imaging required to achieve these approaches. The first, most basic, and cheapest targeting approach is to use scalp measurements to define the localization of the stimulated site. For example, the "5-cm rule" is frequently used to localize the DLPFC by first identifying the motor cortex hot spot as the region which evokes a muscular response in the contralateral hand muscle, and then moving the TMS coil 5 cm rostrally to reach the approximate location of the DLPFC. While these methods are easily applied at a low cost, they have the lowest precision, often missing the desired DLFPC target (Herwig, Padberg, Unger, Spitzer, & Schönfeldt-Lecuona, 2001). Another broadly applied scalp measurement approach used to locate the stimulated site is the international 10–20 system, also called the BEAM approach. This method takes into account individual variability in head size by using percentages of the circumference and distances between anatomical landmarks, such as inion, nasion, and left and right tragus. Under this approach the DLPFC is believed to correspond to the F3 (left-sided) or F4 (right-sided) electrodes (Beam, Borckardt, Reeves, & George, 2009). While this approach more accurately scales the targeting for different head sizes, it still does not take into account interindividual differences in brain organization and might not be accurate enough for precise targeting (Herwig, Satrapi, & Schönfeldt-Lecuona, 2003).

The development of frameless stereotactic neuronavigation systems has allowed for targeting with greater spatial precision that takes into account individual or group physiology. While these systems typically cost tens of thousands of dollars and are optimally used with neuroimaging data, which add additional costs, they represent the current state of the art in targeting to provide much more accurate localization of the stimulated site. This computer-assisted technology registers the location of the participant's head and the TMS coil together (usually through optical tracking) and links these coordinates with the subject's head anatomy, typically acquired through anatomical MRI. As an additional benefit, it is possible to layer on individualized fMRI data, or fMRI data from the literature, the so-called "probabilistic approach," to position the coil on the desired stimulated site. An influential study comparing the behavioral effects of four targeting methods (scalp measurement, probabilistic approach, individual anatomical MRI, and individual functional MRI) found that individualized, functionally guided rTMS with neuronavigation was associated with the largest effects, while scalp measurement localization led to the smallest (Sack et al., 2009). Despite this finding and the potential value of advanced targeting approaches, none of the studies using rTMS to treat PTSD to date have used this approach, potentially leading to reduced efficacy.

Finally the choice of the *coil shape* drastically modifies the electric field entering the brain and, consequently, the rTMS-induced effect. Three main rTMS coil shapes are currently used: circular, figure of 8, and H-coils. While the circular coils have been shown to induce a large nonfocal electric field under the perimeter of the coil, the figure-of-8 coils induce a more focal electric field that maximally stimulates about 1–2 cm^2 of cortex beneath the intersection of the two circular coils. To stimulate deeper areas of the brain, TMS can be applied with H-coils, which induce electric fields that reach as deep as ~4 cm below the scalp; however, there is a depth-focality trade-off, such that no coil can simultaneously prioritize focal and deep stimulation of the brain (Deng, Lisanby, & Peterchev, 2013).

Stimulation waveform parameters: Frequency, number of pulses, and intensity

rTMS involves the application of repeated pulses that are arranged according to a number of parameters, including frequency, intensity, and number of pulses, that describe the waveform parameters. A very common observation in the literature is that rTMS-induced effects are known to be *frequency dependent*, with low frequencies (≤ 1 Hz) being associated with decreased cortical excitability and behavioral inhibition, whereas higher frequencies (≥ 3 Hz) generally lead to increased cortical excitability and behavioral facilitation (for a review, see Fitzgerald, Fountain, & Daskalakis, 2006). According to the neurobiological models of PTSD, the optimal protocol would be to apply high-frequency rTMS over the PFC to reduce its hypoactivation; however, as reviewed later in this chapter, the majority of published studies have actually used low-frequency rTMS under the rationale that these frequencies appear to be safer.

Stimulation intensity, which ultimately reflects the electrical field amplitude entering the participant's brain, strongly impacts rTMS-induced effects. Stimulation intensity can be scaled according to the stimulator output or individualized for each participant according to a physiological response induced by stimulation. By far, the most common approach for scaling stimulation intensity is based on the motor-evoked potential. Under this procedure, single-pulse TMS is applied over the motor cortex, producing an objective, visually observable response in the contralateral hand muscle. The resting motor threshold (rMT) is defined as the lowest stimulation intensity to induce a visible muscle twitch or a motor-evoked potential larger than 50 μV when using electromyogram measurement system, 50% of the time in a finite number of trials (usually 10). Thus rMT reflects an individualized stimulation intensity that represents the amount of energy necessary to induce neuronal depolarization in the motor cortex of a specific participant. When applied over nonmotor areas that do not produce an objective quantifiable response, the intensity is defined as a percentage of the rMT, generally from 80% to 120% rMT. However, while rMT provides individualized information regarding the cortical reactivity of the motor cortex, it does not consider differences in brain physiology and anatomy between motor cortex and other cortical regions within or between individuals. As such, rMT intensity calibration may lead to substantial variation in the desired field strength in targeted brain regions outside the motor cortex. Therefore another practice is to use a fixed intensity, expressed as a percentage of the maximum stimulator output, based on the results from former studies. This method reduces the number of pulses and the overall duration of the experiment. To our knowledge, no studies have investigated the difference between these two methods.

Finally the effects of the *number of pulses* per session or across the whole rTMS treatment remain largely unknown, although a recent metaanalysis investigating the effects of rTMS for treating PTSD suggests that a higher number of pulses may lead to greater symptom improvements (Karsen, Watts, & Holtzheimer, 2014); however, this result needs to be taken cautiously given the small number of studies included in this analysis.

Parameters of acute and maintenance rTMS treatment for PTSD

While a single session of rTMS can induce behavioral effects that outlast the duration of the stimulation train itself, therapeutic applications of rTMS typically involve multiple sessions, over many days, under the logic that these repeated sessions lead to the buildup of cumulative effects. As such, parameters that define the number and spacing of treatment sessions, and maintenance sessions, constitute an important determinant of the effects of therapeutic rTMS. Despite the importance of these parameters, to date, there has been little systematic study of how these factors influence outcomes. rTMS treatment is generally performed with one session a day and many daily sessions spaced over several weeks. This, of course, represents a substantial time investment for both patients and clinicians. Recent studies suggest that "intensified" treatment, that is, applying five daily rTMS sessions, spread over 4 days is safe and might result in faster clinical response when used to treat major depressive disorders (Baeken et al., 2013; Duprat et al., 2016). Such an approach could potentially be effective for PTSD as well, although no studies have investigated this approach for PTSD. Regarding maintenance treatment, very few studies have included follow-up testing to investigate rTMS posteffect duration on symptom improvement. To our knowledge, this has also not been investigated in prior rTMS studies for PTSD.

rTMS for the treatment of PTSD

Between 1998 and 2016, at least 13 studies of rTMS-induced effects on PTSD symptoms were published (Table 1). Outcomes from these studies have been aggregated and compared in five different reviews and/or metaanalyses (Berlim & Van den Eynde, 2014; Clark, Cole, Winter, Williams, & Grammer, 2015; Karsen et al., 2014; Trevizol et al., 2016; Yan, Xie, Zheng, Zou, & Wang, 2017). Collectively, these metaanalyses suggest that rTMS significantly improves PTSD symptoms. For example, Trevizol et al. (2016) metaanalysis not only concluded that *"Active TMS was significantly superior to sham TMS for PTSD symptoms (Hedges' g = 0.74; 95% confidence interval = 0.06–1.42)"* but also noted that *"Meta-regression found no particular influence of any variable on the results."*

Indeed a rather surprising result across these metaanalyses is the lack of significant differences observed between high- and low-frequency stimulation, as both were associated with improved PTSD symptoms. It is largely accepted in the rTMS literature that low-frequency rTMS decreases cortical excitability, whereas higher-frequency stimulation increases excitability.

Table 1 Summary of the parameters used across all TMS studies to treat PTSD.

Authors	n	Frequency	Intensity	Target	Coil shape	Targeting approach	Number of pulses per session	Number of sessions	Sham?
Kozel et al. (2019)	27	1; 10	110% rMT	r-DLPFC	8	BEAM	2400	30	No
Kozel et al. (2018)	62	1	110% rMT	r-DLPFC	8	BEAM	1800	12	Sham coil
Philip et al. (2018)	33	5	NR	l-DLPFC	8	BEAM	3000–4000	33 ± 9	No
Oznur et al. (2014)	20	1	80% rMT	r-DLPFC	8	5-cm rule	600	20	No
Nam, Pae, and Chae (2013)	16	1	100% rMT	r-Prefrontal	8	5-cm rule	1200	15	Tilted coil
Isserles et al. (2013)	30	20	120% rMT	b-mPFC	H-coil	Brainsway Helmet	1680	12	Sham coil
Watts et al. (2012)	20	1	90% rMT	r-DLPFC	8	4 cm + 2 cm	400	10	Sham coil
Boggio et al. (2010)	30	20	80% rMT	r,l-Prefrontal	8	5-cm rule	1600	10	Sham coil
Osuch et al. (2009)	9	1	100% rMT	r-Prefrontal	8	5-cm rule	1800	20	Tilted coil
Cohen et al. (2004)	24	1; 10	80% rMT	r-DLPFC	Circular	5-cm rule	1 Hz: 100 10 Hz: 400	10	Tilted coil
Rosenberg et al. (2002)	12	1; 5	90% rMT	l-Prefrontal	8	4 cm + 2 cm	600	10	No
McCann et al. (1998)	2	1	80% rMT	r-Frontal	8	NE	1200	17; 30	No
Grisaru, Amir, Cohen, and Kaplan (1998)	10	0.3	100% MSO	b-Motor cortex	Angular-shaped	BEAM	300	1	No

The first author of each study is presented with the year of publication; n is the number of subjects per study; l = left, r = right, and b = bilateral.

For example, studies assessing prefrontal rTMS effects on cerebral oxygen perfusion in both local and distant areas find that low- and high-frequency rTMS have opposite effects on cerebral perfusion (Speer et al., 2000). Thus, one would expect the low- and high-frequency stimulation to have opposite effects on symptom changes. One explanation of the lack of frequency dependency in these studies could relate to the differences in study design, particularly the number of pulses administered in the different protocols. Two studies directly comparing the effects of low- and high-frequency rTMS. Rosenberg et al. (2002; 1 Hz vs 5 Hz) did not report any differences. In contrast, Cohen et al. (2004) reported that 10-Hz rTMS was superior to 1 Hz for reducing PTSD symptoms; however, Rosenberg et al. applied 600 pulses for both stimulation frequencies, whereas Cohen et al. applied 1000 pulses at 1 Hz and 4000 pulses at 10 Hz. Thus, if, as reported by Karsen et al. (2014), the number of pulses correlated with the PTSD symptom improvements, then the difference in outcomes could be explained by the number of pulses rather than the frequency itself. A more recent study was conducted to investigate this question by equalizing the number of pulses; however, this study did not find any differences between the two frequencies (Kozel et al., 2019).

An alternative interpretation for this paradoxical finding is that each frequency may affect different elements of neural networks. It has been hypothesized that high frequency applied over the right frontal cortex could activate the hypothalamic-pituitary-adrenal axis to inhibit excessive autonomic responses and therefore suppress amygdala hyperactivation (Cohen et al., 2004). Another interpretation is that stimulation over the right frontal cortex inhibits the left frontal cortex through transcallosal processes, leading to inhibition of the memory retrieval network (Boggio et al., 2010). This assumption is also supported by Rossi et al. (2006) who speculated that rTMS effects could be due to interference with episodic memory. Indeed, when applying high-frequency rTMS over the frontal cortex of healthy subjects during stimulus encoding, performance on the subsequent recognition tasks was impaired (Rossi et al., 2001). This finding suggests that episodic memory is supported by the stimulated frontal region and that improvement associated with high-frequency rTMS could be alternatively explained by a reduction of "abnormal hypermnesia" in PTSD. Since none of the studies included in these metaanalyses acquired imaging before and/or after rTMS, it is difficult to draw conclusions about the neural origins of these effects.

Optimizing rTMS efficacy for PTSD: Development of state-dependent rTMS therapies

Recent research has shown that the effect of rTMS does not simply depend on the stimulation parameters, but instead on the interaction between these parameters with the neural state (Silvanto & Pascual-Leone, 2008). According to this so-called state-dependency assumption, applying rTMS while the targeted brain circuits are actively engaged will lead to the coactivation of the targeted neuronal population and will be more effective by creating "Hebbian-like plasticity." Three studies (Isserles et al., 2013; Kozel et al., 2018; Osuch et al., 2009) have tested this theory in relation to PTSD by combining rTMS with either exposure therapy or cognitive processing therapy. The first double-blind sham-controlled crossover study (Osuch et al., 2009) included nine patients with treatment-resistant PTSD. Patients received 20 sessions of active or sham rTMS at 1 Hz for 30 minutes at 100% rMT over the right DLPFC. Participants were asked to talk about traumatic events for at least 5 minutes. Following this 5-minute period, rTMS was begun, and participants could choose to either keep talking or remain quiet. Contrary to expectations, no differences were found between active and sham rTMS when combined with exposure. When investigating the effect of active rTMS separately, a moderate improvement was found only for the hyperarousal symptoms. The authors concluded that beyond the small sample size and poor sham condition (which utilized a tilted coil at a 45-degree angle), the weak effect could have been due to the low stimulation frequency utilized and hypothesized that higher-frequency stimulation might increase efficacy.

A few years later, Isserles et al. (2013) investigated this assumption with 20-Hz rTMS. Thirty treatment-resistant patients were randomly assigned to three groups: exposure therapy combined with active rTMS, no exposure combined with active rTMS, and exposure therapy combined with sham rTMS. In the "ultrabrief exposure therapy," lasting 30 seconds before rTMS treatment, participants were asked to recall a traumatic memory to elicit a fear response before applying high-frequency rTMS over the mPFC. After 12 sessions of rTMS, only subjects receiving the exposure therapy combined with active rTMS showed improvement on the CAPS. No improvements were found for subjects receiving active rTMS without exposure therapy, or sham rTMS combined with exposure therapy. Interestingly, when looking at physiological data recorded during the treatment, only patients in the active group showed attenuation of their heart rate, a response that has previously been associated with the exposure therapy and shown to correlate with improvement in CAPS scores. This result suggests that stimulation of bilateral mPFC combined with ultrabrief exposure was able to reduce PTSD symptoms, both at behavioral and physiological levels. While this study was well designed and allowed comparison of exposure effects, it remains surprising that active 20-Hz rTMS over the mPFC did not result in any symptom improvement, as this area is highly involved in PTSD. One possible interpretation could be linked to the H-coil utilized for stimulation. As mentioned earlier, while H-coils allow for stimulation of deeper brain structures, they are also much less focal and could have potentially affected other brain structures that counteracted the expected rTMS effect.

Both of the studies noted earlier suffered from a small sample size preventing strong conclusions about the efficacy of state-dependent rTMS. More recently, this approach was performed in a much larger sample size ($n=62$:32 in the active group, 30 in the sham group) (Kozel et al., 2018). Participants received 12 sessions of active or sham 1-Hz rTMS over the right DLPFC for 30 minutes, therefore reproducing the parameters used by Osuch et al. (2009). Instead of experiencing exposure therapy during rTMS application, participants went through cognitive processing therapy immediately after rTMS application and again following a 30-minute break. Clinical improvements were observed on the CAPS for both groups, with significantly greater improvement for the group receiving active rTMS and cognitive processing therapy. Perhaps most important, clinical improvements persisted at 6-month follow-up.

In conclusion, while it remains difficult to compare these three studies, given differences in the stimulation parameters and types of exposures/therapies used, as a group, they do suggest that rTMS combined with therapy is likely to be superior rTMS alone. Importantly the clinical improvement observed in the largest of these studies (Kozel et al., 2018; $n=62$) was sustained for 6 months following treatment, which suggests that the combination of these approaches (i.e., "cognitive paired stimulation") may induce neuronal plasticity. Notably, however, this hypothesis has not been directly tested, as none of these studies used imaging to evaluate rTMS effects. This gap represents a crucial area for future work on rTMS for PTSD.

Investigations of the neural mechanisms underlying response to rTMS

A key objective of many studies using rTMS to treat major depressive disorders has been to develop a better understanding of the neuronal abnormalities associated with depression and predictors of positive treatment outcomes. For example, the inverse correlation between the subgenual

ACC and the DLPFC has been found to predict the response to rTMS treatment (Fox, Buckner, White, Greicius, & Pascual-Leone, 2012). As a result, more recent work has focused on developing innovative targeting approaches based on this finding to individualize rTMS treatment for depression. Unfortunately, this approach has been largely ignored in the field of rTMS for PTSD. To our knowledge, only two studies have reported this information (McCann et al., 1998; Philip et al., 2018). In McCann et al. (1998) pioneering study of PTSD, rTMS treatment was found to reduce hypermetabolism in the right hemisphere where the TMS pulses were applied. More recently, Philip et al. (2018) investigated resting-state connectivity as a potential predictor of rTMS outcomes in 33 patients with comorbid MDD and PTSD receiving 5-Hz rTMS over the left DLPFC, in an open-label fashion. They found that participants' brain state at baseline predicted symptom improvements. Specifically, reduced connectivity between the subgenual ACC and the default mode network and higher connectivity between the amygdala and PFC were both predictive of treatment outcome. Moreover, after 30–40 sessions of rTMS, PTSD symptom reduction was found to be associated with reduced connectivity between the subgenual ACC and the default mode network. A similar effect was also observed between the hippocampus and the salience network. Taken together, these findings (1) provide support for the theoretical assumption that dysfunction in the frontal cortex and amygdala plays a key role in PTSD, (2) that connectivity between these structures also plays a key role, and (3) that rTMS has the ability to modulate the excitability of these regions and the connectivity between them. While these results still need to be replicated and confirmed in additional studies, they indicate that individualizing rTMS parameters (e.g., by targeting regions of the default mode network that are the most negatively correlated with the subgenual ACC) could provide a way to further optimize rTMS efficacy.

Transcranial electrical stimulation for PTSD

Transcranial electrical stimulation techniques use low-voltage electrical currents, typically 1–2 mA, delivered via electrodes placed over subjects' scalp to modulate neuronal function. While there are multiple forms of transcranial electrical stimulation, the general principle is that anode electrodes induce positive currents that propagate into the cortex, while cathode electrodes induce negative currents. Contrary to rTMS, transcranial electric stimulation does not induce neuronal firing, but instead changes the membrane potentials of neurons, which generates polarity-dependent long-term potentiation-like effects and long-term depression-like effects, which outlast the stimulation duration.

An essential aspect of transcranial electric stimulation is the modulatory form of the current alternation, with three typical applications. The most common application is transcranial direct current stimulation (tDCS), where the current is gradually ramped on and then applied continuously for durations that typically last 10–20 minutes, before the current is ramped off. In this case, anodal current is thought to cause depolarization of resting membrane potentials, increasing neuronal excitability that allows for more spontaneous firing, while cathodal current causes hyperpolarization and a decrease in neuron excitability due to the decreased spontaneous cell firing (Nitsche & Paulus, 2000). A second type of transcranial electric stimulation is transcranial alternating current stimulation, where the current is modulated sinusoidally at a chosen frequency, therefore providing a way to entrain ongoing brain oscillations to the transcranial alternating current stimulation resonance frequency (Tavakoli & Yun, 2017). A third approach, referred to as transcranial random noise stimulation, is to apply electrical currents with random amplitude and frequency modulations. While the mechanisms of action of transcranial random noise stimulation are not well understood, it has been proposed that the effects might be attributed to either the repeated opening of sodium channels or the introduction of stochastic noise in the system, which might boost the sensitivity of the neurons, thereby increasing their signal-to-noise ratio (Terney, Chaieb, Moliadze, Antal, & Paulus, 2008).

As with TMS, there are a large number of parameters that influence the neuronal and behavioral effects of transcranial electric stimulation, such as the spatial extent of stimulation and the temporal duration. With regard to the spatial aspects, current diffusion of transcranial electric stimulation is generally quite broad. As such, in practice, the anodal electrode is typically placed over hypoactivated brain region, while the cathode is placed over a neutral location such as the vertex or the contralateral orbital frontal cortex. While there are a large number of temporal parameters, such as the current modulations discussed earlier, to achieve aftereffects, it appears to be necessary to stimulate for at least 3 minutes with the intensity of at least 0.6 mA (Nitsche & Paulus, 2000). While 5–13 minutes of stimulation can produce aftereffects lasting about 1–2 hours (Nitsche & Paulus, 2001), prolonged duration of stimulation can also reverse the expected effects of stimulation. For example, the application of 26 minutes of anodal stimulation resulted in inhibition (Monte-Silva et al., 2013). Therefore stimulation duration constitutes an important factor that contributes to transcranial electric stimulation neuromodulation.

Studies of transcranial direct current stimulation (tDCS) to treat PTSD

At this time, no transcranial electric stimulation techniques have regulatory approval for any clinical indication (though there are ongoing trials), and only tDCS has been tested as a potential treatment for specific aspects of PTSD symptomatology. For example, in 2015, Saunders et al. (2015) investigated the feasibility and efficacy of tDCS to reduce working memory deficits frequently observed among patients with PTSD. Four participants received five sessions of tDCS with the anode electrode placed over left DLPFC (defined by the F3 electrode location) and the cathode over the contralateral supraorbital area. tDCS was applied for 20 minutes, at 1 mA, once a week during 5 weeks of treatment, followed by five additional weeks of a computerized working memory training program. Results from this pilot study showed working memory performance enhancement, as measured by improvements in several tasks such as digit span, memory recognition, and visual memory span and improvement in emotional processing as subjectively reported by participants. These behavioral changes were associated with a normalization of neurophysiological dysregulation, as measured by electroencephalography recordings.

The second feasibility study conducted to test the efficacy of tDCS combined stimulation with an extinction learning process to improve the fear extinction process (van't Wout et al., 2017). As fear extinction is, in theory, driven by top–down control by the vmPFC on the amygdala, anodal tDCS was placed over the vmPFC to increase its metabolism and theoretically improve participants' fear extinction. Twenty-eight patients with PTSD completed a 2-day Pavlovian fear conditioning task. tDCS was applied on the second day for 10 minutes at 2 mA, with the anodal electrode placed over AF3 on the 10–20 grid. Skin conductance was measured as an indication of fear extinction. While results showed that patients who received active tDCS demonstrated moderately better extinction memory, as reflected by a decrease in skin conductance, the effect was not statistically significant. Thus, while this study demonstrated the feasibility and tolerability of this type of approach, it also indicates that a greater duration of tDCS is likely necessary to influence fear-based responses.

More recently the same team investigated the efficacy of tDCS combined with virtual reality exposure (van't Wout-Frank, Shea, Larson, Greenberg, & Philip, 2019). A total of six sessions were performed over 2 weeks, during which 2-mA tDCS was applied for 25 minutes per session, with the anode over AF3 and the cathode over PO8. tDCS application was combined with immersive virtual reality (VR) exposure, during which patients were confronted with driving scenarios in war-zone environments. Results from the 12 patients showed that skin conductance to VR events reduced more quickly over time with active tDCS compared with sham. Importantly, patients in both groups showed significant reduction in PTSD symptoms, and patients who received active tDCS continued showing improvement at the 1-month follow-up assessment. Thus this study demonstrates a promising and innovative approach to reducing PTSD symptoms that combines both stimulation and VR-based exposure sessions.

In summary, while relatively little attention has been given to tDCS as a treatment for PTSD, preliminary results from these three studies suggest that tDCS is feasible and may be associated with positive outcomes—particularly when combined with exposure and/or other psychotherapy approaches. Given these preliminary findings, as well as the low-risk, low cost, and ease-of-use associated with tDCS, it is recommended that larger, more well-powered randomized clinical trials be conducted that combine tDCS with exposure therapy to evaluate the efficacy of tDCS for improving treatment outcomes for patients with PTSD.

Noninvasive and convulsive brain stimulation techniques
Electroconvulsive therapy and magnetic seizure therapy

Electroconvulsive therapy (ECT) and magnetic seizure therapy are clinical procedures performed under general anesthesia during which electrical currents, or magnetic fields, are passed through the brain to intentionally trigger a brief seizure. ECT is FDA approved for the treatment-resistant major depression. While the specific mechanisms of action of these two approaches remain largely unknown, it is often assumed that the seizure convulsions act to resynchronize dysregulated brain activity (McClintock et al., 2014). As reported by a recent review (Singh & Kar, 2017), ECT involves numerous intricate biological processes, including neurophysiological changes that increase cerebral blood flow, neurobiochemical changes that release hormones and neurotransmitters in the brain, and neuroplastic changes that promote neurogenesis or synaptogenesis, which result in changes in brain volume. These changes are associated with positive outcomes in psychiatric disorders. While the use of ECT still remains stigmatized, ECT constitutes the most effective treatment for highly depressed, treatment-resistant patients (Weiner & American Psychiatric Association, 2001). Moreover, since the 1950s, technological developments have been introduced to reduce the side effects associated with ECT, such as the use of unilateral, instead of bilateral stimulation, and shorter pulse width instead of sine wave (Weiner, Rogers, Davidson, & Squire, 1986). As stated in a recent review, the remaining side effects are usually limited to the first 3 days after ECT treatment (Semkovska & McLoughlin, 2010). Magnetic seizure therapy uses magnetic, rather than electrical, stimulation to induce a more focal seizure, leading to reduced side effects (Lisanby, 2002); however, to our knowledge, magnetic seizure therapy has not yet been tested to treat PTSD.

Efficacy of electroconvulsive therapy to treat PTSD
To date, there has been only limited research on the utility of ECT to treat PTSD, including one uncontrolled clinical trial, several retrospective studies, and several case reports (Youssef, McCall, & Andrade, 2017). Accordingly the present review focuses only on the first prospective, uncontrolled trial of ECT for treatment-resistant PTSD to date (Margoob, Ali, & Andrade, 2010) and a large retrospective analysis of ECT for PTSD with comorbid depression (Ahmadi, Moss, Simon, Nemeroff, & Atre-Vaidya, 2016).

Margoob et al. (2010) administered six bilateral ECT sessions to 20 adults with treatment-resistant PTSD, which was defined as failure to respond to a minimum of 4 antidepressants and 12 sessions of cognitive behavioral therapy. As would be expected, this group of treatment-resistant participants reported severe PTSD symptoms at baseline (mean CAPS=90.5; SD=17.3). Seventeen of the 20 participants completed the full course of ECT, and an intent-to-treat analysis demonstrated a statistically significant 34% decrease in CAPS scores from baseline to the end of treatment (M=59.4; SD=25.2; $P<.001$). Notably, much of the improvement in PTSD symptoms was observed by the time of the third ECT session. In addition, improvement in PTSD symptoms was independent of improvement in depression ratings. The investigators further noted that improvements in PTSD severity did not vary as function depression severity. Using a ≥30% decrease in CAPS from baseline to end of treatment to define treatment response, the investigators reported that 70% of participants responded to treatment. While a formal follow-up assessment was not conducted, the majority of participants were followed for 4–6 months by case records and all appeared to maintain clinical gains at follow-up, though no formal ratings of symptomatology were obtained.

Ahmadi et al. (2016) conducted a retrospective, nested case-control study that included 22,164 participants, including 3485 with comorbid

PTSD and depression (92 of whom received ECT and 3393 that did not) and 18,679 participants without PTSD or depression. The Clinical Global Impression scale was used to assess treatment efficacy. Among participants with PTSD and depression, they observed that ECT was associated with significantly better improvement than antidepressants alone (90% vs 50%, $P=.001$). Moreover, during the follow-up period (median$=$8 years), they noted that the rate of suicide among participants with PTSD and depression who were treated with ECT was significantly lower than the rate of suicide observed among participants with PTSD and depression who were treated with antidepressants alone (2.2% vs 5.9%, $P<.05$). Interestingly, ECT was also associated with reduced risk for cardiovascular and all-cause mortality among patients with PTSD and depression.

In sum, while ECT appears to have significant potential for severe, treatment-resistant patients with comorbid PTSD and depression, far more research is needed on this topic, particularly given the significant potential side effects associated with ECT, the lack of control conditions included in prior research on this topic, and the small number of prior studies in this area of research to date. It is notable, however, that no prior studies of ECT have demonstrated a worsening of PTSD symptoms (Youssef et al., 2017), which highlight the potential utility of this approach for patients with PTSD who are unresponsive to other traditional treatment approaches.

Invasive brain stimulation techniques

In addition to the noninvasive techniques described earlier, there has also been growing interest in the use of invasive brain stimulation approaches that involve surgical implantation of electrodes directly into the brain to treat psychiatric disorders. Obviously, such techniques entail markedly different risks and are typically reserved for only the most severely treatment-resistant patients. While such techniques have been associated with positive outcomes for Parkinson's disease, depression, and other neurological disorders, little is known about their efficacy to treat PTSD. As such the following section discusses the developing body of research around invasive brain stimulation techniques with an emphasis on the risks entailed and instances in which these treatment options could potentially merit the greater hazards involved.

Deep brain stimulation

Deep brain stimulation (DBS) consists of surgical implantation of thin electrical leads (1.27 mm in diameter) directly into the brain with millimeter accuracy using stereotactic procedures. These electrical leads are connected to an external pulse generator, generally placed over the chest, which delivers low-voltage electrical current to the electrodes, which, in turn, modify neuronal excitability. Stimulation parameters such as frequency (2–185 Hz), pulse width (60–450 μs), and amplitude (0–10.5V) can be adjusted through a portable handheld device that communicates with the implanted generator by telemetry. Pulse width determines the surface of neural tissue that is affected by the electrical current, amplitude determines the strength of the effect, and frequency determines if the neural tissue is inhibited or facilitated.

While the mechanisms of action underlying DBS remain unclear, it is widely accepted that high-frequency DBS (>100 Hz) inhibits stimulated neural structures, mimicking a surgical resection of the stimulated structure (Montgomery Jr & Gale, 2008). Accordingly, DBS electrodes are implanted on brain structures that are overactive. DBS is currently FDA approved to treat Parkinson's disease and essential tremor. DBS is also used to treat obsessive-compulsive disorders, depression, Alzheimer's disease, anorexia, and Tourette syndrome. DBS is associated with risks due to the surgical intervention itself (e.g., seizure, hemorrhage, and infection) and device-related complications (e.g., lead movements or malfunctions); however, a recent retrospective analysis evaluating the adverse effects of DBS among 728 patients found that the risk of procedure- and hardware-related adverse events was "acceptably low" (Fenoy & Simpson, 2014).

Deep brain stimulation to treat PTSD

Based upon the commonly observed hyperactivity of the basolateral nuclei of the amygdala in PTSD patients, DBS electrodes are typically implanted in the basolateral nuclei to reduce PTSD symptoms. The feasibility and efficacy of DBS were initially tested in rodents prior to use with human patients. Applying electrical shocks to rats in the presence of an object, such as a mini-tennis ball, appears to "traumatize" rats, such that they will tend to bury that object when they are reexposed to it. Notably, this effect lasts at least 28 days after the shocks. As such, it appears to represent a long-lasting and robust animal model of avoidance and hypervigilance symptoms (Mikics, Baranyi, & Haller, 2008).

Langevin, De Salles, Kosoyan, and Krahl (2010) used this model in rats to test the potential efficacy of DBS to modulate PTSD symptoms. The DBS system was implanted in the right basolateral nuclei of the amygdala of the rats and either active or sham stimulation was applied 1 day after receiving the shocks for 7 consecutive days at 160 Hz. After the last day of DBS, the rats were reexposed to the mini-tennis ball. While all the rats in the sham group buried the tennis ball, this behavior occurred significantly less frequently among rats receiving active DBS. These findings suggest that DBS may be effective at decreasing PTSD-like symptoms; however, given the fact that the stimulation was applied immediately after the "trauma," this finding is only applicable to patients with recent traumas. In further studies with rodents, when compared with paroxetine, a selective serotonin reuptake inhibitor that has been shown to be beneficial for PTSD symptoms (Marshall, Beebe, Oldham, & Zaninelli, 2001), only active DBS of the right amygdala was found to reduce burying behavior in traumatized rats (Stidd, Vogelsang, Krahl, Langevin, & Fellous, 2013). Taken together, these findings from the animal literature suggest that DBS has the potential to effectively reduce PTSD symptoms and that it was worthy of further testing in humans.

Since 2014 Koek and colleagues have been conducting a proof-of-concept, phase I study to test the safety, feasibility, and efficacy of DBS for PTSD in humans (Koek et al., 2014). In this ongoing study, six patients with severe, treatment-resistant PTSD are being implanted with a DBS system bilaterally in the basolateral nuclei of the amygdalae. Following surgery, the stimulator is kept off for the first 4 weeks, and electroencephalography is used to monitor for adverse events associated with the stimulation. While the study is still actively recruiting, the initial results from the first included patient have been published (Langevin et al., 2016). Notably the baseline CAPS score for this patient was 119, indicating that the patient was experiencing severe PTSD symptoms despite having received 20 years of pharmacological treatments and psychotherapies. No adverse effects were observed at 1 month postoperative, so long-term DBS was initiated. At 8 months postoperative, the patient's CAPS score had been reduced by 37.8%. The patient also reported an increase in sleep duration, reduction of nightmares, and mood improvement. While this result needs to be interpreted with extreme caution given that it is from a single patient, without a control condition, these preliminary findings do point toward feasibility, safety, and possibly therapeutic effects of DBS to treat severe, treatment-resistant PTSD in humans.

Vagus nerve stimulation

In contrast with DBS, vagus nerve stimulation (VNS) does not aim to directly stimulate brain structures. Rather, it attempts to modulate cranial nerves thought to have broad control of brain stem functions. VNS targets the 10th cranial nerve through implantation of a stimulating electrodes over the vagus nerve and a pulse generator placed in the chest. VNS was the first FDA approved for treatment-resistant epilepsy in the United States in 1997. In the following years, clinician observations of mood improvement associated with VNS—even in the absence of improvement in epileptic symptoms (Elger, Hoppe, Falkai, Rush, & Elger, 2000)—led to additional research on using this approach for mood disorders. Clinical trials have since confirmed that VNS has significant positive effects on mood and anxiety (George et al., 2000), and VNS was approved by the FDA for treatment-resistant depression in 2005.

The vagus nerve extends from the brain stem to the abdominal cavities, passing through the neck and upper chest. It is composed of 20% efferent fibers and 80% afferent fibers that terminate in the nucleus solitary tract, which connects to the locus coeruleus, amygdala, and hypothalamus (Howland, 2014). Its anatomical connections to both cortical and limbic areas, combined with the associated VNS-induced changes in neurotransmitter levels, such as norepinephrine and serotonin, suggest that VNS could also be used to treat PTSD; however, to date, there has been very little research on VNS as a potential treatment for PTSD.

Vagus nerve stimulation to treat PTSD

Peña, Engineer, and McIntyre (2013) evaluated VNS in a rodent model that utilized auditory fear conditioning and found that a single session of VNS significantly enhanced the extinction of conditioned fear in rats. In addition, by comparing rats in which VNS was applied during the exposure, to rats receiving VNS after exposure, the investigators found that only the "paired" condition led to significant reduction in freezing, suggesting that pairing between stimulation and exposure is crucial. Moreover, Peña et al. (2014) demonstrated that the observed enhancement of extinction condition fear response with active VNS was associated with long-term potentiation in the infralimbic mPFC and the basolateral complex of the amygdala, a crucial pathway in PTSD. More recently, Noble et al. (2017) used a single prolonged stress procedure with rats who were stressed and then socially isolated. Fear extinction was performed during 11 days, and rats received active or sham VNS every other day. Rats receiving active VNS showed a significantly higher remission of the evoked freezing behavior than rats receiving sham VNS. As expected a reinstatement of the fear conditioning performed on the 12th day of the procedure increased the freezing behavior for both groups of rats; however, this increase was smaller for the rats receiving active VNS. The authors interpreted these results as demonstrating that VNS may have the potential to facilitate resilience to stress-induced relapses. Importantly, behavioral indices of anxiety, hyperarousal, and avoidance were also largely reduced for the rats receiving active VNS. Taken together, these animal studies suggest that combining VNS with exposure therapy could constitute another avenue to treat patients with treatment-resistant PTSD. To date, however, there has been only one VNS study of PTSD in humans. In this pilot study, George et al. (2008) conducted investigated VNS efficacy with 10 patients with treatment-resistant anxiety disorders, two of which presented with PTSD symptoms. In this open-label trial, VNS was applied at 20 Hz–500 μs pulse width and cycles of 30 seconds on and 5 minutes off during the first 2 weeks and slowly increased over the next 2 weeks. Findings revealed only modest anxiety improvement, with 30% of the patients responding to the acute phase of treatment. Thus, at the present time, there is a lack of evidence to support the use of VNS for PTSD in humans.

Future directions

Invasive and noninvasive brain stimulation approaches both appear to hold substantial promise for the treatment of patients with PTSD—particularly patients who do not respond to traditional first-line psychotherapy and/or pharmaceutical approaches. There are, however, numerous limitations that need to be addressed through future research to increase our understanding of the potential benefits and risks associated with each of the brain stimulation techniques described earlier.

First a pervasive issue across the PTSD brain stimulation literature concerns the very limited sample sizes included in prior studies, which has severely limited statistical power and our ability to draw inferences from prior work. As noted earlier in this chapter, brain stimulation effects are highly sensitive to target and stimulation parameters and subject-specific parameters, such as brain physiology, brain reactivity, mental state during the stimulation, and genetic factors that promote brain plasticity (Cheeran et al., 2008). As such, larger sample sizes are needed to draw strong conclusions that are robust to between-subject variability.

Another important aspect to consider in terms of rTMS concerns placebo effects. Patients receiving brain stimulation are generally treatment-resistant patients who have failed to respond to several pharmaceutical and/or psychotherapeutic treatments. When they initiate new treatment with brain stimulation, it is possible that expectations may lead to placebo effect that can manifest a therapeutic benefit, even in the absence of real neuromodulatory improvements. Despite this very real concern, many of the prior studies in this area have not used a sham-controlled condition to isolate the effects of active brain stimulation. This gap likely constitutes a meaningful bias that hiders strong conclusions from the extant literature. Thus it is critical that future clinical trials in this important area of research utilize believable sham-controlled conditions.

Another area of concern relates to the lack of information currently available about the long-term effects of brain stimulation among patients with PTSD. Thus, while many of the reviewed studies suggest a positive effect of brain stimulation on acute behavioral clinical outcomes, very little is currently known about the stability and long-term effects of brain stimulation within this clinical population. Whereas some of the studies in this area suggest that behavioral improvements may be sustained for months after treatment (e.g., Boggio et al., 2010), others have shown that symptoms relapse (e.g., Watts, Landon, Groft, & Young-Xu, 2012). Maintenance treatment has also been largely ignored in the literature. Should brain stimulation be applied regularly to prevent relapse? Should the maintenance treatment be applied over several weeks, as it is done during the acute phase of the treatment? If so, what is the optimal schedule? The answers to such questions are unknown at the present time but should be carefully considered and addressed in the design of future trials in this area.

Beyond these design aspects the majority of studies described earlier have relied on scalp measurement to define coil location for rTMS studies and the electrodes positions for transcranial electric stimulation studies. While this method is cheap and easy to implement, it is less precise and has been shown to miss the desired target (Herwig et al., 2001, 2003). We suggest that more studies rely on the available technologies (e.g., neuronavigation systems) to ensure accurate coil/electrode positioning to improve the efficacy of brain stimulation. Moreover, with the growth of MRI, fMRI, and connectivity analyses, it is strongly recommended that rTMS be personalized to guide targeting and aid in the choice of the stimulation parameters whenever possible.

Another major limiting factor when using noninvasive brain stimulation techniques is the shallow-depth penetration (~2 cm) that prevents direct stimulation of subcortical structures (e.g., the amygdala) that are believed to be critical for the treatment of PTSD. H-coils have been proposed as a way to stimulate deeper brain areas (~4 cm); however, while appealing as a solution, studies implementing electric field modeling have shown that

H-coils suffer from a lack of focality compared with figure-of-8 coils (Deng et al., 2013). Consequently, this depth-focality trade-off prevents focal stimulation of deeper structures, meaning that deeper stimulation comes at a cost of more broad activation, which could potentially lead to unintended consequences.

More recently, "connectivity-based rTMS" has been proposed as a method to indirectly stimulate deeper brain areas (Addicott et al., 2019; Wang et al., 2014). Indeed, as mentioned previously, while rTMS induces focal effects underneath the TMS coil, remote effects are also observed downstream through transynaptic propagation of the field (Bestmann, Baudewig, Siebner, Rothwell, & Frahm, 2004). Instead of ignoring these effects, connectivity-based rTMS uses them systematically to indirectly modulate deep brain structures by stimulating superficial sites that are functionally connected to deeper targets that are the aim of treatment. This approach has been tested in several studies and has been shown to induce indirect modulation of deeper structures such as the hippocampus (Wang et al., 2014) and the insula (Addicott et al., 2019; Li et al., 2017). Notably, findings from these studies suggest that the expected frequency-dependent heuristic may not apply to indirect/deep stimulation. For example, by comparing the effects of 1-Hz and 10-Hz rTMS, Addicott et al. showed that, instead of leading to opposite effects, both frequencies increased resting-state connectivity between the insula and the targeted postcentral gyrus. While more studies are needed to better understand the effects of TMS on limbic structures, these results constitute a real opportunity to optimize rTMS efficacy for PTSD treatment by modulating amygdala activation without a surgical procedure.

Additionally, as noted earlier, PTSD is explained not only by abnormalities in separate brain structures but also by dysregulation in large-scale brain networks. Communication between, and within, brain networks is modulated by oscillations of brain electrical activity. It has been recently demonstrated that rTMS and transcranial alternating current stimulation can entrain rhythmic endogenous brain oscillations and that rTMS is able to "reset" thalamocortical oscillators, normalize regulation, and facilitate the reemergence of intrinsic cerebral rhythms (Leuchter, Cook, Jin, & Phillips, 2013). Thus another way to potentially optimize rTMS efficacy could be to use electroencephalography recordings to define subject-specific brain rhythms and then apply rTMS at the same frequency, in a closed-loop manner. This approach has already been explored to treat schizophrenia and depression (Arns, Spronk, & Fitzgerald, 2010; Jin et al., 2012) with some promising results; however, this literature is still quite limited, and more studies are needed in this important area of inquiry. Nonetheless, applying personalized stimulation parameters has the potential be another effective means of optimizing rTMS treatments.

References

Addicott, M., Luber, B., Nguyen, D., Palmer, H., Lisanby, S., & Appelbaum, L. (2019). Low and high frequency rTMS effects on resting-state functional connectivity between the postcentral gyrus and the insula. *Brain Connectivity*, *9*(4), 322–328.

Ahmadi, N., Moss, L., Simon, E., Nemeroff, C. B., & Atre-Vaidya, N. (2016). Efficacy and long-term clinical outcome of comorbid posttraumatic stress disorder and major depressive disorder after electroconvulsive therapy. *Depression and Anxiety*, *33*(7), 640–647.

Arns, M., Spronk, D., & Fitzgerald, P. B. (2010). Potential differential effects of 9 Hz rTMS and 10 Hz rTMS in the treatment of depression. *Brain Stimulation: Basic, Translational, and Clinical Research in Neuromodulation*, *3*(2), 124–126.

Baeken, C., Vanderhasselt, M.-A., Remue, J., Herremans, S., Vanderbruggen, N., Zeeuws, D., … De Raedt, R. (2013). Intensive HF-rTMS treatment in refractory medication-resistant unipolar depressed patients. *Journal of Affective Disorders*, *151*(2), 625–631.

Beam, W., Borckardt, J. J., Reeves, S. T., & George, M. S. (2009). An efficient and accurate new method for locating the F3 position for prefrontal TMS applications. *Brain Stimulation*, *2*(1), 50–54.

Berlim, M. T., & Van den Eynde, F. (2014). Repetitive transcranial magnetic stimulation over the dorsolateral prefrontal cortex for treating posttraumatic stress disorder: An exploratory meta-analysis of randomized, double-blind and sham-controlled trials. *The Canadian Journal of Psychiatry*, *59*(9), 487–496.

Bestmann, S., Baudewig, J., Siebner, H. R., Rothwell, J. C., & Frahm, J. (2004). Functional MRI of the immediate impact of transcranial magnetic stimulation on cortical and subcortical motor circuits. *European Journal of Neuroscience*, *19*(7), 1950–1962.

Birn, R. M., Patriat, R., Phillips, M. L., Germain, A., & Herringa, R. J. (2014). Childhood maltreatment and combat posttraumatic stress differentially predict fear-related frontosubcortical connectivity. *Depression and Anxiety*, *31*(10), 880–892.

Boggio, P. S., Rocha, M., Oliveira, M. O., Fecteau, S., Cohen, R. B., Campanhã, C., … Zaghi, S. (2010). Noninvasive brain stimulation with high-frequency and low-intensity repetitive transcranial magnetic stimulation treatment for posttraumatic stress disorder. *The Journal of Clinical Psychiatry*, *71*(8), 992.

Bremner, J. D., Krystal, J. H., Southwick, S. M., & Charney, D. S. (1995). Functional neuroanatomical correlates of the effects of stress on memory. *Journal of Traumatic Stress*, *8*(4), 527–553.

Bremner, J. D., Staib, L. H., Kaloupek, D., Southwick, S. M., Soufer, R., & Charney, D. S. (1999). Neural correlates of exposure to traumatic pictures and sound in Vietnam combat veterans with and without posttraumatic stress disorder: A positron emission tomography study. *Biological Psychiatry*, *45*(7), 806–816.

Britton, J. C., Phan, K. L., Taylor, S. F., Fig, L. M., & Liberzon, I. (2005). Corticolimbic blood flow in posttraumatic stress disorder during script-driven imagery. *Biological Psychiatry*, *57*(8), 832–840.

Carmichael, S. T., & Price, J. L. (1995). Limbic connections of the orbital and medial prefrontal cortex in macaque monkeys. *Journal of Comparative Neurology*, *363*(4), 615–641.

Cheeran, B., Talelli, P., Mori, F., Koch, G., Suppa, A., Edwards, M., … Rothwell, J. C. (2008). A common polymorphism in the brain-derived neurotrophic factor gene (BDNF) modulates human cortical plasticity and the response to rTMS. *The Journal of Physiology*, *586*(23), 5717–5725.

Cisler, J. M., Steele, J. S., Smitherman, S., Lenow, J. K., & Kilts, C. D. (2013). Neural processing correlates of assaultive violence exposure and PTSD symptoms during implicit threat processing: A network-level analysis among adolescent girls. *Psychiatry Research: Neuroimaging, 214*(3), 238–246.

Clark, C., Cole, J., Winter, C., Williams, K., & Grammer, G. (2015). A review of transcranial magnetic stimulation as a treatment for post-traumatic stress disorder. *Current Psychiatry Reports, 17*(10), 83.

Cohen, H., Kaplan, Z., Kotler, M., Kouperman, I., Moisa, R., & Grisaru, N. (2004). Repetitive transcranial magnetic stimulation of the right dorsolateral prefrontal cortex in post-traumatic stress disorder: A double-blind, placebo-controlled study. *American Journal of Psychiatry, 161*(3), 515–524.

Daniels, J. K., McFarlane, A. C., Bluhm, R. L., Moores, K. A., Clark, C. R., Shaw, M. E., … Lanius, R. A. (2010). Switching between executive and default mode networks in posttraumatic stress disorder: Alterations in functional connectivity. *Journal of Psychiatry & Neuroscience, 35*(4), 258.

Deng, Z.-D., Lisanby, S. H., & Peterchev, A. V. (2013). Electric field depth–focality tradeoff in transcranial magnetic stimulation: Simulation comparison of 50 coil designs. *Brain Stimulation, 6*(1), 1–13.

Dosenbach, N. U. F., Fair, D. A., Miezin, F. M., Cohen, A. L., Wenger, K. K., Dosenbach, R. A. T., … Raichle, M. E. (2007). Distinct brain networks for adaptive and stable task control in humans. *Proceedings of the National Academy of Sciences, 104*(26), 11073–11078.

Duprat, R., Desmyter, S., van Heeringen, K., Van den Abbeele, D., Tandt, H., Bakic, J., … Van Autreve, S. (2016). Accelerated intermittent theta burst stimulation treatment in medication-resistant major depression: A fast road to remission? *Journal of Affective Disorders, 200*, 6–14.

Elger, G., Hoppe, C., Falkai, P., Rush, A. J., & Elger, C. E. (2000). Vagus nerve stimulation is associated with mood improvements in epilepsy patients. *Epilepsy Research, 42*(2–3), 203–210.

Etkin, A., & Wager, T. D. (2007). Functional neuroimaging of anxiety: A meta-analysis of emotional processing in PTSD, social anxiety disorder, and specific phobia. *American Journal of Psychiatry, 164*(10), 1476–1488.

Fenoy, A. J., & Simpson, R. K. (2014). Risks of common complications in deep brain stimulation surgery: Management and avoidance. *Journal of Neurosurgery, 120*(1), 132–139.

Fitzgerald, P. B., Fountain, S., & Daskalakis, Z. J. (2006). A comprehensive review of the effects of rTMS on motor cortical excitability and inhibition. *Clinical Neurophysiology, 117*(12), 2584–2596.

Fox, M. D., Buckner, R. L., White, M. P., Greicius, M. D., & Pascual-Leone, A. (2012). Efficacy of transcranial magnetic stimulation targets for depression is related to intrinsic functional connectivity with the subgenual cingulate. *Biological Psychiatry, 72*(7), 595–603.

George, M. S., Sackeim, H. A., Rush, A. J., Marangell, L. B., Nahas, Z., Husain, M. M., … Ballenger, J. C. (2000). Vagus nerve stimulation: A new tool for brain research and therapy. *Biological Psychiatry, 47*(4), 287–295.

George, M. S., Ward, H. E., Jr., Ninan, P. T., Pollack, M., Nahas, Z., Anderson, B., … Ballenger, J. C. (2008). A pilot study of vagus nerve stimulation (VNS) for treatment-resistant anxiety disorders. *Brain Stimulation, 1*(2), 112–121.

Grisaru, N., Amir, M., Cohen, H., & Kaplan, Z. (1998). Effect of transcranial magnetic stimulation in posttraumatic stress disorder: a preliminary study. *Biological Psychiatry, 44*(1), 52–55.

Hayes, J. P., Hayes, S. M., & Mikedis, A. M. (2012). Quantitative meta-analysis of neural activity in posttraumatic stress disorder. *Biology of Mood & Anxiety Disorders, 2*(1), 9.

Herwig, U., Padberg, F., Unger, J., Spitzer, M., & Schönfeldt-Lecuona, C. (2001). Transcranial magnetic stimulation in therapy studies: Examination of the reliability of "standard" coil positioning by neuronavigation. *Biological Psychiatry, 50*(1), 58–61.

Herwig, U., Satrapi, P., & Schönfeldt-Lecuona, C. (2003). Using the international 10-20 EEG system for positioning of transcranial magnetic stimulation. *Brain Topography*, *16*(2), 95–99.

Howland, R. H. (2014). Vagus nerve stimulation. *Current Behavioral Neuroscience Reports*, *1*(2), 64–73.

Isserles, M., Shalev, A. Y., Roth, Y., Peri, T., Kutz, I., Zlotnick, E., & Zangen, A. (2013). Effectiveness of deep transcranial magnetic stimulation combined with a brief exposure procedure in post-traumatic stress disorder—A pilot study. *Brain Stimulation*, *6*(3), 377–383.

Jin, Y., Kemp, A. S., Huang, Y., Thai, T. M., Liu, Z., Xu, W., … Potkin, S. G. (2012). Alpha EEG guided TMS in schizophrenia. *Brain Stimulation*, *5*(4), 560–568.

Karsen, E. F., Watts, B. V., & Holtzheimer, P. E. (2014). Review of the effectiveness of transcranial magnetic stimulation for post-traumatic stress disorder. *Brain Stimulation*, *7*(2), 151–157.

Kim, S. J., Jeong, D.-U., Sim, M. E., Bae, S. C., Chung, A., Kim, M. J., … Lyoo, I. K. (2006). Asymmetrically altered integrity of cingulum bundle in posttraumatic stress disorder. *Neuropsychobiology*, *54*(2), 120–125.

Kitayama, N., Vaccarino, V., Kutner, M., Weiss, P., & Bremner, J. D. (2005). Magnetic resonance imaging (MRI) measurement of hippocampal volume in posttraumatic stress disorder: A meta-analysis. *Journal of Affective Disorders*, *88*(1), 79–86.

Koch, S. B. J., van Zuiden, M., Nawijn, L., Frijling, J. L., Veltman, D. J., & Olff, M. (2016). Aberrant resting-state brain activity in posttraumatic stress disorder: A meta-analysis and systematic review. *Depression and Anxiety*, *33*(7), 592–605.

Koek, R. J., Langevin, J.-P., Krahl, S. E., Kosoyan, H. J., Schwartz, H. N., Chen, J. W. Y., … Sultzer, D. (2014). Deep brain stimulation of the basolateral amygdala for treatment-refractory combat post-traumatic stress disorder (PTSD): Study protocol for a pilot randomized controlled trial with blinded, staggered onset of stimulation. *Trials*, *15*(1), 356.

Koenigs, M., Huey, E. D., Raymont, V., Cheon, B., Solomon, J., Wassermann, E. M., & Grafman, J. (2008). Focal brain damage protects against post-traumatic stress disorder in combat veterans. *Nature Neuroscience*, *11*(2), 232.

Kozel, F. A., Motes, M. A., Didehbani, N., DeLaRosa, B., Bass, C., Schraufnagel, C. D., … Kraut, M. A. (2018). Repetitive TMS to augment cognitive processing therapy in combat veterans of recent conflicts with PTSD: A randomized clinical trial. *Journal of Affective Disorders*, *229*, 506–514.

Kozel, F. A., Van Trees, K., Larson, V., Phillips, S., Hashimie, J., Gadbois, B., … Toyinbo, P. (2019). One hertz versus ten hertz repetitive TMS treatment of PTSD: A randomized clinical trial. *Psychiatry Research*, *273*, 153–162.

Langevin, J.-P., De Salles, A. A. F., Kosoyan, H. P., & Krahl, S. E. (2010). Deep brain stimulation of the amygdala alleviates post-traumatic stress disorder symptoms in a rat model. *Journal of Psychiatric Research*, *44*(16), 1241–1245.

Langevin, J.-P., Koek, R. J., Schwartz, H. N., Chen, J. W. Y., Sultzer, D. L., Mandelkern, M. A., … Krahl, S. E. (2016). Deep brain stimulation of the basolateral amygdala for treatment-refractory posttraumatic stress disorder. *Biological Psychiatry*, *79*(10), e82–e84.

Lanius, R. A., Frewen, P. A., Tursich, M., Jetly, R., & McKinnon, M. C. (2015). Restoring large-scale brain networks in PTSD and related disorders: A proposal for neuroscientifically-informed treatment interventions. *European Journal of Psychotraumatology*, *6*(1), 27313.

Lanius, R. A., Williamson, P. C., Boksman, K., Densmore, M., Gupta, M., Neufeld, R. W. J., … Menon, R. S. (2002). Brain activation during script-driven imagery induced dissociative responses in PTSD: A functional magnetic resonance imaging investigation. *Biological Psychiatry*, *52*(4), 305–311.

Lanius, R. A., Williamson, P. C., Hopper, J., Densmore, M., Boksman, K., Gupta, M. A., … Menon, R. S. (2003). Recall of emotional states in posttraumatic stress disorder: An fMRI investigation. *Biological Psychiatry*, *53*(3), 204–210.

Leuchter, A. F., Cook, I. A., Jin, Y., & Phillips, B. (2013). The relationship between brain oscillatory activity and therapeutic effectiveness of transcranial magnetic stimulation in the treatment of major depressive disorder. *Frontiers in Human Neuroscience*, *7*, 37.

Li, X., Du, L., Sahlem, G. L., Badran, B. W., Henderson, S., & George, M. S. (2017). Repetitive transcranial magnetic stimulation (rTMS) of the dorsolateral prefrontal cortex reduces resting-state insula activity and modulates functional connectivity of the orbitofrontal cortex in cigarette smokers. *Drug and Alcohol Dependence*, *174*, 98–105.

Liberzon, I., Taylor, S. F., Amdur, R., Jung, T. D., Chamberlain, K. R., Minoshima, S., … Fig, L. M. (1999). Brain activation in PTSD in response to trauma-related stimuli. *Biological Psychiatry*, *45*(7), 817–826.

Lisanby, S. H. (2002). Update on magnetic seizure therapy: A novel form of convulsive therapy. *The Journal of ECT*, *18*(4), 182–188.

Logue, M. W., van Rooij, S. J. H., Dennis, E. L., Davis, S. L., Hayes, J. P., Stevens, J. S., … Koch, S. B. J. (2018). Smaller hippocampal volume in posttraumatic stress disorder: A multisite ENIGMA-PGC study: Subcortical volumetry results from posttraumatic stress disorder consortia. *Biological Psychiatry*, *83*(3), 244–253.

Margoob, M. A., Ali, Z., & Andrade, C. (2010). Efficacy of ECT in chronic, severe, antidepressant-and CBT-refractory PTSD: An open, prospective study. *Brain Stimulation*, *3*(1), 28–35.

Marshall, R. D., Beebe, K. L., Oldham, M., & Zaninelli, R. (2001). Efficacy and safety of paroxetine treatment for chronic PTSD: A fixed-dose, placebo-controlled study. *American Journal of Psychiatry*, *158*(12), 1982–1988.

McCann, U. D., Kimbrell, T. A., Morgan, C. M., Anderson, T., Geraci, M., Benson, B. E., … Post, R. M. (1998). Repetitive transcranial magnetic stimulation for posttraumatic stress disorder. *Archives of General Psychiatry*, *55*(3), 276–279.

McClintock, S. M., Choi, J., Deng, Z.-D., Appelbaum, L. G., Krystal, A. D., & Lisanby, S. H. (2014). Multifactorial determinants of the neurocognitive effects of electroconvulsive therapy. *The Journal of ECT*, *30*(2), 165.

Menon, V. (2011). Large-scale brain networks and psychopathology: A unifying triple network model. *Trends in Cognitive Sciences*, *15*(10), 483–506.

Menon, V., & Uddin, L. Q. (2010). Saliency, switching, attention and control: A network model of insula function. *Brain Structure and Function*, *214*(5–6), 655–667.

Mikics, E., Baranyi, J., & Haller, J. (2008). Rats exposed to traumatic stress bury unfamiliar objects—A novel measure of hyper-vigilance in PTSD models? *Physiology & Behavior*, *94*(3), 341–348.

Monte-Silva, K., Kuo, M.-F., Hessenthaler, S., Fresnoza, S., Liebetanz, D., Paulus, W., & Nitsche, M. A. (2013). Induction of late LTP-like plasticity in the human motor cortex by repeated non-invasive brain stimulation. *Brain Stimulation*, *6*(3), 424–432.

Montgomery, E. B., Jr., & Gale, J. T. (2008). Mechanisms of action of deep brain stimulation (DBS). *Neuroscience & Biobehavioral Reviews*, *32*(3), 388–407.

Nam, D.-H., Pae, C.-U., & Chae, J.-H. (2013). Low-frequency, repetitive transcranial magnetic stimulation for the treatment of patients with posttraumatic stress disorder: a double-blind, sham-controlled study. *Clinical Psychopharmacology and Neuroscience*, *11*(2), 96.

Negreira, A. M., & Abdallah, C. G. (2019). A review of fMRI affective processing paradigms used in the neurobiological study of posttraumatic stress disorder. *Chronic Stress*, *3*, 2470547019829035.

Nitsche, M. A., & Paulus, W. (2000). Excitability changes induced in the human motor cortex by weak transcranial direct current stimulation. *The Journal of Physiology*, *527*(3), 633–639.

Nitsche, M. A., & Paulus, W. (2001). Sustained excitability elevations induced by transcranial DC motor cortex stimulation in humans. *Neurology*, *57*(10), 1899–1901.

Noble, L. J., Gonzalez, I. J., Meruva, V. B., Callahan, K. A., Belfort, B. D., Ramanathan, K. R., ... McIntyre, C. K. (2017). Effects of vagus nerve stimulation on extinction of conditioned fear and post-traumatic stress disorder symptoms in rats. *Translational Psychiatry, 7*(8), e1217.

Ohl, F., Michaelis, T., Vollmann-Honsdorf, G. K., Kirschbaum, C., & Fuchs, E. (2000). Effect of chronic psychosocial stress and long-term cortisol treatment on hippocampus-mediated memory and hippocampal volume: A pilot-study in tree shrews. *Psychoneuroendocrinology, 25*(4), 357–363.

Osuch, E. A., Benson, B., Geraci, M., Podell, D., Herscovitch, P., McCann, U. D., & Post, R. M. (2001). Regional cerebral blood flow correlated with flashback intensity in patients with posttraumatic stress disorder. *Biological Psychiatry, 50*(4), 246–253.

Osuch, E. A., Benson, B. E., Luckenbaugh, D. A., Geraci, M., Post, R. M., & McCann, U. (2009). Repetitive TMS combined with exposure therapy for PTSD: A preliminary study. *Journal of Anxiety Disorders, 23*(1), 54–59.

Oznur, T., Akarsu, S., Celik, C., Bolu, A., Ozdemir, B., Akcay, B. D., ... Ozmenler, K. N. (2014). Is transcranial magnetic stimulation effective in treatment-resistant combat related posttraumatic stress disorder. *Neurosciences (Riyadh), 19*(1), 29–32.

Patel, R., Spreng, R. N., Shin, L. M., & Girard, T. A. (2012). Neurocircuitry models of posttraumatic stress disorder and beyond: A meta-analysis of functional neuroimaging studies. *Neuroscience & Biobehavioral Reviews, 36*(9), 2130–2142.

Pavliša, G., Papa, J., Pavić, L., & Pavliša, G. (2006). Bilateral MR volumetry of the amygdala in chronic PTSD patients. *Collegium Antropologicum, 30*(3), 565–568.

Peña, D. F., Childs, J. E., Willett, S., Vital, A., McIntyre, C. K., & Kroener, S. (2014). Vagus nerve stimulation enhances extinction of conditioned fear and modulates plasticity in the pathway from the ventromedial prefrontal cortex to the amygdala. *Frontiers in Behavioral Neuroscience, 8*, 327.

Peña, D. F., Engineer, N. D., & McIntyre, C. K. (2013). Rapid remission of conditioned fear expression with extinction training paired with vagus nerve stimulation. *Biological Psychiatry, 73*(11), 1071–1077.

Philip, N. S., Barredo, J., van't Wout-Frank, M., Tyrka, A. R., Price, L. H., & Carpenter, L. L. (2018). Network mechanisms of clinical response to transcranial magnetic stimulation in posttraumatic stress disorder and major depressive disorder. *Biological Psychiatry, 83*(3), 263–272.

Rauch, S. L., Shin, L. M., & Phelps, E. A. (2006). Neurocircuitry models of posttraumatic stress disorder and extinction: Human neuroimaging research—Past, present, and future. *Biological Psychiatry, 60*(4), 376–382.

Rauch, S. L., Shin, L. M., Segal, E., Pitman, R. K., Carson, M. A., McMullin, K., ... Makris, N. (2003). Selectively reduced regional cortical volumes in post-traumatic stress disorder. *Neuroreport, 14*(7), 913–916.

Rauch, S. L., Whalen, P. J., Shin, L. M., McInerney, S. C., Macklin, M. L., Lasko, N. B., ... Pitman, R. K. (2000). Exaggerated amygdala response to masked facial stimuli in posttraumatic stress disorder: A functional MRI study. *Biological Psychiatry, 47*(9), 769–776.

Rosenberg, P. B., Mehndiratta, R. B., Mehndiratta, Y. P., Wamer, A., Rosse, R. B., & Balish, M. (2002). Repetitive transcranial magnetic stimulation treatment of comorbid posttraumatic stress disorder and major depression. *The Journal of Neuropsychiatry and Clinical Neurosciences, 14*(3), 270–276.

Rossi, S., Cappa, S. F., Babiloni, C., Pasqualetti, P., Miniussi, C., Carducci, F., ... Rossini, P. M. (2001). Prefontal cortex in long-term memory: An "interference" approach using magnetic stimulation. *Nature Neuroscience, 4*(9), 948.

Rossi, S., Cappa, S. F., Ulivelli, M., De Capua, A., Bartalini, S., & Rossini, P. M. (2006). rTMS for PTSD: Induced merciful oblivion or elimination of abnormal hypermnesia? *Behavioural Neurology, 17*(3–4), 195–199.

Sack, A. T., Cohen Kadosh, R., Schuhmann, T., Moerel, M., Walsh, V., & Goebel, R. (2009). Optimizing functional accuracy of TMS in cognitive studies: A comparison of methods. *Journal of Cognitive Neuroscience*, *21*(2), 207–221.

Saunders, N., Downham, R., Turman, B., Kropotov, J., Clark, R., Yumash, R., & Szatmary, A. (2015). Working memory training with tDCS improves behavioral and neurophysiological symptoms in pilot group with post-traumatic stress disorder (PTSD) and with poor working memory. *Neurocase*, *21*(3), 271–278.

Semkovska, M., & McLoughlin, D. M. (2010). Objective cognitive performance associated with electroconvulsive therapy for depression: A systematic review and meta-analysis. *Biological Psychiatry*, *68*(6), 568–577.

Shin, L. M., McNally, R. J., Kosslyn, S. M., Thompson, W. L., Rauch, S. L., Alpert, N. M., ... Pitman, R. K. (1997). A positron emission tomographic study of symptom provocation in PTSD. *Annals of the New York Academy of Sciences*, *821*(1), 521–523.

Shin, L. M., Orr, S. P., Carson, M. A., Rauch, S. L., Macklin, M. L., Lasko, N. B., ... Cannistraro, P. A. (2004). Regional cerebral blood flow in the amygdala and medial prefrontalcortex during traumatic imagery in male and female Vietnam veterans with ptsd. *Archives of General Psychiatry*, *61*(2), 168–176.

Shin, L. M., Whalen, P. J., Pitman, R. K., Bush, G., Macklin, M. L., Lasko, N. B., ... Rauch, S. L. (2001). An fMRI study of anterior cingulate function in posttraumatic stress disorder. *Biological Psychiatry*, *50*(12), 932–942.

Silvanto, J., & Pascual-Leone, A. (2008). State-dependency of transcranial magnetic stimulation. *Brain Topography*, *21*(1), 1.

Simmons, A. N., Norman, S. B., Spadoni, A. D., & Strigo, I. A. (2013). Neurosubstrates of remission following prolonged exposure therapy in veterans with posttraumatic stress disorder. *Psychotherapy and Psychosomatics*, *82*(6), 382–389.

Singh, A., & Kar, S. K. (2017). How electroconvulsive therapy works?: Understanding the neurobiological mechanisms. *Clinical Psychopharmacology and Neuroscience*, *15*(3), 210.

Speer, A. M., Kimbrell, T. A., Wassermann, E. M., Repella, J. D., Willis, M. W., Herscovitch, P., & Post, R. M. (2000). Opposite effects of high and low frequency rTMS on regional brain activity in depressed patients. *Biological Psychiatry*, *48*(12), 1133–1141.

Stidd, D. A., Vogelsang, K., Krahl, S. E., Langevin, J.-P., & Fellous, J.-M. (2013). Amygdala deep brain stimulation is superior to paroxetine treatment in a rat model of posttraumatic stress disorder. *Brain Stimulation*, *6*(6), 837–844.

Tatomir, A., Micu, C., & Crivii, C. (2014). The impact of stress and glucocorticoids on memory. *Clujul Medical*, *87*(1), 3.

Tavakoli, A. V., & Yun, K. (2017). Transcranial alternating current stimulation (tACS) mechanisms and protocols. *Frontiers in Cellular Neuroscience*, *11*, 214.

Terney, D., Chaieb, L., Moliadze, V., Antal, A., & Paulus, W. (2008). Increasing human brain excitability by transcranial high-frequency random noise stimulation. *Journal of Neuroscience*, *28*(52), 14147–14155.

Trevizol, A. P., Barros, M. D., Silva, P. O., Osuch, E., Cordeiro, Q., & Shiozawa, P. (2016). Transcranial magnetic stimulation for posttraumatic stress disorder: An updated systematic review and meta-analysis. *Trends in Psychiatry and Psychotherapy*, *38*(1), 50–55.

Tulving, E., & Markowitsch, H. J. (1998). Episodic and declarative memory: Role of the hippocampus. *Hippocampus*, *8*(3), 198–204.

van't Wout, M., Longo, S. M., Reddy, M. K., Philip, N. S., Bowker, M. T., & Greenberg, B. D. (2017). Transcranial direct current stimulation may modulate extinction memory in posttraumatic stress disorder. *Brain and Behavior*, *7*(5), e00681.

van't Wout-Frank, M., Shea, M. T., Larson, V. C., Greenberg, B. D., & Philip, N. S. (2019). Combined transcranial direct current stimulation with virtual reality exposure for posttraumatic stress disorder: Feasibility and pilot results. *Brain Stimulation*, *12*(1), 41–43.

Wang, J. X., Rogers, L. M., Gross, E. Z., Ryals, A. J., Dokucu, M. E., Brandstatt, K. L., ... Voss, J. L. (2014). Targeted enhancement of cortical-hippocampal brain networks and associative memory. *Science, 345*(6200), 1054–1057.

Watts, B. V., Landon, B., Groft, A., & Young-Xu, Y. (2012). A sham controlled study of repetitive transcranial magnetic stimulation for posttraumatic stress disorder. *Brain Stimulation, 5*(1), 38–43.

Weiner, R. D., & American Psychiatric Association. (2001). *The practice of electroconvulsive therapy: Recommendations for treatment, training, and privileging: a task force report of the American Psychiatric Association.* American Psychiatric Association Publishing Inc.

Weiner, R. D., Rogers, H. J., Davidson, J. R. T., & Squire, L. R. (1986). Effects of stimulus parameters on cognitive side effects. *Annals of the New York Academy of Sciences, 462*(1), 315–325.

Woon, F. L., & Hedges, D. W. (2009). Amygdala volume in adults with posttraumatic stress disorder: A meta-analysis. *The Journal of Neuropsychiatry and Clinical Neurosciences, 21*(1), 5–12.

Yamasue, H., Kasai, K., Iwanami, A., Ohtani, T., Yamada, H., Abe, O., ... Furukawa, S. (2003). Voxel-based analysis of MRI reveals anterior cingulate gray-matter volume reduction in posttraumatic stress disorder due to terrorism. *Proceedings of the National Academy of Sciences, 100*(15), 9039–9043.

Yan, T., Xie, Q., Zheng, Z., Zou, K., & Wang, L. (2017). Different frequency repetitive transcranial magnetic stimulation (rTMS) for posttraumatic stress disorder (PTSD): A systematic review and meta-analysis. *Journal of Psychiatric Research, 89*, 125–135.

Youssef, N. A., McCall, W. V., & Andrade, C. (2017). The role of ECT in posttraumatic stress disorder: A systematic review. *Annals of Clinical Psychiatry, 29*(1), 62–70.

CHAPTER 8

Genetic influences on PTSD

Kaitlin E. Bountress[a], Leslie A. Brick[b,c], Shannon Cusack[a,d], Christina M. Sheerin[a], Nicole R. Nugent[b,e], Ananda B. Amstadter[a]

[a]Virginia Institute for Psychiatry and Behavioral Genetics, Virginia Commonwealth University, Richmond, VA, United States
[b]Department of Psychiatry and Human Behavior, Warren Alpert Medical School of Brown University, Providence, RI, United States
[c]Department of Neurology, Warren Alpert Medical School of Brown University, Providence, RI, United States
[d]Department of Psychology, Virginia Commonwealth University, Richmond, VA, United States
[e]Department of Pediatrics, Warren Alpert Medical School of Brown University, Providence, RI, United States

Traumatic events are common, with 50%–60% of the US adult population exposed to at least one potentially traumatic event during their lifetime (Kessler, Sonnega, Bromet, Hughes, & Nelson, 1995). Despite the high prevalence of trauma, the likelihood of subsequent posttraumatic stress disorder (PTSD) is relatively low, ranging from 7% to 30%, depending on the population and trauma type (Kessler et al., 1995). This considerable discrepancy highlights the need to examine individual differences in responses to trauma, with evidence from extant literature suggesting that genetic influences may play a key role in this etiologic pathway (Sartor et al., 2011; True et al., 1993).

Behavioral genetics of PTSD

Early genetic examinations of PTSD were focused on familial influences, suggesting that PTSD does indeed "run in families" (Lambert, Holzer, & Hasbun, 2014; Leen-Feldner et al., 2013). However, these studies often confound shared genetic and environmental risk factors. Twin studies have since been used in the genetic examination of PTSD to extricate the shared environmental influences from heritable influences. Twin studies to date have estimated that the heritability of PTSD ranges from 30% to 72% (see review; Afifi, Asmundson, Taylor, & Jang, 2010), after controlling for genetic influences on trauma exposure itself (True et al., 1993). Notably, twin studies have also demonstrated that exposure to traumatic events is influenced by genetic factors (i.e., gene-environmental correlation, rGE), with heritability estimates ranging from 20% to 47% depending on the type of trauma and the sample characteristics. Summarizing across the extant literature hints at sex differences, wherein all female samples yield a higher heritability estimate (~72%) as compared with all male samples (~30%) or combined samples (Lyons et al., 1993; Sartor et al., 2011; Stein, Jang, & Livesley, 2002); however, existing studies have lacked the power necessary to formally test for quantitative (i.e., differences in magnitude) or qualitative (i.e., differences in the genetic source) sex effects in the same sample.

Further, twin studies have demonstrated that genetic influences on PTSD are shared with those for other phenotypes. For example, the genetic influences on PTSD have been shown to substantially overlap with those of both major depression (Sartor et al., 2012) and alcohol dependence (Sartor et al., 2011). More recent work has demonstrated that the genetic influences on PTSD also overlap with those of protective factors (i.e., resilience). For example, 59% of the phenotypic correlation between PTSD and resilience (i.e., $r = -.59$) was attributable to a shared genetic factor (Wolf, Miller, et al., 2018). These findings suggest that perhaps genetic examinations of PTSD should focus on a broader spectrum of traumatic stress (Wolf, Miller, et al., 2018). Although extant research has established presence of genetic influences on PTSD, molecular genetic research is needed to identify specific genetic variations associated with the disorder.

Molecular genetic studies

Molecular approaches attempt to identify the specific underlying biological mechanisms that influence the development and maintenance of PTSD. Molecular studies have utilized hypothesis-driven approaches, including candidate gene and candidate gene by environment (cGxE) studies and agnostic approaches (i.e., genome-wide association studies, GWAS).

Candidate gene studies

As is true across the field of human genetics, early research into genetic influences on PTSD focused predominantly on candidate approaches, involving genetic variants selected based on existing theories and findings related to the biological predictors and concomitants of PTSD. Emerging from extant theories involving the role of the stress system in PTSD, many studies have focused on genetic variation in the hypothalamic-pituitary-adrenal (HPA) axis, and the locus coeruleus-noradrenergic system. Although over 100 candidate gene studies have been conducted to date (for a review, Duncan, Cooper, & Shen, 2018; Sheerin, Lind, Bountress, Nugent, & Amstadter, 2017), candidate approaches have faced replication challenges. Lack of replication may be due to numerous factors, such as failure to consider population stratification, inconsistent coverage of genes of interest, and relatively small effects that may be difficult to detect in smaller samples. Metaanalyses have attempted to address sample size limitations, with significant effects for variants in *DRD2* and *SLC6A3* but no overall effect for variation in *COMT* (Li et al., 2016) and significant effects of apolipoprotein E (*APOE*) on combat-related PTSD (Roby, 2017). Metaanalyses of serotonin revealed no main effect for *5-HTTLPR* (Gressier et al., 2013; Navarro-Mateu, Escamez, Koenen, Alonso, & Sanchez-Meca, 2013; Zhao et al., 2017). Metaanalyses of genetic variation in neurotrophins have drawn inconsistent conclusions. Specifically, some have found a significant effect of *BDNF* Val66Met (Bruenig et al., 2016), while others did not (Bountress et al., 2017; Wang, 2015). One potential explanation for inconsistencies observed even in the metaanalysis literature is the possibility that the effects of candidate markers may differ as a function of either untested or underpowered tests of factors such as biological sex or environmental exposures. A metaanalysis of variation in pituitary adenylate cyclase activating polypeptide, *ADCYAP1R1*, reported significant findings in females but not males (Lind et al., 2017).

Candidate gene by environment (cGxE) studies

Given the necessary, but not sufficient, role of trauma exposure in the etiology of PTSD, a natural extension of the candidate gene design adopted by traumatic stress researchers has been cGxE studies. In spite of challenges related to adequate power in tests of interactions (Manuck & McCaffery, 2014; Thomas, 2010), cGxE studies have reported significant interactions between childhood abuse and *FKBP5* variation (Binder et al., 2008; Xie et al., 2010) and between variants in *APOE* and combat exposure (Kimbrel et al., 2015; Lyons et al., 2013). An interaction of social environment (e.g., stressful life events and county-level SES) and serotoninergic variation has also been observed in PTSD cGxE studies (Drevo et al., 2016; Koenen et al., 2009; Zhao et al., 2017). In addition to the analytic challenges related to power to detect an interaction effect, there are concerns about vulnerability to statistical artifacts in cases wherein a dichotomous environment variable (such as yes/no trauma) is used in cGxE research (Eaves, 2006). Additionally, a minority of cGxE findings have been able to be replicated, suggesting that many of the interaction effects detected in recent years may actually be false positives (Duncan & Keller, 2011).

Genome-wide association studies (GWAS)

In contrast to candidate approaches, which are based on extant theory and presumed knowledge of the biology of PTSD, GWAS use an agnostic hypothesis-generating approach. GWAS compare frequencies of millions of common genetic variants dispersed across the genome, correcting for multiple testing of main effects of multiple markers and requiring large samples for adequate power to detect effects. Several GWAS of PTSD have been published (Almli et al., 2015; Ashley-Koch et al., 2015; Guffanti et al., 2013; Kilaru et al., 2016; Logue et al., 2013; Nievergelt et al., 2015; Powers et al., 2016; Stein et al., 2016; Wolf et al., 2014; Xie et al., 2013), with some suggestive evidence of replication in independent cohorts. However, no genetic variations have been independently replicated across GWAS, possibly due to heterogeneity in trauma type, ancestry, sex, and even strategies for phenotyping PTSD. GWAS studies have identified novel variants such as *RORA* (Logue et al., 2013), replicated in a candidate sample (Amstadter et al., 2013), which have permitted new ideas about biological influences on PTSD. Research emerging from this finding has supported the influence of *RORA* on neuroprotection after trauma and distress (Miller, Wolf, Logue, & Baldwin, 2013). Additional information on GWAS of PTSD is summarized in Table 1.

To address the large samples needed for adequately powered GWAS (e.g., estimated 75,000–100,000 cases in depression; Levinson et al., 2014), the Psychiatric Genomic Consortium (PGC) was formed in 2007 and has become the largest psychiatric collaboration to date, including 10 workgroups (e.g., schizophrenia, bipolar disorder, and depression) and over 800 investigators. The PGC-PTSD working group was officially established in 2013 to assemble existing GWAS for the first PGC-PTSD analysis and obtain funding for expanded data collection (Logue, Amstadter, et al., 2015; Logue, Smith, et al., 2015). The first PGC-PTSD data freeze included 11 multiethnic studies with 5000 PTSD cases and 15,000 controls (Duncan, Ratanatharathorn, et al., 2017). Findings supported significant genetic influences on PTSD, with average genetic heritability estimates (h_{SNP}^2) of ~15% in the European subset and with higher heritability observed in females (h_{SNP}^2 of 29%). Furthermore, previous evidence of overlap of polygenic risk scores between PTSD and other psychiatric disorders (e.g., bipolar disorder; Nievergelt et al., 2015) was replicated. The sample size for the second

PGC-PTSD data freeze has increased dramatically, and results of this analysis are forthcoming.

Novel statistical genetic procedures using GWAS data

Although GWAS studies have yielded promising findings, psychiatric phenotypes are complex and highly polygenic, containing small effects across several loci that contribute to genetic variation, calling for novel statistical methods to better understand the genetic architecture and to increase statistical power. Furthermore, given that many psychiatric and physical disorders share overlapping genetic risk, several methods have been developed to assess shared genetic effects between phenotypes (also commonly referred to as genetic correlations, genetic overlap, or coheritability, Duncan, Shen, et al., 2017; Yang, Zeng, Goddard, Wray, & Visscher, 2017). Genetic correlations reveal the extent to which the same genetic factors influence both traits, providing insight into comorbidity and etiology. Driving forces for these statistical and methodological developments include the public availability of genetic data for research, the utility of summary-level GWAS results from published work, and advances in statistical techniques and algorithms, including genotype imputation (Visscher et al., 2017).

Polygenic risk scores

The use of polygenic risk scores (PRSs) allows for the determination of whether additive genetic risk aggregated across numerous genomic loci are associated with a trait of interest. A PRS is typically defined as the sum of the number of alleles associated with a particular trait across several genetic loci (Euesden, Lewis, & O'Reilly, 2015). Often, this score is calculated using GWAS summary statistics from an independent sample (i.e., the discovery or training sample) and is weighted by the effect size of the variant (Lewis & Vassos, 2017). The PRS is interpreted as a single, continuous value indexing the genetic liability for a trait and, once calculated, can be used in subsequent analyses in another sample (i.e., the target or testing sample) to examine its association with a phenotype of interest. The size of the discovery dataset largely impacts the power of the PRS in the target sample (Wray et al., 2014). Key design decision points include whether to sum across all available SNPs or to sum only those that meet a significance threshold (e.g., genome-wide significance, $P < 10^{-8}$). Interestingly, as GWAS sample sizes have increased, the upper tails of the summary test distribution become enriched for true significance (International Schizophrenia Consortium et al.,

Table 1 Genome-/epigenome-wide findings across molecular, DNAm, and expression studies.

Citation	Sample size (% PTSD cases if case control)	Trauma and/or sample type (e.g, interpersonal and accidental)	Ancestry/ethnicity	Outcome/measure (e.g., clinical interview and self-report)	Method used, gene(s) examined, array used	Finding
GWAS						
Ashley-Koch et al. (2015)	1708	Various	Non-Hispanic Black (NHB), NHW	SCID and CAPS	Illumina HumanHap650 Beadchip or Illumina Human1M-Duo Beadchip or Illumina HumanOmni2.5 Beadchip	No genome-wide significance
Chen et al. (2017)	13,474	Mostly combat	EA, AA, Latinx	Abbreviated six item PTSD checklist	Imputed to the 1000 Genomes Project reference panel	EA sample showed one locus for reexperiencing severity at chr. 18, rs2311207; Latinx sample showed one locus for severity of hyperarousal chr10_6953246_D
Duncan et al. (2018)	20,730	PGC—various	EA, AA	Various	Imputed see PGC Pipeline	No genome-wide significant SNPs
Guffanti et al. (2013)	413 (23%) Replication $n=2541$ (23%)	Various	AA	Structured phone interview	Illumina HumanOmniExpress BeadChip	Rs10170218, mapped to the lincRNA AC068718.1 gene, significant at genome-wide level

Kilaru et al. (2016)	3678	Various	AA	TEI, modified PTSD Symptom Scale	Omni-Quad 1M or the Omni Express BeadChip	Significant association of two genes (*NLGN1* and *ZNRD1-AS1*) after multiple test correction in African American Cohort
Ashley-Koch et al. (2015)	1708 (41.70%)	Combat trauma	NHW, AA	Structured clinical interviews including CAPS; also DTS, TLEQ		No genome-wide significant SNPs
Logue et al. (2013)	491 (60.08%) Replication $n=84$ Replication $n=521$	Mostly combat	NHW, AA	Structured clinical interviews including CAPS; also LEC, CTQ	Illumina OMNI 2.5-8 array	Genome-wide significance for rs8042149 (*RORA* gene) in the NHW sample; 2 other *RORA* SNPs were nominally significant in AA samples
Melroy-Greif, Wilhelmsen, Yehuda, and Ehlers (2017)	619	Various	Mexican Americans, American Indians	17 items from Semi-Structured Assessment for the Genetics of Alcoholism	Affymetrix Exome1A chip & Illumina low-coverage whole-genome sequencing (WGS)	No genome-wide significant SNPs
Nievergelt et al. (2015)	940	Combat trauma	AIMs used: EA, AA, Hispanic and Native Americans, and others	Structured clinical interviews including CAPS; also CTQ, LEC	Illumina Human Omni Express Exome array	*PRTFDC1* (rs6482463) was genome-wide significant via metaanalysis and replicated in an independent sample

Continued

Table 1 Genome-/epigenome-wide findings across molecular, DNAm, and expression studies—Cont'd

Citation	Sample size (% PTSD cases if case control)	Trauma and/or sample type (e.g., interpersonal and accidental)	Ancestry/ethnicity	Outcome/measure (e.g., clinical interview and self-report)	Method used, gene(s) examined, array used	Finding
Stein et al. (2016)	774 new soldier study 5916 before/post deployment study	Mostly combat	EA, AA	Screening version of PTSD Checklist and Composite International Diagnostic Interview screen	Illumina OmniExpress + Exome or Ilumina PsychChip	ANKRD55 on chromosome 5 (rs159572; odds ratio [OR], 1.62; 95% CI, 1.37–1.92; $P=2.34\times10^{-8}$) and persisted after adjustment for cumulative trauma exposure (adjusted OR, 1.64; 95% CI, 1.39–1.95; $P=1.18\times10^{-8}$) in the African American samples from the NSS. A genome-wide significant locus was also found in or near ZNF626 on chromosome 19 (rs11085374; OR, 0.77; 95% CI, 0.70–0.85; $P=4.59\times10^{-8}$) in the European American samples from the NSS
Wolf et al. (2014)	484	Veterans and partners	NHW	CAPS	Illumina OMNI 2.5-8 array	No genome-wide significant SNPs
Xie et al. (2013)	EA 1578 (19%) AA 2766 (16%)	Various	NHW, AA	DSM-IV SSADDA	Illumina Omni1-Quad microarray	Genome-wide significance for rs406001 (chromosome 7p12)

Citation	Sample size (% PTSD cases if case control), trauma-exposed control group?	Trauma and/or sample type (e.g., interpersonal and accidental)	Covariates/measured confounds included	Outcome/measure (e.g., clinical interview and self-report)	Method used, gene(s) examined, array used	Finding
DNA methylation						
Hammamieh et al. (2017)	159 (combined training and test sets) 50% Yes	Combat veteran	Cell type and age	CAPS (DSM-IV) diagnosis	450k to examine DMGs	At $P < .05$, 3339 differentially methylated genes (DMGs) 74.4% of which encoded hypermethylated CpGs in cases
Kuan, Waszczuk, Kotov, Clouston, et al. (2017) and Kuan, Waszczuk, Kotov, Marsit, et al. (2017)	473[a] 36% Yes	Civilian; WTC first responders	Cell type, age, smoking status, and race (and also ran with ancestry PCs instead of race)	SCID (DSM-IV) diagnosis	450k	No significant CpG sites at FDR <.05, associated with current PTSD but 7 suggestive CpG sites at .0001. Significant pathways included oxytocin signaling, cholinergic synapse, and inflammatory disease

Continued

Table 1 Genome-/epigenome-wide findings across molecular, DNAm, and expression studies—Cont'd

Citation	Sample size (%PTSD cases if case control), trauma-exposed control group?	Trauma and/or sample type (e.g., interpersonal and accidental)	Covariates/measured confounds included	Outcome/measure (e.g., clinical interview and self-report)	Method used, gene(s) examined, array used	Finding
Martin et al. (2018)	14[a] 57%	Combat military personnel	None reported	PCL-M (DSM-IV)	DNA immunoprecipitation MeDIP-seq array	At an FDR of .10, found changes in 124 regions with 119 genes showing reduced methylation, 8 genes increased methylation associated with PTSD onset between time points (3 months later)
Mehta et al. (2017)	96 (discovery)[a] 50% Yes 115 (replication)[a] N/A Yes	Combat veteran (discovery) Civilian, varied (replication)	Surrogate Variable Analysis (SVA) to determine significant SVA vectors to use as covariates	CAPS (DSM-5) diagnosis (discovery) PSS (replication)	850k (discovery) 450k (replication)	Analyses examined PTSD symptom severity In discovery sample, at 10% FDR (.10), 5 CpGs were significant In replication sample, 3 of 5 CpGs were present in 450k array; of those, 1 CpG significant at $P=.0028$

Mehta et al. (2013)	169 36% Yes	Civilian, varied; separate child and adulthood	Age, sex, ethnicity, and substance abuse	Modified PTSD symptom scale (mPSS) and CAPS in small subset	EWAS[b] 450k	Differential DNAm profiles found between those with and w/o PTSD, but greater differences found comparing those with PTSD but w/ with w/o child abuse (P-value not specified; 10,000 permutations run)
Rutten et al. (2018)	93 (discovery)[a] NA Yes 98 (replication)[a] NA Yes	Military (Dutch service members); combat Military (US marines); combat	DNAm at 1 month before deployment, cell type, ancestry PCs (replication) Excluded participants with DNAm changes in cigarette smoking, alcohol use, and medication	Self-rating inventory for PTSD (SRIP; discovery) CAPS symptom severity (replication)	450k	At FDR of .05 and empirical P-values (100,000 permutations), discovery Sample identified 17 DMPs and 12 DMRs associated with increasing PTSD symptoms Significant DMPs and DMRs tested in replication dataset; 1 DMP and 2 DMRs replicated

Continued

Table 1 Genome-/epigenome-wide findings across molecular, DNAm, and expression studies—Cont'd

Citation	Sample size (% PTSD cases f case control), trauma-exposed control group?	Trauma and/ or sample type (e.g., interpersonal and accidental)	Covariates/ measured confounds included	Outcome/ measure (e.g., clinical interview and self-report)	Method used, gene(s) examined, array used	Finding
Smith et al. (2011)	110 45% Yes	Civilian, varied; grouped cases and controls by those with and without childhood trauma	Sex, age, and batch effects	CAPS diagnosis	27k	FDR .05 found CpG sites in five genes associated with PTSD; also found with PTSD status associated with global methylation ($P=1.4\times10^{-4}$)
Uddin et al. (2010)	100 23% Yes	Civilian, varied	Unknown; stated cases/ controls did not differ with respect to age, sex, race, or mononuclear cell count	PCL cut-off based on DSM-IV symptoms	27k	Averaged methylation levels for each gene if uniquely methylated or unmethylated among cases and controls; significant difference in number of uniquely methylated genes differed ($P<.0001$)

Uddin et al. (2013)	100 23% Yes	Civilian, varied	Age, race, sex, smoking, depression, generalized anxiety disorder, medication use, and mononuclear cell count	PCL cut-off based on DSM-IV and severity score; validated by CAPS in subsample	27k		At an uncorrected P-value <.01, 118 CpG sites associated with PTSD status and 80 CpGs associated with symptom severity. Socioeconomic position moderated the association between methylation and PTSD at 119 CpG sites and symptom severity at 55 CpG sites

Gene expression

Sarapas et al. (2011)	40 (50% with PTSD); Yes	Civilian, exposure to World Trade Center attacks	None mentioned	CAPS		Genome-wide gene Expression, univariate analyses	Using FDR ≤ .01, 25 probe sets were differentially expressed between groups
Segman et al. (2005)	236 (17.4% with PTSD); Yes	Civilian, those entering ER of general hospital	None mentioned	CAPS		Genome-wide gene expression, univariate analyses	656 transcripts were differentially expressed between groups (authors mention FDR was applied, but not what correction rate was)

Continued

Table 1 Genome-/epigenome-wide findings across molecular, DNAm, and expression studies—Cont'd

Citation	Sample size (% PTSD cases if case control), trauma-exposed control group?	Trauma and/or sample type (e.g., interpersonal and accidental)	Covariates/measured confounds included	Outcome/measure (e.g., clinical interview and self-report)	Method used, gene(s) examined, array used	Finding
Yehuda et al. (2009)	40 (50% with PTSD); Yes	Civilian, Exposure to World Trade Center attacks	None mentioned in genome-wide portion of analyses	CAPS	Genome-wide gene expression, univariate analyses	Using Holm-Bonferroni procedure, 16 genes were differentially expressed between groups
Tylee et al. (2015)	50 (50% with PTSD); Yes	Military	Deployment cohort, age, ancestry (Caucasian or not), prior deployment	CAPS	Genome-wide gene expression, univariate analyses	Using FDR ≤.01, 64 probes differentiated groups
Glatt et al. (2013)	50 (50% with PTSD); Yes	Military	Deployment cohort, age, ancestry (Caucasian or not), prior deployment	CAPS	Genome-wide gene expression, univariate analyses	Using FDR <.0001, no genes met statistical significance
Logue, Smith, et al. (2015)	143 (80% with PTSD); Yes	Veterans	Monocytes and lymphocytes	CAPS	Genome-wide gene expression, univariate analyses (+candidate gene expression work)	Using FDR $P<.05$, 41 probes differentiated the groups

Kuan, Waszczuk, Kotov, Clouston, et al. (2017)	201 (61% with PTSD); Yes	Civilian, World Trade Center Responders	Age, race, five cell type proportions (CD8T, CD4T, natural killer, and B-cell monocytes)	PCL	Genome-wide gene expression, univariate analyses (+candidate and pathway analyses)	Using FDR ≤.05, 448 genes were differentially expressed between groups
Breen et al. (2015)	94 (50% with PTSD); Yes	Military/ veterans	None mentioned	CAPS	Gene network analyses	Using FDR <.05, nine modules were associated with PTSD development
Bam et al. (2016)	48 (50% with PTSD); No	Veterans	None mentioned	CAPS, PCL	Gene network analyses	326 genes and 190 miRNAs differentiated groups (FDR not mentioned)

[a] Males only.
[b] Examined the methylation profiles of transcripts that were differentially regulated in the expression analyses (was $n=304$).

2009). Thus PRSs that include more SNPs, using more liberal P-values (e.g., $P<.5$), have explained more variance in phenotypes (Agerbo et al., 2015; Clarke et al., 2015; Genetics of Personality Consortium et al., 2015; Jervis et al., 2015). Other decision points in creating PRSs involve whether to employ clumping methods to select variants in linkage disequilibrium (LD) blocks that are highly associated with the outcome (Marees et al., 2018), how to code variants with positive or negative associations so that effects are not canceled out (Dudbridge, 2013), and how to address strand ambiguity (Chen et al., 2018).

PRS and PTSD

Several recent studies have employed the PRS approach in the context of PTSD, some of which are highlighted in the succeeding text. In the largest published GWAS study of PTSD to date, containing over 20,000 individuals, Duncan et al. (2018) generated PRS to examine genetic overlap of PTSD with discovery datasets for schizophrenia, bipolar, and depression. Among individuals of European ancestry, PRS analyses suggested genetic overlap for PTSD with schizophrenia and bipolar, but not depression. Perhaps due to lower power, no significant effects were observed among individuals of African ancestry. Similar results were observed in a sample of women by Sumner, Duncan, Ratanatharathorn, Roberts, and Koenen (2016) such that common genetic variants for schizophrenia and bipolar disorder index genetic risk, but not depression. Nievergelt et al. (2015) used PRS to demonstrate genetic overlap in a sample of male, trauma-exposed, military cohort of European descent for bipolar disorder, but not for schizophrenia or depression. Thus the PRS literature suggests some, albeit inconsistent, genetic overlap between PTSD and other psychiatric phenotypes.

Several studies have also examined genetic overlap of PTSD and physical health. For example, Wolf et al. (2017) generated PRS indexing obesity risk and found that the association between metabolic syndrome and PTSD severity varied as a function of genetic risk for obesity, but found no evidence for a direct association between genetic risk for obesity and PTSD. Polimanti et al. (2017) constructed PRS of several female anthropomorphic traits (e.g., body mass index, hip circumference, and waist-hip ratio) and found significant associations between six traits and PTSD, with the highest being for waist circumference adjusted for body mass index followed by age-adjusted waist-hip ratio. Thus this emerging literature indicates some genetic overlap between PTSD and obesity, as well as female anthropomorphic traits using PRS.

Genomic relatedness matrix restricted maximum likelihood (GREML)

GREML, implemented in the genome-wide complex trait analysis (GCTA) software tool, provides estimates of marker-based heritability (i.e., as the proportion of variance in a trait that is attributable to additive genetic variance [h_{SNP}^2]) among unrelated individuals (Yang, Lee, et al., 2011; Yang et al., 2013, 2017; Yang, Manolio, et al., 2011). In the GREML approach, genetic relationships (also referred to as kinship) between pairs of individuals are estimated and fit as random effects in a mixed linear model to estimate the variance in a trait that is explained by SNPs. Genetic variance can be further partitioned into separate chromosomes and/or genomic segments (Yang, Manolio, et al., 2011) or based on minor allele frequency and/or LD (Yang et al., 2015). GREML requires individual level genotype and phenotype data, and although it generally requires large sample sizes (i.e., $N > 2300$ to detect $h_{SNP}^2 = 0.4$ at 80% power) for adequately powered studies (Visscher et al., 2014), simulation studies have supported its accuracy over a similar approach, Linkage Disequilibrium Score Regression (LDSC, described in the succeeding text, Ni, Moser, Schizophrenia Working Group of the Psychiatric Genomics Consortium, Wray, & Lee, 2018).

GREML and PTSD

Relatively, few studies have been published using GREML in the context of PTSD. Duncan et al. (2018) conducted the largest GWAS study to date and found overall $h_{SNP-GREML}^2 = 0.12$ ($P = .016$) for both males and females combined. Differences were observed; however, for males and females, such that PTSD was significantly heritable in females ($h_{SNP-GREML-FEMALE}^2 = 0.21$, $P = .019$), but not males ($h_{SNP-GREM-MALE}^2 = 0.08$, $P = .43$). Stein et al. (2016) conducted a GWAS among two cohorts of US army soldiers ($N = 4007$), and while the study found independent individual variants associated with PTSD, they found no significant SNP-heritability estimates ($h_{SNP-GREML}^2 = 0.06$, $P = .10$). Thus additional work in this area is needed to clarify what the heritability estimates of PTSD using GREML might be—particularly examining this question separately for males and females.

Linkage Disequilibrium Score Regression (LDSC, also called LDSR)

LDSR (i.e., as implemented in LDSC) is another approach used to provide an estimate of SNP-based heritability using a regression-based approach for GWAS summary statistics (Bulik-Sullivan et al., 2015). An LD score is the sum of all squared correlations of a variant with other variants in a given region. Thus variants that are associated with several other markers have higher LD scores making them good measures of genetic variation. By regressing the chi-squared statistic from a GWAS on the LD scores of SNPs, usually obtained from a reference panel, an upwardly sloping linear trend is achieved. The value of the intercept in this regression indicates inflation from confounding, while the value of the slope can be used to estimate the SNP-based heritability. While LDSC and GREML both provide estimates of genetic correlation, LDSC does not require individual level genotype data (i.e., summary statistics are used) and can incorporate samples with overlapping individuals. Compared with GREML the computing burden requires far less memory and time; however, it has been shown to have larger standard error estimates (Ni et al., 2018).

LDSC and PTSD

LDSC has been applied to studies investigating the genetic overlap between PTSD and several traits, including both mental and physical health. In their large ($N=20{,}070$) GWAS of PTSD, Duncan et al. (2018) found overall $h_{SNP\text{-}LDSC}^2 = 0.18$ ($P=.003$) for both males and females of European ancestry, combined. Differences were observed; however, for males and females, such that PTSD was significantly heritable in females ($h_{SNP\text{-}LDSC\text{-}FEMALE}^2 = 0.36$, $P=.003$), but not males ($h_{SNP\text{-}GREM\text{-}MALE}^2 = 0.05$, $P=.69$). Significant genetic correlations were found between PTSD and schizophrenia ($r_{G\text{-}LDSC}=.33$, $P=1.3\times10^{-5}$) and depression ($r_{G\text{-}LDSC}=.34$, $P=.006$) but not bipolar disorder, which is notably different from results found using PRS in the same study. Using LDSC estimates from Duncan et al. (2018), Sumner et al. (2017) found a genetic correlation between PTSD and coronary artery disease ($r_{G\text{-}LDSC}=.41$, $P=.01$). Similarly, Lind et al. (2019) found significant genetic correlations between PTSD and several sleep phenotypes, including insomnia ($r_{G\text{-}LDSC}$ range .36–.49, P range 2.6×10^{-13} to 1.2×10^{-5}), oversleeping ($r_{G\text{-}LDSC}=.44$, $P=2.7\times10^{-3}$), undersleeping ($r_{G\text{-}LDSC}=.49$, $P=5.4\times10^{-7}$), ICD-10 sleep diagnosis ($r_{G\text{-}LDSC}=.43$, $P=2.4\times10^{-3}$), sleep duration ($r_{G\text{-}LDSC}=-.23$, $P=1.0\times10^{-4}$), napping during the day ($r_{G\text{-}LDSC}=.21$, $P=2.0\times10^{-4}$), and waking up in the morning ($r_{G\text{-}LDSC}=-.23$, $P=2.8\times10^{-5}$). Thus this emerging literature focusing on LDSC and PTSD suggests heterogeneity in heritability estimates for males and females (stronger in females) and significant genetic correlations with a range of physical and psychological phenotypes.

Mendelian randomization

Mendelian randomization (MR) is a method to examine causal effects of a risk factor on an outcome using variation in known genes as a natural experiment. Specifically, genetic variants with known function serve as proxies for modifiable environmental exposure (i.e., risk factors) as a way to overcome effects from unmeasured confounding and mitigate concerns of reverse causation in observational research (Davies, Holmes, & Davey Smith, 2018). The basic principle is similar to a randomized clinical trial (RCT) in that genetic variants serve as instrumental variables (e.g., variables that can mimic randomization and are independent of confounder) because their alleles have been determined before the exposure or outcome has occurred. This is made possible based on the assumption that individuals are naturally randomized during the random assortment of genetic variants during meiosis, which yields a random distribution of genetic variants in a population (Emdin, Khera, & Kathiresan, 2017; Smith & Ebrahim, 2003). Three core assumptions are critical to MR and selection of genetic instrumental variables (GIV): (1) the GIV is associated with the risk factor, (2) the GIV is not associated with confounders, and (3) the GIV only influences the outcome through the risk factor. Thus MR may be biased if the genetic variants employed are pleiotropic and researchers must take care to ensure that adequate statistical power is achieved to determine the effect of the variant on a risk factor and that no confounding factors can account for the associations between the variant and the outcome independent of the risk factor. To conduct MR the core assumptions must first be assessed, including a direct test of the GIV-exposure association, before carrying out the MR analysis.

MR and PTSD

The use of MR is recently becoming more popular in medical and psychiatric genetics; however, it has not yet been widely applied in the context of PTSD. One study utilized SNPs associated with female waist circumference adjusted for body mass index as instrumental variables and found a 64% decrease in the risk for PTSD per 1 standard deviation increase in adjusted waist circumference (Polimanti et al., 2017). Another study used MR to examine the causal effect of plasma dopamine beta-hydroxylase (DBH), an enzyme involved norepinephrine biosynthesis, on PTSD and found a significant effect on reexperiencing symptoms in a sample of active duty US marines (Mustapic et al., 2014).

Other genomic platforms

Beyond the exciting innovations in molecular genetics that have been developing from statistical innovations using GWAS data, the field is seeing rapid advances in other molecular platforms. Fig. 1 illustrates the number of published methylation and expression studies of PTSD to date. Specifically, in recent years, epigenetic studies, namely, DNA methylation (DNAm) in the form of epigenome-wide association studies (EWAS), have proliferated, as have genome-wide gene expression studies. In human studies, epigenetic changes in peripheral tissue is thought to be driven by processes initiated in the central nervous system and may reflect similar changes in the brain (Zannas, Provencal, & Binder, 2015). As the collection of postmortem brain tissue involves ethical, logistic, and technical confounds (Durrenberger et al., 2010; Hynd, Lewohl, Scott, & Dodd, 2003), using blood tissue has been used as an alternative to applying these methods to the study of PTSD. It should be noted that much of the DNAm and expression sections in the succeeding text fail to control for cell type, which is a notable limitation. In the sections that follow, relevant literature is presented summarizing advances in these areas as applied to PTSD.

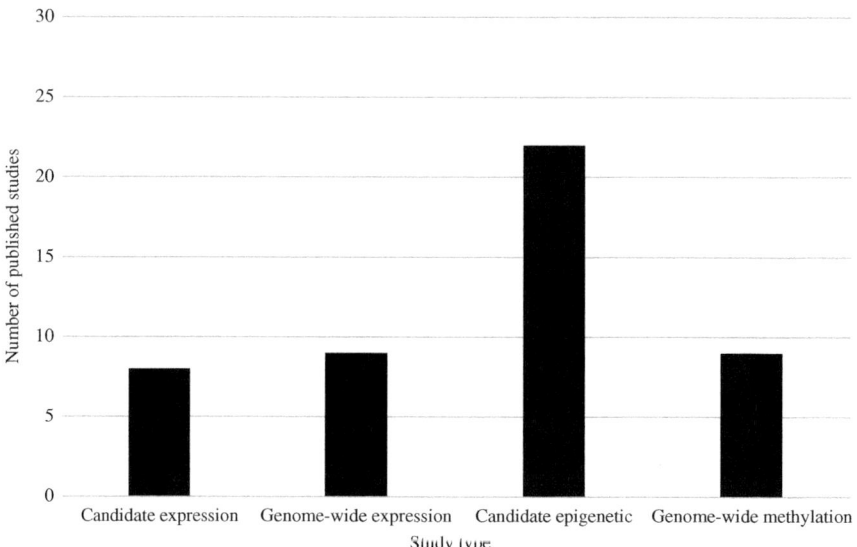

Fig. 1 Depiction of number of studies using an alternative genomic platform (DNAm or gene expression) to date.

DNA methylation

Another important aspect beyond inherited genetic code is the epigenome, through which environmental experiences (including stress) interact with the genome and lead to biological processes (i.e., chemical and protein modifications of chromatin, DNA methylation, histone modifications, and noncoding RNAs; Roth, 2014). These processes impact transcriptional activity without changing the underlying genetic code (Feil & Fraga, 2012). DNAm and epigenetic processes broadly are relevant to trauma and PTSD, as they have been proposed as one way the environment may "get under the skin" (Toyokawa, Uddin, Koenen, & Galea, 2012). Once thought to be stable, it is now known that DNAm is a dynamic process (Teperino, Lempradl, & Pospisilik, 2013; Wu & Zhang, 2014). Most of this dynamic modification occurs at C-phosphate-G (CpG) dinucleotides that cluster in what are known as CpG islands and are highly enriched at or near promoter regions, which links them to transcriptional regulation (Teperino et al., 2013). Methylation status is measured with assays that use a pair of probes to measure the intensities of the methylated and unmethylated alleles at each specific CpG site targeted. Methylation level is generally measured in two ways: The first is a beta-value, the ratio of methylated probe intensity to the overall intensity (sum of methylated and unmethylated probe intensities), which ranges from 0 (fully unmethylated) to 1 (fully methylated) (Du et al., 2010). The second is an *M*-value, which is the log2 ratio of the methylated probe intensity versus unmethylated probe intensity (Du et al., 2010). These values are used in analyses to determine differential methylation status.

Candidate-gene methylation studies

Candidate methylation studies examine differential methylation status at specific probes, or CpG sites, within genes of interest. Differential methylation associated with PTSD has been identified in sites within numerous candidate genes and systems, with over 22 candidate methylation studies conducted to date (see review by Zannas et al., 2015). Common candidates are those within the HPA axis and immune system given the relevance of dysregulation to both of these systems in PTSD. For example, lower methylation levels in the *NR3C1* gene that encodes for the glucocorticoid receptor associated with PTSD risk have been repeatedly demonstrated (e.g., Labonte, Azoulay, Yerko, Turecki, & Brunet, 2014; Yehuda et al., 2015). DNAm in other candidate genes studied in relation to PTSD include other HPA relevant genes studied such *CRHR1* and *FKBP5*, genes relevant for neurotransmitter activity (*SLC6A3* and *SLC6A4*), and genes related to immune regulation such as *IGF2* (Rusiecki et al., 2013). Similar to the limitations of candidate gene studies of sequence variation described in the preceding text, candidate DNAm studies also often suffer from similar limitations, including inconsistent coverage of CpGs, which leads to problems with regard to replication efforts, issues with population stratification, small sample sizes, and limitations in extant knowledge of the pathophysiology of PTSD resulting in insufficient a priori evidence to select genes of interest.

Epigenome-wide methylation studies

EWAS, like GWAS, offer an agnostic approach to examine epigenetic influences and can potentially identify novel candidate genes and biological pathways involved in PTSD. The sample size requirements for EWAS are substantially smaller than for GWAS (e.g., in the hundreds as compared with the thousands; see Rakyan, Down, Balding, & Beck, 2011 for a review of design and power considerations). To date, nine EWAS have been conducted in relation to PTSD (Hammamieh et al., 2017; Kuan, Waszczuk, Kotov, Marsit, et al., 2017; Martin et al., 2018; Mehta et al., 2017, 2013; Rutten et al., 2018; Smith et al., 2011; Uddin et al., 2010, 2013) and are summarized in Table 1. EWAS have suggested the relevance of methylation of genes involved in immune system functioning (e.g., Uddin et al., 2010), and the largest EWAS to date ($n=473$) identified changes related to inflammatory responses and synaptic plasticity (Kuan, Waszczuk, Kotov, Marsit, et al., 2017).

Methylation age

An additional line of research focuses on methylation changes with regard to accelerated DNAm age, based on theory and research that suggest traumatic stress may advance the pace of cellular aging to exceed chronological age, although results have been mixed to date. For example, a recent study using the largest sample to date metaanalyzed nine separate cohorts. This study found evidence for a significant, small association between accelerated DNAm age and childhood trauma and lifetime PTSD severity; however, DNAm was not related to PTSD diagnosis or lifetime trauma exposure (Wolf, Maniates, et al., 2018).

Translating epigenetic findings

DNAm studies have also provided a more mechanistic understanding of how genetic risk and trauma exposure interact. For example, a functional polymorphism in *FKBP5* was associated with multiple stress-related outcomes, including PTSD, in the context of childhood maltreatment, and with decreased DNAm and increased gene expression in glucocorticoid response elements of *FKBP5* (Klengel et al., 2013). As the field develops, the use of translational studies to link DNAm changes with behavior, and relevant endophenotypes will be important. Limited, but promising, work in this area has been conducted to date. In one translational study, higher methylation levels in the catechol-O-methyltransferase (*COMT*) gene were associated with impaired fear inhibition, considered an important marker associated with PTSD (Norrholm et al., 2013).

Considerations and limitations in DNAm

A number of considerations are worth noting in considering the existing literature on DNAm and important factors for moving the field forward. First, many EWAS to date may still be underpowered to detect a signal that survives correction for multiple testing (Daskalakis, Rijal, King, Huckins, & Ressler, 2018). Second, investigations of DNAm must control for cell composition, which has become more common using developed methods (Houseman et al., 2012). Third the timing of trauma exposure in relation to PTSD is important, as at present, the temporal relationship of trauma exposure to DNAm changes and from DNAm changes to development of PTSD remains largely unknown. Of note within the literature is the important confounding effect of smoking, with a recent EWAS finding that DNAm associations with polyvictimization did not remain after controlling for tobacco smoking (Marzi et al., 2018).

Gene expression

In addition to the importance of inherited sequence and methylation influences on risk for PTSD, gene expression studies are also critical in understanding this pathophysiology, as molecular influences on these phenotypes may exert effects via altering gene expression (Ressler et al., 2011; Yehuda et al., 2009). The limited neuronal expression work examining postmortem brains suggests gene expression profile differences between those with PTSD and nontrauma-exposed controls in the dorsolateral prefrontal cortex, which regulates fear-based responses (Su et al., 2008). Additional work examining the postmortem brains of suicide completers with and without a history of child abuse suggests differences in the expression of genes important for myelination (Lutz et al., 2017). Thus there appear to be gene expression differences in the postmortem brains of those with and without trauma history/PTSD.

Candidate gene expression studies

Candidate gene expression studies offer a way to determine whether expression levels may differentiate individuals based on trauma/PTSD, using an a priori test of a specific gene or system of genes. To our knowledge, eight candidate gene expression studies of trauma/PTSD have been conducted. Similar to that of candidate variation studies and candidate DNAm studies, much of this work has focused on the expression of genes implicated in HPA axis functioning (*FKBP5* and *PACAP*) (Kuan, Waszczuk, Kotov, Marsit, et al., 2017; Levy-Gigi, Szabo, Kelemen, & Keri, 2013; Ressler et al., 2011; Yehuda et al., 2009). Some studies have also found differences in genes involved in glucocorticoid signaling (*GR*) from within the HPA axis (Hollifield, Moore, & Yount, 2013; O'Donovan et al., 2011). In addition to genes involved in HPA axis functioning, genes involved in inflammation also appear to differentiate those with and without PTSD (*TNFRSF10B*, *IL10RB*, *IL4R*, *NF-κB*, and *CREB/ATF*) (Hollifield et al., 2013; O'Donovan et al., 2011). Additionally, genes implicated in protein coding (*C9orf84*), normal cell processes (*SOD1*), and cell death (*BBC3*, *CASP2*) (Kuan, Waszczuk, Kotov, Marsit, et al., 2017) have also been associated with PTSD. Another study using this approach found differences between genes involved in glucocorticoid signaling (NR3C1/GR) and on those in the neurotrophin family, important for associative memory formation (Logue, Smith, et al., 2015). Yet another group failed to find any differences between groups; however, this study's sample size was fairly small ($n=50$) (Glatt et al., 2013). Broadly, these findings implicate genes involved in HPA axis, inflammation, memory formation, and cell processes.

Genome-wide gene expression studies

Consistent with the trend in the other molecular platforms, newer studies utilizing gene expression have moved beyond candidate designs to using an agnostic genome-wide approach to examine gene expression. Among the studies using a genome-wide expression approach, nine to our knowledge have conducted univariate analyses aiming to identify expression differences among cases and controls. Results from these studies implicate genes involved in HPA axis functioning (e.g., *FKBP5*; [Kuan, Waszczuk, Kotov, Clouston, et al., 2017; Sarapas et al., 2011; Segman et al., 2005]), signal transduction (e.g., *STAT5B*; [Sarapas et al., 2011; Tylee et al., 2015;

Yehuda et al., 2009]), fear response and memory (e.g., *CHRNA4, DSCAM*; [Bountress et al., under review; Logue, Smith, et al., 2015]), and normal cell processes (e.g., *SOD1*; [Tylee et al., 2015]) in group differences. These studies suggest that the expression of genes important for HPA axis functioning, fear-based responding, and signal transduction is also implicated in PTSD. See Table 1 for more in-depth information about genome-wide gene expression studies.

Network and pathway analyses

Just as novel statistical innovations have been used for GWAS data (e.g., PRS and LDSR), additional statistical approaches to analyzing genome-wide expression arrays have been developed. Two such techniques have been used. First, weighted gene coexpression network analysis (WGCNA) allows for consideration of potential biological interactions between genes in the dataset and provides a systems biology view of gene expression. WGCNA is used to better understand the functional changes observed between the disease and normal transcriptomes. Only two studies to our knowledge have used WGCNA in PTSD case-control studies (Bam et al., 2016; Breen et al., 2015). These studies found overexpression of genes enriched for innate immune response and interferon signaling and differences in the expression of genes enriched for signal transduction.

Second, canonical pathway analysis is used to determine whether genes differentially expressed between groups may be enriched for particular biologic systems and can be used to further characterize expression findings (Kuan, Waszczuk, Kotov, Marsit, et al., 2017). To our knowledge, only one study has applied this methodology to trauma/PTSD, with this group finding evidence for pathways involved in synaptic plasticity, oxytocin signaling, cholinergic synapse, and inflammatory disease pathways (Kuan, Waszczuk, Kotov, Marsit, et al., 2017).

Considerations and limitations in expression studies

A number of limitations within this literature are important to consider. Of the studies comparing those with trauma without PTSD with those with PTSD, most considered the former to not meet criteria for PTSD if they did not meet criteria for current PTSD. This decision may confound important gene expression differences observed between those who previously had been diagnosed with PTSD from those who never met criteria. In other studies, use of a nontrauma-exposed control group is problematic because the molecular genetic risk for these "controls" is unknown, as risk may be masked by a lack of trauma exposure (Amstadter, Sheerin, Lind, & Nugent, 2017).

Unique considerations for PTSD genomics

The field of PTSD genomics is confronted with a number of unique challenges with regard to the design of genomic studies for this complex phenotype, beyond those faced in comparison with other psychiatric disorders. For instance, when attempting to examine differences between "cases" and "controls," great care must be given to how the control group is defined. Specifically, as individuals who have not experienced a trauma may possess "unexpressed vulnerability for PTSD," it is important that the control group have had at least one trauma exposure (Cornelis, Nugent, Amstadter, & Koenen, 2010). In addition to potential differences in trauma "load," there is a need to assess the type and duration of trauma(s) and timing (Chung & Breslau, 2008; Cloitre et al., 2009; Cornelis et al., 2010). To the extent that those with PTSD report more serious traumatic events occurring earlier and lasting for a longer period of time, the presumed genetic effects of PTSD may be in fact be explaining a gene-trauma association (Cornelis et al., 2010).

Prior research finds that a number of factors impact endorsement rates of traumatic events. For instance, the specificity of the question(s) about the event (being too vague or too specific), how data are collected (e.g., in phone vs in person), the specificity of the reference period, and whether the wording references events that ought to be disclosed (e.g., threatened physical harm), impacts endorsement rates (Koss et al., 2007). Similarly, although timesaving, brief screens sometimes result in missed cases (Prins et al., 2016). The use of psychometrically validated full trauma assessment instruments is likely to produce more accurate estimates and thus is recommended to those engaging in trauma research.

Further, more "potent" interpersonal traumatic events tend to be more common among ethnic minority groups (Hatch & Dohrenwend, 2007; Roberts, Gilman, Breslau, Breslau, & Koenen, 2011). As a result the designs of studies with a primary focus on trauma have purposefully attempted to select for individuals enriched for trauma (e.g., Grady Trauma Project; Thompson, Fiorillo, Rothbaum, Ressler, & Michopoulos, 2018). Within such samples, there is a need to control from ancestry markers, with "best practices" involving individuals from genetically similar superpopulations, adjusting for that group's ancestry components, and then metaanalyzing these findings to increase power (Nievergelt et al., 2015).

Another level of complexity with regard to trauma assessment and the diagnosis of PTSD using the Diagnostic and Statistical Manual (DSM) of Mental Disorder version IV differs from DSM-V, which has implications for those attempting to summarize the extant literature. Specifically, DSM-IV requires the event to have prompted a feeling of "fear, helplessness, or horror," while DSM-V does not, resulting—in some samples—in an increase in diagnosed rates of PTSD (Schaal, Koebach, Hinkel, & Elbert, 2015). To date, most of the existing genetic studies of PTSD use DSM-IV (e.g., Duncan, Ratanatharathorn, et al., 2017); thus there is a need to determine whether existing findings on the genetics of PTSD replicate or are different from those using DSM-V.

Studies examining genetic influences on PTSD are also faced with the question of if or how to model potential sex differences. Specifically, behavioral (Sartor et al., 2011; Xian et al., 2000) and molecular genetic research (Duncan, Ratanatharathorn, et al., 2017) converge in suggesting that genetic influences on PTSD are larger among females. Additionally, a recent metaanalysis found that what appeared to be a significant main effect of SNP rs2267735 within gene *ADCYAP1R1* was being driven by a significant effect in females but not males (Lind et al., 2017). These studies indicate that while there may be significant genetic effects on PTSD, there is also significant heterogeneity across sex. Thus sex-specific questions related to the pathoetiology of PTSD ought to be formulated and tested as research in this area continues to move forward.

Finally, as trauma is a nonspecific risk factor for a range of psychiatric disorders, there is concern about what effects, and, in this case, genetic effects might be unique to trauma's signature disorder, PTSD, compared with other co-occurring conditions. For instance, one group found an extremely high degree of genetic overlap between PTSD and depression (Sartor et al., 2012). Another found that there is also moderate genetic overlap between PTSD and alcohol and drug use disorders (Xian et al., 2000). Thus there is certainly a need when conducting genetic studies of PTSD to include measures of comorbid conditions within the modeling framework.

Conclusions and future directions

Family and twin studies using behavioral genetic designs laid the groundwork for molecular studies in the trauma and PTSD arena. While early molecular studies were predominately candidate gene designs, across platforms (e.g., variant, DNAm, and expression), the field has moved beyond candidate gene studies and has embraced larger, agnostic, genome-wide designs across all platforms. Novel biologic mechanisms for PTSD have been discovered via GWAS as we begin to uncover the genetic etiology of PTSD. These efforts to date remain limited by statistical power, but, as noted in the preceding text, with larger collaborative efforts and new funding initiatives, these challenges will be surmounted; however, new challenges are posed, such as those of phenotypic and genomic harmonization across studies. There remains a role for innovative candidate studies, such as those that use sequencing to follow-up hits, or those that characterize gene function following nomination from an agnostic study.

Statistical innovations using GWAS data have yielded key findings, some of which mirror that of the behavioral genetic literature on PTSD. For example, looking across twin studies, the heritability estimate is higher in females than males, and the GCTA findings using GREML mirror this pattern. Similarly, twin studies have highlighted the genetic overlap between PTSD and other psychiatric and physical health phenotypes, and PRS findings are largely consistent with the twin findings. Concordance between the behavioral and molecular literatures is promising and provides validity for these methods. Animal models of PTSD, though not reviewed in the present chapter, will be particularly important in examining the processes by which variants, epigenetic signatures, or expression differences in the brain impact behavior (i.e., fear learning, fear consolidation; see review by Kwapis & Wood, 2014).

The field of epigenetics thus far has focused mainly on DNAm, and the EWAS studies have found numerous genes implicated in PTSD that are worthy of increased study and replication attempts. While DNAm has been the most extensively studied, histone modifications and microRNAs are being increasingly recognized and examined as important epigenetic processes (Roth, 2014). Similarly, methylation quantitative trait loci (mQTLs) have been examined in relation to GWAS findings of PTSD (Almli et al., 2015) and are a promising area of influence. Finally the field of epigenetic inheritance, which examines whether epigenetic alterations can be transmitted to future generations, is gaining interest (Dias, Maddox, Klengel, & Ressler, 2015). With regard to expression studies, both candidate and genome-wide findings indicate that the expression of genes implicated in HPA axis functioning (and specifically, *FKBP5*) are associated with risk for PTSD among those with trauma. Additionally, both candidate and canonical pathway analyses implicate genes involved in inflammation in risk for PTSD, which is consistent with GWAS findings of immune response genes being implicated in risk. Thus these findings are consistent with a larger literature suggesting that genes impacting stress response and inflammatory disease processes are differentially expressed in those with PTSD. Although this work is still in its infancy, this platform has the potential to continue to help clarify the biological mechanisms by which trauma exposure impacts propensity for this debilitating condition.

The field has recently experienced rapid advances in genomic platforms, associated with lowered costs of high throughput arrays across platforms (e.g., GWAS, EWAS, and genome-wide expression). Paired with both commercially available inexpensive, shelf-stable, DNA collection kits, and a growing appreciation for the biologic influences on trauma and PTSD, these advances have resulted in enormous growth of the PTSD genomic literatures, moving it beyond its infancy where it was a mere decade ago. The era of "big data" has generated the need for creation of statistical innovations (e.g., PRS, MR, and WCGNA), and has, perhaps more importantly, espoused large-scale collaborative efforts among investigators (e.g., the PGC) in hopes of achieving statistical power to better understand the genetic architecture of psychiatric phenotypes. This new dawn of team science, big data, statistical innovations, and genomic platforms creates an exciting time for researchers interested in the field of psychiatric genetics. Challenges are also mounting, such as those of how to best harmonize phenotypic data, how to best account for multiple testing, and how to integrate numerous genomic platforms, providing many areas in need of continued research and statistical method development. Further, a critical area in need of development is applying these platforms, and analytic techniques are to better understand the comorbidity of PTSD with other disorders and to leverage the understanding of potential intermediary phenotypes, such as emotion regulation or executive functioning deficits.

References

Afifi, T. O., Asmundson, G. J., Taylor, S., & Jang, K. L. (2010). The role of genes and environment on trauma exposure and posttraumatic stress disorder symptoms: A review of twin studies. *Clinical Psychology Review*, *30*(1), 101–112. https://doi.org/10.1016/j.cpr.2009.10.002.

Agerbo, E., Sullivan, P. F., Vilhjalmsson, B. J., Pedersen, C. B., Mors, O., Borglum, A. D., ... Mortensen, P. B. (2015). Polygenic risk score, parental socioeconomic status, family history of psychiatric disorders, and the risk for schizophrenia: A Danish population-based study and meta-analysis. *JAMA Psychiatry*, https://doi.org/10.1001/jamapsychiatry.2015.0346. [Epub ahead of print].

Almli, L. M., Stevens, J. S., Smith, A. K., Kilaru, V., Meng, Q., Flory, J., ... Ressler, K. J. (2015). A genome-wide identified risk variant for PTSD is a methylation quantitative trait locus and confers decreased cortical activation to fearful faces. *American Journal of Medical Genetics Part B, Neuropsychiatric Genetics*, *168b*(5), 327–336. https://doi.org/10.1002/ajmg.b.32315.

Amstadter, A. B., Sheerin, C., Lind, M., & Nugent, N. R. (2017). Genetic and biological underpinnings and consequences of trauma. In S. N. Gold (Ed.), *APA handbooks in psychology. APA handbook of trauma psychology: Foundations in knowledge* (pp. 443–481). Washington, DC: American Psychological Association.

Amstadter, A. B., Sumner, J. A., Acierno, R., Ruggiero, K. J., Koenen, K. C., Kilpatrick, D. G., ... Gelernter, J. (2013). Support for association of RORA variant and post traumatic stress symptoms in a population-based study of hurricane exposed adults. *Molecular Psychiatry*, *18*(11), 1148–1149. https://doi.org/10.1038/mp.2012.189.

Ashley-Koch, A. E., Garrett, M. E., Gibson, J., Liu, Y., Dennis, M. F., Kimbrel, N. A., ... Hauser, M. A. (2015). Genome-wide association study of posttraumatic stress disorder in a cohort of Iraq–Afghanistan era veterans. *Journal of Affective Disorders*, *184*, 225–234. https://doi.org/10.1016/j.jad.2015.03.049.

Bam, M., Yang, X., Zhou, J., Ginsberg, J. P., Leyden, Q., Nagarkatti, P. S., & Nagarkatti, M. (2016). Evidence for epigenetic regulation of pro-inflammatory cytokines, interleukin-12 and interferon gamma, in peripheral blood mononuclear cells from PTSD Patients. *Journal of Neuroimmune Pharmacology*, *11*(1), 168–181. https://doi.org/10.1007/s11481-015-9643-8.

Binder, E. B., Bradley, R. G., Liu, W., Epstein, M. P., Deveau, T. C., Mercer, K. B., ... Ressler, K. J. (2008). Association of FKBP5 polymorphisms and childhood abuse with risk of posttraumatic stress disorder symptoms in adults. *JAMA*, *299*(11), 1291–1305. https://doi.org/10.1001/jama.299.11.1291.

Bountress, K. E., Bacanu, S. A., Tomko, R. L., Korte, K. J., Hicks, T., Sheerin, C., ... Amstadter, A. B. (2017). The effects of a BDNF Val66Met polymorphism on posttraumatic stress disorder: A meta-analysis. *Neuropsychobiology*, *76*(3), 136–142. https://doi.org/10.1159/000489407.

Bountress, K., Vladimirov, V., McMichael, G., Hardiman, G., Chung, D., Adams, Z., Danielson, C. K., & Amstatder, A. (under review). Gene expression differences between young adults based on trauma history and PTSD.

Breen, M. S., Maihofer, A. X., Glatt, S. J., Tylee, D. S., Chandler, S. D., Tsuang, M. T., ... Woelk, C. H. (2015). Gene networks specific for innate immunity define posttraumatic stress disorder. *Molecular Psychiatry*, *20*(12), 1538–1545. https://doi.org/10.1038/mp.2015.9.

Bruenig, D., Lurie, J., Morris, C. P., Harvey, W., Lawford, B., Young, R. M., & Voisey, J. (2016). A case-control study and meta-analysis reveal BDNF Val66Met is a possible risk factor for PTSD. *Neural Plasticity*, *2016*, 6979435. https://doi.org/10.1155/2016/6979435.

Bulik-Sullivan, B. K., Loh, P. R., Finucane, H. K., Ripke, S., Yang, J., Schizophrenia Working Group of the Psychiatric Genomics Consortium, ... Neale, B. M. (2015). LD Score regression distinguishes confounding from polygenicity in genome-wide association studies. *Nature Genetics*, *47*(3), 291–295. https://doi.org/10.1038/ng.3211.

Chen, C. Y., Stein, M., Ursano, R., Cai, T., Gelernter, J., Heeringa, S., ... Nievergelt, C. (2017). 223. Genome-wide association study of posttraumatic stress disorder symptom domains in two cohorts of United States army soldiers. *Biological Psychiatry*, *81*(10), S91–S92.

Chen, L. M., Yao, N., Garg, E., Zhu, Y., Nguyen, T. T. T., Pokhvisneva, I., ... O'Donnell, K. J. (2018). PRS-on-Spark (PRSoS): A novel, efficient and flexible approach for generating polygenic risk scores. *BMC Bioinformatics*, *19*(1), 295. https://doi.org/10.1186/s12859-018-2289-9.

Chung, H., & Breslau, N. (2008). The latent structure of post-traumatic stress disorder: Tests of invariance by gender and trauma type. *Psychological Medicine, 38*(4), 563–573. https://doi.org/10.1017/s0033291707002589.

Clarke, T. K., Smith, A. H., Gelernter, J., Kranzler, H. R., Farrer, L. A., Hall, L. S., ... McIntosh, A. M. (2015). Polygenic risk for alcohol dependence associates with alcohol consumption, cognitive function and social deprivation in a population-based cohort. *Addiction Biology*, https://doi.org/10.1111/adb.12245. [Epub ahead of print].

Cloitre, M., Stolbach, B. C., Herman, J. L., van der Kolk, B., Pynoos, R., Wang, J., & Petkova, E. (2009). A developmental approach to complex PTSD: Childhood and adult cumulative trauma as predictors of symptom complexity. *Journal of Traumatic Stress, 22*(5), 399–408. https://doi.org/10.1002/jts.20444.

Cornelis, M., Nugent, N. R., Amstadter, A. B., & Koenen, K. C. (2010). Genetics of post-traumatic stress disorder: Review and recommendations for genome-wide association studies. *Current Psychiatry Reports, 12*(4), 313–326.

Daskalakis, N. P., Rijal, C. M., King, C., Huckins, L. M., & Ressler, K. J. (2018). Recent genetics and epigenetics approaches to PTSD. *Current Psychiatry Reports, 20*(5), 30. https://doi.org/10.1007/s11920-018-0898-7.

Davies, N. M., Holmes, M. V., & Davey Smith, G. (2018). Reading Mendelian randomisation studies: A guide, glossary, and checklist for clinicians. *BMJ, 362*, k601. https://doi.org/10.1136/bmj.k601.

Dias, B. G., Maddox, S., Klengel, T., & Ressler, K. J. (2015). Epigenetic mechanisms underlying learning and the inheritance of learned behaviors. *Trends in Neurosciences, 38*(2), 96–107. https://doi.org/10.1016/j.tins.2014.12.003.

Drevo, S., Newman, E., Miller, K. E., Davis, J. L., Craig, C., Sheaff, R. J., ... Bell, K. (2016). The role of social environment and gene interactions on development of posttraumatic stress disorder. *Journal of Articles in Support of the Null Hypothesis, 12*(2).

Du, P., Zhang, X., Huang, C. C., Jafari, N., Kibbe, W. A., Hou, L., & Lin, S. M. (2010). Comparison of Beta-value and M-value methods for quantifying methylation levels by microarray analysis. *BMC Bioinformatics, 11*, 587. https://doi.org/10.1186/1471-2105-11-587.

Dudbridge, F. (2013). Power and predictive accuracy of polygenic risk scores. *PLoS Genetics, 9*(3), e1003348.

Duncan, L. E., Cooper, B. N., & Shen, H. (2018). Robust findings from 25 years of PSTD genetics research. *Current Psychiatry Reports, 20*(12), 115. https://doi.org/10.1007/s11920-018-0980-1.

Duncan, L. E., & Keller, M. C. (2011). A critical review of the first 10 years of candidate gene-by-environment interaction research in psychiatry. *The American Journal of Psychiatry, 168*(10), 1041–1049. https://doi.org/10.1176/appi.ajp.2011.11020191.

Duncan, L. E., Ratanatharathorn, A., Aiello, A. E., Almli, L. M., Amstadter, A. B., Ashley-Koch, A. E., ... Koenen, K. C. (2017). Largest GWAS of PTSD (N=20 070) yields genetic overlap with schizophrenia and sex differences in heritability. *Molecular Psychiatry*, https://doi.org/10.1038/mp.2017.77.

Duncan, L. E., Shen, H., Ballon, J. S., Hardy, K. V., Noordsy, D. L., & Levinson, D. F. (2017). Genetic correlation profile of schizophrenia mirrors epidemiological results and suggests link between polygenic and rare variant (22q11.2) cases of schizophrenia. *Schizophrenia Bulletin*, https://doi.org/10.1093/schbul/sbx174.

Durrenberger, P. F., Fernando, S., Kashefi, S. N., Ferrer, I., Hauw, J. J., Seilhean, D., ... Reynolds, R. (2010). Effects of antemortem and postmortem variables on human brain mRNA quality: A BrainNet Europe study. *Journal of Neuropathology and Experimental Neurology, 69*(1), 70–81. https://doi.org/10.1097/NEN.0b013e3181c7e32f.

Eaves, L. J. (2006). Genotype × environment interaction in psychopathology: Fact or artifact? *Twin Research and Human Genetics, 9*(1), 1–8. https://doi.org/10.1375/183242706776403073.

Emdin, C. A., Khera, A. V., & Kathiresan, S. (2017). Mendelian randomization. *JAMA*, *318*(19), 1925–1926. https://doi.org/10.1001/jama.2017.17219.

Euesden, J., Lewis, C. M., & O'Reilly, P. F. (2015). PRSice: Polygenic Risk Score software. *Bioinformatics*, *31*(9), 1466–1468. https://doi.org/10.1093/bioinformatics/btu848.

Feil, R., & Fraga, M. F. (2012). Epigenetics and the environment: Emerging patterns and implications. *Nature Reviews Genetics*, *13*(2), 97–109. https://doi.org/10.1038/nrg3142.

Genetics of Personality Consortium, de Moor, M. H., van den Berg, S. M., Verweij, K. J., Krueger, R. F., Luciano, M., ... Boomsma, D. I. (2015). Meta-analysis of genome-wide association studies for neuroticism, and the polygenic association with major depressive disorder. *JAMA Psychiatry*, https://doi.org/10.1001/jamapsychiatry.2015.0554. [Epub ahead of print].

Glatt, S. J., Tylee, D. S., Chandler, S. D., Pazol, J., Nievergelt, C. M., Woelk, C. H., ... Tsuang, M. T. (2013). Blood-based gene-expression predictors of PTSD risk and resilience among deployed marines: A pilot study. *American Journal of Medical Genetics Part B, Neuropsychiatric Genetics*, *162b*(4), 313–326. https://doi.org/10.1002/ajmg.b.32167.

Gressier, F., Calati, R., Balestri, M., Marsano, A., Alberti, S., Antypa, N., & Serretti, A. (2013). The 5-HTTLPR polymorphism and posttraumatic stress disorder: A meta-analysis. *Journal of Traumatic Stress*, *26*(6), 645–653. https://doi.org/10.1002/jts.21855.

Guffanti, G., Galea, S., Yan, L., Roberts, A. L., Solovieff, N., Aiello, A. E., ... Koenen, K. C. (2013). Genome-wide association study implicates a novel RNA gene, the lincRNA AC068718.1, as a risk factor for post-traumatic stress disorder in women. *Psychoneuroendocrinology*, *38*(12), 3029–3038. https://doi.org/10.1016/j.psyneuen.2013.08.014.

Hammamieh, R., Chakraborty, N., Gautam, A., Muhie, S., Yang, R., Donohue, D., ... Jett, M. (2017). Whole-genome DNA methylation status associated with clinical PTSD measures of OIF/OEF veterans. *Translational Psychiatry*, *7*(7), e1169. https://doi.org/10.1038/tp.2017.129.

Hatch, S. L., & Dohrenwend, B. P. (2007). Distribution of traumatic and other stressful life events by race/ethnicity, gender, SES and age: A review of the research. *American Journal of Community Psychology*, *40*(3–4), 313–332. https://doi.org/10.1007/s10464-007-9134-z.

Hollifield, M., Moore, D., & Yount, G. (2013). Gene expression analysis in combat veterans with and without posttraumatic stress disorder. *Molecular Medicine Reports*, *8*(1), 238–244. https://doi.org/10.3892/mmr.2013.1475.

Houseman, E. A., Accomando, W. P., Koestler, D. C., Christensen, B. C., Marsit, C. J., Nelson, H. H., ... Kelsey, K. T. (2012). DNA methylation arrays as surrogate measures of cell mixture distribution. *BMC Bioinformatics*, *13*, 86. https://doi.org/10.1186/1471-2105-13-86.

Hynd, M. R., Lewohl, J. M., Scott, H. L., & Dodd, P. R. (2003). Biochemical and molecular studies using human autopsy brain tissue. *Journal of Neurochemistry*, *85*(3), 543–562.

International Schizophrenia Consortium, Purcell, S. M., Wray, N. R., Stone, J. L., Visscher, P. M., O'Donovan, M. C., ... Sklar, P. (2009). Common polygenic variation contributes to risk of schizophrenia and bipolar disorder. *Nature*, *460*(7256), 748–752. https://doi.org/10.1038/nature08185.

Jervis, S., Song, H., Lee, A., Dicks, E., Harrington, P., Baynes, C., ... Antoniou, A. C. (2015). A risk prediction algorithm for ovarian cancer incorporating BRCA1, BRCA2, common alleles and other familial effects. *Journal of Medical Genetics*, https://doi.org/10.1136/jmedgenet-2015-103077. [Epub ahead of print].

Kessler, R. C., Sonnega, A., Bromet, E., Hughes, M., & Nelson, C. B. (1995). Posttraumatic stress disorder in the National Comorbidity Survey. *Archives of General Psychiatry*, *52*(12), 1048–1060.

Kilaru, V., Iyer, S. V., Almli, L. M., Stevens, J. S., Lori, A., Jovanovic, T., ... Ressler, K. J. (2016). Genome-wide gene-based analysis suggests an association between Neuroligin 1 (NLGN1) and post-traumatic stress disorder. *Translational Psychiatry*, *6*, e820. https://doi.org/10.1038/tp.2016.69.

Kimbrel, N. A., Hauser, M. A., Garrett, M., Ashley-Koch, A., Liu, Y., Dennis, M. F., ... Beckham, J. C. (2015). Effect of the APOE allele and combat exposure on PTSD among Iraq/Afghanistan-era veterans. *Depression and Anxiety, 32*(5), 307–315. https://doi.org/10.1002/da.22348.

Klengel, T., Mehta, D., Anacker, C., Rex-Haffner, M., Pruessner, J. C., Pariante, C. M., ... Binder, E. B. (2013). Allele-specific FKBP5 DNA demethylation mediates gene-childhood trauma interactions. *Nature Neuroscience, 16*(1), 33–41. https://doi.org/10.1038/nn.3275.

Koenen, K. C., Aiello, A. E., Bakshis, E., Amstadter, A. B., Ruggiero, K. J., Acierno, R., ... Galea, S. (2009). Modification of the association between serotonin transporter genotype and risk of posttraumatic stress disorder in adults by county-level social environment. *American Journal of Epidemiology, 169*(6), 704–711. https://doi.org/10.1093/aje/kwn397.

Koss, M. P., Abbey, A., Campbell, R., Cook, S., Norris, J., Testa, M., ... White, J. (2007). Revising the SES: A collaborative process to improve assessment of sexual aggression and victimization. *Psychology of Women Quarterly, 31*(4), 357–370.

Kuan, P. F., Waszczuk, M. A., Kotov, R., Clouston, S., Yang, X., Singh, P. K., ... Luft, B. J. (2017). Gene expression associated with PTSD in World Trade Center responders: An RNA sequencing study. *Translational Psychiatry, 7*(12), 1297. https://doi.org/10.1038/s41398-017-0050-1.

Kuan, P. F., Waszczuk, M. A., Kotov, R., Marsit, C. J., Guffanti, G., Gonzalez, A., ... Luft, B. J. (2017). An epigenome-wide DNA methylation study of PTSD and depression in World Trade Center responders. *Translational Psychiatry, 7*(6), e1158. https://doi.org/10.1038/tp.2017.130.

Kwapis, J. L., & Wood, M. A. (2014). Epigenetic mechanisms in fear conditioning: Implications for treating post-traumatic stress disorder. *Trends in Neurosciences, 37*(12), 706–720. https://doi.org/10.1016/j.tins.2014.08.005.

Labonte, B., Azoulay, N., Yerko, V., Turecki, G., & Brunet, A. (2014). Epigenetic modulation of glucocorticoid receptors in posttraumatic stress disorder. *Translational Psychiatry, 4*, e368. https://doi.org/10.1038/tp.2014.3.

Lambert, J. E., Holzer, J., & Hasbun, A. (2014). Association between parents' PTSD severity and children's psychological distress: A meta-analysis. *Journal of Traumatic Stress, 27*(1), 9–17. https://doi.org/10.1002/jts.21891.

Leen-Feldner, E. W., Feldner, M. T., Knapp, A., Bunaciu, L., Blumenthal, H., & Amstadter, A. B. (2013). Offspring psychological and biological correlates of parental posttraumatic stress: Review of the literature and research agenda. *Clinical Psychology Review, 33*(8), 1106–1133. https://doi.org/10.1016/j.cpr.2013.09.001.

Levinson, D. F., Mostafavi, S., Milaneschi, Y., Rivera, M., Ripke, S., Wray, N. R., & Sullivan, P. F. (2014). Genetic studies of major depressive disorder: Why are there no genome-wide association study findings and what can we do about it? *Biological Psychiatry, 76*(7), 510–512. https://doi.org/10.1016/j.biopsych.2014.07.029.

Levy-Gigi, E., Szabo, C., Kelemen, O., & Keri, S. (2013). Association among clinical response, hippocampal volume, and FKBP5 gene expression in individuals with posttraumatic stress disorder receiving cognitive behavioral therapy. *Biological Psychiatry, 74*(11), 793–800. https://doi.org/10.1016/j.biopsych.2013.05.017.

Lewis, C. M., & Vassos, E. (2017). Prospects for using risk scores in polygenic medicine. *Genome Medicine, 9*(1), 96. https://doi.org/10.1186/s13073-017-0489-y.

Li, L., Bao, Y., He, S., Wang, G., Guan, Y., Ma, D., ... Yang, J. (2016). The association between genetic variants in the dopaminergic system and posttraumatic stress disorder: A meta-analysis. *Medicine (Baltimore), 95*(11), e3074. https://doi.org/10.1097/MD.0000000000003074.

Lind, M. J., Brick, L., Gehrman, P. R., Duncan, L., Nugent, N. R., Stein, M. B., ... Amstadter, A. B. (2019). Molecular genetic overlap between posttraumatic stress disorder and sleep phenotypes. *Sleep*, . [in press].

Lind, M. J., Marraccini, M. E., Sheerin, C. M., Bountress, K., Bacanu, S. A., Amstadter, A. B., & Nugent, N. R. (2017). Association of posttraumatic stress disorder with rs2267735 in the ADCYAP1R1 gene: A meta-analysis. *Journal of Traumatic Stress, 30*(4), 389–398. https://doi.org/10.1002/jts.22211.

Logue, M. W., Amstadter, A. B., Baker, D. G., Duncan, L., Koenen, K. C., Liberzon, I., ... Uddin, M. (2015). The psychiatric genomics consortium posttraumatic stress disorder workgroup: Posttraumatic stress disorder enters the age of large-scale genomic collaboration. *Neuropsychopharmacology, 40*(10), 2287–2297. https://doi.org/10.1038/npp.2015.118.

Logue, M. W., Baldwin, C., Guffanti, G., Melista, E., Wolf, E. J., Reardon, A. F., ... Miller, M. W. (2013). A genome-wide association study of post-traumatic stress disorder identifies the retinoid-related orphan receptor alpha (RORA) gene as a significant risk locus. *Molecular Psychiatry, 18*(8), 937–942. https://doi.org/10.1038/mp.2012.113.

Logue, M. W., Smith, A. K., Baldwin, C., Wolf, E., Guffanti, G., Ratanatharathorn, A., ... Miller, M. (2015). An analysis of gene expression in PTSD implicates genes involved in the glucocorticoid receptor pathway and neural responses to stress. *Psychoneuroendocrinology, 57*, 1–13.

Lutz, P. E., Tanti, A., Gasecka, A., Barnett-Burns, S., Kim, J. J., Zhou, Y., ... Turecki, G. (2017). Association of a history of child abuse with impaired myelination in the anterior cingulate cortex: Convergent epigenetic, transcriptional, and morphological evidence. *American Journal of Psychiatry*, https://doi.org/10.1176/appi.ajp.2017.16111286.

Lyons, M. J., Genderson, M., Grant, M. D., Logue, M., Zink, T., McKenzie, R., ... Kremen, W. S. (2013). Gene-environment interaction of Apo E genotype and combat exposure on PTSD. *American Journal of Medical Genetics Part B, Neuropsychiatric Genetics, 162B*(7), 762–769. https://doi.org/10.1002/ajmg.b.32154.

Lyons, M. J., Goldberg, J., Eisen, S. A., True, W., Tsuang, M. T., Meyer, J. M., & Henderson, W. G. (1993). Do genes influence exposure to trauma? A twin study of combat. *American Journal of Medical Genetics, 48*(1), 22–27. https://doi.org/10.1002/ajmg.1320480107.

Manuck, S. B., & McCaffery, J. M. (2014). Gene-environment interaction. *Annual Review of Psychology, 65*, 41–70. https://doi.org/10.1146/annurev-psych-010213-115100.

Marees, A. T., de Kluiver, H., Stringer, S., Vorspan, F., Curis, E., Marie-Claire, C., & Derks, E. M. (2018). A tutorial on conducting genome-wide association studies: Quality control and statistical analysis. *International Journal of Methods in Psychiatric Research, 27*(2), e1608. https://doi.org/10.1002/mpr.1608.

Martin, C., Cho, Y. E., Kim, H., Yun, S., Kanefsky, R., Lee, H., ... Gill, J. (2018). Altered DNA methylation patterns associated with clinically relevant increases in PTSD symptoms and PTSD symptom profiles in military personnel. *Biological Research for Nursing, 20*(3), 352–358. https://doi.org/10.1177/1099800418758951.

Marzi, S. J., Sugden, K., Arseneault, L., Belsky, D. W., Burrage, J., Corcoran, D. L., ... Caspi, A. (2018). Analysis of DNA methylation in young people: Limited evidence for an association between victimization stress and epigenetic variation in blood. *The American Journal of Psychiatry, 175*(6), 517–529. https://doi.org/10.1176/appi.ajp.2017.17060693.

Mehta, D., Bruenig, D., Carrillo-Roa, T., Lawford, B., Harvey, W., Morris, C. P., ... Voisey, J. (2017). Genomewide DNA methylation analysis in combat veterans reveals a novel locus for PTSD. *Acta Psychiatrica Scandinavica, 136*(5), 493–505. https://doi.org/10.1111/acps.12778.

Mehta, D., Klengel, T., Conneely, K. N., Smith, A. K., Altmann, A., Pace, T. W., ... Binder, E. B. (2013). Childhood maltreatment is associated with distinct genomic and epigenetic profiles in posttraumatic stress disorder. *Proceedings of the National Academy of Sciences of the United States of America, 110*(20), 8302–8307. https://doi.org/10.1073/pnas.1217750110.

Melroy-Greif, W. E., Wilhelmsen, K. C., Yehuda, R., & Ehlers, C. L. (2017). Genome-wide association study of post-traumatic stress disorder in two high-risk populations. *Twin Research and Human Genetics, 20*(3), 197–207.

Miller, M. W., Wolf, E. J., Logue, M. W., & Baldwin, C. T. (2013). The retinoid-related orphan receptor alpha (RORA) gene and fear-related psychopathology. *Journal of Affective Disorders, 151*(2), 702–708. https://doi.org/10.1016/j.jad.2013.07.022.

Mustapic, M., Maihofer, A. X., Mahata, M., Chen, Y., Baker, D. G., O'Connor, D. T., & Nievergelt, C. M. (2014). The catecholamine biosynthetic enzyme dopamine beta-hydroxylase (DBH): First genome-wide search positions trait-determining variants acting additively in the proximal promoter. *Human Molecular Genetics, 23*(23), 6375–6384. https://doi.org/10.1093/hmg/ddu332.

Navarro-Mateu, F., Escamez, T., Koenen, K. C., Alonso, J., & Sanchez-Meca, J. (2013). Meta-analyses of the 5-HTTLPR polymorphisms and post-traumatic stress disorder. *PLoS One, 8*(6), e66227. https://doi.org/10.1371/journal.pone.0066227.

Ni, G., Moser, G., Schizophrenia Working Group of the Psychiatric Genomics Consortium, Wray, N. R., & Lee, S. H. (2018). Estimation of genetic correlation via linkage disequilibrium score regression and genomic restricted maximum likelihood. *American Journal of Human Genetics, 102*(6), 1185–1194. https://doi.org/10.1016/j.ajhg.2018.03.021.

Nievergelt, C. M., Maihofer, A. X., Mustapic, M., Yurgil, K. A., Schork, N. J., Miller, M. W., … Baker, D. G. (2015). Genomic predictors of combat stress vulnerability and resilience in U.S. Marines: A genome-wide association study across multiple ancestries implicates PRTFDC1 as a potential PTSD gene. *Psychoneuroendocrinology, 51*, 459–471. https://doi.org/10.1016/j.psyneuen.2014.10.017.

Norrholm, S. D., Jovanovic, T., Smith, A. K., Binder, E., Klengel, T., Conneely, K., … Ressler, K. J. (2013). Differential genetic and epigenetic regulation of catechol-O-methyltransferase is associated with impaired fear inhibition in posttraumatic stress disorder. *Frontiers in Behavioral Neuroscience, 7*, 30. https://doi.org/10.3389/fnbeh.2013.00030.

O'Donovan, A., Sun, B., Cole, S., Rempel, H., Lenoci, M., Pulliam, L., & Neylan, T. (2011). Transcriptional control of monocyte gene expression in post-traumatic stress disorder. *Disease Markers, 30*(2–3), 123–132. https://doi.org/10.3233/dma-2011-0768.

Polimanti, R., Amstadter, A. B., Stein, M. B., Almli, L. M., Baker, D. G., Bierut, L. J., … Psychiatric Genomics Consortium Posttraumatic Stress Disorder Workgroup. (2017). A putative causal relationship between genetically determined female body shape and posttraumatic stress disorder. *Genome Medicine, 9*(1), 99. https://doi.org/10.1186/s13073-017-0491-4.

Powers, A., Almli, L., Smith, A., Lori, A., Leveille, J., Ressler, K. J., … Bradley, B. (2016). A genome-wide association study of emotion dysregulation: Evidence for interleukin 2 receptor alpha. *Journal of Psychiatric Research, 83*, 195–202. https://doi.org/10.1016/j.jpsychires.2016.09.006.

Prins, A., Bovin, M. J., Smolenski, D. J., Marx, B. P., Kimerling, R., Jenkins-Guarnieri, M. A., … Tiet, Q. Q. (2016). The primary care PTSD screen for DSM-5 (PC-PTSD-5): Development and evaluation within a veteran primary care sample. *Journal of General Internal Medicine, 31*(10), 1206–1211. https://doi.org/10.1007/s11606-016-3703-5.

Rakyan, V. K., Down, T. A., Balding, D. J., & Beck, S. (2011). Epigenome-wide association studies for common human diseases. *Nature Reviews. Genetics, 12*(8), 529–541. https://doi.org/10.1038/nrg3000.

Ressler, K. J., Mercer, K. B., Bradley, B., Jovanovic, T., Mahan, A., Kerley, K., … May, V. (2011). Post-traumatic stress disorder is associated with PACAP and the PAC1 receptor. *Nature, 470*(7335), 492–497.

Roberts, A. L., Gilman, S. E., Breslau, J., Breslau, N., & Koenen, K. C. (2011). Race/ethnic differences in exposure to traumatic events, development of post-traumatic stress disorder, and treatment-seeking for post-traumatic stress disorder in the United States. *Psychological Medicine*, *41*(1), 71–83. https://doi.org/10.1017/s0033291710000401.

Roby, Y. (2017). Apolipoprotein E variants and genetic susceptibility to combat-related post-traumatic stress disorder: A meta-analysis. *Psychiatric Genetics*, *27*(4), 121–130. https://doi.org/10.1097/ypg.0000000000000174.

Roth, T. L. (2014). How traumatic experiences leave their signature on the genome: An overview of epigenetic pathways in PTSD. *Frontiers in Psychiatry*, *5*, 93. https://doi.org/10.3389/fpsyt.2014.00093.

Rusiecki, J. A., Byrne, C., Galdzicki, Z., Srikantan, V., Chen, L., Poulin, M., … Baccarelli, A. (2013). PTSD and DNA methylation in select immune function gene promoter regions: A repeated measures case-control study of U.S. military service members. *Frontiers in Psychiatry*, *4*, 56. https://doi.org/10.3389/fpsyt.2013.00056.

Rutten, B. P. F., Vermetten, E., Vinkers, C. H., Ursini, G., Daskalakis, N. P., Pishva, E., … Boks, M. P. M. (2018). Longitudinal analyses of the DNA methylome in deployed military servicemen identify susceptibility loci for post-traumatic stress disorder. *Molecular Psychiatry*, *23*(5), 1145–1156. https://doi.org/10.1038/mp.2017.120.

Sarapas, C., Cai, G., Bierer, L. M., Golier, J. A., Galea, S., Ising, M., … Yehuda, R. (2011). Genetic markers for PTSD risk and resilience among survivors of the World Trade Center attacks. *Disease Markers*, *30*(2), 101–110.

Sartor, C. E., Grant, J. D., Lynskey, M. T., McCutcheon, V. V., Waldron, M., Statham, D. J., … Nelson, E. C. (2012). Common heritable contributions to low-risk trauma, high-risk trauma, posttraumatic stress disorder, and major depression. *Archives of General Psychiatry*, *69*(3), 293–299. https://doi.org/10.1001/archgenpsychiatry.2011.1385.

Sartor, C. E., McCutcheon, V. V., Pommer, N. E., Nelson, E. C., Grant, J. D., Duncan, A. E., … Heath, A. C. (2011). Common genetic and environmental contributions to post-traumatic stress disorder and alcohol dependence in young women. *Psychological Medicine*, *41*(7), 1497–1505. https://doi.org/10.1017/s0033291710002072.

Schaal, S., Koebach, A., Hinkel, H., & Elbert, T. (2015). Posttraumatic stress disorder according to DSM-5 and DSM-IV diagnostic criteria: A comparison in a sample of Congolese ex-combatants. *European Journal of Psychotraumatology*, *6*, 24981. https://doi.org/10.3402/ejpt.v6.24981.

Segman, R. H., Shefi, N., Goltser-Dubner, T., Friedman, N., Kaminski, N., & Shalev, A. Y. (2005). Peripheral blood mononuclear cell gene expression profiles identify emergent post-traumatic stress disorder among trauma survivors. *Molecular Psychiatry*, *10*(5), 500–513.

Sheerin, C. M., Lind, M. J., Bountress, K., Nugent, N. R., & Amstadter, A. B. (2017). The genetics and epigenetics of PTSD: Overview, recent advances, and future directions. *Current Opinion in Psychology*, *14*, 5–11. https://doi.org/10.1016/j.copsyc.2016.09.003.

Smith, A. K., Conneely, K. N., Kilaru, V., Mercer, K. B., Weiss, T. E., Bradley-Davino, B., … Ressler, K. J. (2011). Differential immune system DNA methylation and cytokine regulation in posttraumatic stress disorder. *American Journal of Medical Genetics Part B, Neuropsychiatric Genetics*, *156*, 700–708.

Smith, G. D., & Ebrahim, S. (2003). 'Mendelian randomization': Can genetic epidemiology contribute to understanding environmental determinants of disease? *International Journal of Epidemiology*, *32*(1), 1–22.

Stein, M. B., Chen, C. Y., Ursano, R. J., Cai, T., Gelernter, J., Heeringa, S. G., … Resilience in Servicemembers Collaborators (2016). Genome-wide association studies of posttraumatic stress disorder in 2 cohorts of US army soldiers. *JAMA Psychiatry*, *73*(7), 695–704. https://doi.org/10.1001/jamapsychiatry.2016.0350.

Stein, M. B., Jang, K. L., & Livesley, W. J. (2002). Heritability of social anxiety-related concerns and personality characteristics: A twin study. *The Journal of Nervous and Mental Disease*, *190*(4), 219–224.

Su, Y. A., Wu, J., Zhang, L., Zhang, Q., Su, D. M., He, P., ... Ursano, R. J. (2008). Dysregulated mitochondrial genes and networks with drug targets in postmortem brain of patients with posttraumatic stress disorder (PTSD) revealed by human mitochondria-focused cDNA microarrays. *International Journal of Biological Sciences*, *4*(4), 223–235.

Sumner, J. A., Duncan, L., Ratanatharathorn, A., Roberts, A. L., & Koenen, K. C. (2016). PTSD has shared polygenic contributions with bipolar disorder and schizophrenia in women. *Psychological Medicine*, *46*(3), 669–671. https://doi.org/10.1017/S0033291715002135.

Sumner, J. A., Duncan, L. E., Wolf, E. J., Amstadter, A. B., Baker, D. G., Beckham, J. C., ... Vermetten, E. (2017). Letter to the Editor: Posttraumatic stress disorder has genetic overlap with cardiometabolic traits. *Psychological Medicine*, *47*(11), 2036–2039. https://doi.org/10.1017/S0033291717000733.

Teperino, R., Lempradl, A., & Pospisilik, J. A. (2013). Bridging epigenomics and complex disease: The basics. *Cellular and Molecular Life Sciences*, *70*(9), 1609–1621. https://doi.org/10.1007/s00018-013-1299-z.

Thomas, D. (2010). Gene-environment-wide association studies: Emerging approaches. *Nature Reviews Genetics*, *11*(4), 259–272.

Thompson, N. J., Fiorillo, D., Rothbaum, B. O., Ressler, K. J., & Michopoulos, V. (2018). Coping strategies as mediators in relation to resilience and posttraumatic stress disorder. *Journal of Affective Disorders*, *225*, 153–159. https://doi.org/10.1016/j.jad.2017.08.049.

Toyokawa, S., Uddin, M., Koenen, K. C., & Galea, S. (2012). How does the social environment 'get into the mind'? Epigenetics at the intersection of social and psychiatric epidemiology. *Social Science & Medicine*, *74*(1), 67–74. https://doi.org/10.1016/j.socscimed.2011.09.036.

True, W. R., Rice, J., Eisen, S. A., Heath, A. C., Goldberg, J., Lyons, M. J., & Nowak, J. (1993). A twin study of genetic and environmental contributions to liability for posttraumatic stress symptoms. *Archives of General Psychiatry*, *50*(4), 257–264.

Tylee, D. S., Chandler, S. D., Nievergelt, C. M., Liu, X., Pazol, J., Woelk, C. H., ... Tsuang, M. T. (2015). Blood-based gene-expression biomarkers of post-traumatic stress disorder among deployed marines: A pilot study. *Psychoneuroendocrinology*, *51*, 472–494. https://doi.org/10.1016/j.psyneuen.2014.09.024.

Uddin, M., Aiello, A. E., Wildman, D. E., Koenen, K. C., Pawelec, G., de los Santos, R., ... Galea, S. (2010). Epigenetic and immune function profiles associated with posttraumatic stress disorder. *Proceedings of the National Academy of Sciences of the United States of America*, *107*(20), 9470–9475.

Uddin, M., Galea, S., Chang, S. C., Koenen, K. C., Goldmann, E., Wildman, D. E., & Aiello, A. E. (2013). Epigenetic signatures may explain the relationship between socioeconomic position and risk of mental illness: Preliminary findings from an urban community-based sample. *Biodemography and Social Biology*, *59*(1), 68–84. https://doi.org/10.1080/19485565.2013.774627.

Visscher, P. M., Hemani, G., Vinkhuyzen, A. A., Chen, G. B., Lee, S. H., Wray, N. R., ... Yang, J. (2014). Statistical power to detect genetic (co)variance of complex traits using SNP data in unrelated samples. *PLoS Genetics*, *10*(4), e1004269. https://doi.org/10.1371/journal.pgen.1004269.

Visscher, P. M., Wray, N. R., Zhang, Q., Sklar, P., McCarthy, M. I., Brown, M. A., & Yang, J. (2017). 10 years of GWAS discovery: Biology, function, and translation. *American Journal of Human Genetics*, *101*(1), 5–22. https://doi.org/10.1016/j.ajhg.2017.06.005.

Wang, T. (2015). Does BDNF Val66Met polymorphism confer risk for posttraumatic stress disorder? *Neuropsychobiology*, *71*(3), 149–153. https://doi.org/10.1159/000381352.

Wolf, E. J., Maniates, H., Nugent, N., Maihofer, A. X., Armstrong, D., Ratanatharathorn, A., ... Logue, M. W. (2018). Traumatic stress and accelerated DNA methylation age: A meta-analysis. *Psychoneuroendocrinology*, *92*, 123–134. https://doi.org/10.1016/j.psyneuen.2017.12.007.

Wolf, E. J., Miller, D. R., Logue, M. W., Sumner, J., Stoop, T. B., Leritz, E. C., … Miller, M. W. (2017). Contributions of polygenic risk for obesity to PTSD-related metabolic syndrome and cortical thickness. *Brain, Behavior, and Immunity*, *65*, 328–336. https://doi.org/10.1016/j.bbi.2017.06.001.

Wolf, E. J., Miller, M. W., Sullivan, D. R., Amstadter, A. B., Mitchell, K. S., Goldberg, J., & Magruder, K. M. (2018). A classical twin study of PTSD symptoms and resilience: Evidence for a single spectrum of vulnerability to traumatic stress. *Depression and Anxiety*, *35*(2), 132–139. https://doi.org/10.1002/da.22712.

Wolf, E. J., Rasmusson, A. M., Mitchell, K. S., Logue, M. W., Baldwin, C. T., & Miller, M. W. (2014). A genome-wide association study of clinical symptoms of dissociation in a trauma-exposed sample. *Depression and Anxiety*, *31*(4), 352–360. https://doi.org/10.1002/da.22260.

Wray, N. R., Lee, S. H., Mehta, D., Vinkhuyzen, A. A., Dudbridge, F., & Middeldorp, C. M. (2014). Research review: Polygenic methods and their application to psychiatric traits. *Journal of Child Psychology and Psychiatry*, *55*(10), 1068–1087. https://doi.org/10.1111/jcpp.12295.

Wu, H., & Zhang, Y. (2014). Reversing DNA methylation: Mechanisms, genomics, and biological functions. *Cell*, *156*(1–2), 45–68. https://doi.org/10.1016/j.cell.2013.12.019.

Xian, H., Chantarujikapong, S. I., Scherrer, J. F., Eisen, S. A., Lyons, M. J., Goldberg, J., … True, W. R. (2000). Genetic and environmental influences on posttraumatic stress disorder, alcohol and drug dependence in twin pairs. *Drug and Alcohol Dependence*, *61*(1), 95–102.

Xie, P., Kranzler, H. R., Poling, J., Stein, M. B., Anton, R. F., Farrer, L. A., & Gelernter, J. (2010). Interaction of FKBP5 with childhood adversity on risk for post-traumatic stress disorder. *Neuropsychopharmacology*, *35*(8), 1684–1692. https://doi.org/10.1038/npp.2010.37.

Xie, P., Kranzler, H. R., Yang, C., Zhao, H., Farrer, L. A., & Gelernter, J. (2013). Genome-wide association study identifies new susceptibility loci for posttraumatic stress disorder. *Biological Psychiatry*, *74*(9), 656–663. https://doi.org/10.1016/j.biopsych.2013.04.013.

Yang, J., Bakshi, A., Zhu, Z., Hemani, G., Vinkhuyzen, A. A., Lee, S. H., … Visscher, P. M. (2015). Genetic variance estimation with imputed variants finds negligible missing heritability for human height and body mass index. *Nature Genetics*, *47*(10), 1114–1120. https://doi.org/10.1038/ng.3390.

Yang, J., Lee, S. H., Goddard, M. E., & Visscher, P. M. (2011). GCTA: A tool for genome-wide complex trait analysis. *American Journal of Human Genetics*, *88*(1), 76–82. https://doi.org/10.1016/j.ajhg.2010.11.011. pii: S0002-9297(10)00598-7.

Yang, J., Lee, S. H., Goddard, M. E., & Visscher, P. M. (2013). Genome-wide complex trait analysis (GCTA): Methods, data analyses, and interpretations. *Methods in Molecular Biology*, *1019*, 215–236. https://doi.org/10.1007/978-1-62703-447-0_9.

Yang, J., Manolio, T. A., Pasquale, L. R., Boerwinkle, E., Caporaso, N., Cunningham, J. M., … Visscher, P. M. (2011). Genome partitioning of genetic variation for complex traits using common SNPs. *Nature Genetics*, *43*(6), 519–525. https://doi.org/10.1038/ng.823.

Yang, J., Zeng, J., Goddard, M. E., Wray, N. R., & Visscher, P. M. (2017). Concepts, estimation and interpretation of SNP-based heritability. *Nature Genetics*, *49*(9), 1304–1310. https://doi.org/10.1038/ng.3941.

Yehuda, R., Cai, G., Golier, J. A., Sarapas, C., Galea, S., Ising, M., … Buxbaum, J. D. (2009). Gene expression patterns associated with posttraumatic stress disorder following exposure to the World Trade Center attacks. *Biological Psychiatry*, *66*(7), 708–711. https://doi.org/10.1016/j.biopsych.2009.02.034.

Yehuda, R., Flory, J. D., Bierer, L. M., Henn-Haase, C., Lehrner, A., Desarnaud, F., … Meaney, M. J. (2015). Lower methylation of glucocorticoid receptor gene promoter 1F in peripheral blood of veterans with posttraumatic stress disorder. *Biological Psychiatry*, 77(4), 356–364. https://doi.org/10.1016/j.biopsych.2014.02.006.

Zannas, A. S., Provencal, N., & Binder, E. B. (2015). Epigenetics of posttraumatic stress disorder: Current evidence, challenges, and future directions. *Biological Psychiatry*, 78(5), 327–335. https://doi.org/10.1016/j.biopsych.2015.04.003.

Zhao, M., Yang, J., Wang, W., Ma, J., Zhang, J., Zhao, X., … Yang, Y. (2017). Meta-analysis of the interaction between serotonin transporter promoter variant, stress, and posttraumatic stress disorder. *Scientific Reports*, 7(1), 16532. https://doi.org/10.1038/s41598-017-15168-0.

CHAPTER 9

Psychophysiology of emotional responding in PTSD

Brittney P. Innocente[a], Leah T. Weingast[a], Renie George[a], Seth Davin Norrholm[a,b]

[a]Mental Health Service Line, Atlanta Veterans Administration Health Care System, Decatur, GA, United States
[b]Department of Psychiatry and Behavioral Sciences, Emory University School of Medicine, Atlanta, GA, United States

Introduction

Current PTSD research predominantly utilizes self-report measures to identify and characterize emotion dysregulation and PTSD symptoms. This methodology involves subjective data collection and results in a retrospective study design, rather than a design used to project symptom development. In contrast an increasing number of studies are exploring the usage of psychophysiological measures to provide an objective and internally reliable alternative to assess these items. In addition to the consistent and objective nature of data, psychophysiological techniques can also illuminate biological processes associated with risk and resiliency for PTSD. For example, intermediate phenotypes and biomarkers indicative of PTSD symptomology may contribute to preventative, holistic, and efficacious treatment plans. Intermediate phenotypes can be defined as traits representing fundamental neurobiological mechanisms related to the clinical presentation of an illness (Zuj, Palmer, Hsu, et al., 2016; Zuj, Palmer, Lommen, & Felmingham, 2016). Relatedly, biomarkers are measurable indicators of biological processes, both pathological and healthy, used to identify vulnerability for a disorder and preventative interventions (Briscione, Jovanovic, & Norrholm, 2014; Jovanovic et al., 2011). Laboratory paradigms targeting the autonomic nervous system are currently being employed to identify intermediate phenotypes and biomarkers related to the etiology and maintenance of PTSD symptoms.

Patients suffering from PTSD-induced hyperarousal often experience physical symptoms such as racing heart, breathing difficulties, and increased perspiration, effects that are regulated, in part, by the autonomic nervous system. Simple psychophysiological techniques can be used in translational neuroscience studies to measure these processes by tracking peripheral targets like heart rate (HR), blood pressure (BP), respiratory rate, and skin conductance (for a review see Norrholm & Jovanovic, 2018). In addition to the aforementioned symptoms of hyperarousal, PTSD patients also often experience hypervigilance, which can include enhanced sensory sensitivity exacerbated by one's inability to feel safe. Fear-potentiated startle methodologies encompass both the physical and emotional underpinnings of PTSD signs and symptoms, including hyperarousal and intrusive fear memories. In human studies, exaggerated startle responses are commonly measured as and expressed by the magnitude and frequency of *orbicularis oculi* muscle contractions recorded using electromyography (EMG). Such EMG responses are recorded while conditioned stimuli are paired with aversive unconditioned stimuli and acoustic startle responses are assessed (see fear-potentiated startle; Davis, 1992). Most research performed to date on the psychophysiology of PTSD has largely focused on quantifying elevated activity within cardiovascular, electrodermal, and electromyographic systems in PTSD patients as compared with non-PTSD populations with and without previous trauma exposure. Applied psychophysiological studies, such as those alluded to in the preceding text, often employ Pavlovian fear conditioning paradigms, which appear to have significant predictive value for PTSD symptom development and maintenance (Norrholm & Jovanovic, 2018). These paradigms can be administered efficiently, noninvasively, and by clinicians and researchers alike in an effort to further our understanding of patients' somatic and emotional symptoms in clinical settings.

Psychophysiological indices revealing emotion dysregulation in PTSD
Cardiovascular indices

Cardiovascular activity is most commonly measured via electrocardiographic (ECG) analysis of heart rate, observed changes in blood pressure indices (systolic BP and diastolic BP), or alterations in respiration rate (Cacioppo, Tassinary, & Berntson, 2016). More specifically, heart rate changes evident through electrocardiograms can be expressed as heart rate in beats per minute, the variation in the time interval between heartbeats (interbeat interval), and/or natural variation in heart rate that occurs during each breathing cycle (respiratory sinus arrhythmia, RSA; McCraty & Shaffer, 2015). Cardiovascular measures can be recorded noninvasively through electrode placement on the wrist or chest of a participant, through finger seismogram, or via chest band (ECG, BP, and respiration rate, respectively).

The relationship between fear expression and changes in cardiovascular activity has been extensively observed (Stiedl & Hager, 2017). For example, results tend to show that HR responses generally decrease in the absence of fear expression and increase in the presence of fearful behaviors (Lonsdorf et al., 2017; Lonsdorf & Merz, 2017). A metaanalysis of over 100 studies comparing psychophysiological variables in adults with and without PTSD found that HR variability was robustly correlated with PTSD across all study types (Pole, 2007). PTSD patients display a significantly increased HR response, as compared with non-PTSD trauma survivors, when presented with scripts of their trauma (Michopoulos, Norrholm, & Jovanovic, 2015). In addition to changes in HR response, Wolfe et al. (2000) found higher systolic BP responses in female veterans with PTSD when exposed to standardized trauma cues as compared with a group without PTSD. Buckley and Kaloupek (2001) also conducted a metaanalysis of 34 studies investigating cardiovascular activity as it relates to PTSD diagnosis and found significantly higher HR and BP levels in PTSD+ individuals relative to those without PTSD. Taken together, this expanding body of literature suggests a strong association between altered cardiovascular activity and PTSD symptom expression. Such an association indicates the potential for the clinical utility of various cardiovascular activity measures as a treatment outcome predictor or index.

Electrodermal indices

Electrodermal activity of the skin reflects general states of arousal that are regulated by the activation of the autonomic nervous system through sweat gland stimulation that can, in turn, increase the electrical conductivity of the skin (Jovanovic & Norrholm, 2016). Electrodermal activity is expressed as either skin conductance response or skin conductance level. Skin conductance response refers to the change in moisture level of the skin before and after presentation of a stimulus (phasic), whereas skin conductance level refers to the average level of conductivity created by sweat production during a set time interval (tonic; Jovanovic & Norrholm, 2016). Skin conductance is frequently utilized in human fear conditioning studies due to its ease of use and robust responsivity to the introduction of sensory stimuli. In a typical human study employing skin conductance measures, electrodes filled with isotonic paste are placed either on the hypothenar surface of the palm or on outer phalanges of the middle and index fingers of the nondominant hand. Skin conductance response is relatively slow, with an onset latency of 1–4 s and peak values achieved between 3 and 6 s poststimulus presentation, meaning that skin conductance can be sampled at much lower rates than other autonomic measures (as low as 10 Hz; Jovanovic & Norrholm, 2016).

As one of the most commonly used indicators of physical arousal, electrodermal activity is often utilized in translational PTSD research, and alterations to this type of psychophysiological responding have been consistently associated with PTSD. The aforementioned metaanalysis also reported significant differences in skin conductance responses between individuals with and without PTSD (Pole, 2007). Additionally, several studies have found that, compared with trauma survivors without PTSD, affected patients display higher skin conductance responses when presented with scripts of their trauma (for review, see Michopoulos et al., 2015). However, previous work has also shown that PTSD patients exhibit blunted skin conductance responses when exposed to the threat of shock (Blechert, Michael, Grossman, Lajtman, & Wilhelm, 2007). To date, skin conductance methodologies have proven to be useful complementary tools for the study of human fear and anxiety responses when these data are interpreted within the context of clinical presentations and additional self-report and physiological indices.

Electromyographic indices

Electromyography (EMG) refers to a procedure used to assess the electrical activity produced by skeletal muscles and associated nerve function. PTSD researchers often utilize EMG procedures to measure contraction of the muscles of the face that underlie common human facial expressions, most notably the *orbicularis oculi* (eye blink), the *zygomaticus major* (smile), the *corrugator supercilii* (frown), and the *frontalis* muscles (surprise). It is worth noting that facial expressions are controlled by volitional muscle movements and, as such, they can be voluntarily controlled and thus can be less objective measures than those assessing autonomic responses. As mentioned briefly in a previous section, EMG activity of the *orbicularis oculi* muscle, which mediates eyeblink, can be captured as part of fear-potentiated startle procedures. The rapid onset of startle responses (latency of 20–200 ms) eliminates the possibility of intentional muscle movement and reduces the possible effect of voluntary control (Jovanovic & Norrholm, 2016). Acoustic startle responses are most often elicited through the presentation of brief (~40 ms) bursts of white noise delivered binaurally through earphones. These responses are most readily measured by electrodes placed directly over the *orbicularis* muscle positioned below the pupil and the lateral canthus. Fear-potentiated startle responses are defined as an increase in the frequency of magnitude of the acoustic startle response in the presence of a previously neutral cue that has been repeatedly paired with an aversive outcome (i.e., a conditioned stimulus [CS]–unconditioned stimulus [US] association is made, and a fear memory trace is encoded).

Both baseline acoustic startle and fear-potentiated startle responses have been investigated in varying PTSD populations (Metzger et al., 1999; Morgan, Grillon, Southwick, Davis, & Charney, 1996; Morgan, Grillon, Lubin, & Southwick, 1997; Orr et al., 1995; Shalev et al., 1997) with results supporting the use of these paradigms as laboratory correlates to the clinical etiology and presentation of fear-related PTSD symptoms as described later.

Translational psychophysiological research: Fear conditioning

Overview

Pavlovian threat conditioning, which is rooted in classical conditioning and whose expression is highly conserved phylogenetically across mammals (Pavlov, 1927; Rothbaum & Davis, 2003), has shown great utility by providing researchers with highly translational methods for investigating the dysregulation of fear processing in PTSD (Briscione et al., 2014; Norrholm & Jovanovic, 2011). In typical human fear conditioning paradigms, a neutral stimulus is paired with an aversive unconditioned stimulus (US) that naturally induces an innate physiological fear response (UCR). Through its association with the US, the previously neutral stimulus becomes a conditioned stimulus (CS) and elicits the UCR, even in the absence of the US. The consolidated fear memory, connecting the US and the CS, can persist until the learning parameters change (e.g., repeated pairings of the CS without the US). In the latter example a more recently encoded extinction memory trace can compete with and overcome expression of the original CS–US association. This new memory trace does not eliminate the original CS–US association, but prevents its presentation until environmental or contextual factors change (Norrholm & Jovanovic, 2011).

Threat conditioning has been utilized as a translational tool to model human fear and anxiety disorders, including PTSD. As introduced previously, one successful approach has been the use of paradigms employing *fear-potentiated startle* or the relative increase in the frequency or magnitude of acoustic startle response when a neutral cue is paired with an aversive outcome (Davis, 1992, 1993; Davis, Falls, Campeau, & Kim, 1993).

Early iterations of human fear-potentiated startle paradigms included the use of instructed fear paradigms in which participants were informed that they would receive an aversive US during the experimental procedures. For example, participants may be instructed that one initially neutral stimulus (e.g., a geometric shape or illuminated light) would be paired and presented with a shock and another initially neutral stimulus would not be associated with a shock outcome. In doing so, "threat" and "no-threat" conditions are established during which the acoustic startle response is elicited through the administration of white noise sound "probes." Using this technique in healthy controls, Grillon, Ameli, Woods, Merikangas, and Davis (1991) observed increased acoustic startle responses when the participants anticipated receiving an electric shock as compared with conditions in which they did not expect the aversive shock stimulus, an effect that would be defined as the expression of fear-potentiated startle. This technique was then further applied to a clinical research sample with and without PTSD. For example, Morgan, Grillon, Southwick, Davis, and Charney (1995) assessed fear-potentiated startle responses using the aforementioned instructed fear paradigm and reported that individuals with PTSD exhibited significantly greater startle amplitude while under threat compared with non-PTSD participants. This early work was some of the first of its kind and revealed that PTSD patients appear to experience a disproportionate level of fear in the context of potential threat, even if no actual threat is presented. This suggested that fear dysregulation may be a central feature of PTSD worthy of increased clinical research focus and further refinement of psychophysiological tools such as potentiation of the acoustic startle response.

Translational fear conditioning models
Fear acquisition

Fear acquisition is a putative laboratory model and analogue for the formation of trauma-related fear memories that refers to the initial threat conditioning process where an association forms between a CS and an aversive US. Acquisition often includes two phases, habituation to the conditioned stimuli alone, followed by fear conditioning. During a habituation phase the CSs are presented without any US reinforcement during which investigators can record baseline physiological responses. Fear conditioning consists of the repeated presentation of conditioned stimuli that are reinforced (CS followed by the US, or CS+) or not reinforced (CS followed by the

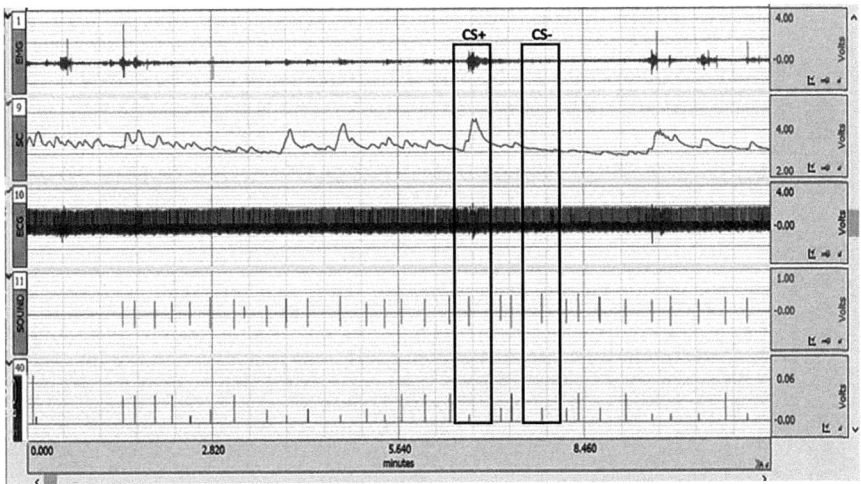

Fig. 1 Representative data acquisition session for acoustic startle (EMG), skin conductance (SC), and heart rate (ECG) during typical fear acquisition phase. SOUND channel provides experimenters with visual confirmation that a startle-eliciting noise burst was delivered.

absence of US, or CS−). In addition, many human conditioning paradigms include the presentation of startle noise probes alone (NA) as a means of maintaining a "rolling" baseline during the session. Current technological advances allow for the concurrent assessment of psychophysiological responses including heart rate, acoustic startle response, respiration, and skin conductance; such concurrent assessment allows researchers to capture fear conditioning-related changes, relative to baseline, in psychophysiological indices in real time (for representative data, see Fig. 1).

As previously introduced, fear conditioning paradigms lend themselves well to translational study across species and allow researchers to move along the translational bridge that spans the space between the preclinical laboratory and the mental health clinic. In fact, one of the most often employed fear-potentiated startle paradigms for studying the acquisition, extinction, and return of fear was developed based on earlier work with rodent models (Norrholm et al., 2006, 2008; Walker & Davis, 2000, 2002). In human fear-potentiated startle models, CSs are often represented by colored shapes on a computer monitor screen and are either paired with a 40-ms, 108-dB noise probe and reinforced by a 250-ms, 140-psi air blast to the larynx (CS + noise probe + air blast; CS+) or paired with only a noise probe (CS + noise probe − air blast; CS−); see Fig. 2 for details on the application of the paradigm. In this widely disseminated and reliable setup, the air blast

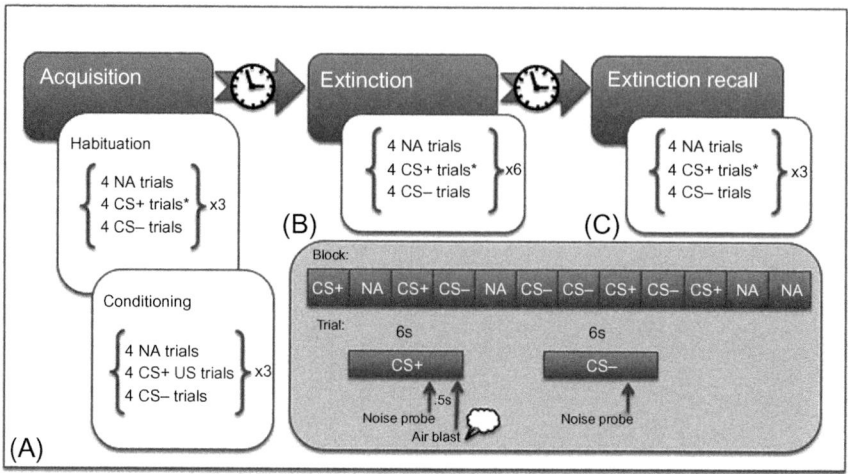

Fig. 2 Schematic outline of experimental (A) fear conditioning, (B) fear extinction, and (C) extinction recall sessions using a fear-potentiated startle paradigm. NA=noise probe alone (40ms, 108dB), synonymous with ITI as shown in similar published paradigms (i.e., Grillon & Morgan III, 1999); CS+=reinforced conditioned stimulus, noise probe; CS−=unreinforced conditioned stimulus, noise probe; US=air blast reinforcement (250ms, 140psi); * CS+ trials are not reinforced with the US during habituation, extinction, or extinction recall.

serves as an unconditioned stimulus (US) and has been shown to elicit robust fear-potentiated startle (Norrholm & Jovanovic, 2011).

There are some aspects of fear conditioning models that are unique to human applications given the obvious complexities and phylogenetic differences between mammalian nervous systems and behaviors. For example, human fear paradigms often include a self-report measure of "fear," "anxiety," or expectancy of the US on a given CS trial. Ratings such as these are typically performed in real time during an experimental session with the use of an input device (e.g., response keypad or visual Likert scale). Participant ratings, such as US-expectancy ratings, help to clarify whether participants can explicitly discriminate between danger (CS+) and safety (CS−) cues, which can help discern dysregulated fear processing at the cognitive level of awareness versus the psychophysiological level of reactivity. As an example of this level of discrimination, one fear conditioning and extinction learning study by Acheson et al. (2015) revealed distinct patterns separating PTSD and anxiety groups from healthy and depressed groups by concurrently collecting EMG responses, self-reported anxiety ratings, and participant US-expectancy ratings (Acheson et al., 2015). The authors reported that active-duty infantry marines and navy corpsmen with PTSD symptoms only displayed larger startle responses to the safety cue (CS−) throughout fear acquisition learning as compared with the psychiatrically healthy control group, who exhibited successful stimulus discrimination and contingency awareness (measured via US-expectancy ratings). Conversely, when examining US-expectancy ratings and postacquisition self-report anxiety ratings, the PTSD group displayed intact contingency awareness and discrimination learning (Acheson et al., 2015). These results demonstrated a dissociation between psychophysiological responses and cognitive awareness responses to danger and safety cues, an observed phenomenon that has been discussed at length in previous work (Norrholm et al., 2008; Warren et al., 2014).

While there is a relatively new clinical distinction being made between the diagnosis of anxiety disorders versus trauma- and stressor-related disorders in the Diagnostic and Statistical Manual of Mental Disorders, Fifth Edition (DSM-5; American Psychiatric Association, 2013), translational neuroscience efforts are underway to determine the degree of overlap versus distinction between disorders in which fear and anxiety lie at the central core of the disease. For example, Jovanovic et al. (2010) reported that impaired fear inhibition was more closely associated with PTSD rather than depression when assessed in groups reporting symptoms of PTSD, major depressive disorder, or a comorbidity with both (Jovanovic et al., 2010). In addition, Acheson et al. (2015) found that, while anxiety, depression, and psychiatrically healthy groups displayed discrimination between danger and safety cues, groups with anxiety-related symptoms showed increased baseline startle and fear-potentiated startle and higher self-reported anxiety to both cue types. These distinct differences suggest a link between more generalized anxiety symptoms and nonspecific exaggerated fear responses, while PTSD symptoms may be more likely to be related to the inability to differentiate between safety and danger cues (Jovanovic et al., 2010). This failure of PTSD populations to differentiate between danger and safety cues has also been established in PTSD treatment-seeking adults, trauma-exposed individuals, and Gulf War Veterans with PTSD (Grillon & Morgan III, 1999; Jovanovic et al., 2010; Peri, Ben-Shakhar, Orr, & Shalev, 2000). The inability to differentiate between safety and danger cues may be specific to PTSD patients and, with further development, could provide clinicians with a complementary predictive and/or diagnostic tool to successfully separate PTSD symptoms from comorbid psychopathology.

While PTSD-specific results from the application of this paradigm have been mixed (see Blechert et al., 2007; Glover et al., 2012), it has been posited that the fear-related elements of experiencing a traumatic event can be conceptualized as a form of fear acquisition learning. In the framework of Pavlovian conditioning, a traumatic event can be viewed as a brief, compressed, and durable form of fear acquisition (Norrholm & Jovanovic, 2018; Rothbaum & Davis, 2003; VanElzakker, Dahlgren, Davis, Dubois, & Shin, 2014). In this conceptualization, exposure to a traumatic event (e.g., a motor vehicle accident) acts as an unlearned US and elicits an unconditioned response (UCR) of intense arousal and fear, while neutral environmental stimuli present during the event (e.g., location, sights, and sounds) become associated with the fear response, rendering them as multimodal CSs. These previously neutral stimuli later possess the properties of conditioned stimuli that can elicit conditioned responses of fear and panic; as a result, reexperiencing symptoms (among other symptoms like hyperarousal) can be conceptualized as enduring conditioned responses (Amstadter, Nugent, & Koenen, 2009; Friedman, 2006). Heightened physiological responses during the aftermath of trauma have been found to predict the likelihood of PTSD development. For example, Bryant, Harvey, Guthrie, and Moulds (2000) found that elevated heart rate during the acute posttraumatic time period predicted the development of PTSD upon 6-month follow-up assessment in motor vehicle accident survivors (Bryant et al., 2000). Similar results have been replicated in firefighters (Guthrie & Bryant, 2006) and assault victims (Griffin, 2008).

Clinically, research such as that described in the previous section could enable clinicians to discern trauma survivors who are most likely to develop PTSD, a distinction that would allow clinicians to target those in greatest need of intervention before PTSD symptoms manifest, thus preventing the development of PTSD rather than relying on treatment alone. Norrholm et al. (2011) found that, compared with participants with lower reexperiencing and hyperarousal symptoms, those with higher symptoms expressed greater fear response to both the danger and safety cues during late acquisition (Norrholm et al., 2011). Identifying dysfunction in safety discrimination early in treatment can help clinicians tailor treatment aims to each patient. For individuals with PTSD and other trauma- and stress-related disorders, fear acquisition provides a model for the development of trauma-related symptoms because of persistent fear responses to once innocuous CSs. However, the fear conditioning process alone fails to explain the majority of individuals who recover after experiencing a traumatic event and do not develop PTSD; these differences may be best explained by extinction learning profiles.

Fear extinction (extinction learning and extinction recall)
Extinction learning
In healthy populations, *fear extinction* occurs after fear acquisition learning has been encoded and refers to a graded reduction in the strength of the association between a previously reinforced CS+ and a US, resulting in the diminution in the expression of a conditioned fear response. The fear extinction process involves actively recalling and establishing a new, competing memory trace by repeatedly presenting the former CS+ in the absence of the US until minimal discrimination exists in the behavioral responses provoked by the CS+ and the CS− (i.e., the CS+ no longer produces the CR; Norrholm & Jovanovic, 2018). Yet, as has been discussed at length in the fear extinction literature, the original fear memory is not erased (see Myers & Davis, 2002). The persistence of the original fear memory has been demonstrated through the return of fear after successful extinction learning (discussed further in following sections). Research strongly supports a model where impaired fear extinction plays a pivotal role in PTSD maintenance (Mineka & Oehlberg, 2008; Shin & Liberzon, 2010; Zuj & Norrholm, 2019). Impaired extinction learning has been exhibited empirically in three primary manners: impaired within-session extinction learning (Norrholm & Jovanovic, 2011), impaired between-session extinction learning (extinction recall; Milad et al., 2009; Shvil et al., 2014), and elevated expression of fear during early extinction (fear load; Briscione et al., 2014; Norrholm et al., 2015).

Within-session extinction learning
Within-session extinction learning refers to the active learning process in which a new association is made, and a previously reinforced CS now predicts the absence of an aversive consequence and, as such, a gradual reduction of conditioned fear expression. Impairments to within-session extinction learning have been reported in studies of traumatized populations with and without PTSD. For example, a study comparing PTSD+, trauma-exposed, and nonexposed healthy control groups revealed that PTSD+ participants showed greater skin conductance and increased heart rate responses to the now nonreinforced CS+ during extinction learning, a trajectory suggesting a tendency for PTSD+ individuals to experience elevated physiological responses and a slower rate of extinguishing conditioned fear (Peri et al., 2000). Norrholm et al. (2011) utilized fear-potentiated startle techniques to assess within-session fear extinction learning in trauma-exposed civilians

with and without PTSD. No differences were found between groups in baseline startle, habituation, or US-expectancy ratings. While both groups displayed significant within-session extinction, the PTSD+ group demonstrated increased fear-potentiated startle to the previously reinforced CS+ throughout the early and middle stages of extinction. These results suggest a relationship between heightened psychophysiological responses and protracted extinction rate in PTSD+ individuals. Supporting this relationship, Acheson et al. (2015) also found that a PTSD+ group of infantry marines and navy corpsmen maintained an elevated level of conditioned fear throughout the extinction session compared with healthy controls despite similar US-expectancy ratings across the groups. These results show impaired within-session extinction of conditioned psychophysiological responses despite an individual's cognitive ability to explicitly discriminate between safety and danger cues, implying a disconnect between the autonomic and cognitive processes associated with PTSD formation and maintenance. Together, these results confirm that PTSD+ populations show higher physiological reactivity during extinction learning compared with controls and an inability to extinguish at the expected rate. It is worth noting that within-session learning impairments relative to PTSD have been largely limited to fear-potentiated startle studies, whereas impaired between-session (extinction recall) deficits have been frequently reported in PTSD study samples assessed with skin conductance methodologies. This is discussed in greater detail by Glover et al. (2011).

In a fear conditioning conceptualization of the fear-related elements of clinical PTSD, if one accepts acquisition to underlie the experience of a traumatic event and the neurobehavioral development of PTSD, then successful fear extinction learning can be considered to mediate the natural reduction of posttraumatic signs and symptoms that the majority of individuals experience immediately following a traumatic event. Impaired extinction learning appears to contribute to the experience of prolonged maintenance of one's initial reaction to a traumatic event in that it either does not or is slow to diminish over time (i.e., the development and persistence of PTSD symptoms; Myers & Davis, 2002; Zuj, Palmer, Lommen, et al., 2016). Impairments in fear extinction have been correlated with and are seen as an underlying mechanism of PTSD-specific fear-related symptoms (Jovanovic & Norrholm, 2016; Zuj, Palmer, Lommen, et al., 2016). For example, reexperiencing symptoms are associated with increased fear-potentiated startle during early and mid-extinction learning (Norrholm et al., 2011). Given the body of literature suggesting an association between impaired fear extinction and the development of PTSD, future research should explore the use of fear extinction as a marker for greater risk of PTSD development after trauma exposure (Acheson et al., 2015; Guthrie & Bryant, 2006; Lommen, Engelhard, Sijbrandij, van den Hout, & Hermans, 2013; Pole et al., 2009). Clinically, it is widely believed that extinction learning processes underlie the foundation upon which exposure-based therapies are based as these treatments include repeated exposure to imagined, actual (in vivo exposure), or simulated fear-related stimuli. Focusing on these fear learning mechanisms allows clinical researchers to explore innovative means for enhancing extinction learning and to identify intrinsic and extrinsic barriers to the persistence of extinction memory or the clinical maintenance of treatment gains. For example, existing extinction research indicates that some early interventions actually inhibit fear extinction, which, in a clinical setting, would result in the development or maintenance of PTSD. Further understanding of the cognitive and psychophysiological processes underlying impaired fear extinction learning may also provide clinicians with an objective measure of potential clinical outcomes for treatment groups utilizing exposure therapies.

New directions: Individual differences in extinction learning

Traditional fear learning research concentrates on the analysis of central tendencies or between-groups mean comparisons of previously traumatized populations with and without PTSD symptom expression. The value of these long-standing analytics has recently come under increased scrutiny due to the risk of overlooking significant individual differences that are displayed during extinction learning (experimentally) and symptom development (experientially), limiting the translational power of the employed models. The heterogeneity of stressor-, fear-, and anxiety-related behaviors has been observed in both rodent and human study samples (Bush, Sotres-Bayon, & LeDoux, 2007; Cavigelli & McClintock, 2003; Duvarci, Bauer, & Pare, 2009; Galatzer-Levy et al., 2017; Galatzer-Levy, Bonanno, Bush, & Ledoux, 2013) supporting the need for further investigation of heterogeneous fear responses during extinction learning specifically.

Emerging translational research has investigated heterogeneous responses in fear extinction using a statistical method termed latent-growth mixture modeling (LGMM) (Muthen, 2004; Weingast, Haas, & Norrholm, 2018) to reveal underlying patterns of change over time, or trajectories. For example, Galatzer-Levy, Ankri, et al. (2013) and Galatzer-Levy, Bonanno, et al. (2013)

reanalyzed data on rodent freezing behavior and identified three heterogeneous patterns of extinction learning: rapid extinction, slow extinction, and a failure to extinguish. Rapid extinction, characterized by rapid and complete extinction of freezing behaviors, was the largest subgroup (57.3%). The second largest class (32.3%), slow extinction, demonstrated an initial protracted decrease during the middle stages of extinction followed by an accelerated decrease in freezing behavior during the final stages. Failure to extinguish, the smallest (10.3%) group, was recognized during the final extinction stages as having no discernible change in freezing behavior over session time. The groups displayed similar fear responses during acquisition, suggesting that differences in fear response patterns are centered in impaired extinction learning (Galatzer-Levy, Ankri, et al., 2013; Galatzer-Levy, Bonanno, et al., 2013). Further exploration of extinction patterns in rodent models will help in the development of translational psychophysiology-based models given that the extinction patterns recognized earlier closely reflect PTSD symptom trajectories established in human research (see Fig. 3) (Galatzer-Levy, Ankri, et al., 2013; Zuj & Norrholm, 2019).

Consistent with the approach used in preclinical rodent models, Galatzer-Levy, Ankri, et al. (2013) and Galatzer-Levy, Bonanno, et al. (2013) investigated PTSD symptom severity across trauma–exposed adults who presented to the emergency room. Following initial data collection and assessment, the authors identified three latent PTSD symptom trajectories that best fit PTSD symptom development and progression: rapid remitting, slow remitting, and nonremitting. Rapid remitting individuals (56%) presented a precipitous decrease in symptoms between 1 and 5 months, depicting the natural course of recovery displayed in larger populations post trauma exposure. Slow remitting individuals (27%) displayed a progressive decrease in symptoms across 15 months and are reflective of individuals who develop PTSD post trauma and eventually recover from the symptoms. Nonremitting individuals (17%) showed persistently elevated symptoms and are analogous to individuals who suffer from chronic PTSD. Similar analytics have been performed with samples recruited from police officers (Galatzer-Levy, Madan, Neylan, Henn-Haase, & Marmar, 2011), US military members (Bonanno, Mancini, et al., 2012), traumatic injury survivors (Bonanno, Kennedy, Galatzer-Levy, Lude, & Elfstrom, 2012; deRoon-Cassini, Mancini, Rusch, & Bonanno, 2010), and college students (Galatzer-Levy & Bonanno, 2013). Identifying heterogeneous trajectories of fear extinction and PTSD symptoms is crucial in developing targeted analytical and treatment methodologies for PTSD development and maintenance. For instance, Galatzer-Levy,

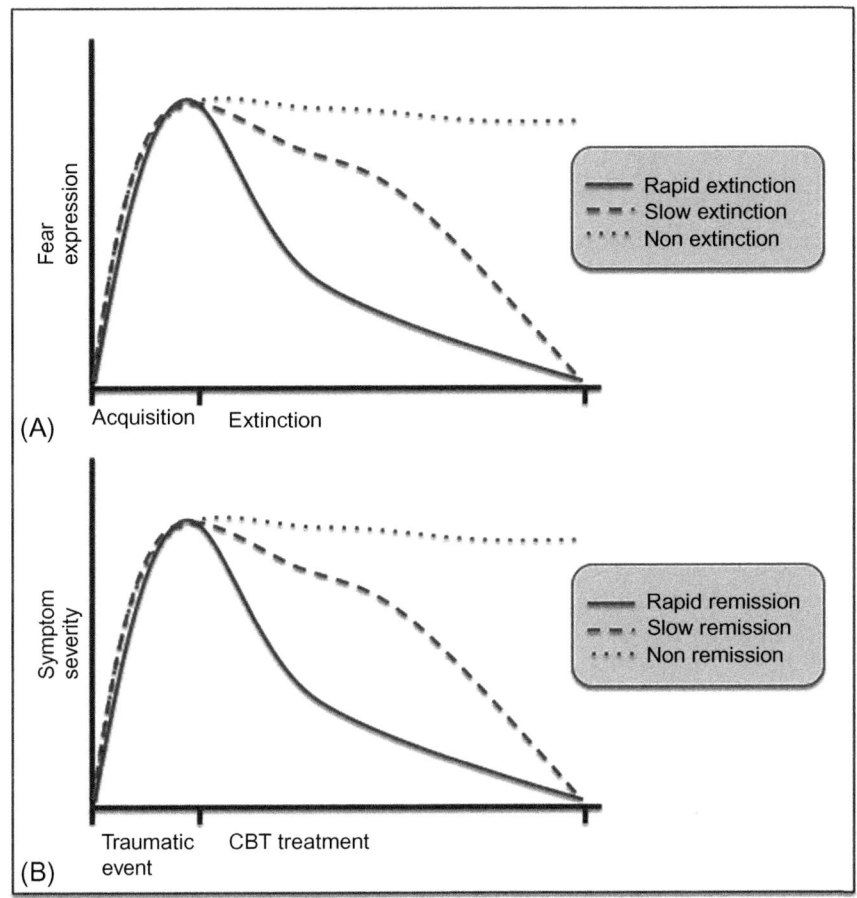

Fig. 3 Illustration of analogous learning processes underlying (A) heterogeneous fear extinction patterns and (B) PTSD symptom recovery trajectories (Zuj & Norrholm, 2019).

Ankri, et al. (2013) and Galatzer-Levy, Bonanno, et al. (2013) examined PTSD symptom trajectories following evidence-based treatment interventions, concluding that cognitive behavioral therapy (CBT) accelerated the recovery of the slow-remitting class only. This implies that identifiable symptom trajectories may prove useful to tailor treatment in a manner such that patient gains are maximized, that is, those in the slow-remitting trajectory may need more sessions of exposure therapy than showing a faster remitting trajectory. Further treatments need to be developed to address specific needs of nonremitting individuals as they presented the most severe hyperarousal and avoidance symptoms in previous work by the Galatzer-Levy group. Increasing our knowledge of the factors that underlie classification into different recovery trajectories clinically or different extinction profiles experimentally may enhance translational models and inform personalized fear- and anxiety-related treatment measures.

Extinction recall

Extinction recall examines the between-session strength and persistence of the extinguished association between the CS+-US (i.e., the extinction memory) formed during extinction learning. Extinction recall does not involve new learning, but rather the ability of an individual to retrieve a previously learned extinction memory. In a laboratory setting, extinction recall with human participants typically takes place 24 h after fear extinction and involves the presentation of the CS+ without air blast reinforcement (US). This process determines whether a conditioned fear response has remained extinguished or has returned (see "Return of fear" in the section). Psychophysiological research in previously traumatized populations with PTSD has shown deficient between-session extinction recall or a reduced ability to retrieve the extinction memory while suppressing the original fear memory (CS-US association; Acheson et al., 2015; Milad et al., 2009; Shvil et al., 2014). When extinction recall is unimpaired, it is comparable with symptom reduction that is maintained after treatment interventions, most commonly prolonged exposure therapy. Conversely, when extinction recall is impaired, it serves as a model for treatment resistance or relapse, as symptom reduction and clinical gains do not persist posttreatment (Zuj & Norrholm, 2019).

Fear load

Fear load, or the overexpression of conditioned fear during early extinction, represents a potential biomarker for PTSD that has been linked to treatment outcomes and symptom severity in some previously traumatized populations. In the previously discussed work by Norrholm et al. (2015, 2011), they found that individuals with PTSD could successfully discriminate between the CS+ and CS− but displayed elevated fear-potentiated startle to the CS+ during late acquisition that extended and increased into early and middle extinction learning (i.e., fear load). However, there were no group differences by late extinction. The authors suggest that fear load is the result of persistent overexcitation of the fear memory being encoded following acquisition and that impaired rates of extinction learning can be predicted by fear load. A follow-up study of highly traumatized individuals found fear load to be significantly correlated with intrusive thoughts and intense physiological reactions to trauma reminders; this relationship remained significant after controlling for demographic data, trauma history, and level of terminal fear acquisition (Norrholm et al., 2015). Further, individuals with the most severe intrusive symptoms, assessed by frequency of fear memories of the traumatic event, had higher fear load than those with less frequent intrusive thoughts (Norrholm et al., 2015). Together, these results indicate that fear load could be utilized as a potential predictor for those who are highly symptomatic and most in need of treatment. Altogether, these results advocate for further investigation of fear load as a potentially informative phenotype in traumatized populations.

Return of fear

Return of fear after extinction learning refers to the reemergence of conditioned but previously extinguished, fear responses that may take place due to the activation of the original fear memory (CS–US association) that outcompetes expression of the extinction memory trace. As mentioned previously, successful extinction learning creates a new memory trace in which the former CS+ predicts the absence of the US in a manner that inhibits, rather than erases, the original conditioned fear response (Myers & Davis, 2002; Norrholm et al., 2008). The coexistence of the original fear memory and the extinction memory is demonstrated through the return of fear after extinction in three principle manifestations: renewal, spontaneous recovery, and reinstatement (see Fig. 4) (Weingast et al., 2018). The existence of both the original fear memory and the extinguished memory can create an ambiguous CS that relies on these procedures to determine which memory trace is exhibited under which circumstances.

Renewal

Renewal describes the return of the conditioned fear response due to a change in the context in which extinction learning took place. In the laboratory, renewal of fear is typically assessed in three variations: ABA renewal, ABC renewal, and AAB renewal (Bouton, 2002). While ABA renewal reflects the reemergence of extinguished fear generated by a return to the original acquisition context, ABC and AAB renewal paradigms assess the return of fear in novel contexts that differ from either the acquisition or extinction learning context or both. The ABA renewal procedure requires fear acquisition to occur in Context A, extinction learning in a second context (Context B), followed by a return to the original acquisition context

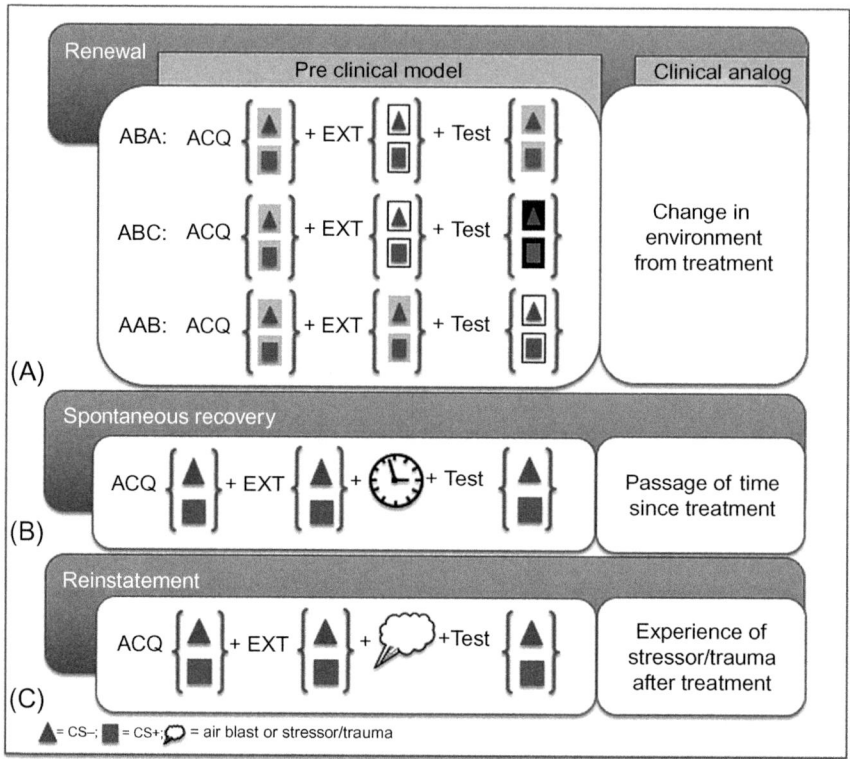

Fig. 4 Preclinical models (or laboratory methods) depicting the return of extinguished fear as well as their clinical analogs. (A) Renewal, or the return of fear in a different context, is displayed by ABA, ABC, and AAB. (B) Spontaneous recovery is the return of fear after the passage of time. (C) Reinstatement is the return of fear with exposure to the US (air blast) postextinction or after experiencing a stressor/trauma posttreatment.

(Context A) for testing fear expression. ABC-renewal procedures administer acquisition in Context A, extinction learning in Context B, and tests of fear response to the extinguished CS in a third context (Context C). Although documented in animal literature, AAB renewal is the most difficult to detect and presents when both acquisition and extinction learning occur in Context A followed by extinction recall testing in a second context (Bouton, 2002; Bouton & Ricker, 1994; Tamai & Nakajima, 2000). Both human and animal studies have consistently demonstrated that the return of extinguished fear can be prompted by changes in context (Bouton & Bolles, 1979; Effting & Kindt, 2007; Thomas, Vurbic, & Novak, 2009; Vansteenwegen et al., 2005; Vervliet, Baeyens, Van den Bergh, & Hermans, 2013). Neumann and Kitlertsirivatana (2010) found a return of US-expectancy ratings in both ABA- and ABC-renewal procedures, although renewal effects were more robust when participants returned to the original acquisition context. These results have clinical implications for extinction-based exposure therapies conducted both in clinical contexts and in vivo environments. The renewal effect has been established after extensive extinction learning and across multiple testing methods (for detailed review, see Bouton, 2002).

As introduced in the previous section, renewal can present in clinical terms as the reemergence of fear-related symptoms prompted by experiences outside the contexts in which exposure therapy occurred, or symptom relapse (Bouton, 2002). As discussed previously, research indicates that, while the original acquisition context is especially prone to induce relapse of symptoms, relapse can also occur in novel, unexpected contexts. Despite the emergence of renewal after treatment, context manipulations have been explored to target mechanisms to prevent renewal of conditioned fear responses, or relapse. Balooch, Neumann, and Boschen (2012) examined the effect of multiple extinction contexts on renewal presentation by comparing the standard ABC-renewal procedure with a modified A(BCD)E variation (Balooch et al., 2012). Based on US-expectancy and startle response, participants who received multiple contexts during extinction learning displayed attenuated renewal to novel contexts. Despite a relatively smaller degree of returned fear response, ABC renewal is considered more clinically relevant than ABA renewal given that the acquisition context(s) may not be clear in clinical situations. The finding that multiple context extinction learning attenuates renewal has major implications for exposure-based therapy techniques that are conceptually based on fear extinction. For instance, Shiban, Pauli, and Muhlberger (2013) examined the use of multiple contexts to attenuate renewal in spider-phobia patients who viewed virtual reality environments containing spiders (Shiban et al., 2013). Based on skin conductance levels and fear ratings, multiple contexts during exposure attenuated the renewal of fear. Together, these results suggest the potential efficacy of treatment taking place across a broad range of settings and the use of virtual reality to expand contexts.

Spontaneous recovery

Spontaneous recovery refers to the return of fear response following the passage of time since the response was extinguished (Pavlov, 1927; Rescorla, 2004). To investigate spontaneous recovery the expression of a conditioned fear response is assessed at a time point following extinction training with little experimental manipulation required save for the nonreinforced presentation of the extinguished CS. The presentation of the previously reinforced CS after extinction learning serves a dual purpose: it allows the researcher to determine if a conditioned fear response has returned through spontaneous recovery or if the extinction memory is recalled (termed extinction recall or between-session extinction). Pavlov (1927) initially established that extinguished conditioned responses would spontaneously recover with the passage of time post extinction. Further investigation of this phenomenon, spanning the next several decades, revealed that, after a 6-day delay, rodents exhibited return of fear levels equal to those present at the end of acquisition (Brooks & Bouton, 1993). Additionally, the extent of rodents' fear recovery at retest is directly proportional to the interval of time since extinction training (Quirk, 2002), resulting in higher levels of CR as time passes. Further, despite the occurrence of spontaneous recovery, the extinction memory remained intact, as evidenced by savings in the rate of reextinction learning (Quirk, 2002). No clear evidence has pointed to a single contributing factor for the expression of spontaneous recovery, implying that it results from a combination of multiple sources (for complete review, see Rescorla, 2004).

Clinically speaking, symptom relapse with regard to intrusive memories or cued emotional and/or physiological reactivity after the successful completion of exposure therapy represents a form of spontaneous recovery. Determining effective strategies to diminish the chance of spontaneous recovery (experimentally) and symptom relapse (experientially) is imperative. Although the timing of extinction relative to acquisition has been identified as a moderating factor for the expression of spontaneous recovery, rodent and human studies have reported inconsistent evidence supporting an "optimal" time interval (Alvarez, Johnson, & Grillon, 2007; Everly Jr. & Mitchell, 2000; Gray & Litz, 2005; Myers, Ressler, & Davis, 2006; Norrholm et al., 2008; Schiller et al., 2010). Further research on the neurobiological underpinnings of and clinical applications for fear extinction may be used to help determine optimal timing between a traumatic event (acquisition) and exposure therapy (extinction) to reduce the likelihood of PTSD symptom relapse (return of conditioned fear) in the future.

Reinstatement

Reinstatement, or the return of a fear response after successful extinction, occurs when the US is unexpectedly presented after extinction, reinforcing the initial CS-US association, even without the original CS+ present (Rescorla & Heth, 1975). This is demonstrated in human studies, after effective acquisition and extinction training, by presenting a small number (typically <5) of unprompted USs (e.g., air blast to the larynx or cutaneous shock), followed by a test session including the previously reinforced CS+ and the CS−, if applicable to the employed paradigm. Reinstatement of the fear response is evident if an elevated physiological response associated with the original CS+ returns. Clinically, reinstatement can underlie the expression of PTSD symptom relapse after the experience of stressful life event (e.g., hardship, loss, and illness).

Similar to renewal, Frohardt, Guarraci, and Bouton (2000) found that context was a critical feature for the expression of reinstatement of a CR (Frohardt et al., 2000). They found that rodents failed to express reinstated fear responses when there was a change between the context of the unsignaled US and the context in which the extinguished CSs were tested. Fear conditioning studies in healthy humans found similar context-dependent results (LaBar & Phelps, 2005; Vervliet, Craske, & Hermans, 2013), signifying the importance of context in reinstatement and return of fear in general. In an effort to separate reinstatement from renewal and to understand the role of a US presentation in reinstatement, Sokol and Lovibond (2012) modified the traditional paradigm to incorporate novel, aversive stimuli (Sokol & Lovibond, 2012). They found that fear response returned after presentation of both the original US and a novel US, suggesting that the unsignaled presence of a stressor is a more significant mechanism of reinstatement than context. Translational conditioning research has shown that reinstatement persists across considerable time intervals after acquisition and extinction training in both rodents and humans (Kindt & Soeter, 2013; Kindt, Soeter, & Vervliet, 2009; Norrholm et al., 2008) deeming it highly relatable to clinical situations where symptom relapse occurs. Several human fear conditioning studies have found that reinstatement procedures cause the return of the CR not only for the original CS+ but also for the CS− as well (Dirikx, Vansteenwegen, Eelen, & Hermans, 2009; Kull, Muller, Blechert, Wilhelm, & Michael, 2012; Norrholm et al., 2006). In contrast to rodent research, these results indicate that factors specific to humans (i.e., self-reported anxiety levels) may generalize the effects of triggered fear return. Due to the notion that the process of reinstatement may underlie symptom relapse triggered by any stress- or fear-inducing life event, even after an extended period of time or successful completion of exposure therapy (Vervliet, Craske, et al., 2013), there remains a compelling interest in further exploring learning and memory facilitators that may increase the clinical effectiveness of exposure therapy and prevention of relapse.

Fear inhibition

Fear inhibition, in general, refers to the reduction in the expression of a conditioned fear response under experimental or experiential conditions in which new learning has occurred and influences responding. In many cases, this involves the transfer of the inhibitory properties of a safety cue on a participant's response to a previously reinforced danger cue (Norrholm & Jovanovic, 2018). Traditionally, fear inhibition was examined using either extinction learning or conditioned inhibition procedures; however, each presented limitations (Myers & Davis, 2002, 2004). As previously discussed, extinction procedures involve a stimulus conditioned to elicit a fear response that is then repeatedly presented in the absence of the US, resulting in diminished fear expression. In extinction learning the CS+ exhibits both excitatory and inhibitory properties making it difficult to determine whether inhibition occurred or whether the stimulus' excitatory properties were diminished. The traditional conditioned inhibition paradigm involves one neutral cue (A) that, when presented alone, is paired with an aversive US (A+) and, when presented in compound with a second neutral cue (B), is not paired with a US (compound BA−; Rescorla, 1969). Through conditioning, the B cue becomes a safety signal that predicts the absence of the US and limits fear response expression when presented with the danger cue, A. After successful conditioning, safety signal (B) is paired with a separately conditioned excitatory stimulus (X); this compound (BX−) is not reinforced. Introduction of a novel conditioned stimuli functions to separate the excitatory and inhibitory properties taken on by the CSs, which allows researchers to test the inhibitory impact of the safety cue. While this model allows for the separation of excitatory (allotted to cue A) and inhibitory (allotted to cue B) properties, the procedure makes the safety signal vulnerable to second-order conditioning (Myers & Davis, 2004). This means that, instead of cue B inhibiting the fear response elicited by cue A, cue A could excite a fear response that conditions cue B to become an additional threat signal.

In addition to the difficulty in independently evaluating extinction versus inhibition, the traditional fear inhibition paradigm is limited by its use of compound cues only in the absence of the US. To address these limitations in human studies, researchers utilize a fear inhibition transfer task called conditional discrimination (AX+/BX−; Jovanovic et al., 2005; Myers & Davis, 2004). This AX+/BX− model, originally based on earlier learning theory experiments (Rescorla & Wagner, 1972; Wagner, Logan, Haberlandt, & Price, 1968), was adapted from rodent models (Myers & Davis, 2004) and applied to humans as a startle paradigm (Jovanovic et al., 2005). In this model, acquisition involves neutral cues A and B that are both paired with a common, third neutral cue (X); the compound AX is reinforced, or paired with the US (AX+), while the compound BX is not reinforced (BX−). Through conditioning, A becomes an excitatory danger cue, predicting the US, while B becomes an inhibitory safety cue, predicting the absence of the US. This model eliminates the pairing of the excitatory cue (A) with the safety cue (B) during conditioning to avoid configurational differences and the potential for second-order conditioning. After conditioning the inhibitory influence of cue B is tested by the nonreinforced pairing of cues A and B alone (AB), which, if successful inhibition occurs, results in decreased fear responding compared with the response when presented with AX+ (Jovanovic et al., 2005). To expose any possible effects of external inhibition, or the unconditioned decrease in CR to a CS when the CS is paired with a novel stimulus (Pavlov, 1927), Jovanovic et al. (2005) included the test compound AC, where C is a novel conditioned stimulus.

In human research, the AX+/BX− model has been applied as a fear-potentiated startle paradigm utilizing colored lights to represent cues A, B, C, and X (Jovanovic et al., 2005). As described earlier in this chapter, white noise probes (40 ms) were used to elicit an acoustic startle response, and air blasts were delivered to the larynx as the US. When applied to healthy controls, this paradigm revealed greater fear-potentiated startle in the presence of AX+ when compared with BX−, signifying successful discrimination between safety and danger cues. Additionally, a decreased startle response was recorded in the presence of the AB− compound as compared with AX+, demonstrating that the inhibitory properties of B successfully transferred learned safety to danger cue A (Jovanovic et al., 2005).

This paradigm is clinically relevant, as an increasing number of studies have found impaired fear inhibition, or safety signal processing, in PTSD populations. One study assessed the relationship between fear inhibition and PTSD symptom severity in Vietnam veterans using the AX+/BX− conditional discrimination model (Jovanovic et al., 2009). A deficit in the ability to transfer learned safety was shown in participants reporting high PTSD symptom severity when compared with individuals with lower symptom severities. In particular, veterans who reported higher re-experiencing and avoidance symptom severities showed more difficulty in transferring learned safety to danger cues even in the absence of reinforcement (Jovanovic et al., 2009). Prior work has also shown that traumatized civilian populations with PTSD display more robust fear-potentiated startle to safety signals, less discrimination between danger and safety cues, and impaired safety transfer compared with PTSD− controls (Jovanovic et al., 2012, 2010). Additionally, PTSD+ participants were able to display contingency awareness for safety and danger cues despite consistently elevated psychophysiological data (Jovanovic et al., 2010). Similar to previous fear learning studies, these results suggest a disconnect between the psychophysiological and cognitive effects of PTSD. Given the high comorbidity between PTSD and major depressive disorder (MDD; Breslau, Davis, Peterson, & Schultz, 2000; Kessler, 2000), Jovanovic et al. (2010) applied the AX+/BX− model to healthy controls, PTSD only, MDD only, and comorbid PTSD+ and MDD groups. Individuals with comorbid PTSD and MDD displayed greater fear-potentiated startle to both the safety cue and during the transfer test when compared with the MDD only group, as well as decreased fear inhibition. These results indicate that impaired fear inhibition may be a specific biomarker for PTSD symptoms. A goal for current prolonged exposure treatment involves the transfer of learned safety to both trauma-related and generalized stimuli, so together these results have immense potential for impact on PTSD treatment models.

Generalization of fear

Fear generalization is the extension of conditioned fear responses to stimuli that are contextually, perceptually, or symbolically similar to the original CS+ (Lissek et al., 2008). In a laboratory setting, fear generalization occurs after acquisition and involves the inclusion of generalization stimuli (GSs) or stimuli similar to the CS+, in addition to the CS− and the CS+; the CS+ may or may not be reinforced (see Fig. 5). If generalization occurs, conditioned fear responses will present for the CS+ and for the GSs. That is to say, if an individual perceives similarities between the conditioned stimuli and novel unconditioned stimuli, the fear memory-conditioned stimulus association can extend to the new stimulus and produce conditioned fear response (Norrholm et al., 2014). Fear generalization is often expressed in gradients centered on the CS+, with sharper inclines indicating less fear generalization to the GS (Dymond, Dunsmoor, Vervliet, Roche, & Hermans, 2015). In the first study to apply fear-potentiated startle to fear generalization, Lissek et al. (2008) used concentric rings that gradually increased in diameter to represent the CS+, CS−, and GSs. Rings on opposite ends of the size range represented the CS+ and CS− stimuli, while the intermediate rings represented the GSs. They found that, in healthy participants, fear response decreased as the GSs became less physically similar to the

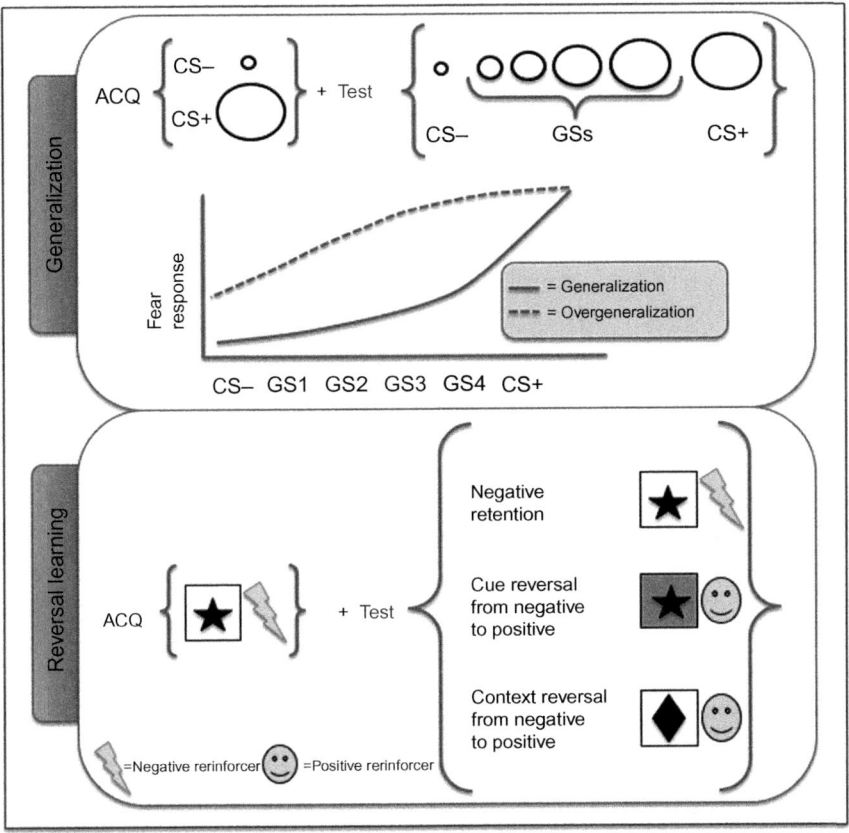

Fig. 5 Schematic illustration of basic laboratory models for stimulus generalization and reversal learning.

CS+ and more physically similar to the CS−, represented by a steeper gradient. This suggests that the more similar the GS is to the CS+ the more fear response is expressed in the presence of the GS. Norrholm et al. (2014) also reported generalization of fear gradients in healthy controls using fear-potentiated startle techniques; however, the authors noted the limitations of this approach depending on the modality of the CSs and GSs (i.e., visual vs auditory; Norrholm et al., 2014). Overgeneralization, represented by a less steep decrease in fear responses, has been increasingly observed in clinical populations, including PTSD (Lissek & Grillon, 2012). Clinically, PTSD patients commonly generalize fear responses to harmless yet similar stimuli that were not present during a traumatic event (Zuj & Norrholm, 2019).

Reversal learning

PTSD can be characterized as a disorder with altered fear processing as a central feature; however, it should not be defined as solely a fear conditioning disorder given its heterogeneous nature and the life domains over which it spans. In fact, there are many lasting cognitive deficits that appear to be linked to PTSD including reversal learning in cue-context associations. Cue-context reversal learning can be defined as changing the expected valence of outcomes after presenting cue and context-related information (Weingast et al., 2018). In other words, if an individual learns that the presentation of a specific cue-context pair predicts an aversive or positive outcome, through reversal learning, one can shift their prediction to expect an opposite outcome in the future (see Fig. 5). To study cues and contexts in isolation, one member of the pair must be kept constant to isolate the reversal of a single dimension. Interestingly, findings from Levy-Gigi and Richter-Levin (2014) suggest that the type of trauma exposure one has experienced may selectively impair reversal of a cue or context that initially predicted a negative outcome but later came to predict a positive outcome (Levy-Gigi & Richter-Levin, 2014). Results from this study suggest that occupations requiring attention to contextual information may result in difficulty reversing the outcome of contexts from dangerous to safe (i.e., firefighters are more attuned to context to determine location of victims in a burning building). Thus it is particularly important to identify individuals at risk for these fear learning deficits so that contextual cue processing skills can be introduced before trauma exposure.

Psychobiological factors that moderate fear processing
Hormones
Stress hormones

The hypothalamic-pituitary-adrenal (HPA) axis is a key biological system regulating the mammalian stress response. A wide range of translational research has implicated this system in identifying potential biomarkers of PTSD including circulating cortisol (human) and corticosterone (rodent models) levels. In the presence of a stressor, the hypothalamus releases corticotropin-releasing hormone (CRH) to the pituitary gland, which in turn releases adrenocorticotrophin-releasing hormone (ACTH) to elicit cortisol release from the adrenal cortex. The activity of this cascade is susceptible to alteration if exposed to early-life (or chronic) traumatic events or stressors and subsequent intergenerational transmission of epigenetic factors affecting steroid metabolism or stress reactivity (Yehuda & Seckl, 2011). While reports on differences between baseline cortisol levels in PTSD and controls have not been consistent, it is important to highlight putative processes that may underlie some of the low cortisol levels that have been observed in PTSD and cortisol reactivity to acute stress (Michopoulos et al., 2015).

It seems counterintuitive that lower levels of a primary stress-related hormone have been observed in individuals diagnosed with PTSD, a trauma- and stressor-related disorder. If individuals are frequently behaving with hypervigilance and are, therefore, subject to chronic stress, shouldn't their circulating cortisol levels be high in comparison with healthy controls? Contrarily, Yehuda and Seckl (2011) highlight a largely metabolic approach to this question stating that trauma exposure, particularly physical or sexual trauma during early life (Meewisse, Reitsma, de Vries, Gersons, & Olff, 2007), causes long-lasting effects on the catabolism of glucocorticoids including cortisol. These cascading effects further diminish the activity of key enzymes, which break down circulating cortisol over time and manifests as increased negative feedback upon the HPA axis and increased target cell sensitivity (i.e., pituitary) to glucocorticoid action. With increased receptor sensitivity the adrenals can secrete less cortisol and still maintain normal blood levels of the hormone, leading to the observed phenotype in PTSD.

Previous research has identified lower cortisol reactivity to acute stressors as a potential biomarker for increased risk for PTSD development. In addition, the use of glucocorticoid administration may be an intervention for the negative cognitive and psychological effects associated with chronically altered cortisol reactivity (Suris, North, Adinoff, Powell, & Greene, 2010). However, there have been conflicting results regarding the effectiveness of the synthetic glucocorticoid, dexamethasone, in facilitating fear extinction (Michopoulos et al., 2017) yet potentially hindering extinction-based exposure therapy (Maples-Keller et al., 2018). That being said, it is important to note that HPA axis activity does not exist in isolation, as other neuroendocrine factors interact with and regulate this system. Neuroendocrine factors interacting with stress responses, such as estradiol, have been extensively characterized as affecting fear learning and extinction processes associated with PTSD development.

Sex hormones
Endogenous factors such as the sex hormones estrogen and progesterone have demonstrated sex-specific putative effects on fear extinction learning and retention (Briscione, Michopoulos, Jovanovic, & Norrholm, 2017;

Glover, Jovanovic, & Norrholm, 2015; Michopoulos et al., 2015). Specifically the cyclical nature of these hormones in women has demonstrated effects on extinction memory across menstrual cycles (Milad et al., 2006; Pace-Schott, Germain, & Milad, 2015a, 2015b). According to results from Glover et al. (2012), higher levels of estrogen may have a *protective* effect against severe PTSD symptoms, as modeled by facilitated fear extinction. These data also showed that women with low estrogen levels and more severe PTSD symptoms expressed a higher fear load and less robust fear extinction. Thus characterizing naturally dynamic neuroendocrine profiles in trauma-exposed populations could contribute to developing novel interventions for PTSD.

Behavioral contributions
Sleep
Sleep disturbance is a core feature of PTSD and is listed in the DSM-5 criteria for PTSD under Criterion E: alterations in arousal and reactivity (Germain, 2013; Zuj, Palmer, Hsu, et al., 2016; Zuj, Palmer, Lommen, et al., 2016). The prevalence of sleep impairment in PTSD patients needs to be taken into consideration when using psychophysiological paradigms targeted at improving fear extinction learning, since numerous studies have shown that reduced sleep is linked to impaired consolidation of fear extinction. For example, a study by Pace-Schott et al. (2009) found that, during a 12-h period after extinction learning and before extinction recall, those that slept during the delay were more likely to generalize the extinguished CS+ to an unextinguished CS+ compared with those engaged in wakefulness during the delay. Zuj, Palmer, Hsu, et al. (2016) and Zuj, Palmer, Lommen, et al. (2016) extended these findings to a clinical population diagnosed with PTSD and found that those with higher PTSD symptom severity displayed poorer extinction learning with increased hours awake. Furthermore, sleep phase cycling may affect extinction learning. This phenomenon is demonstrated by a study by Spoormaker et al. (2010) in which subjects were given a 90-min sleep window between extinction learning and recall. Those that entered into the rapid eye movement (REM) phase of sleep during this sleep window had a reduced skin conductance response to the extinguished stimulus during recall compared with those that did not enter REM (Spoormaker et al., 2010), providing further evidence that memory consolidation during sleep is important to the recall of fear extinction.

The relationship between impaired sleep and fear extinction learning and recall can be seen in the similarities between the neurobiological circuitry involved in extinction learning and recall and memory consolidation during sleep. The key brain regions associated with extinction learning and recall are the amygdala and the ventromedial prefrontal cortex (vmPFC; Pace-Schott et al., 2015a, 2015b; Pace-Schott & Spencer, 2015). These structures also seem to be reactivated during sleep, specifically during REM sleep, and demonstrate repetitive brain activation patterns leading to fear extinction memory consolidation (Pace-Schott et al., 2015a, 2015b). Those who are sleep deprived, especially of the REM phase, do not undergo this reactivation and, therefore, exhibit impaired fear extinction memory (Pace-Schott et al., 2015a, 2015b).

One way to overcome the negative effects of sleep impairment on fear extinction learning is to manipulate the timing of extinction learning paradigms. Pace-Schott et al. (2013) showed that extinction learning was superior in the morning and that extinction recall was more easily generalized from an extinguished to an unextinguished CS+ when compared with later timepoints (Pace-Schott et al., 2013). These observations, in addition to findings by Zuj et al. (2016), provide evidence of a time-of-day effect on performance in psychophysiological fear extinction paradigms. Additionally, there is less homeostatic pressure to sleep in the morning (Zuj et al., 2016), making it an ideal time to provide therapy focused on extinction learning and recall. Identifying the effects of sleep and time-of-day on fear learning and memory may contribute to improving future treatment plans and outcome trajectories for individuals with PTSD.

Frontiers in pharmacotherapy

Fear learning and extinction rely on not only behavioral changes but also neural mechanisms of change. Altered neurotransmitter signaling and activity at receptors involved in learning and memory are crucial for these changes, as substantiated by experiments on brain regions of interest and peripheral agents. These mechanisms may include long-term potentiation (LTP) and long-term depression (LTD), widely accepted theories stating that specific patterns of synaptic activity lead to changes in synaptic strength and transmission (Bliss & Cooke, 2011). Fundamental to these theories, Hebb's rule that "cells that fire together wire together" asserts that learning occurs not only at the contextual, stimulus level but also at the cellular scale. Simultaneous activation of cells from multiple stimuli leads to an increase in receptor expression at the synapse, resulting in an increase in synaptic strength and learning behaviors. These vital processes regulate the release of the excitatory neurotransmitters, such as glutamate, on which

N-methyl-D-aspartate (NMDA) and α-amino-3-hydroxy-5-methyl-4-isoxazolepropionic acid receptors (AMPA) rely to transmit messages (Fitzgerald et al., 2015; Nowak, Bregestovski, Ascher, Herbet, & Prochiantz, 1984).

Scientists have not only explored pharmacological agents aimed at altering CNS signaling involved in fear learning and memory for clinical interventions but also importantly targeted signaling in the peripheral nervous system including the endocrine system, that is, cortisol, and the immune system, that is, glucocorticoids (Altemus, Dhabhar, & Yang, 2006; Jovanovic et al., 2011; Michopoulos et al., 2017). Recent trials of pharmacological agents such as D-cycloserine (DCS), an NMDA agonist, and dexamethasone, a synthetic glucocorticoid agonist, demonstrate the potential for improving outcomes in treatment-resistant patients with disorders related to stress, learning, and memory, such as PTSD.

D-Cycloserine

In 1992 Falls and colleagues utilized a fear-potentiated startle paradigm to suggest that learning to extinguish a conditioned fear was reliant on NMDA receptor activity in areas of the brain related to emotion and fear, such as the amygdala. These findings were substantiated by infusing an NMDA antagonist into the amygdala, which blocked fear extinction (Falls, Miserendino, & Davis, 1992). Due to its effects on the glutamatergic neurotransmitter system, which is the major excitatory signaling pathway, DCS may also act on broader learning and memory processes in an array of neural circuits throughout the cortical and subcortical brain as shown in animal models (Fitzgerald, Seemann, & Maren, 2014). For example, studies in rodent models including Fitzgerald et al. (2014) have suggested that DCS enhances both fear extinction and reconsolidation, or "editing" fear memories, which may help suppress the original conditioned fear memory. Another fear memory process that may be enhanced by DCS administration is fear extinction generalization, as suggested by evidence that rats treated with DCS display less fear responses to a nonextinguished conditioned stimulus in comparison with controls (Ledgerwood, Richardson, & Cranney, 2005). Despite these promising results, there are many limitations and unanswered questions about the potential use of DCS as a fear extinction facilitator. For example, animal studies by Woods and colleagues have hinted that the effects of DCS may be context dependent, as fear renewal was observed after DCS-facilitated extinction once testing resumed in an original conditioning context (Woods & Bouton, 2006).

Although extensive research on DCS in animal models has shown promising fear extinction enhancing potential, bridging the translational divide to human applications has proven more challenging. According to a human study conducted by Klumpers et al. (2012), DCS did not significantly reduce fear response in within-session fear extinction training or fear extinction retention assessed at a later testing session (Klumpers et al., 2012). A recent analysis of cognitive behavioral therapy outcomes by Mataix-Cols et al. (2017) showed that a group given DCS produced a statistically significant advantage in treatment outcomes compared with placebo at post-treatment. However, contrary to reports from Mahan and Ressler (2011) that DCS enhances behavioral therapy in patients with dysregulated fear processing, discrepancies emerged when assessed at follow-up, and insignificant differences in symptom improvement were seen when pretreatment and follow-up were compared in the DCS versus control groups (Mahan & Ressler, 2011; Mataix-Cols et al., 2017). Furthermore, DCS may be an effective agent to enhance the success of treatment outcomes under specific circumstances; however, the number of treatment sessions, dosage and timing of DCS administration, biological predispositions affecting exposure treatment success, and patient compliance may need to be controlled for in order to assess its true potential (Mataix-Cols et al., 2017).

Dexamethasone

Glucocorticoids are hormones that are synthesized and secreted from the adrenal cortex into the blood, which has antiinflammatory effects and influences on learning and memory (Bentz, Michael, de Quervain, & Wilhelm, 2010; van der Velden, 1998). Bentz et al. (2010) assert that glucocorticoids can facilitate consolidation of memory traces involving nonfearful responses to feared situations while also limiting the retrieval of aversive learning outcomes. Targeting the glucocorticoid system the pharmacological agent dexamethasone has been developed for applications to fear extinction in PTSD patients.

Dexamethasone is a synthetic steroid, which acts as a glucocorticoid agonist and can improve fear extinction and retention deficits associated with PTSD when administered prior to or directly after extinction training (Sawamura et al., 2016; Stockhorst & Antov, 2015). It may also regulate gene expression related to glucocorticoid and HPA axis activity during stress responses. Studies suggest that suppression of the HPA axis (resulting in cortisol or corticosterone release) may depend on the dosage of dexamethasone that is administered (Sawamura et al., 2016; Stockhorst & Antov, 2015). For example, results from Sawamura et al. (2016) suggest that a low-dose administration of dexamethasone acts to enhance within-session extinction learning and between-session extinction and retention in rodent models of PTSD. According to Michopoulos et al. (2017), dexamethasone administration led to a reversal of both extinction and stimulus discrimination in individuals with PTSD. In rodent models, dexamethasone and DCS-facilitated reduction of heightened fear responses characteristic of PTSD and suppressed cortisol levels in individuals with PTSD (Jovanovic et al., 2011). These translational findings suggest that dexamethasone may be used for enhancing extinction learning in combination with exposure therapy to optimize treatment; however, recent work from Maples-Keller et al. (2018) suggests that dexamethasone can increase rates of dropout in PTSD populations (Maples-Keller et al., 2018). Further research is certainly warranted to determine the optimal therapeutic conditions under which this drug may be used.

Conclusion

Research in the field of the psychophysiology of emotion and specifically PTSD presents several limitations worth considering. Firstly, laboratory paradigms modeling fear learning have the potential to be mistranslated into clinical applications. Noted heterogeneities in trauma- and stress-related disorders may not always be effectively incorporated into and modeled in study designs, limiting their translational power to improve individual symptom development. For example, pharmacological agents such as dexamethasone, which has been shown to enhance extinction learning and retention in rodent models of PTSD, did not enhance exposure therapy outcomes and was associated with increased treatment dropout (Maples-Keller et al., 2018; Michopoulos et al., 2017; Sawamura et al., 2016). More effective laboratory paradigms modeling fear learning should also incorporate individual special needs unique to veteran populations to develop more effective trauma-focused treatment plans. For example, a large amount of the veteran population is afflicted with eye injuries, which means that the use of a visual fear conditioning paradigm would not be effective for this subset of veterans (Norrholm et al., 2014). Exploring a nonvisual modality in this specific population and keeping in mind which modality would best suit individual patients would allow a larger population of those with PTSD to participate in studies using fear conditioning paradigms.

Future studies could address these limitations and target the return of fear, or clinical relapse through spontaneous recovery, renewal, or reinstatement after one has successfully extinguished a fear memory, thus improving the clinical potential of fear conditioning paradigms. Use of pharmacotherapies aimed at enhancing memory consolidation and retrieval should be explored in conjunction with these paradigms to facilitate fear extinction learning, recall, and extinction. For example, utilizing pharmacological agents such as dexamethasone and D-cycloserine may help decrease fear load, generalization, and reinstatement of fear or improve extinction learning tested using fear learning paradigms.

Measuring psychophysiological responses to these paradigms can help clinicians and researchers pinpoint specific dysregulations in fear processing and directly address the physiological correlates of emotions characteristic of PTSD. These measures can be predictive of treatment trajectories and resistance and can help clinicians develop more personalized treatment courses based on idiosyncratic needs. More so, these paradigms address the heterogeneity of PTSD symptoms through exploring the acquisition of fear, overgeneralization of fear, impairment of fear extinction, and reinstatement of fear, which has applications for understanding the development, maintenance, treatment, and relapse of the fear-related features of PTSD (Zuj & Norrholm, 2019). The use of fear conditioning paradigms and associated psychophysiological data may eventually allow clinicians to identify trauma survivors who are more susceptible to developing PTSD and intervene before the expression or worsening of PTSD symptoms.

References

Acheson, D. T., Geyer, M. A., Baker, D. G., Nievergelt, C. M., Yurgil, K., Risbrough, V. B., & Team, M.-I. (2015). Conditioned fear and extinction learning performance and its association with psychiatric symptoms in active duty marines. *Psychoneuroendocrinology, 51,* 495–505. https://doi.org/10.1016/j.psyneuen.2014.09.030.

Altemus, M., Dhabhar, F. S., & Yang, R. (2006). Immune function in PTSD. *Annals of the New York Academy of Sciences, 1071,* 167–183. https://doi.org/10.1196/annals.1364.013.

Alvarez, R. P., Johnson, L., & Grillon, C. (2007). Contextual-specificity of short-delay extinction in humans: Renewal of fear-potentiated startle in a virtual environment. *Learning & Memory, 14,* 247–253.

American Psychiatric Association. (2013). *Diagnostic and statistical manual of mental disorders, DSM-5* (5th ed.). Arlington, VA: American Psychiatric Association.

Amstadter, A. B., Nugent, N. R., & Koenen, K. C. (2009). Genetics of PTSD: Fear conditioning as a model for future research. *Psychiatric Annals, 39*(6), 358–367.

Balooch, S. B., Neumann, D. L., & Boschen, M. J. (2012). Extinction treatment in multiple contexts attenuates ABC renewal in humans. *Behaviour Research and Therapy, 50*(10), 604–609. https://doi.org/10.1016/j.brat.2012.06.003.

Bentz, D., Michael, T., de Quervain, D. J., & Wilhelm, F. H. (2010). Enhancing exposure therapy for anxiety disorders with glucocorticoids: From basic mechanisms of emotional learning to clinical applications. *Journal of Anxiety Disorders, 24*(2), 223–230. https://doi.org/10.1016/j.janxdis.2009.10.011.

Blechert, J. P., Michael, T. P., Grossman, P. P., Lajtman, M. M., & Wilhelm, F. H. P. (2007). Autonomic and respiratory characteristics of posttraumatic stress disorder and panic disorder. *Psychosomatic Medicine, 69*, 935–943.

Bliss, T. V., & Cooke, S. F. (2011). Long-term potentiation and long-term depression: A clinical perspective. *Clinics (São Paulo, Brazil), 66*(Suppl. 1), 3–17.

Bonanno, G. A., Kennedy, P., Galatzer-Levy, I. R., Lude, P., & Elfstrom, M. L. (2012). Trajectories of resilience, depression, and anxiety following spinal cord injury. *Rehabilitation Psychology, 57*(3), 236–247. https://doi.org/10.1037/a0029256.

Bonanno, G. A., Mancini, A. D., Horton, J. L., Powell, T. M., Leardmann, C. A., Boyko, E. J., … Millennium Cohort Study Team. (2012). Trajectories of trauma symptoms and resilience in deployed U.S. military service members: Prospective cohort study. *The British Journal of Psychiatry, 200*(4), 317–323. https://doi.org/10.1192/bjp.bp.111.096552.

Bouton, M. E. (2002). Context, ambiguity, and unlearning: Sources of relapse after behavioral extinction. *Biological Psychiatry, 52*(10), 976–986.

Bouton, M. E., & Bolles, R. C. (1979). Contextual control of the extinction of conditioned fear. *Learning and Motivation, 10*, 455–466.

Bouton, M. E., & Ricker, S. T. (1994). Renewal of extinguished responding in a second context. *Animal Learning and Behavior, 22*(3), 317–324.

Breslau, N., Davis, G. C., Peterson, E. L., & Schultz, L. R. (2000). A second look at comorbidity in victims of trauma: The posttraumatic stress disorder-major depression connection. *Biological Psychiatry, 48*(9), 902–909.

Briscione, M. A., Jovanovic, T., & Norrholm, S. D. (2014). Conditioned fear associated phenotypes as robust, translational indices of trauma-, stressor-, and anxiety-related behaviors. *Frontiers in Psychiatry, 5*, 88. https://doi.org/10.3389/fpsyt.2014.00088.

Briscione, M. A., Michopoulos, V., Jovanovic, T., & Norrholm, S. D. (2017). Neuroendocrine underpinnings of increased risk for posttraumatic stress disorder in women. *Vitamins and Hormones, 103*, 53–83. https://doi.org/10.1016/bs.vh.2016.08.003.

Brooks, D. C., & Bouton, M. E. (1993). A retrieval cue for extinction attenuates spontaneous recovery. *Journal of Experimental Psychology. Animal Behavior Processes, 19*(1), 77–89.

Bryant, R. A., Harvey, A. G., Guthrie, R. M., & Moulds, M. L. (2000). A prospective study of psychophysiological arousal, acute stress disorder, and posttraumatic stress disorder. *Journal of Abnormal Psychology, 109*(2), 341–344.

Buckley, T. C., & Kaloupek, D. G. (2001). A meta-analytic examination of basal cardiovascular activity in posttraumatic stress disorder. *Psychosomatic Medicine, 63*(4), 585–594.

Bush, D. E. A., Sotres-Bayon, F., & LeDoux, J. E. (2007). Individual differences in fear: Isolating fear reactivity and fear recovery phenotypes. *Journal of Traumatic Stress, 20*(4), 413–422.

Cacioppo, J. T., Tassinary, L. G., & Berntson, G. G. (2016). *Handbook of psychophysiology* (4th ed.). Cambridge, UK: Cambridge University Press.

Cavigelli, S. A., & McClintock, M. K. (2003). Fear of novelty in infant rats predicts adult corticosterone dynamics and an early death. *Proceedings of the National Academy of Sciences of the United States of America, 100*(26), 16131–16136. https://doi.org/10.1073/pnas.2535721100.

Davis, M. (1992). The role of the amygdala in fear-potentiated startle: Implications for animal models of anxiety. *Trends in Pharmacological Sciences, 13*(1), 35–41.

Davis, M. (1993). Pharmacological analysis of fear-potentiated startle. *Brazilian Journal of Medical and Biological Research, 26*, 235–260.

Davis, M., Falls, W. A., Campeau, S., & Kim, M. (1993). Fear-potentiated startle: A neural and pharmacological analysis. *Behavioural Brain Research, 58*(1–2), 175–198.

deRoon-Cassini, T. A., Mancini, A. D., Rusch, M. D., & Bonanno, G. A. (2010). Psychopathology and resilience following traumatic injury: A latent growth mixture model analysis. *Rehabilitation Psychology, 55*(1), 1–11. https://doi.org/10.1037/a0018601.

Dirikx, T., Vansteenwegen, D., Eelen, P., & Hermans, D. (2009). Non-differential return of fear in humans after a reinstatement procedure. *Acta Psychologica, 130*(3), 175–182. https://doi.org/10.1016/j.actpsy.2008.12.002.

Duvarci, S., Bauer, E. P., & Pare, D. (2009). The bed nucleus of the stria terminalis mediates inter-individual variations in anxiety and fear. *The Journal of Neuroscience, 29*(33), 10357–10361. https://doi.org/10.1523/JNEUROSCI.2119-09.2009.

Dymond, S., Dunsmoor, J. E., Vervliet, B., Roche, B., & Hermans, D. (2015). Fear generalization in humans: Systematic review and implications for anxiety disorder research. *Behavior Therapy, 46*(5), 561–582. https://doi.org/10.1016/j.beth.2014.10.001.

Effting, M., & Kindt, M. (2007). Contextual control of human fear associations in a renewal paradigm. *Behaviour Research and Therapy, 45*(9), 2002–2018. https://doi.org/10.1016/j.brat.2007.02.011.

Everly, G. S., Jr., & Mitchell, J. T. (2000). The debriefing "controversy" and crisis intervention: A review of lexical and substantive issues. *International Journal of Emergency Mental Health, 2*(4), 211–225.

Falls, W. A., Miserendino, M. J., & Davis, M. (1992). Extinction of fear-potentiated startle: Blockade by infusion of an NMDA antagonist into the amygdala. *The Journal of Neuroscience, 12*(3), 854–863.

Fitzgerald, P. J., Pinard, C. R., Camp, M. C., Feyder, M., Sah, A., Bergstrom, H. C., … Holmes, A. (2015). Durable fear memories require PSD-95. *Molecular Psychiatry, 20*(7), 913. https://doi.org/10.1038/mp.2015.44.

Fitzgerald, P. J., Seemann, J. R., & Maren, S. (2014). Can fear extinction be enhanced? A review of pharmacological and behavioral findings. *Brain Research Bulletin, 105*, 46–60. https://doi.org/10.1016/j.brainresbull.2013.12.007.

Friedman, M. (2006). *Post-traumatic and acute stress disorders. The latest assessment and treatment strategies* (4th ed.). Kansas City, MO: Dean Psych Press Corporation.

Frohardt, R., Guarraci, F. A., & Bouton, M. E. (2000). The effects of neurotoxic hippocampal lesions on two effects of context following fear extinction. *Behavioral Neuroscience, 114*, 227–240.

Galatzer-Levy, I. R., Andero, R., Sawamura, T., Jovanovic, T., Papini, S., Ressler, K. J., & Norrholm, S. D. (2017). A cross species study of heterogeneity in fear extinction learning in relation to FKBP5 variation and expression: Implications for the acute treatment of posttraumatic stress disorder. *Neuropharmacology, 116*, 188–195. https://doi.org/10.1016/j.neuropharm.2016.12.023.

Galatzer-Levy, I. R., Ankri, Y., Freedman, S., Israeli-Shalev, Y., Roitman, P., Gilad, M., & Shalev, A. (2013). Early PTSD symptom trajectories: Persistence, recovery, and response to treatment: Results from the Jerusalem Trauma Outreach and Prevention Study (J-TOPS). *PLoS One, 8*, 8.

Galatzer-Levy, I. R., & Bonanno, G. A. (2013). Heterogeneous patterns of stress over the four years of college: Associations with anxious attachment and ego-resiliency. *Journal of Personality, 81*(5), 476–486. https://doi.org/10.1111/jopy.12010.

Galatzer-Levy, I. R., Bonanno, G. A., Bush, D. E., & Ledoux, J. E. (2013). Heterogeneity in threat extinction learning: Substantive and methodological considerations for identifying individual difference in response to stress. *Frontiers in Behavioral Neuroscience, 7*, 55. https://doi.org/10.3389/fnbeh.2013.00055.

Galatzer-Levy, I. R., Madan, A., Neylan, T. C., Henn-Haase, C., & Marmar, C. R. (2011). Peritraumatic and trait dissociation differentiate police officers with resilient versus symptomatic trajectories of posttraumatic stress symptoms. *Journal of Traumatic Stress, 24*(5), 557–565. https://doi.org/10.1002/jts.20684.

Germain, A. (2013). Sleep disturbances as the hallmark of PTSD: Where are we now? *The American Journal of Psychiatry, 170*(4), 372–382. https://doi.org/10.1176/appi.ajp.2012.12040432.

Glover, E. M., Jovanovic, T., Mercer, K. B., Kerley, K., Bradley, B., Ressler, K. J., & Norrholm, S. D. (2012). Estrogen levels are associated with extinction deficits in women with posttraumatic stress disorder. *Biological Psychiatry*, *72*(1), 19–24. https://doi.org/10.1016/j.biopsych.2012.02.031.

Glover, E. M., Jovanovic, T., & Norrholm, S. D. (2015). Estrogen and extinction of fear memories: Implications for posttraumatic stress disorder treatment. *Biological Psychiatry*, *78*(3), 178–185. https://doi.org/10.1016/j.biopsych.2015.02.007.

Glover, E. M., Phifer, J. E., Crain, D. F., Norrholm, S. D., Davis, M., Bradley, B., ... Jovanovic, T. (2011). Tools for translational neuroscience: PTSD is associated with heightened fear responses using acoustic startle but not skin conductance measures. *Depression and Anxiety*, https://doi.org/10.1002/da.20880.

Gray, M. J., & Litz, B. T. (2005). Behavioral interventions for recent trauma: Empirically informed practice guidelines. *Behavior Modification*, *29*(1), 189–215. https://doi.org/10.1177/0145445504270884.

Griffin, M. G. (2008). A prospective assessment of auditory startle alterations in rape and physical assault survivors. *Journal of Traumatic Stress*, *21*(1), 91–99.

Grillon, C., Ameli, R., Woods, S. W., Merikangas, K., & Davis, M. (1991). Fear-potentiated startle in humans: Effects of anticipatory anxiety on the acoustic blink reflex. *Psychophysiology*, *28*(5), 588–595.

Grillon, C., Morgan, C. A., III. (1999). Fear-potentiated startle conditioning to explicit and contextual cues in Gulf War veterans with posttraumatic stress disorder. *Journal of Abnormal Psychology*, *108*(1), 134–142.

Guthrie, R. M., & Bryant, R. A. (2006). Extinction learning before trauma and subsequent posttraumatic stress. *Psychosomatic Medicine*, *68*, 307–311.

Jovanovic, T., Ely, T., Fani, N., Glover, E. M., Gutman, D., Tone, E. B., ... Ressler, K. J. (2012). Reduced neural activation during an inhibition task is associated with impaired fear inhibition in a traumatized civilian sample. *Cortex*, https://doi.org/10.1016/j.cortex.2012.08.011.

Jovanovic, T., Keyes, M., Fiallos, A., Myers, K. M., Davis, M., & Duncan, E. (2005). Fear potentiation and fear inhibition in a human fear-potentiated startle paradigm. *Biological Psychiatry*, *57*(12), 1559–1564.

Jovanovic, T., & Norrholm, S. D. (2016). Human psychophysiology and PTSD. In I. Liberzon & K. Ressler (Eds.), *Neurobiology of PTSD: From brain to mind*. New York, NY: Oxford University Press.

Jovanovic, T., Norrholm, S. D., Blanding, N. Q., Davis, M., Duncan, E., Bradley, B., & Ressler, K. J. (2010). Impaired fear inhibition is a biomarker of PTSD but not depression. *Depression and Anxiety*, *27*(3), 244–251.

Jovanovic, T., Norrholm, S. D., Fennell, J. E., Keyes, M., Fiallos, A., Myers, K. M., ... Duncan, E. J. (2009). Posttraumatic stress disorder may be associated with impaired fear inhibition: Relation to symptom severity. *Psychiatry Research*, *167*(1–2), 151–160.

Jovanovic, T., Phifer, J., Sicking, K., Weiss, T., Norrholm, S., Bradley, B., & Ressler, K. (2011). Cortisol suppression by dexamethasone reduces exaggerated fear responses in posttraumatic stress disorder. *Psychoneuroendocrinology*, *36*(10), 1540–1552. https://doi.org/10.1016/j.psyneuen.2011.04.008.

Kessler, R. C. (2000). Posttraumatic stress disorder: The burden to the individual and to society. *The Journal of Clinical Psychiatry*, *61*(Suppl. 5), 4–12 [discussion 13–14].

Kindt, M., & Soeter, M. (2013). Reconsolidation in a human fear conditioning study: A test of extinction as updating mechanism. *Biological Psychology*, *92*, 43–50.

Kindt, M., Soeter, M., & Vervliet, B. (2009). Beyond extinction: Erasing human fear responses and preventing the return of fear. *Nature Neuroscience*, *12*(3), 256–258. https://doi.org/10.1038/nn.2271.

Klumpers, F., Denys, D., Kenemans, J. L., Grillon, C., van der Aart, J., & Baas, J. M. (2012). Testing the effects of Delta9-THC and D-cycloserine on extinction of conditioned fear in humans. *Journal of Psychopharmacology, 26*(4), 471–478. https://doi.org/10.1177/0269881111431624.

Kull, S., Muller, B. H., Blechert, J., Wilhelm, F. H., & Michael, T. (2012). Reinstatement of fear in humans: Autonomic and experiential responses in a differential conditioning paradigm. *Acta Psychologica, 140*(1), 43–49. https://doi.org/10.1016/j.actpsy.2012.02.007.

LaBar, K. S., & Phelps, E. A. (2005). Reinstatement of conditioned fear in humans is context dependent and impaired in amnesia. *Behavioral Neuroscience, 119*(3), 677–686.

Ledgerwood, L., Richardson, R., & Cranney, J. (2005). D-cycloserine facilitates extinction of learned fear: Effects on reacquisition and generalized extinction. *Biological Psychiatry, 57*(8), 841–847.

Levy-Gigi, E., & Richter-Levin, G. (2014). The hidden price of repeated traumatic exposure. *Stress, 17*(4), 343–351. https://doi.org/10.3109/10253890.2014.923397.

Lissek, S., Biggs, A. L., Rabin, S. J., Cornwell, B. R., Alvarez, R. P., Pine, D. S., & Grillon, C. (2008). Generalization of conditioned fear-potentiated startle in humans: Experimental validation and clinical relevance. *Behaviour Research and Therapy, 46*(5), 678–687. pii: S0005-7967(08)00033-8 https://doi.org/10.1016/j.brat.2008.02.005.

Lissek, S., & Grillon, C. (2012). Learning models of PTSD. In J. G. Beck & D. M. Sloan (Eds.), *The Oxford handbook of traumatic stress disorders*. New York, NY: Oxford University Press.

Lommen, M. J., Engelhard, I. M., Sijbrandij, M., van den Hout, M. A., & Hermans, D. (2013). Pre-trauma individual differences in extinction learning predict posttraumatic stress. *Behaviour Research and Therapy, 51*(2), 63–67. https://doi.org/10.1016/j.brat.2012.11.004.

Lonsdorf, T. B., Menz, M. M., Andreatta, M., Fullana, M. A., Golkar, A., Haaker, J., … Merz, C. J. (2017). Don't fear 'fear conditioning': Methodological considerations for the design and analysis of studies on human fear acquisition, extinction, and return of fear. *Neuroscience and Biobehavioral Reviews, 77*, 247–285. https://doi.org/10.1016/j.neubiorev.2017.02.026.

Lonsdorf, T. B., & Merz, C. J. (2017). More than just noise: Inter-individual differences in fear acquisition, extinction and return of fear in humans—Biological, experiential, temperamental factors, and methodological pitfalls. *Neuroscience and Biobehavioral Reviews, 80*, 703–728. https://doi.org/10.1016/j.neubiorev.2017.07.007.

Mahan, A., & Ressler, K. (2011). Fear conditioning, synaptic plasticity and the amygdala: Implications for posttraumatic stress disorder. *Trends in Neurosciences*.

Maples-Keller, J. L., Jovanovic, T., Dunlop, B. W., Rauch, S., Yasinski, C., Michopoulos, V., … Rothbaum, B. O. (2018). When translational neuroscience fails in the clinic: Dexamethasone prior to virtual reality exposure therapy increases drop-out rates. *Journal of Anxiety Disorders*, https://doi.org/10.1016/j.janxdis.2018.10.006.

Mataix-Cols, D., Fernandez de la Cruz, L., Monzani, B., Rosenfield, D., Andersson, E., Perez-Vigil, A., … Thuras, P. (2017). D-Cycloserine augmentation of exposure-based cognitive behavior therapy for anxiety, obsessive-compulsive, and posttraumatic stress disorders: A systematic review and meta-analysis of individual participant data. *JAMA Psychiatry, 74*(5), 501–510. https://doi.org/10.1001/jamapsychiatry.2016.3955.

Metzger, L. J., Orr, S. P., Berry, N. J., Ahern, C. E., Lasko, N. B., & Pitman, R. K. (1999). Physiologic reactivity to startling tones in women with posttraumatic stress disorder. *Journal of Abnormal Psychology, 108*(2), 347–352.

McCraty, R., & Shaffer, F. (2015). Heart rate variability: New perspectives on physiological mechanisms, assessment of self-regulatory capacity, and health risk. *Global Advances in Health and Medicine, 4*(1), 46–61. https://doi.org/10.7453/gahmj.2014.073.

Meewisse, M. L., Reitsma, J. B., de Vries, G. J., Gersons, B. P., & Olff, M. (2007). Cortisol and post-traumatic stress disorder in adults: Systematic review and meta-analysis. *The British Journal of Psychiatry, 191*, 387–392. https://doi.org/10.1192/bjp.bp.106.024877.

Michopoulos, V., Norrholm, S. D., & Jovanovic, T. (2015). Diagnostic biomarkers for post-traumatic stress disorder: Promising horizons from translational neuroscience research. *Biological Psychiatry, 78*(5), 344–353. https://doi.org/10.1016/j.biopsych.2015.01.005.

Michopoulos, V., Norrholm, S. D., Stevens, J. S., Glover, E. M., Rothbaum, B. O., Gillespie, C. F., … Jovanovic, T. (2017). Dexamethasone facilitates fear extinction and safety discrimination in PTSD: A placebo-controlled, double-blind study. *Psychoneuroendocrinology, 83*, 65–71. https://doi.org/10.1016/j.psyneuen.2017.05.023.

Milad, M. R., Goldstein, J. M., Orr, S. P., Wedig, M. M., Klibanski, A., Pitman, R. K., & Rauch, S. L. (2006). Fear conditioning and extinction: Influence of sex and menstrual cycle in healthy humans. *Behavioral Neuroscience, 120*(6), 1196–1203.

Milad, M. R., Pitman, R. K., Ellis, C. B., Gold, A. L., Shin, L. M., Lasko, N. B., … Rauch, S. L. (2009). Neurobiological basis of failure to recall extinction memory in posttraumatic stress disorder. *Biological Psychiatry, 66*(12), 1075–1082.

Mineka, S., & Oehlberg, K. (2008). The relevance of recent developments in classical conditioning to understanding the etiology and maintenance of anxiety disorders. *Acta Psychologica, 127*(3), 567–580.

Morgan, C. A., 3rd, Grillon, C., Southwick, S. M., Davis, M., & Charney, D. S. (1995). Fear-potentiated startle in posttraumatic stress disorder. *Biological Psychiatry, 38*(6), 378–385.

Morgan, C. A., 3rd, Grillon, C., Southwick, S. M., Davis, M., & Charney, D. S. (1996). Exaggerated acoustic startle reflex in Gulf War veterans with posttraumatic stress disorder. *American Journal of Psychiatry, 153*(1), 64–68.

Morgan, C. A., 3rd, Grillon, C., Lubin, H., & Southwick, S. M. (1997). Startle reflex abnormalities in women with sexual assault-related posttraumatic stress disorder. *American Journal of Psychiatry, 154*(8), 1076–1080.

Muthen, B. (2004). Latent variable analysis: Growth mixture modeling and related techniques for longitudinal data. In D. Kaplan (Ed.), *The SAGE handbook of quantitative methodology for the social sciences*. Washington, DC: Sage Research Methods.

Myers, K. M., & Davis, M. (2002). Behavioral and neural analysis of extinction. *Neuron, 36*(4), 567–584.

Myers, K. M., & Davis, M. (2004). AX+, BX- discrimination learning in the fear-potentiated startle paradigm: Possible relevance to inhibitory fear learning in extinction. *Learning & Memory, 11*(4), 464–475.

Myers, K. M., Ressler, K. J., & Davis, M. (2006). Different mechanisms of fear extinction dependent on length of time since fear acquisition. *Learning and Memory, 13*(2), 216–223.

Neumann, D. L., & Kitlertsirivatana, E. (2010). Exposure to a novel context after extinction causes a renewal of extinguished conditioned responses: Implications for the treatment of fear. *Behaviour Research and Therapy, 48*(6), 565–570. https://doi.org/10.1016/j.brat.2010.03.002.

Norrholm, S. D., Glover, E. M., Stevens, J. S., Fani, N., Galatzer-Levy, I. R., Bradley, B., … Jovanovic, T. (2015). Fear load: The psychophysiological over-expression of fear as an intermediate phenotype associated with trauma reactions. *International Journal of Psychophysiology, 98*(2 Pt 2), 270–275. https://doi.org/10.1016/j.ijpsycho.2014.11.005.

Norrholm, S. D., & Jovanovic, T. (2011). Translational fear inhibition models as indicies of trauma-related psychopathology. *Current Psychiatry Reviews, 7*.

Norrholm, S. D., & Jovanovic, T. (2018). Fear processing, psychophysiology, and PTSD. *Harvard Review of Psychiatry, 26*(3), 129–141. https://doi.org/10.1097/HRP.0000000000000189.

Norrholm, S. D., Jovanovic, T., Briscione, M. A., Anderson, K. M., Kwon, C. K., Warren, V. T., … Bradley, B. (2014). Generalization of fear-potentiated startle in the presence of auditory cues: A parametric analysis. *Frontiers in Behavioral Neuroscience, 8*, 361. https://doi.org/10.3389/fnbeh.2014.00361.

Norrholm, S. D., Jovanovic, T., Olin, I. W., Sands, L. A., Karapanou, I., Bradley, B., & Ressler, K. J. (2011). Fear extinction in traumatized civilians with posttraumatic stress disorder: Relation to symptom severity. *Biological Psychiatry, 69*(6), 556–563. pii: S0006-3223(10)00949-2 https://doi.org/10.1016/j.biopsych.2010.09.013.

Norrholm, S. D., Jovanovic, T., Vervliet, B., Myers, K. M., Davis, M., Rothbaum, B. O., & Duncan, E. J. (2006). Conditioned fear extinction and reinstatement in a human fear-potentiated startle paradigm. *Learning & Memory, 13*(6), 681–685.

Norrholm, S. D., Vervliet, B., Jovanovic, T., Boshoven, W., Myers, K. M., Davis, M., … Duncan, E. J. (2008). Timing of extinction relative to acquisition: A parametric analysis of fear extinction in humans. *Behavioral Neuroscience, 122*(5), 1016–1030. https://doi.org/10.1037/a0012604. 2008-13280-002.

Nowak, L., Bregestovski, P., Ascher, P., Herbet, A., & Prochiantz, A. (1984). Magnesium gates glutamate-activated channels in mouse central neurones. *Nature, 307*(5950), 462–465.

Orr, S. P., Lasko, N. B., Shalev, A. Y., & Pitman, R. K. (1995). Physiologic responses to loud tones in Vietnam veterans with posttraumatic stress disorder. *Journal of Abnormal Psychology, 104*(1), 75–82.

Pace-Schott, E. F., Germain, A., & Milad, M. R. (2015a). Effects of sleep on memory for conditioned fear and fear extinction. *Psychological Bulletin, 141*(4), 835–857. https://doi.org/10.1037/bul0000014.

Pace-Schott, E. F., Germain, A., & Milad, M. R. (2015b). Sleep and REM sleep disturbance in the pathophysiology of PTSD: The role of extinction memory. *Biology of Mood & Anxiety Disorders, 5*, 3. https://doi.org/10.1186/s13587-015-0018-9.

Pace-Schott, E. F., Hutcherson, C. A., Bemporad, B., Morgan, A., Kumar, A., Hobson, J. A., & Stickgold, R. (2009). Failure to find executive function deficits following one night's total sleep deprivation in university students under naturalistic conditions. *Behavioral Sleep Medicine, 7*(3), 136–163. https://doi.org/10.1080/15402000902976671.

Pace-Schott, E. F., & Spencer, R. M. (2015). Sleep-dependent memory consolidation in healthy aging and mild cognitive impairment. *Current Topics in Behavioral Neurosciences, 25*, 307–330. https://doi.org/10.1007/7854_2014_300.

Pace-Schott, E. F., Spencer, R. M., Vijayakumar, S., Ahmed, N. A., Verga, P. W., Orr, S. P., … Milad, M. R. (2013). Extinction of conditioned fear is better learned and recalled in the morning than in the evening. *Journal of Psychiatric Research, 47*(11), 1776–1784. https://doi.org/10.1016/j.jpsychires.2013.07.027.

Pavlov, I. P. (1927). *Conditioned reflexes*. Oxford University Press.

Peri, T., Ben-Shakhar, G., Orr, S. P., & Shalev, A. Y. (2000). Psychophysiologic assessment of aversive conditioning in posttraumatic stress disorder. *Biological Psychiatry, 47*(6), 512–519.

Pole, N. (2007). The psychophysiology of posttraumatic stress disorder: A meta-analysis. *Psychological Bulletin, 133*(5), 725–746.

Pole, N., Neylan, T. C., Otte, C., Henn-Hasse, C., Metzler, T. J., & Marmar, C. R. (2009). Prospective prediction of posttraumatic stress disorder symptoms using fear potentiated auditory startle responses. *Biological Psychiatry, 65*(3), 235–240.

Quirk, G. J. (2002). Memory for extinction of conditioned fear is long-lasting and persists following spontaneous recovery. *Learning & Memory, 9*(6), 402–407.

Rescorla, R. A. (1969). Conditioned inhibition of fear resulting from negative CS-US contingencies. *Journal of Comparative and Physiological Psychology, 67*(4), 504–509.

Rescorla, R. A. (2004). Spontaneous recovery. *Learning & Memory, 11*(5), 501–509.

Rescorla, R. A., & Heth, C. D. (1975). Reinstatement of fear to an extinguished conditioned stimulus. *Journal of Experimental Psychology. Animal Behavior Processes, 1*(1), 88–96.

Rescorla, R. A., & Wagner, A. R. (1972). A theory of Pavlovian conditioning: Variations in the effectiveness of reinforcement and nonreinforcement. In A. H. Black & W. F. Prokasy (Eds.), *Classical conditioning II: Current theory and research* (pp. 64–99). New York, NY: Appleton-Century-Crofts.

Rothbaum, B. O., & Davis, M. (2003). Applying learning principles to the treatment of post-trauma reactions. *Annals of the New York Academy of Sciences, 1008*, 112–121.

Sawamura,T., Klengel,T.,Armario,A., Jovanovic,T., Norrholm, S. D., Ressler, K. J., & Andero, R. (2016). Dexamethasone treatment leads to enhanced fear extinction and dynamic *Fkbp*5 regulation in amygdala. *Neuropsychopharmacology, 41*(3), 832–846. https://doi.org/10.1038/npp.2015.210.

Schiller, D., Monfils, M.-H., Raio, C. M., Johnson, D. C., LeDoux, J. E., & Phelps, E. A. (2010). Preventing the return of fear in humans using reconsolidation update mechanisms. *Nature, 463*(7277), 49–53.

Shalev,A.Y., Peri,T., Orr, S. P., Bonne, O., & Pitman, R. K. (1997). Auditory startle responses in help-seeking trauma survivors. *Psychiatry Research, 69*(1), 1–7.

Shiban,Y., Pauli, P., & Muhlberger,A. (2013). Effect of multiple context exposure on renewal in spider phobia. *Behaviour Research and Therapy, 51*(2), 68–74. https://doi.org/10.1016/j.brat.2012.10.007.

Shin, L. M., & Liberzon, I. (2010). The neurocircuitry of fear, stress, and anxiety disorders. *Neuropsychopharmacology, 35*(1), 169–191.

Shvil, E., Sullivan, G. M., Schafer, S., Markowitz, J. C., Campeas, M.,Wager,T. D., … Neria,Y. (2014). Sex differences in extinction recall in posttraumatic stress disorder:A pilot fMRI study. *Neurobiology of Learning and Memory, 113*, 101–108. https://doi.org/10.1016/j.nlm.2014.02.003.

Sokol, N., & Lovibond, P. F. (2012). Cross-US reinstatement of human conditioned fear: Return of old fears or emergence of new ones? *Behaviour Research and Therapy, 50*(5), 313–322. https://doi.org/10.1016/j.brat.2012.02.005.

Spoormaker,V. I., Schroter, M. S., Gleiser, P. M.,Andrade, K. C., Dresler, M.,Wehrle, R., … Czisch, M. (2010). Development of a large-scale functional brain network during human non-rapid eye movement sleep. *The Journal of Neuroscience, 30*(34), 11379–11387. https://doi.org/10.1523/JNEUROSCI.2015-10.2010.

Stiedl, O., & Hager, T. (2017). Cardiovascular conditioning: Neural substrates. In *Reference module in neuroscience and biobehavioral psychology*: Elsevier.

Stockhorst, U., & Antov, M. I. (2015). Modulation of fear extinction by stress, stress hormones and estradiol: A review. *Frontiers in Behavioral Neuroscience, 9*, 359. https://doi.org/10.3389/fnbeh.2015.00359.

Suris, A., North, C., Adinoff, B., Powell, C. M., & Greene, R. (2010). Effects of exogenous glucocorticoid on combat-related PTSD symptoms. *Annals of Clinical Psychiatry, 22*(4), 274–279.

Tamai, N., & Nakajima, S. (2000). Renewal of formerly conditioned fear in rats after extensive extinction training. *International Journal of Comparative Psychology, 13*, 137–147.

Thomas, B. L.,Vurbic, D., & Novak, C. (2009). Extensive extinction in multiple contexts eliminates the renewal of conditioned fear in rats. *Learning and Motivation, 40*(2), 147–159.

van der Velden, V. H. (1998). Glucocorticoids: Mechanisms of action and anti-inflammatory potential in asthma. *Mediators of Inflammation, 7*(4), 229–237. https://doi.org/10.1080/09629359890910.

VanElzakker, M. B., Dahlgren, M. K., Davis, F. C., Dubois, S., & Shin, L. M. (2014). From Pavlov to PTSD:The extinction of conditioned fear in rodents, humans, and anxiety disorders. *Neurobiology of Learning and Memory, 113*, 3–18. https://doi.org/10.1016/j.nlm.2013.11.014.

Vansteenwegen, D., Hermans, D.,Vervliet, B., Francken, G., Beckers,T., Baeyens, F., & Eelen, P. (2005). Return of fear in a human differential conditioning paradigm caused by a return to the original acquisition context. *Behaviour Research and Therapy, 43*(3), 323–336.

Vervliet, B., Baeyens, F.,Van den Bergh, O., & Hermans, D. (2013). Extinction, generalization, and return of fear: A critical review of renewal research in humans. *Biological Psychology, 92*, 51–58.

Vervliet, B., Craske, M. G., & Hermans, D. (2013). Fear extinction and relapse: State of the art. *Annual Review of Clinical Psychology, 9*, 215–248. https://doi.org/10.1146/annurev-clinpsy-050212-185542.

Wagner, A. R., Logan, F. A., Haberlandt, K., & Price, T. (1968). Stimulus selection in animal discrimination learning. *Journal of Experimental Psychology, 76*, 171–180.

Walker, D. L., & Davis, M. (2000). Involvement of NMDA receptors within the amygdala in short- versus long-term memory for fear conditioning as assessed with fear-potentiated startle. *Behavioral Neuroscience, 114*(6), 1019–1033.

Walker, D. L., & Davis, M. (2002). The role of amygdala glutamate receptors in fear learning, fear-potentiated startle, and extinction. *Pharmacology, Biochemistry, and Behavior, 71*(3), 379–392.

Warren, V. T., Anderson, K. M., Kwon, C., Bosshardt, L., Jovanovic, T., Bradley, B., & Norrholm, S. D. (2014). Human fear extinction and return of fear using reconsolidation update mechanisms: The contribution of on-line expectancy ratings. *Neurobiology of Learning and Memory, 113*, 165–173. https://doi.org/10.1016/j.nlm.2013.10.014.

Weingast, L., Haas, H. E., & Norrholm, S. D. (2018). Competing with learned fear: Implications of fear extinction for clinical intervention. In S. M. Powell, R. Gallub, & F. Flach (Eds.), Vol. 38. *Directions in psychiatry*. Hobart, NY: Hatherleigh Company, Ltd.

Wolfe, J., Chrestman, K. R., Ouimette, P. C., Kaloupek, D., Harley, R. M., & Bucsela, M. (2000). Trauma-related psychophysiological reactivity in women exposed to war-zone stress. *Journal of Clinical Psychology, 56*(10), 1371–1379.

Woods, A. M., & Bouton, M. E. (2006). D-cycloserine facilitates extinction but does not eliminate renewal of the conditioned emotional response. *Behavioral Neuroscience, 120*(5), 1159–1162. https://doi.org/10.1037/0735-7044.120.5.1159.

Yehuda, R., & Seckl, J. (2011). Minireview: Stress-related psychiatric disorders with low cortisol levels: A metabolic hypothesis. *Endocrinology, 152*(12), 4496–4503. https://doi.org/10.1210/en.2011-1218.

Zuj, D. V., & Norrholm, S. D. (2019). The clinical applications and practical relevance of human conditioning paradigms for posttraumatic stress disorder. *Progress in Neuro-Psychopharmacology & Biological Psychiatry, 88*, 339–351. https://doi.org/10.1016/j.pnpbp.2018.08.014.

Zuj, D. V., Palmer, M. A., Hsu, C. M., Nicholson, E. L., Cushing, P. J., Gray, K. E., & Felmingham, K. L. (2016). Impaired fear extinction associated with PTSD increases with hours-since-waking. *Depression and Anxiety, 33*(3), 203–210. https://doi.org/10.1002/da.22463.

Zuj, D. V., Palmer, M. A., Lommen, M. J., & Felmingham, K. L. (2016). The centrality of fear extinction in linking risk factors to PTSD: A narrative review. *Neuroscience and Biobehavioral Reviews, 69*, 15–35. https://doi.org/10.1016/j.neubiorev.2016.07.014.

SECTION 3

Difficulties in responding and relating to emotion

CHAPTER 10

Emotion regulation difficulties in PTSD

Matthew T. Tull, Ariana G. Vidaña, James E. Betts
Department of Psychology, University of Toledo, Toledo, OH, United States

There is considerable heterogeneity in the clinical presentation of post-traumatic stress disorder (PTSD). However, at its core, PTSD can be considered a disorder of emotion (Frewen & Lanius, 2006). As other chapters in this book highlight, PTSD is associated with alterations in the activity of areas of the brain (e.g., amygdala and insula) linked with the experience and expression of affect (Lanius, Frewen, Vermetten, & Yehuda, 2010; Sripada et al., 2012). In addition, PTSD is characterized by the experience of frequent and intense negative emotions (Finucane, Dima, Ferreria, & Halvorsen, 2011) and deficits in the experience and expression of positive emotions (Kashdan, Elhai, & Frueh, 2006; Litz & Gray, 2002). Although PTSD has long been considered a disorder of emotion, it has only been in the past two decades that researchers have begun to explore the relation between PTSD and the regulation of emotion. Since 2000, there has been an exponential rise in research on emotion regulation and PTSD (Seligowski, Lee, Bardeen, & Orcutt, 2015) that has largely mirrored the rapid growth of research on emotion regulation in general (Gross, 2015a).

The purpose of this chapter is to provide an overview of the extant literature on emotion regulation and PTSD. This chapter will begin by providing definitions of two prominent models of emotion regulation, followed by a brief discussion as to how these models can be integrated. We will then review representative studies that capture the application of each model of emotion regulation to PTSD. Next, we will discuss a new direction in research on emotion regulation and PTSD, positive emotion regulation difficulties. We will conclude the chapter with a discussion of treatment implications stemming from research on PTSD and emotion regulation.

Definitions of emotion regulation

There are numerous definitions of emotion regulation in the literature, and researchers have yet to come to a consensus as to the most accurate or useful definition (Bloch, Moran, & Kring, 2010; Gratz & Tull, 2010; Gross, 2015a; Tull & Aldao, 2015). It has been suggested that the utility of these definitions of emotion regulation is likely dependent on the particular research or clinical question under investigation (Gratz, Weiss, & Tull, 2015); however, the increase of research in emotion regulation has resulted in the greater refinement of existing definitions, and there is evidence that some prominent models of emotion regulation are beginning to converge. Based on this observation, Tull and Aldao (2015) suggested that current definitions of emotion regulation can be organized according to their focus on the regulation of emotion at the micro or macro level. Micro level approaches emphasize the specific situational, cognitive, or behavioral strategies used in the moment to affect the experience and expression of emotions. Macro level approaches, on the other hand, emphasize the typical or dispositional ways in which individuals understand, regard, and respond to their emotional experiences. Macro level approaches, then, view emotion regulation from the perspective of one's ability or capacity to respond to emotions in an adaptive and effective manner.

An emotion regulation strategy model

The micro approach to emotion regulation is best captured by Gross' (2015a) process model of emotion regulation. This model maps onto the modal model of emotion generation and illustrates the strategies individuals use to alter the trajectory of an emotion at different points in the emotion generative process to facilitate progress toward a desired goal (Gross, 2015a). The different strategies an individual may use to influence emotional experience can be classified according to when they are employed in the emotion generative process: (1) situation selection (i.e., influencing exposure to situations that could generate desirable or undesirable emotions), (2) situation modification (i.e., altering a situation to modify its emotional impact), (3) attentional deployment (i.e., controlling the allocation of attention to modify an emotional response), (4) cognitive change (i.e., changing the evaluation of a situation to influence its emotional impact), and (5) response modulation (i.e., engaging in a behavior to influence some aspect of a generated emotion). According to this model, there are different short- and long-term emotional, cognitive, behavioral, and situational consequences depending on the timing and type of strategy used (Gross, 2015a).

Recently, Gross (2015b) expanded his original model to incorporate the concept of interacting valuation systems, now referring to the model as the extended process model of emotion regulation. This expanded model states that there are three valuation systems (with each unfolding through a process of perception, valuation, and action) involved in emotion regulation distinguished by where they occur in the emotion regulation process. The first stage, identification, involves the detection of an emotion (perception), its evaluation as an experience that requires regulation (valuation), and a decision about whether regulation will begin (action). The second stage, selection, involves the identification of available emotion regulation strategies (perception), evaluation of whether specific strategies will be more or less successful depending on internal and external contextual factors (valuation), and making the choice to use a particular strategy (action). The third and final stage, implementation, involves translating a general emotion regulation strategy into specific behaviors that would be most suitable for that specific situation (perception), evaluating the likely effectiveness or ineffectiveness of specific emotion regulation strategies (valuation), and actually choosing and implementing a specific emotion regulation strategy (action).

An emotion regulation abilities model

The macro approach to emotion regulation is arguably best represented by Gratz and Roemer's (2004) acceptance-based model of emotion regulation. This model was derived from theoretical and empirical literature on the functionality of emotions (Cole, Michel, & Teti, 1994; Ekman & Davidson, 1994; Thompson, 1994) and paradoxical consequences associated with attempts at avoiding or controlling emotions (Hayes, Luoma, Bond, Masuda, & Lillis, 2006). Gratz and Roemer's (2004) model of emotion regulation focuses broadly on adaptive ways of responding to emotions, including the understanding, acceptance, effective use, and modulation of emotions. In evaluating this model, it is important to note that Gratz and Roemer (2004) distinguish the quality of the emotional response (valence, intensity, or reactivity) from responses to emotions, with the former serving only as a vulnerability for emotion regulation difficulties as opposed to an indication of emotion regulation difficulties.

According to this model, adaptive responses are those that aid in the functional use of emotions as information and the pursuit of valued actions and desired goals. Specifically, Gratz and Roemer (2004) conceptualize emotion regulation as a multidimensional construct involving the awareness, understanding, and acceptance of emotions; ability to control impulsive behaviors and engage in goal-directed behaviors when experiencing negative emotions; flexible use of nonavoidant situational appropriate strategies to modulate the intensity and duration of emotional responses to meet individual goals and situational demands; and willingness to experience negative emotions in pursuit of meaningful activities in life. The emphasis on overarching emotion regulation abilities and the contextually dependent nature of adaptive emotion regulation strategies (which can only be evaluated in the context of goals and situational demands; Aldao, 2013; Cole et al., 1994; Thompson, 1994) precludes a focus on the specific emotion regulation strategy used by an individual at any given moment. However, inherent in this model is the idea that emotion regulation strategies focused on the chronic avoidance or control of emotion are maladaptive, whereas adaptive strategies are those that promote awareness, understanding, acceptance, and the functional use of emotions.

Integrating models of emotion regulation

Although Gross' (2015a) and Gratz and Roemer's (2004) models focus on different emotion regulation processes, they can be integrated. Emotion regulation abilities in general would be expected to contribute to the use of more effective emotion regulation strategies in the moment and the flexible use of emotion regulation strategies to meet specific situational demands and individual goals. For example, greater emotional awareness and clarity would provide individuals with a better sense of the most effective emotion regulation strategies for a given situation. Consistent with this idea, Feldman-Barrett, Gross, Christensen, and Benvenuto (2001) found that individuals with greater negative emotion differentiation exhibited the greater ability to employ emotion regulation strategies. Likewise, other studies have found that greater overall emotion regulation abilities are associated with a lower reliance on putatively maladaptive emotion regulation strategies and greater reliance on putatively adaptive emotion regulation strategies (Bardeen & Fergus, 2014; Hofmann & Kashdan, 2010).

Approaching the integration of these models from the other direction, repeated and inflexible use of certain emotion regulation strategies may eventually negatively affect emotion regulation abilities. For example, given the paradoxical consequences of emotional avoidance (Hayes et al., 2006), individuals may experience a reduction in self-efficacy to modulate emotions. Repeated avoidance may also reinforce the belief that certain emotions are dangerous, contributing to a nonacceptance of emotions.

In addition, increased physiological arousal associated with the suppression of emotional expression (Gross & Levenson, 1997) may interfere with emotional clarity and awareness. In sum the relationship between the emotion regulation strategies and abilities models can be conceptualized as bidirectional with both positive and negative feedback loops (Tull & Aldao, 2015), and we can obtain a more complete understanding of the relation between emotion regulation and outcomes (including PTSD) through an integration of the models.

Evaluating emotion regulation strategy and abilities models in the context of PTSD

The emotional experience of individuals with PTSD is often chaotic, unpredictable, and intense. Individuals with PTSD may experience intense fear upon being unexpectedly reminded of a traumatic event, frequent shame and guilt due to intrusive thoughts or self-blame regarding a traumatic event, or sadness as a result of feeling disconnected from others. Individuals may also fluctuate between periods of intense emotional arousal to prolonged episodes of emotional numbing (Horowitz, 1986; Litz & Gray, 2002). These emotional experiences were likely not the typical experiences an individual had prior to the traumatic event and development of PTSD symptoms. As a result, a fear of or unwillingness to experience emotions may develop. Likewise a considerable tax may be placed on an individual's ability to identify specific emotion regulation strategies that can effectively modulate these intense emotional experiences. The experience of intense and rapidly changing emotions may also interfere with emotional clarity and awareness, further hampering an individual's ability to identify the most effective method for responding to an emotion. Consequently, individuals with PTSD may find themselves relying on strategies focused on trying to prevent the experience of emotion (e.g., situational avoidance) or rapidly escape from emotions (e.g., substance use and nonsuicidal self-injury). Although these particular emotion regulation strategies may provide relief in the short term, in the long term, avoidance-/escape-oriented emotion regulation strategies are likely going to contribute to greater disconnection from rewarding activities (e.g., relationships and leisure activities) and increased emotional distress (Tull, Gratz, Salters, & Roemer, 2004). Greater reliance on avoidance- or escape-oriented strategies could also prevent exposure to corrective information, interfering with emotional processing and increasing risk for the development and exacerbation of PTSD symptoms

(Foa & Kozak, 1986). Research on emotion regulation in PTSD from both the emotion regulation strategy and ability perspectives provides support for an association between PTSD and difficulties with emotion regulation.

Emotion regulation strategies and PTSD

Much of the research on emotion regulation strategies in PTSD is cross-sectional in nature, relying heavily on self-report measures of emotion regulation strategy use. Not surprisingly, research in this area shows that individuals with PTSD tend to rely on putatively maladaptive emotion regulation strategies. A recent metaanalysis by Seligowski et al. (2015) examined associations between PTSD and various measures of emotion regulation across 57 studies of college student, community, military, clinical, and substance-dependent samples. Results from this metaanalysis support the hypothesis that PTSD is associated with a greater reliance on putatively maladaptive emotion regulation strategies. Specifically the authors found the largest effects for the specific emotion regulation strategies of rumination, thought suppression, and experiential avoidance. Medium-sized effects were found for expressive suppression and worry. Of note, significant effects were not found for putatively adaptive emotion regulation strategies (acceptance and reappraisal).

In one of the few laboratory-based studies in this area, Tull, Berghoff, Wheeless, Cohen, and Gratz (2018) examined the relation between PTSD symptom severity and the use of specific emotion regulation strategies following exposure to a trauma-related reminder. Specifically, substance-dependent patients exposed to a Criterion A traumatic event listened to a personalized trauma script and then reported on the specific emotion regulation strategies that they used during the script. Interestingly, PTSD symptom severity was positively associated with the use of more putatively adaptive (e.g., distraction) and maladaptive (e.g., emotional avoidance) emotion regulation strategies. In addition, the authors found evidence of indirect effects of PTSD symptom severity on negative affect and cortisol reactivity through both putatively adaptive and maladaptive emotion regulation strategies. These findings speak to the complexity of emotion regulation strategy use among individuals with PTSD. Specifically the findings suggest, that although substance-dependent patients experiencing PTSD symptoms may be more likely to rely on maladaptive emotion regulation strategies, they may also have access to more adaptive emotion regulation strategies as well. It is also possible that, as PTSD symptoms become more severe, individuals may employ a greater number of strategies (potentially insensitive to context) in an attempt to modulate emotional distress.

Research has also demonstrated that the use of particular emotion regulation strategies is prospectively associated with the development of PTSD symptoms. For example, within a sample of individuals who had experienced a motor vehicle accident, dissociation at the time of an accident and persistent dissociation 4 weeks following the accident significantly predicted the severity of PTSD symptoms at 6 months (Murray, Ehlers, & Mayou, 2002). Murray et al. (2002) also found that rumination soon after the accident and 4 weeks postaccident exhibited high-magnitude associations with PTSD symptom severity 6 months after the accident. In a study of undergraduate students exposed to a campus shooting, Kumpla, Orcutt, Bardeen, and Varkovitzky (2011) found that preevent experiential avoidance predicted peritraumatic dissociation and PTSD intrusion and dysphoria symptoms 1 month following the traumatic event. Moreover, peritraumatic dissociation was associated with more severe PTSD symptoms 1 month after the traumatic event and experiential avoidance 1 month after the event was associated with more severe PTSD hyperarousal and dysphoria symptoms 8 months after the traumatic event. Jenness et al. (2016) also found that the use of particular emotion regulation strategies pretraumatic exposure (rumination and low cognitive reappraisal) predicted the onset and severity of PTSD symptoms among adolescents exposed to a terrorist attack. Similarly, Fitzgerald et al. (2018) found that deficits in the ability to engage in cognitive reappraisal (a putatively adaptive emotion regulation strategy) were associated with more PTSD symptoms over the course of 1 year among OEF/OIF/OND combat-exposed veterans. In a recent study, Short, Boffa, Clancy, and Schmidt (2018) used an ecological momentary assessment design to examine prospective associations between the use of putatively maladaptive and adaptive emotion regulation strategies and PTSD symptoms in a sample of individuals with a diagnosis of PTSD recruited from the community and a university setting. Over the course of 8 days, the authors found that the use of putatively maladaptive emotion regulation strategies (e.g., thought suppression, rumination, engaging in impulsive behaviors, and avoidance) in response to a stressor prospectively predicted more severe PTSD symptoms later in the day. Putatively adaptive emotion regulation strategy use (cognitive reappraisal, acceptance, and problem solving) in response to a stressor, however, did not prospectively predict PTSD symptoms later in the day.

Finally, neuroimaging studies provide further evidence of deficits in employing emotion regulation strategies among individuals with PTSD. In general, these studies show that, in the context of provocation paradigms (e.g., exposure to trauma-related or other emotionally evocative cues), individuals with PTSD exhibit hyperactivation of the dorsomedial prefrontal cortex, anterior insula, and amygdala and hypoactivation of the dorsal anterior cingulate cortex—all areas implicated in the experience and regulation of emotion (Taylor & Liberzon, 2007). In addition, several studies have utilized a paradigm where individuals with PTSD are instructed to utilize the specific emotion regulation strategy of cognitive reappraisal when viewing emotionally evocative aversive images. These studies found that, relative to control groups, individuals with PTSD had less recruitment of the dorsolateral prefrontal cortex during cognitive reappraisal (New et al., 2009; Rabinak et al., 2014). Similar to Tull et al. (2018), these findings suggest that individuals with PTSD can employ putatively adaptive emotion regulation strategies (as the dorsolateral prefrontal cortex was activated); however, evidence that the dorsolateral prefrontal cortex was recruited to a lesser extent than individuals without PTSD indicates that they may be less effective in doing so relative to those without PTSD.

Emotion regulation abilities in PTSD

Similar to the literature on emotion regulation strategies and PTSD, much of the research on emotion regulation abilities in PTSD relies on self-report measures, primarily the Difficulties in Emotion Regulation Scale (DERS; Gratz & Roemer, 2004), and is cross-sectional in design. Overall, across multiple studies, there is a large effect-sized relation between PTSD and emotion regulation abilities (Seligowski et al., 2015). In addition, research in this area typically shows deficits in specific emotion regulation abilities among individuals with PTSD. For example, Tull, Barrett, McMillan, and Roemer (2007) found that trauma-exposed college students with (vs without) probable PTSD exhibited greater emotional nonacceptance, more difficulties engaging in goal-directed behaviors when distressed, more difficulties controlling impulsive behaviors when distressed, limited access to emotion regulation strategies perceived as effective, and lower emotional clarity. These findings were replicated by Frewen, Dozois, Neufeld, and Lanius (2012) in a sample of women with current PTSD and women without any history of psychiatric disorders or childhood maltreatment, with the exception that they also found that women with PTSD reported lower emotional awareness than those without PTSD. Moreover, in a sample of African American undergraduate students, Weiss, Tull, Viana, Anestis, and Gratz (2012) found that participants with probable PTSD reported significantly higher levels of emotional nonacceptance, difficulties engaging in goal-directed behavior when distressed, difficulties controlling impulsive behaviors when distressed, and limited access to emotion regulation strategies perceived as effective relative to participants without Criterion A traumatic exposure and Criterion A traumatic exposure but not probable PTSD. Of note, although research in this area has generally relied on undergraduate samples, the aforementioned findings have been replicated in clinical samples, particularly patients in residential substance use disorder treatment (McDermott, Tull, Gratz, Daughters, & Lejuez, 2009; Weiss, Tull, Anestis, & Gratz, 2013) and veterans (Miles, Menefee, Wanner, Tharp, & Kent, 2016).

Research also shows that deficits in emotion regulation abilities may underlie the wide variety of health-compromising and self-destructive behaviors commonly observed among individuals with PTSD. For example, Weiss et al. (2012) found that deficits in emotion regulation abilities explained the association between PTSD and overall involvement in risky behaviors in a sample of substance-dependent patients in residential treatment. Similar evidence was provided by Weiss, Tull, and Gratz (2014) in a sample of women with sexual assault-related PTSD. Likewise, there is evidence that deficits in emotion regulation abilities account for the relationship between PTSD and impulsive aggression in veterans (Miles et al., 2016), alcohol misuse in OEF/OIF/OND veterans (Tripp & McDevitt-Murphy, 2015) and college students (Radomski & Read, 2016), and coping-oriented marijuana use among community-recruited adults (Bonn-Miller, Vujanovic, Boden, & Gross, 2010). In addition, deficits in emotion regulation abilities moderate the association between PTSD and nonsuicidal self-injury, such that the relation between PTSD and nonsuicidal self-injury was stronger among individuals with greater deficits in emotion regulation abilities (Dixon-Gordon, Tull, & Gratz, 2014).

Fewer studies have examined whether emotion regulation abilities prospectively predict the development of PTSD symptoms following a traumatic event; however, the research that has been done in this area supports such a relationship. Bardeen, Kumpula, and Orcutt (2013) assessed emotion regulation abilities and PTSD symptoms before and after a mass shooting on a college campus. Deficits in emotion regulation abilities prior to the event were associated with more severe PTSD symptoms in the immediate aftermath of the shooting and approximately 8 months later. Moreover, within a sample of trauma-exposed individuals recruited from hospital emergency departments, Forbes et al. (2019) found that deficits in emotion regulation abilities assessed soon after the experience of a traumatic event were associated with the development of PTSD 3 months later, even when controlling for initial PTSD symptoms, age, gender, and type of traumatic event experienced.

Regulation of positive emotion in PTSD

Much of the literature on emotion regulation tends to focus on the regulation of negative emotional experiences. However, there is evidence that individuals also regulate positive emotions (Gross, Richards, & John, 2006), and the effective regulation of positive emotions is associated with a number of positive outcomes, such as increased life satisfaction and self-esteem and lower depression (see Tugade & Fredrickson, 2007). On the other hand, individuals can experience difficulties in the regulation of positive emotions (Beblo et al., 2012; Weiss, Gratz, & Lavender, 2015), and these difficulties have been shown to be associated with negative outcomes, such as substance use (Weiss, Forkus, Contractor, & Shick, 2018) and risky sexual behavior (Weiss, Tull, Sullivan, Dixon-Gordon, & Gratz, 2015).

Speaking to the relation between positive emotion regulation difficulties and PTSD in particular, individuals with PTSD may evaluate the experience of physiological arousal as aversive due to the frequent experience of hyperarousal symptoms. Williams, Chambless, and Ahrens (1997) suggest that a tendency to view physiological sensations as aversive may generalize to stimuli that produce similar physiological reactions, such as intense positive emotional experiences. As a result, individuals with PTSD may be motivated to avoid positive emotions. In addition, there is evidence that the experience of positive emotions broadens the visual field of view, whereas negative emotions decrease the visual field of view (Schmitz, De Rosa, & Anderson, 2009). This research would suggest that the experience of positive emotions may counter hypervigilance in PTSD. Although such an outcome may be beneficial in the long term, decreased hypervigilance may initially increase feelings of vulnerability or a lack of preparation for potential threat. Consequently, positive emotions may also be perceived as a threat to safety.

Consistent with these ideas, research has shown that individuals with PTSD do engage in the strategic withholding of positive emotional experiences (Roemer, Litz, Orsillo, & Wagner, 2001). In addition, using a recently developed measure of difficulties in positive emotion regulation, the Difficulties in Emotion Regulation Scale-Positive (DERS-Positive; Weiss et al., 2015), Weiss and colleagues have shown that PTSD is associated with deficits in a number of positive emotion regulation abilities. The DERS-Positive is grounded in the emotion regulation abilities model and assesses deficits in three positive emotion regulation abilities: (1) the nonacceptance of positive emotions, (2) difficulties engaging in goal-directed behaviors when experiencing positive emotions, and (3) difficulties controlling impulsive behaviors when experiencing positive emotions. Difficulties in positive emotion regulation abilities have been found to explain additional variance in PTSD symptom severity above and beyond difficulties in the regulation of negative emotions (Weiss, Nelson, Contractor, & Sullivan, 2019). In addition, greater difficulties in each positive emotion regulation ability corresponded with more severe PTSD symptoms, and individuals with a probable PTSD diagnosis exhibited greater difficulties engaging in goal-directed behaviors and controlling impulsive behaviors when experiencing positive emotions (Weiss, Dixon-Gordon, Peasant, & Sullivan, 2018). This area of research holds promise, and future research on positive emotion regulation difficulties in PTSD may provide us with a more nuanced understanding of emotion dysfunction in PTSD and its role in the myriad of maladaptive behaviors and negative clinical outcomes associated with this disorder.

Implications of emotion regulation research for the treatment of PTSD

Although prominent empirically supported cognitive-behavioral treatments for PTSD (e.g., prolonged exposure and cognitive processing therapy) do not explicitly focus on teaching patients adaptive emotion regulation skills, their ability to reduce the intensity of negative emotions, such as fear, shame, guilt, and anger (Langkaas et al., 2017; Resick, Nishith, Weaver, Astin, & Feuer, 2002; Stapleton, Taylor, & Asmundson, 2006), likely contributes to improved emotion regulation abilities (e.g., increased emotional willingness, improved distress tolerance, and greater emotional clarity) and a reduction in maladaptive emotion regulation strategies (e.g., emotional avoidance and rumination). Engaging in vivo and imaginal exposure could also be viewed as a method for assisting patients in how to navigate multiple stages of the emotion generative process described in Gross' (2015a) process model of emotion regulation. For example, during in vivo and imaginal exposure, patients are instructed to deliberately approach situations/memories that have previously been avoided (situation selection). Patients may also be assisted in controlling the allocation of attention to identify and incorporate new information that is counter to expectations or fears (attentional deployment). Cognitive reappraisal may be used in the context of exposure exercises to address maladaptive cognitions that are maintaining a fear response (cognitive change), and during exposure, patients are instructed to allow their emotional experience without engaging in attempts to avoid, escape, or control their emotions (response modulation).

As evidence grows for the relevance of emotion regulation difficulties for understanding the development and maintenance of PTSD, researchers are also beginning to investigate the role of emotion regulation difficulties in PTSD treatment outcomes and the effect of PTSD treatments on emotion regulation. For example, Cloitre and colleagues have found that incorporating skill training in affective and interpersonal regulation with prolonged exposure (STAIR+PE) is associated with reductions in emotion regulation difficulties among individuals with PTSD as a result of childhood abuse, and these reductions are maintained posttreatment (Cloitre, Koenen, Cohen, & Han, 2002; Cloitre et al., 2010). Written exposure therapy has also been found to be associated with greater reductions in rumination than a waitlist control; however, this between-group difference was only observed 3 months posttreatment (Wisco, Sloan, & Marx, 2013). Jerud, Pruitt, Zoellner, and Feeny (2016) demonstrated that individuals with chronic PTSD and greater levels of emotion regulation difficulty pretreatment exhibited greater improvements in emotion regulation over the course of 10 weeks of prolonged exposure or sertraline. These changes were maintained into the follow-up period. Finally, Boden et al. (2013) showed that residential PTSD treatment for veterans was associated with reductions in expressive suppression (a putatively maladaptive strategy) and increases in cognitive reappraisal (a putatively adaptive strategy). Moreover, the degree of change in expressive suppression predicted PTSD severity at discharge (with greater decreases in expressive suppression associated with greater reductions in PTSD severity).

Conclusion

Research on the role of emotion regulation in the development, maintenance, consequences, and treatment of PTSD has grown rapidly over the past two decades. This body of research clearly supports the argument that PTSD can be characterized primarily by emotion dysregulation (Frewen & Lanius, 2006). Although much research in this area has been done, additional research in this line of inquiry is needed. Specifically, future research would benefit from less reliance on self-report measures of emotion regulation, instead focusing on behavioral and biological indices of emotion regulation. In addition, designs that have more ecological validity (e.g., ecological momentary assessment and daily diary studies) would provide a window into emotion regulation processes (e.g., emotion regulation flexibility and contextual influences on emotion regulation) that are particularly difficult to study in the laboratory. Prospective research is also needed to elucidate the likely bidirectional relation between emotion regulation difficulties and PTSD. Finally, emotion regulation-based interventions for PTSD deserve attention. Given the central role of emotion regulation difficulties in PTSD and many of the maladaptive behaviors observed among individuals with PTSD, interventions that directly target emotion regulation abilities may not only reduce PTSD symptoms but also reduce many of the unhealthy and self-destructive behaviors associated with this disorder.

References

Aldao, A. (2013). The future of emotion regulation research: Capturing context. *Perspectives on Psychological Science, 8*, 155–172.
Bardeen, J. R., & Fergus, T. A. (2014). An examination of the incremental contribution of emotion regulation difficulties to health anxiety beyond specific emotion regulation strategies. *Journal of Anxiety Disorders, 28*, 394–401.
Bardeen, J. R., Kumpula, M. J., & Orcutt, H. K. (2013). Emotion regulation difficulties as a prospective predictor of posttraumatic stress symptoms following a mass shooting. *Journal of Anxiety Disorders, 27*, 188–196.
Beblo, T., Fernando, S., Klocke, S., Griepenstroh, J., Aschenbrenner, S., & Driessen, M. (2012). Increased suppression of negative and positive emotions in major depression. *Journal of Affective Disorders, 141*, 474–479.
Bloch, L., Moran, E. K., & Kring, A. M. (2010). On the need for conceptual and definitional clarity in emotion regulation research on psychopathology. In A. M. Kring & D. M. Sloan (Eds.), *Emotion regulation and psychopathology: A transdiagnostic approach to etiology and treatment* (pp. 88–104). New York, NY: Guilford Press.
Boden, M. T., Westermann, S., McRae, K., Kuo, J., Alvarez, J., Kulkarni, M. R., ... Bonn-Miller, M. O. (2013). Emotion regulation and posttraumatic stress disorder: A prospective investigation. *Journal of Social and Clinical Psychology, 32*, 296–314.
Bonn-Miller, M. O., Vujanovic, A. A., Boden, M. T., & Gross, J. J. (2010). Posttraumatic stress, difficulties in emotion regulation and coping-oriented marijuana use. *Cognitive Behaviour Therapy, 40*, 34–44.
Cloitre, M., Koenen, K. C., Cohen, L. R., & Han, H. (2002). Skills training in affective and interpersonal regulation followed by exposure: A phase-based treatment for PTSD related to childhood abuse. *Journal of Consulting and Clinical Psychology, 70*, 1067–1074.
Cloitre, M., Stovall-McClough, K. C., Nooner, K., Zorbas, P., Cherry, S., Jackson, C. L., ... Petkova, E. (2010). Treatment for PTSD related to childhood abuse: A randomized controlled trial. *American Journal of Psychiatry, 167*, 915–924.
Cole, P. M., Michel, M. K., & Teti, L. O. (1994). The development of emotion regulation and dysregulation: A clinical perspective. *Monographs of the Society for Research in Child Development, 59*, 73–100.

Dixon-Gordon, K. L., Tull, M. T., & Gratz, K. L. (2014). Self-injurious behaviors in post-traumatic stress disorder: An examination of potential moderators. *Journal of Affective Disorders, 166,* 359–367.

Ekman, P. E., & Davidson, R. J. (1994). *The nature of emotion: Fundamental questions.* New York, NY: Oxford University Press.

Feldman-Barrett, L., Gross, J., Christensen, T. C., & Benvenuto, M. (2001). Knowing what you're feeling and knowing what to do about it: Mapping the relation between emotion differentiation and emotion regulation. *Cognition and Emotion, 15,* 713–724.

Finucane, A. M., Dima, A., Ferreria, N., & Halvorsen, M. (2011). Basic emotion profiles in healthy, chronic pain, depressed and PTSD individuals. *Clinical Psychology and Pyschotherapy, 19,* 14–24.

Fitzgerald, J. M., MacNamara, A., Kennedy, A. E., Rabinak, C. A., Rauch, S. A. M., Liberzon, I., & Phan, K. L. (2018). Individual differences in cognitive reappraisal use and emotion regulatory brain function in combat-exposed veterans with and without PTSD. *Depression and Anxiety, 34,* 79–88.

Foa, E. B., & Kozak, M. J. (1986). Emotional processing of fear: Exposure to corrective information. *Psychological Bulletin, 99,* 20–35.

Forbes, C. N., Tull, M. T., Hong, X., Rapaport, D., Kaminski, B., & Wang, X. (2019). *Emotion dysregulation prospectively predicts PTSD symptom severity 3-months after traumatic exposure.* Unpublished data.

Frewen, P. A., Dozois, D. J. A., Neufeld, R. W. J., & Lanius, R. A. (2012). Disturbances of emotional awareness and expression in posttraumatic stress disorder: Meta-mood, emotion regulation, mindfulness, and interference of emotional expressiveness. *Psychological Trauma: Theory, Research, Practice, and Policy, 4,* 152–161.

Frewen, P. A., & Lanius, R. A. (2006). Toward a psychobiology of posttraumatic self-dysregulation: Reexperiencing, hyperarousal, dissociation, and emotional numbing. *Annals of the New York Academy of Sciences, 1071,* 110–124.

Gratz, K. L., & Roemer, L. (2004). Multidimensional assessment of emotion regulation and dysregulation: Development, factor structure, and initial validation of the Difficulties in Emotion Regulation Scale. *Journal of Psychopathology and Behavioral Assessment, 26,* 41–54.

Gratz, K. L., & Tull, M. T. (2010). Emotion regulation as a mechanism of change in acceptance-and mindfulness-based treatments. In R. A. Baer (Ed.), *Assessing mindfulness and acceptance: Illuminating the process of change* (pp. 107–134). Oakland, CA: New Harbinger Publications.

Gratz, K. L., Weiss, N. H., & Tull, M. T. (2015). Examining emotion regulation as an outcome, mechanism, or target of psychological treatments. *Current Opinion in Psychology, 3,* 85–90.

Gross, J. J. (2015a). Emotion regulation: Current status and future prospects. *Psychological Inquiry, 26,* 1–26.

Gross, J. J. (2015b). The extended process model of emotion regulation: Elaborations, applications, and future directions. *Psychological Inquiry, 26,* 130–137.

Gross, J. J., & Levenson, R. W. (1997). Hiding feelings: The acute effects of inhibiting negative and positive emotion. *Journal of Abnormal Psychology, 106,* 95–103.

Gross, J. J., Richards, J. M., & John, O. P. (2006). Emotion regulation in everyday life. In D. K. Snyder, J. A. Simpson, & J. N. Hughes (Eds.), *Emotion regulation in families: Pathways to dysfunction and health* (pp. 13–35). Washington, DC: American Psychological Association.

Hayes, S. C., Luoma, J. B., Bond, F. W., Masuda, A., & Lillis, J. (2006). Acceptance and commitment therapy: Model, processes and outcomes. *Behaviour Research and Therapy, 44,* 1–25.

Hofmann, S. G., & Kashdan, T. B. (2010). The affective style questionnaire: Developmental and psychometric properties. *Journal of Psychopathology and Behavioral Assessment, 32,* 255–263.

Horowitz, M. J. (1986). Stress-response syndromes: A review of posttraumatic and adjustment disorders. *Hospital and Community Psychiatry, 37,* 241–249.

Jenness, J. L., Jager-Hyman, S., Heleniak, C., Beck, A. T., Sheridan, M. A., & McLaughlin, K. A. (2016). Catastrophizing, rumination, and reappraisal prospectively predict adolescent PTSD symptom onset following a terrorist attack. *Depression and Anxiety, 33,* 1039–1047.

Jerud, A. B., Pruitt, L. D., Zoellner, L. A., & Feeny, N. C. (2016). The effects of prolonged exposure and sertraline on emotion regulation in individuals with posttraumatic stress disorder. *Behaviour Research and Therapy, 77*, 62–67.

Kashdan, T. B., Elhai, J. D., & Frueh, C. F. (2006). Anhedonia and emotional numbing in combat veterans with PTSD. *Behaviour Research and Therapy, 44*, 457–467.

Kumpla, M. J., Orcutt, H. K., Bardeen, J. R., & Varkovitzky, R. L. (2011). Peritraumatic dissociation and experiential avoidance as prospective predictors of posttraumatic stress symptoms. *Journal of Abnormal Psychology, 120*, 617–627.

Langkaas, T. F., Hoffart, A., Øktedalen, T., Ulvenes, P. G., Hembree, E. A., & Smucker, M. (2017). Exposure and non-fear emotions: A randomized controlled study of exposure-based and rescripting-based imagery in PTSD treatment. *Behaviour Research and Therapy, 97*, 33–42.

Lanius, R. A., Frewen, P. A., Vermetten, E., & Yehuda, R. (2010). Fear conditioning and early life vulnerabilities: Two distinct pathways of emotional dysregulation and brain dysfunction in PTSD. *European Journal of Pyschotraumatology, 1*, 5467.

Litz, B. T., & Gray, M. J. (2002). Emotional numbing in posttraumatic stress disorder: Current and future research directions. *Australian and New Zealand Journal of Pyschiatry, 36*, 198–204.

McDermott, M. J., Tull, M. T., Gratz, K. L., Daughters, S. B., & Lejuez, C. W. (2009). The role of anxiety sensitivity and difficulties in emotion regulation in posttraumatic stress disorder among crack/cocaine dependent patients in residential substance abuse treatment. *Journal of Anxiety Disorders, 23*, 591–599.

Miles, S. R., Menefee, D. S., Wanner, J., Tharp, A. T., & Kent, T. A. (2016). The relationship between emotion dysregulation and impulsive aggression in veterans with posttraumatic stress disorder symptoms. *Journal of Interpersonal Violence, 31*, 1795–1816.

Murray, J., Ehlers, A., & Mayou, R. A. (2002). Dissociation and post-traumatic stress disorder: Two prospective studies of road traffic accident survivors. *The British Journal of Psychiatry, 180*, 363–368.

New, A. S., Fan, J., Murrough, J. W., Liu, X., Liebman, R. E., Guise, K. G., ... Charney, D. S. (2009). A functional magnetic resonance imaging study of deliberate emotion regulation in resilience and posttraumatic stress disorder. *Biological Psychiatry, 66*, 656–664.

Rabinak, C. A., MacNamara, A., Kennedy, A. E., Angstadt, M., Stein, M. B., Liberzon, I., & Phan, K. L. (2014). Focal and aberrant prefrontal engagement during emotion regulation in veterans with posttraumatic stress disorder. *Depression and Anxiety, 31*, 851–861.

Radomski, S. A., & Read, J. P. (2016). Mechanistic role of emotion regulation in the PTSD and alcohol association. *Traumatology, 22*, 113–121.

Resick, P. A., Nishith, P., Weaver, T. L., Astin, M. C., & Feuer, C. A. (2002). A comparison of cognitive-processing therapy with prolonged exposure and a waiting condition for the treatment of chronic posttraumatic stress disorder in female rape victims. *Journal of Consulting and Clinical Psychology, 70*, 867–879.

Roemer, L., Litz, B. T., Orsillo, S. M., & Wagner, A. W. (2001). A preliminary investigation of the role of strategic withholding of emotions in PTSD. *Journal of Traumatic Stress, 14*, 149–156.

Schmitz, T. W., De Rosa, E., & Anderson, A. K. (2009). Opposing influences of affective state valence on visual cortical encoding. *Journal of Neuroscience, 29*, 7199–7207.

Seligowski, A. V., Lee, D. J., Bardeen, J. R., & Orcutt, H. K. (2015). Emotion regulation and posttraumatic stress symptoms: A meta-analysis. *Cognitive Behaviour Therapy, 44*, 87–102.

Short, N. A., Boffa, J. W., Clancy, K., & Schmidt, N. B. (2018). Effects of emotion regulation strategy use in response to stressors on PTSD symptoms: An ecological momentary assessment study. *Journal of Affective Disorders, 230*, 77–83.

Sripada, R. K., King, A. P., Garfinkel, S. N., Wang, X., Sripada, C. S., Welsh, R. C., & Liberzon, I. (2012). Altered resting-state amygdala functional connectivity in men with posttraumatic stress disorder. *Journal of Psychiatry & Neuroscience, 37*, 241–249.

Stapleton, J., Taylor, S., & Asmundson, G. (2006). Effects of three PTSD treatments on anger and guilt: Exposure therapy, eye movement desensitization and reprocessing, and relaxation training. *Journal of Traumatic Stress, 19,* 369–375.

Taylor, S. F., & Liberzon, I. (2007). Neural correlates of emotion regulation in psychopathology. *Trends in Cognitive Sciences, 11,* 413–418.

Thompson, R. A. (1994). Emotion regulation: A theme in search of definition. *Monographs of the Society for Research in Child Development, 59,* 25–52.

Tripp, J. C., & McDevitt-Murphy, M. E. (2015). Emotion dysregulation facets as mediators of the relationship between PTSD and alcohol misuse. *Addictive Behaviors, 47,* 55–60.

Tugade, M. M., & Fredrickson, B. L. (2007). Regulation of positive emotions: Emotion regulation strategies that promote resilience. *Journal of Happiness Studies, 8,* 311–333.

Tull, M. T., & Aldao, A. (2015). Editorial overview: New directions in the science of emotion regulation. *Current Opinion in Psychology, 3,* iv–x.

Tull, M. T., Barrett, H. M., McMillan, E. S., & Roemer, L. (2007). A preliminary investigation of the relationship between emotion regulation difficulties and posttraumatic stress symptoms. *Behavior Therapy, 38,* 303–313.

Tull, M. T., Berghoff, C. R., Wheeless, L. E., Cohen, R. T., & Gratz, K. L. (2018). PTSD symptom severity and emotion regulation strategy use during trauma cue exposure among patients with substance use disorders: Associations with negative affect, craving, and cortisol reactivity. *Behavior Therapy, 49,* 57–70.

Tull, M. T., Gratz, K. L., Salters, K., & Roemer, L. (2004). The role of experiential avoidance in posttraumatic stress symptoms and symptoms of depression, anxiety, and somatization. *The Journal of Nervous and Mental Disease, 192,* 754–761.

Weiss, N. H., Dixon-Gordon, K. L., Peasant, C., & Sullivan, T. P. (2018). An examination of the role of difficulties regulating positive emotions in posttraumatic stress disorder. *Journal of Traumatic Stress, 31,* 775–780.

Weiss, N. H., Forkus, S. R., Contractor, A. A., & Shick, M. R. (2018). Difficulties regulating positive emotions and alcohol and drug misuse: A path analysis. *Addictive Behaviors, 84,* 45–52.

Weiss, N. H., Gratz, K. L., & Lavender, J. M. (2015). Factor structure and initial validation of a multidimensional measure of difficulties in the regulation of positive emotions: The DERS-positive. *Behavior Modification, 39,* 431–453.

Weiss, N. H., Nelson, R. J., Contractor, A. A., & Sullivan, T. P. (2019). Emotion dysregulation and posttraumatic stress disorder: A test of the incremental role of difficulties regulating positive emotions. *Anxiety, Stress, & Coping, An International Journal, 32,* 443–456.

Weiss, N. H., Tull, M. T., Anestis, M. D., & Gratz, K. L. (2013). The relative and unique contributions of emotion dysregulation and impulsivity to posttraumatic stress disorder among substance dependent inpatients. *Drug and Alcohol Dependence, 128,* 45–51.

Weiss, N. H., Tull, M. T., & Gratz, K. L. (2014). A preliminary experimental examination of the effect of emotion dysregulation and impulsivity on risky behaviors among women with sexual assault-related posttraumatic stress disorder. *Behavior Modification, 38,* 914–939.

Weiss, N. H., Tull, M. T., Sullivan, T. P., Dixon-Gordon, K. L., & Gratz, K. L. (2015). Posttraumatic stress disorder symptoms and risky behaviors among trauma-exposed inpatients with substance dependence: The influence of negative and positive urgency. *Drug and Alcohol Dependence, 155,* 147–153.

Weiss, N. H., Tull, M. T., Viana, A. G., Anestis, M. D., & Gratz, K. L. (2012). Impulsive behaviors as an emotion regulation strategy: Examining associations between PTSD, emotion dysregulation, and impulsive behaviors among substance dependent inpatients. *Journal of Anxiety Disorders, 26,* 453–458.

Williams, K. E., Chambless, D. L., & Ahrens, A. (1997). Are emotions frightening? An extension of the fear of fear construct. *Behaviour Research and Therapy, 35,* 239–248.

Wisco, B. E., Sloan, D. M., & Marx, B. P. (2013). Cognitive emotion regulation and written exposure therapy for posttraumatic stress disorder. *Clinical Psychological Science, 1,* 435–442.

CHAPTER 11

The regulatory role of attention in PTSD from an information processing perspective

Joseph R. Bardeen
Department of Psychology, Auburn University, Auburn, AL, United States

Introduction: Overview of attentional deployment as a form of emotion regulation

Information processing models of emotion regulation suggest that the ability to deploy attention in the service of altering one's emotional experience (i.e., attentional deployment, also described as attentional control) is essential for maintaining psychological well-being (Gross, 1998). Moreover, threat-related attentional impairments predict general emotion dysregulation and increased use of maladaptive emotion regulation strategies (Bardeen & Daniel, 2017a; Bardeen, Daniel, Hinnant, & Orcutt, 2017). In Gross's (2015) process model of emotion regulation, attentional deployment is one of five points in the emotion generative process at which emotions can be regulated. The four other points in Gross's (2015) model include situation selection, situation modification, cognitive change, and response modulation (Gross & Thompson, 2007). Of these points in the emotion generative process, cognitive change (e.g., cognitive reappraisal) and response modulation (e.g., expressive suppression) have received considerable attention in the extant literature, while regulatory processes occurring earlier in the temporal chain, such as attentional deployment, have received relatively little consideration. This is unfortunate because these earlier processes are thought to require less cognitive resources to enact and, thus, may be more easily altered through psychological intervention (Renna et al., 2018).

Attentional control is distinct from other regulation strategies in that it is present from infancy onward (Rothbart, Sheese, & Posner, 2007). For example, children who disengage and shift their attention from appetitive stimuli (e.g., candy) are more successful at delaying gratification (Mischel,

Shoda, & Rodriguez, 1989). The beneficial effects of attentional control have been trumpeted in the extant literature because evidence from a number of studies suggests that attentional control protects those who are vulnerable to maladaptive psychological outcomes, such as posttraumatic stress (Bardeen, Fergus, & Orcutt, 2015), from experiencing such outcomes (Fergus, Bardeen, & Orcutt, 2012; Jones, Fazio, & Vasey, 2012; Richey, Keough, & Schmidt, 2012).

It may be especially important to consider the regulatory role of attentional deployment in the pathogenesis of posttraumatic stress disorder (PTSD) because attention-related abnormalities and deficits have been observed among those with PTSD (Scott et al., 2015). Moreover, hypervigilance toward threatening information is a hallmark symptom of the disorder (*Diagnostic and Statistical Manual of Mental Disorders* [DSM-5]; American Psychiatric Association, 2013). Because hypervigilance is a hallmark symptom of PTSD, there has been an emphasis on understanding the attentional processing of threat- and trauma-related information among those with PTSD, with the idea that hypervigilance and facilitated threat detection are synonymous. Following from this logic, individuals with PTSD (vs those without) should preferentially process threat- and trauma-related stimuli (i.e., attentional bias to threat [ABT]), and according to theory, this ABT should promote the development and maintenance of PTSD symptoms (e.g., Beck, Emery, & Greenberg, 1985; Williams, Watts, MacLeod, & Mathews, 1997). Specifically, prolonged attentional engagement with threat information is thought to result in protracted states of distress that decrease the cognitive resources that are available for the emotional processing of threat- and trauma-related information (Foa & Kozak, 1986), thus resulting in the development and/or maintenance of PTSD symptoms.

Findings regarding the degree to which individuals with PTSD exhibit ABT have been mixed, with some suggesting that PTSD-related ABT is a robust phenomenon (e.g., Buckley, Galovski, Blanchard, & Hickling, 2003; Constans, 2005) and other work suggesting that the results used to support PTSD-related ABT are weak at best. In fact a considerable number of null findings have been reported, and failure to replicate significant findings is fairly common in this literature (e.g., Kimble, Frueh, & Marks, 2009; Van Bockstaele et al., 2014). Discrepancies in the extant literature may be the result of failing to consider the role that top-down attentional processes (i.e., controlled, effortful, and goal directed) play in regulating bottom-up reactivity (i.e., sensory driven and automatic).

Information processing in PTSD: The fear network

Information processing models of PTSD were developed from a framework consistent with a classic cognitive network model of memory. Specifically the network (a "fear network") is made up of interconnected representations of trauma-related memories. Emotional processing theory (Foa & Kozak, 1986), the most prominent information processing theory of PTSD, presupposes automatic threat processing. In this model a network of interconnected trauma memory nodes is hypothesized to be linked by a process of spreading activation. Memory nodes represent a variety of features related to a traumatic event (e.g., stimulus information, physiological and emotional reactions, and response information; Foa, Huppert, & Cahill, 2006; Foa & Kozak, 1986). The triggering of one node in the network, presumably by threat-relevant stimuli in one's environment, causes spreading activation across the network and a subsequent triggering of all other nodes, thus resulting in a response consistent with the physiological symptoms of PTSD. Moreover, repeated activation of the network is thought to increase autonomic fear responses, which in turn results in a state of hypervigilance and the preferential processing of threat-relevant stimuli (i.e., facilitated engagement).

Unlike emotional processing theory (Foa & Kozak, 1986), which implies that there is a single representation of a traumatic event in one's memory, dual-representation theory proposes a model in which traumatic memories are encoded in two systems (Brewin, Dalgleish, & Joseph, 1996). Although these systems typically act in parallel, one system may take precedence over the other. In the "verbally accessible memory" (VAM) system, memories are stored in narrative form and are integrated into the larger narrative of one's history (Brewin et al., 1996). These are memories that can be retrieved through active efforts because they contain information that the individual has consciously processed and stored in long-term memory. In contrast the "situationally accessible memory" (SAM) system contains information that may not have been processed using higher-level perceptual processes (Brewin et al., 1996). The SAM system does not use verbal narratives; memories are stored as a collection of sights, sounds, smells, bodily sensations, etc. Because SAM system memories have not been consciously processed, the individual does not have voluntary control over their conscious retrieval. Thus the SAM system is associated with processes that seem to be reflexive or more automatic, such as trauma reexperiencing (e.g., flashbacks) and bodily sensations experienced at the time of the traumatic event (i.e., autonomic nervous system arousal and hypervigilance). According to Brewin et al. (1996), it is necessary to integrate SAM system memories into the VAM system for one to recover from PTSD.

Ehlers and Clark (2000) proposed a similar two-system model in which trauma memories are poorly integrated into an autobiographical memory system among individuals with PTSD. The autobiographical memory system is similar to Brewin et al.'s (1996) VAM system. Ehlers and Clark (2000) also proposed an associative memory system, similar to Brewin et al.'s (1996) SAM system, in which information is processed preconsciously. The associative memory system is sensory driven and primes the individual to automatically react to trauma reminders when threat stimuli are present. Although these multisystem models are very similar, Ehlers and Clark (2000) suggested that the encoding of information in the associative memory system at the time of the traumatic event is a risk factor for the development of PTSD, whereas dual-representation theory (Brewin et al., 1996) suggests that SAM system encoding of trauma information is only harmful if SAM system encoding greatly outweighs VAM system encoding (Brewin & Holmes, 2003). Additionally, Ehlers and Clark's (2000) cognitive model of PTSD is more complex, placing greater emphasis and providing more conceptual detail on the appraisal of trauma sequelae and maladaptive cognitive and behavioral coping strategies.

The fear network model presented in emotional processing theory suggests that exposure therapy is effective in treating PTSD because pathological associations in the fear network are weakened when exposure to threat stimuli is coupled with disconfirmation of feared outcomes, a process of unlearning (Foa et al., 2006). However, evidence suggests that extinction is not a matter of unlearning, but rather new learning; that is, extinction does not occur through a disintegration of associations in the network, but by establishing new associations that inhibit pathological responses to conditioned stimuli (i.e., inhibitory learning; Craske, Treanor, Conway, Zbozinek, & Vervilet, 2014; Knowles & Olatunji, 2019; McNally, 2007). Thus recovery from PTSD may occur through the creation of new associations in the network that inhibit previously dominant pathways. Although the single level of representation proposed in emotional processing theory has provided us with an important conceptualization of bottom-up threat processing in PTSD (Foa & Kozak, 1986), it does not account for the regulation of bottom-up reactivity through the use of top-down resources. As such, multiple representation models such as dual-representation theory (Brewin et al., 1996) and Ehlers and Clark's (2000) cognitive model, which account for bottom-up and top-down processing, may provide a more accurate conceptualization of the processing of trauma and threat information.

Cognitive network conceptualizations of PTSD provide a framework from which to understand the role of attentional processes and their influence on threat processing in PTSD. Much like network models of PTSD, conceptualizations of ABT in PTSD vary in the degree to which bottom-up and top-down processing are emphasized. Bottom-up processing is emphasized in a number of theories (e.g., Beck & Clark, 1997; Litz & Keane, 1989; Williams et al., 1997), while dual-process models of threat processing in PTSD assert that two systems (bottom-up and top-down) interact to differentially impact the expression of threat biases (e.g., Corbetta & Shulman, 2002; Eysenck & Derakshan, 2011).

Attentional bias to threat: Bottom-up, top-down, or both?

The majority of theories that describe the etiology of fear-related pathology, including PTSD, implicate ABT in the development and maintenance of these pathological presentations (Weierich, Treat, & Hollingworth, 2008). ABT is most often conceptualized as being reflexive, automatic, and occurring outside of conscious awareness (Constans, 2005; Yiend, 2010). This conceptualization suggests that individuals with relatively higher levels of anxiety/fear exhibit faster detection of threat stimuli (i.e., facilitated engagement). Although faster detection of threat may be extremely beneficial in some contexts (e.g., identifying weapons in a combat zone), repeated and prolonged attention toward stimuli that are objectively safe (e.g., weapons in movies and magazines) is thought to maintain hyperarousal and negative affective states, resulting in greater functional impairment and increased vulnerability for the development of fear-related pathology (Constans, 2005).

ABT has also been defined more broadly (e.g., "differential attentional allocation toward threatening stimuli relative to neutral stimuli," Cisler & Koster, 2010). This type of definition does not presuppose facilitated engagement as the attentional process through which individuals with higher levels of anxiety/fear attend to threat stimuli for a greater duration of time. Instead, it allows for the possibility that other attentional processes may account for prolonged attention to threat (i.e., deficits in strategic processing that result in difficulty disengaging from threat stimuli). In other words, this type of definition accounts for automatic (i.e., capacity-free processing occurring without awareness) and strategic information processing (i.e., limited capacity, controllable, and implies awareness).

Vigilance-avoidance versus attention maintenance

Two seemingly contradictory hypotheses are frequently cited to explain evidence of fear-related ABT: vigilance-avoidance and attention maintenance. The vigilance-avoidance hypothesis of ABT presupposes reflexive orienting of attention toward threat (i.e., facilitated engagement), which, in turn, increases sympathetic nervous system arousal and facilitates rapid responding. According to this hypothesis, following initial facilitated detection, fearful individuals quickly shift attention away from threat stimuli to reduce negative affect and the physiological arousal that is provoked by initially attending to the stimulus (i.e., avoidance). Habitually disengaging and shifting attention from threat is considered a maladaptive avoidance strategy that may alleviate physiological and emotional distress in the short term but may maintain trauma-related fear and PTSD symptoms over time by failing to provide the opportunity for new learning. The prominence of the vigilance-avoidance hypothesis in the ABT-PTSD literature (compared with the attention maintenance hypothesis) may be a function of its intuitive appeal, as hypervigilance for, and avoidance of, trauma-related stimuli are central to the symptom profile of PSTD (*DSM-5*; APA, 2013).

In contrast to the vigilance-avoidance hypothesis, the attention maintenance hypothesis does not presuppose faster orienting toward threat stimuli; instead, once threat stimuli are identified, it is more difficult for the fearful individual to disengage and shift attention away from such stimuli (Weierich et al., 2008). Thus the mechanism underlying ABT is not facilitated engagement (i.e., bottom-up), but difficulty disengaging from threat stimuli due to a relative deficit in the ability to regulate attention deployment (i.e., top-down). Further, according to the attention maintenance hypothesis, prolonged emotional distress and the subsequent development of fear-related pathology are thought to be a function of prolonged attending to threat stimuli. This hypothesis is consistent with evidence that prolonged attentional engagement with perceived threat, due to disengagement difficulties, maintains negative affective states (Bardeen & Read, 2010; Compton, 2000).

At first glance, these two hypotheses appear incompatible. The vigilance-avoidance hypothesis explains ABT by focusing on individual differences in bottom-up reactivity, whereas the attention maintenance hypothesis emphasizes deficits in top-down regulatory abilities as the mechanism by which differences in ABT are observed. The apparent discrepancy between these two hypotheses appears to be a function of focusing on only one of the two primary attentional streams (i.e., bottom-up or top-down) in conceptualizing ABT rather than considering that these two systems may both contribute to ABT.

Dual-process models of ABT: Considering the role of attentional control

In contrast to theories that focus on one specific attentional system in accounting for ABT, dual-process theories consider the combined influence of bottom-up and top-down attentional systems on ABT. More specifically, dual-process theories of ABT suggest that anxiety/fear potentiate the bottom-up system, increasing the likelihood that those with higher levels of anxiety/fear will identify potential threat stimuli quicker than those with relatively lower levels of anxiety/fear (Corbetta & Shulman, 2002; Eysenck, Derakshan, Santos, & Calvo, 2007; Eysenck & Derakshan, 2011; Metcalfe & Mischel, 1999). Additionally, anxiety/fear simultaneously reduce one's ability to use top-down attentional control to decrease the impact of this bottom-up reactivity. Corbetta and Shulman (2002) hypothesized that the bottom-up system is stimulus driven, designed for rapid responding to salient and self-relevant stimuli, whereas the top-down system is influenced by current goals, knowledge, and expectations. Similarly, Metcalfe and Mischel (1999) hypothesized that the bottom-up system (i.e., the "hot" system) is specialized for immediate responding, whereas the top-down system (i.e., the "cool" system) is specialized for reflective emotion regulation and control of impulsive tendencies (Metcalfe & Mischel, 1999).

Attentional control theory suggests that the balance between these two attentional systems is disrupted when high levels of anxiety/fear potentiate bottom-up reactivity and impair top-down attentional control (i.e., Eysenck & Derakshan, 2011; Eysenck et al., 2007). However, some evidence suggests that the potentiation of stimulus-driven attention among those with fear-related distress may be the result of enhanced bottom-up reactivity for salient or novel stimuli in general rather than being specific to stimuli that are trauma or threat related (Esterman et al., 2013; Sarapas, Weinberg, Langenecker, & Shankman, 2017). Individuals with PTSD may detect all salient cues more quickly because of the trauma-related perception that (a) the probability of threat being present is high and (b) failing to detect threat quickly will result in a catastrophic outcome (i.e., probability overestimation; White, McManus, & Ehlers, 2008). These individuals may experience a state of tonic alertness that is conceptually consistent with PTSD-related hypervigilance. Results showing that fear/anxiety-related bottom-up reactivity is stimulus nonspecific are also consistent with the fear generalization that is observed among those with PTSD. Instead of trauma-specific fear and related avoidance, individuals with PTSD exhibit fear responding to, and avoidance of, contexts that are unpredictable and in response to stimuli that are seemingly unrelated to one's traumatic event(s) (Dymond, Dunsmoor, Vervliet, Roche, & Hermans, 2015; Jovanovic & Ressler, 2010; Lissek, 2012).

To consider the role of attentional control and whether or not it is impaired or dysregulated among those with higher levels of fear/anxiety, it may be important to break this construct down into its component parts. There are three primary cognitive processes that have been identified as central to attentional control: (a) inhibition of prepotent task-irrelevant stimuli and associated responses, (b) flexibly orienting attention toward task-relevant stimuli (i.e., shifting), and (c) updating working memory (Eysenck et al., 2007; Miyake, Friedman, Emerson, Witzki, & Howerter, 2000). In support of attentional control theory (Eysenck et al., 2007), anxiety/fear has been shown to impair the inhibition and shifting functions of attentional control (Graydon & Eysenck, 1989; Lavie, Hirst, de Fockert, & Viding, 2004). Thus individuals in a heightened state of anxiety/fear will be less successful at tasks in which these attentional control processes are needed. Furthermore, when task-irrelevant stimuli (e.g., trauma-relevant stimuli) increase participant distress, these stimuli will be attended to for a greater length of time as a result of anxiety/fear's detrimental influence on inhibition and shifting. Thus deficits in attentional control processes that limit the ability of the individual to disengage attention from trauma- and threat-related stimuli may differentiate those who experience mild posttraumatic stress symptoms that remit in the acute aftermath of a traumatic event from those who go on to develop more severe, chronic posttraumatic stress symptomatology (Aupperle, Melrose, Stein, & Paulus, 2012).

It is important to consider that the bottom-up influence of anxiety/fear on the stimulus-driven system can be reduced, and perhaps eliminated, among those with fear-related pathology by increasing effort, using more resources, or by limiting oneself to contexts in which trauma and threat stimuli are less salient (Eysenck & Derakshan, 2011). Specifically, among individuals with PTSD, those with relatively better attentional control may be able to disengage and shift attention from perceived threat by drawing on reserve attentional control resources through active effort, while those with relatively poorer attentional control abilities should theoretically have a more difficult time disengaging from threat stimuli because they lack the requisite resources to do so. Thus, while anxiety/fear impairs the inhibition and switching functions of attentional control, individuals who have an abundance of attentional control may still use this resource to regulate bottom-up reactivity. Environmental factors may make it more or less difficult to use attentional control in this manner.

It has been suggested that competition between bottom-up and top-down attentional systems is affected not only by one's level of anxiety/fear but also by factors such as the duration of exposure to the threat, threat saliency or intensity (stimulus valence), and competition with other stimuli for processing resources (Bishop, 2008). For example, among individuals with PTSD, those with relatively better (vs relatively worse) attentional control abilities may be able to use these abilities to downregulate bottom-up reactivity when threat salience is at low to moderate levels. When threat salience is high (e.g., revisiting the site of the index trauma), however, the influence of fear on attentional control resources may be so intense that top-down regulation of bottom-up reactivity is improbable (e.g., disengaging and shifting attention from trauma stimuli). Evidence from studies in which EEG and fMRI have been employed has shown that (a) anxiety/fear impairs prefrontal brain regions associated with attentional control and (b) this impairment can be reversed in individuals with greater anxiety/fear-related pathology to compensate for potential impairments in task demands by drawing on available top-down resources (Rosler, Heil, & Roder, 1997; Wager, Jonides, & Reading, 2004).

Some have suggested that the distinction between top-down and bottom-up threat processing (automatic vs controlled) is one that is rarely clear (Todd, Cunningham, Anderson, & Thompson, 2012; Verbruggen, 2016). The line between automatic and controlled threat processing is sometimes blurry because sensory systems may be "pretuned" so that specific categories of stimuli are preferentially processed in an attempt to modulate emotional distress (Todd et al., 2012), especially when the presence of trauma and/or threat cues are predictable. For example, a survivor of a traumatic car accident may predict the occurrence of the feared stimulus (e.g., a motor vehicle) in a given situation (e.g., grocery store parking lot). In preparation for exposure to the feared stimulus, the trauma survivor proactively reconfigures his or her bias settings (i.e., prioritizes the use of top-down attentional control) so that attention is quickly diverted from the feared stimulus to stimuli from a neutral category. This in turn results in the downregulation of emotional distress in the short term. This has been described elsewhere as a "prepared reflex;" once proactive adjustments in top-down control have been made, the inhibitory response can be activated easily by relevant stimuli (Verbruggen, 2016).

In the aforementioned example the ability to predict the presence of the feared stimulus is important; the trauma survivor would not be prepared to shift attention away from a feared stimulus that appears in an unusual context (e.g., a car parked on the roof of a house). However, the more frequently that one uses top-down attentional control to inhibit bottom-up reactivity, the faster that inhibitory effect becomes (Koole, Webb, & Sheeran, 2015). Consistent with this proposition, evidence suggests that it takes as little as 150 ms for top-down attentional processes to influence the processing of threat information (Bardeen & Orcutt, 2011; Shomstein, Kravitz, & Behrmann, 2012). In addition, bottom-up activation is more likely to trigger top-down inhibition in individuals with better developed attention control, with inhibition becoming an almost automatic response to bottom-up activation (Metcalfe & Mischel, 1999). As such, the use of attentional control as a trauma-related regulatory mechanism may become habitual, no longer requiring active deliberation after prolonged use of the once effortful strategy. In summary, among individuals with PTSD, those with relatively better attentional control and repeated experience employing this top-down resource to inhibit bottom-up reactivity may be better able to disengage and shift attention from threat information than individuals with PTSD and poorly developed attentional control. Before we consider whether deploying attention away from threat stimuli in this manner is adaptive, let us first discuss commonly used methods of assessing ABT.

Measuring ABT in PTSD

Although a number of different stimulus-response paradigms have been used to assess PTSD-related ABT (e.g., rapid serial visual presentation [e.g., Olatunji, Armstrong, McHugo, & Zald, 2013], visual search [e.g., Pineles, Shipherd, Mostoufi, Abramovitz, & Yovel, 2009]), the modified Stroop task has received the most attention in this literature. In the traditional Stroop task, participants view a list of words and are asked to name the color in which each word is printed as quickly as possible while ignoring the meaning of the word. This task becomes more difficult when the color of the ink differs from the meaning of the word (e.g., the word "blue" written in green ink). Similarly, in the version of the task that has been modified to assess ABT, participants are asked to name the color of a series of one-word stimuli as quickly as possible while paying no attention to the word's meaning. Each word comes from a specific category (e.g., trauma, positive, and neutral). Slower responding to specific word categories is thought to occur when attention is briefly captured by the potency of word meaning, thus disrupting the processing of color information.

Although some have suggested that the evidence of ABT among those with PTSD is "strong" and "abundant," with the majority of studies examining PTSD-related ABT via the modified Stroop showing slowed responding of color naming trauma-relevant words among those with PTSD (Constans, 2005), a review of peer-reviewed journal articles and dissertation abstracts suggests otherwise. Specifically, Kimble et al. (2009) conducted a review of studies in which the modified Stroop task was used to examine ABT in trauma-exposed adults (with and without PTSD). Slowed responding to threat words among individuals with PTSD was only reported in 8% of dissertation abstracts and 44% of peer-reviewed journal articles. Before taking these results as evidence of an absence of ABT among those with PTSD, it is important to consider what the modified Stroop task actually measures. The task was originally developed to assess bottom-up facilitated engagement, but some evidence suggests that response times on the modified Stroop task represent individual differences in the top-down processing of threat information (McKenna & Sharma, 2004; Phaf & Kan, 2007; Weierich et al., 2008).

The modified Stroop task is typically presented in one of two formats: (a) a random format in which all word types are mixed together and presented at random and a (b) blocked format in which each category of word stimuli is presented together. The blocked format provides a measure of the combined effect of the bottom-up and top-down components of attention, whereas the random format only provides a measure of bottom-up ABT (McKenna & Sharma, 2004). In the traditional Stroop paradigm, slowed responding occurs when there is a conflict between two processes that are relatively automatic (i.e., word meaning and color naming). In contrast, in the modified Stroop task, color naming, a more automatic process, competes with the decoding of trauma and threat word meanings (Phaf & Kan, 2007). If the decoding of these words occurred at an automatic, preattentive level, then the modified Stroop task might be an appropriate measure of bottom-up ABT when the random format is employed. Contrary to the expectation of automaticity that the original task is founded on, evidence of the modified Stroop effect has only been observed when the blocked format has been used. This pattern of findings suggests that observed effects of PTSD-related ABT using this task are likely the result of differences in top-down attentional processes (Bar-Haim, Lamy, Pergamin, Bakermans-Kranenburg, & van IJzandoorn, 2007; McKenna & Sharma, 2004).

The dot-probe task is another stimulus-response task used to examine PTSD-related ABT (e.g., Bardeen & Orcutt, 2011; Bryant & Harvey, 1997; Iacoviello et al., 2014; Wald et al., 2011). At the beginning of each trial of the task, a fixation cross appears in the center of the screen and is then replaced by two stimuli presented side by side. The stimuli remain on the screen for a specified duration of time (e.g., typically a stimulus-onset asynchrony of 500 ms), and then a dot appears on the screen, replacing one of the two pictures. To provide the researcher with a snapshot of the participant's attention allocation at that point in time, the participant presses a button as quickly as possible that corresponds to the relative position of the dot on the screen. If the participant has a bias for attending to trauma or threat stimuli, she/he should respond faster when the dot appears in the spatial position previously held by these types of stimuli.

Among the studies in which this task has been used to examine PTSD-related ABT, design decisions have limited the degree to which study findings help to elucidate the nature of threat processing. Specifically the dot-probe task can provide temporal snapshots of attention allocation when multiple stimulus presentation durations are used. However, using multiple presentation durations requires substantially lengthening task duration, thus increasing concerns regarding the impact of participant fatigue on responding. As a result, in the large majority of studies in which the task has been employed, stimuli are typically only presented for 500 ms (Bar-Haim et al., 2007; Bryant & Harvey, 1997; Elsesser, Satory, & Tackenberg, 2004; Fani, Jovanovic, et al., 2012; Fani, Tone, et al., 2012; Iacoviello et al., 2014; Schäfer et al., 2016; Sipos, Bar-Haim, Abend, Adler, & Bliese, 2014; Yuval, Zvielli, & Bernstein, 2017). This presentation duration allows for multiple shifts in attention (Mogg & Bradley, 1998), thus precluding inferences regarding the degree to which ABT, when it is observed and is primarily bottom-up, top-down, or a combination of the two.

Findings from dot-probe studies in which PTSD-related ABT is examined have been mixed. Differences between studies in dot-probe design and administration may account for some of the ambiguity of findings in this area of research. It is unclear from these findings not only whether those with PTSD exhibit ABT but also whether PTSD-related abnormalities in threat processing (a) take the form of facilitated engagement, avoidance, or disengagement difficulties; (b) are observed across a variety of stimuli (i.e., word, face, and images); and (c) are trauma specific or threat general. For example, in one of the first studies to use the dot-probe to assess PTSD-related ABT, Bryant and Harvey (1997) found that participants with PTSD (vs those with subclinical PTSD or low anxiety) exhibited ABT for mild, but not strong threat words. Bryant and Harvey (1997) concluded that study findings suggested greater hypervigilance toward threat among individuals with PTSD (i.e., facilitated engagement); however, as described earlier, the use of a 500 ms stimulus presentation duration precludes this conclusion. Additional modifications to the dot-probe task that was used in this study may have increased error variance and contributed to the unexpected finding of a PTSD-related bias to "mild," but not "strong" threat words. Some of these modifications include simultaneous presentation of word stimuli and response options and the use of word stimuli that require greater semantic processing (Pineles et al., 2009) and are more prone to subjective familiarity and frequency of use than pictorial stimuli (Bradley et al., 1997). In studies in which the same stimulus presentation duration used by Bryant and Harvey (1997) was employed (i.e., 500 ms), support for PTSD-related ABT (i.e., faster responding when the dot replaces threat stimuli compared with neutral stimuli; Fani, Jovanovic, et al., 2012; Fani, Tone, et al., 2012) and avoidance of threat stimuli among those with PTSD (i.e., faster responding when the dot replaced neutral stimuli compared with threat stimuli; Bar-Haim et al., 2007; Sipos et al., 2014) have been observed. In contrast to Bryant and Harvey (1997), Dalgleish et al. (2003) did not find ABT for threat words among participants with PTSD versus those with depression, generalized anxiety disorder, or with no diagnosis. The use of a relatively long presentation duration (1500 ms) may explain these discrepant findings.

Despite the fact that ABT is conceptualized as a vulnerability factor for the development of fear-related pathology (Weierich et al., 2008) and PTSD more specifically (Constans, 2005), relatively few studies have been conducted to clarify the temporal relations among ABT, trauma, and PTSD. In fact, Van Bockstaele et al. (2014) identified the issue of temporality (i.e., ABT preceding the onset of fear-related pathology) as a "major gap in the current state of the literature." Findings from the few longitudinal studies that have employed stimulus-response paradigms in this area are mixed. There is evidence in favor of both avoidance of threat (Wald et al., 2011) and general attention dysregulation (both toward and away from threat; Schäfer et al., 2016) prospectively predicting higher posttraumatic symptoms. In contrast, some evidence suggests that ABT develops in response to trauma exposure rather than being a pretrauma vulnerability factor (Iacoviello et al., 2014). These equivocal findings may be the result of the same methodological limitations of stimulus-response paradigms (modified Stroop and dot-probe) described earlier.

Another potential limitation in this area of research, which may account for discrepant findings, is the use of traditional ABT scores that are calculated by aggregating task response times. Take the dot-probe task for example; ABT is typically calculated by subtracting mean response latencies on trials where the probe replaces the threat image from mean response latencies on trials where the probe replaces the neutral image in neutral-threat pairings (Frewen, Dozois, Joanisse, & Neufeld, 2008). Negative bias scores indicate greater attention deployment toward neutral stimuli, and positive scores indicate attention deployment toward threat stimuli. Importantly, these difference scores represent ABT as a static signal (i.e., bias toward or away from threat at a constant rate over time), thus failing to take into account the temporal dynamics of ABT (i.e., a succession of shifts toward and away from threat stimuli; Zvielli, Bernstein, & Koster, 2015). Use of the traditional static method of calculating ABT may be responsible for (a) the notoriously poor reliability of these scores (Schmukle, 2005; Staugaard & Rosenberg, 2011; Waechter, Nelson, Wright, Hyatt, & Oakman, 2014), (b) difficulty replicating significant findings (e.g., Fani, Jovanovic, et al., 2012; Fani, Tone, et al., 2012), and (c) the considerable number of null findings and small magnitude effects in this area of research (Van Bockstaele et al., 2014).

To remedy these issues, response latencies from these same tasks can be used to calculate a score that represents threat-related attention dysregulation and accounts for within-subject variability of ABT, both toward and away from threat stimuli (i.e., attention bias variability; Zvielli et al., 2015). Although this method is relatively new, preliminary evidence suggests that individuals with PTSD exhibit significantly greater attention bias variability in response to trauma-specific (Yuval et al., 2017) and general-threat (Bardeen, Tull, Daniel, Evenden, & Stevens, 2016; Iacoviello et al., 2014; Naim et al., 2015) stimuli compared with trauma control and healthy participants. Additionally, in contrast to the poor reliability observed for traditional attentional bias scores, acceptable reliability coefficients have been reported for attention bias variability scores (e.g., Schäfer et al., 2016; Zvielli et al., 2015). However, this approach does not necessarily remedy all of the limitations of stimulus-response tasks. For example, the majority of these preliminary studies have employed the same stimulus presentation duration (i.e., 500ms) as the large majority of dot-probe studies described earlier. As described, the use of this specific presentation duration allows one to shift attention multiple times before a response is made (Mogg & Bradley, 1998), and thus inferences regarding bottom-up reflexive orienting cannot be made through the use of this approach.

Eye-tracking technology, which has a number of benefits over stimulus-response tasks for assessing PTSD-related ABT, has been used in relatively few studies in this line of research. Unlike stimulus-response tasks that use button press to make inferences about covert attention, with eye tracking, the construct of interest (overt attention) is directly assessed via eye movements. As such, eye-tracking indices are less susceptible to alternate explanations than attentional bias scores from stimulus-response tasks. Additionally, indices of attentional bias obtained via eye tracking (e.g., proportion of viewing time on threat vs neutral stimuli) have been shown to have adequate reliability (Bardeen & Daniel, 2018; Waechter et al., 2014). There is, however, one notable exception to this rule. To assess facilitated engagement, researchers often calculate a score that incorporates the participants' first fixations to the stimuli that are presented (e.g., proportion of first fixations to threat vs neutral stimuli). These scores tend to have unacceptably low reliability (Waechter et al., 2014). Fortunately, unlike the covert assessment of ABT at 500 ms, overt assessment of ABT (via eye tracking) in the first 500 ms of stimulus presentation has been identified as a valid measure of facilitated engagement (Armstrong & Olatunji, 2012) and, importantly, exhibits significantly better reliability than attentional bias scores from stimulus-response paradigms (e.g., Bardeen & Daniel, 2018). In addition to producing indices of ABT that have adequate reliability, another benefit of using eye-tracking technology is that eye movements are recorded almost continuously. It is common for researchers to use eye-tracking equipment that takes a measurement every 16.67 ms, and sampling frequency can be increased even further (Armstrong & Olatunji, 2012). Thus, unlike stimulus-response tasks, eye tracking is well suited for distinguishing different components of attention on each trial (i.e., facilitated engagement, avoidance, and disengagement difficulties).

To date, free-viewing tasks have been used in all of the published studies in which eye tracking has been employed to examine PTSD-related ABT (Armstrong, Bilsky, Zhao, & Olatunji, 2013; Bardeen & Daniel, 2017b; Beevers, Lee, Wells, Ellis, & Telch, 2011; Felmingham, Rennie, Manor, & Bryant, 2011; Kimble, Fleming, Bandy, Kim, & Zambetti, 2010; Lee & Lee, 2012). For free-viewing tasks, ABT is not measured based on task performance (e.g., speed of responding and identifying a predefined target). Instead, participants are instructed to view the stimuli that appear on the computer screen naturally, as if they were watching television. Much like the dot-probe and modified Stroop literature, there have been considerable differences in task design between eye-tracking studies. In terms of stimuli, photographs (Bardeen & Daniel, 2017b; Kimble et al., 2010; Lee & Lee, 2012), faces (Armstrong et al., 2013; Beevers et al., 2011), and words (Felmingham et al., 2011) have been used. Additionally, considerable variability in stimulus presentation duration has been observed (i.e., from 1000 to 30,000 ms; Beevers et al., 2011; Felmingham et al., 2011, respectively). Despite substantial between-study variability in task design, one finding has been observed fairly consistently—individuals with PTSD (or relatively higher posttraumatic symptoms in some cases) spend significantly more time looking at negatively valenced stimuli than trauma control participants (i.e., attention maintenance; Armstrong et al., 2013; Bardeen & Daniel, 2017b; Kimble et al., 2010, Lee & Lee, 2012). The only study in which the results suggested facilitated engagement to both trauma-specific and threat-general stimuli was also the only study in which word stimuli were used and the measure of facilitation that was used is known for having unacceptably low reliability (i.e., number of initial fixations; Felmingham et al., 2011). One additional finding is of note. Beevers et al. (2011) asked US soldiers to complete a free-viewing task, using face stimuli with facial expressions (i.e., happy, sad, fearful, and neutral), prior to being deployed. Participants were assessed a second time for posttraumatic stress symptomatology after 3 months of being deployed to Iraq. Interestingly, soldiers who spent relatively less time attending to fearful faces at baseline (i.e., avoidance) were at greater risk for developing PTSD after being deployed compared with those who spent more time attending to fearful faces at baseline.

Attentional control as a trauma-related regulatory mechanism

The distress-buffering effects of attentional control have been reported in relation to a wide variety of maladaptive psychological outcomes, and these protective effects also apply to those who are vulnerable to such outcomes (e.g., Armstrong, Zald, & Olatunji, 2011; Eisenberg, Fabes, Guthrie, & Reiser, 2000; Fergus et al., 2012; Jones et al., 2012; Richey et al., 2012). Nonetheless, relatively few published studies have examined attention control in the context of trauma and posttraumatic stress. Of those that have, the majority suggest that attentional control can be used as a trauma-related regulatory mechanism. For example, Bardeen and Read (2010) found that trauma-exposed adult participants with better attentional control recovered significantly faster from trauma retelling (i.e., verbalization of the first-person account of their most traumatic event) induced negative affect than individuals with relatively worse attentional control. In a cross-sectional study, Bardeen and Fergus (2016) found that attentional control moderated the relationship between three PTSD-related vulnerability factors (i.e., emotional distress intolerance, anxiety sensitivity, and experiential avoidance) and posttraumatic stress symptoms, such that the relationship between these vulnerability factors and posttraumatic stress symptoms was significantly stronger among those with relatively worse attentional control. The authors concluded that attentional control may protect those with PTSD-related vulnerabilities from developing the disorder following trauma exposure. Preliminary longitudinal findings have similarly suggested the regulatory value of attentional control. Bardeen et al. (2015) found that those with relatively better attentional control, measured prior to a potentially traumatic event, reported relatively lower posttraumatic stress symptoms following that event in comparison with those that had relatively worse attentional control pretrauma.

Bardeen and Orcutt (2011) considered the regulatory role of attentional control at a more proximal level (i.e., information processing) in an attempt to clarify mixed findings in the PTSD/ABT literature that may be a function of failing to account for the impact of top-down attentional processes on bottom-up reactivity. Specifically, they conducted a laboratory study in which participants completed a modified dot-probe task to assess ABT and a battery of self-report measures, which included a self-report measure of attentional control. They found that, among participants with relatively higher posttraumatic stress symptoms, those with better attentional control disengaged and shifted attention from threat to neutral stimuli, whereas those with relatively worse attentional control maintained attention on threat stimuli (pictorial stimuli that were rated high on arousal and negative valence [IAPS images]; Lang, Bradley, & Cuthbert, 1999). This effect remained significant even after statistically controlling for state levels of anxious arousal. The authors hypothesized that the use of attentional control to disengage and shift attention from threat stimuli among those with higher posttraumatic stress symptoms may help to downregulate emotional distress and sympathetic nervous system arousal in the short term (Bardeen & Orcutt, 2011). Consistent with this proposition, shifting attention to safe or novel stimuli has been shown to reduce negative affect (Harman, Rothbart, & Posner, 1997; Nolen-Hoeksema & Morrow, 1993), and mild distraction has been shown to help expedite fear reduction during exposure therapy (Johnstone & Page, 2004). The authors also acknowledged that consistently shifting attention away from threat stimuli may be perceived as maladaptive avoidance but offered the possibility that using attentional control in this manner may reduce the likelihood that one uses even less adaptive regulatory strategies that are known to maintain and exacerbate posttraumatic stress (e.g., substance use and physical escape). By reducing the intensity of the emotional experience, the individual may be able to remain in the environment in which the fear-provoking stimuli are present, thus increasing the likelihood that inhibitory learning will occur and symptoms will be reduced.

Before considering longitudinal research, which is necessary to determine the adaptive value of using attentional control to regulate the emotional distress that is associated with viewing threat stimuli among those with higher posttraumatic stress, one other significant limitation in this line of research should be addressed. Attentional control is often assessed via self-report (Derryberry & Reed, 2002). As described earlier, evidence suggests that attentional control processes can be enacted in a fraction of a second (e.g., Shomstein et al., 2012), and thus it may be difficult to accurately report on cognitive processes that occur so rapidly that they go unnoticed on a moment-by-moment basis. In fact, some have suggested that self-report measures of attentional control may be a better indicator of beliefs about attentional control rather than providing an index of attentional control abilities (Spada, Georgiou, & Wells, 2010). Moreover, self-reported attentional control has failed to correlate with behavioral measures of attentional control processes (e.g., working memory and inhibitory ability) in some studies (Quigley, Wright, Dobson, & Sears, 2017).

To address the limitation of using self-report to assess attentional control while also replicating the moderation effect observed by Bardeen and Orcutt (2011), Bardeen et al. (2016) incorporated a behavioral measure of attentional control and a dot-probe task into their laboratory study. Additionally, ABT scores were calculated using the trial-level bias method to improve score reliability and account for the temporal dynamics of ABT (Zvielli et al., 2015). As predicted, attentional control (measured via a behavioral task) moderated the association between PTSD status and ABT, such that, among those with PTSD, those with relatively worse attentional control exhibited significantly greater ABT (Bardeen et al., 2016). This effect remained significant even after statistically controlling for variability on trials with only neutral content, thus increasing confidence that the observed effect was specific to threat and not a function of general variability in responding.

Preliminary longitudinal findings suggest that pretrauma attentional control may serve as a protective factor against developing PTSD in the acute aftermath of a trauma (Bardeen et al., 2015). Evidence also suggests that those with PTSD and relatively better attentional control disengage and shift attention from threat to neutral stimuli, while those with PTSD and relatively worse attentional control maintain attention on threat stimuli (Bardeen & Orcutt, 2011; Bardeen et al., 2016). Bardeen and Daniel (2017b) conducted a longitudinal study to determine whether deploying attention away from threat is adaptive over a prolonged period (i.e., 6 months). As described by Bardeen and Daniel (2017b), two hypotheses were plausible. First, temporarily disengaging attention from threat stimuli and refocusing elsewhere may serve to downregulate sympathetic nervous system arousal and associated emotional distress. By reducing the intensity of the emotional experience, the individual may be able to remain in the environment in which the fear-provoking stimuli are present, rather than using more extreme avoidance strategies, such as physical escape, which do not provide opportunities for new learning and allow for fear extinction. Although the intensity of the uncomfortable emotion is reduced, it is still experienced. Second the chronic inflexible use of attentional control processes to avoid threat information may reduce emotional arousal in the short term but maintain and perhaps exacerbate posttraumatic stress symptoms over time by failing to provide the opportunity for new learning.

To test these hypotheses, trauma-exposed adults participated in a laboratory session in which they completed self-report measures (e.g., attentional control), behavioral tasks that separately assessed the primary components of attentional control, and a free-viewing task in which eye movements were recorded in response to threat and neutral stimuli over the course of 60 trials (Bardeen & Daniel, 2017b). Eye-tracking technology was used to ensure reliability of ABT scores and to provide a more direct, overt measure of attention allocation. The stimulus set that was presented during the free-viewing task (i.e., threat-neutral pairings) had been used in previous research (Bardeen, 2015), and the format of presentation was similar to Bardeen and Orcutt (2011) and Bardeen et al. (2016). Specifically, for each trial, two images (threat-neutral pairings) appeared side by side on the screen for 3000 ms, and participants were free to pay as much or as little attention to each image as they desired; a response was not required. To examine within-trial variability in ABT, dwell time (i.e., proportion of time attending to threat versus neutral) was calculated in 500 ms epochs (i.e., 0–500, 501–1000, 1001–1500, 1501–2000, 2001–2500, and 2501–3000 ms).

Additionally, pupil response was assessed and served as an indicator of emotional arousal in response to task stimuli (Bradley, Miccoli, Escrig, & Lang, 2008). Based on the hypothesis that shifting attention from threat to neutral stimuli downregulates sympathetic nervous system arousal, Bardeen and Daniel (2017b) hypothesized that, among participants with relatively higher posttraumatic stress symptoms, those with better attentional control would exhibit attenuated pupillary response during the free-viewing task compared with those with worse attentional control. Finally, participants completed a battery of self-report measures online, including a measure of posttraumatic stress symptoms, 6 months after completing the laboratory session.

Consistent with previous research (Bardeen & Orcutt, 2011), self-reported attentional control moderated the relationship between posttraumatic stress symptoms and ABT, such that, among those with relatively higher posttraumatic stress symptoms, those with better attentional control disengaged from threat and shifted attention to the neutral stimulus, while those with worse attentional control maintained attention on threat. Importantly, this finding was replicated using behavioral measures of attentional control. Specifically, inhibitory ability appeared to be driving the moderation effect. This effect was most pronounced at a relatively later stage of information processing (1500–3000 ms), but not at a stage of processing indicative of facilitated engagement (i.e., 0–500 ms). This might help explain the apparent discrepancy between the vigilance-avoidance and attention maintenance hypotheses. While a difference in initial reflexive orienting toward threat stimuli based on posttraumatic stress symptomatology was not observed in this study, findings suggest that, among those with higher posttraumatic stress symptoms, those with relatively worse inhibitory ability lack the requisite resources to disengage attention from threat stimuli and refocus elsewhere, thus resulting in attention maintenance. In contrast, those with higher posttraumatic stress symptoms and relatively better inhibitory ability quickly disengaged attention from threat and maintained attentional focus on neutral stimuli once disengagement had occurred (i.e., threat avoidance). Additionally, results of this study suggest that this pattern of avoidance, among those with higher posttraumatic stress symptoms and relatively better attentional control, leads to short-term relief from emotional distress (i.e., downregulation of sympathetic nervous system arousal), as this group exhibited significantly lower pupillary reactivity to threat stimuli in comparison with those with higher posttraumatic stress symptoms and relatively worse attentional control.

Results from the longitudinal portion of the study suggest that using attentional control to shift attention away from threat stimuli and downregulate emotional arousal maintains and perhaps exacerbates posttraumatic stress symptoms over a 6-month period. It is important to keep in mind that the maladaptive nature of this effect is specific to participants who had relatively higher posttraumatic stress symptoms to begin with. However, Beevers et al. (2011) found that soldiers who exhibited the same pattern of threat avoidance, albeit to fearful faces, were at greater risk for developing PTSD after being deployed compared with those who attended to fearful faces more often at the initial assessment. To understand the effect observed by Bardeen and Daniel (2017b), it may important to distinguish ability (i.e., abilities underlying attention control: inhibition, shifting, and working memory updating) from the effortful application of ability (i.e., attentional deployment). As described, information processing theories of emotion regulation suggest that the flexible use of attentional control is important for maintaining psychological well-being (Gross, 2015). The pattern of threat-related avoidance exhibited in this study by those with higher posttraumatic symptoms and relatively better attentional control suggests a rigid fear-based approach to reducing emotional arousal. In contrast, *flexible* use of attentional control suggests a willingness to experience fluctuations in emotions and affective states. As identified in the *DSM-5* (APA, 2013), the use of avoidance to provide short-term relief from aversive internal experiences (i.e., trauma-related bodily sensations, emotions, memories, and thoughts) is a symptom of PTSD. As suggested by the results of Bardeen and Daniel's (2017b) study, avoidance of aversive internal experiences (i.e., experiential avoidance) that are associated with a traumatic event, and threat stimuli in general (i.e., fear generalization), may reduce trauma-related distress in the short term, thus reinforcing the likelihood that attentional avoidance of threat stimuli will continue. As this regulatory approach is used more often (i.e., "pretuning"), it should become easier to enact, eventually becoming an almost automatic reflex to specific stimulus categories (i.e., "prepared reflex"; Verbruggen, 2016). Over time the chronic use of attentional avoidance prevents disconfirmation of faulty threat appraisals, thus resulting in the continued use of maladaptive avoidance behaviors and the maintenance of posttraumatic stress.

The findings of Bardeen and Daniel (2017b) highlight a broader issue regarding the linear assumption that is often made regarding the nature of constructs that represent top-down regulation. That is, it is typically assumed that the more you have (e.g., attentional control, inhibitory control, and self-regulation/control), the better off you are (Tangney, Baumeister, & Boone, 2004). In a chapter on "Self-regulation: Self-control," Peterson and Seligman (2004) write, "We have focused on the positive aspects of self-regulation and self-control and have not discussed the possible drawbacks of self-control. This omission arises from our belief that there is no true disadvantage of having too much self-control" (p. 515). The results from Bardeen and Daniel (2017b) suggest that attentional control, like any other tool of self-regulation, can be used in such a way as to result in maladaptive outcomes. Consistent with this proposition, examinations of the role of maladaptive overcontrol, which requires a high level of top-down inhibitory ability, in the development of psychopathology (e.g., anorexia nervosa) are increasing. Additionally, interventions have recently been developed that aim to relax inhibitory control and increase flexibility among individuals suffering from overcontrol-related pathology (e.g., Lynch, 2018).

The problem of maladaptive overcontrol has also received some attention in the emotion regulation literature. As described by Gross (2015), some individuals are prone to "delayed stopping" in the process of emotion regulation. That is, they continue to employ an emotion regulation strategy well beyond the point at which regulation is needed, thus resulting in overtaxed cognitive resources and a wide variety of maladaptive outcomes (e.g., inhibiting emotional expression across contexts, which is sometimes adaptive [business meeting], and other times maladaptive [in the presence of family]). While it is important to consider individual differences in top-down attentional control abilities to understand the role of attention in the development and maintenance of trauma-related distress, more emphasis should be placed on how and when these abilities are used. Specifically, it seems important to assess whether individuals employ top-down attention with chronic rigidity across contexts or flexibly based on the demands of a given situation. Some evidence suggests that the ability to flexibly shift between attentional avoidance and engagement based on threat intensity is psychologically healthy (i.e., avoiding in high threat situations and engaging in low-moderate threat situations; Sheppes, Scheibe, Suri, & Gross, 2011).

Also of note, Bardeen and Daniel (2017b) found that the combination of high pupillary reactivity and high attentional control buffered the effect of posttraumatic stress symptoms at baseline on posttraumatic stress symptoms 6 months later. This finding suggests that the combination of ability (i.e., attentional control) and willingness to experience short-term emotional distress may be protective. This finding is consistent with a functional-contextual perspective in which the chronic and rigid avoidance of unwanted internal experiences reduces short-term emotional distress but paradoxically exacerbates emotional distress over prolonged periods of time (Hayes, Luoma, Bond, Masuda, & Lillis, 2006). This finding is also consistent with empirical evidence that those with outcome-specific vulnerabilities and greater willingness to experience uncomfortable emotions report greater short-term emotional distress in response to negative mood induction but relatively less long-term maladaptive outcomes (Bardeen, 2015). In contrast, those with outcome-specific vulnerabilities and greater avoidance of uncomfortable emotions report less short-term distress in response to negative mood induction but greater maladaptive long-term outcomes.

Attention bias modification

A great deal of effort has been spent researching emotion regulation strategies that are employed at later stages of the emotion generative process (Gross & Thompson, 2007), and this research has been used to support the use of resource-intensive interventions that focus on altering emotions after they are more fully developed (e.g., cognitive therapy/restructuring). However, it may be easier and less resource intensive to target regulatory processes that occur earlier in the emotion generative process before an emotion has completely unfolded (e.g., attentional deployment and situation modification). Attention bias modification (ABM), a computer-delivered treatment for anxiety- and fear-related pathology, was developed to train attention (implicitly) away from threat and toward nonthreat stimuli (i.e., threat avoidance). ABM training is conducted most often using a modified version of the dot-probe task. Specifically, in ABM, the probe appears in the spatial position previously held by the neutral stimuli on the majority of task trials (e.g., 80%–100% of the time; Schoorl, Putman, & Van Der Does, 2013).

Relatively few randomized control trials (RCTs) have been conducted in which ABM has been used to treat individuals with PTSD (Badura-Brack et al., 2015; Kuckertz et al., 2014; Schoorl et al., 2013). In these studies, and in the application of ABM in the broader fear/anxiety literature, the control condition typically consists of completion of the standard dot-probe task; the probe appears equally in place of both types of stimuli (i.e., neutral and threat). Using this approach, Schoorl et al. (2013) found that both conditions (ABM and the standard dot probe) were equally effective in reducing posttraumatic stress symptoms among outpatient participants with chronic PTSD. However, neither intervention performed better than the use of a placebo pill, and neither intervention altered ABT. Schoorl et al. (2013) concluded that ABM is not an effective treatment for PTSD.

In another RCT, Kuckertz et al. (2014) assigned patients with PTSD who were receiving frontline treatments for the disorder (i.e., prolonged exposure or cognitive processing therapy) and pharmacological intervention, to receive ABM or the standard dot-probe task as an adjunct. Although PTSD symptoms decreased in both conditions, the decrease was significantly larger for participants in the ABM condition compared with control. Interestingly, symptom reductions were largest among participants who exhibited avoidance of threat, rather than ABT, at baseline. This finding runs counter to the fundamental assumption of ABM that ABT plays a causal role in the development of PTSD and, thus, can be reduced through the use of ABM to reduce PTSD symptoms. To explain this finding, Kuckertz et al. (2014) hypothesized that "it may be the case that individuals who initially present with a bias away from threat possess a cognitive strength that is more easily maximized through attention training away from threat, relative to individuals who have a preexisting difficulty attending away from threat" (p. 33). Evidence suggests that, among individuals with PTSD, those with relatively better attentional control exhibit threat avoidance rather than ABT (Bardeen & Daniel, 2017b; Bardeen & Orcutt, 2011; Bardeen et al., 2016). Following from this evidence, Kuckertz et al. (2014) may have been corrected that patients in their study who exhibited attentional avoidance at baseline had a relative strength in cognitive ability (e.g., attentional control). However, it seems unlikely that ABM would somehow draw on this ability to alleviate PTSD symptoms. Specifically, it is not clear why individuals with PTSD who already exhibit threat avoidance would benefit from an intervention designed to enhance threat avoidance, especially given evidence that using attentional control to avoid threat maintains PTSD symptoms (Bardeen & Daniel, 2017b). It may be that the relative difference in cognitive ability at baseline interacted with one of the many components of the primary treatment that the patients were receiving (e.g., exposure, cognitive processing, and pharmacological intervention) to enhance symptom reduction, but this hypothesis would need to be explored in future research.

Finally, perhaps one of the most unusual findings if one assumes that ABM works by training attention away from threat comes from a set of RCTs in which participants (veterans of Israel's Defense Force and the US military) were randomly assigned to receive ABM or the standard dot-probe task (Badura-Brack et al., 2015). In contrast to Kuckertz et al. (2014), participants assigned to the control condition exhibited significantly greater reductions in PTSD symptoms compared with those assigned to the ABM condition. Additionally, those who completed the standard dot probe (typically used as a control condition) exhibited significant reductions in attention bias variability, whereas those in the active ABM condition did not (Badura-Brack et al., 2015). Badura-Brack et al. (2015) explained these findings by suggesting that the standard dot-probe task, with the probe appearing equally in place of both types of stimuli, helps train attention toward task success rather than away from threat stimuli, thus promoting flexibility of attentional control rather than threat avoidance.

Many more ABM studies have been conducted with people suffering from other forms of fear/anxiety-related pathology. Unfortunately the results of these studies do not provide clarity surrounding issue of ABM efficacy. Although some metaanalytic evidence suggests small- to medium-sized effects for the use of ABM for treating anxiety and related disorders (e.g., Hakamata et al., 2010), Cristea, Kok, and Cuijpers (2015) found that previously significant effects became nonsignificant when (a) adjusting for publication bias, (b) accounting for extreme outliers, and (c) confining analyses to patient samples. These equivocal findings are less perplexing when one considers that the basic assumption of ABM, which individuals with fear- and anxiety-related pathology (i.e., PTSD) preferentially process threat, appears to be incorrect. Results from a number of studies (Bardeen & Daniel, 2017b; Bardeen & Orcutt, 2011; Bardeen et al., 2016; Derryberry & Reed, 2002; Ho, Yeung, & Mak, 2017) suggest that the assessment of top-down attentional control processes is important when designing studies that seek to understand threat processing in PTSD and related disorders.

Although standard ABM may not be suitable for individuals with PTSD and relatively better attentional control, this approach may have value for reducing PTSD symptoms among those that lack the ability to maintain threat disengagement (i.e., PTSD and lower attentional control). In contrast, those that exhibit chronic and inflexible threat avoidance (i.e., PTSD and higher attentional control) may be better served by participating in interventions that promote flexibility of attentional control (e.g., the attention training component of emotion regulation therapy: Renna et al., 2018; the attention training technique: Wells, 1990; see Fergus & Bardeen, 2016, for a review). Additionally, treatments that have been shown to be effective in increasing willingness to stay in contact with uncomfortable emotions and related internal experiences may be beneficial for individuals who exhibit rigid threat avoidance (e.g., acceptance and commitment therapy: Hayes et al., 2006; mindfulness-based stress reduction: Kabat-Zinn, 1990).

References

American Psychiatric Association. (2013). *Diagnostic and statistical manual of mental disorders* (5th ed.). Arlington, VA: American Psychiatric Publishing.

Armstrong, T., Bilsky, S. A., Zhao, M., & Olatunji, O. (2013). Dwelling on potential threat cues: An eye movement marker for combat-related PTSD. *Depression and Anxiety, 30*, 497–502.

Armstrong, T., & Olatunji, B. O. (2012). Eye tracking of attention in the affective disorders: A meta-analytic review and synthesis. *Clinical Psychology Review, 32*, 704–723.

Armstrong, T., Zald, D. H., & Olatunji, B. O. (2011). Attentional control in OCD and GAD: Specificity and associations with core cognitive symptoms. *Behaviour Research and Therapy, 49*, 756–762.

Aupperle, R. L., Melrose, A. J., Stein, M. B., & Paulus, M. P. (2012). Executive function and PTSD: Disengaging from trauma. *Neuropharmacology, 62*, 686–694.

Badura-Brack, A. S., Naim, R., Ryan, T. J., Levy, O., Abend, R., Khanna, M. M., ... Bar-Haim, Y. (2015). Effect of attention training on attention bias variability and PTSD symptoms: Randomized controlled trials in Israeli and U.S. combat Veterans. *The American Journal of Psychiatry, 172*, 1233–1241.

Bardeen, J. R. (2015). Short-term pain for long-term gain: The role of experiential avoidance in the relation between anxiety sensitivity and emotional distress. *Journal of Anxiety Disorders, 30*, 113–119.

Bardeen, J. R., & Daniel, T. A. (2017a). An eye-tracking examination of emotion regulation difficulties, cognitive reappraisal, and expressive suppression as predictors of attentional bias and pupillary reactivity to threat stimuli. *Cognitive Therapy & Research, 41*, 853–866.

Bardeen, J. R., & Daniel, T. A. (2017b). A longitudinal examination of the role of attentional control in the relationship between posttraumatic stress and threat-related attentional bias: An eye-tracking study. *Behaviour Research and Therapy, 99*, 67–77.

Bardeen, J. R., & Daniel, T. A. (2018). Anxiety sensitivity and attentional bias to threat interact to prospectively predict anxiety. *Cognitive Behaviour Therapy, 47*, 482–494.

Bardeen, J. R., Daniel, T. A., Hinnant, J. B., & Orcutt, H. K. (2017). Emotion dysregulation and threat-related attention bias variability. *Motivation and Emotion, 41*, 402–409.

Bardeen, J. R., & Fergus, T. A. (2016). Emotional distress intolerance, experiential avoidance, and anxiety sensitivity: The buffering effect of attentional control on associations with posttraumatic stress symptoms. *Journal of Psychopathology and Behavioral Assessment, 38*, 320–329.

Bardeen, J. R., Fergus, T. A., & Orcutt, H. K. (2015). Attentional control as a prospective predictor of posttraumatic stress symptomatology. *Personality and Individual Differences, 81*, 124–128.

Bardeen, J. R., & Orcutt, H. K. (2011). Attentional control as a moderator of the relationship between posttraumatic stress symptoms and attentional threat bias. *Journal of Anxiety Disorders, 25*, 1008–1018.

Bardeen, J. R., & Read, J. P. (2010). Attentional control, trauma, and affect regulation: A preliminary investigation. *Traumatology, 16*, 11–18.

Bardeen, J. R., Tull, M. T., Daniel, T. A., Evenden, J., & Stevens, E. N. (2016). A preliminary investigation of the time course of attention bias variability in posttraumatic stress disorder: The moderating role of attentional control. *Behaviour Change, 33*, 94–111.

Bar-Haim, Y., Lamy, D., Pergamin, L., Bakermans-Kranenburg, M. J., & van IJzandoorn, M. H. (2007). Threat-related attentional bias in anxious and nonanxious individuals: A meta-analytic study. *Psychological Bulletin, 133*, 1–24.

Beck, A. T., & Clark, D. A. (1997). An information processing model an anxiety: Automatic and strategic processes. *Behavior Research and Therapy, 35*, 49–58.

Beck, A. T., Emery, G., & Greenberg, R. (1985). *Anxiety disorders and phobias: A cognitive perspective*. New York, NY: Basic Books.

Beevers, C. G., Lee, H. J., Wells, T. T., Ellis, A. J., & Telch, M. J. (2011). Association of predeployment gaze bias for emotion stimuli with later symptoms of PTSD and depression in soldiers deployed in Iraq. *American Journal of Psychiatry, 168*, 735–741.

Bishop, S. J. (2008). Neural mechanisms underlying selective attention to threat. *Annals of the New York Academy of Sciences, 1129*, 141–152.

Bradley, M. M., Miccoli, L., Escrig, M. A., & Lang, P. J. (2008). The pupil as a measure of emotional arousal and autonomic activation. *Psychophysiology, 45*, 602–607.

Bradley, B. P., Mogg, K., Millar, N., Bonham-Carter, C., Fergusson, E., ... Parr, M. (1997). Attentional biases for emotional faces. *Cognition and Emotion, 11*, 25–42.

Brewin, C. R., Dalgleish, T., & Joseph, S. (1996). A dual-representation theory of posttraumatic stress disorder. *Psychological Review, 106*, 670–686.

Brewin, C. R., & Holmes, E. A. (2003). Psychological theories of posttraumatic stress disorder. *Clinical Psychology Review, 23*, 339–376.

Bryant, R. A., & Harvey, A. G. (1997). Attentional bias in posttraumatic stress disorder. *Journal of Traumatic Stress, 10*, 635–644.

Buckley, T. C., Galovski, T., Blanchard, E. B., & Hickling, E. J. (2003). Is the emotional Stroop paradigm sensitive to malingering? A between-groups study with professional actors and actual trauma survivors. *Journal of Traumatic Stress, 16*, 59–66.

Cisler, J. M., & Koster, E. H. W. (2010). Mechanisms of attentional biases towards threat in anxiety disorders. *Clinical Psychology Review, 30*, 203–216.

Compton, J. R. (2000). Ability to disengage attention predicts negative affect. *Cognition and Emotion, 14*, 401–415.

Constans, J. I. (2005). Information-processing bias in PTSD. In J. J. Vasterling & C. R. Brewin (Eds.), *Neuropsychology of PTSD: Biological, cognitive, and clinical perspectives* (pp. 105–130). New York, NY: The Guilford Press.

Corbetta, M., & Shulman, G. L. (2002). Control of goal-directed behavior and stimulus-driven attention in the brain. *Nature Reviews Neuroscience, 3*, 201–215.

Craske, M. G., Treanor, M., Conway, C. C., Zbozinek, T., & Vervliet, B. (2014). Maximizing exposure therapy: An inhibitory learning approach. *Behaviour Research and Therapy, 58*, 10–23.

Cristea, I. A., Kok, R. N., & Cuijpers, P. (2015). Efficacy of cognitive bias modification interventions in anxiety and depression: Meta-analysis. *The British Journal of Psychiatry, 206*, 7–16.

Dalgleish, T., Taghavi, R., Neshat-Doost, H., Moradi, A., Canterbury, R., & Yule, W. (2003). Patterns for processing bias for emotional information across clinical disorders: A comparison of attention, memory, and prospective cognition in children and adolescents with depression, generalized anxiety, and posttraumatic stress disorder. *Journal of Clinical Child and Adolescent Psychology, 32*, 10–21.

Derryberry, D., & Reed, M. A. (2002). Anxiety-related attentional biases and their regulation by attentional control. *Journal of Abnormal Psychology, 111*, 225–236.

Dymond, S., Dunsmoor, J. E., Vervliet, B., Roche, B., & Hermans, D. (2015). Fear generalization in humans: Systematic review and implications for anxiety disorder research. *Behavior Therapy, 46*, 561–582.

Ehlers, A., & Clark, D. M. (2000). A cognitive model of posttraumatic stress disorder. *Behaviour Research and Therapy, 38*, 319–345.

Eisenberg, N., Fabes, R. A., Guthrie, I. K., & Reiser, M. (2000). Dispositional emotionality and regulation: Their role in predicting quality of social functioning. *Journal of Personality Social Psychology, 78*, 136–157.

Elsesser, K., Satory, G., & Tackenberg, A. (2004). Initial symptoms and reactions to trauma-related stimuli and the development of posttraumatic stress disorder. *Depression and Anxiety, 21*, 61–70.

Esterman, M., DeGutis, J., Mercado, R., Rosenblatt, A., Vasterling, J. J., Milberg, W., & McGlinchey, R. (2013). Stress-related psychological symptoms are associated with increased attentional capture by visually salient distractors. *Journal of the International Neuropsychological Society, 19,* 835–840.

Eysenck, M. W., & Derakshan, N. (2011). New perspectives in attentional control theory. *Personality and Individual Differences, 50,* 955–960.

Eysenck, M. W., Derakshan, N., Santos, R., & Calvo, M. G. (2007). Anxiety and cognitive performance: Attentional control theory. *Emotion, 7,* 336–353.

Fani, N., Jovanovic, T., Ely, T. D., Bradley, B., Gutman, D., Tone, E. B., & Ressler, K. J. (2012). Neural correlates of attention bias to threat in posttraumatic stress disorder. *Biological Psychology, 90,* 134–142.

Fani, N., Tone, E. B., Phifer, J., Norrholm, S. D., Bradley, B., Ressler, K. J., … Jovanovic, T. (2012). Attention bias toward threat is associated with exaggerated fear expression and impaired extinction in PTSD. *Psychological Medicine, 42,* 533–543.

Felmingham, K. L., Rennie, C., Manor, B., & Bryant, R. A. (2011). Eye tracking and physiological reactivity to threatening stimuli in posttraumatic stress disorder. *Journal of Anxiety Disorders, 25,* 668–673.

Fergus, T. A., & Bardeen, J. R. (2016). The attention training technique: A review of a neurobehavioral therapy for anxiety and related disorders. *Cognitive and Behavioral Practice, 23,* 502–516.

Fergus, T. A., Bardeen, J. R., & Orcutt, H. K. (2012). Attentional control moderates the relationship between activation of the cognitive attentional syndrome and symptoms of psychopathology. *Personality and Individual Differences, 53*(3), 213–217.

Foa, E. B., Huppert, J. D., & Cahill, S. P. (2006). Emotional processing theory: An update. In B. O. Rothbaum (Ed.), *Pathological anxiety: Emotional processing in etiology and treatment* (pp. 3–24). New York, NY: Guilford Press.

Foa, E. B., & Kozak, M. J. (1986). Emotional processing of fear: Exposure to corrective information. *Psychological Bulletin, 99,* 20–35.

Frewen, P. A., Dozois, D. J. A., Joanisse, M. F., & Neufeld, R. W. J. (2008). Selective attention to threat versus reward: Meta-analysis and neural-network modeling of the dot-probe task. *Clinical Psychology Review, 28,* 307–337.

Graydon, J., & Eysenck, M. W. (1989). Distraction and cognitive performance. *European Journal of Cognitive Psychology, 1,* 161–179.

Gross, J. J. (1998). The emerging field of emotion regulation: An integrative review. *Review of General Psychology, 2,* 271–299.

Gross, J. J. (2015). Emotion regulation: Current status and future prospects. *Psychological Inquiry, 26,* 1–26.

Gross, J. J., & Thompson, R. A. (2007). Emotion regulation: Conceptual foundations. In J. J. Gross (Ed.), *Handbook of emotion regulation* (pp. 3–24). New York, NY: Guilford Press.

Hakamata, Y., Lissek, S., Bar-Haim, Y., Britton, J. C., Fox, N. A., Leibenluft, E., … Pine, D. S. (2010). Attention bias modification treatment: A meta-analysis toward the establishment of novel treatment for anxiety. *Biological Psychiatry, 68,* 982–990.

Harman, C., Rothbart, M. K., & Posner, M. I. (1997). Distress and attention interactions in early infancy. *Motivation and Emotion, 21,* 27–43.

Hayes, S. C., Luoma, J. B., Bond, F. W., Masuda, A., & Lillis, J. (2006). Acceptance and commitment therapy: Model, processes and outcomes. *Behaviour Research and Therapy, 44,* 1–25.

Ho, S. M. Y., Yeung, D., & Mak, C. W. Y. (2017). The interaction effect of attentional bias and attentional control on dispositional anxiety among adolescents. *British Journal of Psychology, 108,* 564–582.

Iacoviello, B. M., Wu, G., Abend, R., Murrough, J. W., Feder, A., Fruchter, E., … Charney, D. S. (2014). Attention bias variability and symptoms of posttraumatic stress disorder. *Journal of Traumatic Stress, 27,* 232–239.

Johnstone, K. A., & Page, A. C. (2004). Attention to phobic stimuli during exposure: The effect of distraction on anxiety reduction, self-efficacy and perceived control. *Behaviour Research and Therapy, 42,* 249–275.

Jones, C. R., Fazio, R. H., & Vasey, M. W. (2012). Attentional control buffers the effect of public-speaking anxiety on performance. *Social Psychological and Personality Science, 3,* 556–561.

Jovanovic, T., & Ressler, K. J. (2010). How the neurocircuitry and genetics of fear inhibition may inform our understanding of PTSD. *American Journal of Psychiatry, 167,* 648–662.

Kabat-Zinn, J. (1990). *Full catastrophe living: Using the wisdom of your body and mind to face stress, pain, and illness.* New York, NY: Delacourt.

Kimble, M. O., Fleming, K., Bandy, C., Kim, J., & Zambetti, A. (2010). Eye tracking and visual attention to threatening stimuli in veterans of the Iraq war. *Journal of Anxiety Disorders, 24,* 293–299.

Kimble, M. O., Frueh, B. C., & Marks, L. (2009). Does the modified Stroop effect exist in PTSD? Evidence from dissertation abstracts and the peer reviewed literature. *Journal of Anxiety Disorders, 23,* 650–655.

Knowles, K. A., & Olatunji, B. O. (2019). Enhancing inhibitory learning: The utility of variability in exposure. *Cognitive and Behavioral Practice, 26,* 186–200.

Koole, S. L., Webb, T. L., & Sheeran, P. L. (2015). Implicit emotion regulation: Feeling better without knowing why. *Current Opinion in Clinical Psychology, 3,* 6–10.

Kuckertz, J. M., Amir, N., Boffa, J. W., Warren, C. K., Rindt, S. E. M., Norman, S., ... McLay, R. (2014). The effectiveness of an attention bias modification program as an adjunctive treatment for post-traumatic stress disorder. *Behaviour Research and Therapy, 63,* 25–35.

Lang, P. J., Bradley, M. M., & Cuthbert, B. N. (1999). *International affective picture system (IAPS): Technical manual and affective ratings.* Gainesville, FL: The Center for Research in Psychophysiology, University of Florida.

Lavie, N., Hirst, A., de Fockert, J. W., & Viding, E. (2004). Load theory of selective attention and cognitive control. *Journal of Experimental Psychology: General, 133,* 339–354.

Lee, J., & Lee, J. (2012). Attentional bias to violent images in survivors of dating violence. *Cognition and Emotion, 26,* 1124–1133.

Lissek, S. (2012). Toward an account of clinical anxiety predicated on basic, neurally-mapped mechanisms of Pavlovian fear-learning: The case for conditioned overgeneralization. *Depression and Anxiety, 29,* 257–263.

Litz, B. T., & Keane, T. M. (1989). Information processing in anxiety disorders: Application to the understanding of post-traumatic stress disorder. *Clinical Psychology Review, 9,* 243–257.

Lynch, T. R. (2018). *Radically open dialectical behavior therapy: Theory and practice for treating disorders of overcontrol.* Oakland, CA: New Harbinger.

McKenna, F. P., & Sharma, D. (2004). Reversing the emotional Stroop effect reveals that it is not what it seems: The role of fast and slow components. *Journal of Experimental Psychology, 30,* 382–392.

McNally, R. J. (2007). Mechanisms of exposure therapy: How neuroscience can improve psychological treatments for anxiety disorders. *Clinical Psychology Review, 27,* 750–759.

Metcalfe, J., & Mischel, W. (1999). A hot/cool-system analysis of delay of gratification: Dynamics of willpower. *Psychological Review, 106,* 3–18.

Mischel, W. S., Shoda, Y., & Rodriguez, M. L. (1989). Delay of gratification in children. *Science, 244,* 933–938.

Miyake, A., Friedman, N. P., Emerson, M. J., Witzki, A. H., & Howerter, A. (2000). The unity of diversity of executive functions and contributions to complex "frontal lobe" tasks: A latent variable analysis. *Cognitive Psychology, 41,* 49–100.

Mogg, K., & Bradley, B. P. (1998). A cognitive-motivational analysis of anxiety. *Behaviour Research and Therapy, 36,* 809–848.

Naim, R., Abend, R., Wald, I., et al. (2015). Threat-related attention bias variability and post-traumatic stress. *American Journal of Psychiatry, 172,* 1242–1250.

Nolen-Hoeksema, S., & Morrow, J. (1993). Effects of rumination and distraction on naturally occurring depressed mood. *Cognition & Emotion, 7,* 561–570.

Olatunji, B. O., Armstrong, T., McHugo, M., & Zald, D. H. (2013). Heightened attentional capture to threat in veterans. *Journal of Abnormal Psychology, 122*, 397–405.

Peterson, C., & Seligman, M. E. (2004). *Character strengths and virtues: A handbook and classification*. Oxford: Oxford University Press.

Phaf, R. H., & Kan, K. J. (2007). The automaticity of emotional Stroop: A meta-analysis. *Journal of Behavior Therapy and Experimental Psychiatry, 38*, 184–199.

Pineles, S. L., Shipherd, J. C., Mostoufi, S. M., Abramovitz, S. M., & Yovel, I. (2009). Attentional biases in PTSD: More evidence for interference. *Behaviour Research and Therapy, 47*, 1050–1057.

Quigley, L., Wright, C. A., Dobson, K. S., & Sears, C. R. (2017). Measuring attentional control ability or beliefs? Evaluation of the factor structure and convergent validity of the Attentional Control Scale. *Journal of Psychopathology and Behavioral Assessment, 39*, 742–754.

Renna, M. E., Seeley, S. H., Heimberg, R. G., Etkin, A., Fresco, D. M., & Mennin, D. S. (2018). Increased attention regulation from emotion regulation therapy for generalized anxiety disorder. *Cognitive Therapy and Research, 42*, 121–134.

Richey, J. A., Keough, M. E., & Schmidt, N. B. (2012). Attentional control moderates fearful responding to a 35% CO_2 challenge. *Behavior Therapy, 43*, 285–299.

Rosler, F., Heil, M., & Roder, B. (1997). Slow negative brain potentials as reflections of specific modular resources of cognition. *Biological Psychology, 45*, 109–141.

Rothbart, M. K., Sheese, B. E., & Posner, M. I. (2007). Executive attention and effortful control: Linking temperament, brain networks, and genes. *Child Development Perspectives, 1*, 2–7.

Sarapas, C., Weinberg, A., Langenecker, S. A., & Shankman, S. A. (2017). Relationships among attention networks and physiological responding to threat. *Brain and Cognition, 111*, 63–72.

Schäfer, J., Bernstein, A., Zvielli, A., Höfler, M., Wittchen, H.-U., & Schönfeld, S. (2016). Attentional bias temporal dynamics predict posttraumatic stress symptoms: A prospective–longitudinal study among soldiers. *Depression and Anxiety, 33*, 630–639.

Schmukle, S. C. (2005). Unreliability of the dot probe task. *European Journal of Personality, 19*, 595–605.

Schoorl, M., Putman, P., & Van Der Does, W. (2013). Attentional bias modification in posttraumatic stress disorder: A randomized controlled trial. *Psychother Psychosom, 82*, 99–105.

Scott, J. C., Matt, G. E., Wrocklage, K. M., Crnich, C., Jordan, J., Southwick, S. M., ... Schweinsburg, B. C. (2015). A quantitative meta-analysis of neurocognitive functioning in posttraumatic stress disorder. *Psychological Bulletin, 141*, 105–140.

Sheppes, G., Scheibe, S., Suri, G., & Gross, J. J. (2011). Emotion-regulation choice. *Psychological Science*, (11), 1391–1396.

Shomstein, S., Kravitz, D. J., & Behrmann, M. (2012). Attentional control: Temporal relationships within the fronto-parietal network. *Neuropsychologia, 50*, 1202–1210.

Sipos, M. L., Bar-Haim, Y., Abend, R., Adler, A. B., & Bliese, P. D. (2014). Postdeployment threat-related attention bias interacts with combat exposure to account for PTSD and anxiety symptoms in soldiers. *Depression and Anxiety, 31*, 124–129.

Spada, M. M., Georgiou, G. A., & Wells, A. (2010). The relationship among metacognitions, attentional control, and state anxiety. *Cognitive Behaviour Therapy, 39*, 64–71.

Staugaard, S. R., & Rosenberg, N. K. (2011). Processing of emotional faces in social phobia. *Mental Illness, 3*, 14–20.

Tangney, J., Baumeister, R., & Boone, A. (2004). High self-control predicts good adjustment, less pathology, better grades, and interpersonal success. *Journal of Personality, 72*, 271–324.

Todd, R. M., Cunningham, W. A., Anderson, A. K., & Thompson, E. (2012). Affect-biased attention as emotion regulation. *Trends in Cognitive Science, 16*, 365–372.

Van Bockstaele, B., Verschuere, B., Tibboel, H., De Houwer, J., Crombez, G., & Koster, E. H. (2014). A review of current evidence for the causal impact of attentional bias on fear and anxiety. *Psychological Bulletin, 140,* 682–721.

Verbruggen, F. (2016). Executive control of actions across time and space. *Current Directions in Psychological Science, 25,* 399–404.

Waechter, S., Nelson, A. L., Wright, C., Hyatt, A., & Oakman, J. (2014). Measuring attentional bias to threat: Reliability of dot probe and eye movement indices. *Cognitive Therapy and Research, 38,* 313–333.

Wager, T. D., Jonides, J., & Reading, S. (2004). Neuroimaging studies of shifting attention: A meta-analysis. *NeuroImage, 22,* 1679–1693.

Wald, I., Schechner, T., Bitton, S., Holoshitz, Y., Charney, D. S., Muller, D., … Bar-Haim, Y. (2011). Attention bias away from threat during life threatening danger predicts PTSD symptoms at one-year follow-up. *Depression and Anxiety, 28,* 406–411.

Weierich, M. R., Treat, T. A., & Hollingworth, A. (2008). Theories and measurement of visual attentional processing in anxiety. *Cognition & Emotion, 22,* 985–1018.

Wells, A. (1990). Panic disorder in association with relaxation induced anxiety: An attentional training approach to treatment. *Behavior Therapy, 21,* 273–280.

White, M., McManus, F., & Ehlers, A. (2008). An investigation of whether patients with post-traumatic stress disorder overestimate the probability and cost of future negative events. *Journal of Anxiety Disorders, 22,* 1244–1254.

Williams, J. M. G., Watts, F. N., MacLeod, C., & Mathews, A. (1997). *Cognitive psychology and the emotional disorders* (2nd ed.). New York, NY: Wiley.

Yiend, J. (2010). The effects of emotion on attention: A review of attentional processing of emotional information. *Cognition & Emotion, 24,* 3–47.

Yuval, K., Zvielli, A., & Bernstein, A. (2017). Attentional bias dynamics and posttraumatic stress in survivors of violent conflict and atrocities: New directions in clinical psychological science of refugee mental health. *Clinical Psychological Science, 5,* 64–73.

Zvielli, A., Bernstein, A., & Koster, E. H. W. (2015). Temporal dynamics of attentional bias. *Clinical Psychological Science,* (5), 772–788.

CHAPTER 12
Distress tolerance in PTSD

Anka A. Vujanovic, Maya Zegel
Department of Psychology, University of Houston, Houston, TX, United States

Introduction

Distress tolerance (DT) is a cognitive-behavioral factor that has received increasing attention in the trauma field due to its potential theoretical and clinical relevance to the etiology, maintenance, and treatment of posttraumatic stress disorder (PTSD), which will include subclinical PTSD symptoms for the purposes of this chapter (Vujanovic, Bernstein, & Litz, 2011; Vujanovic, Litz, & Farris, 2015). DT is defined as the perceived or actual ability to tolerate negative or aversive emotional or physical states (Leyro, Zvolensky, & Bernstein, 2010). The construct of DT can be conceptualized as a risk or resilience factor, as a clinical intervention target for PTSD and related conditions, or as an active ingredient for adjunctive skill-based treatments for established evidence-based interventions for PTSD.

The increased interest in DT is driven largely by its potential as a target for intervention in clinical problems when emotional dysregulation is prominent (e.g., Linehan, 1993). First a substantial proportion of individuals with PTSD do not seek treatment, drop out of treatment prematurely, and refuse treatment or do not manifest significant improvements in well-established evidence-based treatments (Imel, Laska, Jakupcak, & Simpson, 2013; Schottenbauer, Glass, Arnkoff, Tendick, & Gray, 2008). Second, individuals with PTSD may not feel "ready" to engage in trauma-focused treatment and may be referred to skill-based group treatment programs as a precursor or adjunct intervention; DT is a promising factor with significant clinical implications for such adjunctive interventions. Third, DT skills may serve as a beneficial preventive or early intervention for PTSD for populations with a high probability of exposure to intense or chronic potentially traumatic events (PTE), such as first responders or military personnel. Fourth, DT may be clinically relevant to the well-established associations between PTSD and suicidal ideation and behavior (Anestis, Tull, Bagge, & Gratz, 2012; Boffa, Short, Gibby, Stentz, & Schmidt, 2018; Viana, Woodward, Raines, Hanna, & Zvolensky, 2018; Vujanovic, Bakhshaie, Martin, Reddy, & Anestis, 2017; Vujanovic, Berenz, & Bakhshaie, 2018). Finally, DT may have clinical relevance to cooccurring PTSD and substance use disorders (SUD), a highly prevalent comorbidity with limited evidence-based intervention options (e.g., Roberts, Roberts, Jones, & Bisson, 2015), and/or PTSD and other comorbidities. Indeed, DT may function as a mechanism of treatment change, a predictor of treatment outcome, or a treatment component for either suicide interventions or PTSD/SUD treatment programs.

The goal of this chapter is to elucidate the construct of DT and its potential utility in conceptualizing, preventing, and treating PTSD (and related conditions) to stimulate further scholarly and clinical thought in this domain. The most salient research relevant to exposure to PTE and DT is reviewed. Suicidal ideation and behavior as well as substance use/SUD are examined as clinically pertinent conditions with growing empirical relevance to the DT-PTSD literature. The clinical implications of DT for the prevention and treatment of PTSD and related conditions are also discussed. Finally, several future research directions with the potential of informing both clinical and research efforts are suggested.

Theoretical framework

Although original conceptualizations of the DT construct suggested its trait-like stability over time and across contexts, emerging empirical and theoretical literature suggests that it may be context sensitive (e.g., Leyro et al., 2010) and amenable to change via intervention (e.g., Linehan, 1993). Thus several theoretical pathways have been posited to explain the potential associations between DT and PTSD (Vujanovic et al., 2015; Vujanovic, Bernstein, & Litz, 2011), which may help to inform the rationale for its continued study in the context of traumatic stress.

First, low levels of DT might predispose an individual to developing PTSD following exposure to PTE due to insufficient perceived or actual capacities to experience trauma-relevant distress. Generally, avoidant coping behaviors may be particularly negatively reinforcing for individuals with lower DT and may significantly impair healing and recovery due to missed opportunities to learn that trauma-related distress can be tolerated and regulated. Conversely, individuals with higher tolerance of emotional distress may be less likely to react with avoidance to PTSD symptoms following PTE, or such individuals may perceive their tolerance as greater for such emotional experiences, thus enabling them to cope more adaptively with trauma-related distress.

Second, DT levels may change or fluctuate as a function of exposure to PTE. It is important to note that a majority of individuals who experience a PTE do not develop PTSD (e.g., Kilpatrick et al., 2013). However, PTE experiences are ripe preconditions for changes in the perception of the ability to tolerate distress. For some a PTE might diminish a previously held perceived or actual ability to tolerate psychological or physical (e.g., racing heart and shakiness) distress. For others, living through a PTE might serve to bolster their previously held belief about or actual capacity to tolerate emotional distress, as evidenced by their ability to survive a previously unthinkable event.

Third, DT and PTSD symptoms may relate bidirectionally or transactionally following exposure to a PTE. For example, as an individual's PTSD symptoms increase or decrease in the aftermath of trauma, DT may follow suit. Fourth, it is possible that DT might operate differentially depending on the type of PTE endured (e.g., sexual assault and military combat), the severity or chronicity of the PTE, one's previous history of trauma or stress, and other salient factors separate from but related to the PTE and one's ability to recover adaptively from it (e.g., substance use and social support).

Finally, distress overtolerance may contribute to self-harming and self-handicapping behaviors, such as suicidality, in the aftermath of PTE (Anestis, Tull, et al., 2012; Lynch & Mizon, 2011). For example, distress overtolerance may be exhibited in individuals with an especially elevated propensity to persist in the face of adversity despite (a) high levels of distress, (b) evidence that the desired goal may not be achieved, or (c) indication that the effort of persistence may result in negative effects. Distress overtolerance should be distinguished from heightened DT, a theoretically adaptive state, by its extremity. This may be exemplified by individuals who maintain involvement in damaging or tumultuous relationships despite chronically enduring the negative effects and with evidence that those relationships will remain constant or deteriorate further. Another example of distress overtolerance is in the ability of certain individuals to tolerate the great emotional and physical distress or pain associated with nonsuicidal self-injury (NSSI) or suicidal behavior so much that it puts them at greater risk of severe suicidal acts (Anestis, Tull, et al., 2012).

Defining and measuring DT

A significant challenge in the DT literature is the operationalization and measurement of the construct. Due to the preponderance of self-report and behavioral indices of DT (Leyro et al., 2010), studies often include certain measures to the exclusion of others, precluding our ability to generalize and synthesize findings across indices. DT measures tend to be characterized by assessment modality (i.e., self-report vs behavioral assessment) and type of distress (i.e., physical vs psychological/emotional). Regardless of the type of distress, self-report DT measures tend to be moderately correlated (Bernstein, Zvolensky, Vujanovic, & Moos, 2009; McHugh et al., 2011; McHugh & Otto, 2012), and behavioral measures tend to be moderately correlated (Marshall-Berenz, Vujanovic, Bon-Miller, Bernstein, & Zvolensky, 2010; McHugh et al., 2011). However, self-report and behavioral measures are *typically* not significantly correlated (Anestis et al., 2012; Marshall-Berenz et al., 2010; McHugh et al., 2011). Given that measures of perceived (i.e., self-report) and behavioral DT are not necessarily assessing a single overarching DT construct, efforts are needed to better define the nature of these constructs and possible similarities and differences in their relations to PTSD symptomatology.

Notably the majority of the research on DT in the traumatic stress field has utilized self-report measures of DT. As discussed earlier, fewer than 10 studies to date have utilized multimodal DT measures, evaluated concurrently, in examining associations with PTSD. To better contextualize the literature reviewed in this chapter, a brief overview of DT measures used in the trauma-relevant literature is offered in Tables 1 and 2. A comprehensive overview of all available DT measures is beyond the scope of this chapter (see Leyro et al., 2010).

DT and PTSD symptoms

Associations between DT and PTSD symptoms have been evaluated across various populations, including community adults, undergraduate students, psychiatric inpatient adults, outpatient treatment-seeking adults, adolescents, and substance-using populations. The relevant literature on PTE or PTSD and DT is reviewed in the succeeding text by population type. Separate sections are devoted to discussing the emerging empirical and theoretical applications to suicidal ideation and behavior as well as substance use. See Table 3 for an overview of studies.

Table 1 Self-report measures of distress tolerance employed in trauma-related studies to date.

Measure	# Items	Format	Distress tolerance facet	Sample question
Distress Tolerance Scale (DTS; Simons & Gaher, 2005)	15	5-Point Likert-style scale (1 "strongly agree" to 5 "strongly disagree")	Perceived ability to withstand negative emotional states	See in the succeeding text
DTS: Tolerance	3		DTS subscale: perceived ability to tolerate emotion	"Feeling distressed or upset is unbearable to me"
DTS: Absorption	3		DTS subscale: level of attention absorbed by negative emotion and related functional interference	"My feelings of distress are so intense that they completely take over"
DTS: Appraisal	6		DTS subscale: assessment of emotional situation as acceptable	"My feelings of distress or being upset are not acceptable"
DTS: Regulation	3		DTS subscale: ability to self-regulate emotion	"I'll do anything to avoid feeling distressed or upset"
Discomfort Intolerance Scale (DIS; Schmidt, Richey, & Fitzpatrick, 2006)	5	7-Point Likert-style scale (0 "Not at all like me" to 6 "Extremely like me")	Perceived ability to withstand uncomfortable physical states	"I take extreme measures to avoid feeling physically uncomfortable"

Table 2 Behavioral distress tolerance tasks employed in trauma-relevant studies to date.

Tasks	Description	Distress tolerance facet
Mirror Tracing Persistence Task (MTPT; Quinn, Brandon, & Copeland, 1996)	This computer-based task requires participants to use a mouse to trace objects on the screen, as if viewing them through a mirror. Loud buzzer sounds when mouse goes beyond the lines. Distress tolerance is measured as the length of time (number of seconds) that participants persist in the task	Psychological distress (frustration) tolerance
Breath-Holding Task (Hajek, Belcher, & Stapleton, 1987)	This behavioral task requires participants to hold their breath for as long as possible. Distress tolerance is measured as the length of time that participants are able to hold their breath (Daughters, Lejuez, Kahler, Strong, & Brown, 2005; Hajek et al., 1987)	Physical discomfort tolerance
Behavioral Indicator of Resiliency to Distress (BIRD; Daughters et al., 2009; Lejuez, Daughters, Danielson, & Ruggiero, 2006)	This computer-based task requires participants to visually track and click on numbers associated with a green dot. Participants earn points for each time they click the correct number prior to the green dot disappearing. The task lasts 5 min, but participants can self-terminate by clicking a button on the computer screen. Lower distress tolerance is indexed by shorter latency to discontinue	Psychological distress (frustration) tolerance
Paced Auditory Serial Addition Task (Gronwall, 1977) Paced Auditory Serial Addition Task—Computerized Version (PASAT-C; Lejuez, Kahler, & Brown, 2003).	This computer-based task requires participants, over three levels (increasing in difficulty), to add numbers by continually summing the two most recently presented digits. Incorrect or missed responses result in a loud buzzing sound, and the latency between number of presentations decreases as level of difficulty increases. On the third level the participant can self-terminate the task—lower distress tolerance is indexed by shorter latency to discontinue the task	Psychological distress (frustration) tolerance

Table 3 Articles cited within this chapter examining distress tolerance among samples exposed to PTE.

In-text citation	Distress tolerance measure(s)	Trauma exposure measure(s)	PTSD symptom measure(s)
Anestis, Tull, et al. (2012)	PASAT-C	PTSD Criterion A trauma exposure (CAPS-IV)	CAPS-IV
Banducci, Bujarski, Bonn-Miller, Patel, and Connolly (2016)	DTS	MINI (*DSM-IV*)	PCL-S
Banducci, Connolly, Vujanovic, Alvarez, and Bonn-Miller (2017)	DTS	Self-reported trauma type(s)	PCL
Banducci, Lejuez, Dougherty, and MacPherson (2017)	BIRD	CTQ: childhood emotional abuse	N/A
Bardeen and Fergus (2016)	DTS	LEC-5	PCL-5
Bartlett et al. (2018)	DTS	LEC-5	PCL-5
Batchelder et al. (2017)	DTS	Davidson Trauma Scale	Davidson Trauma Scale
Berenz et al. (2018a)	DTS B-H Task PASAT-C	TLEQ	PCL-5
Berenz, Vujanovic, Coffey, and Zvolensky (2012)	B-H Task	LEC	CAPS-IV
Berenz et al. (2018b)	MTPT B-H Task DTS	LEC-5 CTQ	PCL-5
Boffa et al. (2018)	DTS	SCID-5	PCL-C
Chowdhury et al. (2018)	DTS DIS B-H task PASAT-C	TLEQ	N/A
Cohen, Danielson, Adams, and Ruggiero (2016)	BIRD	Participants were selected based on exposure to a particular natural disaster	The National Survey of Adolescents— Replication PTSD Module
Contractor, Weiss, Dranger, Ruggero, and Armour (2017)	DTS	SLESQ	PCL-5
Danielson, Ruggiero, Daughters, and Lejuez (2010)	BIRD	Self-reported trauma count and type(s)	UCLA PTSD RI

Continued

Table 3 Articles cited within this chapter examining distress tolerance among samples exposed to PTE—cont'd

In-text citation	Distress tolerance measure(s)	Trauma exposure measure(s)	PTSD symptom measure(s)
Duranceau, Fetzner, and Carleton (2014)	DTS	List of traumatic experiences	PCL-C
Erwin et al. (2018)	DTS	SLESQ	PCL-5
Farris, Vujanovic, Hogan, Schmidt, and Zvolensky (2014)	B-H Task	PDS	PDS
Fergus and Bardeen (2016)	DTS	LEC-5	PCL-5
Fetzner, Peluso, and Asmundson (2014)	DTS	TLEQ	PCL-C
Gerber et al. (2018)	DTS	SLESQ	PCL-5
Hashoul-Andary et al. (2016)	DTS	Carmel Trauma Questionnaire PDS	PDS
Himmerich and Orcutt (2019)	DTS	LEC-5	PCL-5
Holliday, Pedersen, and Leventhal (2016)	DTS	N/A	PCL-5
Hyland et al. (2017)	DTS	LEC-5 ITQ (Version 1.2)	PCL-5
Kraemer, Luberto, and McLeish (2013)	DTS	N/A	IDAS Traumatic Intrusions and Panic subscales
Marshall-Berenz et al. (2010)	DTS DIS MTPT B-H Task	LEC	CAPS-IV
Marshall-Berenz, Vujanovic, and Macpherson (2011)	DTS	LEC	CAPS-IV
Marshall-Berenz, Vujanovic, and Zvolensky (2011)	DTS	LEC	CAPS-IV
Pinciotti, Seligowski, and Orcutt (2017)	DTS	TLEQ	PSDS
Potter, Vujanovic, Marshall-Berenz, Bernstein, and Bonn-Miller (2011)	DTS	PDS	PDS
Powers et al. (2016)	DTS	PDS	PSS-I PDS

Table 3 Articles cited within this chapter examining distress tolerance among samples exposed to PTE—cont'd

In-text citation	Distress tolerance measure(s)	Trauma exposure measure(s)	PTSD symptom measure(s)
Tull, Gratz, Coffey, Weiss, and McDermott (2013)	PASAT-C	PTSD Criterion A trauma exposure (CAPS-IV)	CAPS-IV
Viana et al. (2018)	PASAT-C	PTSD Criterion A trauma exposure (CPSS)	CPSS
Vinci, Mota, Berenz, and Connolly (2016)	DTS	Self-report of "most distressing trauma"	PCL-S
Vujanovic, Bakhshaie, et al. (2017)	DTS	LEC-5	PCL-5
Vujanovic, Berenz, and Bakhshaie (2018)	DTS MTPT B-H Task	LEC-5	N/A
Vujanovic, Bonn-Miller, Potter, Marshall, and Zvolensky (2011)	DTS	PDS	PDS
Vujanovic, Dutcher, and Berenz (2017)	DTS MTPT B-H Task	LEC-5	PCL-5
Vujanovic et al. (2013)	DTS	PDS	PDS
Vujanovic, Marshall-Berenz, and Zvolensky (2011)	DTS	PDS	PDS
Vujanovic et al. (2018)	DTS MTPT B-H Task PASAT-C	LEC-5	CAPS-5

Note: B-H Task, Breath-Holding Task; BIRD, Behavioral Indicator of Resiliency to Distress; CAPS, Clinician-Administered PTSD Scale; CPSS, Childhood PTSD Symptom Scale; CTQ, Childhood Trauma Questionnaire; DIS, Discomfort Intolerance Scale; DTS, Distress Tolerance Scale; IDAS, Inventory of Depression and Anxiety Symptoms; ITQ, International Trauma Questionnaire; LEC-5, Life Events Checklist for DSM-5; MINI, Mini-International Neuropsychiatric Interview; MTPT, Mirror Tracing Persistence Task; PASAT-C, Paced Auditory Serial Addition Task—Computerized Version; PCL, Posttraumatic Stress Disorder Checklist for DSM-IV; PCL-5, Posttraumatic Stress Disorder Checklist for DSM-5; PCL-C, Posttraumatic Stress Disorder Checklist for DSM-IV (Civilian Version); PCL-S, Posttraumatic Stress Disorder Checklist for DSM-IV-TR (specific); PDS, Posttraumatic Diagnostic Scale; PSDS, PTSD Screening and Diagnostic Scale; PSS-I, PTSD Symptom Scale—Interview Version; SCID-5, Structured Clinical Interview for DSM-5 Disorders; SLESQ, Stressful Life Events Screening Questionnaire; TLEQ, Traumatic Life Events Questionnaire; UCLA PTSD RI, UCLA PTSD Reaction Index for DSM-IV.

Community adults

Broadly, among community samples endorsing exposure to PTE, negative (inverse) associations between DT and PTSD symptoms have been documented, such that lower DT is related to higher PTSD symptom severity and vice versa. One of the first studies in this domain focused on a sample of community-recruited adults exposed to PTE ($N=140$, 51% women). Perceived DT was incrementally and inversely associated with global PTSD symptom severity and severity of each of the *Diagnostic and Statistical Manual of Mental Disorders, Fourth Edition* (*DSM-IV*; American Psychiatric Association, 2000) PTSD symptom clusters, above and beyond the variance accounted for by number of PTE experienced and negative affectivity (Vujanovic, Bonn-Miller, et al., 2011). An extension of this work was conducted among a sample of adults exposed to PTE ($N=122$, 79.5% women) recruited from across North America via online social media (Fetzner et al., 2014). Participants completed online surveys. Partially consistent with findings from Vujanovic, Bonn-Miller, et al. (2011), DT was significantly incrementally associated with *DSM-IV* PTSD reexperiencing and avoidance symptom clusters only, after accounting for sex, number of years since the index trauma, number of types of PTE, and depressive symptoms. Discrepancies in results between the studies may be attributed to differences in sampling methodology, types of covariates included, and/or measures of PTE exposure and PTSD symptomatology (all self-report indices) implemented. Further work is needed to clarify these relations.

A multimethod study of DT in relation to PTSD symptoms among community adults exposed to PTE ($N=81$, 63.1% women) implemented two self-report measures and two behavioral indices of DT (Marshall-Berenz et al., 2010). Only the Distress Tolerance Scale (DTS; see Table 1) emerged as a significant incremental inverse predictor of PTSD symptoms. This study highlighted the relevance of *perceived* tolerance of negative emotional states in terms of PTSD symptom severity. Together, these preliminary studies support the premise that lower levels of DT are related to heightened PTSD symptom severity among community adults exposed to PTE, manifesting mostly subclinical levels of PTSD.

Notably, there is minimal research on DT among community survivors of *specific* types of trauma. At the time of publication, only one published study examined the association between DT and PTSD symptom severity among sexual trauma survivors, specifically. In a national sample of female survivors of sexual trauma ($N=101$), DT was significantly negatively correlated with PTSD symptom severity and symptom cluster severity (Fergus & Bardeen, 2016). While these findings are consistent with extant research among community adults, further research is needed, particularly among traditionally underrepresented groups, including racial/ethnic minorities and male survivors of sexual trauma.

Interactive effects

Several studies have examined the interactive effects of DT and panic-relevant symptoms or related cognitive-affective factors in terms of their impact on the manifestation of PTSD symptom severity. A series of studies was theoretically based on extant research indicating that exposure to PTE is related to panic vulnerability (i.e., heightened arousability; Brown & McNiff, 2009) and that individuals who experience PTE often experience panic symptomology (e.g., Nixon & Bryant, 2003). One of the first such studies was based on a community sample of adults exposed to various types of PTE ($N=91$, 62.6% women; Marshall-Berenz, Vujanovic, & Zvolensky, 2011). Self-reported DT and nonclinical panic attack history, defined as a past 2-year history of unexpected panic attacks not meeting diagnostic criteria for panic disorder, interacted to significantly relate to *DSM-IV* PTSD hyperarousal symptoms only (Marshall-Berenz, Vujanovic, & Zvolensky, 2011). Synergistically, lower levels of DT and a history of nonclinical panic attacks were related to the highest levels of PTSD hyperarousal symptoms. A follow-up study was conducted to examine the interactive effect of breath-holding duration and anxiety sensitivity (AS) in terms of PTSD symptom severity ($N=88$, 63.6% women; Berenz et al., 2012). AS is a cognitive factor defined as the fear of anxiety-related sensations (McNally, 2002) and studied extensively as a risk and maintenance factor for panic disorder and PTSD (e.g., Marshall, Miles, & Stewart, 2010; Schmidt, Zvolensky, & Maner, 2006). Results indicated that lower physical DT (shorter breath-holding durations) was a significant moderator of the association between higher levels of AS and greater PTSD avoidance symptom severity. Interestingly, no effects emerged for other PTSD symptom clusters.

Negative affect intensity, or the temperamental dimension reflecting individual differences in the activation and intensity of negative emotional responses (Larsen & Diener, 1987), has also been examined as a relevant factor that may interplay with DT in terms of PTSD symptom presentation. Specifically, one study examined the interaction of perceived (self-report) DT and self-reported negative affect intensity among adults exposed to PTE ($N=190$, 52.6% women; Vujanovic et al., 2013). Here, negative affect intensity was found to moderate (exacerbate) the effect of DT on PTSD symptom severity and specifically the severity of emotional numbing PTSD symptoms. Low levels of DT in the context of elevated levels of negative affect intensity were associated with the greatest levels of PTSD symptoms.

A recent study extended this literature to attentional control, defined as the ability to self-regulate externally triggered emotional reactions by executive control of one's attention, a factor with theoretical potential to mitigate the distressing effects of trauma (Bardeen & Fergus, 2016). A large, national online survey of adults with a history of PTE exposure ($N=903$, 67.6% women) demonstrated self-reported attentional control to be a moderator of the association between self-reported emotional DT and PTSD symptoms, such that greater attentional control decreased the strength of this association (Bardeen & Fergus, 2016). This investigation suggests that heightened attentional control may attenuate the association between low DT and PTSD symptoms.

Longitudinal studies

A significant criticism of the DT-PTSD literature is the dearth of published longitudinal studies. One of the few longitudinal studies measured distress at three time points in a sample of adults ($N=151$, 76% women) who had experienced a natural disaster (Hashoul-Andary et al., 2016). All participants had recently and simultaneously endured the same type of trauma, and all were recruited from the same community. Participants were considered for the study if they had been impacted by the disaster and felt fearful, horrified, or helpless at the time, thus meeting *DSM-IV* Criterion A for PTSD. Approximately 83% of participants were retained to the 6-month follow-up. Higher levels of emotional avoidance and distress 30-day postdisaster exposure were associated with greater distress and lower perceived (self-report) DT at the 3-month follow-up. Furthermore, lower perceived DT at the 3-month follow-up was subsequently predictive of greater distress posttrauma, measured at the 6-month follow-up (Hashoul-Andary et al., 2016). These results suggest that the level of DT *pretrauma* or immediately *posttrauma* may predict long-term mental health outcomes. In addition, this study elucidates the potential clinical relevance of DT interventions for those exposed to PTE who are at risk of developing more severe PTSD symptoms.

Undergraduate students

Consistent associations between DT and PTSD symptoms have been documented among undergraduate students. For example, in a sample of 322 PTE-exposed undergraduate students (72% women), self-reported DT was significantly negatively correlated with self-reported symptom severity of the *DSM-IV* PTSD reexperiencing, avoidance, and hyperarousal symptom clusters (Pinciotti et al., 2017). This investigation replicates and extends to undergraduate students past work among community adults exposed to PTE (Fergus & Bardeen, 2016; Vujanovic, Bonn-Miller, et al., 2011). Given that the study did not collect data on the duration of time elapsed since trauma exposure, this presents a fruitful potential avenue for future research.

In addition to documenting relations between DT and PTSD symptoms in undergraduate students, it is important to understand associations between specific types of trauma and DT in this population. A recent study examined the association between types of childhood trauma and self-reported and behavioral DT among undergraduate students ($N=320$, 75% women; Berenz et al., 2018a). Greater frequency of childhood physical abuse was related to increased perceived DT; however, witnessing family violence more frequently was linked to lower behavioral DT, as indexed by the Breath-Holding Task and not the Mirror-Tracing Persistence Task-Computerized Version (MTPT-C; Quinn et al., 1996). It is possible that those who have endured childhood physical abuse may have a greater perceived ability to withstand distress, while those who experienced family violence as children may associate physical arousal symptoms with fearful or angry emotions. The type and frequency of childhood trauma experienced can have a differential effect on both behavioral and perceived DT. Given that childhood is composed of multiple critical development periods, age at the time of the trauma and the duration and frequency of the trauma should be accounted for in future research. As corroborated by other studies, the specificity of trauma type may be important in understanding DT-PTSD associations.

Another study sought to investigate the interactive effect of DT and AS-physical concerns (i.e., fear of anxiety-related physical sensations, specifically) in terms of PTSD symptoms among undergraduate students (Kraemer et al., 2013). Among this sample of students ($N=416$, 72.1% women), PTSD symptoms were defined as "traumatic intrusions" and measured by the self-report inventory of depression and anxiety symptoms (IDAS; Watson et al., 2007). This study found a significant incremental relation between DT and traumatic intrusions, after covarying for negative affectivity; however, no evidence of an interactive effect between DT and AS-physical concerns emerged regarding "traumatic intrusions." Notably, this investigation did not sample adults with PTE, specifically, which likely influenced findings; these undergraduate participants may or may not have had a history of exposure to PTE.

Few studies to date have examined personality differences that may affect the association between PTE exposure and DT. One study of undergraduate students with a history of at least one PTE ($N=440$, 71.4% women) found that lower perceived DT was associated with higher levels of neuroticism and lower levels of conscientiousness, when controlling for sex and trauma load (Chowdhury et al., 2018). The associations between these personality factors and other measures of DT, including the DIS, were not significant. However, higher levels of extraversion were significantly related to greater DT, as measured by the DIS and Paced Auditory Serial Addition Task (PASAT-C; Lejuez et al., 2003). Openness was positively correlated with behavioral DT, indexed by the Breath-Holding Task. Consistent with prior research, DT and trauma load were significantly negatively correlated in this sample. Notably, this study did not report associations between DT and the examined personality factors with PTSD symptoms. Further research is needed to elucidate the effect of personality factors on DT and resiliency after exposure to PTE.

Psychiatric inpatient adults

Three published studies to date have evaluated DT-PTSD associations in general acute-care psychiatric inpatients. Vujanovic, Dutcher, and Berenz (2017) conducted a multimodal examination of DT and PTSD symptoms in a sample of acute-care psychiatric inpatients with a history of trauma exposure ($N=103$, 41.7% women). This investigation found that, as expected, self-reported DT was negatively associated with PTSD symptom severity, after accounting for trauma load, substance use, gender, race, and subjective social status (i.e., rating of perceived social standing on 10-point scale) as covariates (Vujanovic, Dutcher, & Berenz, 2017). Additionally, higher levels of behavioral DT, as measured by the Breath-Holding Task, were related to more severe PTSD arousal symptoms. Interestingly, this study did not find any gender differences in perceived DT, yet women demonstrated lower levels of behavioral DT on the Breath-Holding Task and MTPT-C. It was also noted that women reported greater PTSD symptom severity, specifically PTSD avoidance symptoms, which could influence this gender comparison.

As previously mentioned (e.g., Berenz et al., 2018a), DT-PTSD associations can vary by trauma type. A recent study examined self-reported (i.e., perceived) and behavioral DT in a sample of psychiatric inpatients who reported histories of childhood trauma ($N=87$, 40.2% women; Berenz et al., 2018b). Greater perceived DT was related to higher levels of physical abuse and emotional neglect. However, *lower* perceived DT was associated with greater *emotional* abuse, and lower *behavioral* DT was linked to higher levels of *physical* neglect. More research is needed to understand the developmental associations between specific types of trauma and behavioral or perceived DT, as such relations can have a significant impact on risk or resilience processes for various types of psychopathology, including PTSD.

Outpatient treatment-seeking adults

A single exposure to a traumatic event is sufficient for an individual to develop PTSD. However, researchers have shown that, among treatment-seeking adults ($N=123$, 68.3% women), the number of PTE types was significantly inversely correlated with self-reported (perceived) DT (Gerber et al., 2018). Additionally, all DTS subscales, except DTS tolerance, were significantly associated with number of types of PTE. This finding is consistent with the broader literature, which supports significant negative associations between number of traumas or number of trauma types and DT (Berenz et al., 2018a, 2018b; Erwin et al., 2018). Additional studies have examined DT as it relates to PTSD symptoms in treatment-seeking adults (Contractor et al., 2017). A clinical sample ($N=123$, 68.3% women) drawn from a Midwestern community health center was sorted by the endorsement of the PTSD "reckless and self-destructive behavior" criterion (E2) at a clinically significant level. Those who endorsed PTSD criterion E2 reported lower scores on the absorption and regulation subscales of the DTS, as compared with those who did not endorse E2 (see Table 1 for subscale descriptions). More research is needed to clarify distinctions between DTS subscales and aspects of emotion regulation.

Extensions of this work have proposed more complex models explicating the potential interplay of DT and other theoretically pertinent factors in relation to PTSD symptomatology. For example, one study conducted with treatment-seeking adults ($N=119$, 68.1% women) at a Midwestern community health center found the regulation subscale of the DTS was inversely associated with each of the *DSM-5* PTSD symptom clusters, when controlling for depression and number of PTE (Erwin et al., 2018). A secondary analysis indicated counterfactual rumination was a significant mediator of the association between global DT and the PTSD avoidance and intrusion symptom clusters. In this study, counterfactual thinking (CFT) was defined as the act of mentally playing out various scenarios for situations, regardless of whether those situations had already occurred or not. For individuals who have experienced a PTE, CFT can be beneficial or detrimental, depending on the level of perseveration and the positive or negative frame the individual has placed on a newly imagined scenario. Participants for this study were drawn from a sample of first-time treatment-seeking individuals with a history of PTE, possibly limiting generalizability.

Notably, changes in DT and PTSD have been documented in the context of two residential PTSD programs for military veterans ($N_1 = 53$, 82.7% men and $N_2 = 33$, 97% men; Banducci, Connolly, et al., 2017). These treatment program facilities were located in different geographic regions, and one program was designed to treat comorbid PTSD/SUD. Despite the differences between facilities, results showed that veterans with the greatest increases in DT from pre- to posttreatment were found to have the lowest PTSD symptom severity upon being discharged. Causality cannot be inferred from the study design, but this research highlights that (a) DT levels may change during PTSD treatment and (b) changes in DT are related to changes in PTSD symptoms in the context of treatments.

Adolescents

Three studies to date have examined the relevance of DT in terms of PTSD symptom severity among youth exposed to PTE. A pilot study among 24 youth exposed to PTE (12 girls and 12 boys) documented large effect sizes for the association between behavioral DT, as indexed via the Behavioral Indicator of Resiliency to Distress (BIRD) task, and PTSD avoidance symptoms in girls as compared with boys (Danielson et al., 2010). Specifically, PTE-exposed girls with low DT, as compared with girls with high DT, reported higher levels of PTSD avoidance symptoms. This effect was not found for boys, suggesting that relations between DT and PTSD among youth may be gender specific in some contexts. Among a sample of adolescents exposed to a natural disaster ($N = 352$, 55.4% girls), those with decreased behavioral DT, as indexed by the BIRD, in conjunction with low levels of perceived social support, had the highest risk for developing PTSD symptoms when assessed at 12 and 20 months post trauma (Cohen et al., 2016). Given the potential variability of outcomes based upon trauma type, future research is needed to assess adolescent DT in the context of different forms of potentially traumatic events.

Finally a longitudinal study annually assessed adolescents between the ages of 10–14 ($N=244$, 45% girls) over the course of 5 years (Banducci, Lejuez, et al., 2017). Results demonstrated that baseline behavioral DT was related to the greatest levels of anxiety every year for those with higher levels of childhood emotional abuse (Banducci, Lejuez, et al., 2017). Notably, this study did not directly assess whether participants experienced PTSD Criterion A events; only childhood emotional abuse was queried. Extant research on DT among community adolescents exclusively employed the BIRD, suggesting a dearth of well-established measures for self-reported adolescent DT.

Applications to suicidal ideation and behavior

PTSD is associated positively with both suicidal ideation and suicide attempts (e.g., Belik, Cox, Stein, Asmundson, & Sareen, 2007; Krysinska & Lester, 2010; Nock et al., 2014; Nock, Prinstein, & Sterba, 2009). Indeed, PTSD may be among few disorders that can independently differentiate suicidal ideation from suicide attempts (Bernal et al., 2007; Bryan, Grove, & Kimbrel, 2017; Wilcox, Storr, & Breslau, 2009). Given that the association between PTSD and suicidal ideation and behavior is well-established, an emerging literature has sought to understand the role that DT may play in this association to inform suicide prevention and intervention efforts.

Only one study to date examined DT as a potential mechanism of change in the context of interventions designed to target suicide risk among adults with PTSD (Boffa et al., 2018). Within a sample of treatment-seeking adults with PTSD and elevated suicide risk ($N=54$, 61.1% women), participants were randomized into one of four treatment conditions: anxiety condition (i.e., Cognitive AS Treatment and Cognitive Bias Modification Program [CBM] for AS), mood condition (i.e., computerized mood intervention and CBM-Mood Interpretation Bias), and repeated contact control condition (i.e., repeated in-person assessments). While there were no significant differences between treatment groups regarding change in perceived DT from pre- to postintervention, change in perceived DT was predictive of PTSD symptom severity at 3-month follow-up. Change in perceived DT thus significantly mediated the effect of treatment condition group on PTSD symptom severity at 3-month follow-up, suggesting long-term treatment implications for DT-based interventions. Future research similar to this work is essential, as DT may be a pertinent mechanism of change, related to reductions in PTSD symptoms and in suicide risk interventions for populations with PTSD.

The emergent contextual literature suggests differential associations between self-reported DT versus behaviorally indexed DT with regard to the association between PTSD symptomatology and suicidal behavior (Anestis, Tull, et al., 2012; Bartlett et al., 2018; Bender, Anestis, Anestis, Gordon, & Joiner, 2012; Vujanovic, Bakhshaie, et al., 2017). *Higher* levels of behaviorally indexed DT (i.e., PASAT-C) significantly moderate—or exacerbate—the association between PTSD symptom severity and number of past suicide attempts in adult substance users in residential treatment ($N=164$; 44.5% women; Anestis, Tull, et al., 2012). Heightened behavioral DT has also been shown to intensify the association between PTSD and past medically attended suicide attempts, above and beyond gender, age, history of NSSI, and the presence of mood disorder, anxiety disorder, cocaine dependence, and borderline personality disorder (Anestis, Tull, et al., 2012). Conversely, among firefighters ($N=765$; 6% women), *lower* levels of *self-reported* (perceived) DT exacerbate associations between PTSD symptom severity and suicide risk—defined by suicidal ideation and behavior, after controlling for trauma load, depressive symptom severity, and sociodemographic factors (i.e., gender, race, education, and age; Bartlett et al., 2018).

Furthermore, among acute-care psychiatric inpatients ($N=105$, 52.2% women), PTSD symptom severity indirectly affected suicidality (i.e., ideation, intent, or behavior) as a basis for hospitalization and suicidal desire severity through perceived DT (Vujanovic, Bakhshaie, et al., 2017), after controlling for gender, number of trauma types, number of psychiatric diagnoses, and substance use. A complementary study among a sample of adolescent inpatients with a history of PTE in acute psychiatric care ($N=50$, 52% girls) evaluated associations between emotional clarity, NSSI, DT, and suicidal behavior. Among hospitalized adolescents, increased NSSI behaviors were strongly associated with a greater number of previous suicide attempts. In this sample, lower behavioral DT, indexed via the PASAT-C, significantly exacerbated the association between emotional clarity and NSSI, when controlling for prior suicide attempts (Viana et al., 2018).

Among psychiatric inpatients exposed to PTE ($N=102$, 44.1% women), various indices of DT (DTS, MTPT-C, and Breath-Holding) were examined multimodally and concurrently in relation to number of past suicide attempts, self-reported suicidal ideation severity, and suicidality as reason for the current hospital admission (Vujanovic, Berenz, & Bakhshaie, 2018). After covarying for gender, number of psychiatric diagnoses, number of drug classes used, and number of PTE, MTPT-C duration was negatively associated with number of past suicide attempts, while perceived DT was negatively associated with suicidal ideation severity and suicidality as reason for the current admission. Unfortunately, patients' self-report was used to derive the "suicidality as reason for current admission" variable, which was ultimately composed of either ideation or behavior that was severe enough to lead to hospitalization. This methodological limitation impeded conclusions pertaining to specificity of findings regarding suicidal ideation versus behavior. Notably, at the bivariate level, trauma load (i.e., number of PTE types experienced) was significantly negatively associated with perceived (vs behavioral) DT and positively with suicidal ideation severity only. This is consistent with past work (Gerber et al., 2018), suggesting inverse associations between number of PTE and DT, and other work documenting associations between number of PTE and suicidal ideation (Afifi et al., 2016; Kimerling, Makin-Byrd, Louzon, Ignacio, & McCarthy, 2016; Krysinska, Lester, & Martin, 2009; McMahon et al., 2018).

Taken together, it may be that lower levels of DT are associated with suicidal ideation/desire but that this association is potentially mediated by NSSI or other painful and provocative experiences (e.g., exposure to PTE), which in turn account for suicidal behaviors or attempts (Anestis, Bender, Selby, Ribeiro, & Joiner, 2011; Anestis, Gratz, Bagge, & Tull, 2012; Anestis & Joiner, 2012; Anestis, Knorr, Tull, Lavender, & Gratz, 2013; Anestis, Pennings, Lavender, Tull, & Gratz, 2013; Anestis, Tull, et al., 2012; Vujanovic, Berenz, & Bakhshaie, 2018). Furthermore, individuals low in perceived DT experience life stressors as being more unmanageable than those high in DT, thereby leading to greater suicidal ideation as a means of escape. Recent work suggests that perhaps it is not any DT facet in isolation but a physical pain/emotional DT combination, indexed via higher levels of persistence through painful *and* distressing tasks, that might amplify the association between current suicidal ideation and lifetime suicide attempts (Anestis & Capron, 2016). Future studies of suicidality among populations exposed to PTE or those with PTSD might benefit from enhanced and separate assessment of suicidal ideation and behavior to allow examination of the DT facets that are differentially related to ideation versus behavior and the role of PTSD symptoms in those associations.

Overall, extant literature has documented the association between DT and suicidality among those exposed to PTE; however, minimal research has been conducted comparing men and women, inpatients and outpatients, military personnel and civilians, differences in professions, and variability across cultures, to name a few. Furthermore, there is a dearth of studies examining associations between PTE and/or PTSD, DT, and suicidal ideation and behavior among children or adolescents, leaving a large gap in developmental perspectives of these phenomena. It is necessary for future studies to examine the association between DT and suicidality among various populations, given the contextual importance of demographics, culture, and environment.

Applications to substance use

The comorbidity of PTSD and SUD is complex and highly prevalent (e.g., McCauley, Killeen, Gros, Brady, & Back, 2012), as it is estimated that one in two individuals with either disorder will meet criteria for the other disorder (Vujanovic & Back, in press). The comorbidity is challenging, difficult to treat, and marked by a more costly and chronic clinical course, when compared with either disorder alone (e.g., McCauley et al., 2012; Mills, Teesson, Ross, & Peters, 2006; Schäfer & Najavits, 2007). DT has presented a clinically relevant factor with promise in terms of better understanding the PTSD/SUD comorbidity and improving extant evidence-based interventions.

Community adults

The first published work directly citing DT as a pertinent construct of interest for PTSD populations was a theoretical review of factors with potential to influence smoking relapse among individuals with PTSD (Cook, McFall, Calhoun, & Beckham, 2007). Here, it was postulated that DT might be a significant factor related to smoking abstinence among smokers with PTSD, with lower levels of DT, specifically, giving rise to the increased risk of lapse or relapse among smokers with PTSD attempting cessation (Cook et al., 2007). In a follow-up empirical study, Vujanovic, Marshall, Gibson, and Zvolensky (2010) found significantly higher levels of self-reported physical discomfort tolerance among smokers with PTSD ($N=123$, 62% women) relative to smokers with nonclinical panic attacks only and those without current Axis I psychopathology per the *DSM-IV*.

Research also has examined associations between DT and PTSD symptoms in other substance-using populations. For example, a recent study examined such relations in cocaine-dependent adults with a history of PTE ($N=138$, 81% men) (Vujanovic, Rathnayaka, Amador, & Schmitz, 2016). In this sample, DT was significantly negatively associated with PTSD symptom severity, after controlling for gender, past-month cocaine-use severity, depression symptoms, and trauma load. Another study examined the association of DT, PTSD symptoms, and alcohol consumption among a community-based sample of participants exposed to PTE ($N=146$, 81%

women) who self-identified as using alcohol to cope with anxiety symptoms (Duranceau et al., 2014). Participants completed an online survey, which included measures of exposure to PTE and PTSD symptom severity, alcohol consumption, and DT. Although there was no direct effect of DT on alcohol use, this study found an indirect effect of DT on alcohol consumption via the hyperarousal cluster of PTSD symptoms. The authors concluded that, among individuals exposed to PTE, those with low DT who manifest PTSD hyperarousal symptoms may feel a particular need to consume alcohol, perhaps to avoid negative affect.

Associations between DT and PTSD symptoms also have been explored with regard to coping-oriented motives for substance use. In this context, DT has emerged as a potential explanatory factor (i.e., statistical mediator) in several studies. For example, one study found that DT partially mediated the association of PTSD symptoms and coping-oriented alcohol use among a community sample of adults reporting exposure to PTE ($N=83$, 63.8% women; Vujanovic, Marshall-Berenz, & Zvolensky, 2011). Additionally, in a sample of adults exposed to PTE who reported alcohol use ($N=86$, 64.3% women), DT partially mediated the association between impulsivity and alcohol use coping motives, after controlling for the variance explained by PTSD symptom severity and alcohol use problems (Marshall-Berenz, Vujanovic, & Macpherson, 2011). Furthermore, among community-recruited marijuana users with a history of PTE ($N=142$, 46.5% women), DT partially mediated the association between PTSD symptom severity and coping motives for marijuana use (Potter et al., 2011). Taken together, these studies indicate that DT may at least partially account for the relations between PTSD symptoms and coping-oriented substance use, such that lower DT might aid in the explanation (at least in part) of the associations between PTSD symptom severity—or related factors, such as impulsivity—and substance use to cope with negative affective states among adults exposed to PTE. In other words, among individuals exposed to PTE, PTSD symptoms may lead to coping-oriented substance use, due to lower levels of DT. Notably, however, all studies employed a cross-sectional design, limiting any inferences about temporal relations or causality among variables.

Undergraduate students

Among a sample of undergraduate students reporting a history of PTE ($N=318$, 56% men), perceived DT was significantly inversely correlated with PTSD symptom severity (Himmerich & Orcutt, 2019). DT was also significantly negatively associated with alcohol use and positive alcohol expectancies.

These associations remained true when controlling for gender, age, race, and education. Lastly, DT moderated the indirect effect of PTSD symptoms on alcohol use through negative changes in cognition and affect, possibly suggesting that trauma-related negative alterations in cognitions and mood are more salient for individuals with higher perceived abilities to withstand negative emotional states. It is important to note that 23.1% of participants screened positive for PTSD diagnosis via self-report (Himmerich & Orcutt, 2019). Further research is needed to understand the role of DT in the well-established association between PTE or PTSD symptoms and alcohol use in undergraduate students.

Treatment-seeking adults

There is also some evidence that DT may impact SUD treatment outcomes among adults with PTSD. Specifically, in a sample of substance-dependent patients enrolled in a residential SUD treatment program ($N=214$, 63% men), a significant three-way interactive effect emerged whereby men (relative to women) with a current diagnosis of PTSD (relative to no PTSD) and lower DT (shorter durations on the PASAT-C) completed a significantly lower proportion of SUD treatment (Tull et al., 2013). Interestingly, in this study, the main effects of PTSD, lower DT, and male gender were not significantly related to SUD treatment completion, suggesting more complex associations among PTSD, DT, and SUD treatment (Tull et al., 2013).

More recently, research was conducted among a sample of treatment-seeking, low-income, inner-city adults ($N=58$, 49.1% women), assessing cue reactivity and DT multimodally (Vujanovic, Wardle, et al., 2018). Participants in this sample met criteria for current substance dependence (per the *DSM-IV*), endorsed at least four current PTSD symptoms (per the *DSM-5*; American Psychiatric Association, 2013), and had a history of PTE exposure. Consistent with the literature, perceived DT was significantly negatively correlated with interview-based PTSD symptom severity. In the presence of trauma-related script-driven imagery cues, participants with lower behaviorally indexed psychological DT (MTPT-C duration) reported significantly greater levels of craving/urges to use substances, when controlling for gender, PTSD symptom severity, and number of SUD diagnoses. In addition, lower DTS scores predicted lower levels of self-reported control/safety ratings in response to substance-related script-driven imagery cues. These results support the self-medication model of substance use, such that those with lower DT may use substances to mitigate the distressing effects of exposure to trauma-related cues and may feel less safe or in control when exposed to substance cues.

Furthermore, among a sample of smokers exposed to PTE recruited for a smoking cessation treatment study ($N=137$, 48.2% women), the effect of DT was examined in terms self-reported PTSD symptom severity (Farris et al., 2014). Results indicated that DT was not directly associated with PTSD symptom severity. However, among smokers who were high in AS, lower DT (i.e., shorter breath-holding duration) was associated with the most severe PTSD hyperarousal symptoms. A randomized clinical trial (Powers et al., 2016) is currently being conducted to evaluate the efficacy of a novel, integrated smoking cessation treatment for adult smokers with PTSD ($N=80$). This study will examine DT, AS, and PTSD symptom severity associations longitudinally in the context of the clinical trial and at follow-up.

Overall, DT appears to be a factor of clinical significance to PTSD/SUD associations. DT is related to coping-oriented substance use and may thus be a risk or protective factor for the development of SUD among individuals exposed to PTE or clinical PTSD populations. Also, DT may function as a predictor of PTSD/SUD treatment outcomes or as a mechanism of change in treatment. Further work is needed, using longitudinal and experimental methodologies and among varied populations, before more definitive conclusions can be drawn.

Military veterans

A large proportion of extant research on DT in the context of PTSD/SUD has been conducted among military veterans. Consistent with findings across populations, perceived DT is significantly negatively correlated PTSD in military veterans. Among young adult veterans ($N=783$, 83.3% men; Holliday et al., 2016) with probable PTSD, as defined by a score of 33 or more on the PTSD Checklist (PCL-5; Bovin et al., 2015), negative associations between DT and PTSD were documented. In the same study, perceived DT was inversely associated with alcohol use severity. When separated by gender, these correlations remained significant; however, the strength of the correlations was weaker for women, possibly due to a smaller sample size. Interestingly, DT was a significant mediator in the association between probable PTSD symptomatology and problematic alcohol use, when controlling for gender, race, and ethnicity. Similarly, among a sample of male veterans ($N=75$) in a PTSD/SUD treatment program at a southern VA facility (Vinci et al., 2016), perceived DT was negatively related to PTSD symptom severity, even when accounting for depression and substance use severity. Specifically, low perceived DT was significantly associated with higher PTSD intrusion and arousal symptoms. Thus, it was conjectured that individuals with a lower perceived ability to cope with distressing PTSD symptoms, including memories of trauma and arousal, may be more inclined to use substances to numb or avoid this distress. Furthermore, among a sample of veterans in a PTSD/SUD treatment program at a southern VA facility ($N=70$, 95.5% men; Banducci et al., 2016), perceived DT was significantly negatively correlated with both PTSD symptom severity. In this study, lower DT also was significantly associated with greater substance cravings in response to trauma cues. Further understanding of trauma-related cues and subsequent craving will inform more effective treatment for cooccurring PTSD and SUD.

A unique study examined the association between condomless sex and PTSD symptom severity, accounting for DT and substance use, among a population of men who have sex with men (MSM) with a history of childhood sexual abuse (CSA; $N=288$; Batchelder et al., 2017). Participants with substance dependence reported a higher proportion of condomless sex for all levels of PTSD symptom severity; however, among participants without substance dependence, greater PTSD symptom severity was associated with a higher proportion of condomless sex compared with those with lower PTSD symptom severity. Among participants with greater PTSD symptom severity, both high and low levels of perceived DT were related to condomless sex. This suggests a potential parallel to the DT-suicide literature, such that both low and high levels of DT may be related to higher risk-taking and self-destructive behaviors. Additional replications of this work are necessary to disentangle whether DT-based interventions for high-risk sexual behaviors among PTSD/SUD populations are relevant.

Limitations and future directions

Despite mounting interest in the role of DT in the context of traumatic experience, the extant research is limited in several key ways that should be addressed by future work to promote a more clinically meaningful understanding of the DT construct. Most studies to date have used cross-sectional designs and self-report methodologies. Future studies using multimodal approaches to examine the association between DT and PTSD symptoms are needed. More studies utilizing interview-based (diagnostic) measures of PTSD, such as the Clinician-Administered PTSD Scale (Weathers et al., 2013), are needed to assess the association between DT and PTSD. Interview-based indices of PTSD are less subject to various self-report biases and share less method variance with paper-and-pencil questionnaires. Research has focused mostly on adults reporting exposure to PTE with varying, often subclinical, levels of PTSD, thus limiting our ability to generalize to clinical populations with PTSD. Future work focused on delineating the relevance of DT to specific types of PTE and to clinical populations with PTSD, for example, might be fruitful.

Further research is necessary to understand associations between DT and specific PTSD symptom criteria. Some studies in this chapter have found DT to be significantly associated with certain clusters of PTSD criteria but not others (Erwin et al., 2018; Fetzner et al., 2014; Pinciotti et al., 2017; Vinci et al., 2016; Vujanovic, Bonn-Miller, et al., 2011). Furthermore, in defining complex PTSD (CPTSD) criteria, perceived DT was evaluated with regard to disturbances in self-organization (DSO), composed of three factors including affective dysregulation, negative self-concept, and disturbed relationships (Hyland et al., 2017). Specifically, among adults who reported PTSD Criterion A trauma exposure and either a prior PTSD diagnosis or a positive PTSD diagnostic screen, DSO (but not PTSD) was negatively associated with DT, such that those with greater DSO symptoms had lower levels of DT (and vice versa).

Currently, no intervention or prevention studies have been published delineating the outcomes of a DT intervention in a population exposed to PTE nor any describing the efficacy of a DT intervention on PTSD symptoms. Intervention or prevention studies targeting DT are necessary to more directly test the clinical significance of the construct. Studies utilizing experimental and prospective methodologies are needed to better establish the temporal relations between DT and symptoms of PTSD or other psychological disturbances among individuals exposed to PTE. Neurobiological and genetic studies relevant to DT in the context of exposure to PTE, PTSD, or PTSD/SUD symptoms are also warranted.

In clinical contexts, DT is targeted directly through several treatment programs, including dialectical behavior therapy (DBT; Linehan, 1993), unified treatment for emotional disorders (Allen, McHugh, & Barlow, 2008), acceptance-based emotion regulation group therapy (Gratz & Gunderson, 2006), and acceptance and commitment therapy (Hayes, Wilson, Gifford, Follette, & Strosahl, 1996). Notably, DT may be targeted secondarily in other interventions, including mindfulness-based therapies, exercise interventions, and exposure-based or cognitive-emotional processing therapies. Individuals undergoing such treatments might increase their awareness of cognitive and emotional experiences; tolerance of physiological sensations; or ability to understand and confront trauma-related cognitions, memories, and emotions. In so doing, it is plausible that individuals will increase their ability to tolerate negative emotional states. Relevant clinical research studies might integrate assessments of DT to allow for the examination of this factor as a potential mechanism of change in the context of treatment. Alternatively, it is interesting to consider that DT interventions could be offered prior to evidence-based treatments for PTSD, which may be useful in improving treatment engagement and completion. Furthermore, DT may be a significant factor to consider in terms of prevention and early intervention efforts for populations exposed to PTE, such as military personnel or first responders. Similarly, providing DT skill training in an indicated prevention framework for individuals experiencing early subsyndromal levels of PTSD in the aftermath of a PTE might reduce incidence of psychopathology.

Furthermore, studies with more diverse populations are necessary to increase external validity of this literature. For example, greater research should be conducted among children and youth, among specific types of PTE (e.g., sexual assault), among more varied types of SUD patients, and among demographically diverse populations as DT may manifest differently across cultures. Relatedly the role of gender in the context of relations between DT and PTSD symptomatology should be further investigated, as extant work suggests a potential moderating role of gender. There remains a dearth of research among male survivors of sexual trauma and female veterans, for example. The literature could also benefit from more research focused upon first responders (e.g., police and firefighters) or other populations chronically exposed to trauma, as emerging research is suggesting that such populations may report greater levels of perceived DT as compared with community samples (Bartlett et al., 2018). This may indicate that individuals with higher levels of DT self-select into such stressful occupations, that such occupations demand a higher perceived level of emotional tolerance and control, or that mental health stigma may moderate reports of DT and mental health symptoms.

In research contexts, it is important to utilize multimodal assessments of DT to promote a more comprehensive understanding of the construct. Most of the research to date is limited in that studies implemented only one or two measures that might assess specific aspects of DT, thus curtailing our ability to generalize to other aspects of the construct. It is plausible that individuals suffering from PTSD symptoms may perceive their capacity for DT as lower than their actual capacity due to maladaptive, trauma-related alterations in cognitions about self-efficacy, for example.

Of note, DT can be assessed through additional self-report and behavioral measures that have not yet been employed in the traumatic stress literature. For example, there are at least two recently developed self-report measures of relevance, including the Distress Intolerance Index (DII; McHugh & Otto, 2012) and the Distress Overtolerance Scale (DOS; Gorey, Rojas, & Bornovalova, 2018). Both measures demonstrate good psychometric properties based upon preliminary studies. To the best of our knowledge, no studies of samples exposed to PTE have implemented the DII or DOS to date, presenting an area for empirical growth in this literature. Both measures have been evaluated in terms of relations to substance use behavior, suggesting relevance to PTSD/SUD populations, as well. Further, additional measures of physical discomfort tolerance, such as the Cold Pressor Task (Leyro et al., 2010) or pain algometers, may be relevant to this literature but have not yet been evaluated in samples reporting PTE. Such indices may help clarify distinctions between perceived and actual DT and potential associations between pain tolerance and PTSD/SUD or PTSD-suicide associations.

References

Afifi, T. O., Taillieu, T., Zamorski, M. A., Turner, S., Cheung, K., & Sareen, J. (2016). Association of child abuse exposure with suicidal ideation, suicide plans, and suicide attempts in military personnel and the general population in Canada. *JAMA Psychiatry, 73*(3), 229–238. https://doi.org/10.1001/jamapsychiatry.2015.2732.

Allen, L. B., McHugh, R. K., & Barlow, D. H. (2008). Emotional disorders: A unified protocol. In D. H. Barlow (Ed.), *Clinical handbook of psychology disorders: A step-by-step treatment manual.* (4th ed.)(pp. 216–249). New York: Guilford Press.

American Psychiatric Association. (2000). *Diagnostic and statistical manual of mental disorders: DSM-IV-TR.* Text Revision(4th ed.). Washington, DC: American Psychiatric Association.

American Psychiatric Association. (2013). *Diagnostic and statistical manual of mental disorders: DSM-5* (5th ed.). Washington, DC: American Psychiatric Association.

Anestis, M. D., Bender, T. W., Selby, E. A., Ribeiro, J. D., & Joiner, T. E. (2011). Sex and emotion in the acquired capability for suicide. *Archives of Suicide Research, 15*(2), 172–182. https://doi.org/10.1080/13811118.2011.566058.

Anestis, M. D., & Capron, D. W. (2016). The associations between state veteran population rates, handgun legislation, and statewide suicide rates. *Journal of Psychiatric Research, 74*, 30–34. https://doi.org/10.1016/j.jpsychires.2015.12.014.

Anestis, M. D., Gratz, K. L., Bagge, C. L., & Tull, M. T. (2012). The interactive role of distress tolerance and borderline personality disorder in suicide attempts among substance users in residential treatment. *Comprehensive Psychiatry, 53*(8), 1208–1216. https://doi.org/10.1016/j.comppsych.2012.04.004.

Anestis, M. D., & Joiner, T. E. (2012). Behaviorally-indexed distress tolerance and suicidality. *Journal of Psychiatric Research, 46*(6), 703–707. https://doi.org/10.1016/j.jpsychires.2012.02.015.

Anestis, M. D., Knorr, A. C., Tull, M. T., Lavender, J. M., & Gratz, K. L. (2013). The importance of high distress tolerance in the relationship between nonsuicidal self-injury and suicide potential. *Suicide and Life-threatening Behavior, 43*(6), 663–675. https://doi.org/10.1111/sltb.12048.

Anestis, M. D., Lavender, J. M., Marshall-Berenz, E. C., Gratz, K. L., Tull, M. T., & Joiner, T. E. (2012). Evaluating distress tolerance measures: Interrelations and associations with impulsive behaviors. *Cognitive Therapy and Research, 36*, 593–602.

Anestis, M. D., Pennings, S. M., Lavender, J. M., Tull, M. T., & Gratz, K. L. (2013). Low distress tolerance as an indirect risk factor for suicidal behavior: Considering the explanatory role of non-suicidal self-injury. *Comprehensive Psychiatry, 54*(7), 996–1002. https://doi.org/10.1016/j.comppsych.2013.04.005.

Anestis, M. D., Tull, M. T., Bagge, C. L., & Gratz, K. L. (2012). The moderating role of distress tolerance in the relationship between posttraumatic stress disorder symptom clusters and suicidal behavior among trauma exposed substance users in residential treatment. *Archives of Suicide Research, 16*, 198–2011. https://doi.org/10.1080/13811118.2012.695269.

Banducci, A. N., Bujarski, S. J., Bonn-Miller, M. O., Patel, A., & Connolly, K. M. (2016). The impact of intolerance of emotional distress and uncertainty on veterans with co-occurring PTSD and substance use disorders. *Journal of Anxiety Disorders, 41*, 73–81. https://doi.org/10.1016/j.janxdis.2016.03.003.

Banducci, A. N., Connolly, K. M., Vujanovic, A. A., Alvarez, J., & Bonn-Miller, M. O. (2017). The impact of changes in distress tolerance on PTSD symptom severity post-treatment among veterans in residential trauma treatment. *Journal of Anxiety Disorders, 47*, 99–105. https://doi.org/10.1016/j.janxdis.2017.01.004.

Banducci, A. N., Lejuez, C. W., Dougherty, L. R., & MacPherson, L. (2017). A prospective examination of the relations between emotional abuse and anxiety: moderation by distress tolerance. *Prevention Science, 18*(1), 20–30. https://doi.org/10.1007/s11121-016-0691-y.

Bardeen, J. R., & Fergus, T. A. (2016). Emotional distress intolerance, experiential avoidance, and anxiety sensitivity: The buffering effect of attentional control on associations with posttraumatic stress symptoms. *Journal of Psychopathology and Behavioral Assessment, 38*(2), 320–329. https://doi.org/10.1007/s10862-015-9522-x.

Bartlett, B. A., Jardin, C., Martin, C., Tran, J. K., Buser, S., Anestis, M. D., & Vujanovic, A. A. (2018). Posttraumatic stress and suicidality among firefighters: The moderating role of distress tolerance. *Cognitive Therapy and Research, 42*(4), 483–496. https://doi.org/10.1007/s10608-018-9892-y.

Batchelder, A. W., Ehlinger, P. P., Boroughs, M. S., Shipherd, J. C., Safren, S. A., Ironson, G. H., & O'Cleirigh, C. (2017). Psychological and behavioral moderators of the relationship between trauma severity and HIV transmission risk behavior among MSM with a history of childhood sexual abuse. *Journal of Behavioral Medicine, 40*(5), 794–802. https://doi.org/10.1007/s10865-017-9848-9.

Belik, S. L., Cox, B. J., Stein, M. B., Asmundson, G. J., & Sareen, J. (2007). Traumatic events and suicidal behavior: Results from a national mental health survey. *Journal of Nervous and Mental Disease, 195*(4), 342–349. https://doi.org/10.1097/01.nmd.0b013e318060a869.

Bender, T. W., Anestis, M. D., Anestis, J. C., Gordon, K. H., & Joiner, T. E. (2012). Affective and behavioral paths toward the acquired capacity for suicide. *Journal of Social and Clinical Psychology, 31*(1), 81–100.

Berenz, E. C., Vujanovic, A., Rappaport, L. M., Kevorkian, S., Gonzalez, R. E., Chowdhury, N., … Amstadter, A. (2018a). A multimodal study of childhood trauma and distress tolerance in young adulthood. *Journal of Aggression, Maltreatment & Trauma, 27*(7), 795–810. https://doi.org/10.1080/10926771.2017.1382636.

Berenz, E. C., Vujanovic, A. A., Coffey, S. F., & Zvolensky, M. J. (2012). Anxiety sensitivity and breath-holding duration in relation to PTSD symptom severity among trauma exposed adults. *Journal of Anxiety Disorders, 26*(1), 134–139.

Berenz, E. C., Vujanovic, A. A., Rappaport, L., Kevorkian, S., Gonzalez, R. E., Chowdhury, N., … Amstadter, A. B. (2018b). Childhood trauma and distress tolerance in a trauma-exposed acute-care psychiatric inpatient sample. *Psychological Trauma: Theory, Research, Practice and Policy, 10*(3), 368–375. https://doi.org/10.1037/tra0000300.

Bernal, M., Haro, J. M., Bernert, S., Brugha, T., de Graaf, R., Bruffaerts, R., … Investigators, E. M. (2007). Risk factors for suicidality in Europe: Results from the ESEMED study. *Journal of Affective Disorders, 101*(1–3), 27–34. https://doi.org/10.1016/j.jad.2006.09.018.

Bernstein, A., Zvolensky, M. J., Vujanovic, A. A., & Moos, R. (2009). Integrating anxiety sensitivity, distress tolerance, and discomfort intolerance: A hierarchical model of affect sensitivity and tolerance. *Behavior Therapy, 40*, 291–301.

Boffa, J. W., Short, N. A., Gibby, B. A., Stentz, L. A., & Schmidt, N. B. (2018). Distress tolerance as a mechanism of PTSD symptom change: Evidence for mediation in a treatment-seeking sample. *Psychiatry Research, 267*, 400–408. https://doi.org/10.1016/j.psychres.2018.03.085.

Bovin, M. J., Marx, B. P., Weathers, F. W., Gallagher, M. W., Rodriguez, P., Schurr, P. P., & Keane, T. M. (2015). Psychometric properties of the PTSD Checklist for *Diagnostic and Statistical Manual of Mental Disorders—Fifth Edition* (PCL-5) in veterans. *Psychological Assessment, 28*, 1379–1391.

Brown, T. A., & McNiff, J. (2009). Specificity of autonomic arousal to DSM-IV panic disorder and posttraumatic stress disorder. *Behaviour Research and Therapy, 47*(6), 487–493. https://doi.org/10.1016/j.brat.2009.02.016.

Bryan, C. J., Grove, J. L., & Kimbrel, N. A. (2017). Theory-driven models of self-directed violence among individuals with PTSD. *Current Opinion in Psychology, 14*, 12–17. https://doi.org/10.1016/j.copsyc.2016.09.007.

Chowdhury, N., Kevorkian, S., Hawn, S. E., Amstadter, A. B., Dick, D., Kendler, K. S., & Berenz, E. C. (2018). Associations between personality and distress tolerance among trauma-exposed young adults. *Personality and Individual Differences, 120*, 166–170. https://doi.org/10.1016/j.paid.2017.08.041.

Cohen, J. R., Danielson, C. K., Adams, Z. W., & Ruggiero, K. J. (2016). Distress tolerance and social support in adolescence: Predicting risk for internalizing and externalizing symptoms following a natural disaster. *Journal of Psychopathology and Behavioral Assessment, 38*(4), 538–546. https://doi.org/10.1007/s10862-016-9545-y.

Contractor, A. A., Weiss, N. H., Dranger, P., Ruggero, C., & Armour, C. (2017). PTSD's risky behavior criterion: Relation with DSM-5 PTSD symptom clusters and psychopathology. *Psychiatry Research, 252*, 215–222. https://doi.org/10.1016/j.psychres.2017.03.008.

Cook, J. W., McFall, M. E., Calhoun, P. S., & Beckham, J. C. (2007). Posttraumatic stress disorder and smoking relapse: A theoretical model. *Journal of Traumatic Stress, 20*(6), 989–998. https://doi.org/10.1002/jts.20275.

Danielson, C. K., Ruggiero, K. J., Daughters, S. B., & Lejuez, C. W. (2010). Distress tolerance, risk-taking propensity, and PTSD symptoms in trauma-exposed youth: Pilot study. *The Behavior Therapist, 33*(2), 28–34.

Daughters, S. B., Lejuez, C. W., Kahler, C. W., Strong, D. R., & Brown, R. A. (2005). Psychological distress tolerance and duration of most recent abstinence attempt among residential treatment-seeking substance abusers. *Psychology of Addictive Behaviors, 19*(2), 208–211. https://doi.org/10.1037/0893-164X.19.2.208.

Daughters, S. B., Reynolds, E. K., MacPherson, L., Kahler, W. W., Danielson, C. K., Zvolensky, M. J., & Lejuez, C. W. (2009). Negative reinforcement and early adolescent externalizing and internalizing symptoms: The moderating role of gender and ethnicity. *Behaviour Research and Therapy, 47*, 198–205.

Duranceau, S., Fetzner, M. G., & Carleton, R. N. (2014). Low distress tolerance and hyperarousal posttraumatic stress disorder symptoms: A pathway to alcohol use? *Cognitive Therapy and Research, 38*(3), 280–290.

Erwin, M. C., Mitchell, M. A., Contractor, A. A., Dranger, P., Charak, R., & Elhai, J. D. (2018). The relationship between distress tolerance regulation, counterfactual rumination, and PTSD symptom clusters. *Comprehensive Psychiatry, 82*, 133–140. https://doi.org/10.1016/j.comppsych.2018.01.012.

Farris, S. G., Vujanovic, A. A., Hogan, J., Schmidt, N. B., & Zvolensky, M. J. (2014). Main and interactive effects of anxiety sensitivity and physical distress intolerance with regard to PTSD symptoms among trauma-exposed smokers. *Journal of Trauma & Dissociation, 15*(3), 254–270. https://doi.org/10.1080/15299732.2013.834862.

Fergus, T. A., & Bardeen, J. R. (2016). Main and interactive effects of mental contamination and tolerance of negative emotions in relation to posttraumatic stress symptoms following sexual trauma. *Journal of Psychopathology and Behavioral Assessment, 38*(2), 274–283. https://doi.org/10.1007/s10862-015-9511-0.

Fetzner, M. G., Peluso, D. L., & Asmundson, G. J. G. (2014). Tolerating distress after trauma: Differential associations between distress tolerance and posttraumatic stress symptoms. *Journal of Psychopathology and Behavioral Assessment, 36*(3), 475–484.

Gerber, M. M., Frankfurt, S. B., Contractor, A. A., Oudshoorn, K., Dranger, P., & Brown, L. A. (2018). Influence of multiple traumatic event types on mental health outcomes: Does count matter? *Journal of Psychopathology and Behavioral Assessment*, https://doi.org/10.1007/s10862-018-9682-6.

Gorey, C. M., Rojas, E. A., & Bornovalova, M. A. (2018). More of a good thing is not always better: Validation of a distress overtolerance measure. *Assessment, 25*(4), 446–457. https://doi.org/10.1177/1073191116654218.

Gratz, K. L., & Gunderson, J. G. (2006). Preliminary data on acceptance-based emotion regulation group intervention for deliberate self-harm among women with borderline personality disorder. *Behavior Therapy, 37*(1), 25–35.

Gronwall, D. M. A. (1977). Paced auditory serial-addition task: A measure of recovery from concussion. *Perceptual and Motor Skills, 44*, 367–373.

Hajek, P., Belcher, M., & Stapleton, J. (1987). Breath-holding endurance as a predictor of success in smoking cessation. *Addictive Behaviors, 12*(3), 285–288.

Hashoul-Andary, R., Assayag-Nitzan, Y., Yuval, K., Aderka, I. M., Litz, B., & Bernstein, A. (2016). A longitudinal study of emotional distress intolerance and psychopathology following exposure to a potentially traumatic event in a community sample. *Cognitive Therapy and Research, 40*(1), 1–13. https://doi.org/10.1007/s10608-015-9730-4.

Hayes, S. C., Wilson, K. G., Gifford, E. V., Follette, V. M., & Strosahl, K. (1996). Experiential avoidance and behavioral disorders: A functional dimensional approach to diagnosis and treatment. *Journal of Consulting and Clinical Psychology, 64*(6), 1152–1168. https://doi.org/10.1037/0022-006X.64.6.1152.

Himmerich, S., & Orcutt, H. (2019). Alcohol expectancies and distress tolerance: Potential mechanisms in the relationship between posttraumatic stress and alcohol use. *Personality and Individual Differences, 137*, 39–44. https://doi.org/10.1016/j.paid.2018.08.004.

Holliday, S. B., Pedersen, E. R., & Leventhal, A. M. (2016). Depression, posttraumatic stress, and alcohol misuse in young adult veterans: The transdiagnostic role of distress tolerance. *Drug and Alcohol Dependence, 161*, 348–355. https://doi.org/10.1016/j.drugalcdep.2016.02.030.

Hyland, P., Shevlin, M., Brewin, C. R., Cloitre, M., Downes, A. J., Jumbe, S., … Roberts, N. P. (2017). Validation of post-traumatic stress disorder (PTSD) and complex PTSD using the International Trauma Questionnaire. *Acta Psychiatrica Scandinavica, 136*(3), 313–322. https://doi.org/10.1111/acps.12771.

Imel, Z. E., Laska, K., Jakupcak, M., & Simpson, T. L. (2013). Meta-analysis of dropout in treatments for posttraumatic stress disorder. *Journal of Consulting and Clinical Psychology, 81*(3), 394–404. https://doi.org/10.1037/a0031474.supp (Supplemental).

Kilpatrick, D. G., Resnick, H. S., Milanak, M. E., Miller, M. W., Keyes, K. M., & Friedman, M. J. (2013). National estimates of exposure to traumatic events and PTSD prevalence using DSM-IV and DSM-5 criteria. *Journal of Traumatic Stress*, *26*(5), 537–547. https://doi.org/10.1002/jts.21848.

Kimerling, R., Makin-Byrd, K., Louzon, S., Ignacio, R.V., & McCarthy, J. F. (2016). Military sexual trauma and suicide mortality. *American Journal of Preventive Medicine*, *50*(6), 684–691. https://doi.org/10.1016/j.amepre.2015.10.019.

Kraemer, K. M., Luberto, C. M., & McLeish, A. C. (2013). The moderating role of distress tolerance in the association between anxiety sensitivity physical concerns and panic and PTSD-related re-experiencing symptoms. *Anxiety, Stress, and Coping*, *26*(3), 330–342.

Krysinska, K., & Lester, D. (2010). Post-traumatic stress disorder and suicide risk: A systematic review. *Archives of Suicide Research*, *14*(1), 1–23. https://doi.org/10.1080/13811110903478997.

Krysinska, K., Lester, D., & Martin, G. (2009). Suicidal behavior after a traumatic event. *Journal of Trauma Nursing*, *16*(2), 103–110.

Larsen, R. J., & Diener, E. (1987). Affect intensity as an individual difference characteristic: A review. *Journal of Research in Personality*, *21*(1), 1–39.

Lejuez, C. W., Daughters, S. B., Danielson, C. K., & Ruggiero, K. J. (2006). *The Behavioral Indictor of Resiliency to Distress (BIRD)*. Unpublished Manuscript.

Lejuez, C. W., Kahler, C. W., & Brown, R. A. (2003). A modified computer version of the Paced Auditory Serial Addition Task (PASAT) as a laboratory-based stressor. *The Behavior Therapist*, *26*(4), 290–293.

Leyro, T. M., Zvolensky, M. J., & Bernstein, A. (2010). Distress tolerance and psychopathological symptoms and disorders: A review of the empirical literature among adults. *Psychological Bulletin*, *136*(4), 576–600. https://doi.org/10.1037/a0019712.

Linehan, M. M. (1993). *Skills training manual for treating borderline personality disorder*. New York, NY: Guilford Press.

Lynch, T. R., & Mizon, G. A. (2011). Distress overtolerance and distress intolerance: A behavioral perspective. In M. J. Zvolensky, A. Bernstein, & A. A. Vujanovic (Eds.), *Distress tolerance: Theory, research, and clinical applications* (pp. 52–79). New York, NY: Guilford Press.

Marshall, G. N., Miles, J. N., & Stewart, S. H. (2010). Anxiety sensitivity and PTSD symptom severity are reciprocally related: Evidence from a longitudinal study of physical trauma survivors. *Journal of Abnormal Psychology*, *119*(1), 143–150. https://doi.org/10.1037/a0018009.

Marshall-Berenz, E. C., Vujanovic, A. A., Bon-Miller, M. O., Bernstein, A., & Zvolensky, M. J. (2010). Multimethod study of distress tolerance and PTSD symptom severity in a trauma-exposed community sample. *Journal of Traumatic Stress*, *23*(5), 623–630. https://doi.org/10.1002/jts.20568.

Marshall-Berenz, E. C., Vujanovic, A. A., & Macpherson, L. (2011). Impulsivity and alcohol use coping motives in a trauma-exposed sample: The mediating role of distress tolerance. *Personality and Individual Differences*, *50*(5), 588–592. https://doi.org/10.1016/j.paid.2010.11.033.

Marshall-Berenz, E. C., Vujanovic, A. A., & Zvolensky, M. J. (2011). Main and interactive effects of a nonclinical panic attack history and distress tolerance in relation to PTSD symptom severity. *Journal of Anxiety Disorders*, *25*(2), 185–191. https://doi.org/10.1016/j.janxdis.2010.09.001.

McCauley, J. L., Killeen, T., Gros, D. F., Brady, K. T., & Back, S. E. (2012). Posttraumatic stress disorder and co-occurring substance use disorders: Advances in assessment and treatment. *Clinical Psychology: A Publication of the Division of Clinical Psychology of the American Psychological Association*, *19*(3), https://doi.org/10.1111/cpsp.12006.

McHugh, R. K., Daughters, S. B., Lejuez, C. W., Murray, H. W., Hearon, B. A., Gorka, S. M., & Otto, M. W. (2011). Shared variance among self-report and behavioral measures of distress intolerance. *Cognitive Therapy and Research, 35*(3), 266–275.

McHugh, R. K., & Otto, M. W. (2012). Refining the measurement of distress intolerance. *Behavior Therapy, 43*, 641–651.

McMahon, K., Hoertel, N., Olfson, M., Wall, M., Wang, S., & Blanco, C. (2018). Childhood maltreatment and impulsivity as predictors of interpersonal violence, self-injury and suicide attempts: A national study. *Psychiatry Research, 269*, 386–393. https://doi.org/10.1016/j.psychres.2018.08.059.

McNally, R. J. (2002). Anxiety sensitivity and panic disorder. *Biological Psychiatry, 52*(10), 938–946. https://doi.org/10.1016/S0006-3223(02)01475-0.

Mills, K. L., Teesson, M., Ross, J., & Peters, L. (2006). Trauma, PTSD, and substance use disorders: Findings from the Australian National Survey of Mental Health and Well-Being. *The American Journal of Psychiatry, 163*(4), 651–658. https://doi.org/10.1176/appi.ajp.163.4.652.

Nixon, R. D., & Bryant, R. A. (2003). Peritraumatic and persistent panic attacks in acute stress disorder. *Behaviour Research and Therapy, 41*(10), 1237–1242.

Nock, M. K., Prinstein, M. J., & Sterba, S. K. (2009). Revealing the form and function of self-injurious thoughts and behaviors: A real-time ecological assessment study among adolescents and young adults. *Journal of Abnormal Psychology, 118*(4), 816–827.

Nock, M. K., Stein, M. B., Heeringa, S. G., Ursano, R. J., Colpe, L. J., Fullerton, C. S., … Army, S. C. (2014). Prevalence and correlates of suicidal behavior among soldiers: Results from the Army Study to Assess Risk and Resilience in Servicemembers (Army STARRS). *JAMA Psychiatry, 71*(5), 514–522. https://doi.org/10.1001/jamapsychiatry.2014.30.

Pinciotti, C. M., Seligowski, A. V., & Orcutt, H. K. (2017). Psychometric properties of the PACT Scale and relations with symptoms of PTSD. *Psychological Trauma: Theory, Research, Practice and Policy, 9*(3), 362–369. https://doi.org/10.1037/tra0000206.

Potter, C. M., Vujanovic, A. A., Marshall-Berenz, E. C., Bernstein, A., & Bonn-Miller, M. O. (2011). Posttraumatic stress and marijuana use coping motives: The mediating role of distress tolerance. *Journal of Anxiety Disorders, 25*(3), 437–443.

Powers, M. B., Kauffman, B. Y., Kleinsasser, A. L., Lee-Furman, E., Smits, J. A., Zvolensky, M. J., & Rosenfield, D. (2016). Efficacy of smoking cessation therapy alone or integrated with prolonged exposure therapy for smokers with PTSD: Study protocol for a randomized controlled trial. *Contemporary Clinical Trials, 50*, 213–221. https://doi.org/10.1016/j.cct.2016.08.012.

Quinn, E. P., Brandon, T. H., & Copeland, A. L. (1996). Is task persistence related to smoking and substance abuse? The application of learned industriousness theory to addictive behaviors. *Experimental and Clinical Psychopharmacology, 4*(2), 186–190.

Roberts, N. P., Roberts, P. A., Jones, N., & Bisson, J. I. (2015). Psychological interventions for post-traumatic stress disorder and comorbid substance use disorder: A systematic review and meta-analysis. *Clinical Psychology Review, 38*, 25–38. https://doi.org/10.1016/j.cpr.2015.02.007.

Schäfer, I., & Najavits, L. M. (2007). Clinical challenges in the treatment of patients with posttraumatic stress disorder and substance abuse. *Current Opinion in Psychiatry, 20*(6), 614–618. https://doi.org/10.1097/YCO.0b013e3282f0ffd9.

Schmidt, N. B., Richey, J. A., & Fitzpatrick, K. K. (2006). Discomfort intolerance: Development of a construct and measure relevant to panic disorder. *Journal of Anxiety Disorders, 20*(3), 263–280.

Schmidt, N. B., Zvolensky, M. J., & Maner, J. K. (2006). Anxiety sensitivity: Prospective prediction of panic attacks and Axis I pathology. *Journal of Psychiatric Research, 40*(8), 691–699. https://doi.org/10.1016/j.jpsychires.2006.07.009.

Schottenbauer, M. A., Glass, C. R., Arnkoff, D. B., Tendick, V., & Gray, S. H. (2008). Nonresponse and dropout rates in outcome studies on PTSD: Review and methodological considerations. *Psychiatry: Interpersonal and Biological Processes*, *71*(2), 134–168.

Simons, J. S., & Gaher, R. M. (2005). The distress tolerance scale: Development and validation of a self-report measure. *Motivation and Emotion*, *29*(2), 83–102.

Tull, M. T., Gratz, K. L., Coffey, S. F., Weiss, N. H., & McDermott, M. J. (2013). Examining the interactive effect of posttraumatic stress disorder, distress tolerance, and gender on residential substance use disorder treatment retention. *Psychology of Addictive Behaviors*, *27*(3), 763–773.

Viana, A. G., Woodward, E. C., Raines, E. M., Hanna, A. E., & Zvolensky, M. J. (2018). The role of emotional clarity and distress tolerance in deliberate self-harm in a sample of trauma-exposed inpatient adolescents at risk for suicide. *General Hospital Psychiatry*, *50*, 119–124. https://doi.org/10.1016/j.genhosppsych.2017.10.009.

Vinci, C., Mota, N., Berenz, E., & Connolly, K. (2016). Examination of the relationship between PTSD and distress tolerance in a sample of male veterans with comorbid substance use disorders. *Military Psychology*, *28*(2), 104–114. https://doi.org/10.1037/mil0000100.

Vujanovic, A. A., & Back, S. E. (in press). PTSD and substance use disorders: A clinical overview. In A. A. Vujanovic & S. E. Back (Eds.), *Posttraumatic stress and substance use disorders: A comprehensive clinical handbook*. New York: Routledge.

Vujanovic, A. A., Bakhshaie, J., Martin, C., Reddy, M. K., & Anestis, M. D. (2017). Posttraumatic stress and distress tolerance: Associations with suicidality in acute-care psychiatric inpatients. *Journal of Nervous and Mental Disease*, *205*(7), 531–541. https://doi.org/10.1097/NMD.0000000000000690.

Vujanovic, A. A., Berenz, E. C., & Bakhshaie, J. (2018). Multimodal examination of distress tolerance and suicidality in acute-care psychiatric inpatients. *Journal of Experimental Psychopathology*, *8*(4), https://doi.org/10.5127/jep.059416.

Vujanovic, A. A., Bernstein, A., & Litz, B. T. (2011). Traumatic stress. In A. A. Vujanovic, A. Bernstein, M. J. Zvolensky, A. Bernstein, & A. A. Vujanovic (Eds.), *Distress tolerance: Theory, research, and clinical applications* (pp. 126–148). New York, NY: Guilford Press.

Vujanovic, A. A., Bonn-Miller, M. O., Potter, C. M., Marshall, E. C., & Zvolensky, M. J. (2011). An evaluation of the relation between distress tolerance and posttraumatic stress within a trauma-exposed sample. *Journal of Psychopathology and Behavioral Assessment*, *33*(1), 129–135.

Vujanovic, A. A., Dutcher, C. D., & Berenz, E. C. (2017). Multimodal examination of distress tolerance and posttraumatic stress disorder symptoms in acute-care psychiatric inpatients. *Journal of Anxiety Disorders*, *48*, 45–53. https://doi.org/10.1016/j.janxdis.2016.08.005.

Vujanovic, A. A., Hart, A. S., Potter, C. M., Berenz, E. C., Niles, B., & Bernstein, A. (2013). Main and interactive effects of distress tolerance and negative affect intensity in relation to PTSD symptoms among trauma-exposed adults. *Journal of Psychopathology and Behavioral Assessment*, *35*(2), 235–243.

Vujanovic, A. A., Litz, B. T., & Farris, S. G. (2015). Distress tolerance as risk and maintenance factor for PTSD: Empirical and clinical implications. In C. R. Martin, V. R. Preedy, & V. B. Patel (Eds.), *Comprehensive guide to post-traumatic stress disorder*. London: Springer International Publishing.

Vujanovic, A. A., Marshall, E. C., Gibson, L. E., & Zvolensky, M. J. (2010). Cognitive–affective characteristics of smokers with and without posttraumatic stress disorder and panic psychopathology. *Addictive Behaviors*, *35*(5), 419–425.

Vujanovic, A. A., Marshall-Berenz, E. C., & Zvolensky, M. J. (2011). Posttraumatic stress and alcohol use motives: A test of the incremental and mediating role of distress tolerance. *Journal of Cognitive Psychotherapy*, *25*(2), 130–141.

Vujanovic, A. A., Rathnayaka, N., Amador, C. D., & Schmitz, J. M. (2016). Distress tolerance: Associations with posttraumatic stress disorder symptoms among trauma-exposed, cocaine-dependent adults. *Behavior Modification*, *40*(1–2), 120–143. https://doi.org/10.1177/0145445515621490.

Vujanovic, A. A., Wardle, M. C., Bakhshaie, J., Smith, L. J., Green, C. E., Lane, S. D., & Schmitz, J. M. (2018). Distress tolerance: Associations with trauma and substance cue reactivity in low-income, inner-city adults with substance use disorders and posttraumatic stress. *Psychology of Addictive Behaviors, 32*(3), 264–276. https://doi.org/10.1037/adb0000362.

Watson, D., O'Hara, M. W., Simms, L. J., Kotov, R., Chmielewski, M., McDade-Montez, E. A., … Stuart, S. (2007). Development and validation of the inventory of depression and anxiety symptoms (IDAS). *Psychological Assessment, 19*(3), 253–268. https://doi.org/10.1037/1040-3590/19.3.253.supp (Supplemental).

Weathers, F. W., Blake, D. D., Schnurr, P. P., Kaloupek, D. G., Marx, B. P., & Keane, T. M. (2013). *The clinician-administered PTSD scale for DSM-5 (CAPS-5).* Retrieved from Interview available from the National Center for PTSD atwww.ptsd.va.gov.

Wilcox, H. C., Storr, C. L., & Breslau, N. (2009). Posttraumatic stress disorder and suicide attempts in a community sample of urban american young adults. *Archives of General Psychiatry, 66*(3), 305–311.

CHAPTER 13

Emotional granularity in PTSD

Michael K. Suvak[a], Regina M. Musicaro[a], Hilary Hodgdon[a,b]
[a]Psychology Department, Suffolk University, Boston, MA, United States
[b]The Trauma Center at Justice Resource Institute, Brookline, MA, United States

Introduction: Emotional granularity and emotion differentiation

Emotional granularity and emotion differentiation are terms used to refer to the same theoretical construct, individual differences in one's ability to make fine-grained distinctions among similar emotional states (Kashdan et al., 2015; Smidt & Suvak, 2015). However, the use of these terms tends to correspond to slightly different operational definitions of the construct. Both are operationalized by summarizing correlations among the use of emotion terms across several occasions. One common methodology for investigating these constructs is experiential sampling, sometimes referred to as ecological momentary assessment. In a typical experiential sampling study, participants are given a device that prompts them to report on their emotions several times throughout the day across many days as they go about their normal lives. At each assessment, participants are asked to rate the extent to which they are experiencing several different emotional states (happy, sad, angry, fearful, etc.). If a person exhibits a perfect correlation ($r = 1.00$) between two similarly valenced states with different arousal levels, such as angry and fearful or happy and ecstatic, then they are not making any distinction between these states in their self-reports of emotional experience; in other words, they are exhibiting low granularity.

Lisa Feldman Barrett (e.g., Barrett, 2004) coined the term emotional granularity and developed a method to quantify granularity. Barrett's model was built upon the affective circumplex model (Russell, 1980), perhaps the most empirically supported dimensional model of emotion. This model purports that emotional phenomena (e.g., emotional states, self-reports of emotion, conceptual representations of emotion, facial expressions of emotion, and emotional stimuli) can be adequately described using two dimensions, valence and arousal (e.g., Russell, 2003; Russell & Barrett, 1999).

The valence dimension captures hedonic tone, or the amount of pleasure or displeasure, of a specific emotional state or event, while arousal refers to felt activation (activated or deactivated) of an emotional phenomenon. A large body of empirical work has established that, on a nomothetic level, the circumplex model accounts for emotional phenomena quite well (Russell & Barrett, 1999). However, Barrett developed the concept of emotional granularity based on her work documenting meaningful individual differences in how the circumplex model accounts for self-reports of emotional experience (Barrett, 1995). Fig. 1 depicts individual differences in self-reports of emotion.

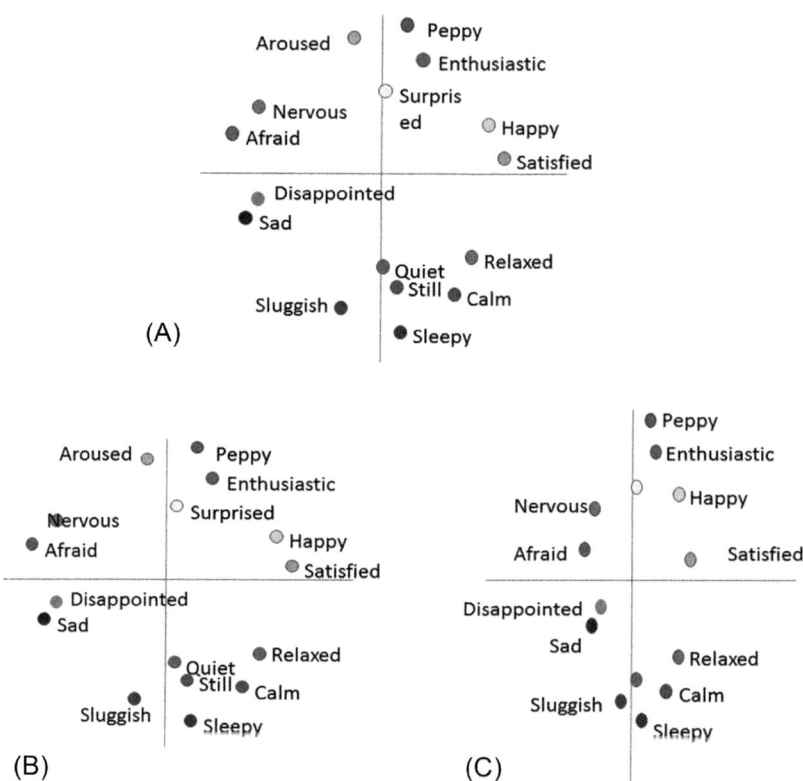

Fig. 1 Variations in the structure of self-report emotion. Part (A) depicts equal weighing of valence and arousal, (B) depicts an emphasis on valence relative to arousal, and (C) depicts an emphasis on arousal relative to valence; x-axis – valence; y axis = arousal. Adapted from Barrett, L. F. (1995). Valence focus and arousal focus: Individual differences in the structure of affective experience. Journal of Personality and Social Psychology 69, 153–166.

Fig. 1A shows the typical depiction of the circumplex model, a circular structure with equal weight given to valence (horizontal axis) and arousal (vertical arousal). Fig. 1B and C depicts variations in the circular structure, with the structure in Fig. 1B depicting someone who emphasizes valence more than arousal in self-reports of affect and Fig. 1C depicting the structure of someone who emphasizes arousal more than valence. To operationalize emotional granularity, Barrett (1995) developed two individual difference variables: valence focus and arousal focus. Valence focus refers to the degree to which one emphasizes valence in representations of emotions, and arousal focus refers to the degree to which one emphasizes arousal in representations of emotion. Being emotionally granular requires adequate levels of both valence and arousal focus (similar to the structure depicted in Fig. 1A). Quantitative estimates are derived by determining the degree to which valence and arousal account for the correlations among emotion terms across reports (see Barrett, 2004, for more details).

Operationalizing emotion differentiation involves summarizing correlations among emotion terms in a slightly different way. Quantifying emotion differentiation typically involves computing intraclass correlations (ICCs) to represent the average correlation across many types of emotions. The most commonly computed differentiation variable is negative emotion differentiation, which essentially represents the average correlation among all negatively valenced emotional states (e.g., Erbas et al., 2018). Less often, positive emotion differentiation, essentially the average correlation among all positively valence emotional states, is computed.

In sum, emotional granularity and emotion differentiation are essentially the same construct on a theoretical level, operationalized via two different methods to summarize correlations among the use of emotional labels across a variety of situations or contexts. Unfortunately, no research exists that compares the predictive power of these different operational definitions, a worthy avenue for future investigations. The following sections apply to both emotional granularity and emotion differentiation; however, we primarily (but not exclusively) focus on emotional granularity for two reasons: (1) emotional granularity inherently focuses on mechanisms, as the degree to which valence and arousal are incorporated into representations of emotion (i.e., valence and arousal focus) are mechanisms that contribute to overall level of emotional granularity, and (2) an emotional granularity framework allows for the examination of different patterns or types of low granularity, for instance, low granularity due high valence focus relative to arousal focus (see Fig. 1B) or low granularity due to high arousal focus relative to valence focus (see Fig. 1C).

Emotional granularity, related constructs, and PTSD

Over the past few decades, several related, yet distinct, constructs have been introduced in an attempt to understand and operationalize the richness and sophistication of one's emotional life (Smidt & Suvak, 2015). Emotional complexity is a broad umbrella term that encapsulates a number of dispositions and processes including emotional granularity, emotion differentiation, alexithymia emotional clarity, and emotional awareness (Kang & Shaver, 2004; Lindquist & Barrett, 2008). We are aware of no studies that have examined PTSD from an emotional granularity perspective; however, a substantial body of empirical research has established associations between PTSD and emotional complexity constructs similar to emotional granularity, in particular emotional clarity and alexithymia.

Emotional clarity

The term meta-mood experience refers to the ongoing process in which individuals "continually reflect upon their feelings by monitoring, evaluating, and regulating them" (Salovey, Mayer, Goldman, Turvey, & Palfai, 1995, p. 127). An important characteristic of one's meta-mood experience is emotional clarity, or the degree to which individuals can understand and identify their emotions. While the empirical association between emotional clarity and emotional granularity remain to be explicated, theoretically they are similar. Thus you would expect at least a moderate association between the two.

Several studies have used the clarity subscale of the Difficulties in Emotion Regulation scale (DERS; Gratz & Roemer, 2004) to examine the relationship between PTSD and clarity (e.g., Bornovalova, Ouimette, Crawford, & Levy, 2009; Doolan, Bryant, Liddell, & Nickerson, 2017; Tull, Barrett, McMillan, & Roemer, 2007; Tull, Weiss, Adams, & Gratz, 2012; Viana et al., 2018; Weiss, Tull, Anestis, & Gratz, 2013). The DERS has six subscales that measure: (1) goals (i.e., how emotional dysregulation interferes with ability to achieve personal goals, (2) nonacceptance (of one's own emotions), (3) impulse (i.e., behavioral control while feeling upset), (4) strategies (i.e., beliefs about one's ability to effectively manage emotions), (5) awareness (i.e., how much attention is given to feelings), and (6) clarity (i.e., inability to differentiate between emotional states). Research has consistently documented a relationship between posttraumatic stress symptom severity and all of the DERS subscales, with the exception of the awareness subscale (e.g., Tripp, McDevitt-Murphy, Avery, & Bracken, 2015;

Tull et al., 2007; Weiss et al., 2013). Short, Norr, Mathes, Oglesby, and Schmidt (2016) applied structural equation modeling to data from a trauma-exposed community sample to examine the association between PTSD symptom clusters and DERS subscales. On a bivariate level the emotional clarity subscale was associated with all PTSD symptom clusters. However, when controlling for neuroticism and all other DERS subscales, emotional clarity was only associated with the numbing symptoms of PTSD. Emotional clarity may also account for high levels of co-occurrence between PTSD and other psychosocial functioning problems such as comorbid substance abuse. Adults with comorbid substance use disorder and PTSD admitted to a residential treatment facility reported lower levels of clarity as assessed by the DERS clarity subscale compared with those with substance use disorder alone (Weiss et al., 2013). Another study showed that low levels of emotional clarity partially accounted for the association between PTSD symptoms and substance abuse in men but not women who had been admitted to an inpatient drug and alcohol abuse treatment center (Bornovalova et al., 2009).

Alexithymia

Alexithymia refers to deficits in recognizing, labeling, and communicating emotional states and consists of the following criteria: (a) difficulty identifying and describing subjective feelings and differentiating between feelings and the physical sensations of emotional arousal, (b) trouble verbally expressing emotion, (c) limited imagination, and (d) an externally oriented cognitive style with very limited thought directed toward internal reality (Kooiman, Spinhoven, & Trijsburg, 2002; Timoney & Holder, 2013). Barrett (2017, p. 107) described alexithymia as having an "impoverished conceptual system for emotion." Individuals reporting high levels of alexithymia are described as having an "absence of words" for emotions or may experience a deficit in propositional knowledge of emotions (Pond Jr. et al., 2012). Individuals endorsing high levels of alexithymia generally report less intense experiences of emotions and often use fewer emotion words to describe their emotional state. Those with alexithymia have poor clarity in identifying their feelings (Lischetzke, Angelova, & Eid, 2011). Erbas, Ceulemans, Lee Pe, Koval, and Kuppens (2014) provided initial evidence for an association between alexithymia and emotion differentiation, showing that negative emotion differentiation was negatively associated with the difficulty describing feelings and difficulty identifying feelings, but not the externally oriented thinking, subscales of the Toronto Alexithymia Scale (Bagby, Parker, & Taylor, 1994), the most widely used self-report measure of alexithymia.

A relationship between PTSD and alexithymia has been well documented by a number of empirical studies. Frewen, Dozois, Neufeld, and Lanius (2008) conducted a meta-analysis of all studies that reported alexithymia scores in PTSD samples up to July 2007 that included 12 studies and 1095 participants diagnosed with PTSD and a control sample of 460 participants without PTSD. A large effect size ($d=0.80$) emerged in the comparison between PTSD and control participants. Significant PTSD-alexithymia associations have been reported in a several subsequent studies including multiple populations such as Turkish university students (Balaban et al., 2012), people who suffer from asthma attacks (Chung, Rudd, & Wall, 2012), people with epileptic seizure disorder (Chung & Allen, 2013; Chung, Allen, & Dennis, 2013), adolescents (Chen & Chung, 2016), inpatient alcohol-dependent men (Evren, Dalbudak, Durkaya, Cetin, & Evren, 2010), paramedics (Halpern, Maunder, Schwartz, & Gurevich, 2012), military veterans (Kušević, Ćusa, Babić, & Marčinko, 2015; O'Brien, Gaher, Pope, & Smiley, 2008; Polusny, Dickinson, Murdoch, & Thuras, 2008), stroke survivors (Wang, Chung, Hyland, & Bahkeit, 2011), heart attack survivors (Gao, Zhao, Li, & Cao, 2015), and people with somatoform disorder (Kienle et al., 2017). Thus many empirical studies have demonstrated that the PTSD-alexithymia relationship is strong and robust.

Recent research has also begun to elucidate mechanisms that might underlie the PTSD-alexithymia association. Several studies found that alexithymia was associated with a higher severity of dissociative symptoms in individuals with PTSD (Hetzel-Riggin & Meads, 2016; Powers, Fani, Cross, Ressler, & Bradley, 2016; Terock et al., 2016). Avoidance/numbing symptoms of PTSD (Eichhorn, Brähler, Franz, Friedrich, & Glaesmer, 2014) and a denial coping style (Gaher, O'Brien, Smiley, & Hahn, 2016) have been found to mediate the PTSD-alexithymia association. Alexithymia has also been shown to predict comorbidity with other psychiatric disorders among individuals diagnosed with PTSD (Chen & Chung, 2016) and strongly predicted suicide attempts in veterans of the Croatian war (Kušević et al., 2015). One study found that alexithymia predicted sexual revictimization in adulthood for survivors of childhood victimization (Bell & Naugle, 2008). Further solidifying this association, clinical trials have found that alexithymia improves when PTSD symptoms improve, suggesting that alexithymia is an important target for trauma-informed treatments (Berke et al., 2017; Classen, Muller, Field, Clark, & Stern, 2017).

We contend that the documented associations between alexithymia and emotional clarity, two constructs closely related to emotional granularity, and PTSD provide a strong rationale for future research examining PTSD from a granularity perspective. It is important to note that emotional clarity and alexithymia are typically assessed using global self-report measures asking participants to rate the extent to which they generally experience emotions clearly, while, as described earlier, emotional granularity and emotion differentiation are assessed by having participants report on their current emotional state at several distinct instances, with granularity or differentiation scores computed by summarizing correlations among self-reported emotions across several situations. Research has shown that self-reports of emotional experience are influenced by two factors: episodic information about one's current emotional state that is anchored in a person's experience and semantic decontextualized beliefs that a person has about him or herself (Robinson & Clore, 2002a). Robinson and Clore (2002b) showed that, when people are asked to report on their emotional experience over short time frames (momentary assessment or a few days), their reports are more anchored in episodic information and less anchored in semantic beliefs than when people are asked to report on their emotional experience over longer time frames (e.g., in general or over a period of weeks or months), which tend to be more anchored in semantic beliefs than episodic information. Therefore, although emotional granularity and emotion differentiation are typically conceptualized as traits, they are assessed via momentary self-reports of emotions across several occasions; thus granularity and differentiation likely tap episodic processes (i.e., information anchored in one's experience) more so than self-report measures of clarity and alexithymia, which ask people to aggregate their experiences over a longer time period and are likely more anchored in one's beliefs about him or herself. It is quite possible that both episodic characteristics of people's emotional experience and the beliefs they have about themselves regarding their emotionality contribute to PTSD. Future research examining PTSD from a granularity perspective should include both types of assessments (momentary assessment that taps episodic information and global self-reports that tap beliefs) to provide a comprehensive account of emotional processes associated with PTSD. Now, let us continue to build the case that examining PTSD from an emotional granularity perspective may help further elucidate the emotional processes that contribute to this debilitating condition.

PTSD and emotional granularity: A preliminary road map

We created Fig. 2 to depict several ways that emotional granularity may be related to PTSD, including mechanisms that (1) specify emotional granularity as a variable that might be directly impacted by trauma exposure, and/or (2)

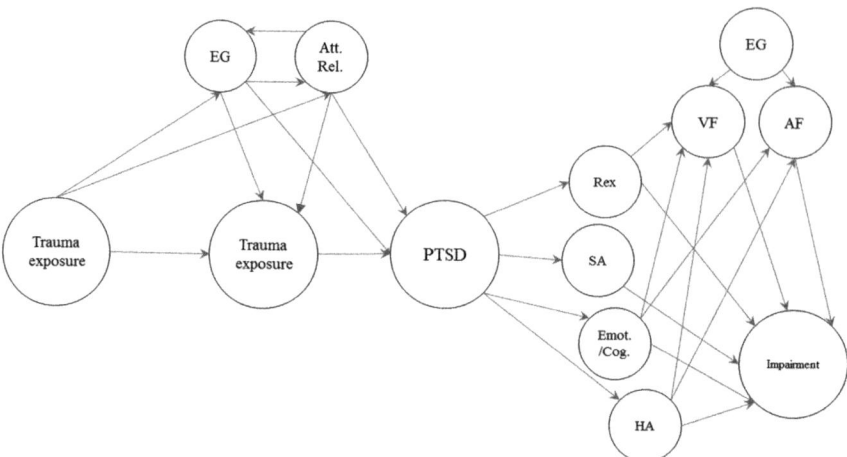

Fig. 2 Map of potential links between PTSD and emotional granularity. *EG*, emotional granularity; *Att. Rel.*, attachment relationships; *PTSD*, posttraumatic stress disorder; *Rex*, reexperiencing; *SA*, strategical avoidance; *Emot./Cog.*, emotion cognition (negative alterations in cognitions and mood); *HA*, hyperarousal; *VF*, valence focus; *AF*, arousal focus.

impact the likelihood of being exposed to trauma and/or (3) PTSD moderating the relationship between exposure and the subsequent development of PTSD, and/or (4) account for (i.e., mediate), (5) or moderate the relationship between PTSD and functional impairment.

Several testable hypotheses can be derived from the model. Of course, all of these paths require empirical examination, and we are certain that empirical research will warrant substantial revisions to this model. We offer this as an initial map (think of a map of the United States circa 1776) that empirical research can revise and improve over time (moving toward a satellite driven Google map that can zoom in and out circa 2018).

Close/attachment relationships, emotional granularity, and trauma exposure

Let us start on the left side of the map (i.e., Fig. 2), which depicts processes that occur prior to or immediately after traumatic event exposure, before the development of PTSD symptoms (Criterion F of DSM-5 PTSD specifies that symptoms must last for more than 1 month). Attachment/close relationships appear on this area of the map in addition to emotional granularity. There is widespread agreement that the quality of the bond between an infant and his or her caregivers strongly influences the blueprint of one's emotional life. Kernberg's (1967, 1975, 1976) object-relation theory purported that the capacity to experience a broad range of complex and nuanced emotions develops parallel with and influences the capacity to develop mature and healthy object relations. Healthy development is characterized by increasingly more sophisticated and balanced representations of the self and others, and these representations are linked to more nuanced and complex representations of affect. A disruption in the development of more nuanced and complex affect leads to pathology. For instance, Kernberg's (1967, 1975, 1976) description of borderline pathology included broad, undifferentiated anxiety and unadulterated rage, emotional states that are crude and unrefined.

Shaver, Mikulincer, and colleagues (e.g., Mikulincer, Shaver, & Pereg, 2003; Shaver & Mikulincer, 2002) have proposed an empirically based model illustrating how attachment relationships profoundly impact the development of emotion generation, regulation, and expression. Mikulincer, Shaver, and Solomon (2015) integrated the impact of trauma into this model in a review of theory and research that demonstrates how trauma exposure can impact attachment relationships and emotional capacities. They reviewed research demonstrating that individual differences in attachment anxiety and avoidance are associated with how individuals manage stress and destress. High levels of attachment anxiety and avoidance have been shown to be associated with doubts about one's safety and lovability, and people with high levels of attachment anxiety and avoidance tend to evoke what Mikulincer et al. (2015) refer to as "hyperactivating" or "deactivating" of the attachment behavioral system in an effort to cope with threats, stress, and destress. These strategies likely interfere with the mobilization of social resources that facilitate recovery following exposure to traumatic events and may contribute to the hypervigilance (i.e., diminished feelings of safety and security) and arousal and emotion regulation problems associated with PTSD (see later).

Charuvastra and Cloitre (2008) also reviewed empirical studies demonstrating both the acute and chronic biological effects of the parent-child dynamic with regard to safety and threat exposure. Infant stress-regulatory responses are substantially dependent on the infants' relationship to their primary caretaker. Several studies have also identified insecure attachment, particularly attachment anxiety, as a predictor of revictimization (Bockers, Roepke, Michael, Renneberg, & Knaevelsrud, 2014; Brenner & Ben-Amitay, 2015; Reese-Weber & Smith, 2011), with impaired risk recognition identified as likely mediator of this relationship (Bockers et al., 2014). Although speculative due to a lack of empirical research, it is quite possible that being able to make nuanced distinctions among emotional states (i.e., emotional granularity) could enhance risk identification. Future research is needed to empirically elucidate what is likely a dynamic process involving reciprocal relationships among the quality of close/attachment relationships, emotional experience and capacities (including emotional granularity), stress exposure, and reactions to stress exposures. While attachment relationships provide the context for the initial development of emotional granularity, being able to make distinctions among similar emotional states likely facilitates the development of additional close relationships (hence the double/bidirectional arrows between emotional granularity and attachment in Fig. 2). Furthermore, both attachment/close relationships and emotional granularity are likely influenced by previous exposure to traumatic events, and both likely impact the likelihood of being exposed to subsequent traumatic events.

Emotional granularity and coping with traumatic stressors: Context probably matters

An interesting unexplored question is whether high levels of granularity represents a resource for dealing with trauma exposures in the context in which the trauma occurs. Emotional granularity is almost always conceptualized as a trait-like construct that is positively associated with healthy psychosocial functioning. However, like most psychological constructs, context probably matters necessitating the examination of trait-state/situation interactions. For instance, in an environment in which a person is frequently exposed to potentially traumatic events (such as a war zone, or chronic abuse by a care giver, or maybe the environments navigated by our very distant ancestors) making fine-grain distinctions among emotional states could possibly be harmful, not helpful. In these situations, it may be more adaptive to be able to quickly identify and respond to threats present in the environment, with no nuance necessary. Bonanno and Burton (2013) identified sensitivity to context and the availability of a diverse repertoire of regulatory capacities and strategies as key components to what they term as regulatory flexibility, or the ability to match regulation strategies with contextual demands. It is quite likely that the same applies to emotional granularity, that is, emotional granularity is likely adaptive in complex social environments where one needs to effectively respond to subtle emotional cues, but is not adaptive in environments in which serious and dangerous threats are often encountered. Possessing what might be called "emotional granularity flexibility," or possessing the ability to make fine-grain distinctions among emotional states when necessary, possessing the ability to identify and respond to threats in a quick and crude manner when necessary, and possessing the ability to match level of granularity with the level of granularity demanded by the situation would provide one with a complete repertoire of skills allowing them to function effectively across a range of situations. One notable gap in research on emotional granularity and emotion differentiation is studies that examine the role of situational factors. We are aware of only one study that has considered state or situational factors in an empirical examination of emotion differentiation. Erbas et al. (2018) showed that stress was negatively associated with emotion differentiation, and analyses conducted on the day-to-day level showed that stress prospectively negatively predicted emotion differentiation, but emotional granularity did not prospectively predict stress. However, this study did not address the question of whether exhibiting high levels of emotional granularity is differentially adaptive across situations, a question that should be addressed by future research. In a sample of trauma-exposed patients receiving substance use disorder treatment in a residential facility, Schoenleber, Berghoff, Gratz, and Tull (2018) showed that associations among anger, anxiety, and self-conscious emotions were stronger when patients were listening to individualized trauma scripts compared with when they were listening to neutral control scripts, suggesting that granularity/differentiation may have been lower during the trauma scripts than during the neutral scripts. It is also quite likely that the degree to which emotional granularity facilitates adaptive responses varies across different types of trauma exposures. For instance, it is hard to think of ways that granularity might be helpful for a soldier confronting the enemy in war. However, it may be helpful for the victim of ongoing abuse perpetrated by a caregiver to make fine-grained distinctions regarding the emotional state of the perpetrator. Understanding the role of context and situational factors on the adaptiveness of emotional granularity represents a notable gap in research establishing the nature and importance of emotional granularity, and examining the possible differential impact of emotional granularity across different context in which one confronts traumatic events will help address this gap.

Emotional granularity as a moderator between trauma exposure and PTSD

Research has shown that immediately following exposure it is quite common for individuals to exhibit substantial levels of psychological distress that decline over time, and PTSD is often considered a failure of recovery (e.g., Gillihan, Cahill, & Foa, 2014; McFarlane, 2000; Yehuda & LeDoux,

2007). Emotion regulation has been identified as a factor that influences the course of psychological distress following exposure to a traumatic event and is the target of early interventions administered developed to be delivered shortly after exposure to a traumatic event to prevent the development of PTSD (e.g., Fine, Achituv, Etkin, Merin, & Shalev, 2018), and both longitudinal and laboratory research has substantiated the role of emotion regulation in recovery following exposure to a traumatic event (e.g., Bardeen, Kumpula, & Orcutt, 2013; Shepherd & Wild, 2014).

One of the potential benefits of high levels of emotional granularity (or emotion differentiation) is the effective implementation of emotion regulation, or as Barrett puts it "preciseness leads to efficiency (Barrett, 2017, p. 121)." One of the functions of emotions is to provide information that guides decision-making and behavior (e.g., Schwarz, 1990), and having access to more precise and nuanced information should facilitate effective and efficient emotion regulation. Barrett, Gross, Christensen, and Benvenuto (2001) provided the first empirical link between emotion differentiation and emotion regulation, and subsequent studies have showed that high levels of emotional granularity/emotion differentiation mitigate the relationship between intense emotional negative states and/or psychological symptoms and maladaptive behaviors. Three studies showed that emotion differentiation played an important role in the relationship between anger and aggressive behaviors, with this association being significantly weaker for individuals exhibiting higher levels of emotion differentiation relative to those exhibiting lower levels (Pond Jr. et al., 2012). Negative emotion differentiation has also been found to moderate the association between intensity of negative emotions and alcohol consumption, with this relationship significantly weaker for participants exhibiting high emotion differentiation compared with those exhibiting low emotion differentiation. Zaki, Coifman, Rafaeli, Berenson, and Downey (2013) also showed that emotion differentiation moderated the relationship between rumination and the urge to engage in nonsuicidal self-injury. Together, this evidence suggests that emotional granularity may provide a resource to help those who have been exposed to a traumatic event to effectively cope with the stress associated with exposure to a traumatic event and facilitate the reduction of posttraumatic stress symptoms that many experience, perhaps preventing the development of PTSD. Up until this point, we have been primarily focusing on processes and constructs that may provide a link between emotional granularity and PTSD prior to the development of PTSD symptoms. Now, let us turn to potential connections between emotional granularity and PTSD symptoms.

PTSD symptoms and emotional granularity

PTSD is a heterogeneous disorder. Informed by factor analysis studies, the DSM-5 specification of PTSD includes four separate clusters of symptoms (Criteria B: Reexperiencing, Criteria C: Strategic Avoidance, Criteria D: Negative Alterations in Cognitions and Mood, Hyperarousal, Criteria E: Hyperarousal/Hyperreactivity). Galatzer-Levy and Bryant (2013) demonstrated that 636,120 symptom combinations can produce a DSM-5 PTSD diagnosis. Each of the DSM-5 PTSD symptom clusters includes symptoms that involve problematic emotional processing including unpleasant feelings (e.g., Criterion B4: Emotional distress after exposure to traumatic reminders, Criterion D4: Persistent negative emotional state, and Criterion D7: Inability to experience positive emotions), emotional cognitions (e.g., Criterion D2: Overly negative thoughts and assumptions about oneself or the world and Criterion D3: Exaggerated blame of self or others for causing the trauma), and emotional behaviors (e.g., Criterion C2: Avoidance of trauma reminders and Criterion E1: Irritable behavior). While early learning-based models of PTSD conceptualized PTSD as a disorder of the fear circuit (e.g., Shin & Handwerger, 2009), more contemporary accounts have stressed that PTSD involves numerous emotional states including shame, guilt, anger, and disgust (McLean & Foa, 2017; Resick & Miller, 2009), which influenced multiple changes from DSM-IV to DSM-5, including PTSD no longer being classified as an anxiety disorder. While a comprehensive model parsimoniously accounting for the various emotional processes associated with PTSD has yet to be developed, most models of emotion and PTSD make a common observation, that emotional processes associated with PTSD involve both hyper- (e.g., becoming upset/physiologically aroused in response to trauma cues, hypervigilance, and exaggerated startle response) and hyporeactivity (e.g., anhedonia, the lack of interest, and detachment from others) (e.g., Frewen & Lanius, 2006; Litz, 1992; Litz & Gray, 2002). A comprehensive review of models of emotion of PTSD is beyond the scope of the current chapter. Here, we focus on symptoms, processes, and models that provide the most direct link between PTSD and emotional granularity, starting with PTSD as a disorder of arousal dysregulation.

PTSD and arousal focus

To meet criteria for DSM-5 PTSD, one has to endorse at least two of the following hyperarousal symptoms (Criterion E): (E1) Irritability or aggression, (E2) Risky or destructive behavior, (E3) Hypervigilance, (E4) Heightened startle reaction, (E5) Difficulty concentrating, or (E5) Difficulty sleeping. While the DSM-5 lumps all of these symptoms together, confirmatory factor analysis studies have suggested that hypervigilance and exaggerated startle are hyperarousal symptoms specifically associated with PTSD, with the other Criteria E symptoms being more indicative of nonspecific distress (Simms, Watson, & Doebbeling, 2002; Yufik & Simms, 2010) or dysphoric arousal (Armour, Műllerová, & Elhai, 2016). Regardless of the exact configuration, hyperarousal symptoms are central to all models of PTSD, and longitudinal research has identified hyperarousal symptoms as the cluster most predictive of subsequent levels of other symptom clusters (Marshall, Schell, Glynn, & Shetty, 2006; Schell, Marshall, & Jaycox, 2004).

The importance of hyperarousal symptoms to PTSD has contributed to conceptualizations of PTSD as a disorder of arousal dysregulation. Frewen and Lanius (2006) noted that PTSD is characterized by both an inability to manage or downregulate arousal levels, particularly after being exposed to reminders of the trauma or after experiencing symptoms such as flashbacks or nightmares, and, at other times, an inability to upregulate their level of arousal leading to hypoarousal symptoms such as anhedonia, loss of interest, and detachment from others. In addition to an individual diagnosed with PTSD vacillating between hypo- and hyperaroused states, there appears to be individual differences in the manifestation of PTSD, with some individuals exhibiting more hyperarousal reactions relative to hypoarousal and others exhibiting the opposite pattern. Based on neuroimaging findings, Lanius et al. (2010) specified two subtypes of PTSD, one characterized by undermodulation of arousal and predominantly by hyperarousal and intrusions and the other characterized by dissociation. Lanius et al. (2010) reviewed psychophysiological and neural studies showing that a majority of individuals diagnosed with PTSD (approximately 70% according to their estimate) showed an "undermodulation" pattern of reactivity to characterized by an increase in heart rate when remembering their traumatic event, with a substantial minority (30%) exhibiting an "overmodulation" pattern of reactivity showing no increase in heart rate when remembering their traumatic event.

Before making a connection between PTSD and emotional granularity via arousal, we will discuss one important variable in the relationship between PTSD and arousal, anxiety sensitivity. Anxiety sensitivity is a trait-like cognitive characteristic that is defined as the fear of anxiety-related bodily sensations (McNally, 2002). Anxiety sensitivity has been found to be associated with all anxiety disorders; however, anxiety sensitivity levels are more elevated in PTSD relative to all other anxiety disorders with the exception of panic disorder (Taylor, 2003). Longitudinal studies have substantiated the association between PTSD and anxiety sensitivity. Marshall, Miles, and Stewart (2010) assessed PTSD symptoms and ASI within days of a traumatic physical injury that resulted in a hospital visit, at 6-month follow-up, and at 12-month follow-up. Cross lagged panel analysis indicated a reciprocal relationship between anxiety sensitivity and PTSD symptoms across time; that is, PTSD predicted subsequent changes in anxiety sensitivity and vice versa. Again using data from individuals ($n=677$) who went to the hospital because of a traumatic injury, Zahradnik, Stewart, Marshall, Schell, and Jaycox (2009) showed that anxiety sensitivity and alexithymia both significantly independently predicted PTSD symptoms within 1 month of the injury, with effect size estimates indicating a larger association between anxiety sensitivity and PTSD relative to the association between alexithymia and PTSD. In sum, there is strong evidence suggesting that the tendency to interpret the bodily cues of arousal as dangerous plays a role in the development and manifestation of PTSD symptoms.

A plethora of empirical evidence has established arousal dysregulation as a core component of PTSD. As the brief review earlier indicates, PTSD is associated with periods of both hyper- and hypoarousal, with meaningful variation across individuals in the prominent pattern of arousal (i.e., undermodulators characterized more by hyperarousal responses relative to overmodulators who display more hypoarousal). Furthermore, empirical research has documented a robust relationship between anxiety sensitivity, or the tendency to interpret physical responding of arousal as dangerous, and PTSD. We contend that examining PTSD from an emotional granularity perspective can contribute to the ongoing process of developing a more specific, accurate, and comprehensive understanding of the role of arousal in PTSD. Empirical research is needed to explicate the precise nature of the relationship between emotional granularity and PTSD. Let us consider a few possibilities. Intuitively, since arousal is a key feature of PTSD, it is reasonable to hypothesize a positive correlation between the arousal focus dimension and PTSD. However, we hypothesize the opposite. Arousal focus refers to the degree to which one incorporates arousal into representations of emotions. Conceptually, arousal focus is distinct from the frequency and intensity that a person experiences arousal. However, processes that contribute to dysfunctional arousal (too high, too frequent, too low, not frequent enough, vacillation between hypo- and hyperaroused states, fear of bodily arousal cues, etc.) likely impair one's ability to effectively incorporate information about arousal into their representations of emotion; thus, we predict a negative association between PTSD and the arousal focus.

PTSD and valence focus

PTSD may also be associated with valence focus. Again, valence focus refers to the degree to which individuals incorporate information about hedonic tone (unpleasantness-pleasantness) into representations of emotion. Behavioral research consistently finds that individuals with PTSD demonstrate an attentional bias toward threat cues (see Buckley, Blanchard, & Neill, 2000 for a review; see Bar-Haim, Lamy, Pergamin, Bakermans-Kranenburg, & van IJzendoorn, 2007 for a meta-analysis), either due to enhanced detection of or difficulty disengaging from such cues. A number of studies using a modified Stroop paradigm show that individuals with PTSD demonstrate an interference effect to trauma-related, but not neutral, words, in that they are slower to color name high-threat words (Beck, Freeman, Shipherd, Hamblen, & Lackner, 2001; Bryant & Harvey, 1995; Cassiday, McNally, & Zeitlin, 1992; Foa, Feske, Murdock, Kozak, & McCarthy, 1991; Harvey, Bryant, & Rapee, 1996; Kaspi, McNally, & Amir, 1995; McNally, Kaspi, Riemann, & Zeitlin, 1990). Research using a visual search task showed difficulty disengaging from threat-related words, but did not support facilitated attention toward threat words, among combat veterans with high versus low levels of PTSD (Pineles, Shipherd, Welch, & Yovel, 2007). Studies utilizing the dot-probe paradigm show that individuals with PTSD demonstrate an attentional bias toward threat cues (angry faces) in comparison with traumatized controls (Fani et al., 2012).

The research demonstrating attentional bias to threat cues in PTSD is also supported by brain imaging methods (e.g., fMRI, PET, and SPECT) that have shown increased amygdala activation in response to threatening stimuli among individuals with PTSD (see Etkin & Wager, 2007 for a meta-analysis and McNally, 2006 for a review) including trauma-related script-driven imagery (Shin et al., 1997; Shin, Orr, Carson, et al., 2004), sounds reminiscent of combat situations (Liberzon et al., 1999), trauma-related words (Protopopescu et al., 2005), and fearful facial expressions (Bryant et al., 2008; Felmingham et al., 2010; Rauch et al., 2000; Shin, Wright, Cannistraro, et al., 2005; Williams et al., 2006). Individuals with PTSD show exaggerated amygdala reactivity during both conscious (Shin et al., 2005; Williams et al., 2006) and unconscious (i.e., masked facial stimuli; Bryant et al., 2008; Felmingham et al., 2010; Rauch et al., 2000) processing of fearful faces, suggesting that the neural substrates of attention to threat are reactive to threat cues regardless of whether the person is aware that the threat is present or not. Additionally, research examining amygdala reactivity to masked fearful, happy, and neutral facial expressions among healthy controls and individuals with anxiety disorders (including panic DO, specific phobia, and PTSD) demonstrated increased amygdala response among individuals with anxiety disorders to both happy and fearful faces (Killgore et al., 2014). It is not clear whether these findings are due to an increased attention to valence (although happy is pleasant and fearful is negative, they both involve valence), or alternatively, amygdala reactivity may be driven by the emotional arousal depicted in the face rather than the valence. Research conducted from an emotional granularity perspective may help clarify the processes involved. We offer two tentative hypotheses. First, as mentioned earlier, we predict that the arousal dysregulation associated with PTSD will interfere with the ability to adequately include information about arousal; therefore we hypothesis an inverse relationship between PTSD and arousal focus. Second, based on research demonstrating attentional biases toward threatening information, we hypothesize a positive association between PTSD and valence focus.

Psychological constructionism, PTSD, and emotional granularity

In addition to developing and producing initial support for the construct validity of emotional granularity, Lisa Feldman Barrett (e.g., Barrett, 2009a, 2009b, 2017) has developed a psychological constructionist model of emotion to capture the considerable variability of emotional states within (i.e., instances of fear for any individual varies considerable across situations; see Russell, 2003) and between (e.g., people differ in how they experience fear) individuals. According to her model, mental states (including emotions) are being constantly constructed via several biological and psychological processes. Barrett (e.g., Barrett, 2009a, 2017) often uses a cooking analogy to describe her model, with the biological and psychological processes representing different ingredients that are mixed together to produce a mental state, and the oven representing the context in which these ingredients meld into the final product. Important ingredients include core affect, physiological changes, memory, attention, knowledge (e.g., emotion concepts), and language.

To account for the heterogeneity in the expression of PTSD, Suvak and Barrett (2011) provided a psychological constructionist account of the brain processes that contribute to PTSD. They purported that PTSD arises from dysfunction in basic mechanisms (i.e., biological and psychological ingredients) related to salience and attention, hyperarousal, working memory, and long-term memory. A particular individual's manifestation of PTSD is a function of (a) the amount and quality of basic ingredients involved (i.e., which basic mechanisms are involved and the nature of the dysfunction in these mechanisms), (b) the interaction among these ingredients and other ingredients that produce the mental states of PTSD, and (c) the context in which these basic ingredients are mixed together.

The exact psychological ingredients that contribute to emotional granularity have not been empirically investigated. However, it is likely that there is considerable overlap between the ingredients that contribute to PTSD and the ingredients that contribute to emotional granularity. Given the substantial heterogeneity in the expression of PTSD (see earlier), it is also likely that the ingredients associated with PTSD vary considerably across different manifestations of PTSD. The heterogeneity of PTSD may obscure overall associations between PTSD and emotional granularity. Therefore, future research on the association between PTSD and emotional granularity should include focus on identifying common ingredients contributing to these constructs.

Working memory is part of what Barrett (2009a) referred to as the attentional matrix, or different sources of attention that can change the rate of neural firing to focus attention and recruit necessary cognitive and behavioral resources in a given situation, a basic psychological ingredient that helps construct emotional experience. Working memory is an important executive function that controls attention and helps us negotiate the interplay of automatic and controlled processes by both maintaining multiple sources of information in an active state and manipulating this information to complete complex cognitive tasks (Barrett, Tugade, & Engle, 2004). While the relationship between emotional granularity and working memory/controlled attention has not been adequately empirically investigated, studies documenting associations between mindfulness and emotion differentiation provide an indirect link (Fogarty et al., 2015; Hill & Updegraff, 2012; Tong & Keng, 2017 & Van der Gucht et al., 2018). Van der Gucht et al. (2018) recently showed that a mindfulness intervention involving training participants to focus their attention on inner mental states and remain focused on them regardless of how unpleasant they were without judging or attempting to change them leads to significant increases in negative emotion differentiation. They also showed that changes in mindfulness skills accounted for subsequent changes in negative emotion differentiation, even when controlling for levels of negative affect. Mindfulness training has also been shown to increase working memory capacity (Mrazek, Franklin, Phillips, Baird, & Schooler, 2013).

There is a wide body of research indicating that working memory processes are disrupted among adults with PTSD (see Polak, Witteveen, Reitsma, & Olff, 2012 and Aupperle, Melrose, Stein, & Paulus, 2012 for reviews), thought to arise from an overtaxing of cognitive resources that occurs when one is attempting to engage in complex cognitive tasks while simultaneously having to attend to and set aside distressing, intrusive trauma-related thoughts and memories (Schweizer & Dalgleish, 2011). A meta-analysis of 60 studies examining data from over 4000 individuals comparing those with PTSD, trauma-exposed comparison subjects, and healthy subjects found a significant working memory deficits in PTSD, with a medium effect size ($d=-0.50$; Scott et al., 2015). One study using an emotional working memory task, requiring adults with and without PTSD to remember lists of neutral words while processing either neutral sentences or sentences of trauma-related thoughts, showed that individuals with PTSD showed working memory deficits in an emotionally laden context but not in a neutral context (Schweizer & Dalgleish, 2011). Thus, disruptions in working memory exhibited by individuals in PTSD represent a potential link between PTSD and emotional granularity that warrants further investigation.

Core affect is another important ingredient in psychological constructionist models of emotion. Russell (2003) defined core affect as "A neurophysiological state that is consciously accessible as a simple, nonreflective feeling that is an integral blend of hedonic (pleasure-displeasure) and arousal (sleepy-activated) values (p. 147)." In other words, core affect is how you feel at any given moment, and it is always present even if it is often characterized by neutral valence and deactivated levels of arousal. Core affect can be experienced as a general mood, or it can be tied to a certain object or event contributing to an emotional episode.

The core affect of most mental disorders, in particular anxiety and mood disorders, is characterized by high levels of negative affectivity or high levels of nonspecific emotional distress (captured by symptoms such as difficulty sleeping, irritability, and difficulty concentrating). Confirmatory factor analysis studies have demonstrated that these symptoms of nonspecific emotional distress (referred to as dysphoria or negative affectivity) linked PTSD to major depressive disorder and generalized anxiety disorders, two diagnosis that often co-occur with PTSD following exposures to traumatic event (Byllesby, Charak, Durham, Wang, & Elhai, 2016; Grant, Beck, Marques,

Palyo, & Clapp, 2008). Undifferentiated negative affect (sometimes referred to as negative affectivity, neuroticism, dysphoria, etc.) has been identified as a transdiagnostic construct associated with a variety of mental disorders (e.g., Krueger & Markon, 2006).

The tendency to experience nonspecific distress or negative affectivity has also been shown to be inversely related to emotion differentiation (e.g., Erbas et al., 2014), and as cited earlier, in one of the few studies that have longitudinally investigated emotion differentiation, Erbas et al., 2018) found that negative emotion differentiation varied significantly within individuals across time and that from day to day stress levels prospectively predicted negative motion differentiation, but negative emotion differentiation did not prospectively predict stress. Thus one possible link between PTSD and emotional granularity might be the tendency to experience nonspecific emotional distress, which may not be unique to PTSD but rather a transdiagnostic process common to several mental disorders characterized by high levels of negative affectivity. A comprehensive review of common psychological and biological ingredients that may link between PTSD and emotional granularity is beyond the scope of this chapter, but we want to note that adopting a psychological constructionist perspective may help elucidate the connection between PTSD and the ability to make fine-grained nuanced distinction among emotional states.

Conclusion

We hope that we provided a compelling argument that examining PTSD from an emotional granularity perspective can assist future research efforts to further refine the field's understanding of the emotional processes that contribute to and result from PTSD. Future research on the connection between PTSD and emotional granularity will also help build the nomological network of support for the construct validity of emotional granularity. As reviewed earlier, several studies have demonstrated a relationship between being able to make fine-grained associations among similar emotional states and a variety of psychosocial functioning variables including multiple different types of psychopathology with those showing higher levels of emotional granularity or emotion differentiation tending to exhibiting higher levels of functioning and lower levels of pathology relative to those who exhibit lower levels of granularity or differentiation. However, more research is needed to further elucidate the precise nature of the relationship between emotional granularity and psychopathology. No longitudinal research exists that addresses temporal precedent. As depicted in Fig. 2, deficits in emotional granularity may represent a diathesis increasing the likelihood that an individual will experience chronic psychological distress following exposure to trauma, and/or the psychological distress experienced after being exposed to traumatic stress might lead to diminished emotional granularity. The specificity in the relationship between PTSD and other forms of psychopathology also warrants further investigation. It may be that low levels of emotional granularity or emotion differentiation represent a nonspecific factor associated with psychological distress, or more interestingly, it may be that different manifestations of emotional granularity are associated with different dimensions and manifestations of psychopathology. As depicted in the right side of Fig. 2, it is possible that different symptoms of PTSD are associated with different aspects of emotional granularity. Perhaps, reexperiencing symptoms are related positively to valence focus, while hyperarousal symptoms are inversely related to arousal focus. Emotional granularity's role may include mitigating the relationship between PTSD symptoms and functional impairment. Right now, because of a lack of empirical data, this is all speculation, but exploring these possibilities will contribute to the understanding of both PTSD and emotional granularity.

As reviewed earlier, a sizable and growing body of research has documented negative associations between emotional granularity (and emotion differentiation) and several types of psychopathology and associated maladaptive behaviors and coping strategies, and we argued that future research will reveal a negative association between emotional granularity and PTSD. Of course an association does not imply causality, but while we await longitudinal and experimental research that can address temporal precedence and causality, it may be prudent to incorporate strategies to increase emotional granularity into interventions, including interventions for PTSD. In this last section, we will consider some strategies that might increase emotional granularity.

A recent review of neuroimaging studies examining the effect of psychotherapy indicated that areas implicated in coding semantic representations play an important role in enhanced emotion regulation facilitated by therapy (Messina, Sambin, Beschoner, & Viviani, 2016). Studies have also documented a relationship between the structure of one's emotional knowledge anchored in semantic knowledge or language and emotional granularity (Barrett, 2004; Suvak & Barrett, 2011) indicating that high levels of emotional granularity are partly due to possessing a nuanced cognitive map of emotions. Therefore, helping clients develop more nuanced and precise language to describe their emotional experiences may contribute to increased efficacy of interventions for emotional disorders. Work needs to be done to figure out how to most effectively and efficiently refine mental representations of emotion. Some skill-based treatments such as dialectical behavior therapy (DBT; Linehan, 1993) include modules that attempt to increase semantic emotion knowledge. Incorporating these strategies into trauma-focused and PTSD interventions, especially strategies that help clients clearly differentiate the nature and role of valence and arousal, may facilitate PTSD symptom reduction. In her recent book, Lisa Feldman Barrett (2017) argues developing one's conceptual understanding of emotions or "becoming more emotionally intelligent (p. 179)" will help a person more effectively and efficiently manage emotions. She recommended activities such as taking trips to new places, trying new foods, watching moves, and particularly activities that increase one's vocabulary (such as learning a foreign language or reading books out of one's comfort range). Therefore encouraging clients to participate in activities that will further develop their semantic emotion knowledge may increase granularity and, in turn, reduce symptoms.

In addition to developing conceptual understanding of emotion and related processes, interoception represents a potential target of intervention to increase emotional granularity. Interoception is a person's ability to sense bodily signals such as temperature, pain, touch, hunger, thirst, muscular sensations, and feelings in the viscera (Craig, 2002), which is a system that is hypothesized to be integrated with emotional awareness, as evidenced by joint activation of the anterior insula cortex and anterior cingulate cortex implicated in bodily interoception, during studies measuring emotional experiences (Craig, 2009). In two separate studies, Barrett et al. (2004) showed that individuals high in arousal focus exhibited heightened interoceptive sensitivity, as measured by their performance on a heartbeat detection task, suggesting that the degree to which individuals access internal bodily cues impacts the specificity of their emotional experience. Mindfulness and biofeedback training represents two potential ways to increase granularity by increasing interoceptive ability. In a randomized controlled trial, Mehling et al. (2018) found improvement on three subscales of the Multidimensional Assessment of Interoceptive Awareness (MAIA), emotional awareness, self-regulation, and body listening, in war veterans with PTSD participating in a 12-week mindfulness/exercise intervention compared with veterans assigned to a waitlist control condition. There is a small, but growing, body of research demonstrating that biofeedback (Petta, 2017; Reyes, 2014) and particularly neurofeedback (e.g., van der Kolk et al., 2016) can significantly reduce PTSD symptoms on its or in combination with other empirically supported PTSD treatments. Although not yet empirically investigated, it is possible that increased emotional granularity is an important mechanism of action in bio-/neurofeedback PTSD symptom reduction and could possibly be a mechanism underlying currently used empirically supported treatments.

References

Armour, C., Műllerová, J., & Elhai, J. D. (2016). A systematic literature review of PTSD's latent structure in the diagnostic and statistical manual of mental disorders: DSM-IV to DSM-5. *Clinical Psychology Review*, *44*, 60–74.

Aupperle, R. L., Melrose, A. J., Stein, M. B., & Paulus, M. P. (2012). Executive function and PTSD: Disengaging from trauma. *Neuropharmacology*, *62*(2), 686–694. https://doi.org/10.1016/j.neuropharm.2011.02.008.

Bagby, R. M., Parker, J. D. A., & Taylor, G. J. (1994). The twenty-item Toronto alexithymia scale-I. Item selection and cross-validation of the factor structure. *Journal of Psychosomatic Research*, *38*(1), 23–32. https://doi.org/10.1016/0022-3999(94)90005-1.

Balaban, H., Semiz, M., Şentürk, İ. A., Kavakçı, Ö., Çınar, Z., Dikici, A., & Topaktaş, S. (2012). Migraine prevalence, alexithymia, and post-traumatic stress disorder among medical students in Turkey. *The Journal of Headache and Pain*, *13*(6), 459–467. https://doi-org.ezproxysuf.flo.org/10.1007/s10194-012-0452-7.

Bardeen, J. R., Kumpula, M. J., & Orcutt, H. K. (2013). Emotion regulation difficulties as a prospective predictor of posttraumatic stress symptoms following a mass shooting. *Journal of Anxiety Disorders*, *27*(2), 188–196. https://doi.org/10.1016/j.janxdis.2013.01.003.

Bar-Haim, Y., Lamy, D., Pergamin, L., Bakermans-Kranenburg, M. J., & van IJzendoorn, M. H. (2007). Threat-related attentional bias in anxious and nonanxious individuals: A meta-analytic study. *Psychological Bulletin*, *133*, 1–24.

Barrett, L. F. (1995). Valence focus and arousal focus: Individual differences in the structure of affective experience. *Journal of Personality and Social Psychology*, *69*, 153–166.

Barrett, L. F. (2004). Feelings or words? Understanding the content in self-report ratings of experienced emotion. *Journal of Personality and Social Psychology*, *87*(2), 266–281. https://doi-org.ezproxysuf.flo.org/10.1037/0022-3514.87.2.266.

Barrett, L. F. (2009a). The future of psychology: Connecting mind to brain. *Perspectives in Psychological Science*, *4*, 326–339.

Barrett, L. F. (2009b). Variety is the spice of life: A psychological construction approach to understanding variability in emotion. *Cognition and Emotion*, *23*(7), 1284–1306.

Barrett, L. F. (2017). *How emotions are made: The secret life of the brain*. New York, NY: Houghton Mifflin Harcourt.

Barrett, L. F., Gross, J., Christensen, T. C., & Benvenuto, M. (2001). Knowing what you're feeling and knowing what to do about it: Mapping the relation between emotion differentiation and emotion regulation. *Cognition and Emotion*, *15*(6), 713–724. https://doi-org.ezproxysuf.flo.org/10.1080/02699930143000239.

Barrett, L. F., Tugade, M. M., & Engle, R. W. (2004). Individual differences in working memory capacity and dual-process theories of the mind. *Psychological Bulletin*, *130*(4), 553–573. https://doi-org.ezproxysuf.flo.org/10.1037/0033-2909.130.4.553.

Beck, J. G., Freeman, J. B., Shipherd, J. C., Hamblen, J. L., & Lackner, J. M. (2001). Specificity of Stroop interference in patients with pain and PTSD. *Journal of Abnormal Psychology*, *110*(4), 536.

Bell, K. M., & Naugle, A. E. (2008). The role of emotion recognition skills in adult sexual revictimization. *The Journal of Behavior Analysis of Offender and Victim Treatment and Prevention*, *1*(4), 93–118. https://doi-org.ezproxysuf.flo.org/10.1037/h0100459.

Berke, D. S., Macdonald, A., Poole, G. M., Portnoy, G. A., McSheffrey, S., Creech, S. K., & Taft, C. T. (2017). Optimizing trauma-informed intervention for intimate partner violence in veterans: The role of alexithymia. *Behaviour Research and Therapy*, *97*, 222–229. https://doi-org.ezproxysuf.flo.org/10.1016/j.brat.2017.08.007.

Bockers, E., Roepke, S., Michael, L., Renneberg, B., & Knaevelsrud, C. (2014). Risk recognition, attachment anxiety, self-efficacy, and state dissociation predict revictimization. *PLoS One*, *9*(9). Retrieved from http://ezproxysuf.flo.org/login?url=http://search.ebscohost.com/login.aspx?direct=true&db=psyh&AN=2014-40445-001&site=ehost-live.

Bonanno, G. A., & Burton, C. L. (2013). Regulatory flexibility: An individual differences perspective on coping and emotion regulation. *Perspectives on Psychological Science*, *8*(6), 591–612.

Bornovalova, M. A., Ouimette, P., Crawford, A. V., & Levy, R. (2009). Testing gender effects on the mechanisms explaining the association between post-traumatic stress symptoms and substance use frequency. *Addictive Behaviors*, *34*(8), 685–692. https://doi-org.ezproxysuf.flo.org/10.1016/j.addbeh.2009.04.005.

Brenner, I., & Ben-Amitay, G. (2015). Sexual revictimization: The impact of attachment anxiety, accumulated trauma, and response to childhood sexual abuse disclosure. *Violence and Victims*, *30*(1), 49–65.

Bryant, R. A., & Harvey, A. G. (1995). Processing threatening information in posttraumatic stress disorder. *Journal of Abnormal Psychology*, *104*, 537–541.

Bryant, R. A., Kemp, A. H., Felmingham, K. L., Liddell, B., Olivieri, G., Peduto, A., … Williams, L. M. (2008). Enhanced amygdala and medial prefrontal activation during nonconscious processing of fear in posttraumatic stress disorder: An fMRI study. *Human Brain Mapping*, *29*, 517–523. https://doi.org/10.1002/hbm.20415.

Buckley, T. C., Blanchard, E. B., & Neill, W. T. (2000). Information processing and PTSD: A review of the empirical literature. *Clinical Psychology Review*, *20*, 1041–1065. https://doi.org/10.1016/S0272-7358(99)00030-6.

Byllesby, B. M., Charak, R., Durham, T. A., Wang, X., & Elhai, J. D. (2016). The underlying role of negative affect in the association between PTSD, major depressive disorder, and generalized anxiety disorder. *Journal of Psychopathology and Behavioral Assessment*, *38*(4), 655–665. https://doi-org.ezproxysuf.flo.org/10.1007/s10862-016-9555-9.

Cassiday, K. L., McNally, R., & Zeitlin, S. B. (1992). Cognitive processing of trauma cues in rape victims with post-traumatic stress disorder. *Cognitive Therapy and Research*, *16*, 283–295.

Charuvastra, A., & Cloitre, M. (2008). Social bonds and posttraumatic stress disorder. *Annual Review of Psychology*, *59*, 301–328. https://doi-org.ezproxysuf.flo.org/10.1146/annurev.psych.58.110405.085650.

Chen, Z. S., & Chung, M. C. (2016). The relationship between gender, posttraumatic stress disorder from past trauma, alexithymia and psychiatric co-morbidity in Chinese adolescents: A moderated mediational analysis. *Psychiatric Quarterly*, *87*(4), 689–701. https://doi-org.ezproxysuf.flo.org/10.1007/s11126-016-9419-1.

Chung, M. C., & Allen, R. D. (2013). Alexithymia and posttraumatic stress disorder following epileptic seizure. *Psychiatric Quarterly*, *84*(3), 271–285. https://doi-org.ezproxysuf.flo.org/10.1007/s11126-012-9243-1.

Chung, M. C., Allen, R. D., & Dennis, I. (2013). The impact of self-efficacy, alexithymia and multiple traumas on posttraumatic stress disorder and psychiatric co-morbidity following epileptic seizures: A moderated mediation analysis. *Psychiatry Research*, *210*(3), 1033–1041. https://doi-org.ezproxysuf.flo.org/10.1016/j.psychres.2013.07.041.

Chung, M. C., Rudd, H., & Wall, N. (2012). Posttraumatic stress disorder following asthma attack (post-asthma attack PTSD) and psychiatric co-morbidity: The impact of alexithymia and coping. *Psychiatry Research*, *197*(3), 246–252.

Classen, C. C., Muller, R. T., Field, N. P., Clark, C. S., & Stern, E.-M. (2017). A naturalistic study of a brief treatment program for survivors of complex trauma. *Journal of Trauma & Dissociation, 18*(5), 720–734. https://doi-org.ezproxysuf.flo.org/10.1080/15299732.2017.1289492.

Craig, A. D. (2002). How do you feel? Interoception: The sense of the physiological condition of the body. *Nature Reviews Neuroscience, 3*, 655–666. https://doi.org/10.1038/nrn894.

Craig, A. D. (2009). How do you feel—Now? The anterior insula and human awareness. *Nature Reviews Neuroscience, 10*, 59–70. https://doi.org/10.1038/nrn2555.

Doolan, E. L., Bryant, R. A., Liddell, B. J., & Nickerson, A. (2017). The conceptualization of emotion regulation difficulties, and its association with posttraumatic stress symptoms in traumatized refugees. *Journal of Anxiety Disorders, 50*, 7–14. https://doi-org.ezproxysuf.flo.org/10.1016/j.janxdis.2017.04.005.

Eichhorn, S., Brähler, E., Franz, M., Friedrich, M., & Glaesmer, H. (2014). Traumatic experiences, alexithymia, and posttraumatic symptomatology: A cross-sectional population-based study in Germany. *European Journal of Psychotraumatology, 5*, Retrieved from: http://ezproxysuf.flo.org/login?url=http://search.ebscohost.com/login.aspx?direct=true&db=psyh&AN=2015-02809-001&site=ehost-live.

Erbas, Y., Ceulemans, E., Kalokerinos, E. K., Houben, M., Koval, P., Pe, M. L., & Kuppens, P. (2018). Why I don't always know what I'm feeling: The role of stress in within-person fluctuations in emotion differentiation. *Journal of Personality and Social Psychology, 115*(2), 179–191. https://doi-org.ezproxysuf.flo.org/10.1037/pspa0000126.supp. [Supplemental].

Erbas, Y., Ceulemans, E., Lee Pe, M., Koval, P., & Kuppens, P. (2014). Negative emotion differentiation: Its personality and well-being correlates and a comparison of different assessment methods. *Cognition and Emotion, 28*(7), 1196–1213.

Etkin, A., & Wager, T. D. (2007). Functional neuroimaging of anxiety: A meta-analysis of emotional processing in PTSD, social anxiety disorder, and specific phobia. *American Journal of Psychiatry, 164*, 1476–1488.

Evren, C., Dalbudak, E., Durkaya, M., Cetin, R., & Evren, B. (2010). Interaction of life quality with alexithymia, temperament and character in male alcohol-dependent inpatients. *Drug and Alcohol Review, 29*(2), 177–183.

Fani, N., Tone, E. B., Phifer, J., Norrholm, S. D., Bradley, B., Ressler, K. J., ... Jovanovic, T. (2012). Attention bias toward threat is associated with exaggerated fear expression and impaired extinction in PTSD. *Psychological Medicine, 42*(3), 533–543.

Felmingham, K., Williams, L. M., Kemp, A. H., Liddell, B., Falconer, E., Peduto, A., & Bryant, R. (2010). Neural responses to masked fear faces: Sex differences and trauma exposure in posttraumatic stress disorder. *Journal of Abnormal Psychology, 119*, 241–247.

Fine, N. B., Achituv, M., Etkin, A., Merin, O., & Shalev, A. Y. (2018). Evaluating web-based cognitive-affective remediation in recent trauma survivors: Study rationale and protocol. *European Journal of Psychotraumatology, 9*(1), https://doi.org/10.1080/20008198.2018.1442602.

Foa, E. B., Feske, U., Murdock, T. B., Kozak, M. J., & McCarthy, P. R. (1991). Processing of threat-related information in rape victims. *Journal of Abnormal Psychology, 100*, 156–162.

Fogarty, F. A., Lu, L. M., Sollers, J. J., Krivoschekov, S. G., Booth, R. J., & Consedine, N. S. (2015). Why it pays to be mindful: Trait mindfulness predicts physiological recovery from emotional stress and greater differentiation among negative emotions. *Mindfulness, 6*(2), 175–185. https://doi.org/10.1007/s12671-013-0242-6.

Frewen, P. A., Dozois, D. J. A., Neufeld, R. W. J., & Lanius, R. A. (2008). Meta-analysis of alexithymia in posttraumatic stress disorder. *Journal of Traumatic Stress, 21*(2), 243–246. https://doi-org.ezproxysuf.flo.org/10.1002/jts.20320.

Frewen, P. A., & Lanius, R. A. (2006). Toward a psychobiology of posttraumatic self-dysregulation: Reexperiencing, hyperarousal, dissociation, and emotional numbing. In R. Yehuda & R. Yehuda (Eds.), *Psychobiology of Posttraumatic Stress Disorders: A Decade of Progress* (pp. 110–124). Malden: Blackwell Publishing.

Gaher, R. M., O'Brien, C., Smiley, P., & Hahn, A. M. (2016). Alexithymia, coping styles and traumatic stress symptoms in a sample of veterans who experienced military sexual trauma. *Stress and Health: Journal of the International Society for the Investigation of Stress, 32*(1), 55–62. https://doi-org.ezproxysuf.flo.org/10.1002/smi.2578.

Galatzer-Levy, I. R., & Bryant, R. A. (2013). 636,120 ways to have posttraumatic stress disorder. *Perspectives on Psychological Science, 8*(6), 651–662.

Gao, W., Zhao, J., Li, Y., & Cao, F. (2015). Post-traumatic stress disorder symptoms in first-time myocardial infarction patients: Roles of attachment and alexithymia. *Journal of Advanced Nursing, 71*(11), 2575–2584. https://doi-org.ezproxysuf.flo.org/10.1111/jan.12726.

Gillihan, S. J., Cahill, S. P., & Foa, E. B. (2014). Psychological theories of PTSD. In M. J. Friedman, T. M. Keane, P. A. Resick, M. J. Friedman, T. M. Keane, & P. A. Resick (Eds.), *Handbook of PTSD: Science and practice* (pp. 166–184). New York, NY, USA: Guilford Press.

Grant, D. M., Beck, J. G., Marques, L., Palyo, S. A., & Clapp, J. D. (2008). The structure of distress following trauma: Posttraumatic stress disorder, major depressive disorder, and generalized anxiety disorder. *Journal of Abnormal Psychology, 117*(3), 662–672. https://doi.org/10.1037/a0012591.

Gratz, K. L., & Roemer, L. (2004). Multidimensional assessment of emotion regulation and dysregulation: Development, factor structure, and initial validation of the difficulties in emotion regulation scale. *Journal of Psychopathology and Behavioral Assessment, 26*, 41–54. doihttps://doi.org/10.1023/B:JOBA.0000007455.08539.94.

Halpern, J., Maunder, R. G., Schwartz, B., & Gurevich, M. (2012). Identifying, describing, and expressing emotions after critical incidents in paramedics. *Journal of Traumatic Stress, 25*(1), 111–114. https://doi-org.ezproxysuf.flo.org/10.1002/jts.21662.

Harvey, A. G., Bryant, R. A., & Rapee, R. M. (1996). Preconscious processing of threat in posttraumatic stress disorder. *Cognitive Therapy and Research, 20*, 613–623.

Hetzel-Riggin, M. D., & Meads, C. L. (2016). Interrelationships among three avoidant coping styles and their relationship to trauma, peritraumatic distress, and posttraumatic stress disorder. *Journal of Nervous and Mental Disease, 204*(2), 123–131. https://doi-org.ezproxysuf.flo.org/10.1097/NMD.0000000000000434.

Hill, C. L. M., & Updegraff, J. A. (2012). Mindfulness and its relationship to emotional regulation. *Emotion, 12*(1), 81–90. https://doi.org/10.1037/a0026355.

Kang, S., & Shaver, P. R. (2004). Individual differences in emotional complexity: Their possible psychological implications. *Journal of Personality, 72*, 687–726.

Kashdan, T. B., Barrett, L. F., & McKnight, P. E. (2015). Unpacking emotion differentiation: Transforming unpleasant experience by perceiving distinctions in negativity. *Current Directions in Psychological Science, 24*(1), 10–16. https://doi.org/10.1177/0963721414550708.

Kaspi, S. P., McNally, R. J., & Amir, N. (1995). Cognitive processing of emotional information in posttraumatic stress disorder. *Cognitive Therapy and Research, 19*, 433–444.

Kernberg, O. F. (1967). Borderline personality organization. *Journal of the American Psychoanalysis Association, 15*, 641–685.

Kernberg, O. F. (1975). *Borderline conditions and pathological narcissism.* New York: Jason Aronson, Inc..

Kernberg, O. F. (1976). *Object relations theory and clinical psychoanalysis.* New York: Jason Aronson, Inc..

Kienle, J., Rockstroh, B., Bohus, M., Fiess, J., Huffziger, S., & Steffen-Klatt, A. (2017). Somatoform dissociation and posttraumatic stress syndrome—Two sides of the same medal? A comparison of symptom profiles, trauma history and altered affect regulation between patients with functional neurological symptoms and patients with PTSD. *BMC Psychiatry, 17*, Retrieved from: http://ezproxysuf.flo.org/login?url=http://search.ebscohost.com/login.aspx?direct=true&db=psyh&AN=2017-30634-001&site=ehost-live.

Killgore, W. D., Britton, J. C., Schwab, Z. J., Price, L. M., Weiner, M. R., Gold, A. L., ... Rauch, S. L. (2014). Cortico-limbic responses to masked affective faces across PTSD, panic disorder, and specific phobia. *Depression and Anxiety*, *31*(2), 150–159.

Kooiman, C. G., Spinhoven, P., & Trijsburg, R. W. (2002). The assessment of alexithymia: A critical review of the literature and a psychometric study of the Toronto Alexithymia Scale-20. *Journal of Psychosomatic Research*, *53*(6), 1083–1090.

Krueger, R. F., & Markon, K. E. (2006). Reinterpreting comorbidity: A model-based approach to understanding and classifying psychopathology. *Annual Review of Clinical Psychology*, *2*, 111–133.

Kušević, Z., Ćusa, B. V., Babić, G., & Marčinko, D. (2015). Could alexithymia predict suicide attempts—A study of Croatian war veterans with post-traumatic stress disorder. *Psychiatria Danubina*, *27*(4), 420–423. Retrieved from: http://ezproxysuf.flo.org/login?url=http://search.ebscohost.com/login.aspx?direct=true&db=psyh&AN=2015-54669-012&site=ehost-live.

Lanius, R. A., Vermetten, E., Loewenstein, R. J., Brand, B., Schmahl, C., Bremner, J. D., & Spiegel, D. (2010). Emotion modulation in PTSD: Clinical and neurobiological evidence for a dissociative subtype. *The American Journal of Psychiatry*, *167*(6), 640–647. https://doi-org.ezproxysuf.flo.org/10.1176/appi.ajp.2009.09081168.

Liberzon, I., Taylor, S. F., Amdur, R., Jung, T. D., Chamberlain, K. R., Minoshima, S., ... Fig, L. M. (1999). Brain activation in PTSD in response to trauma-related stimuli. *Biological Psychiatry*, *45*, 817–826. https://doi.org/10.1016/S0006-3223(98)00246-7.

Lindquist, K., & Barrett, L. F. (2008). Emotional complexity. In M. Lewis, J. M. Haviland-Jones, & L. F. Barrett (Eds.), *The handbook of emotion*. (3rd ed., pp. 513–530). New York: Guilford.

Linehan, M. M. (1993). *Cognitive-behavioral treatment of borderline personality disorder*. New York: Guilford Press.

Lischetzke, T., Angelova, R., & Eid, M. (2011). Validating an indirect measure of clarity of feelings: Evidence from laboratory and naturalistic settings. *Psychological Assessment*, *23*, 447–455.

Litz, B. T. (1992). Emotional numbing in combat-related post-traumatic stress disorder: A critical review and reformulation. *Clinical Psychology Review*, *12*(4), 417–432. https://doi.org/10.1016/0272-7358(92)90125-R.

Litz, B. T., & Gray, M. J. (2002). Emotional numbing in posttraumatic stress disorder: Current and future research directions. *Australian and New Zealand Journal of Psychiatry*, *36*(2), 198–204. https://doi.org/10.1046/j.1440-1614.2002.01002.x.

Marshall, G. N., Miles, J. V., & Stewart, S. H. (2010). Anxiety sensitivity and PTSD symptom severity are reciprocally related: Evidence from a longitudinal study of physical trauma survivors. *Journal of Abnormal Psychology*, *119*(1), 143–150. https://doi.org/10.1037/a0018009.

Marshall, G. N., Schell, T. L., Glynn, S. M., & Shetty, V. (2006). The role of hyperarousal in the manifestation of posttraumatic psychological distress following injury. *Journal of Abnormal Psychology*, *115*(3), 624–628. https://doi.org/10.1037/0021-843X.115.3.624.

McFarlane, A. C. (2000). Posttraumatic stress disorder: A model of the longitudinal course and the role of the risk factors. *The Journal of Clinical Psychiatry*, *61*(Suppl. 5), 15–23.

McLean, C. P., & Foa, E. B. (2017). Emotions and emotion regulation in posttraumatic stress disorder. *Current Opinion in Psychology*, *14*, 72–77. https://doi-org.ezproxysuf.flo.org/10.1016/j.copsyc.2016.10.006.

McNally, R. J. (2002). Anxiety sensitivity and panic disorder. *Biological Psychiatry*, *52*(10), 938–946. https://doi-org.ezproxysuf.flo.org/10.1016/S0006-3223(02)01475-0.

McNally, R. J. (2006). Cognitive abnormalities in post-traumatic stress disorder. *Trends in Cognitive Sciences*, *10*, 271–277. https://doi.org/10.1016/j.tics.2006.04.007.

McNally, R. J., Kaspi, S. P., Riemann, B. C., & Zeitlin, S. B. (1990). Selective processing of threat cues in posttraumatic stress disorder. *Journal of Abnormal Psychology*, *99*, 398–402.

Mehling, W. E., Chesney, M. A., Metzler, T. J., Goldstein, L. A., Maguen, S., Geronimo, C., ... Neylan, T. C. (2018). A 12-week integrative exercise program improves self-reported mindfulness and interoceptive awareness in war veterans with posttraumatic stress symptoms. *Journal of Clinical Psychology, 74*(4), 554–565.

Messina, I., Sambin, M., Beschoner, P., & Viviani, R. (2016). Changing views of emotion regulation and neurobiological models of the mechanism of action of psychotherapy. *Cognitive, Affective, & Behavioral Neuroscience, 16*, 571–587. https://doi.org/10.3758/s13415-016-0440-5.

Mikulincer, M., Shaver, P. R., & Pereg, D. (2003). Attachment theory and affect regulation: The dynamics, development, and cognitive consequences of attachment-related strategies. *Motivation and Emotion, 27*(2), 77–102.

Mikulincer, M., Shaver, P. R., & Solomon, Z. (2015). An attachment perspective on traumatic and posttraumatic reactions. In M. Safir, H. Wallach, & A. Rizzo (Eds.), *Future Directions in Post-Traumatic Stress Disorder* (pp. 79–96). Boston, MA: Springer.

Mrazek, M. D., Franklin, M. S., Phillips, D. T., Baird, B., & Schooler, J. W. (2013). Mindfulness training improves working memory capacity and GRE performance while reducing mind wandering. *Psychological Science, 24*(5), 776–781. https://doi-org.ezproxysuf.flo.org/10.1177/0956797612459659.

O'Brien, C., Gaher, R. M., Pope, C., & Smiley, P. (2008). Difficulty identifying feelings predicts the persistence of trauma symptoms in a sample of veterans who experienced military sexual trauma. *Journal of Nervous and Mental Disease, 196*(3), 252–255. https://doi-org.ezproxysuf.flo.org/10.1097/NMD.0b013e318166397d.

Petta, L. M. (2017). Resonance frequency breathing biofeedback to reduce symptoms of subthreshold PTSD with an Air Force Special Tactics operator: A case study. *Applied Psychophysiology and Biofeedback, 42*(2), 139–146.

Pineles, S. L., Shipherd, J. C., Welch, L. P., & Yovel, I. (2007). The role of attentional biases in PTSD: Is it interference or facilitation? *Behaviour Research and Therapy, 45*, 1903–1913. https://doi.org/10.1016/j.brat.2006.08.021.

Polak, A. R., Witteveen, A. B., Reitsma, J. B., & Olff, M. (2012). The role of executive function in posttraumatic stress disorder: A systemic review. *Journal of Affective Disorders, 141*, 11–21. https://doi.org/10.1016/j.jad.2012.01.001.

Polusny, M. A., Dickinson, K. A., Murdoch, M., & Thuras, P. (2008). The role of cumulative sexual trauma and difficulties identifying feelings in understanding female veterans' physical health outcomes. *General Hospital Psychiatry, 30*(2), 162–170. https://doi-org.ezproxysuf.flo.org/10.1016/j.genhosppsych.2007.11.006.

Pond, R. S., Jr., Kashdan, T. B., DeWall, C. N., Savostyanova, A., Lambert, N. M., & Fincham, F. D. (2012). Emotion differentiation moderates aggressive tendencies in angry people: A daily diary analysis. *Emotion, 12*, 326–337. https://doi.org/10.1037/a0025762.

Powers, A., Fani, N., Cross, D., Ressler, K. J., & Bradley, B. (2016). Childhood trauma, PTSD, and psychosis: Findings from a highly traumatized, minority sample. *Child Abuse & Neglect, 58*, 111–118. https://doi-org.ezproxysuf.flo.org/10.1016/j.chiabu.2016.06.015.

Protopopescu, X., Pan, H., Tuescher, O., Cloitre, M., Goldstein, M., Engelien, W., ... Stern, E. (2005). Differential time courses and specificity of amygdala activity in posttraumatic stress disorder subjects and normal control subjects. *Biological Psychiatry, 57*, 464–473. https://doi.org/10.1016/j.biopsych.2004.12.026.

Rauch, S. L., Whalen, P. J., Shin, L. M., McInerney, S. C., Macklin, M. L., Lasko, N. B., ... Pitman, R. K. (2000). Exaggerated amygdala response to masked facial stimuli in posttraumatic stress disorder: A functional MRI study. *Biological Psychiatry, 47*, 769–776. https://doi.org/10.1016/S0006-3223(00)00828-3.

Reese-Weber, M., & Smith, D. M. (2011). Outcomes of child sexual abuse as predictors of later sexual victimization. *Journal of Interpersonal Violence, 26*(9), 1884–1905.

Resick, P. A., & Miller, M. W. (2009). Posttraumatic stress disorder: Anxiety or traumatic stress disorder? *Journal of Traumatic Stress, 22*(5), 384–390. https://doi.org/10.1002/jts.20437.

Reyes, F. J. (2014). Implementing heart rate variability biofeedback groups for veterans with posttraumatic stress disorder. *Biofeedback*, *42*(4), 137–142.
Robinson, M. D., & Clore, G. L. (2002a). Belief and feeling: Evidence for an accessibility model of emotional self-report. *Psychological Bulletin*, *128*(6), 934–960. https://doi-org. ezproxysuf.flo.org/10.1037/0033-2909.128.6.934.
Robinson, M. D., & Clore, G. L. (2002b). Episodic and semantic knowledge in emotional self-report: Evidence for two judgment processes. *Journal of Personality and Social Psychology*, *83*(1), 198–215. https://doiorg.ezproxysuf.flo.org/10.1037/0022-3514.83.1.198.
Russell, J. A. (1980). A circumplex model of affect. *Journal of Personality and Social Psychology*, *39*, 1161–1178.
Russell, J. A. (2003). Core affect and the psychological construction of emotion. *Psychological Review*, *110*(1), 145.
Russell, J. A., & Barrett, L. F. (1999). Core affect, prototypical emotional episodes, and other things called emotion: Dissecting the elephant. *Journal of Personality and Social Psychology*, *76*(5), 805–819. https://doi-org.ezproxysuf.flo.org/10.1037/0022-3514.76.5.805.
Salovey, P., Mayer, J. D., Goldman, S. L., Turvey, C., & Palfai, T. (1995). Emotional attention, clarity, and repair: Exploring emotional intelligence using the Trait Meta-Mood Scale. In J. Pennebaker (Ed.), *Emotion, Disclosure, and Health*. Washington, DC: American Psychological Association.
Schell, T. L., Marshall, G. N., & Jaycox, L. H. (2004). All symptoms are not created equal: The prominent role of hyperarousal in the natural course of posttraumatic psychological distress. *Journal of Abnormal Psychology*, *113*(2), 189–197. https://doi. org/10.1037/0021-843X.113.2.189.
Schoenleber, M., Berghoff, C. R., Gratz, K. L., & Tull, M. T. (2018). Emotional lability and affective synchrony in posttraumatic stress disorder pathology. *Journal of Anxiety Disorders*, *53*, 68–75.
Schwarz, N. (1990). Feelings as information: Informational and motivational functions of affective states. In E. T. Higgins & R. Sorentino (Eds.), Vol. 2. *Handbook of motivation and cognition: Foundations of social behavior* (pp. 527–561). New York, NY: Guilford Press.
Schweizer, S., & Dalgleish, T. (2011). Emotional working memory capacity in posttraumatic stress disorder (PTSD). *Behaviour Research and Therapy*, *49*, 498–504. https://doi. org/10.1016/j.brat.2011.05.007.
Scott, J. C., Matt, G. E., Wrocklage, K. M., Crnich, C., Jordan, J., Southwick, S. M., Krystal, J. H., et al. (2015). A quantitative meta-analysis of neurocognitive functioning in posttraumatic stress disorder. *Psychological Bulletin*, *141*, 105–140. https://doi.org/10.1037/a0038039.
Shaver, P. R., & Mikulincer, M. (2002). Attachment-related psychodynamics. *Attachment & Human Development*, *4*(2), 133–161.
Shepherd, L., & Wild, J. (2014). Emotion regulation, physiological arousal and PTSD symptoms in trauma-exposed individuals. *Journal of Behavior Therapy and Experimental Psychiatry*, *45*(3), 360–367.
Shin, L. M., & Handwerger, K. (2009). Is posttraumatic stress disorder a stress induced fear circuitry disorder? *Journal of Traumatic Stress*, *22*, 409–415. https://doi.org/10.1002/jts.20442.
Shin, L. M., McNally, R. J., Kosslyn, S. M., Thomson, W. L., Rauch, S. L., Alpert, N. M., … Pitman, R. K. (1997). A positron emission tomographic study of symptom provocation in PTSD. *Annals of the New York Academy of Sciences*, *821*, 521–523. https://doi. org/10.1111/j.1749-6632.1997.tb48320.x.
Shin, L. M., Orr, S. P., Carson, M. A., et al. (2004). Regional cerebral blood flow in the amygdala and medial prefrontal cortex during traumatic imagery in male and female Vietnam veterans with PTSD. *Archives of General Psychiatry*, *61*, 168–176. https://doi. org/10.1001/archpsyc.61.2.168.
Shin, L. M., Wright, C. I., Cannistraro, P. A., et al. (2005). A functional magnetic resonance imaging study of amygdala and medial prefrontal cortex responses to overtly presented fearful faces in posttraumatic stress disorder. *Archives of General Psychiatry*, *62*, 273–281. https://doi.org/10.1001/archpsyc.62.3.273.

Short, N. A., Norr, A. M., Mathes, B. M., Oglesby, M. E., & Schmidt, N. B. (2016). An examination of the specific associations between facets of difficulties in emotion regulation and posttraumatic stress symptom clusters. *Cognitive Therapy and Research, 40*(6), 783–791. https://doi-org.ezproxysuf.flo.org/10.1007/s10608-016-9787-8.

Simms, L. J., Watson, D., & Doebbeling, B. N. (2002). Confirmatory factor analyses of posttraumatic stress symptoms in deployed and nondeployed veterans of the Gulf War. *Journal of Abnormal Psychology, 111*(4), 637–647.

Smidt, K. E., & Suvak, M. K. (2015). A brief, but nuanced, review of emotional granularity and emotion differentiation research. *Current Opinion in Psychology, 3*, 48–51. https://doi.org/10.1016/j.copsyc.2015.02.007.

Suvak, M. K., & Barrett, L. F. (2011). Considering PTSD from the perspective of brain processes: A psychological construction approach. *Journal of Traumatic Stress, 24*(1), 3–24. https://doi-org.ezproxysuf.flo.org/10.1002/jts.20618.

Taylor, S. (2003). Anxiety sensitivity and its implications for understanding and treating PTSD. *Journal of Cognitive Psychotherapy, 17*(2), 179–186. https://doi.org/10.1891/jcop.17.2.179.57431.

Terock, J., Van der Auwera, S., Janowitz, D., Spitzer, C., Barnow, S., Miertsch, M., … Grabe, H.-J. (2016). From childhood trauma to adult dissociation: The role of PTSD and alexithymia. *Psychopathology, 49*(5), 374–382. https://doi-org.ezproxysuf.flo.org/10.1159/000449004.

Timoney, L. R., & Holder, M. D. (2013). *Emotional Processing Deficits and Happiness: Assessing the Measurement, Correlates, and Well-Being of People With Alexithymia*. New York, NY: Springer Science + Business Media. https://doi-org.ezproxysuf.flo.org/10.1007/978-94-007-7177-2.

Tong, E. M. W., & Keng, S.-L. (2017). The relationship between mindfulness and negative emotion differentiation: A test of multiple mediation pathways. *Mindfulness, 8*(4), 933–942. https://doi.org/10.1007/s12671-016-0669-7.

Tripp, J. C., McDevitt-Murphy, M. E., Avery, M. L., & Bracken, K. L. (2015). PTSD symptoms, emotion dysregulation, and alcohol-related consequences among college students with a trauma history. *Journal of Dual Diagnosis, 11*(2), 107–117. https://doi-org.ezproxysuf.flo.org/10.1080/15504263.2015.1025013.

Tull, M. T., Barrett, H. M., McMillan, E. S., & Roemer, L. (2007). A preliminary investigation of the relationship between emotion regulation difficulties and posttraumatic stress symptoms. *Behavior Therapy, 38*(3), 303–313. https://doi-org.ezproxysuf.flo.org/10.1016/j.beth.2006.10.001.

Tull, M. T., Weiss, N. H., Adams, C. E., & Gratz, K. L. (2012). The contribution of emotion regulation difficulties to risky sexual behavior within a sample of patients in residential substance abuse treatment. *Addictive Behaviors, 37*(10), 1084–1092. https://doi-org.ezproxysuf.flo.org/10.1016/j.addbeh.2012.05.001.

Van der Gucht, K., Dejonckheere, E., Erbas, Y., Takano, K., Vandemoortele, M., Maex, E., … Kuppens, P. (2018). An experience sampling study examining the potential impact of a mindfulness-based intervention on emotion differentiation. *Emotion*, https://doi.org/10.1037/emo0000406.

van der Kolk, B. A., Hodgdon, H., Gapen, M., Musicaro, R., Suvak, M. K., Hamlin, E., & Spinazzola, J. (2016). A randomized controlled study of neurofeedback for chronic PTSD. *PLoS One, 11*(12).

Viana, A. G., Hanna, A. E., Woodward, E. C., Raines, E. M., Paulus, D. J., Berenz, E. C., & Zvolensky, M. J. (2018). Emotional clarity, anxiety sensitivity, and PTSD symptoms among trauma-exposed inpatient adolescents. *Child Psychiatry and Human Development, 49*(1), 146–154. https://doi-org.ezproxysuf.flo.org/10.1007/s10578-017-0736-x.

Wang, X., Chung, M. C., Hyland, M. E., & Bahkeit, M. (2011). Posttraumatic stress disorder and psychiatric co-morbidity following stroke: The role of alexithymia. *Psychiatry Research, 188*(1), 51–57. https://doi-org.ezproxysuf.flo.org/10.1016/j.psychres.2010.10.002.

Weiss, N. H., Tull, M. T., Anestis, M. D., & Gratz, K. L. (2013). The relative and unique contributions of emotion dysregulation and impulsivity to posttraumatic stress disorder among substance dependent inpatients. *Drug and Alcohol Dependence, 128*(1–2), 45–51. https://doi-org.ezproxysuf.flo.org/10.1016/j.drugalcdep.2012.07.017.

Williams, L. M., Kemp, A. H., Felmingham, K., Barton, M., Olivieri, G., Peduto, A., … Bryant, R. A. (2006). Trauma modulates amygdala and medial prefrontal responses to consciously attended fear. *NeuroImage, 29*, 347–357. https://doi.org/10.1016/j.neuroimage.2005.03.047.

Yehuda, R., & LeDoux, J. (2007). Response variation following trauma: A translational neuroscience approach to understanding PTSD. *Neuron, 56*(1), 19–32.

Yufik, T., & Simms, L. J. (2010). A meta-analytic investigation of the structure of posttraumatic stress disorder symptoms. *Journal of Abnormal Psychology, 119*(4), 764–776.

Zahradnik, M., Stewart, S. H., Marshall, G. N., Schell, T. L., & Jaycox, L. H. (2009). Anxiety sensitivity and aspects of alexithymia are independently and uniquely associated with posttraumatic distress. *Journal of Traumatic Stress, 22*(2), 131–138. https://doi.org/10.1002/jts.20397.

Zaki, L. F., Coifman, K. G., Rafaeli, E., Berenson, K. R., & Downey, G. (2013). Emotion differentiation as a protective factor against nonsuicidal self-injury in borderline personality disorder. *Behavior Therapy, 44*(3), 529–540.

CHAPTER 14
Experiential avoidance and PTSD

Holly K. Orcutt, Anthony N. Reffi, Robyn A. Ellis
Northern Illinois University, DeKalb, IL, United States

What is experiential avoidance?

Experiential avoidance (EA) is the natural tendency of humans to avoid threatening or uncomfortable experiences. The universality of this instinct can be attributed to its adaptiveness. Evolutionarily, taking actionable steps toward reducing fear or anxiety meant increasing chances of survival and thus the propagation of this behavioral response. Conversely the failure to respond to these emotions appropriately likely increased the chances of serious injury or death. For instance, a healthy fear of heights might deter otherwise risk-taking people from dangerous, high-elevation conditions, thereby allowing them to ultimately pass down this cautious disposition. However, within the context of nonthreatening, objectively safe situations, this instinct to avoid discomfort does not necessarily entail a corresponding decrease in danger and instead can paradoxically invite more pain and suffering into experience. This is the problem of EA.

EA reflects an aversion to sitting with distressing present-moment experience *and* subsequent efforts to eliminate or reduce this distress, even when doing so is antithetical to one's values (Hayes, Wilson, Gifford, Follette, & Strosahl, 1996). The source of this distress could include thoughts, emotions, physical sensations, memories, behavioral tendencies and impulses, people, or situations (Boulanger, Hayes, & Pistorello, 2010). Given EA's function of downregulating distress, many researchers have endeavored to delineate the boundaries between EA, emotion regulation, and specific regulatory strategies (e.g., Boulanger et al., 2010; Chapman, Dixon-Gordon, & Walters, 2011; Chawla & Ostafin, 2007; Kashdan, Barrios, Forsyth, & Steger, 2006).

Distinguishing experiential avoidance from emotion regulation

EA has been conceptualized as falling within the broader construct of emotion regulation (e.g., Sheppes, Suri, & Gross, 2015). More precisely, EA drives emotion regulation efforts by overestimating the costs of experiencing or maintaining an emotional state (e.g., "I cannot handle the anxiety"), coupled with an overvaluation of the effectiveness of avoidant strategies (Sheppes et al., 2015). Subsequently, avoidant strategies (e.g., distraction)—the perceived benefits of which are informed by prior experiences of negative reinforcement—are readily accessible, chosen, and implemented to reduce distress arising from unwanted private experiences. Such manifestations of experiential avoidance are varied.

In its most observable forms, EA may include substance use; binge eating; risky sexual behavior; physical avoidance of people, places, and situations; self-harm; and suicide attempts (Blackledge, 2004; Chapman et al., 2011; Chawla & Ostafin, 2007). An example of overt EA is illustrated when combat veterans elect to take a seat facing the exit, as this functions to reduce (i.e., avoid) the anxiety associated with having their back to the door. However, EA also operates through covert, private acts such as thought, emotion, and expressive suppression; worry; rumination; and distraction (Blackledge, 2004). For example, ruminative thoughts focused on "what if" questions that attempt to "undo" a trauma divert attention away from facing the reality of what occurred, in turn impeding processing (Elwood, Hahn, Olatunji, & Williams, 2009). However, EA is less a matter of the avoidant *forms* of emotion regulation than it is the *function* of such regulatory attempts. Put simply, EA differs from emotion regulation insofar as EA serves as the function of emotion regulation (Boulanger et al., 2010).

There is empirical basis for this distinction between EA and emotion regulation. In a daily diary study, EA emerged as a more robust predictor of daily psychological well-being and social anxiety compared with emotion suppression and cognitive reappraisal (Kashdan et al., 2006). Further, EA fully or partially mediated the effects of suppression and reappraisal on these outcomes, respectively. Along with other studies (e.g., Tull & Gratz, 2008), this highlights EA's role as a mechanism connecting emotion regulation with psychological outcomes—specific regulatory strategies can become problematic when employed for the purposes of EA. For example, a woman whose husband is deployed in the US army may become distressed over his safety when she sees coverage of the war on the news. She may choose to suppress the urge to cry in front of her daughter to prevent causing her worry. Moreover, she may continue to suppress this urge even while alone because she "cannot deal" with experiencing those emotions. Notably, emotion regulation is evidenced in both scenarios, yet EA is present only in the latter. The first situation is not because of an unwillingness to feel sad or the perceived need to control it and as such would not be considered EA, whereas this is the exact motivation for her suppressing her tears while alone. This raises the question of how EA develops.

The role of language

According to relational frame theory (RFT; Hayes, Barnes-Holmes, & Roche, 2001), EA is rooted in our ability to use language (Blackledge, 2004; Blackledge & Hayes, 2001; Hayes et al., 1996; Törneke, 2010). Without this higher-order process, EA could not exist. To illustrate this point, consider a child repeatedly hearing the name of her uncle as he enters the room (e.g., "Hi, John!"). The child sees her uncle as others provide contextual cues by speaking with him and using his name. Eventually the child learns a relation between these two stimuli (seeing uncle ⇨ "John"). Importantly the child will also spontaneously *derive* (or infer) the inverse relation ("John" ⇨ seeing uncle), meaning a new relation is formed without explicit training (i.e., not learned through experience), and the child will now expect to see John after hearing his name (e.g., Törneke, 2010). For the child the two stimuli have been rendered *equivalent*—she looks at her uncle and understands that he *is* John (seeing uncle ⇔ "John") and now pictures a mental image of her uncle in response to the name "John" (whether it is heard, spoken, thought, etc.). To understand how this ability is tied to language, contrast this example with another of training a dog a similar relation.

If the dog repeatedly hears someone call out "John" as its owner returns home, this will be followed by seeing its owner and the subsequent reinforcement of the dog's excitement. The dog will now get excited and run to the door to greet its owner when it hears his name being called (hear "John" ⇨ see owner). However, the dog does not derive the inverse relationship (see owner ⇨ hear "John"), such that we would not expect it to think of the name "John" when it sees its owner. Unlike the child, we do not believe that the dog understands that its owner *is* John, only that hearing "John" precedes seeing its owner. This and similar observations have been made in past research (e.g., Devany, Hayes, & Nelson, 1986; Heagle & Rehfeldt, 2006; McHugh, Barnes-Holmes, & Barnes-Holmes, 2004; Sidman & Tailby, 1982) and have been observed only in humans (Dugdale & Lowe, 2000; Hayes, 1989; Hayes et al., 2001). These processes form the basis for language development (Devany et al., 1986; Hayes et al., 2001) and have consequences for our behavior and engagement in EA.

Due to our ability to derive new relationships between stimuli, we are also able to transform *stimulus functions* (Törneke, 2010). A stimulus' function refers to how that stimulus influences an organism's behavior (Törneke, 2010). In the previous example of the dog that runs to the door to greet its owner, "John" is merely a sound that signals the initiation of this behavior due to prior learning (hearing "John" ⇨ seeing owner). In this case the stimulus "John" carries the function of approaching the front door, though this function can be transformed. For instance, if John became an abusive owner, the function of "John" would be transformed to become avoidance. Extending this clinically, consider a young boy whose father abuses him. Before the abuse began the stimulus function of his father would have been approach (based on prior receipt of love, affection, and warmth). However, the abuse invokes negative reactions (e.g., fear, pain, and racing heart) and accordingly has the function of avoidance. Through conditioning the boy will learn that seeing his father precedes being abused (seeing father ⇨ abuse), such that seeing his father triggers these same abuse reactions, effectively *transforming* the function of seeing his father from that of approach to avoidance. The use of language allows for an exponential amount of transformations of stimulus functions, because they can be born out of derived relations not only direct experience.

As described previously the spontaneously derived, inverse relation means that the experience of abuse reactions now triggers mental images of his father (seeing father ⇔ abuse). Similarly the word "father" represents the boy's father and vice versa (the word "father" ⇔ the boy's father). As a result a relation is derived between the word "father" and the abuse, despite this word never being paired with the abuse (see image in the succeeding text). Consequently, this equivalency transfers the function of the abuse (i.e., avoidance) to the word "father," thus producing the same behavioral response (e.g., Dougher, Augustson, Markham, Greenway, & Wulfert, 1994; Dougher, Hamilton, Fink, & Harrington, 2007). This might lead the boy to avoid talking about his father because it produces feelings similar to the abuse.

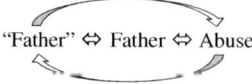
"Father" ⇔ Father ⇔ Abuse

Notably, humans do not simply endure these reactions but use language to *evaluate* what they mean (Ehlers & Clark, 2000; Hayes et al., 1996). These appraisals are similarly informed by prior experiences. Since the boy learned that abuse ⇨ racing heart, then he can again derive the inverse and thus equivalency relation, racing heart ⇔ abuse, in turn prompting the thought, "The abuse is happening again." Together with cognitive fusion—the belief that thoughts are equivalent to reality—this thought is perceived as the abuse happening all over again. Such fusion with private experience blurs the boundaries between these experiences and what they represent, leading to negatively appraising them and their implications (e.g., "I should not be feeling this way"). Together, these appraisals further motivate EA, casting downstream effects on psychological functioning and well-being.

Summary

Regardless of form, EA functions to escape or avoid unpleasant, uncomfortable, or distressing private experiences. Language permits an infinite variety of stimuli to be labeled as unwanted, even those that are not directly experienced (e.g., a therapist *hearing* details about the abuse). Together with cognitive fusion, language dictates what ought to be avoided (e.g., accelerated heart rate). In the short term, such efforts to avoid do in fact provide relief, thereby reinforcing and maintaining EA. Over time, these *verbal rules*[a] (e.g., "I cannot handle hearing about child abuse") promote EA that in turn reinforces these evaluations, as EA prevents the corrective experience of what would have otherwise occurred (e.g., habituation). In other words, avoiding an unwanted private experience maintains the belief in the feared outcome (e.g., "I will lose control of my emotions") because doing so precludes an opportunity to test whether that belief is true (Blackledge, 2004). Overall, there is ample evidence that this process is maladaptive, as EA is seen across psychopathology (Chawla & Ostafin, 2007; Spinhoven, Drost, de Rooij, van Hemert, & Penninx, 2014). Although this seemingly suggests that EA is a transdiagnostic *symptom*, it may be more useful to infer psychopathology as the *consequence* of rigid EA (Blackledge & Hayes, 2001; Hayes et al., 1996). Though it is certainly not unique to PTSD, EA is nonetheless intimately linked with the onset, maintenance, and remittance of this condition.

Experiential avoidance and posttraumatic stress disorder

A link between EA and PTSD symptoms (PTSS) is well established, particularly cross sectionally (see, e.g., Seligowski, Lee, Bardeen, & Orcutt, 2015 for a metaanalytic review). What is less clear, however, is what it might mean when we say that EA and PTSD are related. To start, if we imagine both constructs as overlapping circles in a Venn diagram, the margins

---

[a] Not unlike "stuck points" as discussed in cognitive processing therapy (CPT; Resick, Monson, & Chard, 2017).

are fuzzy at best. How is EA distinct or similar to what Spinhoven et al. (2014) refer to as "bordering psychological constructs" such as worry and anxiety sensitivity? These bordering constructs can be expanded to include alexithymia, rumination, thought suppression, and avoidance coping. With regard to PTSD, controversy exists at present as to the diagnostic features, as evidenced by the discrepancies between DSM-5, the current ICD 10, and the proposed changes to ICD 11 with such discrepancies resulting in different diagnostic classifications (Wisco et al., 2017). Further, more than 600,000 possible symptom permutations reveal very distinct presentations of PTSD in DSM-5 (Galatzer-Levy & Bryant, 2013).

Aside from the fuzziness of both EA and PTSD, unpacking this relationship is further complicated by the fact that EA focuses on avoidance of negative private experiences while a key diagnostic feature of PTSD is avoidance of negative private experiences. This conceptual overlap raises a critical and complex question as to whether observed relationships between PTSS and EA, particularly cross sectionally, reflect a "double counting" of avoidance across both constructs rather than a substantive relationship.

Cross-sectional designs are limited in the ability to parse out the relationship between EA and PTSD. Longitudinal designs, albeit still correlational, allow for temporal sequencing that may inform our understanding of this relationship. Prior to reviewing the extant longitudinal research, however, it is useful to start by exploring possible models for the relation between EA and PTSD.

Models of experiential avoidance and posttraumatic stress disorder

Spinhoven et al. (2014), in their longitudinal investigation of the relation between EA and the development of anxiety and depressive disorders in a longitudinal sample of adults in the Netherlands, adopted a useful classification structure employed by Klein, Kotov, and Bufferd (2011). Klein et al. (2011) presented seven possible models for the relationship between personality and mood disorders. In thinking about EA and PTSD, the following three models were identified as most relevant and will be reviewed briefly: common cause, predisposition, and pathoplasticity.

First the *common cause* model views EA and PTSD as having similar causal influences without specifying that one has a causal influence on the other. Second the *predisposition* model posits that EA plays a causal role in the development of PTSD and that the etiologic processes that underlie EA differ from those that underlie PTSD. The predisposition model is consistent with a diathesis-stress model and involves mediating and moderating relationships among risk factors (Klein et al., 2011). Third the *pathoplasticity* model is similar to the predisposition model in viewing EA as having a causal influence on PTSD. However, rather than viewing EA as contributing to the onset of PTSD, the pathoplasticity model posits that EA would causally influence the expression of the disorder after onset. Specifically, EA would influence the symptomatology, course, and response to treatment among individuals with PTSS.

Longitudinal evidence
Pretrauma EA
Evidence in support of the predisposition model would seem to require assessment of EA prior to the index trauma to examine whether EA plays a potentially causal role in the development of PTSD. In our review of the literature, we identified two data sets that have an EA (or bordering construct) assessment prior to an index trauma. The first data set, the Northern Illinois University (NIU) Trauma Study, is from our research team and involves a sample of college women who were enrolled in a longitudinal study at the time that a mass shooting occurred on our campus in 2008. We were in the unique position to have preshooting assessments of EA, trauma history, and PTSS. Two publications from this data set are relevant to the question of whether EA may play a role as a predisposition to PTSD (Kumpula, Orcutt, Bardeen, & Varkovitzky, 2011; Orcutt, Bonanno, Hannan, & Miron, 2014).

Using three timepoints (preshooting, approximately 1 month post shooting, and approximately 8 months post shooting) and a cross lagged panel design, we asked whether EA prospectively predicted the four DSM-IV PTSD symptom clusters proposed by the Simms/Watson dysphoria model (Simms, Watson, & Doebbeling, 2002) in 532 college women (Kumpula et al., 2011). The cross lagged panel design included the four clusters of PTSS and EA at all three timepoints with the inclusion of peritraumatic dissociation and a shooting exposure variable at the second timepoint. Results indicate that preshooting EA significantly predicted intrusions and dysphoria 1 month post shooting and that EA 1 month post shooting significantly predicted dysphoria and hyperarousal symptoms 8 months later. Preshooting PTSS did not predict increased EA at the latter two timepoints. In this case, we were able to look specifically at whether the relationship between EA and PTSS was being carried by the avoidance cluster. We found that preshooting EA was marginally predictive of PTSD avoidance symptoms 1 month post shooting and EA was not predictive of PTSD avoidance symptoms 8 months post shooting. The strongest and most consistent relations between EA and PTSD symptoms were with the dysphoria cluster.

In Orcutt et al. (2014), we utilized all seven timepoints (including the preshooting timepoint) to examine the trajectory of PTSS over approximately 30 months following the shooting. Growth mixture modeling suggested the presence of four classes in our sample—a minimal impact-resilience class (65.2% of the sample) with very little PTSS except a slight increase immediately following the shooting; a high impact-recovery class (25.0%), which included slightly higher level of PTSS preshooting and a dramatic increase immediately post shooting and then equally dramatic decrease in symptoms by the timepoint 6 months later; a moderate impact–moderate symptom class (7.9%), which reported moderately high PTSS at each timepoint with an increase immediately post shooting; and finally a chronic dysfunction class (1.8%), which included endorsement of approximately 12 PTSS at each timepoint, which did not increase following the shooting. Preshooting EA was examined as a potential predictor of class membership and was found to be significantly related to membership in the chronic dysfunction class as compared with those in both the resilient and recovery classes. In interpreting this finding, it is important to consider that the chronic dysfunction class was the smallest group and included individuals who were reporting elevated levels of PTSS preshooting. Thus, in this particular class, the causal direction is less clear. Stronger evidence for the predisposition model would have included a relationship between preshooting EA and the other three classes who were less symptomatic preshooting.

Data from the NIU Trauma Study provide some evidence in support of the predisposition model, particularly the cross lagged model across three timepoints broken out by PTSD cluster. The trajectory analyses over seven timepoints are less clear-cut in support of the predisposition model. Importantly the timepoint between the preshooting assessment and the shooting was quite variable because of the nature of the longitudinal study in progress when the shooting occurred. Although the effect sizes are small, these data suggest that EA may confer risk for PTSD following trauma exposure.

The second known data set with a pretrauma assessment of an EA-related construct was in response to a disaster on campus. Gil (2005) had preattack measures of avoidance coping on 81 students exposed to a terrorist explosion. Students had been assessed 2 weeks prior to the explosion and were then assessed 1 month and 6 months post attack. Trait coping was assessed before the attack, state coping was assessed 1 month post attack, and PTSS were assessed 6 months post attack. Trait and state avoidance coping both predicted PTSS 6 months post attack. PTSD symptoms were not assessed before the attack; however, so it is unknown how many participants might have had existing PTSD symptoms.

Posttrauma EA

Although lacking pretrauma assessment, Ehlers, Mayou, and Bryant (1998) found that negative interpretations of intrusive recollections from a car accident and thought suppression significantly predicted PTSS 3 months and 1 year post accident in a large longitudinal study of individuals presenting to the emergency room (ER) ($N = 781$ 1 year post accident). In this case the assessment of EA factors occurred prior to the development of PTSD but after the index trauma; however, both the negative interpretations of intrusive recollections and the thought suppression are accident specific as well, so preaccident assessment would not have been possible.

In another study employing an ER design, Thompson, Fiorillo, Rothbaum, Ressler, and Michopoulos (2018) assessed individuals experiencing a range of trauma exposures. Participants were first assessed in the ER within a few hours of the trauma exposure and then assessed 1, 3, and 6 months after the index trauma ($n = 164$ at 6 months). Multiple avoidant coping strategies were assessed at baseline (in the ER) and were found to be positively associated with PTSD over time. PTSD was not assessed at baseline and thus was not used as a control for later analyses.

A review of additional longitudinal studies examining PTSS and EA (and bordering constructs such as avoidance coping, rumination, and thought suppression) identified studies that are consistent with the pathoplasticity model. These studies examine whether EA is relevant to the symptom expression, course of the disorder, and response to treatment. With regard to EA, specifically, Marx and Sloan (2005) examined 185 students in an introductory psychology course who had reported trauma exposure and were subsequently assessed 4 weeks later. A randomly selected subset of 70 participants also completed follow-up questionnaires an additional 4 weeks later. In this sample, limited by low to moderate level of PTSS severity, EA significantly predicted PTSS severity at both 4 and 8 weeks later, controlling for baseline levels of PTSS severity.

In a sample of 236 veterans who served in post 9/11 conflict in Iraq and Afghanistan, Meyer and colleagues (Meyer, La Bash, et al., 2018) found that Time 1 assessment of EA (post deployment) significantly predicted severity of PTSD symptoms 1 year later. The predictive power of EA on subsequent PTSD severity remained significant after controlling for a number of variables including Time 1 levels of PTSD avoidance symptoms, neuroticism, combat exposure, perceived threat, and peritraumatic dissociation.

Rumination, viewed as a cognitive avoidance strategy, was examined in 73 participants who had been assaulted within the 3 months prior to the Time 1 assessment (Michael, Halligan, Clark, & Ehlers, 2007). Of these participants, 37% were positive for a PTSD diagnosis. Rumination at Time 1, as measured by an interview, was predictive of PTSD severity 6 months later, controlling for Time 1 PTSD diagnostic status.

The remaining identified longitudinal studies examine avoidance coping and PTSS across a range of samples. In all cases, coping and PTSS were assessed following trauma exposure. Avoidance coping at the initial timepoint was found to predict later PTSD severity in nontreatment-seeking military reservists following the Gulf War (Benotsch et al., 2000). In a second Gulf War sample, Sharkansky et al. (2000) examined the ratio of avoidance coping to approach coping strategies used during the conflict and found that, although it did significantly predict PTSS cross sectionally at Time 1, it did not significantly predict Time 2 PTSS after controlling for other model variables. In a large study of adolescents who experienced the 2008 Wenchuan earthquake, An, Fu, Wu, Lin, and Zhang (2013) employed a cross lagged panel model with three timepoints (1 year, 1.5 years, and 2 years post earthquake) and three variables at each timepoint (neuroticism, avoidance coping, and PTSS). PTSS predicted increased use of avoidant coping across time and, whereas avoidant coping at Time 1 did not predict PTSD symptoms at Time 2, avoidant coping at Time 2 did significantly predict PTSS at Time 3. In a study of 255 Israeli soldiers at two timepoints (1 year and 2 years post 1982 Lebanon War), Solomon, Mikulincer, and Flum (1989) found that PTSS at Time 1 was associated with increased avoidance coping strategies. Additionally, Solomon et al. examined intervening stressful life events between Time 1 and Time 2 and found that the negative life events were most impactful on PTSS at Time 2 among those individuals using avoidance coping strategies (specifically emotion-focused coping). Finally, Tiet et al. (2006) examined the impact of behavioral versus avoidance coping strategies among 178 veterans with chronic PTSD. Cognitive avoidance (but not behavioral) assessed at baseline significantly predicted PTSS 10 months later (controlling for baseline PTSS), and baseline PTSS predicted increased use of behavioral avoidance (but not cognitive) coping strategies.

EA and treatment response
Several studies have examined how avoidant coping may impact response to PTSD treatment. Leiner, Kearns, Jackson, Astin, and Rothbaum (2012)

investigated how PTSD outcomes were predicted by pretreatment avoidant coping (e.g., avoiding thoughts about the rape). Women with primary diagnoses of PTSD who had been raped within the past 3 months were randomized to receive either prolonged exposure (PE; Foa, Hembree, & Rothbaum, 2007) or eye movement desensitization and reprocessing (EMDR; Shapiro, 2001). After controlling for initial PTSD symptoms, higher levels of pretreatment avoidance predicted less PTSD symptom severity posttreatment. These relationships remained consistent even after excluding the PTSD avoidance symptoms from posttreatment scores to account for potential overlap with avoidant coping. Moreover, Leiner and colleagues (2012) found that greater levels of pretreatment avoidant coping increased the likelihood of a meaningful treatment response. In other words, women with more avoidant coping styles prior to beginning treatment experienced the largest benefits posttreatment. Consequently, those with less reliance on avoidance pretreatment did not experience significant gains in PTSD symptom recovery. If avoidant coping is not pervasive, we might not expect that efforts toward further reducing avoidance would be effective and confer an improvement in symptoms (Leiner et al., 2012). Unfortunately, we cannot extrapolate from these conclusions whether EA will continue to influence the presentation of PTSD beyond an immediate posttreatment assessment. Moreover, the EA PTSD relationship appears more complicated when dealing with treatment-resistant, chronic PTSD.

To examine the links between PTSD and general avoidance coping (i.e., not trauma event specific—denial, behavioral disengagement, and substance use) both within and outside the context of treatment, another study measured these variables at intake, discharge, and 4 months after treatment (Badour, Blonigen, Boden, Feldner, & Bonn-Miller, 2012). Treatment was administered through a residential PTSD program for mostly male veterans with chronic PTSD who had previously not responded to outpatient treatment. In this study, veterans received group therapy that predominantly followed a cognitive behavioral framework. In short,[b] greater use of avoidance at intake predicted more severe PTSS at discharge, which in turn predicted greater use of avoidance at follow-up. These findings are inconsistent with the previous study that found greater pretreatment avoidance predicts less posttreatment PTSS severity (cf. Leiner et al., 2012) and might

---

[b] Controlling for the stability of PTSD symptom severity (and length of stay), pretreatment avoidant coping positively predicted PTSD symptom severity at discharge, but avoidant coping at discharge did not predict PTSD at follow-up. Controlling for the stability of avoidance coping, PTSD at intake was not associated with avoidance at discharge, though PTSD at discharge was associated with greater avoidance at follow-up (Badour et al., 2012).

reflect the chronicity of PTSD in Badour and colleagues' sample. Whereas the women in Leiner et al.'s study may have had more room to improve, this sample did not exhibit substantial treatment gains[c] (Badour et al., 2012). Without pronounced improvements in avoidance or PTSD, the finding that higher pretreatment avoidance predicted more severe PTSD symptoms at discharge seems to suggest that preexisting avoidance may have maintained symptom severity. This may also reflect differences in treatment modality. Perhaps, there would have been greater reduction in avoidance and thus PTSD symptoms, if veterans received an evidence-based psychotherapy for PTSD (EBP; e.g., PE, CPT, and EMDR), as was seen by Leiner and colleagues. Finally, elevated PTSS severity at discharge predicted greater use of avoidance at follow-up. This is consistent with previous research outside of treatment contexts (e.g., Benotsch et al., 2000) and fits with evidence that maladaptive regulatory strategies (e.g., avoidance) are linked with more psychopathology than adaptive ones (e.g., acceptance [Aldao, Nolen-Hoeksema, & Schweizer, 2010]).

In a similar, difficult-to-treat veteran sample, change in avoidant coping was examined during cognitive behavioral-oriented groups (Boden, Bonn-Miller, Vujanovic, & Drescher, 2012). Although there was a significant reduction in PTSS severity, there was no statistically significant difference in pre- to posttreatment avoidant coping. Nonetheless the authors did find that avoidant coping had lessened, and these changes were important for outcomes. Specifically, decreases in avoidant coping between intake and discharge predicted lower PTSS severity at discharge (adjusting for initial PTSS at intake, length of stay, and trauma severity). These relationships remained consistent when analyzing how change in coping predicted each PTSD symptom cluster, suggesting these findings are not attributable to overlap between PTSD avoidance/numbing symptoms and avoidant coping. Overall, findings offer support for a pathoplastic model of EA by demonstrating its influence on treatment outcomes, thereby highlighting the relevance of reducing EA to facilitate treatment success.

---

[c] See Table 1 in Badour et al. (2012) for average PTSD and avoidance scores across timepoints. Overall, PTSD symptom severity significantly lessened from intake to follow-up. Specifically, there was a decrease in symptoms from intake to discharge, but an increase in symptoms was seen from discharge to follow-up. This is consistent with the treatment-resistant nature of this sample's PTSD. Avoidance coping also significantly decreased from intake to follow-up. Although no differences were found between intake and discharge, avoidance decreased from discharge to follow-up (Badour et al., 2012).

Thus, with regard to the pathoplasticity model, EA and bordering constructs, particularly avoidant coping strategies, appear to be associated with prospective prediction of PTSS, suggesting that EA following trauma exposure confers a vulnerability to the development and/or worsening of PTSD symptoms. Importantly, there is also some evidence to suggest some bidirectional relationships in which avoidance may increase symptoms and that the symptoms may increase avoidance.

Summary

As noted earlier the construct fuzziness and definitional overlap between the avoidance in EA and PTSS complicate our ability to speak definitively to the nature of the relationship between EA and PTSD. However, the extant longitudinal evidence, although quite limited with regard to addressing whether EA predisposes the development of PTSD following trauma exposure, is consistent with EA as a risk factor for the development of PTSD. It is very difficult to obtain longitudinal data in which EA and trauma history/PTSS are obtained before an index trauma, and the existing studies to do so have been in response to unanticipated disasters occurring when relevant data are already in hand. Further, there is consistent evidence in the literature for a pathoplastic model between EA (and bordering constructs) and PTSS, with several studies having documented how treatment success is impacted by both preexisting (e.g., Badour et al., 2012; Leiner et al., 2012) and changing levels of EA (e.g., Boden et al., 2012), although discrepant findings would suggest these variables remain bidirectionally related (Boden et al., 2013). These findings are especially meaningful because they persisted beyond other baseline characteristics, such as initial PTSS severity, age, length of treatment, and trauma severity. Nonetheless, what is largely missing from the posttraumatic literature regarding the EA PTSD relationship is a discussion of *why* EA is consistently related to PTSD and *how* the bidirectional relationship may function.

Possible mechanisms of the EA PTSD relationship

As noted previously, EA has been implicated in other forms of psychopathology (e.g., depression, borderline personality disorder, and generalized anxiety disorder [e.g., see Chawla & Ostafin, 2007, and Spinhoven et al., 2014, for reviews]). As such the extant literature investigating these disorders' relationship to EA may offer insight into potential mechanisms that drive the EA PTSD relationship and offer evidence for the three proposed models (i.e., common cause, predisposition, and pathoplasticity).

Inflexible use of experiential avoidance

EA may not be maladaptive in and of itself. Rather, EA appears to promote PTSD when it is applied inflexibly and across situations. The bereavement literature implicates EA as an adaptive component to processing grief after the loss of a loved one but, when used exclusively or excessively, can lead to complicated grief reactions (Shear, 2010). Bowlby (1980) suggests that in the wake of a death the emotional pain is reconciled over time through the dynamic integration of the painful memory with positive memories of the deceased individual, wherein avoidance plays a role in the oscillation between approaching and avoiding the painful emotions regarding the death. This oscillation is driven by the *flexible* use of EA as the individual approaches the painful emotions to the degree they can manage and then utilizes EA for emotional reprieve. In theory the more an individual utilizes EA flexibly, the less often they will need to use EA overall. As individuals oscillate between processing and avoidance, they process more of the death, ultimately leading to less painful emotions and therefore less need to engage in EA. In contrast, maladaptive avoidance is characterized by a more rigid, stable utilization of EA to avoid experiencing the painful emotions. Those reliant on avoidance to cope with their painful emotions may not process the painful memory and therefore not experience reductions in painful emotions, further maintaining their reliance on EA to cope (Shear, 2010). The relationship between EA and bereavement processing may parallel processing in the wake of a traumatic experience and offer a potential mechanism for the role of EA in natural recovery and the development of PTSD.

Similar to bereavement, traumatic exposure can bring about strong, painful emotions and memories that are difficult to reconcile with previous experiences. Additionally, bereavement is associated with feelings of anger and self-blame (Shear, 2010), emotions also considered natural in the wake of a trauma (e.g., Ehlers & Clark, 2000). Given the adaptive role of avoidance in processing bereavement, EA may function similarly in the processing of traumatic events, within the context of both natural recovery and the development of PTSD. When an individual is exposed to trauma, they may engage in more EA due to the presence of posttraumatic symptoms (e.g., intrusive memories and nightmares), but those who naturally recover may oscillate between processing and avoidance. Over time, those who oscillate facilitate integration of new information, and the emotional intensity may subside, leading to recovery; the maladaptive reliance on EA in those who go on to develop PTSD may inhibit this process. This is consistent with the broader emotion regulation and coping literature on the importance of flexibility (e.g., Bonanno & Burton, 2013).

There is empirical evidence that may support this hypothesis. For instance, Orcutt et al. (2014) findings, as previously described, potentially suggest that those who have high levels of EA prior to traumatic exposure may continue to rely on EA in a maladaptive manner to cope with the trauma, hindering natural recovery, whereas those with lower levels of EA pretrauma may not rely on EA rigidly and naturally recover. As suggested previously, this study may offer some tentative evidence for the predisposition model of the EA PTSD relationship.

Experiential avoidance and subjective arousal

This propensity to avoid unpleasant thoughts and feelings may be due to a greater sensitivity to the experience of those emotions. Engagement in EA has been linked to increases in PTSS (e.g., Tiet et al., 2006), severity of symptoms (e.g., Badour et al., 2012), and increases in the incidence of specific symptom clusters of PTSD, such as reexperiencing (e.g., Macrae, Bodenhausen, Milne, & Ford, 1997), hyperarousal, and further avoidance (e.g., Shenk, Putnam, Rausch, Peugh, & Noll, 2014). Further, other research suggests that it is not the presence of intrusions but how individuals relate to their inner experiences that influence the etiology of PTSD (Shenk et al., 2014). This may be one mechanism by which EA and PTSD are mutually reinforcing; as one engages in EA to suppress or avoid unpleasant intrusions, it in turn increases the incidence of such intrusions (i.e., increasing severity of the PTSD symptom cluster; see Chapman et al., 2011 for review), therefore driving the individual to engage in more EA to cope with the increased intrusions, and the cycle restarts.

One study found that individuals who reported engaging in high levels of EA endorsed greater subjective experiences of anxiety and distress compared with individuals who reported low engagement in EA, but no significant differences were found in their physiological arousal (Feldner, Zvolensky, Eifert, & Spira, 2003). This study suggests that there may be differences in the interpretation and sensitivity to negative emotions based on engagement in EA, that is, individuals high in EA interpret some physiological arousal as more intense than those with low EA, which in turn leads them to report more subjective distress. In the context of PTSD, these differences may also inform why EA increases the severity of PTSS; the development of PTSD may increase physiological responding (e.g., hyperarousal), which may lead to further engagement in EA due to the heightened distress interpretation in those high in EA, thus maintaining, if not increasing, the PTSS. This proposed mechanism also may provide further evidence for the pathoplasticity model, suggesting that EA influences the expression, or severity of the PTSS, not necessarily the onset of the symptoms themselves.

Another study found that EA moderated the sensitivity of the behavioral inhibition system (BIS; i.e., the neural motivational system to avoid punishment and aversive stimuli) as it relates to PTSS, more specifically that PTSS was only predicted by increased BIS activation when EA engagement was high, but not when EA was low (Pickett, Bardeen, & Orcutt, 2011). Other research has suggested that anxiety sensitivity is also highly related to PTSD (e.g., Simpson, Jakupcak, & Luterek, 2006). Although a review of the anxiety sensitivity literature is outside the scope of this chapter, in conjunction with the BIS sensitivity findings and those regarding the subjective versus physiological measurement of distress and EA, it appears that EA confers risk for PTSD through interactions with other vulnerabilities (e.g., BIS sensitivity and anxiety sensitivity) when EA utilization is high, but potentially not when EA utilization is low (e.g., Bardeen, 2015; Bardeen, Fergus, & Orcutt, 2013; Bardeen, Tull, Stevens, & Gratz, 2015; Cobb, Lancaster, Meyer, Lee, & Telch, 2017; Naifeh, Tull, & Gratz, 2012). These findings seemingly support both the common cause model, such that EA and PTSD have similar causal influences (e.g., BIS sensitivity and anxiety sensitivity), and the pathoplasticity model, such that EA appears to affect PTSS with higher engagement once the PTSS have already emerged.

Experiential avoidance and rumination

Another potential mechanism by which EA and PTSD are mutually reinforced can be understood through recent research on rumination. Rumination, traditionally thought of as an approach behavior, may serve as an avoidant coping strategy due to the repetitive nature of the thoughts that may inhibit emotional processing (Ehring, Frank, & Ehlers, 2008; Michael et al., 2007, Smith & Alloy, 2009). Engagement in rumination shortly after a trauma (i.e., 2 weeks) has been shown to predict PTSD symptom severity at 6 months, controlling for symptom severity at 2 weeks, with similar results being found for rumination at 1 month post trauma (Ehring et al., 2008). Additionally, rumination has been found to be associated with increased frequency of intrusions in those with PTSD (Michael et al., 2007). In the same study, those in the PTSD group also reported greater reliance on rumination to alleviate their trauma-related intrusions (Michael et al., 2007), thus highlighting the mutually reinforcing relationship between intrusions and rumination.

Although not developed specifically for PTSD, Newman and Llera's (2011) contrast avoidance model may help elucidate how rumination is involved in the maintenance of PTSD through inhibition of emotional processing. The contrast avoidance model proposes that, to avoid negative emotional shifts (knowns as contrasts) occurring after an unpleasant event, individuals may use worry to create a sustained negative emotional state to protect against experiencing emotional shifts from positive to negative. Rumination appears to function similarly to worry (Newman & Llera, 2011). Given that worry has been empirically shown to increase negative affect, both subjectively and physiologically (see Borkovec, Alcaine, & Behar, 2004, for review), it may function to prolong the negative emotionality so that the individual does not have to experience the shift from positive to negative emotions. Since emotional processing requires reactivity (both physiological and subjective) to a feared stimulus for the fear structure to be activated and habituation to occur, this process is inhibited due to prolonged negative emotionality prohibiting further reactivity to the stimulus, and thus blocking the emotions from being fully processed (see Newman & Llera, 2011, for review).

Other research has suggested that the reliance on worry in the context of contrast avoidance may lead to perseveration on threatening stimuli (i.e., hypervigilance [Wells & Matthews, 1994] and hyperarousal [Marshall et al., 2006]) due to the maintenance of the conditioned fear response associated with cognitive rehearsal (Davey & Matchett, 1994). Bardeen et al. (2013) empirically tested the contrast avoidance model as it related to PTSS in a cross-sectional sample, finding that worry positively predicted PTSS scores and specifically symptoms of hyperarousal. Taken together, this literature suggests that chronic worry or rumination influences the severity and maintenance of PTSS by inhibiting emotional processing through prolonged emotional arousal (i.e., hyperarousal); this inhibition may lead to increased hypervigilance and greater frequency of intrusions, ultimately resulting in greater engagement in EA (e.g., rumination) to cope with the increased symptomatology. In the context of the three proposed models of EA and PTSD, this proposed mechanism appears to support the pathoplasticity model, such that chronic rumination and worry increase the severity and maintain PTSS, therefore modifying the expression of the symptoms once they have already emerged.

Summary

The PTSD literature has presented strong empirical evidence for a bidirectional relationship between engagement in EA and PTSD symptoms (e.g., Badour et al., 2012) but has contributed less in terms of the mechanisms behind this mutually reinforcing relationship. The extant literature on EA in other forms of psychopathology offers several potential mechanisms by which EA may function in PTSD including mechanisms in the development of PTSD and the function of EA in the maintenance of PTSD symptoms over time. These mechanisms provide further support for the three proposed models of the EA PTSD relationship, with the most support for the pathoplasticity model, suggesting that the mechanisms of the relationship may modify the expression of the PTSD symptoms rather than directly cause the symptoms.

Malleability of experiential avoidance

Given its demonstrable consequences, cultivating patients' ability to disengage from rigid adherence to EA is a key concern for trauma clinicians. Research has shown that EA is reduced more efficiently following PE (Zang et al., 2017). In a randomized controlled trial, adults with PTSD and comorbid alcohol dependence received either a psychotropic medication (naltrexone) or a placebo with or without PE; all received concurrent supportive counseling on alcohol use (Zang et al., 2017). From baseline to 6-month follow-up, all participants (regardless of condition) demonstrated reduced avoidant coping (e.g., "I've been refusing to believe that it has happened"). Notably, participants who received PE evidenced faster decreases in avoidance than those in supportive counseling only, and at 6 months a trend emerged in the direction of lower avoidance following PE compared with no PE. Additionally, greater PTSS reduction was associated with a faster rate of decrease in avoidance across groups, yet a more marked effect was seen for those receiving PE compared with supportive counseling alone. Further, those who received PE demonstrated a higher ratio of adaptive-to-avoidant coping posttreatment, indicating more use of adaptive relative to avoidant coping over time (i.e., greater coping flexibility; see Bonanno & Burton, 2013). These findings are unsurprising given that PE actively challenges avoidant behaviors through safe and direct confrontation with feared stimuli.

Changes in EA and the subsequent effect on treatment outcomes were explored among male veterans (mostly combat exposed) enrolled in a residential treatment program for PTSD (Boden et al., 2013). Use of expressive suppression and cognitive reappraisal[d] were assessed in relation to PTSS at intake and discharge. Expressive suppression is a specific emotion regulation strategy employed to inhibit the outward display of emotion (Gross & John, 2003) and is considered a form of EA (Chawla & Ostafin, 2007; Hayes et al., 2004; Kashdan et al., 2006). Though there were a variety of treatment groups (e.g., cognitive therapy and recreation therapy), the primary trauma-focused treatment that veterans received was group CPT (with the written trauma account). Engagement in expressive suppression at intake and discharge was associated with greater PTSS severity at both timepoints. Moreover, use of expressive suppression lessened significantly throughout treatment. These reductions in expressive suppression predicted less PTSS severity at discharge and less symptom severity among the avoidance and numbing clusters specifically (after adjusting for initial PTSS severity, length of stay, and age). These findings indicate that EA—in the form of expressive suppression—is amenable to therapeutic intervention. Consistent with a pathoplasticity model, decreased use of this type of EA predicted less PTSS severity. Interestingly, however, change in PTSS severity was also found to predict less expressive suppression, casting more confusion onto the directionality of the relationship between PTSD and EA.

In a recent uncontrolled pilot study (Meyer, Walser, et al., 2018), Meyer and colleagues examined the effectiveness of a 12-session acceptance and commitment therapy protocol for comorbid PTSD and alcohol use disorders in a sample of 43 veterans (88.4% male). Results indicated that participants demonstrated significant reductions in PTSD and alcohol use symptoms post treatment and at 3-month follow-up. Further, in most cases, reductions in EA were associated with reductions in symptoms. Although lacking randomization and a control condition, Meyer et al.'s findings are promising with regard to a linkage between EA reductions and PTSD symptom reductions. Thus, taken together, these studies demonstrate the effective reduction of EA and the subsequent impact on treatment outcomes, thus validating the rationale shared by EBPs for PTSD of targeting trauma-related EA.

Avoidant coping and PTSD outcomes were investigated once more among male veterans with comorbid PTSD and substance use disorders (SUD) randomly assigned to Seeking Safety (SS; Najavits, 2002) or treatment as usual, which fostered and sought to maintain abstinence

---

[d] A review of emotion regulation variables that do not appear to measure experiential avoidance will not be discussed in this chapter. For a discussion of emotion regulation difficulties as they relate to PTSD, please see Chapter 11.

(Boden et al., 2014). Regardless of treatment condition, avoidant coping significantly decreased from treatment start to finish, and these changes were maintained over the course of 9 months. However, treatment dose was not a significant predictor of avoidant coping, and avoidant coping did not influence rate of change of PTSS severity (or alcohol use), despite covarying with these variables across timepoints. Said differently the relationship between avoidance and PTSD and SUD was similar at baseline, posttreatment, and follow-up assessments, suggesting that veterans continued utilizing avoidant strategies to cope with symptoms. Thus, although SS aims to build more active coping styles (e.g., problem solving and positive reappraisal) while simultaneously reducing avoidant ones, these skills may not directly transfer to how patients respond to trauma-related stimuli. By not explicitly attending to patients' traumas (Najavits, 2002), avoidance of trauma-related cues may be reinforced, thus preserving their emotional potency and ultimately promoting continued avoidance (e.g., Boden et al., 2014).

Summary

EA appears to be malleable and amenable to intervention. Differences across study findings in EA malleability appear to hinge on treatment modality. EBPs for PTSD all share components antithetical to avoidance, such as gradually confronting the feared stimulus or generating alternative beliefs that empower and encourage patients to approach rather than continue avoiding. Without this overt intention to reduce EA, it may persist undetected and remain resistant to change. For instance, markers of EA were significantly reduced following CPT (Boden et al., 2013) and PE (Zang et al., 2017) but not a cognitive behavioral group for chronic PTSD (Boden et al., 2012). Acceptance and commitment therapy for comorbid PTSD/alcohol use disorders is intended to reduce EA, and results indicated that reductions in EA were associated with reductions in symptoms (Meyer, Walser, et al., 2018). Although EA significantly decreased following SS, this change was no more significant than treatment as usual, was not affected by treatment dose, and had no bearing on the rate of change in PTSD symptoms (Boden et al., 2014). Thus EBPs for PTSD seem to uniquely influence EA, with similarly distinct corollaries for treatment response.

Conclusions

Humans want to avoid emotional pain; thus EA represents an evolutionarily adaptive paradox; the more an individual uses EA to cope with their emotional pain, the more they may increase the very pain they are trying to avoid. RFT posits that this propensity to avoid unpleasant emotions may be based in our language abilities; the uniquely human ability to derive relationships and transfer responses across stimuli without direct experience is at the core of EA (Hayes et al., 2001). Individuals who have experienced trauma may be even further incentivized to avoid their emotional distress due to the natural development of PTSS in the wake of trauma exposure.

There is a large body of literature supporting the relationship between EA and PTSD, although the direction of this relationship is much less clear. The research seemingly points to a bidirectional relationship between EA and PTSD, with the majority of the limited longitudinal research pointing to a pathoplastic relationship, such that EA causally exerts influence over the expression and maintenance of PTSS once they have developed. This conclusion may be a consequence of the necessary assessment of EA prior to the exposure to a trauma to evaluate a predispositional model; to the authors' knowledge, there are only two data sets that meet this criterion, thus severely limiting the confidence in which a statement of the nature of the relationship between EA and PTSD can be made. Further, there is a dearth of research examining the potential mechanisms underlying the bi-directionality of EA and PTSD. The applicable extant literature from other psychopathologies associated with EA, in conjunction with the limited research regarding the mechanisms behind the EA PTSD relationship, also offers further support for a pathoplastic model, wherein greater utilization of EA may increase severity and contribute to the maintenance of PTSD symptomatology over time.

Clinical implications

Of specific clinical importance, EA appears to be malleable, especially when directly challenged, and may have important implications in regard to reducing PTSD symptoms (e.g., Zang et al., 2017), but the temporal influence of reductions in EA versus PTSD symptoms remains unclear (Boden et al., 2013). Additionally, Leiner et al. (2012) found that those with more baseline avoidant coping stand to benefit the most from PE or EMDR. Given that PE and EMDR both require approaching trauma-related content, these treatments may lend themselves more to individuals who rely on avoidant coping. In contrast, following a cognitive behavioral group, pretreatment EA was associated with greater PTSS severity post-treatment, which was related to more EA 4 months later (Badour et al., 2012). Albeit a difficult-to-treat sample, conclusions from aforementioned studies suggest outcomes may have been different following completion of EBPs. Collectively, these findings point to the importance of keeping EA a vital target of intervention, perhaps especially when faced with the challenge of chronic PTSD.

Limitations and future directions

Despite substantial research conducted on the relationship between EA and PTSD symptoms, there are significant limitations to the literature as it stands. There are issues with regard to the measurement of EA as a construct. One of the most widely used measures of EA, the AAQ in its original form (Hayes et al., 2004), demonstrated issues with internal consistency and reliability across different populations (e.g., Bond et al., 2011). As such the second version of this measure (i.e., AAQ-II; Bond et al., 2011) appeared to improve upon the reliability of the AAQ. However, the construct validity (e.g., Wolgast, 2014) and discriminant validity with negative emotionality (e.g., Gámez, Chmielewski, Kotov, Ruggero, & Watson, 2011) of the AAQ-II have been critiqued. Utilized significantly less often the Multidimensional Experiential Avoidance Questionnaire (MEAQ; Gámez et al., 2011) and its short version, Brief Experiential Avoidance Questionnaire (BEAQ; Gámez et al., 2014), appear to offer strong psychometric properties (e.g., Rochefort, Baldwin, & Chmielewski, 2018), and it will be important to increase the frequency of the use of these instruments. Another measurement limitation in the EA PTSD literature is the underutilization of ecological momentary assessment (EMA) data collection methods, which increase ecological validity through moment, or short-term retrospective reporting, of how individuals interact with their emotions, thus potentially offering a more accurate assessment of engagement in EA than traditional retrospective reporting; this method of data collection has been utilized to study EA in the context of other types of psychopathology (e.g., Machell, Goodman, & Kashdan, 2015) but is largely missing from the EA PTSD literature. Further, though challenging to obtain, the limited number of studies that include a pre-trauma measurement of EA significantly restricts the conclusions that can be drawn in terms of causal relationships between EA and PTSD. Relatedly the literature mostly supports a bidirectional relationship between EA and PTSD, but this consensus could be a by-product of the measurement limitations and the conceptual overlap between the construct of EA and the symptom cluster of avoidance in PTSD; more specific measurement and prospective designs could further clarify the directionality of the relationship between EA and PTSD.

Additionally, as discussed previously, the literature on EA and PTSD is largely lacking a discussion of the mechanisms by which EA and PTSD are related. The literature could benefit from future research investigating potentially common mechanisms of EA across psychopathology (e.g., increased sensitivity to stress) to further our understanding of how EA functions in the context of PTSD. Considering the relationship between sleep disturbances and PTSD (e.g., Pickett, Barbaro, & Mello, 2016), one potential avenue of future work may include investigating the role of sleep in EA, as sleep is highly related to emotions (Kahn, Sheppes, & Sadeh, 2013) and emotion regulation ability (Yoo, Hu, Gujar, Jolesz, & Walker, 2007). Further, identifying more specific mechanisms may facilitate our understanding of how to modify EA when it is utilized maladaptively. On the whole, there is not much understanding of the malleability of EA, which may be attributed to limited measurement of EA outside the context of the PTSD symptom cluster in trauma-focused treatment. Future research may benefit from more specific, intentional measurement of EA in the context of PTSD treatment to increase our understanding of the temporal relationship between changes in EA and PTSD symptomatology.

Finally, one of the largest limitations in need of further exploration is the construct overlap and definitional bleeding between EA, emotion regulation, PTSD avoidance symptom cluster, and different "bordering constructs" of EA. There is little consensus in the literature of how to operationalize EA (e.g., what strategies are and are not considered EA versus "bordering constructs" [e.g., Chawla and Ostafin, 2007]), limiting conclusions that can be drawn across studies. This chapter, in line with Boulanger et al. (2010), offers a definition of EA and delineation between the constructs of EA and emotion regulation with empirical backing but would benefit from further empirical testing. The relationship between EA and PTSD has strong empirical support but is far from understood. The literature could benefit from further exploration to shed light on this evolutionarily driven desire to avoid our painful emotions.

References

Aldao, A., Nolen-Hoeksema, S., & Schweizer, S. (2010). Emotion-regulation strategies across psychopathology: A meta-analytic review. *Clinical Psychology Review, 30*(2), 217–237. https://doi.org/10.1016/j.cpr.2009.11.004.

An, Y., Fu, F., Wu, X., Lin, C., & Zhang, Y. (2013). Longitudinal relationships between neuroticism, avoidant coping, and posttraumatic stress disorder symptoms in adolescents following the 2008 Wenchuan earthquake in China. *Journal of Loss and Trauma, 18*(6), 556–571. https://doi.org/10.1080/15325024.2012.719351.

Badour, C. L., Blonigen, D. M., Boden, M. T., Feldner, M. T., & Bonn-Miller, M. O. (2012). A longitudinal test of the bi-directional relations between avoidance coping and PTSD severity during and after PTSD treatment. *Behaviour Research and Therapy, 50*(10), 610–616. https://doi.org/10.1016/j.brat.2012.06.006.

Bardeen, J. R. (2015). Short-term pain for long-term gain: The role of experiential avoidance in the relation between anxiety sensitivity and emotional distress. *Journal of Anxiety Disorders, 30*, 113–119. https://doi.org/10.1016/j.janxdis.2014.12.013.

Bardeen, J. R., Fergus, T. A., & Orcutt, H. K. (2013). Experiential avoidance as a moderator of the relationship between anxiety sensitivity and perceived stress. *Behavior Therapy, 44*(3), 459–469. https://doi.org/10.1016/j.beth.2013.04.001.

Bardeen, J. R., Tull, M. T., Stevens, E. N., & Gratz, K. L. (2015). Further investigation of the association between anxiety sensitivity and posttraumatic stress disorder: Examining the influence of emotional avoidance. *Journal of Contextual Behavioral Science, 4*(3), 163–169. https://doi.org/10.1016/j.jcbs.2015.05.002.

Benotsch, E. G., Brailey, K., Vasterling, J. J., Uddo, M., Constans, J. I., & Sutker, P. B. (2000). War zone stress, personal and environmental resources, and PTSD symptoms in gulf war veterans: A longitudinal perspective. *Journal of Abnormal Psychology, 109*(2), 205–213.

Blackledge, J. T. (2004). Functional contextual processes in posttraumatic stress. *International Journal of Psychology and Psychological Therapy, 4*(3), 443–468.

Blackledge, J. T., & Hayes, S. C. (2001). Emotion regulation in acceptance and commitment therapy. *Journal of Clinical Psychology, 57*, 243–255.

Boden, M., Kimerling, R., Kulkarni, M., Bonn-Miller, M. O., Weaver, C., & Trafton, J. (2014). Coping among military veterans with PTSD in substance use disorder treatment. *Journal of Substance Abuse Treatment, 47*(2), 160–167. https://doi.org/10.1016/j.jsat.2014.03.006.

Boden, M. T., Bonn-Miller, M. O., Vujanovic, A. A., & Drescher, K. D. (2012). A prospective investigation of changes in avoidant and active coping and posttraumatic stress disorder symptoms among military veteran. *Journal of Psychopathology and Behavioral Assessment, 34*(4), 433–439. https://doi.org/10.1007/s10862-012-9293-6.

Boden, M. T., Westermann, S., McRae, K., Kuo, J., Alvarez, J., Kulkarni, M. R., ... Bonn-Miller, M. O. (2013). Emotion regulation and posttraumatic stress disorder: A prospective investigation. *Journal of Social and Clinical Psychology, 32*(3), 296–314. https://doi.org/10.1521/jscp.2013.32.3.296.

Bonanno, G. A., & Burton, C. L. (2013). Regulatory flexibility: An individual differences perspective on coping and emotion regulation. *Perspectives on Psychological Science, 8*(6), 591–612. https://doi.org/10.1177/1745691613504116.

Bond, F. W., Hayes, S. C., Baer, R. A., Carpenter, K. M., Guenole, N., Orcutt, H. K., ... Zettle, R. D. (2011). Preliminary psychometric properties of the acceptance and action questionnaire–II: A revised measure of psychological inflexibility and experiential avoidance. *Behavior Therapy, 42*(4), 676–688. https://doi.org/10.1016/j.beth.2011.03.007.

Borkovec, T. D., Alcaine, O. M., & Behar, E. (2004). Avoidance theory of worry and generalized anxiety disorder. In R. Heimberg, C. Turk, & D. Mennin (Eds.), *Generalized anxiety disorder: Advances in research and practice*. New York, NY: Guilford Press.

Boulanger, J. L., Hayes, S. C., & Pistorello, J. (2010). Experiential avoidance as a functional contextual concept. In *Emotion regulation and psychopathology: A transdiagnostic approach to etiology and treatment*. New York, NY: Guilford Press. (pp. 107–136).

Bowlby, J. (1980). *Attachment and loss*. New York, NY: Basic Books.

Chapman, A. L., Dixon-Gordon, K. L., & Walters, K. N. (2011). Experiential avoidance and emotion regulation in borderline personality disorder. *Journal of Rational-Emotive & Cognitive-Behavior Therapy, 29*(1), 35–52. https://doi.org/10.1007/s10942-011-0124-6.

Chawla, N., & Ostafin, B. (2007). Experiential avoidance as a functional dimensional approach to psychopathology: An empirical review. *Journal of Clinical Psychology, 63*(9), 871–890. https://doi.org/10.1002/jclp.20400.

Cobb, A. R., Lancaster, C. L., Meyer, E. C., Lee, H.-J., & Telch, M. J. (2017). Pre-deployment trait anxiety, anxiety sensitivity and experiential avoidance predict war-zone stress-evoked psychopathology. *Journal of Contextual Behavioral Science*, *6*(3), 276–287. https://doi.org/10.1016/j.jcbs.2017.05.002.

Davey, G. C., & Matchett, G. (1994). Unconditioned stimulus rehearsal and the retention and enhancement of differential "fear" conditioning: Effects of trait and state anxiety. *Journal of Abnormal Psychology*, *103*(4), 708–718.

Devany, J. M., Hayes, S. C., & Nelson, R. O. (1986). Equivalence class formation in language-able and language-disabled children. *Journal of the Experimental Analysis of Behavior*, *46*(3), 243–257. https://doi.org/10.1901/jeab.1986.46-243.

Dougher, M. J., Augustson, E., Markham, M. R., Greenway, D. E., & Wulfert, E. (1994). The transfer of respondent eliciting and extinction functions through stimulus equivalence classes. *Journal of the Experimental Analysis of Behavior*, *62*(3), 331–351.

Dougher, M. J., Hamilton, D. A., Fink, B. C., & Harrington, J. (2007). Transformation of the discriminative and eliciting functions of generalized relational stimuli. *Journal of the Experimental Analysis of Behavior*, *88*(2), 179–197. https://doi.org/10.1901/jeab.2007.45-05.

Dugdale, N., & Lowe, C. F. (2000). Testing for symmetry in the conditional discriminations of language-trained chimpanzees. *Journal of the Experimental Analysis of Behavior*, *73*(1), 5–22. https://doi.org/10.1901/jeab.2000.73-5.

Ehlers, A., & Clark, D. M. (2000). A cognitive model of posttraumatic stress disorder. *Behaviour Research and Therapy*, *38*(4), 319–345.

Ehlers, A., Mayou, R. A., & Bryant, B. (1998). Psychological predictors of chronic posttraumatic stress disorder after motor vehicle accidents. *Journal of Abnormal Psychology*, *107*(3), 508–519.

Ehring, T., Frank, S., & Ehlers, A. (2008). The role of rumination and reduced concreteness in the maintenance of posttraumatic stress disorder and depression following trauma. *Cognitive Therapy and Research*, *32*(4), 488–506. https://doi.org/10.1007/s10608-006-9089-7.

Elwood, L. S., Hahn, K. S., Olatunji, B. O., & Williams, N. L. (2009). Cognitive vulnerabilities to the development of PTSD: A review of four vulnerabilities and the proposal of an integrative vulnerability model. *Clinical Psychology Review*, *29*(1), 87–100. https://doi.org/10.1016/j.cpr.2008.10.002.

Feldner, M., Zvolensky, M., Eifert, G., & Spira, A. (2003). Emotional avoidance: An experimental test of individual differences and response suppression using biological challenge. *Behaviour Research and Therapy*, *41*(4), 403–411. https://doi.org/10.1016/S0005-7967(02)00020-7.

Foa, E. B., Hembree, E., & Rothbaum, B. (2007). *Prolonged exposure therapy for PTSD: Therapist guide: Emotional processing of traumatic experiences*. Oxford University Press. https://doi.org/10.1093/med:psych/9780195308501.001.0001.

Galatzer-Levy, I. R., & Bryant, R. A. (2013). 636,120 Ways to have posttraumatic stress disorder. *Perspectives on Psychological Science*, *8*(6), 651–662. https://doi.org/10.1177/1745691613504115.

Gámez, W., Chmielewski, M., Kotov, R., Ruggero, C., Suzuki, N., & Watson, D. (2014). The brief experiential avoidance questionnaire: Development and initial validation. *Psychological Assessment*, *26*(1), 35–45. https://doi.org/10.1037/a0034473.

Gámez, W., Chmielewski, M., Kotov, R., Ruggero, C., & Watson, D. (2011). Development of a measure of experiential avoidance: The multidimensional experiential avoidance questionnaire. *Psychological Assessment*, *23*(3), 692–713. https://doi.org/10.1037/a0023242.

Gil, S. (2005). Pre-traumatic personality as a predictor of post-traumatic stress disorder among undergraduate students exposed to a terrorist attack: A prospective study in Israel. *Personality and Individual Differences*, *39*(4), 819–827. https://doi.org/10.1016/j.paid.2005.03.007.

Gross, J. J., & John, O. P. (2003). Individual differences in two emotion regulation processes: Implications for affect, relationships, and well-being. *Journal of Personality and Social Psychology*, *85*(2), 348–362.

Hayes, S. C. (1989). Nonhumans have not yet shown stimulus equivalence. *Journal of the Experimental Analysis of Behavior*, *51*(3), 385–392. https://doi.org/10.1901/jeab.1989.51-385.

Hayes, S. C., Barnes-Holmes, D., & Roche, B. (Eds.), (2001). *Relational frame theory: A post-Skinnerian account of human language and cognition*. New York, NY: Kluwer Academic/Plenum Publishers.

Hayes, S. C., Strosahl, K., Wilson, K. G., Bissett, R. T., Pistorello, J., Toarmino, D., ... McCurry, S. M. (2004). Measuring experiential avoidance: A preliminary test of a working model. *The Psychological Record*, *54*(4), 553–578. https://doi.org/10.1007/BF03395492.

Hayes, S. C., Wilson, K. G., Gifford, E. V., Follette, V. M., & Strosahl, K. (1996). Experiential avoidance and behavioral disorders: A functional dimensional approach to diagnosis and treatment. *Journal of Consulting and Clinical Psychology*, *64*(6), 1152–1168. https://doi.org/10.1037/0022-006X.64.6.1152.

Heagle, A. I., & Rehfeldt, R. A. (2006). Teaching perspective-taking skills to typically developing children through derived relational responding. *Journal of Early and Intensive Behavior Intervention*, *3*(1), 1–34. https://doi.org/10.1037/h0100321.

Kahn, M., Sheppes, G., & Sadeh, A. (2013). Sleep and emotions: Bidirectional links and underlying mechanisms. *International Journal of Psychophysiology*, *89*(2), 218–228. https://doi.org/10.1016/j.ijpsycho.2013.05.010.

Kashdan, T. B., Barrios, V., Forsyth, J. P., & Steger, M. F. (2006). Experiential avoidance as a generalized psychological vulnerability: Comparisons with coping and emotion regulation strategies. *Behaviour Research and Therapy*, *44*(9), 1301–1320. https://doi.org/10.1016/j.brat.2005.10.003.

Klein, D. N., Kotov, R., & Bufferd, S. J. (2011). Personality and depression: Explanatory models and review of the evidence. *Annual Review of Clinical Psychology*, *7*(1), 269–295. https://doi.org/10.1146/annurev-clinpsy-032210-104540.

Kumpula, M. J., Orcutt, H. K., Bardeen, J. R., & Varkovitzky, R. L. (2011). Peritraumatic dissociation and experiential avoidance as prospective predictors of posttraumatic stress symptoms. *Journal of Abnormal Psychology*, *120*(3), 617–627. https://doi.org/10.1037/a0023927.

Leiner, A. S., Kearns, M. C., Jackson, J. L., Astin, M. C., & Rothbaum, B. O. (2012). Avoidant coping and treatment outcome in rape-related posttraumatic stress disorder. *Journal of Consulting and Clinical Psychology*, *80*(2), 317–321. https://doi.org/10.1037/a0026814.

Machell, K. A., Goodman, F. R., & Kashdan, T. B. (2015). Experiential avoidance and well-being: A daily diary analysis. *Cognition and Emotion*, *29*(2), 351–359. https://doi.org/10.1080/02699931.2014.911143.

Macrae, C. N., Bodenhausen, G. V., Milne, A. B., & Ford, R. L. (1997). On regulation of recollection: The intentional forgetting of stereotypical memories. *Journal of Personality and Social Psychology*, *72*(4), 709–719. https://doi.org/10.1037/0022-3514.72.4.709.

Marshall, R. D., Turner, J. B., Lewis-Fernandez, R., Koenan, K., Neria, Y., & Dohrenwend, B. P. (2006). Symptom patterns associated with chronic PTSD in male veterans: New findings from the National Vietnam Veterans Readjustment Study. *The Journal of Nervous and Mental Disease*, *194*(4), 275–278. https://doi.org/10.1097/01.nmd.0000207363.25750.56.

Marx, B. P., & Sloan, D. M. (2005). Peritraumatic dissociation and experiential avoidance as predictors of posttraumatic stress symptomatology. *Behaviour Research and Therapy*, *43*(5), 569–583. https://doi.org/10.1016/j.brat.2004.04.004.

McHugh, L., Barnes-Holmes, Y., & Barnes-Holmes, D. (2004). Perspective-taking as relational responding: A developmental profile. *The Psychological Record*, *54*(1), 115–144. https://doi.org/10.1007/BF03395465.

Meyer, E. C., La Bash, H., DeBeer, B. B., Kimbrel, N. A., Gulliver, S. B., & Morisette, S. B. (2018). Psychological inflexibility predicts PTSD symptom severity in war veterans after accounting for established PTSD risk factors and personality. *Psychological Trauma Theory Research Practice and Policy*, https://doi.org/10.1037/tra0000358. Advance online publication.

Meyer, E. C., Walser, R., Hermann, B., La Bash, H., DeBeer, B. B., Morisette, S. B., ... Schnurr, P. P. (2018). Acceptance and commitment therapy for co-occurring posttraumatic stress disorder and alcohol use disorders in veterans: Pilot treatment outcomes. *Journal of Traumatic Stress*, *31*, 781–789.

Michael, T., Halligan, S. L., Clark, D. M., & Ehlers, A. (2007). Rumination in posttraumatic stress disorder. *Depression and Anxiety*, *24*(5), 307–317. https://doi.org/10.1002/da.20228.

Naifeh, J. A., Tull, M. T., & Gratz, K. L. (2012). Anxiety sensitivity, emotional avoidance, and PTSD symptom severity among crack/cocaine dependent patients in residential treatment. *Cognitive Therapy and Research*, *36*(3), 247–257. https://doi.org/10.1007/s10608-010-9337-8.

Najavits, L. M. (2002). *Seeking safety: A treatment manual for PTSD and substance abuse*. New York, NY: Guilford Press.

Newman, M. G., & Llera, S. J. (2011). A novel theory of experiential avoidance in generalized anxiety disorder: A review and synthesis of research supporting a contrast avoidance model of worry. *Clinical Psychology Review*, *31*(3), 371–382. https://doi.org/10.1016/j.cpr.2011.01.008.

Orcutt, H. K., Bonanno, G. A., Hannan, S. M., & Miron, L. R. (2014). Prospective trajectories of posttraumatic stress in college women following a campus mass shooting: Symptom trajectories after mass shooting. *Journal of Traumatic Stress*, *27*(3), 249–256. https://doi.org/10.1002/jts.21914.

Pickett, S. M., Barbaro, N., & Mello, D. (2016). The relationship between subjective sleep disturbance, sleep quality, and emotion regulation difficulties in a sample of college students reporting trauma exposure. *Psychological Trauma Theory Research Practice and Policy*, *8*(1), 25–33. https://doi.org/10.1037/tra0000064.

Pickett, S. M., Bardeen, J. R., & Orcutt, H. K. (2011). Experiential avoidance as a moderator of the relationship between behavioral inhibition system sensitivity and posttraumatic stress symptoms. *Journal of Anxiety Disorders*, *25*(8), 1038–1045. https://doi.org/10.1016/j.janxdis.2011.06.013.

Resick, P. A., Monson, C. M., & Chard, K. M. (2017). *Cognitive processing therapy for PTSD: A comprehensive manual*. New York, NY: Guilford Press.

Rochefort, C., Baldwin, A. S., & Chmielewski, M. (2018). Experiential avoidance: An examination of the construct validity of the AAQ-II and MEAQ. *Behavior Therapy*, *49*(3), 435–449. https://doi.org/10.1016/j.beth.2017.08.008.

Seligowski, A. V., Lee, D. J., Bardeen, J. R., & Orcutt, H. K. (2015). Emotion regulation and posttraumatic stress symptoms: A meta-analysis. *Cognitive Behaviour Therapy*, *44*(2), 87–102. https://doi.org/10.1080/16506073.2014.980753.

Shapiro, F. (2001). *Eye movement desensitization and reprocessing: Basic principles, protocols, and procedures* (2nd ed.). New York, NY: Guilford Press.

Sharkansky, E. J., King, D. W., King, L. A., Wolfe, J., Erickson, D. J., & Stokes, L. R. (2000). Coping with gulf war combat stress: Mediating and moderating effects. *Journal of Abnormal Psychology*, *109*(2), 188–197.

Shear, M. K. (2010). Exploring the role of experiential avoidance from the perspective of attachment theory and the dual process model. *OMEGA-Journal of Death and Dying*, *61*(4), 357–369. https://doi.org/10.2190/OM.61.4.f.

Shenk, C. E., Putnam, F. W., Rausch, J. R., Peugh, J. L., & Noll, J. G. (2014). A longitudinal study of several potential mediators of the relationship between child maltreatment and posttraumatic stress disorder symptoms. *Development and Psychopathology*, *26*(01), 81–91. https://doi.org/10.1017/S0954579413000916.

Sheppes, G., Suri, G., & Gross, J. J. (2015). Emotion regulation and psychopathology. *Annual Review of Clinical Psychology*, *11*, 379–405. https://doi.org/10.1146/annurev-clinpsy-032814-112739.

Sidman, M., & Tailby, W. (1982). Conditional discrimination vs. matching to sample: An expansion of the testing paradigm. *Journal of the Experimental Analysis of Behavior*, *37*(1), 5–22. https://doi.org/10.1901/jeab.1982.37-5.

Simms, L. J., Watson, D., & Doebbeling, B. N. (2002). Confirmatory factor analyses of posttraumatic stress symptoms in deployed and non-deployed veterans of the Gulf War. *Journal of Abnormal Psychology, 111*, 637–647.

Simpson, T., Jakupcak, M., & Luterek, J. A. (2006). Fear and avoidance of internal experiences among patients with substance use disorders and PTSD: The centrality of anxiety sensitivity. *Journal of Traumatic Stress, 19*(4), 481–491. https://doi.org/10.1002/jts.20128.

Smith, J. M., & Alloy, L. B. (2009). A roadmap to rumination: A review of the definition, assessment, and conceptualization of this multifaceted construct. *Clinical Psychology Review, 29*(2), 116–128. https://doi.org/10.1016/j.cpr.2008.10.003.

Solomon, Z., Mikulincer, M., & Flum, H. (1989). The implications of life events and social integration in the course of combat-related post-traumatic stress disorder. *Social Psychiatry and Psychiatric Epidemiology, 24*(1), 41–48. https://doi.org/10.1007/BF01788199.

Spinhoven, P., Drost, J., de Rooij, M., van Hemert, A. M., & Penninx, B. W. (2014). A longitudinal study of experiential avoidance in emotional disorders. *Behavior Therapy, 45*(6), 840–850. https://doi.org/10.1016/j.beth.2014.07.001.

Thompson, N. J., Fiorillo, D., Rothbaum, B. O., Ressler, K. J., & Michopoulos, V. (2018). Coping strategies as mediators in relation to resilience and posttraumatic stress disorder. *Journal of Affective Disorders, 225*, 153–159. https://doi.org/10.1016/j.jad.2017.08.049.

Tiet, Q. Q., Rosen, C., Cavella, S., Moos, R. H., Finney, J. W., & Yesavage, J. (2006). Coping, symptoms, and functioning outcomes of patients with posttraumatic stress disorder. *Journal of Traumatic Stress, 19*(6), 799–811. https://doi.org/10.1002/jts.20185.

Törneke, N. (2010). *Learning RFT: An introduction to relational frame theory and its clinical applications*. Oakland, CA: Context Press.

Tull, M. T., & Gratz, K. L. (2008). Further examination of the relationship between anxiety sensitivity and depression: The mediating role of experiential avoidance and difficulties engaging in goal-directed behavior when distressed. *Journal of Anxiety Disorders, 22*(2), 199–210. https://doi.org/10.1016/j.janxdis.2007.03.005.

Wells, A., & Matthews, G. (1994). *Attention and emotion: A clinical perspective*. Hillsdale, NJ: Lawrence Erlbaum Associates, Inc.

Wisco, B. E., Marx, B. P., Miller, M. W., Wolf, E. J., Krystal, J. H., Southwick, S. M., & Pietrzak, R. H. (2017). A comparison of ICD-11 and DSM criteria for posttraumatic stress disorder in two national samples of U.S. military veterans. *Journal of Affective Disorders, 223*, 17–19. https://doi.org/10.1016/j.jad.2017.07.006.

Wolgast, M. (2014). What does the acceptance and action questionnaire (AAQ-II) really measure? *Behavior Therapy, 45*(6), 831–839. https://doi.org/10.1016/j.beth.2014.07.002.

Yoo, S.-S., Hu, P. T., Gujar, N., Jolesz, F. A., & Walker, M. P. (2007). A deficit in the ability to form new human memories without sleep. *Nature Neuroscience, 10*(3), 385–392. https://doi.org/10.1038/nn1851.

Zang, Y., Yu, J., Chazin, D., Asnaani, A., Zandberg, L. J., & Foa, E. B. (2017). Changes in coping behavior in a randomized controlled trial of concurrent treatment for PTSD and alcohol dependence. *Behaviour Research and Therapy, 90*, 9–15. https://doi.org/10.1016/j.brat.2016.11.013.

CHAPTER 15

Emotion-driven impulsivity in PTSD

Nicole H. Weiss[a], Shannon R. Forkus[a], Svetlana Goncharenko[a], Ateka A. Contractor[b]

[a]Department of Psychology, University of Rhode Island, Kingston, RI, United States
[b]Department of Psychology, University of North Texas, Denton, TX, United States

Posttraumatic stress disorder and emotion-driven impulsivity

Research in the past decade has linked posttraumatic stress disorder (PTSD) to a wide range of deficits in emotion regulation (e.g., Ehring & Quack, 2010; McDermott, Tull, Gratz, Daughters, & Lejuez, 2009; Tull, Barrett, McMillan, & Roemer, 2007; Weiss, Tull, Anestis, & Gratz, 2013). Emotion-driven impulsivity is one such deficit that has recently gained increasing promise as an explanatory factor in the etiology and maintenance of PTSD (e.g., Contractor, Caldas, Weiss, & Armour, 2018; Roley, Contractor, Weiss, Armour, & Elhai, 2017; Weiss, Tull, Anestis, et al., 2013) and cooccurring health outcomes (e.g., substance use and aggressive behavior [Contractor, Weiss, Tull, & Elhai, 2017; Weiss, Connolly, Gratz, & Tull, 2017; Weiss, Tull, Sullivan, Dixon-Gordon, & Gratz, 2015; Weiss, Tull, Viana, Anestis, & Gratz, 2012]). The aim of the current chapter is to synthesize theoretical and empirical research on the relation of emotion-driven impulsivity to PTSD to inform future research and clinical practice.

Conceptualizing emotion-driven impulsivity

The importance of emotion has long been recognized in preparing the body for action (Frijda, 1986; Lang, 1994), and this process is thought to be fundamentally adaptive (Depue, 1996). However, while emotions may motivate an individual to take important steps to modify their current situation (Gray, 1987), intense emotional experiences may prevent access to available cognitive resources (Inzlicht & Schmeichel, 2012; Muraven & Baumeister, 2000) or interfere with rational decision-making (Bechara, 2004, 2005;

Bechara, Damasio, & Damasio, 2000; Loewenstein & Lerner, 2003; Shiv, Loewenstein, & Bechara, 2005), increasing the likelihood of rash or impulsive action. Alternatively, individuals may not always be able to satisfy the need that initially triggered the emotion and thus may engage in alternative actions to diminish the intensity of an emotional experience or to elicit reward (Larsen, 2000); such behaviors may take the form of rash or impulsive behavior that function to bring about immediate negative or positive reinforcement.

Consistent with these premises, an extensive body of literature has linked intense emotion states to one's propensity to engage in rash and impulsive behavior (for reviews, see Weiss, Sullivan, & Tull, 2015; Weiss, Tull, & Sullivan, 2015). For instance, daily diary studies indicate that individuals are more likely to engage in alcohol use (Swendsen et al., 2000), binge/purging (Smyth et al., 2007), nonsuicidal self-injury (Nock, Prinstein, & Sterba, 2009), and HIV/sexual risk behaviors (Mustanski, 2007) on days when they experience intense emotions. Moreover, these investigations found that rash and impulsive behavior resulted in a decrease (albeit temporary) in emotional intensity. As such, and consistent with principles of negative reinforcement (Baker, Piper, McCarthy, Majeskie, & Fiore, 2004), engagement in rash and impulsive action following intense emotions may increase the likelihood of such action in similar emotional contexts in the future (Fischer, Smith, Spillane, & Cyders, 2005).

Emotion-driven impulsivity is one construct that has recently gained growing attention in literature examining the link between intense emotion states and rash action. Our understanding of emotion-driven impulsivity is best informed by research on the UPPS-P Impulsive Behavior Scale assessed traits of negative and positive urgency (Cyders et al., 2007; Whiteside & Lynam, 2001). Negative and positive urgency are defined as the tendency to act rashly in response to intense negative or positive emotional experiences, respectively, although the work of several other theorists has also contributed to the conceptualization and measurement of emotion-driven impulsivity (e.g., Difficulties in Emotion Regulation Scale [Gratz & Roemer, 2004], Difficulties in Emotion Regulation Scale – Positive [Weiss, Gratz, & Lavender, 2015], and Risky Behavior Questionnaire [Weiss, Tull, Dixon-Gordon, & Gratz, 2018]).

The construct of urgency was first described by Whiteside and Lynam (2001) who, in an attempt to delimit the facets underlying impulsivity, conducted a factor analysis of several widely used impulsivity measures. Results of Whiteside and Lynam (2001), as well as subsequent factor analyses

(Cyders et al., 2007; Smith et al., 2007; Whiteside, Lynam, Miller, & Reynolds, 2005), indicate that the urgency trait(s) is part of a larger impulsivity model that also includes sensation seeking (i.e., the tendency to enjoy and pursue exciting activities and an openness to trying new experiences), lack of perseverance (i.e., an inability to focus or follow through on difficult or boring tasks), and lack of premeditation (i.e., a failure to reflect on the consequences of an act before engaging in that act). Factor analysis of the urgency traits in relation to personality factors suggests that they are linked to high levels of neuroticism and low levels of agreeableness and conscientiousness (Cyders & Smith, 2008), consistent with evidence that the urgency traits are analogous to the impulsiveness facet of neuroticism proposed by Costa and McCrae (1995). These findings suggest that individuals with elevated levels of the urgency traits may be emotionally unstable and less likely to inhibit impulsive behaviors and cooperate with others. Further, empirical reviews (Berg, Latzman, Bliwise, & Lilienfeld, 2015; Coskunpinar, Dir, & Cyders, 2013; Cyders & Smith, 2007, 2008) indicate that the urgency trait generally demonstrates the greatest concurrent and prospective predictive validity in clinically relevant outcomes, including both psychopathology and rash and impulsive behavior. Based on these findings, theorists (Cyders & Smith, 2008; Selby, Anestis, & Joiner, 2008) have proposed that the urgency trait(s) may serve as a strong indicator of one's likelihood of engaging in rash or impulsive action to modulate intense emotional experiences.

Emotion-driven impulsivity and PTSD

Not surprisingly, most of the theoretical and empirical literature on emotion-driven impulsivity has focused on its relation to impulsive behaviors (for reviews, see Berg et al., 2015; Coskunpinar et al., 2013; Cyders & Smith, 2007, 2008), suggesting that higher levels of urgency are associated with greater overall impulsive behaviors (Weiss, Tull, Sullivan, Dixon-Gordon, et al., 2015) and the specific impulsive behaviors of alcohol use (for a review, see Coskunpinar et al., 2013), drug use (Verdejo-García, Bechara, Recknor, & Pérez-García, 2007), disordered eating (for a review, see Fischer, Smith, & Cyders, 2008), aggressive behaviors (Miller, Flory, Lynam, & Leukefeld, 2003; Weiss et al., 2017), problematic gambling (Smith et al., 2007), compulsive buying (Billieux, Rochat, Rebetez, & Van der Linden, 2008), HIV/sexual risk behavior (Deckman & DeWall, 2011; Zapolski, Cyders, & Smith, 2009), and problematic smartphone use (Billieux, Van der Linden, & Rochat, 2008; Contractor et al., 2017). Nonetheless, there is growing evidence for a relation between emotion-driven impulsivity and psychopathology (e.g., borderline personality features, depression, and anxiety sensitivity [Berg et al., 2015; Peters, Upton, & Baer, 2013; Smith, Guller, & Zapolski, 2013; Weitzman, McHugh, & Otto, 2011]), including PTSD (Contractor et al., 2018; Ehring & Quack, 2010; McDermott et al., 2009; Roley et al., 2017; Tull, Barrett, et al., 2007; Weiss, Tull, Davis, et al., 2012; Weiss, Tull, Lavender, & Gratz, 2013; Weiss, Tull, Sullivan, Dixon-Gordon, et al., 2015).

In the following sections, we will first describe psychological mechanisms underlying emotion-driven impulsivity and how this may inform our understanding of emotion-driven impulsivity in PTSD. We will then identify biological mechanisms (i.e., neurotransmitters, brain pathways, and gene polymorphisms) that have been shown to underlie both PTSD and emotion-driven impulsivity separately and therefore may explain their relation. Following this, we will provide a comprehensive review of empirical investigations of the association between PTSD and emotion-driven impulsivity. Then, we will describe important avenues for future research, specifically the exploration of positive emotion-driven impulsivity in PTSD, biological mechanisms underlying the PTSD–emotion-driven impulsivity relation, and the directionality of the association between PTSD and emotion-driven impulsivity. Finally, we will present clinical implications of the literature examining PTSD and emotion-driven impulsivity.

Psychological mechanisms

Researchers have described several key psychological mechanisms underlying the urgency traits, most notably the following: (1) a reduced focus on the long-term consequences of behavior and (2) a poor capacity for prepotent response inhibition (Billieux, Gay, Rochat, & Van der Linden, 2010; Cyders & Coskunpinar, 2012). Regarding the former, Cyders and Smith (2008) proposed that the urgency traits are associated with a heightened focus on immediate needs, such as the desire to obtain relief from aversive emotions without consideration of long-term consequences. Consistent with the assertion, several studies (Billieux et al., 2010; Cyders & Coskunpinar, 2012; Dolan, Bechara, & Nathan, 2008; Xiao et al., 2009) have found support for an association between the urgency trait(s) and performance on laboratory-based behavioral tasks, such as the Iowa gambling task (IGT [Bechara, Damasio, Damasio, & Anderson, 1994]) and the two-choice impulsivity paradigm (Dougherty, Mathias, Marsh, & Jagar, 2005), that require participants to forego immediate gratification to acquire long-term rewards. Moreover, in one such study (Billieux et al., 2010), the negative urgency trait was found to be associated with more marked disturbance on the second half of the IGT in particular, which requires effortful conscious processes, such as modification to working memory, response inhibition, and set-shifting (compared with performance on the first half of the IGT, which involves making choices without any conscious knowledge of reinforcement contingencies).

Regarding the latter mechanism, the ability to deliberately control or suppress an automatic response has been acknowledged as central to rash and impulsive action (Logan, Schachar, & Tannock, 1997) and has been proposed to underlie the urgency trait(s) (Bechara & Van Der Linden, 2005). However, mixed findings have been found with regard to the urgency-response inhibition relation on go/no-go and stop-start tasks (Verbruggen & Logan, 2008), laboratory-based behavioral assessments of response inhibition, with one study finding a significant positive relation between the negative trait and the number of commission errors (i.e., false alarms) and others detecting no significant relation between the urgency traits and task performance (Billieux et al., 2010; Cyders & Coskunpinar, 2012; Perales, Verdejo-García, Moya, Lozano, & Pérez-García, 2009). Interestingly, results of Billieux et al. (2010) provide support for a more complex model, finding that a low capacity to inhibit a prepotent response in an emotional (but not neutral) condition predicted a proneness to act without forethought in the context of risky decision-making, which ultimately led to heightened levels of the negative urgency trait and, then, rash and impulsive action. Indeed, several investigations have found that it is more difficult to inhibit a prepotent response for an emotional stimulus than it is for a neutral stimulus (Schulz et al., 2007; Verbruggen & Logan, 2008), and an emotional context is central to the rash and impulsive action seen in the urgency traits (Cyders & Smith, 2007, 2008).

The presence of PTSD may exacerbate these deficits in decision-making related to emotion-driven impulsivity. Individuals with PTSD display dysfunction in both bottom-up (e.g., perception) and top-down (e.g., appraisal) emotional processing (Zoladz & Diamond, 2013). For instance, individuals with PTSD are more likely to (a) exhibit an emotional response after the presentation of an emotionally evocative cue (Litz, Orsillo, Kaloupek, & Weathers, 2000; Orsillo, Batten, Plumb, Luterek, & Roessner, 2004), (b) experience a higher amplitude of emotional response to such a cue (Liberzon et al., 1999; Tull, Jakupcak, McFadden, & Roemer, 2007), and (c) exhibit a slower return to baseline levels of emotions following the experience of intense emotions (Perry, 1994).

There is also substantial evidence for maladaptive cognitive appraisals in PTSD (Ehlers & Clark, 2000), including toward their emotional experiences. For instance, research indicates that individuals with PTSD are more likely to take an evaluative and judgmental stance toward their negative (Tull, Barrett, et al., 2007; Weiss, Tull, Anestis, et al., 2013) and positive (Weiss, Darosh, et al., 2018; Weiss, Dixon-Gordon, Peasant, & Sullivan, 2018) emotional experiences, judging them to be undesirable, overwhelming, and/or frightening. Such secondary emotional responses may relate to the fact that emotions are experienced as more intense, reactive, and long-lasting; individuals with PTSD may feel as though their emotional experiences are uncontrollable or unpredictable. Alternatively, emotions may be more likely to elicit secondary emotional reactions among individuals with PTSD because of their association with traumatic experiences. As an example, through stimulus generalization, fear of physiological arousal originally associated with trauma cues may expand to the experience of arousal more broadly, including that associated with negative and positive emotions (Roemer, Litz, Orsillo, & Wagner, 2001).

Notably, these deficits in top-down and bottom-up emotional processing among individuals with PTSD may interfere with decision-making (Bechara, 2004, 2005; Bechara et al., 2000; Ciccarelli, Griffiths, Nigro, & Cosenza, 2017; Heilman, Crişan, Houser, Miclea, & Miu, 2010; Loewenstein & Lerner, 2003; Shiv et al., 2005), increasing the likelihood of emotion-driven impulsivity. For instance, to obtain relief from intense, reactive, and/or prolonged emotional experiences, individuals with PTSD may engage in rash or impulsive action without giving much consideration to the long-term consequences of their behavior. Further, they may be less successful in inhibiting responses that are situationally ineffective and/or that interfere with goal-directed behavior, particularly in emotional contexts. For instance, Billieux et al. (2010) found evidence for impaired performance on the stop-start task following the presentation of pictures of human faces expressing joy. Consistent with the previous literature, individuals with PTSD have been shown to demonstrate deficits in both delay discounting (Engelmann, Maciuba, Vaughan, Paulus, & Dunlop, 2013) and response inhibition (Carrion, Garrett, Menon, Weems, & Reiss, 2008; Swick, Honzel, Larsen, Ashley, & Justus, 2012; Wu et al., 2010).

Additional factors that may influence decision-making (and, thus, emotion-driven impulsivity) among individuals with PTSD warrant mention. Deficits in attentional processing among individuals with PTSD (e.g., attentional threat bias and attentional control [Bardeen, Fergus, &

Orcutt, 2015; Bardeen & Orcutt, 2011]) may make it more difficult to control impulses for rash or impulsive behavior when faced with cues for these behaviors (Tull, McDermott, Gratz, Coffey, & Lejuez, 2011). Indeed, attentional processing deficits (e.g., narrowed attention) have been implicated in emotion-driven impulsivity (Gable & Harmon-Jones, 2008). Additionally, both PTSD (Clark et al., 2003; Weber et al., 2005) and emotion-driven impulsivity (Gunn & Finn, 2015; Romer et al., 2011) have been found to be associated with impairments in working memory. Among individuals with PTSD, distractibility (related to working memory deficits) may interfere with goal-directed behavior, which may explain its relation to emotion-driven impulsivity. Finally, higher levels of emotion-driven impulsivity among individuals with PTSD may be the result of changes in motivation for self-regulation. According to Inzlicht and Schmeichel (2012), the down-regulation of intense emotions is an effortful and not immediately rewarding process. Thus, as the need for regulation persists, individuals may begin to experience a shift in motivation from the regulation of emotion toward acquiring more immediately rewarding experiences. This shift coincides with an increased allocation of attention toward cues that signal more immediate gratification and reward (vs the need for self-regulation), increasing the likelihood of rash and impulsive action. The impaired regulatory control capacity observed among individuals with PTSD (Frewen & Lanius, 2006) may make efforts to inhibit impulsive responding in the context of intense emotions particularly difficult.

Biological mechanisms
Neurotransmitters

Serotonin and dopamine activities have been associated with key processes that underlie both PTSD and emotion-driven impulsivity, separately, and thus may be critical to understanding emotion-driven impulsivity among individuals with PTSD. The serotonin system has been implicated in both affect and impulse modulation (Frankle et al., 2005), and various studies have found that alterations in serotonin activity are associated with changes in emotional functioning (Knutson et al., 1998; Vaswani, Linda, & Ramesh, 2003; Whittington et al., 2004) and with behavioral and motivational systems (Lesch & Merschdorf, 2000; Montoya, Terburg, Bos, & Van Honk, 2012). Regarding the relation of PTSD to serotonin, many of the symptoms associated with PTSD (e.g., mood changes, sleep disturbance, and destructive behavior) are directly and indirectly affected by serotonin (Monti &

Jantos, 2008; Ryding, Lindström, & Träskman-Bendz, 2008). Furthermore, the efficacy of pharmaceutical interventions aimed at increasing levels of available selective serotonin reuptake inhibitors (SSRIs) has shown to successfully reduce many of the symptoms of PTSD (Asnis, Kohn, Henderson, & Brown, 2004; Seedat et al., 2004), suggesting a potential role of serotonin in PTSD. While not as well understood, serotonin also appears to play a role in emotion-driven impulsivity. Heightened levels of negative and positive affect (Zald & Depue, 2001) and rash and impulsive behavior (Harrison, Everitt, & Robbins, 1997) are related to lower levels of serotonin. Further, animal models provide support for the relation between serotonin and mood-based rash action (Yates, Darna, Gipson, Dwoskin, & Bardo, 2015). These findings converge with our understanding of the serotonin system and the functional role of SSRIs to increase serotonin, which then reduces impulsivity (Rinne, van den Brink, Wouters, & van Dyck, 2002).

Studies examining the role of dopamine, primarily implicated in reward and reinforcement (Wise, 2008), have also found evidence of altered dopaminergic systems in individuals with PTSD. Specifically, individuals with PTSD have been shown to have increased levels of dopamine (Spivak et al., 1999), and greater levels of dopamine have been associated with more severe PTSD symptomatology (Yehuda, Southwick, Giller, Ma, & Mason, 1992). Further, elevated levels of dopamine have been associated with emotionality and impulsivity (Friedel, 2004), suggesting the role of this neurotransmitter in emotion-driven impulsivity. Interestingly, there is evidence that the interaction between serotonin and dopamine systems may be particularly important to emotion-driven impulsivity, as serotonin has been shown to exert an inhibitory influence on dopamine activity in certain regions of the brain, particularly those that are believed to play a role in approach behaviors (Spoont, 1992). Some researchers have theorized that the inhibitory influence of serotonin on dopamine may deter behaviors that are inconsistent with long-term goals (Davidson et al., 2003). Thus low levels of serotonin might increase rash and impulsive action by not exerting this inhibitory influence on dopamine.

Brain pathways

The areas of the brain most strongly implicated in emotion-based impulsivity include the amygdala, long acknowledged for its role in emotional processing, including its critical role in directing attention to emotionally salient stimuli, most notably that associated with negative affect (Davidson, 2003), and sectors of the frontal cortex, including the orbitofrontal cortex (OFC), responsible for modulation of emotion-based reactivity

(Davidson, 2003), and the ventromedial prefrontal cortex (VM PFC [Bechara, 2005]), theorized to facilitate affect-guided planning (Bechara et al., 2000) and subjective reward value (Eiler, Dzemidzic, Case, Considine, & Kareken, 2012; Kareken et al., 2010). Activation of the amygdala during negative emotion maintenance and reappraisal has been positively related to the negative urgency trait (Albein-Urios et al., 2013). Further, emotional cue-induced activation in the right lateral OFC and left amygdala is related to heightened levels of negative urgency trait and risk-taking in emotionally salient contexts through the negative urgency trait (Cyders et al., 2014b). In addition, the negative urgency trait has been shown to be related to neural responses to alcohol cues in the VM PFC, suggesting that emotion-driven impulsivity may alter craving and brain regions involved in reward (Cyders et al., 2014a). Notably, these brain regions have also been implicated in PTSD (Koenigs & Grafman, 2009; Myers-Schulz & Koenigs, 2012; Shin et al., 1999, 2004, 2005; Shin, Rauch, & Pitman, 2006), providing support for common underlying mechanisms.

Gene polymorphisms

Despite evidence of genetic influence in the etiology of PTSD among molecular studies, a precise genetic association has not yet been identified (for a review, see Skelton, Ressler, Norrholm, Jovanovic, & Bradley-Davino, 2012). Nonetheless, several specific candidate genes have been found to impact risk or resilience for PTSD, including those associated with the HPA axis (FKBP5 [Binder et al., 2008; Xie et al., 2010]), noradrenergic system (NPY and DBH [Lappalainen et al., 2002; Mustapić et al., 2007]), and frontal system (e.g., DRD2 and SLC6A4 [Comings, Muhleman, & Gysin, 1996; Kolassa et al., 2010]). The vast majority of this research points to the role of genes associated with serotonin and dopamine production in the development, maintenance, and/or exacerbation of PTSD (Comings et al., 1991, 1996; Dragan & Oniszczenko, 2009; Voisey et al., 2009; Young et al., 2002). Similarly, research on emotion-driven impulsivity shows analogous associations, providing evidence for variably in the serotonin transporter gene (5HTTLPR) and dopamine receptor genes DRD2, DRD3, and DRD4 (Cyders & Smith, 2008; Ficks & Waldman, 2014; Grant, Odlaug, & Chamberlain, 2016; Lejuez et al., 2010).

Empirical evidence

Over the past decade a growing body of empirical work has begun to document an association of PTSD to emotion-driven impulsivity

(Contractor et al., 2017, 2018; Dutra & Sadeh, 2017; Ehring & Quack, 2010; Gaher et al., 2014; Hahn, Tirabassi, Simons, & Simons, 2015; McDermott et al., 2009; Mirhashem et al., 2017; Price, Bell, & Lilly, 2014; Roley et al., 2017; Tull, Barrett, et al., 2007; Weiss et al., 2017; Weiss, Tull, Anestis, et al., 2013; Weiss, Tull, Davis, et al., 2012; Weiss, Tull, Lavender, et al., 2013; Weiss, Tull, Sullivan, Dixon-Gordon, et al., 2015; Weiss, Tull, Viana, et al., 2012). Research in this area provides support for higher levels of emotion-driven impulsivity among individuals with (vs without) a diagnosis of PTSD (e.g., Tull, Barrett, et al., 2007; Weiss, Tull, Anestis, et al., 2013; Weiss, Tull, Lavender, et al., 2013). Moreover, it suggests that greater severity of PTSD symptoms is related to higher levels of emotion-driven impulsivity among trauma-exposed individuals (e.g., Ehring & Quack, 2010; Roley et al., 2017; Weiss, Tull, Sullivan, Dixon-Gordon, et al., 2015). Subsequent investigations in this area have sought to clarify the association of PTSD to emotion-driven impulsivity.

Several researchers have explored the unique contribution of emotion-driven impulsivity to PTSD relative to other clinically relevant factors. Weiss, Tull, Anestis, et al. (2013) examined the roles of the UPPS impulsivity dimensions in PTSD status (present/absent), finding that the negative urgency trait was a significant and unique predictor of the presence of PTSD, above and beyond the other UPPS impulsivity traits. Likewise, Weiss, Tull, Lavender, et al. (2013) found that the dimension of emotion dysregulation characterized by emotion-driven impulsivity (i.e., difficulties controlling impulsive behaviors when experiencing distress) was uniquely associated with the presence of PTSD above and beyond the other emotion dysregulation facets. Further, Roley et al. (2017) examined the contributions of the UPPS impulsivity dimensions to PTSD severity; the negative urgency trait was the most predictive of PTSD severity, as it was the only facet to demonstrate associations with all of PTSD's symptom clusters. Finally, Contractor et al. (2018) used latent profile analysis to explore patterns of PTSD symptom severity and UPPS impulsivity dimensions. The negative urgency trait was found to be the most distinguishing impulsivity facet, with the three identified classes being characterized with high, moderate, and low levels of this impulsivity dimension parallel to PTSD severity levels. Overall, these findings highlight the role of emotion-driven impulsivity in PTSD diagnosis and symptom severity.

Recent investigations have also examined the role of traumatic experiences in the association between PTSD and emotion-driven impulsivity. Weiss, Tull, Davis, et al. (2012) found higher levels of emotion-driven impulsivity among trauma-exposed African Americans with PTSD compared with African Americans without Criterion A traumatic exposure and those with Criterion A traumatic exposure but no PTSD, underscoring the relevance of emotion-driven impulsivity to PTSD per se versus simply Criterion A traumatic exposure. In another study, Weiss, Tull, Lavender, et al. (2013) found that emotion-driven impulsivity mediated the relation of childhood physical and emotional abuse to PTSD in adulthood. Lastly, Price, Connor, and Allen (2017) provided initial support for the moderating role of childhood abuse/maltreatment in the association between the positive urgency trait and PTSD, such that this relation was only significant at lower levels of childhood emotional abuse and neglect. These previous findings present a mixed picture with regard to the role of trauma in the link between PTSD and emotion-driven impulsivity, suggesting the need for additional research in this area.

Not surprisingly the vast majority of empirical work in this area has examined PTSD and emotion-driven impulsivity in relation to rash and impulsive behavior. Several such studies have examined the mediating role of emotion-driven impulsivity in the association between PTSD and rash and impulsive behavior (a relation that has been well established in the literature [for a review, see Tull, Weiss, & McDermott, 2015]). Weiss, Tull, Sullivan, Dixon-Gordon, et al. (2015) found that the urgency traits mediated the relation of PTSD symptom severity to overall risky behaviors. With regard to specific impulsive behaviors, Weiss et al. (2017) found that the negative urgency trait was the only UPPS dimension that explained the association of PTSD status to physical aggression, whereas Contractor et al. (2017) reported that the negative urgency trait was the only UPPS dimension that accounted for the PTSD severity–problematic smartphone use relation. Extending these findings, Dutra and Sadeh (2017) found that PTSD symptom severity predicted externalizing behavior through the negative urgency trait and that psychological flexibility moderated this association, such that the relation between the negative urgency trait and externalizing behavior was attenuated at high levels of psychological flexibility.

Other studies have examined the potential underlying role of PTSD in the association between emotion-driven impulsivity and rash and impulsive behavior. For instance, Hahn et al. (2015) found that PTSD symptom severity mediated the association between the negative urgency trait and alcohol-related problems. Similarly, Mirhashem et al. (2017) demonstrated that the negative urgency trait was related to substance-related problems through PTSD symptom severity. Finally, in an experience sampling study, the effect of the negative urgency trait on alcohol use and problems was found to be indirect through an association with PTSD symptom severity (Gaher et al., 2014). In conclusion, initial evidence supports the underlying roles of (a) emotion-driven impulsivity in the association between PTSD and rash and impulsive behavior and (b) PTSD in the relation of emotion-driven impulsivity to rash and impulsive behavior. Future research would benefit from examining the directionality of these associations.

Important avenues for future research
Positive emotions

The vast majority of research on PTSD and emotion-driven impulsivity has focused on rash action in the context of negative emotions. Nonetheless, there is growing evidence to suggest that individuals may be prone to rash and impulsive behavior when experiencing positive emotions (for measures of positive emotion-driven impulsivity, see Cyders et al., 2007; Weiss, Gratz, et al., 2015; Weiss, Tull, et al., 2018). Indeed, positive emotion-driven impulsivity has been found to be associated with overall risky behaviors (Weiss, Tull, Sullivan, Dixon-Gordon, et al., 2015) and a wide range of specific risky behaviors (for reviews, see Berg et al., 2015; Coskunpinar et al., 2013; Cyders & Smith, 2007, 2008), including substance use (Cyders et al., 2010; Weiss, Forkus, Contractor, & Schick, 2018), HIV/sexual risk (Weiss, Tull, et al., 2018; Zapolski et al., 2009), nonsuicidal self-injury (Dir, Karyadi, & Cyders, 2013; Weiss, Tull, et al., 2018), disordered eating (Cyders et al., 2007; Dir et al., 2013), and aggressive behavior (Grimaldi, Napper, & LaBrie, 2014; Miller, Zeichner, & Wilson, 2012).

Recent research also suggests that positive emotion-driven impulsivity relates to PTSD. Individuals with PTSD have been shown to have higher levels of positive emotion-driven impulsivity than those without PTSD (Weiss, Dixon-Gordon, et al., 2018). Further, results of Weiss, Tull, Sullivan, Dixon-Gordon, et al. (2015) and Weiss, Dixon-Gordon, et al. (2018) provide support for higher levels of positive emotion-driven impulsivity among individuals with more severe PTSD symptoms. Finally, using latent profile analysis, Weiss, Darosh, et al. (2018) found that women in a class identified by higher levels of positive emotion-driven impulsivity reported more severe PTSD symptoms. One explanation for the previous findings is that individuals with PTSD exhibit physiological arousal in response to positive emotions (Litz et al., 2000), which may be experienced as aversive because of the relation of arousal to trauma-related symptoms (Taylor, Koch, & McNally, 1992). In turn, these individuals may engage in rash and impulsive behavior to escape arousal, consistent with tension reduction theory (Conger, 1956). Alternatively, positive emotions may impair behavioral control, consistent with evidence to suggest that positive emotional states increase distractibility (Dreisbach & Goschke, 2004) and narrow attention (Gable & Harmon-Jones, 2008) and lead to less discriminative use of information (Forgas, 1992). Future research is needed to better understand the functional association between PTSD and positive emotion-driven impulsivity.

Biological mechanisms

As was described earlier, research suggests overlap in the specific brain pathways, neurotransmitters, and genes identified in PTSD and emotion-driven impulsivity. However, we are not aware of any studies that have examined the direct contribution of these—or other—biological mechanisms to the emotion-driven impulsivity observed among individuals with PTSD. This is a critical next step. Indeed, because emotional experiences are multifaceted (Gratz & Roemer, 2004; Weiss, Gratz, et al., 2015) and PTSD may result in a lack of coherence between verbal and nonverbal expressions of emotions (Negrao, Bonanno, Noll, Putnam, & Trickett, 2005), research relying on subjective report alone may provide an incomplete picture of the role of emotion-driven impulsivity in PTSD.

Directionality

Research examining the relation between PTSD and emotion-driven impulsivity has relied on cross-sectional and correlational designs, precluding determination of the exact nature and direction of this association. For instance, it is unclear whether (a) higher levels of emotion-driven impulsivity are related to an increased risk of PTSD following trauma; (b) post trauma, the presence of PTSD contributes to the development of emotion-driven impulsivity; or (c) the association of PTSD to emotion-driven impulsivity is bidirectional. At least conceptually, it makes sense that each of these explanations may be plausible for any given individual. For instance, childhood trauma has been posited to interfere with the development of adaptive strategies for modulating emotional experiences (Cicchetti & Howes, 1991), and consistent with this assertion, adults with a history of childhood trauma have been shown to utilize more emotionally avoidant strategies (Weiss, Peasant, & Sullivan, 2019), which may take the form of rash or impulsive action. These findings suggest that, for some individuals, emotion-driven impulsivity may emerge after the development of PTSD. Conversely, for others, emotion-driven impulsivity may result in a higher level of vulnerability to PTSD. Specifically, higher levels of emotion-driven impulsivity may repeatedly place individuals in dangerous situations and thus at higher risk for trauma, consistent with the compulsive exposure hypothesis (Cottler, Compton, Mager, Spitznagel, & Janca, 1992). Finally, PTSD symptoms and emotion regulation difficulties (including emotion-driven impulsivity) have been reported to reciprocally influence one another from pretrauma to post trauma (Bardeen, Kumpula, & Orcutt, 2013), suggesting a potential bidirectional association between PTSD and emotion-driven impulsivity.

Clinical implications

Research on the link between PTSD and emotion-driven impulsivity has important implications for working with trauma-exposed individuals. These findings may inform the assessment of individuals at risk for PTSD or exhibiting PTSD symptoms. Specifically, our review of the literature suggests that evaluation of emotion-driven impulsivity may identify individuals at risk for displaying PTSD symptoms post trauma. Further, among individuals with PTSD, emotion-driven impulsivity may identify risk for more severe PTSD symptomatology and cooccurring psychopathology (e.g., borderline personality features) and impulsive behaviors (e.g., substance use and HIV/sexual risk). Moreover, examination of emotion-driven impulsivity may aid in identifying treatment targets for individuals with PTSD. Evidence of heightened levels of emotion-driven impulsivity among patients with PTSD may suggest the potential utility of teaching skills for tolerating intense emotions without acting impulsively. For example, distress tolerance skills (Bornovalova, Gratz, Daughters, Hunt, & Lejuez, 2012; Brown et al., 2008; Linehan, 1993) may facilitate behavioral control in the context of intense emotions by redirecting attention to nonemotional stimuli and promoting more adaptive actions in the face of emotional arousal. Likewise, treatments that emphasize emotional acceptance and willingness (Gratz, Tull, & Levy, 2014; Hayes, Strosahl, & Wilson, 1999) may reduce urgency stemming from intense emotions by decreasing secondary emotional responses and increasing tolerance for previously avoided emotions. Indeed, dialectical behavior therapy prolonged exposure (DBT PE), which targets PTSD and emotion-driven impulsivity as part of a more comprehensive treatment, has been found to reduce PTSD symptoms and impulsive behaviors (Harned, Korslund, Foa, & Linehan, 2012; Harned, Korslund, & Linehan, 2014). The DBT PE protocol is implemented concurrently with standard DBT (i.e., individual therapy, group skills training, phone consultation, and therapist consultation team meeting) once patients have attained control over higher-priority treatment targets (e.g., life-threatening behaviors) with standard DBT. Specifically, patients receive individual therapy (either once [90 minutes of DBT PE and 30 minutes of DBT] or twice [one 90-minute DBT PE session and one 60-minute DBT session] weekly sessions), group DBT skills training, and phone consultation. DBT PE integrates DBT strategies into in vivo and imaginal exposure, with the goal of reducing risk for negative reactions during or as a result of exposure and addressing the needs of patients with BPD. The duration of DBT PE is determined by a particular patient's PTSD symptoms and treatment goals.

Acknowledgment

Work on this paper by the first author (NHW) was supported by National Institute on Drug Abuse grant K23DA039327.

References

Albein-Urios, N., Verdejo-Román, J., Soriano-Mas, C., Asensio, S., Martínez-González, J. M., & Verdejo-García, A. (2013). Cocaine users with comorbid Cluster B personality disorders show dysfunctional brain activation and connectivity in the emotional regulation networks during negative emotion maintenance and reappraisal. *European Neuropsychopharmacology, 23*, 1698–1707.

Asnis, G. M., Kohn, S. R., Henderson, M., & Brown, N. L. (2004). SSRIs versus non-SSRIs in post-traumatic stress disorder. *Drugs, 64*, 383–404.

Baker, T. B., Piper, M. E., McCarthy, D. E., Majeskie, M. R., & Fiore, M. C. (2004). Addiction motivation reformulated: An affective processing model of negative reinforcement. *Psychological Review, 111*, 33–51.

Bardeen, J. R., Fergus, T. A., & Orcutt, H. K. (2015). Attentional control as a prospective predictor of posttraumatic stress symptomatology. *Personality and Individual Differences, 81*, 124–128.

Bardeen, J. R., Kumpula, M. J., & Orcutt, H. K. (2013). Emotion regulation difficulties as a prospective predictor of posttraumatic stress symptoms following a mass shooting. *Journal of Anxiety Disorders, 27*, 188–196.

Bardeen, J. R., & Orcutt, H. K. (2011). Attentional control as a moderator of the relationship between posttraumatic stress symptoms and attentional threat bias. *Journal of Anxiety Disorders, 25*, 1008–1018.

Bechara, A. (2004). The role of emotion in decision-making: Evidence from neurological patients with orbitofrontal damage. *Brain and Cognition, 55*, 30–40.

Bechara, A. (2005). Decision making, impulse control and loss of willpower to resist drugs: A neurocognitive perspective. *Nature Neuroscience, 8*, 1458–1463.

Bechara, A., Damasio, A. R., Damasio, H., & Anderson, S. W. (1994). Insensitivity to future consequences following damage to human prefrontal cortex. *Cognition, 50*, 7–15.

Bechara, A., Damasio, H., & Damasio, A. R. (2000). Emotion, decision making and the orbitofrontal cortex. *Cerebral Cortex, 10*, 295–307.

Bechara, A., & Van Der Linden, M. (2005). Decision-making and impulse control after frontal lobe injuries. *Current Opinion in Neurology, 18*, 734–739.

Berg, J. M., Latzman, R. D., Bliwise, N. G., & Lilienfeld, S. O. (2015). Parsing the heterogeneity of impulsivity: A meta-analytic review of the behavioral implications of the UPPS for psychopathology. *Psychological Assessment, 27*, 1129–1146.

Billieux, J., Gay, P., Rochat, L., & Van der Linden, M. (2010). The role of urgency and its underlying psychological mechanisms in problematic behaviours. *Behavior Research and Therapy, 48*, 1085–1096.

Billieux, J., Rochat, L., Rebetez, M. M. L., & Van der Linden, M. (2008). Are all facets of impulsivity related to self-reported compulsive buying behavior? *Personality and Individual Differences, 44*, 1432–1442.

Billieux, J., Van der Linden, M., & Rochat, L. (2008). The role of impulsivity in actual and problematic use of the mobile phone. *Applied Cognitive Psychology, 22*, 1195–1210.

Binder, E. B., Bradley, R. G., Liu, W., Epstein, M. P., Deveau, T. C., Mercer, K. B., ... Nemeroff, C. B. (2008). Association of FKBP5 polymorphisms and childhood abuse with risk of posttraumatic stress disorder symptoms in adults. *JAMA, 299*, 1291–1305.

Bornovalova, M. A., Gratz, K. L., Daughters, S. B., Hunt, E. D., & Lejuez, C. W. (2012). Initial RCT of a distress tolerance treatment for individuals with substance use disorders. *Drug and Alcohol Dependence, 122*, 70–76.

Brown, R. A., Palm, K. M., Strong, D. R., Lejuez, C. W., Kahler, C. W., Zvolensky, M. J., ... Gifford, E. V. (2008). Distress tolerance treatment for early-lapse smokers: Rationale, program description, and preliminary findings. *Behavior Modification, 32*, 302–332.

Carrion, V. G., Garrett, A., Menon, V., Weems, C. F., & Reiss, A. L. (2008). Posttraumatic stress symptoms and brain function during a response-inhibition task: An fMRI study in youth. *Depression and Anxiety, 25*, 514–526.

Ciccarelli, M., Griffiths, M. D., Nigro, G., & Cosenza, M. (2017). Decision making, cognitive distortions and emotional distress: A comparison between pathological gamblers and healthy controls. *Journal of Behavior Therapy and Experimental Psychiatry, 54*, 204–210.

Cicchetti, D., & Howes, P. W. (1991). Developmental psychopathology in the context of the family: Illustrations from the study of child maltreatment. *Canadian Journal of Behavioural Science, 23*, 257–281.

Clark, C. R., McFarlane, A. C., Morris, P., Weber, D. L., Sonkkilla, C., Shaw, M., ... Egan, G. F. (2003). Cerebral function in posttraumatic stress disorder during verbal working memory updating: A positron emission tomography study. *Biological Psychiatry, 53*, 474–481.

Comings, D. E., Comings, B. G., Muhleman, D., Dietz, G., Shahbahrami, B., Tast, D., ... Kovacs, B. W. (1991). The dopamine D2 receptor locus as a modifying gene in neuropsychiatric disorders. *JAMA, 266*, 1793–1800.

Comings, D. E., Muhleman, D., & Gysin, R. (1996). Dopamine D2 receptor (DRD2) gene and susceptibility to posttraumatic stress disorder: A study and replication. *Biological Psychiatry, 40*, 368–372.

Conger, J. J. (1956). Reinforcement theory and the dynamics of alcoholism. *Quarterly Journal of Studies on Alcohol, 17*, 296–305.

Contractor, A. A., Caldas, S., Weiss, N. H., & Armour, C. (2018). Examination of the heterogeneity in PTSD and impulsivity facets: A latent profile analysis. *Personality and Individual Differences, 125*, 1–9.

Contractor, A. A., Weiss, N. H., Tull, M. T., & Elhai, J. D. (2017). PTSD's relation with problematic smartphone use: Mediating role of impulsivity. *Computers in Human Behavior, 75*, 177–183.

Coskunpinar, A., Dir, A. L., & Cyders, M. A. (2013). Multidimensionality in impulsivity and alcohol use: A meta-analysis using the UPPS model of impulsivity. *Alcoholism: Clinical and Experimental Research, 37*, 1441–1450.

Costa, T. P., & McCrae, R. R. (1995). Domains and facets: Hierarchical personality assessment using the Revised NEO Personality Inventory. *Journal of Personality Assessment, 64*, 21–50.

Cottler, L. B., Compton, W. M., Mager, D., Spitznagel, E. L., & Janca, A. (1992). Posttraumatic stress disorder among substance users from the general population. *American Journal of Psychiatry, 149*, 664–670.

Cyders, M. A., & Coskunpinar, A. (2012). The relationship between self-report and lab task conceptualizations of impulsivity. *Journal of Research in Personality, 46*, 121–124.

Cyders, M. A., Dzemidzic, M., Eiler, W. J., Coskunpinar, A., Karyadi, K., & Kareken, D. A. (2014a). Negative urgency and ventromedial prefrontal cortex responses to alcohol cues: fMRI evidence of emotion-based impulsivity. *Alcoholism: Clinical and Experimental Research, 38*, 409–417.

Cyders, M. A., Dzemidzic, M., Eiler, W. J., Coskunpinar, A., Karyadi, K. A., & Kareken, D. A. (2014b). Negative urgency mediates the relationship between amygdala and orbitofrontal cortex activation to negative emotional stimuli and general risk-taking. *Cerebral Cortex, 25*, 4094–4102.

Cyders, M. A., & Smith, G. T. (2007). Mood-based rash action and its components: Positive and negative urgency. *Personality and Individual Differences, 43*, 839–850.

Cyders, M. A., & Smith, G. T. (2008). Emotion-based dispositions to rash action: Positive and negative urgency. *Psychological Bulletin, 134*, 807–828.

Cyders, M. A., Smith, G. T., Spillane, N. S., Fischer, S., Annus, A. M., & Peterson, C. (2007). Integration of impulsivity and positive mood to predict risky behavior: Development and validation of a measure of positive urgency. *Psychological Assessment, 19*, 107–118.

Cyders, M. A., Zapolski, T. C. B., Combs, J. L., Settles, R. F., Fillmore, M. T., & Smith, G. T. (2010). Experimental effect of positive urgency on negative outcomes from risk taking and on increased alcohol consumption. *Psychology of Addictive Behaviors, 24*, 367–375.

Davidson, R. J. (2003). Darwin and the neural bases of emotion and affective style. *Annals of the New York Academy of Sciences, 1000*, 316–336.

Davidson, R. J., Kabat-Zinn, J., Schumacher, J., Rosenkranz, M., Muller, D., Santorelli, S. F., … Sheridan, J. F. (2003). Alterations in brain and immune function produced by mindfulness meditation. *Psychosomatic Medicine, 65*, 564–570.

Deckman, T., & DeWall, C. N. (2011). Negative urgency and risky sexual behaviors: A clarification of the relationship between impulsivity and risky sexual behavior. *Personality and Individual Differences, 51*, 674–678.

Depue, R. A. (1996). *A neurobiological framework for the structure of personality and emotion: Implications for personality disorders.* New York, NY: Guilford Press.

Dir, A. L., Karyadi, K., & Cyders, M. A. (2013). The uniqueness of negative urgency as a common risk factor for self-harm behaviors, alcohol consumption, and eating problems. *Addictive Behaviors, 38*, 2158–2162.

Dolan, S. L., Bechara, A., & Nathan, P. E. (2008). Executive dysfunction as a risk marker for substance abuse: The role of impulsive personality traits. *Behavioral Sciences & the Law, 26*, 799–822.

Dougherty, D. M., Mathias, C. W., Marsh, D. M., & Jagar, A. A. (2005). Laboratory behavioral measures of impulsivity. *Behavior Research Methods, 37*, 82–90.

Dragan, W. L., & Oniszczenko, W. (2009). The association between dopamine D4 receptor exon III polymorphism and intensity of PTSD symptoms among flood survivors. *Anxiety, Stress, and Coping, 22*, 483–495.

Dreisbach, G., & Goschke, T. (2004). How positive affect modulates cognitive control: Reduced perseveration at the cost of increased distractibility. *Journal of Experimental Psychology: Learning, Memory, and Cognition, 30*, 343–353.

Dutra, S. J., & Sadeh, N. (2017). Psychological flexibility mitigates effects of PTSD symptoms and negative urgency on aggressive behavior in trauma-exposed veterans. *Personality Disorders, Theory, Research, and Treatment, 9*, 315–323.

Ehlers, A., & Clark, D. M. (2000). A cognitive model of posttraumatic stress disorder. *Behavior Research and Therapy, 38*, 319–345.

Ehring, T., & Quack, D. (2010). Emotion regulation difficulties in trauma survivors: The role of trauma type and PTSD symptom severity. *Behavior Therapy, 41*, 587–598.

Eiler, W. J. A., Dzemidzic, M., Case, K. R., Considine, R. V., & Kareken, D. A. (2012). Correlation between ventromedial prefrontal cortex activation to food aromas and cue-driven eating: An fMRI study. *Chemosensory Perception, 5*, 27–36.

Engelmann, J. B., Maciuba, B., Vaughan, C., Paulus, M. P., & Dunlop, B. W. (2013). Posttraumatic stress disorder increases sensitivity to long term losses among patients with major depressive disorder. *PLoS One, 8*, e78292.

Ficks, C. A., & Waldman, I. D. (2014). Candidate genes for aggression and antisocial behavior: A meta-analysis of association studies of the 5HTTLPR and MAOA-uVNTR. *Behavior Genetics, 44*, 427–444.

Fischer, S., Smith, G. T., & Cyders, M. A. (2008). Another look at impulsivity: A meta-analytic review comparing specific dispositions to rash action in their relationship to bulimic symptoms. *Clinical Psychology Review, 28*, 1413–1425.

Fischer, S., Smith, G. T., Spillane, N., & Cyders, M. A. (2005). Urgency: Individual differences in reaction to mood and implications for addictive behaviors. In A. V. Clark (Ed.), *The Psychology of Mood* (pp. 85–107). New York, NY: Nova Science Publishers.

Forgas, J. P. (1992). Mood and the perception of unusual people: Affective asymmetry in memory and social judgments. *European Journal of Social Psychology, 22*, 531–547.

Frankle, W. G., Lombardo, I., New, A. S., Goodman, M., Talbot, P. S., Huang, Y., … Abi-Dargham, A. (2005). Brain serotonin transporter distribution in subjects with impulsive aggressivity: A positron emission study with [11C] McN 5652. *American Journal of Psychiatry, 162*, 915–923.

Frewen, P. A., & Lanius, R. A. (2006). Toward a psychobiology of posttraumatic self-dysregulation. *Annals of the New York Academy of Sciences, 1071*, 110–124.

Friedel, R. O. (2004). Dopamine dysfunction in borderline personality disorder: A hypothesis. *Neuropsychopharmacology, 29*, 1029–1039.

Frijda, N. H. (1986). *The emotions*. New York, NY: Cambridge University Press.

Gable, P. A., & Harmon-Jones, E. (2008). Approach-motivated positive affect reduces breadth of attention. *Psychological Science, 19*, 476–482.

Gaher, R. M., Simons, J. S., Hahn, A. M., Hofman, N. L., Hansen, J., & Buchkoski, J. (2014). An experience sampling study of PTSD and alcohol-related problems. *Psychology of Addictive Behaviors, 28*, 1013–1025.

Grant, J. E., Odlaug, B. L., & Chamberlain, S. R. (2016). Neural and psychological underpinnings of gambling disorder: A review. *Progress in Neuro-Psychopharmacology and Biological Psychiatry, 65*, 188–193.

Gratz, K. L., & Roemer, L. (2004). Multidimensional assessment of emotion regulation and dysregulation: Development, factor structure, and initial validation of the difficulties in emotion regulation scale. *Journal of Psychopathology and Behavioral Assessment, 26*, 41–54.

Gratz, K. L., Tull, M. T., & Levy, R. (2014). Randomized controlled trial and uncontrolled 9-month follow-up of an adjunctive emotion regulation group therapy for deliberate self-harm among women with borderline personality disorder. *Psychological Medicine, 11*, 2099–2112.

Gray, J. A. (1987). *The psychology of fear and stress*. New York, NY: Cambridge University Press.

Grimaldi, E. M., Napper, L. E., & LaBrie, J. W. (2014). Relational aggression, positive urgency and negative urgency: Predicting alcohol use and consequences among college students. *Psychology of Addictive Behaviors, 28*, 893–898.

Gunn, R. L., & Finn, P. R. (2015). Applying a dual process model of self-regulation: The association between executive working memory capacity, negative urgency, and negative mood induction on pre-potent response inhibition. *Personality and Individual Differences, 75*, 210–215.

Hahn, A. M., Tirabassi, C. K., Simons, R. M., & Simons, J. S. (2015). Military sexual trauma, combat exposure, and negative urgency as independent predictors of PTSD and subsequent alcohol problems among OEF/OIF veterans. *Psychological Services*, *12*, 378–383.

Harned, M. S., Korslund, K. E., Foa, E. B., & Linehan, M. M. (2012). Treating PTSD in suicidal and self-injuring women with borderline personality disorder: Development and preliminary evaluation of a Dialectical Behavior Therapy Prolonged Exposure Protocol. *Behavior Research and Therapy*, *50*, 381–386.

Harned, M. S., Korslund, K. E., & Linehan, M. M. (2014). A pilot randomized controlled trial of Dialectical Behavior Therapy with and without the Dialectical Behavior Therapy Prolonged Exposure protocol for suicidal and self-injuring women with borderline personality disorder and PTSD. *Behavior Research and Therapy*, *55*, 7–17.

Harrison, A. A., Everitt, B. J., & Robbins, T. W. (1997). Central 5-HT depletion enhances impulsive responding without affecting the accuracy of attentional performance: interactions with dopaminergic mechanisms. *Psychopharmacology*, *133*, 329–342.

Hayes, S. C., Strosahl, K. D., & Wilson, K. G. (1999). *Acceptance and commitment therapy: An experiential approach to behavior change*. New York, NY: Guilford Press.

Heilman, R. M., Crişan, L. G., Houser, D., Miclea, M., & Miu, A. C. (2010). Emotion regulation and decision making under risk and uncertainty. *Emotion*, *10*, 257–265.

Inzlicht, M., & Schmeichel, B. J. (2012). What is ego depletion? Toward a mechanistic revision of the resource model of self-control. *Perspectives on Psychological Science*, *7*, 450–463.

Kareken, D. A., Liang, T., Wetherill, L., Dzemidzic, M., Bragulat, V., Cox, C., … Foroud, T. (2010). A polymorphism in GABRA2 is associated with the medial frontal response to alcohol cues in an fMRI study. *Alcoholism: Clinical and Experimental Research*, *34*, 2169–2178.

Knutson, B., Wolkowitz, O. M., Cole, S. W., Chan, T., Moore, E. A., Johnson, R. C., … Reus, V. I. (1998). Selective alteration of personality and social behavior by serotonergic intervention. *American Journal of Psychiatry*, *155*, 373–379.

Koenigs, M., & Grafman, J. (2009). Posttraumatic stress disorder: The role of medial prefrontal cortex and amygdala. *The Neuroscientist*, *15*, 540–548.

Kolassa, I., Ertl, V., Eckart, C., Glöckner, F., Kolassa, S., Papassotiropoulos, A., … Elbert, T. (2010). Association study of trauma load and SLC6A4 promoter polymorphism in posttraumatic stress disorder: Evidence from survivors of the Rwandan genocide. *Journal of Clinical Psychiatry*, *71*, 543–547.

Lang, P. J. (1994). *The motivational organization of emotion: Affect-reflex connections*. Hillsdale, NJ: Erlbaum.

Lappalainen, J., Kranzler, H. R., Malison, R., Price, L. H., Van Dyck, C., Rosenheck, R. A., … Krystal, J. (2002). A functional neuropeptide Y Leu7Pro polymorphism associated with alcohol dependence in a large population sample from the United States. *Archives of General Psychiatry*, *59*, 825–831.

Larsen, R. J. (2000). Toward a science of mood regulation. *Psychological Inquiry*, *11*, 129–141.

Lejuez, C. W., Magidson, J. F., Mitchell, S. H., Sinha, R., Stevens, M. C., & De Wit, H. (2010). Behavioral and biological indicators of impulsivity in the development of alcohol use, problems, and disorders. *Alcoholism: Clinical and Experimental Research*, *34*, 1334–1345.

Lesch, K. P., & Merschdorf, U. (2000). Impulsivity, aggression, and serotonin: A molecular psychobiological perspective. *Behavioral Sciences & the Law*, *18*, 581–604.

Liberzon, I., Taylor, S. F., Amdur, R., Jung, T. D., Chamberlain, K. R., Minoshima, S., … Fig, L. M. (1999). Brain activation in PTSD in response to trauma-related stimuli. *Biological Psychiatry*, *45*, 817–826.

Linehan, M. M. (1993). *Cognitive behavioral treatment of borderline personality disorder*. New York, NY: Guilford Press.

Litz, B. T., Orsillo, S. M., Kaloupek, D., & Weathers, F. (2000). Emotional processing in posttraumatic stress disorder. *Journal of Abnormal Psychology*, *109*, 26–39.

Loewenstein, G., & Lerner, J. S. (2003). *The role of affect in decision making.* New York, NY: Oxford University Press.

Logan, G. D., Schachar, R. J., & Tannock, R. (1997). Impulsivity and inhibitory control. *Psychological Science, 8,* 60–64.

McDermott, M. J., Tull, M. T., Gratz, K. L., Daughters, S. B., & Lejuez, C. W. (2009). The role of anxiety sensitivity and difficulties in emotion regulation in posttraumatic stress disorder among crack/cocaine dependent patients in residential substance abuse treatment. *Journal of Anxiety Disorders, 23,* 591–599.

Miller, J., Flory, K., Lynam, D., & Leukefeld, C. (2003). A test of the four-factor model of impulsivity-related traits. *Personality and Individual Differences, 34,* 1403–1418.

Miller, J., Zeichner, A., & Wilson, L. F. (2012). Personality correlates of aggression: Evidence from measures of the five-factor model, UPPS model of impulsivity, and BIS/BAS. *Journal of Interpersonal Violence, 27,* 2903–2919.

Mirhashem, R., Allen, H. C., Adams, Z. W., van Stolk-Cooke, K., Legrand, A., & Price, M. (2017). The intervening role of urgency on the association between childhood maltreatment, PTSD, and substance-related problems. *Addictive Behaviors, 69,* 98–103.

Monti, J. M., & Jantos, H. (2008). The roles of dopamine and serotonin, and of their receptors, in regulating sleep and waking. *Progress in Brain Research, 172,* 625–646.

Montoya, E. R., Terburg, D., Bos, P. A., & Van Honk, J. (2012). Testosterone, cortisol, and serotonin as key regulators of social aggression: A review and theoretical perspective. *Motivation and Emotion, 36,* 65–73.

Muraven, M., & Baumeister, R. F. (2000). Self-regulation and depletion of limited resources: Does self-control resemble a muscle? *Psychological Bulletin, 126,* 247–259.

Mustanski, B. (2007). The influence of state and trait affect on HIV risk behaviors: A daily diary study of MSM. *Health Psychology, 26,* 618–626.

Mustapić, M., Pivac, N., Kozarić-Kovačić, D., Deželjin, M., Cubells, J. F., & Mück-Šeler, D. (2007). Dopamine beta-hydroxylase (DBH) activity and -1021C/T polymorphism of DBH gene in combat-related post-traumatic stress disorder. *American Journal of Medical Genetics Part B: Neuropsychiatric Genetics, 144,* 1087–1089.

Myers-Schulz, B., & Koenigs, M. (2012). Functional anatomy of ventromedial prefrontal cortex: Implications for mood and anxiety disorders. *Molecular Psychiatry, 17,* 132–141.

Negrao, C., Bonanno, G. A., Noll, J. G., Putnam, F. W., & Trickett, P. K. (2005). Shame, humiliation, and childhood sexual abuse: Distinct contributions and emotional coherence. *Child Maltreatment, 10,* 350–363.

Nock, M. K., Prinstein, M. J., & Sterba, S. K. (2009). Revealing the form and function of self-injurious thoughts and behaviors: A real-time ecological assessment study among adolescents and young adults. *Journal of Abnormal Psychology, 118,* 816–827.

Orsillo, S. M., Batten, S. V., Plumb, J. C., Luterek, J. A., & Roessner, B. M. (2004). An experimental study of emotional responding in women with posttraumatic stress disorder related to interpersonal violence. *Journal of Traumatic Stress, 17,* 241–248.

Perales, J. C., Verdejo-García, A., Moya, M., Lozano, Ó., & Pérez-García, M. (2009). Bright and dark sides of impulsivity: Performance of women with high and low trait impulsivity on neuropsychological tasks. *Journal of Clinical and Experimental Neuropsychology, 31,* 927–944.

Perry, B. D. (1994). *Neurobiological sequelea of childhood trauma: PTSD in children.* Washington, DC: American Psychiatric Association.

Peters, J. R., Upton, B. T., & Baer, R. A. (2013). Relationships between facets of impulsivity and borderline personality features. *Journal of Personality Disorders, 27,* 547–552.

Price, M., Connor, J. P., & Allen, H. C. (2017). The moderating effect of childhood maltreatment on the relations among PTSD symptoms, positive urgency, and negative urgency. *Journal of Traumatic Stress, 30,* 432–437.

Price, R. K., Bell, K. M., & Lilly, M. M. (2014). The interactive effects of PTSD, emotion regulation, and anger management strategies on female-perpetrated IPV. *Violence and Victims, 29*, 907–926.

Rinne, T., van den Brink, W., Wouters, L., & van Dyck, R. (2002). SSRI treatment of borderline personality disorder: A randomized, placebo-controlled clinical trial for female patients with borderline personality disorder. *American Journal of Psychiatry, 159*, 2048–2054.

Roemer, L., Litz, B. T., Orsillo, S. M., & Wagner, A. W. (2001). A preliminary investigation of the role of strategic withholding of emotions in PTSD. *Journal of Traumatic Stress, 14*, 149–156.

Roley, M. E., Contractor, A. A., Weiss, N. H., Armour, C., & Elhai, J. D. (2017). Impulsivity facets' predictive relations with DSM-5 PTSD symptom clusters. *Psychological Trauma Theory Research Practice and Policy, 9*, 76–79.

Romer, D., Betancourt, L. M., Brodsky, N. L., Giannetta, J. M., Yang, W., & Hurt, H. (2011). Does adolescent risk taking imply weak executive function? A prospective study of relations between working memory performance, impulsivity, and risk taking in early adolescence. *Developmental Science, 14*, 1119–1133.

Ryding, E., Lindström, M., & Träskman-Bendz, L. (2008). The role of dopamine and serotonin in suicidal behaviour and aggression. *Progress in Brain Research, 172*, 307–315.

Schulz, K. P., Fan, J., Magidina, O., Marks, D. J., Hahn, B., & Halperin, J. M. (2007). Does the emotional go/no-go task really measure behavioral inhibition? Convergence with measures on a non-emotional analog. *Archives of Clinical Neuropsychology, 22*, 151–160.

Seedat, S., Warwick, J., van Heerden, B., Hugo, C., Zungu-Dirwayi, N., Van Kradenburg, J., & Stein, D. J. (2004). Single photon emission computed tomography in posttraumatic stress disorder before and after treatment with a selective serotonin reuptake inhibitor. *Journal of Affective Disorders, 80*, 45–53.

Selby, E. A., Anestis, M. D., & Joiner, T. E. (2008). Understanding the relationship between emotional and behavioral dysregulation: Emotional cascades. *Behavior Research and Therapy, 46*, 593–611.

Shin, L. M., McNally, R. J., Kosslyn, S. M., Thompson, W. L., Rauch, S. L., Alpert, N. M., ... Pitman, R. K. (1999). Regional cerebral blood flow during script-driven imagery in childhood sexual abuse-related PTSD: A PET investigation. *American Journal of Psychiatry, 156*, 575–584.

Shin, L. M., Orr, S. P., Carson, M. A., Rauch, S. L., Macklin, M. L., Lasko, N. B., ... Cannistraro, P. A. (2004). Regional cerebral blood flow in the amygdala and medial prefrontalcortex during traumatic imagery in male and female vietnam veterans with PTSD. *Archives of General Psychiatry, 61*, 168–176.

Shin, L. M., Rauch, S. L., & Pitman, R. K. (2006). Amygdala, medial prefrontal cortex, and hippocampal function in PTSD. *Annals of the New York Academy of Sciences, 1071*, 67–79.

Shin, L. M., Wright, C. I., Cannistraro, P. A., Wedig, M. M., McMullin, K., Martis, B., ... Krangel, T. S. (2005). A functional magnetic resonance imaging study of amygdala and medial prefrontal cortex responses to overtly presented fearful faces in posttraumatic stress disorder. *Archives of General Psychiatry, 62*, 273–281.

Shiv, B., Loewenstein, G., & Bechara, A. (2005). The dark side of emotion in decision-making: When individuals with decreased emotional reactions make more advantageous decisions. *Cognitive Brain Research, 23*, 85–92.

Skelton, K., Ressler, K. J., Norrholm, S. D., Jovanovic, T., & Bradley-Davino, B. (2012). PTSD and gene variants: New pathways and new thinking. *Neuropharmacology, 62*, 628–637.

Smith, G. T., Fischer, S., Cyders, M. A., Annus, A. M., Spillane, N. S., & McCarthy, D. M. (2007). On the validity and utility of discriminating among impulsivity-like traits. *Assessment, 14*, 155–170.

Smith, G. T., Guller, L., & Zapolski, T. C. B. (2013). A comparison of two models of urgency: Urgency predicts both rash action and depression in youth. *Clinical Psychological Science*, (3), 266–275.

Smyth, J. M., Wonderlich, S. A., Heron, K. E., Sliwinski, M. J., Crosby, R. D., Mitchell, J. E., & Engel, S. G. (2007). Daily and momentary mood and stress are associated with binge eating and vomiting in bulimia nervosa patients in the natural environment. *Journal of Consulting and Clinical Psychology*, 75, 629–638.

Spivak, B., Vered, Y., Graff, E., Blum, I., Mester, R., & Weizman, A. (1999). Low platelet-poor plasma concentrations of serotonin in patients with combat-related posttraumatic stress disorder. *Biological Psychiatry*, 45, 840–845.

Spoont, M. R. (1992). Modulatory role of serotonin in neural information processing: Implications for human psychopathology. *Psychology Bulletin*, 112, 330–350.

Swendsen, J. D., Tennen, H., Carney, M. A., Affleck, G., Willard, A., & Hromi, A. (2000). Mood and alcohol consumption: An experience sampling test of the self-medication hypothesis. *Journal of Abnormal Psychology*, 109, 198–204.

Swick, D., Honzel, N., Larsen, J., Ashley, V., & Justus, T. (2012). Impaired response inhibition in veterans with post-traumatic stress disorder and mild traumatic brain injury. *Journal of the International Neuropsychological Society*, 18, 917–926.

Taylor, S., Koch, W. J., & McNally, R. J. (1992). How does anxiety sensitivity vary across the anxiety disorders? *Journal of Anxiety Disorders*, 6, 249–259.

Tull, M. T., Barrett, H. M., McMillan, E. S., & Roemer, L. (2007). A preliminary investigation of the relationship between emotion regulation difficulties and posttraumatic stress symptoms. *Behavior Therapy*, 38, 303–313.

Tull, M. T., Jakupcak, M., McFadden, M. E., & Roemer, L. (2007). The role of negative affect intensity and the fear of emotions in posttraumatic stress symptom severity among victims of childhood interpersonal violence. *Journal of Nervous and Mental Disorders*, 195, 580–587.

Tull, M. T., McDermott, M. J., Gratz, K. L., Coffey, S. F., & Lejuez, C. W. (2011). Cocaine-related attentional bias following trauma cue exposure among cocaine dependent inpatients with and without post-traumatic stress disorder. *Addiction*, 106, 1810–1818.

Tull, M. T., Weiss, N. H., & McDermott, M. J. (2015). Posttraumatic stress disorder and impulsive and risky behavior: An overview and discussion of potential mechanisms. In C. R. Martin, V. R. Preedy, & V. B. Patel (Eds.), *Comprehensive Guide to Post-traumatic Stress Disorders* (pp. 803–816). New York, NY: Springer.

Vaswani, M., Linda, F. K., & Ramesh, S. (2003). Role of selective serotonin reuptake inhibitors in psychiatric disorders: A comprehensive review. *Progress in Neuro-Psychopharmacology and Biological Psychiatry*, 27, 85–102.

Verbruggen, F., & Logan, G. D. (2008). Response inhibition in the stop-signal paradigm. *Trends in Cognitive Sciences*, 12, 418–424.

Verdejo-García, A., Bechara, A., Recknor, E. C., & Pérez-García, M. (2007). Negative emotion-driven impulsivity predicts substance dependence problems. *Drug and Alcohol Dependence*, 91, 213–219.

Voisey, J., Swagell, C. D., Hughes, I. P., Morris, C. P., van Daal, A., Noble, E. P., ... Lawford, B. R. (2009). The DRD2 gene 957C> T polymorphism is associated with posttraumatic stress disorder in war veterans. *Depression and Anxiety*, 26, 28–33.

Weber, D. L., Clark, C. R., McFarlane, A. C., Moores, K. A., Morris, P., & Egan, G. F. (2005). Abnormal frontal and parietal activity during working memory updating in post-traumatic stress disorder. *Psychiatry Research: Neuroimaging*, 140, 27–44.

Weiss, N. H., Connolly, K. M., Gratz, K. L., & Tull, M. T. (2017). The role of impulsivity dimensions in the relation between probable posttraumatic stress disorder and aggressive behavior among substance users. *Journal of Dual Diagnosis*, 13, 109–118.

Weiss, N. H., Darosh, A., Contractor, A., Forkus, S. R., Dixon-Gordon, K. L., & Sullivan, T. P. (2018). Heterogeneity in emotion regulation difficulties among women victims of domestic violence: A latent profile analysis. *Journal of Affective Disorders, 239*, 192–200.

Weiss, N. H., Dixon-Gordon, K. L., Peasant, C., & Sullivan, T. P. (2018). An examination of the role of difficulties regulating positive emotions in posttraumatic stress disorder. *Journal of Traumatic Stress, 31*(5), 775–780.

Weiss, N. H., Forkus, S. R., Contractor, A. A., & Schick, M. R. (2018). Difficulties regulating positive emotions and alcohol and drug misuse: A path analysis. *Addictive Behaviors, 84*, 45–52.

Weiss, N. H., Gratz, K. L., & Lavender, J. (2015). Factor structure and initial validation of a multidimensional measure of difficulties in the regulation of positive emotions: The DERS-Positive. *Behavior Modification, 39*, 431–453.

Weiss, N. H., Peasant, C., & Sullivan, T. P. (2019). Avoidant coping as a moderator of the association between childhood abuse types and HIV/sexual risk behaviors. *Child Maltreatment, 24*(1), 26–35.

Weiss, N. H., Sullivan, T. P., & Tull, M. T. (2015). Explicating the role of emotion dysregulation in risky behaviors: A review and synthesis of the literature with directions for future research and clinical practice. *Current Opinion in Psychology, 3*, 22–29.

Weiss, N. H., Tull, M. T., Anestis, M. D., & Gratz, K. L. (2013). The relative and unique contributions of emotion dysregulation and impulsivity to posttraumatic stress disorder among substance dependent inpatients. *Drug and Alcohol Dependence, 128*, 45–51.

Weiss, N. H., Tull, M. T., Davis, L. T., Dehon, E. E., Fulton, J. J., & Gratz, K. L. (2012). Examining the association between emotion regulation difficulties and probable posttraumatic stress disorder within a sample of African Americans. *Cognitive Behaviour Therapy, 41*, 5–14.

Weiss, N. H., Tull, M. T., Dixon-Gordon, K. L., & Gratz, K. L. (2018). Assessing the negative and positive emotion-dependent nature of risky behaviors among substance dependent patients. *Assessment, 25*, 702–715.

Weiss, N. H., Tull, M. T., Lavender, J., & Gratz, K. L. (2013). Role of emotion dysregulation in the relationship between childhood abuse and probable PTSD in a sample of substance abusers. *Child Abuse and Neglect, 37*, 944–954.

Weiss, N. H., Tull, M. T., & Sullivan, T. P. (2015). Emotion dysregulation and risky, self-destructive, and health compromising behaviors: A review of the literature. In M. L. Bryant (Ed.), *Handbook on Emotion Regulation: Processes, Cognitive Effects and Social Consequences* (pp. 37–56). Hauppauge, NY: Nova Science Publishers.

Weiss, N. H., Tull, M. T., Sullivan, T. P., Dixon-Gordon, K. L., & Gratz, K. L. (2015). Posttraumatic stress disorder symptoms and risky behaviors among trauma-exposed inpatients with substance dependence: The influence of negative and positive urgency. *Drug and Alcohol Dependence, 155*, 147–153.

Weiss, N. H., Tull, M. T., Viana, A. G., Anestis, M. D., & Gratz, K. L. (2012). Impulsive behaviors as an emotion regulation strategy: Examining associations between PTSD, emotion dysregulation, and impulsive behaviors among substance dependent inpatients. *Journal of Anxiety Disorders, 26*, 453–458.

Weitzman, M. L., McHugh, R. K., & Otto, M. W. (2011). The association between affect amplification and urgency. *Depression and Anxiety, 28*, 1105–1110.

Whiteside, S. P., & Lynam, D. R. (2001). The five factor model and impulsivity: Using a structural model of personality to understand impulsivity. *Personality and Individual Differences, 30*, 669–689.

Whiteside, S. P., Lynam, D. R., Miller, J. D., & Reynolds, S. K. (2005). Validation of the UPPS Impulsive Behaviour Scale: A four-factor model of impulsivity. *European Journal of Personality, 19*, 559–574.

Whittington, C. J., Kendall, T., Fonagy, P., Cottrell, D., Cotgrove, A., & Boddington, E. (2004). Selective serotonin reuptake inhibitors in childhood depression: Systematic review of published versus unpublished data. *The Lancet, 363,* 1341–1345.

Wise, R. A. (2008). Dopamine and reward: The anhedonia hypothesis 30 years on. *Neurotoxicity Research, 14,* 169–183.

Wu, J., Ge, Y., Shi, Z., Duan, X., Wang, L., Sun, X., & Zhang, K. (2010). Response inhibition in adolescent earthquake survivors with and without posttraumatic stress disorder: A combined behavioral and ERP study. *Neuroscience Letters, 486,* 117–121.

Xiao, L., Bechara, A., Grenard, L. J., Stacy, W. A., Palmer, P., Wei, Y., … Johnson, C. A. (2009). Affective decision-making predictive of Chinese adolescent drinking behaviors. *Journal of the International Neuropsychological Society, 15,* 547–557.

Xie, P., Kranzler, H. R., Poling, J., Stein, M. B., Anton, R. F., Farrer, L. A., & Gelernter, J. (2010). Interaction of FKBP5 with childhood adversity on risk for post-traumatic stress disorder. *Neuropsychopharmacology, 35,* 1684–1692.

Yates, J. R., Darna, M., Gipson, C. D., Dwoskin, L. P., & Bardo, M. T. (2015). Dissociable roles of dopamine and serotonin transporter function in a rat model of negative urgency. *Behavioural Brain Research, 291,* 201–208.

Yehuda, R., Southwick, S. M., Giller, E. L., Ma, X., & Mason, J. W. (1992). Urinary catecholamine excretion and severity of PTSD symptoms in Vietnam combat veterans. *Journal of Nervous and Mental Disease, 180,* 321–325.

Young, R. M., Lawford, B. R., Noble, E. P., Kann, B., Wilkie, A., Ritchie, T., … Shadforth, S. (2002). Harmful drinking in military veterans with post-traumatic stress disorder: Association with the D2 dopamine receptor A1 allele. *Alcohol and Alcoholism, 37,* 451–456.

Zald, D. H., & Depue, R. A. (2001). Serotonergic functioning correlates with positive and negative affect in psychiatrically healthy males. *Personality and Individual Differences, 30,* 71–86.

Zapolski, T. C. B., Cyders, M. A., & Smith, G. T. (2009). Positive urgency predicts illegal drug use and risky sexual behavior. *Psychology of Addictive Behaviors, 23,* 348–354.

Zoladz, P. R., & Diamond, D. M. (2013). Current status on behavioral and biological markers of PTSD: A search for clarity in a conflicting literature. *Neuroscience & Biobehavioral Reviews, 37,* 860–895.

SECTION 4

Treatment and cultural considerations

CHAPTER 16

Prolonged Exposure for PTSD: Impact on emotions

Katie A. Ragsdale[a], Lauren B. McSweeney[a], Sheila A.M. Rauch[a,b]
[a]Emory University School of Medicine, Atlanta, GA, United States
[b]Atlanta VA Healthcare System, Decatur, GA, United States

Prolonged Exposure (PE)

Prolonged Exposure (PE [Foa, Hembree, & Rothbaum, 2007]) therapy is an evidence-based trauma-focused treatment for posttraumatic stress disorder (PTSD) with extensive efficacy and flexibility across patient presentations (Rauch, Eftekhari, & Ruzek, 2012). PE is one of the most studied interventions for PTSD and has been used with many complex and comorbid patient presentations, including serious mental illness/psychosis (van Minnen, Arntz, & Keijsers, 2002), borderline personality disorder (Harned, Korslund, Foa, & Linehan, 2012), alcohol and substance use disorders (Norman et al., 2015), traumatic brain injury (Ragsdale & Voss Horrell, 2016; Sripada et al., 2013), suicidality (Cox et al., 2016), dissociation (Hagenaars, van Minnen, & Hoogduin, 2010), and depression (Jayawickreme et al., 2014). Generally, research indicates that even severely comorbid patients with PTSD can benefit from standard PE (Van Minnen, Harned, Zoellner, & Mills, 2012). Based on its robust outcomes and flexible and individualized procedures, PE has been used across many populations and settings with effectiveness (Rauch, Cigrang, Austern, & Evans, 2017; Rauch & Rothbaum, 2016). PE is truly robust in its simplicity that has proven to increase its applicability across people, languages, and settings.

PE is typically conducted weekly within an outpatient setting over the course of approximately 10 sessions, though recent intensive treatment programs (e.g., daily PE for 2 weeks) have shown similar effectiveness with reduced dropout (Beidel, Frueh, Neer, & Lejuez, 2017; Harvey et al., 2017; Rauch, Post, et al., 2017; Yasinski, Sherrill, Maples-Keller, Rauch, & Rothbaum, 2017). The treatment protocol includes imaginal exposure to an index trauma, in vivo exposure to real-life stimuli that are avoided and/or

Table 1 PE treatment components.

Psychoeducation	Discussion about the development of PTSD, the role of avoidance in maintaining PTSD, and the role of exposure in reducing PTSD symptoms for the long term
In vivo exposure	Deliberate exposure to real-life situations that are objectively safe and avoided or endured with distress. In vivo exposures begin with situations that evoked moderate distress (e.g., SUDS=50)
Imaginal exposure	Deliberate therapist assisted exposure to the trauma memory. Patients recount their memory aloud, in first person and present tense language. Patients are encouraged to fully experience all emotions, cognitions, and sensations that are present
Emotional processing	Open-ended and nondirective discussion of cognitions and emotions related to the event, including discussion of how the experience influenced beliefs about the self and world. Processing is often considered part of imaginal and in vivo exposure rather than its own component

endured with distress and emotional processing of the memory (e.g., examination of cognitions and emotions related to the event and how the experience influences beliefs about the self and world). See Table 1 for a table of PE treatment components. During exposures, patients self-report their subjective level of distress using the Subjective Units of Distress Scale (SUDS; 0–100 [Wolpe & Lazarus, 1966]) to monitor emotional activation. The goals of PE include reduction of physiological and psychological distress related to the traumatic event and trauma-related stimuli, decreased avoidance, and more helpful and/or accurate appraisal of the event and its meaning.

For example, a patient with PTSD related to a combat trauma would engage in repeated imaginal exposures during sessions to the memory of the traumatic event (e.g., a firefight during a combat mission), recounting the memory aloud in first person and present tense with their eyes closed. Imaginal exposures would last approximately 45 min and would be audio recorded and listened to for homework. The clinician would solicit the patient's SUDS every 5 min to track emotional engagement and reduction; the patient would record their pre, post, and peak SUDS for the same purpose during daily homework exposures. The goal of imaginal exposure is a decline in emotional and physiological arousal to the memory (in this case, of the combat trauma), where the patient no longer experiences significant distress when thinking or talking about the memory. While trauma memories would likely never be regarded as pleasant, being reminded of the event should no longer elicit disabling distress.

Following imaginal exposures in session, emotional processing would target thoughts and feelings that occurred during the event or imaginal exposure and general beliefs that were created or reinforced by the event. Emotional processing of a combat trauma includes discussion of the experience of imaginal and in vivo exposure and may include exploration of feelings of guilt and/or cognitions of self-blame. Other examples may include feelings of weakness and/or cognitions related to what it means to have developed PTSD. The goals of emotional processing include the development of a helpful and accurate appraisal of the event (e.g., it was not my fault), modification of unhelpful cognitions in general (e.g., having a mental health diagnosis does not make me weak), and amelioration of negative emotions (e.g., reduced sadness and guilt).

Finally, in vivo exposures would be delineated within a hierarchy (see Fig. 1) and could include any stimulus present during the trauma that is presently avoided or endured with distress and is objectively safe (e.g., fireworks that remind the patient of gunfire or situations that induce vulnerability, such as sitting in the middle of a restaurant while unknown individuals are behind the patient). Patients complete in vivo exposures for homework, also recording their pre, post, and peak SUDS experienced during the in vivo exposure. Patients are instructed to engage in the in vivo exposure for approximately 45 min or until their SUDS decreases by half. Patients are also encouraged to refrain from engaging in safety behaviors or behaviors that artificially reduce their anxiety or distress (e.g., deep breathing and visually scanning for threat). The goal of in vivo exposures includes accurate appraisal of the safety or tolerability of the stimulus and the ability to live typically without feeling the need to avoid or endure stimuli related to the traumatic event.

PE is recommended as a first-line PTSD treatment in clinical practice guidelines (Courtois et al., 2017; Department of Veterans Affairs & Department of Defense, 2017) and demonstrates a large effect size compared with control conditions ($ES = 1.08$) in the treatment of PTSD following a wide range of traumatic events (e.g., sexual assault and combat trauma; Powers, Halpern, Ferenschak, Gillihan, & Foa, 2010). Recent meta-analysis found that exposure, including PE specifically, is the PTSD treatment with the strongest support (Cusack et al., 2016). PE has been widely disseminated within the Veterans Administration and the Department of Defense (Karlin et al., 2010), increasing its accessibility, particularly to

In vivo hierarchy

Name: Joe
Date: July 20, 2018

Item		SUDS (Start)	SUDS (End)
1.	Intimacy with wife	90	___
2.	Talking about rape	90	___
3.	Locker room at gym	85	___
4.	Hanging out with male friends	80	___
5.	Sleeping without a light on	80	___
6.	Sleeping naked	75	___
7.	Being in a room with the door closed	75	___
8.	Large crowds (e.g., concerts)	70	___
9.	Going to a family barbeque	70	___
10.	Talking to male co-workers	65	___
11.	Taking a bath	65	___
12.	Sitting with back to strangers	60	___
13.	Non sexual contact with wife	60	___
14.	Going to the grocery store	50	___
15.	Being home alone	45	___
15.	Taking a shower (door unlocked)	35	___

Fig. 1 In vivo hierarchy.

veterans, and increasing the number of trained providers. While it is well accepted clinical knowledge that PE is effective at treating PTSD, the mechanisms of action underlying its effectiveness are a prominent target of recent research as the field of psychotherapy research focuses on *how* treatments work (e.g., Rauch & Liberzon, 2016).

Emotional processing theory

The foundation of PE is emotional processing theory (EPT [Foa & Kozak, 1986]). While initial versions of EPT focused almost exclusively on fear and anxiety, later developments of the theory have evolved to include *all* types of emotion and affective processes (Rauch & Foa, 2006). Originally, EPT posited that fear was represented as a structure or memory network in the brain, which serves as a blueprint for behavior; that is, when the fear structure is activated, it prescribes avoidance of the feared stimulus. Fear structures can be adaptive/helpful (e.g., a bear is approaching, I must escape) or maladaptive/unhelpful, whereby fear structures erroneously motivate escape or avoidance of innocuous stimuli (e.g., smelling a particular cologne that reminds a sexual assault survivor of the perpetrator and prescribes escape of an objectively safe situation). In PTSD a fear structure related to a traumatic event includes information about stimuli present at the time of the trauma: sensory cues (e.g., sights, sounds, smells, and tastes), cognitive interpretations (e.g., it was my fault), emotional responses (e.g., horror, helplessness, and fear), physiological activation (e.g., increased heartrate or perspiration), and other internal and external factors present during memory encoding. When the fear structure is activated by encountering an associated stimulus, the program of escape is also activated. If patients escape the presence of an objectively safe stimulus, the fear network and its subsequent avoidance are reinforced, and PTSD is maintained (Rauch & Foa, 2006).

While the aforementioned example focuses on fear, one can consider any type of emotion structure as working similarly. For example, PTSD marked by a primary emotional response of sadness would be represented by a memory structure that largely elicits sadness. Sadness structures can also be adaptive (e.g., encouraging empathy between individuals and eliciting reactions that can facilitate social connection and support), they can also be unhelpful. When this sadness structure is related to a traumatic event that is avoided, the sad affective responses may be triggered in situations where sadness is not appropriate or required and may present as more intense than the situation calls for. As such an unhelpful sadness structure can also lead to avoidance of a trauma memory and/or trauma-related stimuli in an effort to avoid the sad affective response, thereby reinforcing and maintaining avoidance and, thus, PTSD.

To modify an unhelpful emotion structure, Foa and Kozak (1986) argue the emotion network must be "processed." This processing first requires activation of the emotion network. Within PE, activation occurs with planned exposure (e.g., exposure to the memory of the index event or an associated stimulus). Activation can be measured by assessing a patient's physiological and/or emotional response (e.g., increased skin conductance or increased SUDS). The second requirement for processing is that the new information is integrated into the emotion structure. As patients stay with the memory or related stimulus without escaping, new information that is not compatible with the emotion network becomes apparent. Such information may include awareness of the ability to tolerate negative affect (e.g., I can handle the sadness) that negative affect reduces with time (e.g., the sadness declines if I do not avoid it) and that feared negative outcomes do not occur (e.g., I did not go crazy or stay sad forever). New information also includes altered interpretation or perception of the memory as integration of the context of the event occurs. For instance, when the memory is approached rather than avoided, patients become aware of the totality of the event opposed to focusing on avoiding the most distressing parts, which can lead to shifts in cognitions (e.g., I engaged in the firefight because I was in danger; I did everything I could).

Foa and Kozak (1986) propose the mechanisms underlying EPT that include emotional engagement (e.g., emotion activation), reduction of negative affect (previously called habituation), and modification of erroneous cognitions (e.g., encoding true harm probability and/or the ability to tolerate distress). Notably, we have chosen to use reduction in affect opposed to habituation, the term used by Foa and Kozak (1986), given recent advancement in theory and neuroscience research (see Rauch & Liberzon, 2016 for review and additional discussion in the succeeding text). To summarize EPT, activation of the emotion network allows for reduction of the physiological and emotional reaction, which facilitates shifts in the interpretation of the meaning of the emotion structure. In other words, patients learn that the previously avoided stimuli are indeed safe and that emotions reduce and are tolerable, thereby encoding extinction learning and extinguishing the negative emotional response.

More recently, EPT theorists have suggested that exposure therapy involves the formation of new structures that override the pathological fear network (Foa & McNally, 1996), opposed to a weakening of the original fear structure (Foa & Kozak, 1986). Further, researchers have recently argued that strengthening inhibitory learning, or the encoding of these new nonthreat associations (i.e., the new structure), should outweigh the importance of emotion reduction (Craske et al., 2008; Craske, Treanor, Conway, Zbozinek, & Vervliet, 2014). While the role of reduction of negative affect in exposure therapy outcomes is indeed mixed (Rupp, Doebler, Ehring, & Vossbeck-Elsebusch, 2017), between-session reduction may be particularly important for PTSD treatment outcome specifically (Sripada & Rauch, 2015); in general, support for the importance of between-session reduction of negative affect within PE is more consistent (e.g., Bluett, Zoellner, & Feeny, 2014; Jaycox, Foa, & Morral, 1998; Sripada & Rauch, 2015). Similarly, while violation of harm expectancies (i.e., expecting an aversive outcome that does not occur; an indication of inhibitory learning) does take place during PE, one study found it was not related to PTSD symptom reduction, whereas reduction of negative affect was (de Kleine, Hendriks, Becker, Broekman, & van Minnen, 2017).

While PE is a manualized program of exposure therapy, the mechanisms of action underlying the treatment may be somewhat unique compared with basic fear extinction in general and/or somewhat different than the mechanisms underlying other effective anxiety disorder treatments (e.g., exposure therapy for social anxiety disorder). For instance, within PE, reinforcement of inhibitory learning may take place largely within processing, where implicit learning is made explicit byway of reflective listening and nondirective discussion. As such, reduction of affect may remain an important mechanism of PE specifically, as PE modifies *all* emotions and is not simply a program of fear extinction.

The influence of PE on specific emotions

While research has yet to fully elucidate the mechanisms underlying successful PE, it is well established that PE effectively treats PTSD (Cusack et al., 2016; Powers et al., 2010). As noted previously, EPT originally focused on fear processing and fear extinction (Foa & Kozak, 1986); as such, fear and anxiety may be the most apparent emotions that are influenced by PE. In fact, fear was originally conceptualized as the central emotional response to trauma (e.g., Kilpatrick, Resick, & Veronen, 1981) and earlier versions of the *Diagnostic and Statistical Manual of Mental Disorders* (DSM) classified PTSD as an anxiety disorder with fear being the primary and defining response. In the most recent DSM (*DSM-5*; American Psychiatric Association, 2013), PTSD is now categorized under "trauma- and stress-related disorders" and no longer requires the DSM-IV-TR (American Psychiatric Association, 2013) A2 criterion of fear, helplessness, or horror in response to trauma. As such, while the reduction or extinction of fear and anxiety is perhaps the most apparent emotion modification that occurs within the context of PE, recent conceptualization changes, and newer research indicates that PE successfully modifies a wide range of emotional responses. Indeed, Powers et al. (2010) indicated that patients who receive PE outperform those in control conditions on secondary outcome measures (e.g., measures of general anxiety and depression; Hedges's $g=0.41$), indicating PE's ability to modify emotions outside of fear and anxiety.

Recently, discussion of nonfear-based conceptualizations of PTSD has emerged, particularly related to PE's ability to effectively treat these nonfear-based responses to trauma. For example, moral injury or "perpetrating, failing to prevent, or bearing witness to acts that transgress deeply held moral beliefs and expectations" (Litz et al., 2009) has been discussed as one particular sequelae that may follow trauma. Litz et al. (2009) posit that moral injury may lead to specific emotions of guilt and shame, which may not be fully ameliorated by extinction (i.e., PE). Conversely, Smith, Duax, and Rauch (2013) provide a framework for how PE specifically addresses moral injury and subsequent emotional reactions (i.e., guilt and shame), noting that emotionally processing morally injurious events allows for reduction of diverse trauma-related emotions and alteration of beliefs that support guilt and shame. In a secondary analysis of the influence of guilt on treatment outcome, Clifton, Feeny, and Zoellner (2017) found that those reporting higher levels of guilt at pretreatment showed better outcomes with PE than those with lower guilt. In addition, self-reported guilt was reduced with PE (Clifton et al., 2017).

Indeed a recent RCT by Langkaas et al. (2017) examined exposure to nonfear emotions (e.g., anger, trauma-related guilt, and shame) by randomly assigning 65 patients to 10 weeks of either rescripting-based imagery (IR [Smucker, Dancu, Foa, & Niederee, 1995]) or PE. Notably, it has been suggested that IR is superior at reducing nonfear-based emotions (Arntz, Tiesema, & Kindt, 2007). Results indicated that PE and IR showed similar reductions in nonfear emotions (e.g., anger, trauma-related guilt, and shame). Another study looked at PE and eye movement desensitization and reprocessing (EMDR [Shapiro, 2001]) and replicated reductions in both guilt and anger (Stapleton, Taylor, & Asmundson, 2006). Further, research has indicated that PE was effective in ameliorating nonfear emotions (e.g., anger) compared with waitlist controls and that pretreatment anger did not hinder treatment outcome (van Minnen et al., 2002). Regarding reductions in trauma-related guilt, Resick, Nishith, Weaver, Astin, and Feuer (2002) conducted a comparison of PE and cognitive processing therapy (CPT [Resick, Monson, & Chard, 2016]) to a control group. Results indicated that both PE and CPT exhibited reduction in trauma-related guilt in comparison with a control group. Taken together, these results indicate that active ingredients in PE (imaginal exposure, in vivo exposure, and emotional processing) not only are limited to reduction and extinction of fear but also can include nonfear-based emotions (e.g., guilt, shame, and anger).

In addition to guilt, shame, and anger, PE shows robust efficacy in modifying depression and sadness. Notably, depression symptom reduction is an important outcome in PE. This is notable given the overlapping symptoms between depression and PTSD (e.g. impaired concentration, sleep difficulties, and loss of ability to feel emotions) and the common comorbidity between depression and PTSD. In fact, results from two epidemiological studies of adults (National Comorbidity Survey [NCS] [Kessler, Sonnega, Bronet, Hughes, & Nelson, 1995]) and young adults (the Michigan Study of Young Adults [Breslau, Roth, Rosenthal, & Andreski, 1995]) indicated that patients with PTSD are 2–8 times as likely to meet criteria for major depressive disorder than persons who do not have PTSD and 3–11 times as likely as those who have never experienced psychological trauma (Breslau, Davis, Peterson, & Schultz, 2000). In the NCS, 48% of men and 49% of women who had PTSD also had major depressive episodes (Kessler et al., 1995). Further, according to Breslau et al. (2000), individuals with major depressive disorder are twice as likely as those not diagnosed with major depression to experience psychological trauma and more than three times as likely to develop PTSD compared with those without major depression. Therefore it appears that experiencing psychological trauma may either follow from or lead to major depression, particularly if the onset of PTSD occurs soon after psychological trauma (Breslau et al., 2000).

As indicated in Powers et al. (2010) metaanalytic review, PE is effective at reducing depression symptoms. For example, female survivors of sexual assault with a primary diagnosis of PTSD and comorbid depression experienced decreases in depressive symptomatology during the course of PE (Resick et al., 2002). Specifically, at pretreatment, the rate of comorbid depression among sexual assault survivors who received PE was 47.5%. At 9-month follow-up, the rate dropped to 29.5% (Resick et al., 2002), and at long-term follow-up (6 years), the rate had dropped to 6.9% (Resick et al., 2012). PE has also demonstrated effectiveness at reducing depression symptoms within combat- and terror-related PTSD (Nacasch et al., 2011), as well as in intensive outpatient treatment utilizing daily PE (Rauch, Post, et al., 2017; Yasinski et al., 2017).

To understand how PE may target and ameliorate nonfear-based emotions, it is useful to refer back to the EPT. During the course of PE, nonfear-based emotions (e.g., anger, guilt, sadness, and shame) are elicited through emotional engagement with the trauma memory (i.e., exposure to all emotions present at the time of the trauma) and examined during emotional processing. This may occur through discussion of all thoughts and emotions that arise during the imaginal exposure and discussion of all thoughts and emotions that occurred during the traumatic event. For instance, processing may center on guilt, focusing on the context of the event that may have led to certain behaviors, such as active combat where the survivor may have had to kill someone to save others or himself/herself. Once the actions are placed in the trauma context, there is room for therapeutic growth. This growth often occurs through recognition of previously disregarded or underemphasized elements of the trauma context. Take for instance a soldier who has called himself a monster for killing a child. Once this solider recognizes his actions occurred in the context of war, where the child had a bomb strapped to her chest and was thrown in front of his convoy, the soldier's guilt and self-blame can reduce. Often, if the survivor sees the full circumstance and all that occurred, he/she can think about actions in a new light that allows for less judgment and more validation of the experience. That growth may come through realization of reason for the actions or even in realizing the complexity of the situation. Accepting the totality of the event allows for examination of how he/she may make amends for events and behaviors that continues to cause guilt and other unhelpful emotions and beliefs.

Rauch and Foa (2006) argue that emotional engagement with all elements of the trauma structure, including all nonfear-based emotions, is a necessary component of successful PE. As such, reduction of guilt, anger, or other emotions may not have separate mechanisms of action within PE; emotional engagement and reduction of fear- and nonfear-based emotions may exhibit similar treatment trajectories and mechanisms. In a secondary analysis examining anger over the course of PE, researchers found that anger significantly reduced in patients receiving PE (Cahill, Rauch, Hembree, & Foa, 2003). Another clinical trial using PE found that anger did not impede outcome and was reduced with PE (Clifton et al., 2017). In addition, this study found that, with a standard course of PE, those patients who reported clinically significant anger at pretreatment were reduced to the within normal limits of reported anger at post treatment. Notably, this study did not provide any additional specific intervention for anger; however, therapists were directed to validate anger (as they validate all emotional reactions related to the trauma), process the meaning of the anger at the time of the trauma and now, and consider the role of anger in avoidance of other emotions. Specifically, anger can be a powerful emotion that trauma survivors use to avoid more vulnerable and uncomfortable emotions (e.g., Clifton et al., 2017). Moving quickly to anger when reminded of the trauma can become a habit that serves to avoid other important aspects of the trauma and prevent fully processing the memory. Given the impact of imaginal exposure on anger, some have even proposed exposure as a primary treatment for anger-related disorders (Grodnitzky & Tafrate, 2000).

In addition to direct reduction in negative affect, modification of unhelpful cognitions can also lead to reduction in trauma-related negative emotions in PTSD. During the course of PE, alterations in negative thoughts about the self and the world have been associated with symptom reduction (Foa & Rauch, 2004; Hagenaars et al., 2010). For example, Zalta et al. (2015) found that, among female assault survivors who engaged in 10 weekly sessions of PE, reductions in PTSD-related cognitions lead to subsequent reductions in PTSD symptoms. Similarly, Kumpula et al. (2017) and McLean, Su, and Foa (2015) demonstrated that change in negative posttraumatic cognitions preceded change in PTSD symptoms during the course of PE.

Given the importance of cognitions in the development and maintenance of PTSD, studies have examined whether the addition of cognitive restructuring (CR) enhances PE outcomes (Marks, Lovell, Noshirvani, Livanou, & Thrasher, 1998), above and beyond typical exposure and emotional processing. Within PTSD treatment, CR is used to label, identify, and modify trauma-related schemas via interpretation of unhelpful cognitions (Bryant, Moulds, Guthrie, Dang, & Nixon, 2003). Studies that have combined CR with imaginal exposure have yielded mixed results. For example, Bryant et al.'s (2003) study demonstrated that the addition of CR to imaginal exposure enhanced treatment gains as measured by decreased avoidance, depression, and catastrophic cognitions. Notably, however, the full PE model was not utilized (i.e., emotional processing and in vivo exposure were not included in the imaginal exposure treatment). Alternatively a large body of evidence (Foa et al., 2005; Marks et al., 1998; Paunovic & Ost, 2001) has demonstrated that the addition of CR does not improve treatment outcomes when the full PE model is utilized. In fact, recent studies have found that cognitive restructuring in PE is not necessary to reduce trauma-related cognitions (Foa & Rauch, 2004; Hagenaars et al., 2010), indicating treatment gains are related to mechanisms underlying PE, not CR.

In line with the latter body of literature, Smith and colleagues (2013) suggest that the flexibility of PE allows for reduction of a wide range of emotions. Repeated exposures and subsequent processing allows the patient to gain a new perspective without formal cognitive restructuring procedures. When the full memory, including all the emotional and cognitive responses, is activated, updated information that is incompatible with the trauma memory can be incorporated into the memory structure (Sripada & Rauch, 2015). As such, imaginal exposure and emotional processing (in other words, the standard PE protocol) appear to be sufficient to ameliorate negative emotions.

Finally, difficulties with emotion regulation may be one mechanism that underlies the etiology of PTSD (e.g., Etkin & Wager, 2007) and thus may be affected by effective treatment. While PE does not deliberately target emotion dysregulation, research indicates that PE improves regulation of emotion (Jerud, Pruitt, Zoellner, & Feeny, 2016; Jerud, Zoellner, Pruitt, & Feeny, 2014). Similarly, it appears that patients with PTSD comorbid with borderline personality disorder (BPD), a disorder marked by emotion dysregulation, can also benefit from PE. For example, Harned et al. (2012) examined a 13.5-week course of PE/dialectal behavior therapy (DBT [Linehan, 1993]) with 13 recently suicidal and/or self-injuring women with BPD. Results indicated that patients with BPD showed large and significant improvements in PTSD (60% remission) at post treatment, suggesting that patients with BPD can benefit from PE.

It has been posited that emotion regulation may improve by way of inhibitory learning and distress tolerance within exposures (Craske et al., 2008), suggesting that PE mechanisms improve emotion dysregulation. Indeed, emotional activation does entail tolerating distress, allowing for alternative cognitive interpretations related to a perceived ability to handle and/or manage strong emotion. For instance, the ability to remain engaged in an exposure despite experiencing high levels of anxiety allows a patient to remain engaged in goal-directed behavior (i.e., completing PE) despite significant negative emotion. For example, a patient with a primary emotional response of fear and anxiety would experience significantly elevated fear and anxiety while engaged in an exposure. Their fear structure would prescribe avoidance (i.e., stop thinking about this memory, leave this situation); however, within PE, providers would encourage the patient to remain engaged in the exposure, allowing emotions to be fully present however distressing they may be. As such the patient would be practicing distress tolerance in service of processing the trauma memory and treating their PTSD. Notably the by-product of this distress tolerance is actually a reduction of the distress (i.e., a reduction of the emotional reaction tied to the trauma memory structure).

Case study

Joe is a 33-year-old black male who presented to a psychology clinic with primary complaints of nightmares and anxiety. Given his presenting complaints, Dr. Stevens, a licensed clinical psychologist, administered a general psychosocial interview and the Clinician-Administered PTSD Scale for DSM-5 (CAPS-5; Weathers et al., 2017). Upon further assessment, Joe disclosed that he was an Operation Iraqi Freedom veteran and noted that symptoms began during his military service. He stated he was married (4 years in total), had a 1-year-old daughter, and worked as a security guard for a local hospital. On the CAPS-5, Joe disclosed a Criterion A trauma of being sexually assaulted by a higher-ranking service member while on deployment. Joe met criteria for PTSD and was offered psychoeducation about different treatment options. Joe ultimately selected PE and was referred to a trained provider in the clinic.

Prior to starting treatment, Dr. Davis, his treating psychologist, had Joe complete the posttraumatic stress disorder checklist for DSM-5 (PCL-5; Blevins, Weathers, Davis, Witte, & Domino, 2015) to obtain a baseline self-report score of PTSD symptoms. Joe completed this measure each week for the purpose of fostering measurement-based care. Dr. Davis utilized Joe's scores each week to ensure PE was effectively treating his complaints and to assist in the decision-making process of when to terminate treatment. The first few sessions of outpatient treatment included psychoeducation on PE, including the rationale for imaginal and in vivo exposure. Through collaborative discussion, Joe was able to identify that his avoidance of the memory and associated triggers ultimately perpetuated his PTSD and interfered with his relationships. He noted his wife complained of his reluctance to leave the house and of his hesitancy to engage in sexual intimacy. He stated he also avoided getting close to coworkers and had been described as short-tempered, irritable, and standoffish at work. Joe reported that his wife was generally aware of his trauma history and was understanding and supportive of his decision to engage in treatment.

Once Joe had an understanding of the PE rationale, he worked with his psychologist, Dr. Davis, to create an in vivo hierarchy (see Fig. 1). Together, they selected situations that Joe avoided or tolerated with distress, estimating his peak SUDS for each activity. These situations and stimuli were selected only if they were objectively safe; that is, exposures targeted stimuli that were avoided due to unhelpful emotion structures. Joe was given instruction to engage in these exposures for approximately 45 min while ensuring he did not engage in any safety behaviors. Joe was able to verbalize his typical safety behaviors, noting that he scans the environment for threat, attempts to push his anxiety away, or escapes the environment entirely. Joe began engaging in in vivo exposures for homework following session two. He and Dr. Davis selected a couple exposures he could engage in that would elicit moderate levels (i.e., SUDS of 50–60) of distressing emotion. The purpose of initially prescribing moderate in vivo exposures was to ensure Joe experienced effective and appropriate learning; in other words, that Joe learned he could tolerate exposures and subsequent distress, that the stimulus or environment was objectively safe, and that distress would reduce without escape.

At session three, Joe began imaginal exposure to the memory of his sexual assault. First, Dr. Davis reiterated the imaginal exposure rationale (e.g., disentangling the strong negative emotional response the memory automatically elicits from the memory itself, learning that thinking about the memory is not the same as experiencing the trauma, and gaining confidence that he can tolerate distressing emotion) to ensure that Joe understood and was willing to fully engage in exposure. Joe was reminded to abstain from all safety behaviors and encouraged to fully experience any emotions and/or physiological sensations that may arise without attempting to avoid. Joe and Dr. Davis then confirmed parameters of the memory (i.e., helpful start and stop points of the memory) obtained during assessment and early session discussion to guide exposure. Dr. Davis informed Joe they would likely engage in multiple repetitions of the memory, pending available time. Dr. Davis then assessed Joe's baseline SUDS prior to engaging in the imaginal exposure (see Fig. 2) and instructed him to start his audio recording for later homework. Joe then closed his eyes and began to recount the memory in first person and present tense language.

During the first imaginal exposure, Joe became quite emotionally activated. He displayed significant anxiety (i.e., bouncing his legs and wringing his hands) and reported elevated SUDS throughout (see Fig. 2). Dr. Davis frequently encouraged him (e.g., "You're doing great, stay with your emotions."), assisted him with fully engaging ("What are you feeling in your body? What emotions are present?"), and assessed his SUDS every 5 min. Once Joe completed three repetitions (which took approximately 40 min), Dr. Davis invited Joe to open his eyes and return to the room. He then asked an open-ended, nondirective question: "What was that like for you?". Joe noted that he felt significant anger throughout and admitted that he wanted to run out of the room during most of the exposure. Processing focused on reinforcing Joe's willingness, troubleshooting signs of avoidance (e.g., safety behaviors noted throughout), and a discussion of prominent emotions and cognitions Joe experienced. Joe predominately discussed the anger he felt directed at his perpetrator. Dr. Davis examined these experiences with Joe, largely reflecting what Joe was reporting. For homework, Joe was instructed

Therapist imaginal exposure recording form

Name: Joe   **Date**: July 27, 2018

Exposure: # 1   **Session**: # 3

Description of target event: *Sexually assaulted by a higher-ranking service member while on deployment*

Time	Suds	Notes:
Beginning	70	
5 min	70	
10 min	75	Bouncing his legs
15 min	75	Wringing his hands
20 min	60	
25 min	65	"I'm so angry! How could I let this happen?"
30 min	65	"I should have fought him off!"
35 min	55	Appears somewhat less tense
40 min	50	Leg bouncing has stopped
45 min	—	
50 min	—	
55 min	—	
60 min	—	

Fig. 2 Session 3 imaginal exposure.

to listen to his imaginal exposure recording daily while fully experiencing any emotions that may arise and to document his pre, post, and peak SUDS on his homework sheet. He and Dr. Davis then selected a few in vivo exposures to practice over the next week, including a return to the in vivo exposures that he began last week.

Subsequent sessions followed a similar structure: setting an agenda, reviewing homework (in vivo and imaginal exposures), and engaging in imaginal exposure and processing. Homework review included exploration of emotions experienced, SUDS reported, and learning experienced (i.e., "What are you noticing about your ability to go out to eat with your wife? What happens to your distress as you remain in the situation?") during both in vivo and imaginal homework exposures. During sessions, imaginal exposures eventually focused only on hot spots or the most distressing and emotional segments of the trauma memory. Two hot spots were identified. The worst was the moment when he was raped, which quickly reduced over two sessions. The second hot spot was when his attacker first pushed the gun into his ribs. This hot spot also required two sessions of focused imaginal exposure. During exposures, Dr. Davis continued to encourage full immersion in the memory (i.e., instruction to experience all aspects of the memory including all sensations, emotions, and cognitions) and frequently provided reinforcement and praise. Notably, while Joe initially expressed prominent anger, as he allowed himself to feel this anger and explore his emotions and cognitions during processing, Joe began to become aware of other emotions and cognitions related to this event (e.g., disgust, vulnerability, and sadness).

As sessions progressed, Joe learned he could immerse himself in strong emotion without "going crazy" and realized that he was supported nonjudgmentally by Dr. Davis, even with Dr. Davis knowing all the details of the traumatic event. As such, Joe became more comfortable with vulnerability and more open to emotion. As a result of this learning and increased willingness (continually fostered throughout exposure and processing), Joe's anger slowly became less prominent as sadness began to emerge. In fact, Joe allowed himself to cry during his fourth imaginal exposure. Following, during processing, Joe disclosed he believed he was weak and "less of a man" as a result of the assault. He described feelings of disgust and identified his anger as an understandable response to mask or avoid other distressing emotions. He and Dr. Davis explored these evolved emotions and cognitions, agreeing to continue allowing the sadness and disgust to present during subsequent imaginal exposures.

As sessions progressed, Joe initially experienced increased sadness and disgust, even becoming nauseous at times. Dr. Davis encouraged Joe to remain with whatever emotion or sensation that presented, knowing these strong unpleasant reactions would reduce with continued exposure and willingness. Processing continued to explore these emotions and the cognitions related to feelings of weakness and what it means to be sexually assaulted. As Joe engaged in exposure, he began to place the event into its full context. A key piece in processing was Joe's epiphany that the perpetrator had a firearm, whereas Joe was unarmed. He began to conceptualize his inability "to fight him off" as a potentially lifesaving decision. This realization impacted Joe's feelings of guilt and self-blame, as Joe began to consider he was fully coerced and did the best he could to protect himself, given the circumstances.

Throughout treatment, Joe continued in vivo exposures, which worked to modify his beliefs about the safety of the world. During homework review at session eight, Joe disclosed he no longer emotionally responded to typical living (e.g., taking his daughter to a new playground and going to a work function) as if he were back in a warzone being sexually assaulted. He stated that he no longer experienced physiological anxiety (e.g., racing heart, increased sweat, and body tension) or a strong urge to escape when he engaged in situations he used to avoid (e.g., grocery shopping or having dinner at a crowded restaurant). He described a continued willingness to allow negative emotion to arise, which he reported was less intense and dissipated more quickly than when he initially began exposures. Through in vivo exposure, Joe was able to discriminate what was objectively safe and began to reengage with activities that he used to enjoy, including increased intimacy with his wife and the ability to take his young daughter to crowded parks and museums. Joe's coworkers even noticed his more relaxed demeanor and began to invite him to social events, activities that Joe now felt were safe and even enjoyable.

By session 11, Joe was reporting minimal activation during his memory as indicated by SUDS peaking at or below 25 (see Fig. 3). While a few situations on his hierarchy still elicited a mild to moderate level of distress (e.g., SUDS of 40), Joe indicated that he no longer avoided any internal or external stimuli and was willing to engage in all in vivo exposures on his hierarchy. Furthermore, Joe's scores on the PCL-5 had fallen below the threshold suggestive of PTSD. Dr. Davis and Joe collaboratively discussed termination. Joe was confident he could transition out of therapy and engage in his own in vivo exposures as needed. Dr. Davis was confident Joe had a solid understanding of the treatment rationale and would make decisions based on exposure, not avoidance, in service of maintaining treatment gains. Together, they agreed to one final session to revisit the entire memory during the final imaginal exposure. During the final session, Joe again did not report SUDS above 25 throughout, and his PCL-5 score remained low. Joe and Dr. Davis agreed to terminate, planning a check-in session in 1 month to ensure Joe continued to utilize skills learned in PE (e.g., continued exposure and discontinuation of avoidance and safety behaviors).

In summary, Joe completed 12 sessions of PE focused on a sexual assault. He initially presented with significant anxiety, avoidance, and anger. Except for work, he largely isolated at home and felt his marriage was failing.

Therapist imaginal exposure recording form

Name: Joe **Date**: September 21, 2018

Exposure: # 9 **Session**: # 11

Description of target event: *Sexually assaulted by a higher-ranking service member while on deployment*

Time	Suds	Notes:
Beginning	25	"I feel pretty calm, I'm not anxious"
5 min	25	Appears relaxed, no psychomotor activity
10 min	20	"There is nothing I could have done"
15 min	20	
20 min	20	Yawning
25 min	15	
30 min	15	"I did the best I could, It's not my fault"
35 min	—	
40 min	—	
45 min	—	
50 min	—	
55 min	—	
60 min	—	

Fig. 3 Session 11 imaginal exposure.

Through imaginal exposure, Joe was able to process the strong distressing emotions, which included a natural progression from anger to sadness. Joe's processing of his trauma history did not mean he accepted that sexual assault was acceptable or that the memory no longer elicited any negative emotion. The processing of the event meant he considered the event in its entirety and was able to think about the event without debilitating negative emotion. Further, processing meant the memory, and the emotions originally associated with the memory no longer controlled Joe's life. Largely during processing, Joe also reported significant cognitive shifts. He no longer blamed himself for the assault or believed the assault defined his masculinity or self-worth. Through in vivo exposure, Joe was able to learn the world was generally safe or, at least, not to be lived as if he were in a warzone being sexually assaulted. This newfound ability to discriminate safety and threat meant Joe could reengage in life, which had a positive impact on his marriage, work, and ultimately his own happiness. Indeed, he reported that PE had been transformative for him and that he had taken his life back from PTSD.

Impact of culture on PE on emotion

While little research has examined how culture impacts emotion modification within PE specifically, PE has been used with many thousands of patients across race/ethnicities and even throughout the world (Foa, Gillihan, & Bryant, 2013) with excellent results. Indeed the manual has been translated into over 10 languages, and PE trainers have engaged with trauma survivors throughout the world, often moving at the request of foreign governments following traumatic events. For a full review of PE dissemination, the reader is referred to Foa et al. (2013). PE has a strong presence in the Netherlands, the United Kingdom, Australia, Israel, and elsewhere and is growing in other countries. For example, following the triple disaster in Japan in 2011, Drs. Rauch, Tuerk, Yoder, and Hall led a team to establish a training network for PE in Japan that expanded the small group of PE providers who already worked in Japan (Yoder, Tuerk, Rauch, & Hall, 2012). The network of providers and trainers that arose from this effort remains today.

Of note, in a PE treatment sample of veterans with combat-related PTSD, race was not a predictor of treatment completion, nor was it meaningfully associated with treatment outcome or slope of treatment outcome over time (Tuerk et al., 2011). This is consistent with other studies showing excellent outcomes across ethnic and racial groups (Foa et al., 2005; Foa et al., 2018). Nonetheless, PE therapists, as with all psychotherapy, must meet the patient where he/she is and must be culturally sensitive and individualized to each distinct patient. For instance, basic knowledge of military culture may be important for treatment of military-related PTSD (Coll, Weiss, & Yarvis, 2011), and consideration of cultural mistrust and/or racism and discrimination may be important when working with African Americans (Williams et al., 2014). Indeed, PE allows the therapist the opportunity to let the patient lead on cultural issues as the patient provides the content. It is up to the therapist to disclose those cultural matters that they may not be aware of and let the patient provide their expertise in the session. As an example of this, for therapists who have not served in the military, it is good to get some basic training in military culture, but it can go a long way for the therapist to let the patient know they are not an expert on the patient's military experience. The therapist can explain that they know how to treat PTSD but will rely on the patient to discuss their military experiences and how that impacts their views of the trauma and the self. A similar stance of openness to learn from the patient can be effective with other cultural issues as well. However, in the end, ensuring that the patient has the support they desire is important. If additional consultation or referral is needed, that can also be an option. Regardless, cultural factors should not exclude patients from receiving this first-line PTSD treatment, nor should providers or patients expect attenuated results due to cultural factors. PE has proven a robust treatment across settings and cultures.

One factor that has contributed to the success of dissemination of PE across cultures is the individualized nature of the protocol. The PE therapist is always working with the patient who is present in the office and attuning to the motivations and values of that individual and their culture to get the most out of treatment. Understanding and exploring with the patient how his/her cultural beliefs and values influence the trauma and their thoughts regarding their behavior during and after the trauma are a part of processing and one reason that the reflective style of PE processing can be especially flexible. Therapists are following the patient without judgment. Exploring meaning without a specific destination is required for the patient to benefit from the therapy. Each patient provides their own path to wellness, and the therapist's job is to help shine a light on a path that will work for that person.

Limitations and future research

Limitations of this body of literature lie largely in the novelty of studying PE's effect on nonfear-based emotions. Theory and research have historically considered PTSD to be a disorder of fear and anxiety, and thus PE has historically centered on the activation and habituation of fear and anxiety. Newer research and clinical insights gained over 30 years of practicing PE indicate that PE effectively treats diverse emotional reactions by way of reducing negative affect and modifying unhelpful cognitions that support distressing affective states. While we now know that PE is effective at ameliorating a wide range of distressing emotions (e.g., sadness and anger), there is minimal understanding of *how* PE modifies those specific emotions. For instance, it is unclear whether sadness resolves through the same mechanism(s) as guilt or as fear.

As noted previously the mechanisms of PE have been conceptualized using EPT and the learning theory (Rauch & Liberzon, 2016). Within EPT, extinction is conceptualized as relearning (Rauch & Liberzon, 2016). During PE, extinction occurs when new inhibitory associations are formed in place of the fear associations and is indicated by a reduction in subjective fear response to the avoided memory and its reminders. This relearning is facilitated through the process of contextualization or learning to discriminate between safety and threat cues, depending on the context in which they occur (Maren, Phan, & Liberzon, 2013). In addition, emotional engagement during imaginal and in vivo exposures is one of the theoretically proposed mechanisms of action in PE (Rauch & Foa, 2006) and has been shown to be important to the process of inhibitory learning (Culver, Stoyanova, & Craske, 2012). Emotional processing occurs in tandem with extinction; relearning is marked by increased sense of competence, reduced sense that the world as dangerous, and reduction in social and emotional withdrawal (Rauch & Liberzon, 2016; Rothbaum, Astin, & Marsteller, 2005). Although changes in emotion and trauma-related beliefs are not central to learning models, this learning of increased competence to cope with negative affect (e.g., sadness and anger) and reduced sense of a dangerous world can be viewed as a form of inhibitory learning (Rauch & Liberzon, 2016). As such, extant literature could benefit from research that further elucidates the specific mechanisms of action underlying PE, including elucidation of specific mechanisms that modify emotions specifically.

As discussed, EPT and PE have historically understood treatment of PTSD as fear extinction by way of repeated habituation. However, newer conceptualizations focused on inhibitory learning suggest that distress tolerance and consolidation and retrievability of new learning may also be at work (e.g., Craske et al., 2008). While emotions other than fear do decline over time and may extinguish, inhibitory learning explains a cognitive process that may also occur. For example, if a patient prominently experiences shame related to their trauma and expects others to blame them and/or think negatively about them, learning that talking about the trauma does not lead to those predicted outcomes would likely decrease the shame reaction elicited by the trauma memory. Understanding the specific mechanisms underlying distress reduction in PTSD treatment could lead to more targeted interventions.

Further, it is unclear whether there are certain emotions that do or do not contribute to the development of PTSD and, similarly, the treatment of PTSD through PE. It is unclear if there is a common trajectory of symptom development (i.e., which symptoms present first) and/or symptom reduction during PE (i.e., which symptoms, or symptom clusters, decrease first). Complicating this line of research is the reliance of treatment outcome studies on a symptom severity total score (i.e., reduction in CAPS-5 [Weathers et al., 2017] or PCL-5 [Blevins et al., 2015] scores) rather than symptom clusters scores. The notion that PTSD is a homogeneous disorder discounts the heterogeneity of the symptom clusters and inhibits more nuanced research. Given the change in the diagnostic nomenclature, factor analytic studies have yet to determine the underlying symptom structure of the DSM-5 criteria; research that elucidates this may improve future research examining symptoms and symptom clusters and thus provides further insight on the mechanisms of emotion modification within PE.

A growing wave of research focuses on identifying psychological and biological factors involved in treatment change in PE, which is an important research emphasis. Specifically, despite PE being a robust and effective intervention that can be used with many comorbidities and other issues, only about 50% of patients complete PE when provided in clinical care. Rates of dropout are not specific to PE, but plague all PTSD treatments including both medication and psychotherapy. Indeed, since PTSD is characterized by avoidance, expecting people suffering from PTSD to return to a clinic on a weekly basis may be inconsistent with the nature of the disorder. To reduce dropout and improve outcomes, research has focused on identification of predictors of slow or low response to PE. For instance, augmentation or modifications can be made to speed response and prevent dropout that can occur when people are experiencing less gains.

Also in an effort to address low retention in PTSD treatment, PE researchers have been looking for new models of administration of PE to provide the right treatment to the right patient in the right setting. These new models are varied. PE for primary care is one new mode of delivery, where a behavioral health professional provides a brief version of PE in four to six 30-min sessions. This model has shown significant change over time and superior change to a minimal attention control condition. Indeed, 50% of those with PTSD at pretreatment no longer met criteria for a PTSD diagnosis at 8 weeks (Cigrang et al., 2017). PE has also demonstrated effectiveness when provided over telemental health (Gros, Yoder, Tuerk, Lozano, & Acierno, 2011). Finally, many groups have developed intensive outpatient models of care that are showing excellent results (e.g., Blount, Cigrang, Foa, Ford, & Peterson, 2014; Harvey et al., 2017, 2018). These models adhere to EPT but provide PE each day for 2–3 weeks. These intensive models allow for the therapy to fit the needs and lifestyles of the patient and allow treatment choice while retaining the efficacy of the treatment.

Finally, there is minimal research on how culture impacts emotion modification in PE. However, as noted previously, PE has proven to be a robust treatment across cultures. Nonetheless, future research could incorporate how different emotions (e.g., guilt and shame) are experienced and displayed by different cultures and how these emotions reduce through PE.

Conclusion

In summary, research indicates that PE is effective at treating PTSD, regardless of the predominant emotional reaction produced by the event and its subsequent cognitive appraisals. PE has long been conceptualized as fear extinction and has substantial research support indicating its ability to facilitate reduction of anxiety and fear responses. Newer research and a huge body of clinical experience gathered over 30 years of PE use suggest that the underlying mechanisms of PE are also effective at treating diverse trauma-related sequelae, including moral injury reactions, guilt, shame, anger, sadness, and even emotion dysregulation. Importantly, standard PE appears to be sufficient at targeting these emotional reactions; that is, additional treatment components (e.g., CR) do not appear necessary. Modification of emotion likely takes place within both exposure and emotional processing, where reduction of distress and altered interpretations of events lead to more accurate and helpful emotions. As such, PE should be considered a first-line treatment for PTSD effective in addressing a wide range of trauma-related emotions. In addition, research examining more closely the specifics of emotion related to PTSD development and treatment response is warranted to clarify and further improve efficacy of PTSD treatment. For instance, examining more specific emotions and their patterns of response in PE may help to elucidate whether the patterns of change and mechanisms involved are the same across emotions or specific to certain emotions. Better understanding of the mechanisms involved in change in PE may help to improve the treatment generally and speed reduction of emotions most relevant to treatment change.

Acknowledgments and disclosures

The views expressed in this article presentation are solely those of the author(s) and do not reflect an endorsement by or the official policy of the Department of Veterans Affairs, Department of Defense, or the US Government, or the official views of the National Institutes of Health. Dr. Rauch receives support from Wounded Warrior Project (WWP), Department of Veterans Affairs (VA), National Institute of Health (NIH), Woodruff Foundation, and Department of Defense (DOD). Dr. Rauch receives royalties from Oxford University Press. Dr. Ragsdale and Dr. McSweeney have nothing to disclose.

References

American Psychiatric Association. (2013). *Diagnostic and statistical manual of mental disorders—DSM-5* (5th ed.). Washington, DC: American Psychiatric Association.

Arntz, A., Tiesema, M., & Kindt, M. (2007). Treatment of PTSD: A comparison of imaginal exposure with and without imagery rescripting. *Journal of Behavior Therapy and Experimental Psychiatry*, *38*(4), 345–370. https://doi.org/10.1016/J.JBTEP.2007.10.006.

Beidel, D. C., Frueh, B. C., Neer, S. M., & Lejuez, C. W. (2017). The efficacy of trauma management therapy: A controlled pilot investigation of a three-week intensive outpatient program for combat-related PTSD. *Journal of Anxiety Disorders*, *50*, 23–32. https://doi.org/10.1016/j.janxdis.2017.05.001.

Blevins, C. A., Weathers, F. W., Davis, M. T., Witte, T. K., & Domino, J. L. (2015). The posttraumatic stress disorder checklist for DSM-5 (PCL-5): Development and initial psychometric evaluation. *Journal of Traumatic Stress*, *28*(6), 489–498. https://doi.org/10.1002/jts.22059.

Blount, T. H., Cigrang, J. A., Foa, E. B., Ford, H. L., & Peterson, A. L. (2014). Intensive outpatient prolonged exposure for combat related PTSD: A case study. *Cognitive and Behavioral Practice*, *21*(1), 89–96.

Bluett, E. J., Zoellner, L. A., & Feeny, N. C. (2014). Does change in distress matter? Mechanisms of change in prolonged exposure for PTSD. *Journal of Behavior Therapy and Experimental Psychiatry*, *45*(1), 97–104. https://doi.org/10.1016/j.jbtep.2013.09.003.

Breslau, N., Davis, G. C., Peterson, E. L., & Schultz, L. R. (2000). A second look at comorbidity in victims of trauma: The posttraumatic stress disorder-major depression connection. *Biological Psychiatry*, *48*(9), 902–909. https://doi.org/10.1016/S0006-3223(00)00933-1.

Breslau, N., Roth, T., Rosenthal, L., & Andreski, P. (1995). Sleep disturbance and psychiatric disorders. *Biological Psychiatry*, *37*(9), 655–656.

Bryant, R. A., Moulds, M. L., Guthrie, R. M., Dang, S. T., & Nixon, R. D. V. (2003). Imaginal exposure alone and imaginal exposure with cognitive restructuring in treatment of posttraumatic stress disorder. *Journal of Consulting and Clinical Psychology*, *71*(4), 706–712. https://doi.org/10.1037/0022-006X.71.4.706.

Cahill, S. P., Rauch, S. A., Hembree, E. A., & Foa, E. B. (2003). Effect of cognitive-behavioral treatments for PTSD on anger. *Journal of Cognitive Psychotherapy*, *17*(2), 113–131. https://doi.org/10.1891/jcop.17.2.113.57434.

Cigrang, J. A., Rauch, S. A., Mintz, J., Brundige, A. R., Mitchell, J. A., Najera, E., … Peterson, A. L. (2017). Moving effective treatment for posttraumatic stress disorder to primary care: A randomized controlled trial with active duty military. *Families, Systems & Health*, *35*(4), 450–462. https://doi.org/10.1037/fsh0000315.

Clifton, E. G., Feeny, N. C., & Zoellner, L. A. (2017). Anger and guilt in treatment for chronic posttraumatic stress disorder. *Journal of Behavior Therapy and Experimental Psychiatry*, *54*, 9–16. https://doi.org/10.1016/j.jbtep.2016.05.003.

Coll, J. E., Weiss, E. L., & Yarvis, J. S. (2011). No one leaves unchanged: Insights for civilian mental health care professionals into the military experience and culture. *Social Work in Health Care*, *50*(7), 487–500. https://doi.org/10.1080/00981389.2010.528727.

Courtois, C. A., Sonis, J., Fairbank, J. A., Friedman, M., Jones, R., Roberts, J., & Schulz, P. (2017). *Clinical practice guideline for the treatment of posttraumatic stress disorder (PTSD) in adults*. American Psychological Association. Retrieved from https://www.apa.org/about/offices/directorates/guidelines/ptsd.pdf.

Cox, K. S., Mouilso, E. R., Venners, M. R., Defever, M. E., Duvivier, L., Rauch, S. A. M., … Tuerk, P. W. (2016). Reducing suicidal ideation through evidence-based treatment for posttraumatic stress disorder. *Journal of Psychiatric Research*, *80*, 59–63. https://doi.org/10.1016/j.jpsychires.2016.05.011.

Craske, M. G., Kircanski, K., Zelikowsky, M., Mystkowski, J., Chowdhury, N., & Baker, A. (2008). Optimizing inhibitory learning during exposure therapy. *Behaviour Research and Therapy*, *46*(1), 5–27. https://doi.org/10.1016/j.brat.2007.10.003.

Craske, M. G., Treanor, M., Conway, C. C., Zbozinek, T., & Vervliet, B. (2014). Maximizing exposure therapy: An inhibitory learning approach. *Behaviour Research and Therapy*, *58*, 10–23. https://doi.org/10.1016/j.brat.2014.04.006.

Culver, N. C., Stoyanova, M., & Craske, M. G. (2012). Emotional variability and sustained arousal during exposure. *Journal of Behavior Therapy and Experimental Psychiatry*, *43*(2), 787–793. https://doi.org/10.1016/J.JBTEP.2011.10.009.

Cusack, K., Jonas, D. E., Forneris, C. A., Wines, C., Sonis, J., Middleton, J. C., ... Gaynes, B. N. (2016). Psychological treatments for adults with posttraumatic stress disorder: A systematic review and meta-analysis. *Clinical Psychology Review*, *43*(290), 128–141. https://doi.org/10.1016/j.cpr.2015.10.003.

de Kleine, R. A., Hendriks, L., Becker, E. S., Broekman, T. G., & van Minnen, A. (2017). Harm expectancy violation during exposure therapy for posttraumatic stress disorder. *Journal of Anxiety Disorders*, *49*(June 2016), 48–52. https://doi.org/10.1016/j.janxdis.2017.03.008.

Department of Veterans Affairs & Department of Defense. (2017). *VA/DoD clinical practice guidelines for management of post-traumatic stress*. Department of Veterans Affairs & Department of Defense. Retrieved from http://www.healthquality.va.gov/guidelines/MH/ptsd/cpgPTSDFULL201011612c.pdf.

Etkin, A., & Wager, T. D. (2007). Functional neuroimaging of anxiety: A meta-analysis of emotional processing in PTSD, social anxiety disorder, and specific phobia. *The American Journal of Psychiatry*, *164*(10), 1476–1488. https://doi.org/10.1176/appi.ajp.2007.07030504.

Foa, E. B., Gillihan, S. J., & Bryant, R. A. (2013). Challenges and successes in dissemination of evidence-based treatments for posttraumatic stress. *Psychological Science in the Public Interest*, *14*(2), 65–111. https://doi.org/10.1177/1529100612468841.

Foa, E. B., Hembree, E. A., Cahill, S. P., Rauch, S. A. M., Riggs, D. S., Feeny, N. C., & Yadin, E. (2005). Randomized trial of prolonged exposure for posttraumatic stress disorder with and without cognitive restructuring: Outcome at academic and community clinics. *Journal of Consulting and Clinical Psychology*, *73*(5), 953–964. https://doi.org/10.1037/0022-006X.73.5.953.

Foa, E. B., Hembree, E. A., & Rothbaum, B. O. (2007). *Prolonged exposure therapy for PTSD: Emotional processing of traumatic experiences*. New York, NY: Oxford University Press.

Foa, E. B., & Kozak, M. J. (1986). Emotional processing of fear: Exposure to correct information. *Psychological Bulletin*, *99*(1), 20–35.

Foa, E. B., McLean, C. P., Zang, Y., Rosenfield, D., Yadin, E., Yarvis, J. S., ... Peterson, A. L. (2018). Effect of prolonged exposure therapy delivered over 2 weeks vs 8 weeks vs present-centered therapy on PTSD symptom severity in military personnel: A randomized clinical trial. *JAMA: Journal of the American Medical Association*, *319*(4), 354–364.

Foa, E. B., & McNally, R. J. (1996). Mechanisms of change in exposure therapy. In M. Rapee (Ed.), *Current controversies in the anxiety disorders* (pp. 329–343). New York, NY: The Guilford Press.

Foa, E. B., & Rauch, S. A. M. (2004). Cognitive changes during prolonged exposure versus prolonged exposure plus cognitive restructuring in female assault survivors with posttraumatic stress disorder. *Journal of Consulting and Clinical Psychology*, *72*(5), 879–884.

Grodnitzky, G. R., & Tafrate, R. C. (2000). Imaginal exposure for anger reduction in adult outpatients: A pilot study. *Journal of Behavior Therapy and Experimental Psychiatry*, *31*(3–4), 259–279. https://doi.org/10.1016/S0005-7916(01)00010-6.

Gros, D. F., Yoder, M., Tuerk, P. W., Lozano, B. E., & Acierno, R. (2011). Exposure therapy for PTSD delivered to veterans via telehealth: Predictors of treatment completion and outcome and comparison to treatment delivered in person. *Behavior Therapy*, *42*, 276–283.

Hagenaars, M. A., van Minnen, A., & Hoogduin, K. A. L. (2010). The impact of dissociation and depression on the efficacy of prolonged exposure treatment for PTSD. *Behaviour Research and Therapy*, *48*(1), 19–27. https://doi.org/10.1016/j.brat.2009.09.001.

Harned, M. S., Korslund, K. E., Foa, E. B., & Linehan, M. M. (2012). Treating PTSD in suicidal and self-injuring women with borderline personality disorder: Development and preliminary evaluation of a dialectical behavior therapy prolonged exposure protocol. *Behaviour Research and Therapy*, *50*(6), 381–386. https://doi.org/10.1016/j.brat.2012.02.011.

Harvey, M. M., Petersen, T. J., Sager, J. C., Makhija-Graham, N. J., Wright, E. C., Clark, E. L., & Simon, N. M. (2018). An intensive outpatient program for veterans with posttraumatic stress disorder and traumatic brain injury. *Cognitive and Behavioral Practice*, https://doi.org/10.1016/j.cbpra.2018.07.003.

Harvey, M. M., Rauch, S. A. M., Zalta, A. K., Sornborger, J., Pollack, M. H., Rothbaum, B. O., … Simon, N. M. (2017). Intensive treatment models to address posttraumatic stress among post-9/11 warriors: The warrior care network. *Focus*, *15*(4), 378–383. https://doi.org/10.1176/appi.focus.20170022.

Jayawickreme, N., Cahill, S. P., Riggs, D. S., Rauch, S. A. M., Resick, P. A., Rothbaum, B. O., & Foa, E. B. (2014). Primum non nocere (first do no harm): Symptom worsening and improvement in female assault victims after prolonged exposure for PTSD. *Depression and Anxiety*, *31*(5), 412–419. https://doi.org/10.1002/da.22225.

Jaycox, L. H., Foa, E. B., & Morral, A. R. (1998). Influence of emotional engagement and habituation on exposure therapy for PTSD. *Journal of Consulting and Clinical Psychology*, *66*(1), 185–192. https://doi.org/10.1037/0022-006X.66.1.185.

Jerud, A. B., Pruitt, L. D., Zoellner, L. A., & Feeny, N. C. (2016). The effects of prolonged exposure and sertraline on emotion regulation in individuals with posttraumatic stress disorder. *Behaviour Research and Therapy*, *77*, 62–67. https://doi.org/10.1016/j.brat.2015.12.002.

Jerud, A. B., Zoellner, L. A., Pruitt, L. D., & Feeny, N. C. (2014). Changes in emotion regulation in adults with and without a history of childhood abuse following posttraumatic stress disorder treatment. *Journal of Consulting and Clinical Psychology*, *82*(4), 721–730. https://doi.org/10.1037/a0036520.

Karlin, B. E., Ruzek, J. I., Chard, K. M., Eftekhari, A., Monson, C. M., Hembree, E. A., … Foa, E. B. (2010). Dissemination of evidence-based psychological treatments for posttraumatic stress disorder in the Veterans Health Administration. *Journal of Traumatic Stress*, *23*(6), 663–673. https://doi.org/10.1002/jts.

Kessler, R. C., Sonnega, A., Bronet, E., Hughes, M., & Nelson, C. B. (1995). Posttraumatic stress disorder in the national comorbidity survey. *Archive of General Psychiatry*, *52*(October), 1048–1060. https://doi.org/10.1002/1099-1298(200011/12)10.

Kilpatrick, D. G., Resick, P. A., & Veronen, L. J. (1981). Effects of a rape experience: A longitudinal study. *Journal of Social Issues*, *37*(4), 105–122. https://doi.org/10.1111/j.1540-4560.1981.tb01073.x.

Kumpula, M. J., Pentel, K. Z., Foa, E. B., LeBlanc, N. J., Bui, E., McSweeney, L. B., … Rauch, S. A. M. (2017). Temporal sequencing of change in posttraumatic cognitions and PTSD symptom reduction during prolonged exposure therapy. *Behavior Therapy*, *48*(2), 156–165. https://doi.org/10.1016/j.beth.2016.02.008.

Langkaas, T. F., Hoffart, A., Øktedalen, T., Ulvenes, P. G., Hembree, E. A., & Smucker, M. (2017). Exposure and non-fear emotions: A randomized controlled study of exposure-based and rescripting-based imagery in PTSD treatment. *Behaviour Research and Therapy*, *97*, 33–42. https://doi.org/10.1016/J.BRAT.2017.06.007.

Linehan, M. M. (1993). *Diagnosis and treatment of mental disorders. Cognitive-behavioral treatment of borderline personality disorder*. New York, NY: Guilford Press.

Litz, B. T., Stein, N., Delaney, E., Lebowitz, L., Nash, W. P., Silva, C., & Maguen, S. (2009). Moral injury and moral repair in war veterans: A preliminary model and intervention strategy. *Clinical Psychology Review*, *29*(8), 695–706. https://doi.org/10.1016/j.cpr.2009.07.003.

Maren, S., Phan, K. L., & Liberzon, I. (2013). The contextual brain: Implications for fear conditioning, extinction and psychopathology. *Nature Reviews Neuroscience, 14*(6), 417–428. https://doi.org/10.1038/nrn3492.

Marks, I., Lovell, K., Noshirvani, H., Livanou, M., & Thrasher, S. (1998). Treatment of posttraumatic stress disorder by exposure and/or cognitive restructuring. *Archives of General Psychiatry, 55*, 317–325.

McLean, C. P., Su, Y. J., & Foa, E. B. (2015). Mechanisms of symptom reduction in a combined treatment for comorbid posttraumatic stress disorder and alcohol dependence. *Journal of Consulting and Clinical Psychology, 83*(3), 655–661. https://doi.org/10.1037/ccp0000024.

Nacasch, N., Foa, E. B., Huppert, J. D., Tzur, D., Fostick, L., Dinstein, Y., ... Zohar, J. (2011). Prolonged exposure therapy for combat- and terror-related posttraumatic stress disorder: A randomized control comparison with treatment as usual. *Journal of Clinical Psychiatry, 72*(9), 1174–1180. https://doi.org/10.4088/JCP.09m05682blu.

Norman, S. B., Davis, B. C., Colvonen, P. J., Haller, M., Myers, U. S., Trim, R. S., ... Robinson, S. K. (2015). Prolonged exposure with veterans in a residential substance use treatment program. *Cognitive and Behavioral Practice, 23*(2), 162–172. https://doi.org/10.1016/j.cbpra.2015.08.002.

Paunovic, N., & Ost, L.-G. (2001). Cognitive-behavior therapy vs exposure therapy in the treatment of PTSD in refugees. *Behaviour Research and Therapy, 39*(10), 1183–1197.

Powers, M. B., Halpern, J. M., Ferenschak, M. P., Gillihan, S. J., & Foa, E. B. (2010). A meta-analytic review of prolonged exposure for posttraumatic stress disorder. *Clinical Psychology Review, 30*(6), 635–641. https://doi.org/10.1016/j.cpr.2010.04.007.

Ragsdale, K. A., & Voss Horrell, S. C. (2016). Effectiveness of prolonged exposure and cognitive processing therapy for U.S. veterans with a history of traumatic brain injury. *Journal of Traumatic Stress, 29*, 474–477.

Rauch, S. A. M., Cigrang, J., Austern, D., & Evans, A. (2017). Expanding the reach of effective PTSD treatment into primary care: Prolonged exposure for primary care. *Focus, 15*(4), 406–410.

Rauch, S. A. M., Eftekhari, A., & Ruzek, J. I. (2012). Review of exposure therapy: A gold standard for PTSD treatment. *Journal of Rehabilitation Research and Development, 49*(5), 679–688.

Rauch, S. A. M., & Foa, E. (2006). Emotional processing theory (EPT) and exposure therapy for PTSD. *Journal of Contemporary Psychotherapy, 36*(2), 61–65. https://doi.org/10.1007/s10879-006-9008-y.

Rauch, S. A. M., & Liberzon, I. (2016). Mechanisms of action in psychotherapy. In I. Liberzon & K. J. Ressler (Eds.), *Neurobiology of PTSD: From brain to mind*. New York, NY: Oxford University Press.

Rauch, S. A. M., Post, L. M., Yasinski, C. M., Sherrill, A. M., Maples-Keller, J. L., Breazeale, K., & Rothbaum, B. O. (2017). Healing the invisible wounds of war. In *Symposium Presented at the International Society for Traumatic Stress Studies 33rd Annual Meeting, Chicago, IL*.

Rauch, S. A. M., & Rothbaum, B. O. (2016). Innovations in exposure therapy for PTSD treatment. *Practice Innovations, 1*(3), 189.

Resick, P. A., Bovin, M. J., Calloway, A. L., Dick, A. M., King, M. W., Mitchell, K. S., ... Wolf, E. J. (2012). A critical evaluation of the complex PTSD literature: Implications for DSM-5. *Journal of Traumatic Stress, 25*(3), 241–251.

Resick, P. A., Monson, C. M., & Chard, K. M. (2016). *Cognitive processing therapy for PTSD: A comprehensive manual*. New York, NY: The Guilford Press.

Resick, P. A., Nishith, P., Weaver, T. L., Astin, M. C., & Feuer, C. A. (2002). A comparison of cognitive-processing therapy with prolonged exposure and a waiting condition for the treatment of chronic posttraumatic stress disorder in female rape victims. *Journal of Consulting and Clinical Psychology, 70*(4), 867–879. https://doi.org/10.1037/0022-006X.70.4.867.

Rothbaum, B. O., Astin, M. C., & Marsteller, F. (2005). Prolonged exposure versus eye movement desensitization and reprocessing (EMDR) for PTSD rape victims. *Journal of Traumatic Stress, 18*(6), 607–616. https://doi.org/10.1002/jts.20069.

Rupp, C., Doebler, P., Ehring, T., & Vossbeck-Elsebusch, A. N. (2017). Emotional processing theory put to test: A meta-analysis on the association between process and outcome measures in exposure therapy. *Clinical Psychology & Psychotherapy*, *24*(3), 697–711. https://doi.org/10.1002/cpp.2039.

Shapiro, F. (2001). *Eye movement desensitization and reprocessing: Basic principles, protocols, and procedures*. New York, NY: The Guilford Press.

Smith, E. R., Duax, J. M., & Rauch, S. A. M. (2013). Perceived perpetration during traumatic events: Clinical suggestions from experts in prolonged exposure therapy. *Cognitive and Behavioral Practice*, *20*(4), 461–470. https://doi.org/10.1016/j.cbpra.2012.12.002.

Smucker, M. R., Dancu, C., Foa, E. B., & Niederee, J. L. (1995). Imagery rescripting: A new treatment for survivors of childhood sexual abuse suffering from posttraumatic stress. *Journal of Cognitive Psychotherapy*, *9*(1), 3–17. Retrieved from http://www.ingentaconnect.com/content/springer/jcogp/1995/00000009/00000001/art00001.

Sripada, R. K., & Rauch, S. A. M. (2015). Between-session and within-session habituation in prolonged exposure therapy for posttraumatic stress disorder: A hierarchical linear modeling approach. *Journal of Anxiety Disorders*, *30*, 81–87. https://doi.org/10.1016/j.janxdis.2015.01.002.

Sripada, R. K., Rauch, S. A. M., Tuerk, P. W., Smith, E., Defever, A. M., Mayer, R. A., … Venners, M. (2013). Mild traumatic brain injury and treatment response in prolonged exposure for PTSD. *Journal of Traumatic Stress*, *26*(3), 369–375.

Stapleton, J., Taylor, S., & Asmundson, G. (2006). Effects of three PTSD treatments on anger and guilt: Exposure therapy, eye movement desensitization and reprocessing, and relaxation training. *Journal of Traumatic Stress*, *19*(1), 369–375.

Tuerk, P. W., Yoder, M., Grubaugh, A., Myrick, H., Hamner, M., & Acierno, R. (2011). Prolonged exposure therapy for combat-related posttraumatic stress disorder: An examination of treatment effectiveness for veterans of the wars in Afghanistan and Iraq. *Journal of Anxiety Disorders*, *25*(3), 397–403. https://doi.org/10.1016/j.janxdis.2010.11.002.

van Minnen, A., Arntz, A., & Keijsers, G. P. (2002). Prolonged exposure in patients with chronic PTSD: Predictors of treatment outcome and dropout. *Behaviour Research and Therapy*, *40*(4), 439–457. https://doi.org/10.1016/S0005-7967(01)00024-9.

Van Minnen, A., Harned, M. S., Zoellner, L., & Mills, K. (2012). Examining potential contraindications for prolonged exposure therapy for PTSD. *European Journal of Psychotraumatology*, *3*, https://doi.org/10.3402/ejpt.v3i0.18805.

Weathers, F. W., Bovin, M. J., Lee, D. J., Sloan, D. M., Schnurr, P. P., Kaloupek, D. G., … Marx, B. P. (2017). The clinician-administered PTSD scale for DSM–5 (CAPS-5): Development and initial psychometric evaluation in military veterans. *Psychological Assessment*, *28*(6), 489–498. https://doi.org/10.1037/pas0000486.

Williams, M., Malcoun, E., Sawyer, B., Davis, D., Nouri, L., & Bruce, S. (2014). Cultural adaptations of prolonged exposure therapy for treatment and prevention of posttraumatic stress disorder in African Americans. *Behavioral Science*, *4*(2), 102–124. https://doi.org/10.3390/bs4020102.

Wolpe, J., & Lazarus, A. A. (1966). *Behavior therapy techniques*. New York, NY: Pergamon.

Yasinski, C. M., Sherrill, A. M., Maples-Keller, J. L., Rauch, S. A. M., & Rothbaum, B. O. (2017). Intensive outpatient prolonged exposure for PTSD in post 9/11 veterans and service-members: Program structure and preliminary outcomes of the Emory healthcare veterans program. *Trauma Psychology News*, Winter, 1–4.

Yoder, M., Tuerk, P., Rauch, S. A. M., & Hall, B. (2012). Ecological collaborations: Mental health response to the Great East Japan earthquake. *International Perspective in Victimology*, *62*(2), 81.

Zalta, A. K., Gillihan, S. J., Fisher, A. J., Mintz, J., Mclean, C. P., Yehuda, R., & Foa, E. B. (2015). Change in negative cognitions associated with PTSD predicts symptom reduction in prolonged exposure. *Journal of Consulting and Clinical Psychology*, *82*(1), 171–175. https://doi.org/10.1037/a0034735.Change.

CHAPTER 17

Emotion in cognitive processing therapy for PTSD

Colleen Martin, Laura Stayton, Kathleen M. Chard
Cincinnati VA Medical Center, Cincinnati, OH, United States

Emotion in cognitive processing therapy

Overview of cognitive processing therapy

Cognitive processing therapy (CPT) was created by Resick and Schnicke (1993) for survivors of sexual trauma to treat posttraumatic stress disorder (PTSD), based on the assumption that certain posttraumatic cognitions maintain PTSD symptoms. Over the years, CPT has evolved to be one of the gold-standard, evidence-based psychotherapies for survivors of sexual trauma and other events, such as combat or natural disasters (Resick, Monson, & Chard, 2010). CPT is widely disseminated throughout various treatment centers, most notably within the veterans affairs healthcare system (VA; Chard, Ricksecker, Healy, Karlin, & Resick, 2012; Karlin et al., 2010). Largely grounded in cognitive theory, the goal of CPT is to modify maladaptive, trauma-related cognitions into more realistic thoughts, ultimately changing the emotional intensity and/or valence attached to the original thoughts. CPT is a manualized treatment that consists of 7–15 60-min sessions with a focus on trauma-related beliefs (assimilated beliefs) in the first half of therapy and a focus on exaggerated beliefs about the self, others, and the world (overaccommodated) in the second half of therapy. According to the cognitive behavioral theoretical underpinnings of CPT, the modification of these maladaptive cognitions that are associated with reexperiencing symptoms and emotional distress will lead to a reduction in PTSD symptomatology and other comorbid disorders (e.g., depressive symptoms). Throughout treatment, Socratic dialogue is used to challenge cognitions that are maintaining PTSD symptoms (i.e., "stuck points") to facilitate the recovery process through adopting more balanced beliefs and engaging in natural emotional processing. For example, posttraumatic cognitions regarding self-blame and other blame are common in PTSD and frequently lead to feelings of guilt and shame. When individuals have the opportunity to critically examine their thoughts stemming from their perceived role in the cause of the traumatic event, they may in turn feel less guilty and/or ashamed, lessening PTSD symptomatology.

In CPT, therapists outline the distinction between natural and manufactured emotions. Natural emotions are those thought to be experienced "naturally" in a situation based on the facts. These responses are often "hard wired" and may even provide helpful information to the individual. For example, if a person loses a loved one, they are likely to feel the natural emotion of sadness. Manufactured emotions, on the other hand, are emotions that stem from our beliefs or interpretations of the situation. For example, if in the same situation someone tells themselves, "I didn't do enough to save the person from dying," they are likely to then feel the manufactured emotion of guilt. Through psychoeducation, CPT therapists explain to patients that, while natural emotions usually run their course and resolve over time, manufactured emotions persist for as long as the belief is entertained and they can include emotions across the spectrum. The goal in CPT is to reduce painful manufactured emotions by adapting more balanced beliefs and perspectives that take into account all available information. Through Socratic questioning and challenging, patients are able to process natural emotion that may have been avoiding and reduce painful manufactured emotions through adoption of a new perspective.

Research on natural and manufactured emotions

In the literature, natural and manufactured emotions are sometimes described as primary and secondary emotions, respectively. Primary (natural) emotions occur during, and as a direct consequence of, the traumatic experience and are universal in nature, whereas secondary (manufactured) emotions are based on the cognitive appraisals made following trauma. One may interpret the cause and/or consequences of the event to oneself, which can lead to an array of secondary emotions that are a product of how the individual interprets the event (e.g., guilt and shame). These cognitive appraisals influence the onset and maintenance of secondary emotions and the accuracy of the memory of the event (Gross, 2002). Secondary emotions have been associated with a slower recovery process from PTSD, as these emotions remain if the appraisals underlying them are maintained (Brewin & Holmes, 2003). As Ehlers and Clark (2000) described in their work on emotion dysregulation, symptoms of PTSD will persist due to the presence of secondary emotions such as shame, guilt, and fear, as the individual maintains an ongoing sense of threat.

In a community sample, reappraisals following a stressor were examined to determine if more accurate and positive reappraisals would be associated with less frequent intrusive symptoms. They found that, when reappraisals were used following an emotionally stressful event, those individuals who exhibited more adaptive appraisals experienced less intrusive memories of the stressor (Woud, Holmes, Postma, Dalgleish, & Mackintosh, 2012). Additionally, the presence of secondary emotions following maladaptive cognitive appraisals of a trauma was associated with three symptom clusters of PTSD in a sample of survivors of interpersonal trauma (Semb, Henningsson, Fransson, & Sundbom, 2009). This finding demonstrates the impact of cognitions on secondary emotions and, in turn, symptoms of PTSD. Given this relationship, CPT targets cognitions with the aim of changing emotions.

Trials of CPT have demonstrated its efficacy in reducing PTSD symptom severity in a variety of trauma-exposed samples. Randomized control trials of CPT have shown that CPT results in PTSD symptom relief with individuals with histories of interpersonal violence, childhood sexual abuse, and military-related trauma (e.g., Chard, 2005; Monson et al., 2006; Resick, Nishith, Weaver, Astin, & Feuer, 2002; Resick et al., 2008). To date, several trials of CPT have been conducted in active-duty military and veteran samples, with results indicating that CPT effectively reduces PTSD symptomatology (Forbes et al., 2012; Monson et al., 2006; Resick et al., 2015; Suris, Link-Malcolm, Chard, Ahn, & North, 2013). Based on the results of these stringent trials, CPT continues to be one of the top two recommended treatments for PTSD based on the VA/DoD guidelines for the treatment of PTSD (Department of Veterans Affairs and Department of Defense, 2017). The RCTs examining the effectiveness of CPT in treating PTSD have also shown moderate to large effect size reductions in dysfunctional posttraumatic cognitions and large effect size reductions in both PTSD and depression symptomatology (Monson et al., 2006; Resick et al., 2008; Resick et al., 2002). Research on the effectiveness of CPT in trauma-exposed populations has shown that CPT results in reductions in comorbid symptoms, such as depression and suicidal ideation. Specifically, several studies have found that depressive symptoms and suicidal ideation decrease following engagement in CPT for active-duty samples, survivors of sexual trauma, and veteran samples (e.g., Resick, Monson, et al., 2017; Resick, Wachen, et al., 2017; Bryan et al., 2016; Gradus, Suvak, Wisco, Marx, & Resick, 2013; Bryan et al., 2018; Cox et al., 2016; Forbes et al., 2012; Suris et al., 2013; Resick et al., 2015; Chard, 2005). These findings are likely due to CPT's ability to target maladaptive cognitions that are inherent in other associated psychiatric conditions.

Specific emotional changes in CPT
Theoretical rationale
There is a significant body of empirical work demonstrating changes in emotional states following CPT treatment; however, several theoretical frameworks also help clarify how CPT influences the intensity and/or nature of emotions associated with PTSD. From a cognitive theoretical perspective, one's interpretation of a traumatic event can ultimately lead to dysfunctional or maladaptive cognitions that influence emotional experiences. CPT has shown to effectively reduce PTSD symptoms by identifying and modifying maladaptive cognitions about the traumatic event(s) and about the individual, others, and the world (e.g., Holliday, Holder, & Suris, 2018; Scher, Suvak, & Resick, 2017). This framework has received substantial empirical support as many studies have found that changes in posttraumatic cognitions lead to not only decreased PTSD symptomatology but also decreases in other emotions commonly associated with trauma (e.g., Schumm, Dickstein, Walter, Owens, & Chard, 2015). As previously discussed, assimilated beliefs are targeted first in treatment to address any unrealistic beliefs about the cause of the traumatic event. This includes a focus on retrospective emotions about the trauma (e.g., guilt, shame, and sadness), as these are thought to be inclined to change through modification of cognitions (Resick et al., 2008). Following this phase of treatment, over-accommodated beliefs often related to safety, trust, power/control, esteem, and intimacy are examined and modified to create more balanced beliefs (Resick et al., 2002). If an individual can modify these cognitions through CPT, it follows that the associated emotions will either change or lessen in intensity (Resick & Schnicke, 1993; Schumm et al., 2015). For example, if an individual has maladaptive beliefs about safety that have led to heightened levels of fear (e.g., I am never safe), therapists are able to help the individual challenge these beliefs to create more balanced cognitions that result in less intense fear responses. Similarly, when an individual can challenge unrealistic interpretations about the cause of the traumatic event, they are likely to experience a decrease in guilt and/or shame. As explained in the CPT protocol, manufactured emotions such as guilt and shame continue if the maladaptive cognitions behind the emotion are maintained.

From an emotion regulation perspective, difficulty with effectively managing emotions has been associated with increased PTSD symptom severity (Ehring & Quack, 2010; McDermott, Tull, Gratz, Daughters, & Lejuez, 2009; Short, Norr, Mathes, Oglesby, & Schmidt, 2016). In CPT, individuals are asked to identify and rate the intensity of emotions associated with various stuck points after receiving psychoeducation on the complexity of emotional experiences. By routinely identifying emotions that accompany these stuck points through worksheets and emotional processing, the individual increases their awareness and acceptance of their emotions and thereby decreases PTSD symptomatology (e.g., Berke et al., 2017). Many times, individuals will have stuck points about emotional expression itself that can also be addressed and challenged through CPT worksheets (e.g., If I show emotions, I will lose control). When more realistic beliefs about emotions are developed and solidified, the individual increases their ability to effectively tolerate and manage the emotions.

Within the CPT framework, emotions can also inform treatment providers of underlying core beliefs that have not been addressed over the course of treatment. After processing and challenging stuck points, emotional intensity can act as a "thermometer" to determine whether there may be additional stuck points rooted in core beliefs that should be given attention. In general, guilt, shame, anger, fear, and sadness are some of the most commonly observed emotional changes in CPT treatment.

Empirical support
Guilt

In the current literature, several studies have examined changes in guilt over the course of CPT treatment among various samples, including veterans, trauma-exposed community samples, and active-duty personnel (e.g., Monson et al., 2006; Resick et al., 2002; Resick et al., 2008; Resick, Monson, et al., 2017; Resick, Wachen, et al., 2017). Guilt is an emotion that is frequently seen in individuals with PTSD, and it typically develops out of one's interpretation of responsibility for the traumatic event (e.g., Bryan, Morrow, Etienne, & Ray-Sannerud, 2013; Tangney, Stuewig, & Mashek, 2007). It is influenced by hindsight bias, perceived responsibility for a negative outcome, perceived violation of values, and perceived insufficient justification for actions (Kubany & Watson, 2003). When individuals possess beliefs about the traumatic event that fall into these different categories (e.g., "I should have fought back" or "I should have stopped the attack"), they are likely to experience trauma-related guilt (Bannister, Colvonen, Angkaw, & Norman, 2018; Tangney et al., 2007). CPT has been successful at reducing trauma-related guilt by modifying guilt cognitions and thereby reducing the intensity of the emotion in various populations (e.g., Held, Klassen, Brennan, & Zalta, 2017; Paul et al., 2014; Resick et al., 2002).

Most of the literature examining emotional changes in CPT has been with trauma-exposed community samples. An early study of emotional changes in CPT was conducted with a sample of women with sexual trauma histories (Resick et al., 2002). The authors of this study examined changes in PTSD, depression, and guilt in individuals enrolled in CPT or PE, and they found that those individuals in CPT, as compared with the PE condition, had significant decreases in PTSD, depression, and guilt at posttreatment, 3-month, and 9-month follow-up time points. Following both treatments, participants endorsed significantly lower levels of global guilt and wrongdoing than those in the minimal attention condition (MA) group; however, those enrolled in the CPT condition had significantly lower levels of hindsight bias and lack of justification scores than those in the other two conditions. This is noteworthy, as this is evidence that CPT may be better suited than other treatments to address these specific guilt conditions since it provides more focused intervention for these thoughts (Resick et al., 2002). Guilt has often been associated with depression in the literature, so there have been questions about whether CPT is specifically targeting guilt or if the decreases in depression play a role in decreased guilt. Even in trauma-exposed samples with comorbid PTSD and depression, greater proportions of individuals in CPT, as compared with PE, had clinically significant decreases on measures of guilt (Nishith, Nixon, & Resick, 2005). This finding suggests that CPT alters certain trauma-related guilt conditions, even in the presence of comorbid depression.

When CPT was examined as part of a dismantling study comparing it with CPT without a trauma account and a written account-only condition, guilt cognitions significantly decreased across all three forms of the treatment in a sample of women with histories of interpersonal violence (Resick et al., 2008). In a more recent sample of women who experienced sexual assault, trauma-related guilt significantly decreased across CPT and continued to decrease at long-term follow-up 5–10 years after the completion of CPT (Larsen, Fleming, & Resick, 2018). Therefore not only CPT reduces guilt in the short term, but also changes are maintained up to 10 years later. Guilt was also examined in a sample of men and women with sexual trauma histories who were engaged in CPT, and the authors found that guilt exhibited rapid initial decreases at the beginning of treatment, followed by a slower decrease (Galovski, Blain, Chappuis, & Fletcher, 2013). In fact, women showed more rapid decreases than men on global guilt and guilt cognitions, suggesting that men and women have different experiences with interpretations of guilt in CPT; however, guilt significantly decreased across treatment for both men and women.

In addition to trauma-exposed community samples, guilt has been examined in veteran samples engaged in CPT. A controlled trial comparing CPT with a wait-list condition in a sample of veterans was conducted to determine the effectiveness of CPT in reducing primary symptoms of PTSD and secondary symptoms of guilt, depression, anxiety, and affect functioning. They found that guilt-related distress significantly decreased in the CPT condition versus the wait-list condition (Monson et al., 2006). In addition to reductions in guilt, participants also exhibited improvements in affect control and alexithymia over the course of CPT. These findings highlight the ability of CPT to address guilt in a variety of trauma-exposed samples (Kubany & Watson, 2003). Shame often goes hand in hand with guilt in trauma-exposed samples. Therefore research has also explored the impact of CPT on feelings of shame.

Shame

When one is exposed to a traumatic event that goes against deeply held values, the individual may engage in thought processes that involve negative evaluations of themselves, eventually leading to feelings of shame. This commonly occurs in individuals with interpersonal trauma histories, and it has been related to the onset and maintenance of PTSD (Andrews, Brewin, Rose, & Kirk, 2000; Andrews, Brewin, Stewart, Philpott, & Hejdenberg, 2009). In fact, shame has been more strongly associated with PTSD than guilt in several studies (e.g., Beck et al., 2011; Dorahy et al., 2013; Pineles, Street, & Koenen, 2006). Studies have shown that, when individuals experience shame, they are more motivated to escape and withdraw to conceal aspects of their identity (Tangney et al., 2007). Over the course of CPT, beliefs underlying shame (e.g., "I am a monster, I am damaged") are examined and modified to create more functional beliefs. When shame was examined in the Resick et al.'s (2008) study, women with histories of interpersonal traumas exhibited significant improvements in shame and cognitive distortions associated with shame at posttreatment.

Although there have been limited empirical literature specifically examining changes in shame during CPT, theoretical models point to several posttraumatic cognitions that lead to the development of shame in trauma survivors. The emotional experience of shame involves negative appraisals of the self as being inadequate, inferior, and powerless following a traumatic event. Within CPT, maladaptive posttraumatic cognitions such as these are modified to reduce the intensity of shame, which has been frequently associated with the development and maintenance of PTSD (Ehlers & Clark,

2000; Resick & Schnicke, 1993). When guilt and shame were examined in a sample of veterans from Operation Iraqi Freedom (OIF)/Operation Enduring Freedom (OEF), shame was more strongly associated with PTSD symptom severity than guilt (Bannister et al., 2018). In another veteran sample in a residential treatment program, changes in self-blame and negative appraisals about the self were associated with a reduction in PTSD symptoms (Schumm et al., 2015).

Anger

Another commonly observed emotion in individuals with PTSD is anger. Several studies have demonstrated the ability of CPT to reduce anger levels during treatment (e.g., Resick et al., 2008; Resick, Nishith, & Griffin, 2003). As previously discussed, anger can be a manufactured and natural emotion, depending on the cognitions that may be driving the anger. For example, if an individual has thoughts that are balanced and realistic about the traumatic event, anger may in fact be a natural emotional reaction; however, if the anger is fueled by thoughts that are maladaptive and serve to maintain the anger (e.g., "It shouldn't be this way; They should have done something different") for an indefinite period, it is likely a manufactured emotion. Studies have shown that anger is predictive of initial PTSD development and that anger at pretreatment has been associated with less successful treatment outcomes (e.g., Forbes et al., 2008; Meffert et al., 2008).

In a sample of men and women with sexual assault histories, Galovski et al. (2013) examined sex differences in the expression and progression of anger over the course of CPT. They found that, over the course of treatment, changes in anger were evidenced by an initial rapid decline and followed by a slower rate of decline as treatment progressed. The authors suggest that, as individuals in this study began to increase their emotional control of anger, they experienced significant gains in treatment. Regarding sex differences, men exhibited higher levels of anger at pretreatment, and women experienced more rapid decreases in anger and irritability over the course of treatment. Although men and women had different initial changes in anger, the maintenance of the gains following treatment was similar.

Anger has been implicated in treatment response and outcome, with several studies finding that pretreatment anger was associated with worse outcomes in exposure-based therapies (e.g., Foa, Riggs, Massie, & Yarczower, 1995; Pitman et al., 1991); however, other studies have not found any negative treatment effects of anger (e.g., Cahill, Rauch, Hembree, & Foa, 2003; van Minnen, Arntz, & Keijsers, 2002). When anger was examined in a sample of women with histories of rape randomized to either CPT or PE for PTSD, those with higher trait anger who were enrolled in CPT had a lower dropout rate than those in the PE condition (Rizvi, Vogt, & Resick, 2009). The authors suggest that individuals with higher pretreatment levels of trait anger may be better served by CPT due to its ability to retain those individuals in treatment, potentially due to the modification of beliefs about anger. Interestingly, one study with outpatient veterans of wars in Iraq and Afghanistan examined how fear of emotions and losing control over emotions influenced treatment outcome (Miles, Smith, Maieritsch, & Ahearn, 2015). These authors found that fear of anger was associated with decreased levels of PTSD symptom severity following engagement in CPT. Specifically, their findings suggest that a heightened fear of one's anger may motivate the individual to reduce their anger symptoms to avoid some of the negative consequences of anger. In this way, CPT can assist individuals in gaining a better understanding of their anger by developing more balanced and adaptive cognitions about the origin of their anger and their ability to effectively cope with it. In addition to fear of emotions, CPT has also been shown to target more general emotion of fear.

Fear

Fear is a prominent component in the development of PTSD, given that it is often one of the most common emotions present during, and immediately following, a traumatic event. Emotional processing theory suggests that a fear structure is created following a trauma and this structure may include inaccurate beliefs and meanings about the trauma itself (Scher et al., 2017). In CPT, this fear structure is activated when processing and discussing beliefs about the trauma and beliefs that exist within this structure are challenged to create more balanced cognitions. Many times, beliefs about one's safety influence how they function in the world and the level of fear and/or anxiety they experience in certain situations (e.g., crowds, standing in lines, and traffic). In CPT+A (with the trauma account), the habituation of the fear response is achieved through the repeated processing of the thoughts and emotions associated with the traumatic event (Resick et al., 2008). Following CPT treatment in a trauma-exposed sample, anxiety significantly decreased at posttreatment and maintained this decrease at follow-up (Resick et al., 2008). Additionally, this same finding held in a sample of veterans in which general anxiety significantly decreased across CPT along with several other secondary symptoms (Monson et al., 2006).

Fear and anxiety can also be present for individuals when thinking about the actual experience of certain emotions, as unrealistic beliefs about losing control over emotions can increase these fears. For example, in a sample of Vietnam veterans, cognitive changes in the fear of experiencing emotions was associated with changes in PTSD symptom severity (Price, Monson, Callahan, & Rodriguez, 2006). When these participants developed realistic beliefs about emotions, it likely decreased the urge to avoid the emotions, thus decreasing PTSD symptom severity. In another sample of veterans enrolled in CPT in an outpatient PTSD clinic, fear of anxiety was associated with higher rates of dropout from treatment, potentially as a method to avoid the experience of the emotion during treatment (Miles et al., 2015). Significant reductions in anxiety were also seen in a sample of veterans randomly assigned to CPT as opposed to treatment as usual (Lloyd et al., 2014). As individuals with PTSD begin to integrate their newly balanced beliefs into their daily lives, the emotional reactivity in certain situations (e.g., crowds) will begin to subside as they find evidence that they are not in danger.

Depression

Finally, depression is one of the most frequent comorbid conditions with PTSD in trauma-exposed samples, and many beliefs underlying PTSD overlap with certain aspects of depression as well (e.g., Breslau, Davis, Peterson, & Schultz, 2000). In addition to changes in guilt, shame, anger, and fear, numerous studies have found that depression symptoms decrease across CPT. These findings have held in samples ranging from military veterans with combat-related trauma (Forbes et al., 2012), veterans with military sexual trauma (Suris et al., 2013), active-duty military members (Resick et al., 2015), child sexual abuse survivors (Chard, 2005), and victims of interpersonal violence (Resick et al., 2008). More specifically, studies have compared CPT with other treatments and found that it results in greater reduction in depressive symptoms.

In a sample of child sexual abuse survivors, depressive symptoms were significantly reduced following a combined group and individual CPT format. Specifically, 79% of the participants in this combined and individual CPT condition evidenced good end-state depressive functioning as defined by a Beck Depression Inventory score of 10 or less (Chard, 2005). Moreover, the effects of treatment on depression were maintained at both 3- and 12-month follow-up time points. One hypothesis is that CPT worked equally well for reducing both PTSD and depression in this sample because of the shared cognitive techniques that have proven successful in reducing both conditions in other studies (e.g., DeRubeis, Gelfand, Tang, & Simons, 1999). When CPT was examined with a trauma-exposed community sample in a dismantling study conducted by Resick et al. (2008), self-report and clinician-rated depressive symptoms significantly decreased in both CPT with and without the trauma account. In a longitudinal examination of female sexual assault survivors, changes in trauma-related beliefs resulted in maintained changes in depression and PTSD at 5–10 year follow-up time points (Iverson, King, Cunningham, & Resick, 2015).

In addition to community samples, veterans with depressive symptoms have also shown to benefit from engagement in CPT. In a randomized control trial examining the influence of CPT on PTSD symptomatology, Monson and colleagues (2006) found treatment to result in significantly reduced levels of depression symptoms at post treatment. Veterans with varying levels of PTSD symptom severity in both residential (Alvarez et al., 2011; Walter, Buckley, Simpson, & Chard, 2014) and outpatient treatment programs (Jeffreys et al., 2014) have also demonstrated significant reductions in depression following a course of CPT. Veterans enrolled in a community-delivered CPT for military-related PTSD exhibited increased reductions in depressions, as well as other posttraumatic symptoms, as compared with a treatment as usual condition (Forbes et al., 2012). This finding also held true for veterans with histories of military sexual trauma, such that depression symptoms decreased over the course of CPT (Suris et al., 2013). In another veteran sample enrolled in a 3-week intensive outpatient program for PTSD that included a combination of CPT and psychoeducational groups, depression symptoms significantly decreased over the treatment. Additionally, the authors found that reductions in posttraumatic cognitions predicted the decreases in both PTSD and depression (Zalta et al., 2018). When comparing the effectiveness of group with individual CPT+A in a sample of veterans, depression symptoms significantly decreased in both formats of CPT+A. Through hierarchical linear modeling analyses, the data showed that depression symptoms decreased significantly more in the individual therapy format than in group format (Lamp, Avallone, Maieritsch, Buchholz, & Rauch, 2018). In this particular study, white veterans exhibited more significant reductions in depression symptoms than African American veterans, highlighting the need for increased attention on of the influence of culture on belief systems for African American veterans.

Lastly, changes in depression have been examined across CPT in active-duty military personnel samples. In a study comparing CPT with present-centered therapy (PCT), those individuals who participated in CPT experienced greater reductions in depression symptoms as opposed to those enrolled in PCT (Resick et al., 2015). Taken together, these studies suggest that CPT is effective in reducing symptoms of depression in addition to PTSD.

Emotion regulation

In addition to altering specific emotion states, CPT has also been shown to address difficulties with general emotion regulation in a variety of samples. In the literature, difficulty with identifying and describing emotions has been associated with increased PTSD symptom severity, likely due to avoidance of internal experiences when feeling overwhelmed by emotions (Berke et al., 2017). Despite higher symptom severity, individuals with heightened difficulties in emotion regulation (i.e., borderline personality traits) have been no more likely to drop out of CPT than those with low difficulties (Clarke, Rizvi, & Resick, 2008), suggesting that CPT is well tolerated even for those with emotion regulation difficulties. Additionally, those with high levels of emotion regulation difficulties (as compared with low), and more severe PTSD and depression, actually exhibited more improvements during CPT. In a sample of veterans in a residential treatment program, Walter, Bolte, Owens, and Chard (2012) found no significant differences in PTSD treatment gains between those with and without personality disorders. Those with personality disorders also exhibited greater improvement in posttreatment depression following CPT treatment. The finding that CPT addresses emotion regulation difficulties in trauma-exposed samples also held true for women with military sexual trauma-related PTSD with comorbid borderline personality disorder. In this study, those with and without a borderline personality disorder diagnosis benefitted equally well from CPT, suggesting that CPT has the capacity to affect change on emotion dysregulation through cognitive restructuring (Holder, Holliday, Pai, & Suris, 2017). In another sample of veterans with PTSD, those that engaged in more balanced cognitive reappraisals during attempts to regulate emotion had less activation in the amygdala (Fitzgerald et al., 2017). This lends evidence to the effectiveness of actively monitoring cognitive appraisals in the regulation of emotion.

Mechanisms of action in CPT

Several possible mechanisms of action have been identified within CPT. Given the underlying theory of CPT, strong support has been found for change in cognitions as the mechanism by which CPT reduces symptoms and emotions. Numerous studies have found that changes in cognitions precede changes in PTSD symptoms across CPT (Gallagher & Resick, 2012; Holliday et al., 2018; Iverson et al., 2015; Schumm et al., 2015). Schumm et al. (2015) examined the temporality of changes in posttraumatic cognitions, PTSD symptom, and depressive symptoms in a group of veterans receiving CPT in a residential treatment program. They found that changes in posttraumatic cognitions preceded changes in PTSD symptoms (Schumm et al., 2015). More specifically, changes in negative beliefs about the self preceded changes in depression. Subsequently, changes in depression preceded changes in self-blame cognitions and PTSD symptoms (Schumm et al., 2015). These findings suggest that changes in posttraumatic cognitions may be the mechanism by which PTSD symptoms change in CPT. In addition, changes in cognitions preceded changes in mood (i.e., depression) suggesting, as theorized, that CPT directly impacts changes in emotion.

In addition to trauma-related cognitions, studies have examined the impact of specific types of cognitions, such as hopelessness, as a potential mediator of changes in symptoms across CPT. Gallagher and Resick (2012) examined a sample of female sexual assault survivors and found that changes in hopelessness mediated changes in PTSD symptoms for individuals receiving CPT, but not PE. This suggests that changes in hopelessness may be a unique mechanism of CPT. Similarly, Gilman, Schumm, and Chard (2012) examined reports of hope in veterans completing CPT in a residential PTSD program and found that higher hope at mid-treatment predicted greater changes in PTSD and depressive symptoms at post treatment (Gilman et al., 2012). This demonstrates further evidence of the impact of CPT on emotion and subsequent symptom change.

Not only are trauma-related cognitions correlated with PTSD symptom change immediately following treatment, but also long-term follow-up studies found similar patterns of results (Scher et al., 2017). Scher and colleagues assessed changes in cognitions and PTSD symptoms in a sample of female victims of sexual assault during treatment and at 3- and 9-month follow-up periods, as well as 5- and 10-year follow-up periods. They found that trauma-related cognitions continued to predict PTSD symptoms up to 10 years following treatment completion (Scher et al., 2017). Similarly, Iverson et al. (2015) followed a sample of sexual assault survivors during and after treatment and found that changes in trauma-related beliefs between treatment completion and 5- to 10-year follow-up predicted changes in PTSD and depression during the same time period. These studies together suggest that trauma-related cognitions are not only a mechanism of action during PTSD treatment but also a mechanism for maintenance of treatment gains.

Biological mechanisms

Given the wealth of evidence that CPT results in cognitive changes, which, in turn, lead to reductions in symptoms of PTSD and depression, researchers have attempted to investigate the biological pathway by which these changes occur. Understanding the biological impact of cognitive changes also helps us to understand the mechanism by which cognitive changes result in emotional changes. It is theorized that, by altering one's cognitions, this may result in increased cognitive flexibility and, in turn, activate the prefrontal cortex while downregulating the amygdala (Resick, Monson, et al., 2017; Resick, Wachen, et al., 2017), both of which are systems known to be implicated in individuals with PTSD. While no studies to date have examined these changes within CPT specifically, studies have examined biological changes across cognitive behavioral treatments (CBT) for PTSD more broadly.

One study assessed adolescent females who had been diagnosed with PTSD related to sexual or physical assault pretreatment and post treatment with trauma-focused CBT (Cisler et al., 2016). They conducted fMRI tests on participants while they engaged in a cognitive reappraisal task. They found that those with greater symptom reductions across treatment had greater suppression of the amygdala–insula functional connectivity (FC) (Cisler et al., 2016). Interestingly the changes in amygdala to insula FC that scaled with PTSD symptom reduction also scaled with improvements in emotion regulation. This suggests that CBT for PTSD may directly impact the FC between these brain areas, thereby producing changes in both PTSD symptoms and emotion regulation.

Another study conducted fMRI tests on women who experienced domestic violence and were participating in cognitive therapy for battered women (CTT-BW; Aupperle et al., 2013). During the study, participants were presented with images containing positive or negative affective content and activation during this task was examined. Women who received CTT-BW showed enhanced anterior cingulate cortex (ACC) activation and decreased amygdala-insula activation during anticipation of image presentation and decreased dorsolateral prefrontal cortex and amygdala response during image presentation (Aupperle et al., 2013). These results suggest that cognitive therapy may enhance patients' ability to prepare for emotional events and subsequently lead to a decreased reactivity during such events. Taken together, these studies support the hypothesis that CBT and cognitive therapies for PTSD result in neural changes that likely account for self-reported changes in emotions and PTSD symptoms. While this research has not directly examined CPT, future research may expand current findings to include the influence of biological changes on emotions across CPT.

Limitations and future directions

Although an extensive body of work exists that highlights the effectiveness of CPT in addressing emotional processes, there are several limitations to note. There is a noticeable lack of empirical evidence that examines shame over the course of CPT. Many studies focus on blame-related cognitions, which may in turn lead to emotional responses of shame; however, the effects of CPT on shame have rarely been examined (e.g., Bannister et al., 2018). This may be due in part to outdated measures of shame and/or the limitations that exist in differentiating shame and guilt. Additionally, there is a reliance on primarily self-report measures for the secondary emotional outcomes in CPT. Given that self-report measures are inherently subjective, this may lead to biased reports of symptomatology following CPT.

Also missing from the current literature is a thorough examination of the effects of CPT on emotion regulation in trauma-exposed populations. It is unclear how CPT specifically influences aspects of emotion regulation (e.g., awareness of emotion, describing emotion, and having strategies to cope with emotions) and how, in turn, increased emotion regulation predicts decreases in PTSD symptomatology. Relatedly, there is limited research on cross-lag models of the mechanisms of emotional change (i.e., changes in dysfunctional cognitions) in CPT. Some studies have found changes in depressive cognitions to precede changes in PTSD (e.g., Schumm et al., 2015); however, it is unclear how other emotions function in relation to decreased PTSD symptomatology.

Future research would benefit from addressing these limitations by expanding the existing knowledge base of CPT's ability to address emotional processing following traumatic events. It is critical to have a better understanding of the construct of shame in relation to guilt and to have a better method of operationalizing this construct as well. Researchers may also be interested in examining how emotions function in CPT using samples with different types of traumatic events (e.g., childhood abuse, sexual trauma, and combat trauma). It could be that the effects of more interpersonal trauma types may differentially affect emotions and how they are processed. CPT's influence on different aspects of emotion regulation is another important line of inquiry. Given that emotion dysregulation frequently accompanies PTSD presentations, it is important to investigate changes in difficulties with emotion regulation as a potential mechanism of change in PTSD symptomatology. Additionally, using the most appropriate and up-to-date measures of secondary outcomes (e.g., shame, guilt, and depression) in future studies is recommended. For example, many studies examined in this chapter used outdated measures of shame; therefore measures such as the trauma-related shame inventory (Øktedalen, Hoffart, & Langkaas, 2015) may be a useful tool for future studies. Whenever possible, it is also important to use clinician-administered instruments to assess symptomatology, in addition to self-report measures, to ensure a more accurate assessment. Related to potential mechanisms of change, future research should seek to use more advanced statistical methods to determine how various changes in emotions influence changes in cognitions and vice versa. For example, based on recent studies examining shame and guilt, cross-lag statistical models may help elucidate whether decreases in trauma-related guilt precede decreases in shame. Other studies have examined the pattern of change in depressive cognitions and PTSD symptoms (e.g., Schumm et al., 2015); however, further replication is needed in samples with different characteristics. Another important line of research will be to expand research on biological mechanisms of change from CBT to CPT, specifically. To do so, studies involving biomarkers of PTSD symptoms will be needed to monitor how symptoms change across CPT for individuals with certain biological presentations. Finally, future research lines may examine how changes in emotion through CPT influence rates of dropout in trauma-exposed populations.

References

Alvarez, J., McLean, C., Harris, A. H., Rosen, C. S., Ruzek, J. I., & Kimerling, R. (2011). The comparative effectiveness of cognitive processing therapy for male veterans treated in a VHA posttraumatic stress disorder residential rehabilitation program. *Journal of Consulting and Clinical Psychology, 79*, 590–599. https://doi.org/10.1037/a0024466.

Andrews, B., Brewin, C. R., Rose, S., & Kirk, M. (2000). Predicting PTSD symptoms in victims of violent crime: The role of shame, anger, and childhood abuse. *Journal of Abnormal Psychology, 109*, 69–73. https://doi.org/10.1037/0021-843X.109.1.69.

Andrews, B., Brewin, C. R., Stewart, L., Philpott, R., & Hejdenberg, J. (2009). Comparison of immediate-onset and delayed-onset posttraumatic stress disorder in military veterans. *Journal of Abnormal Psychology, 118*, 767–777. https://doi.org/10.1037/a0017203.

Aupperle, R. L., Allard, C. B., Simmons, A. N., Flagan, T., Thorp, S. R., Norman, S. B., … Stein, M. B. (2013). Neural responses during emotional processing before and after cognitive therapy for battered women. *Psychiatry Research: Neuroimaging, 214*, 48–55. https://doi.org/10.1016/j.pscychresns.2013.05.001.

Bannister, J. A., Colvonen, P. J., Angkaw, A. C., & Norman, S. B. (2018). Differential relationships of guilt and shame on posttraumatic stress disorder among veterans. *Psychological Trauma Theory Research Practice and Policy*. https://doi.org/10.1037/tra0000392.

Beck, J. G., McNiff, J., Clapp, J. D., Olsen, S. A., Avery, M. L., & Hagewood, J. H. (2011). Exploring negative emotion in women experiencing intimate partner violence: Shame, guilt, and PTSD. *Behavior Therapy, 42*, 740–750. https://doi.org/10.1016/j.beth.2011.04.001.

Berke, D. S., Macdonald, A., Poole, G. M., Portnoy, G., McSheffrey, S., Creech, S. K., & Taft, C. T. (2017). Optimizing trauma-informed intervention for intimate partner violence in veterans: The role of alexithymia. *Behaviour Research and Therapy, 97*, 222–229. https://doi.org/10.1016/j.brat.2017.08.007.

Breslau, N., Davis, G. C., Peterson, E. L., & Schultz, L. R. (2000). A second look at comorbidity in victims of trauma: The posttraumatic stress disorder–major depression connection. *Biological Psychiatry, 48*, 902–909.

Brewin, C. R., & Holmes, E. A. (2003). Psychological theories of posttraumatic stress disorder. *Clinical Psychology Review, 23*, 339–376.

Bryan, C. J., Clemans, T. A., Hernandez, A., Mintz, J., Peterson, A. L., Yarvis, J. S., … Resick, P. A. (2016). Evaluating potential iatrogenic suicide risk in trauma-focused group cognitive behavioral therapy for the treatment of PTSD in active-duty military personnel. *Depression and Anxiety, 33*, 549–557. https://doi.org/10.1002/da.22456.

Bryan, C. J., Leifker, F. R., Rozek, D. C., Bryan, A. O., Reynolds, M. L., Oakley, D. N., & Roberge, E. (2018). Examining the effectiveness of an intensive, 2-week treatment program for military personnel and veterans with PTSD: Results of a pilot, open-label, prospective cohort trial. *Journal of Clinical Psychology*, Advanced online publication. https://doi.org/10.1002/jclp.22651.

Bryan, C. J., Morrow, C. E., Etienne, N., & Ray-Sannerud, B. (2013). Guilt, shame, and suicidal ideation in a military outpatient clinical sample. *Depression and Anxiety, 30*, 55–60. https://doi.org/10.1002/da.22002.

Cahill, S. P., Rauch, S. A., Hembree, E. A., & Foa, E. B. (2003). Effect of cognitive behavioral treatments for PTSD on anger. *Journal of Cognitive Psychotherapy: An International Quarterly, 17*, 113–131.

Chard, K. M. (2005). An evaluation of cognitive processing therapy for the treatment of posttraumatic stress disorder related to childhood sexual abuse. *Journal of Consulting and Clinical Psychology, 75*, 965–971. https://doi.org/10.1037/0022-006X.73.5.965.

Chard, K. M., Ricksecker, E. G., Healy, E. T., Karlin, B. C., & Resick, P. A. (2012). Dissemination and experience with cognitive processing therapy. *Journal of Rehabilitation Research and Development, 49*, 667–678.

Cisler, J. M., Sigel, B. A., Steele, J. S., Smitherman, S., Vanderzee, K., Pemberton, J., … Kilts, C. D. (2016). Changes in functional connectivity of the amygdala during cognitive reappraisal predict symptoms reduction during trauma-focused cognitive-behavioral therapy among adolescent girls with post-traumatic stress disorder. *Psychological Medicine, 46*, 3013–3023. https://doi.org/10.1017/S0033291716001847.

Clarke, S. B., Rizvi, S. L., & Resick, P. A. (2008). Borderline personality characteristics and treatment outcome in cognitive-behavioral treatments for PTSD in female rape victims. *Behavior Therapy, 39*, 72–78.

Cox, K. S., Mouilso, E. R., Venners, M. R., Defever, M. E., Duvivier, L., Rauch, A. M., … Tuerk, P. W. (2016). Reducing suicidal ideation through evidence-based treatment for posttraumatic stress disorder. *Journal of Psychiatric Research, 80*, 59–63. https://doi.org/10.1016/j.jpsychires.2016.05.011.

Department of Veterans Affairs and Department of Defense. (2017). *VA/DoD clinical practice guideline for the management of post-traumatic stress disorder and acute stress disorder.* Washington, DC: Department of Veterans Affairs and Department of Defense.

DeRubeis, R. J., Gelfand, L. A., Tang, T. Z., & Simons, A. (1999). Medications versus cognitive behavioral therapy for severely depressed outpatients: Meta-analysis for four randomized comparisons. *American Journal of Psychiatry, 156,* 1007–1013.

Dorahy, M. J., Corry, M., Shannon, M., Webb, K., McDermott, B., Ryan, M., & Dyer, K. F. (2013). Complex trauma and intimate relationships: The impact of shame, guilt and dissociation. *Journal of Affective Disorders, 147,* 72–79. https://doi.org/10.1016/j.jad.2012.10.010.

Ehlers, A., & Clark, D. M. (2000). A cognitive model of posttraumatic stress disorder. *Behaviour Research and Therapy, 38,* 319–345.

Ehring, T., & Quack, D. (2010). Emotion regulation difficulties in trauma survivors: The role of trauma type and PTSD symptom severity. *Behavior Therapy, 41,* 587–598.

Fitzgerald, J. M., MacNamara, A., Kennedy, A. E., Rabinak, C. A., Rauch, S. A., Liberzon, I., ... Phan, K. L. (2017). Individual differences in cognitive reappraisal use and emotion regulatory brain function in combat-exposed veterans with and without PTSD. *Depression and Anxiety, 34,* 79–88. https://doi.org/10.1002/da.22551.

Foa, E. B., Riggs, D. S., Massie, E. D., & Yarczower, M. (1995). The impact of fear activation and anger on the exposure treatment for posttraumatic stress disorder. *Behavior Therapy, 26,* 487–499.

Forbes, D., Lloyd, D., Nixon, R. D., Elliott, P., Varker, T., Perry, D., ... Creamer, M. (2012). A multisite randomized controlled effectiveness trial of cognitive processing therapy for military-related posttraumatic stress disorder. *Journal of Anxiety Disorders, 26,* 442–452.

Forbes, D., Parslow, R., Creamer, M., Allen, N., McHugh, T., & Hopwood, M. (2008). Mechanisms of anger and treatment outcome in combat veterans with posttraumatic stress disorder. *Journal of Traumatic Stress, 212,* 142–149. https://doi.org/10.1002/jts.20315.

Gallagher, M. W., & Resick, P. A. (2012). Mechanisms of change in cognitive processing therapy and prolonged exposure therapy for PTSD: Preliminary evidence for the differential effects of hopelessness and habituation. *Cognitive Therapy and Research, 36,* 750–755. https://doi.org/10.1007/s10608-011-9423-6.

Galovski, T. E., Blain, L. M., Chappuis, C., & Fletcher, T. (2013). Sex differences in recovery from PTSD in male and female interpersonal assault survivors. *Behaviour Research and Therapy, 51,* 247–255. https://doi.org/10.1016/j.brat.2013.02.002.

Gilman, R., Schumm, J., & Chard, K. M. (2012). Hope as a change mechanism in the treatment of posttraumatic stress disorder. *Psychological Trauma Theory Research Practice and Policy, 4,* 270–277. https://doi.org/10.1037/a0024252.

Gradus, J. L., Suvak, M. K., Wisco, B. E., Marx, B. P., & Resick, P. A. (2013). Treatment of posttraumatic stress disorder reduces suicidal ideation. *Depression and Anxiety, 30,* 1046–1053.

Gross, J. J. (2002). Emotion regulation: Affective, cognitive, and social consequences. *Psychophysiology, 39,* 281–291. https://doi.org/10.1017/S0048577201393198.

Held, P., Klassen, B. J., Brennan, M. B., & Zalta, A. K. (2017). Using prolonged exposure and cognitive processing therapy to treat veterans with moral injury-based PTSD: Two case examples. *Cognitive and Behavioral Practice,* Advance online publication. https://doi.org/10.1016/j.cbpra.2017.09.003.

Holder, N., Holliday, R., Pai, A., & Suris, A. (2017). Role of borderline personality disorder in the treatment of military sexual trauma-related posttraumatic stress disorder with cognitive processing therapy. *Behavioral Medicine, 43,* 184–190. https://doi.org/10.1080/08964289.2016.1276430

Holliday, R., Holder, N., & Suris, A. (2018). Reductions in self-blame cognitions predict PTSD improvements with cognitive processing therapy for military sexual trauma-related PTSD. *Psychiatry Research, 263,* 181–184. https://doi.org/10.1016/j.psychres.2018.03.007.

Iverson, K. M., King, M. W., Cunningham, K. C., & Resick, P. A. (2015). Rape survivors' trauma-related beliefs before and after cognitive processing therapy: Associations with PTSD and depression symptoms. *Behaviour Research and Therapy, 66*, 49–55. https://doi.org/10.1016/j.brat.2015.01.002.

Jeffreys, M. D., Reinfeld, C., Nair, P. V., Garcia, H. A., Mata-Galan, E., & Rentz, T. O. (2014). Evaluating treatment of posttraumatic stress disorder with cognitive processing therapy and prolonged exposure therapy in a VHA specialty clinic. *Journal of Anxiety Disorders, 28*, 108–114. https://doi.org/10.1016/j.janxdis.2013.04.010.

Karlin, B. E., Ruzek, J. I., Chard, K. M., Eftekhari, A., Monson, C. M., Hembree, E. A., … Foa, E. B. (2010). Dissemination of evidence-based psychological treatments for post-traumatic stress disorder in the Veterans Health Administration. *Journal of Traumatic Stress, 23*, 663–673. https://doi.org/10.1002/jts.20588.

Kubany, E. S., & Watson, S. B. (2003). Guilt: Elaboration of a multidimensional model. *The Psychological Record, 53*, 51–90.

Lamp, K. E., Avallone, K. M., Maieritsch, K. P., Buchholz, K. R., & Rauch, S. A. M. (2018). Individual and group cognitive processing therapy: Effectiveness across two Veterans Affairs posttraumatic stress disorder treatment clinics. *Psychological Trauma Theory Research Practice and Policy*, Advanced online publication. https://doi.org/10.1037/tra0000370.

Larsen, S. E., Fleming, C. J. E., & Resick, P. A. (2018). Residual symptoms following empirically supported treatment for PTSD. *Psychological Trauma Theory Research Practice and Policy, 11*, 207–215. Advanced online publication.

Lloyd, D., Nixon, R. D. V., Varker, T., Elliott, P., Perry, D., Bryant, R. A., … Forbes, D. (2014). Comorbidity in the prediction of cognitive processing therapy treatment outcomes for combat-related posttraumatic stress disorder. *Journal of Anxiety Disorders, 28*, 237–240. https://doi.org/10.1016/j.janxdis.2013.12.002.

McDermott, M. J., Tull, M. T., Gratz, K. L., Daughters, S. B., & Lejuez, C. W. (2009). The role of anxiety sensitivity and difficulties in emotion regulation in posttraumatic stress disorder among crack/cocaine dependent patients in residential substance abuse treatment. *Journal of Anxiety Disorders, 23*, 591–599.

Meffert, S. M., Metzler, T. J., Henn-Haase, C., McCaslin, S., Inslicht, S., Chemtob, C., … Marmar, C. R. (2008). A prospective study of trait anger and PTSD symptoms in police. *Journal of Traumatic Stress, 21*, 410–416. https://doi.org/10.1002/jts.20350.

Miles, S. R., Smith, T. L., Maieritsch, K. P., & Ahearn, E. P. (2015). Fear of losing emotional control is associated with cognitive processing therapy outcomes in U.S. Veterans of Afghanistan and Iraq. *Journal of Traumatic Stress, 28*, 475–479. https://doi.org/10.1002/jts.22036.

Monson, C. M., Schnurr, P. P., Resick, P. A., Friedman, M. J., Young-Xu, Y., & Stevens, S. P. (2006). Cognitive processing therapy for veterans with military-related posttraumatic stress disorder. *Journal of Consulting and Clinical Psychology, 74*, 898–907.

Nishith, P., Nixon, R. D. V., & Resick, P. A. (2005). Resolution of trauma-related guilt following treatment of PTSD in female rape victims: A result of cognitive processing therapy targeting comorbid depression? *Journal of Affective Disorders, 86*, 259–265. https://doi.org/10.1016/j.jad.2005.02.013.

Øktedalen, T., Hoffart, A., & Langkaas, T. F. (2015). Trauma-related shame and guilt as time-varying predictors of posttraumatic stress disorder symptoms during imagery exposure and imagery rescripting—A randomized controlled trial. *Psychotherapy Research, 25*, 518–532.

Paul, L. A., Gros, D. F., Strachan, M., Worsham, G., Foa, E. B., & Acierno, R. (2014). Prolonged exposure for guilt and shame in a veteran of operation Iraqi freedom. *American Journal of Psychotherapy, 68*, 277–286.

Pineles, S. L., Street, A. E., & Koenen, K. C. (2006). The differential relationships of shame-proneness and guilt-proneness to psychological and somatization symptoms. *Journal of Social and Clinical Psychology, 25*, 688–704. https://doi.org/10.1521/jscp.2006.25.6.688.

Pitman, R. K., Altman, B., Greenwald, E., Longpre, R. E., Macklin, M. L., ... Poire, R. E. (1991). Psychiatric complications during flooding therapy for posttraumatic stress disorder. *Journal of Clinical Psychiatry, 52,* 17–20.

Price, J. L., Monson, C. M., Callahan, K., & Rodriguez, B. F. (2006). The role of emotional functioning in military-related PTSD and its treatment. *Journal of Anxiety Disorders, 20,* 661–674. https://doi.org/10.1016/j.janxdis.2005.04.004.

Resick, P. A., Galovski, T. E., Uhlmansiek, M. O., Scher, C. D., Clum, G. A., & Young-Xu, Y. (2008). A randomized clinical trial to dismantle components of cognitive processing therapy for posttraumatic stress disorder in female victims of interpersonal violence. *Journal of Consulting and Clinical Psychology, 76,* 243–258. https://doi.org/10.1037/0022-006X.76.2.243.

Resick, P. A., Monson, C. M., & Chard, K. M. (2010). *Cognitive processing therapy: Veteran/military version: Therapist's manual.* Washington, DC: Department of Veterans Affairs.

Resick, P. A., Monson, C. M., & Chard, K. M. (2017). *Cognitive processing therapy for PTSD: A comprehensive manual.* New York, NY: Guilford Press.

Resick, P. A., Nishith, P., & Griffin, M. G. (2003). How well does cognitive–behavioral therapy treat symptoms of complex PTSD? An examination of child sexual abuse survivors within a clinical trial. *CNS Spectrums, 8,* 340–355.

Resick, P. A., Nishith, P., Weaver, T. L., Astin, M. C., & Feuer, C. A. (2002). A comparison of cognitive-processing therapy with prolonged exposure and a waiting condition for the treatment of chronic posttraumatic stress disorder in female rape victims. *Journal of Consulting and Clinical Psychology, 70,* 867–879. https://doi.org/10.1037/0022-006X.70.4.867.

Resick, P. A., & Schnicke, M. K. (1993). *Cognitive processing therapy for rape victims: A treatment manual.* Newbury Park, CA: Sage.

Resick, P. A., Wachen, J. S., Dondanville, K. A., Pruiksma, K. E., Yarvis, J. S., Peterson, A. L., ... Mintz, J. (2017). Effect of group vs individual cognitive processing therapy in active-duty military seeking treatment for posttraumatic stress disorder: A randomized clinical trial. *JAMA Psychiatry, 74,* 28–36. https://doi.org/10.1001/jamapsychiatry.2016.2729.

Resick, P. A., Wachen, J. S., Mintz, J., Young-McCaughan, S., Roache, J. D., Borah, A. M., ... Peterson, A. L. (2015). A randomized clinical trial of group cognitive processing therapy compared with group present-centered therapy for PTSD among active duty military personnel. *Journal of Consulting and Clinical Psychology, 83,* 1058–1068. https://doi.org/10.1037/ccp0000016.

Rizvi, S. L., Vogt, D. S., & Resick, P. A. (2009). Cognitive and affective predictors of treatment outcome in cognitive processing therapy and prolonged exposure for posttraumatic stress disorder. *Behavior Research & Therapy, 47,* 737–743. https://doi.org/10.1016/j.brat.2009.06.003.

Scher, C. D., Suvak, M. K., & Resick, P. A. (2017). Trauma cognitions are related to symptoms up to 10 years after cognitive behavioral treatment for posttraumatic stress disorder. *Psychological Trauma Theory Research Practice and Policy, 9,* 750–757. https://doi.org/10.1037/tra0000258.

Schumm, J. A., Dickstein, B. D., Walter, K. H., Owens, G. P., & Chard, K. M. (2015). Changes in posttraumatic cognitions predict change in posttraumatic stress disorder symptoms during cognitive processing therapy. *Journal of Consulting and Clinical Psychology, 83,* 1161–1166. https://doi.org/10.1037/ccp0000040.

Semb, O., Henningsson, M., Fransson, P., & Sundbom, E. (2009). Trauma-related symptoms after violent crime: The role of risk factors before, during, and eight months after victimization. *The Open Psychology Journal, 2,* 77–88.

Short, N. A., Norr, A. M., Mathes, B. M., Oglesby, M. E., & Schmidt, N. B. (2016). An examination of the specific associations between facets of difficulties in emotion regulation and posttraumatic stress symptom clusters. *Cognitive Therapy and Research, 40,* 783–791.

Suris, A., Link-Malcolm, J., Chard, K., Ahn, C., & North, C. (2013). A randomized clinical trial of cognitive processing therapy for veterans with PTSD related to military sexual trauma. *Journal of Traumatic Stress, 26*, 28–37.

Tangney, J. P., Stuewig, J., & Mashek, D. J. (2007). Moral emotions and moral behavior. *Annual Review of Psychology, 58*, 345–372. https://doi.org/10.1146/annurev.psych.56.091103.070145.

van Minnen, A., Arntz, A., & Keijsers, G. P. J. (2002). Prolonged exposure in patients with chronic PTSD: Predictors of treatment outcome and dropout. *Behaviour Research and Therapy, 40*, 439–457.

Walter, K. H., Bolte, T. A., Owens, G. P., & Chard, K. M. (2012). The impact of personality disorders on treatment outcome for veterans in a posttraumatic stress disorder residential treatment program. *Cognitive Therapy and Research, 36*, 576–584.

Walter, K. H., Buckley, A., Simpson, J. M., & Chard, K. M. (2014). Residential PTSD treatment for female veterans with military sexual trauma: does a history of childhood sexual abuse influence outcome? *Journal of Interpersonal Violence, 29*, 971–986. https://doi.org/10.1177/0886260513506055.

Woud, M. L., Holmes, E. A., Postma, P., Dalgleish, T., & Mackintosh, B. (2012). Ameliorating intrusive memories of distressing experiences using computerized reappraisal training. *Emotion, 12*, 778–784. https://doi.org/10.1037/a0024992.

Zalta, A. K., Held, P., Smith, K. L., Klassen, B. J., Lofgreen, A. M., Normand, P. S., … Karnik, N. S. (2018). Evaluating patterns and predictors of symptom change during a three-week intensive outpatient treatment for veterans with PTSD. *BMC Psychiatry, 18*, 242. Advanced online publication.

CHAPTER 18

Skills Training in Affective and Interpersonal Regulation (STAIR) Narrative Therapy: Making meaning while learning skills

Kile M. Ortigo[a,b], Ashley Bauer[a], Marylene Cloitre[a,c]
[a]National Center for Posttraumatic Stress Disorder—Dissemination & Training Division, VA Palo Alto Health Care System, Palo Alto, CA, United States
[b]Center for Existential Exploration, Palo Alto, CA, United States
[c]Stanford University Department of Psychiatry and Behavioral Sciences, Stanford, CA, United States

Despite, historically speaking, a relatively recent entrance into official mental health vernacular (in *DSM-III*; American Psychiatric Association [APA], 1980), posttraumatic stress disorder (PTSD) and related posttrauma phenomena have been around in some form since the ancient Greeks (and likely well before) (Shay, 1994/2010). With such a long history, it is no surprise that several approaches to healing trauma have been applied throughout history, from shamanistic rituals to modern cognitive behavioral interventions. The tools we have today, though, afford us ways to test systematically which approaches have better efficacy and effectiveness. There are several evidence-based therapies currently available (see treatment guidelines, ISTSS, 2018). The dominant models for trauma treatment over the past 20 years have been cognitive behavioral therapies. These therapies focus predominantly on cognitive models and/or behavioral fear conditioning to explain the development and maintenance of PTSD as well as the mechanisms of action underlying its resolution. Cognitive therapies view PTSD essentially as a disorder of maladaptive cognitions or schema generated by the traumas, and recovery from PTSD results from reappraisal and adjustment of trauma-generated beliefs. Exposure therapies, originally derived from classical behaviorism, assume that PTSD is essentially a conditioned fear response that can be resolved via imaginal or in vivo exposure to the fear-generating stimuli under safe conditions.

Nevertheless, new treatment models continue to emerge as complements to the earlier cognitive behavioral traditions and seek to address more directly certain reactions common to some, if not many, people who experience a trauma. In this chapter, we focus on one such new approach, a two-module treatment, skills training in affective and interpersonal regulation (STAIR) Narrative Therapy (Cloitre, Cohen, & Koenen, 2006; Cloitre, Cohen, Ortigo, Koenen, & Jackson, 2020). STAIR Narrative Therapy views trauma as creating a loss of internal and external resources. It highlights specifically childhood trauma as a resource loss of immense proportion because of its adverse effects on psychological development. Childhood abuse has been demonstrated to undermine the development of healthy emotion regulation and relational and social capacities and contribute to the loss of a healthy, positive, and coherent sense of self (see Cook et al., 2005). STAIR Narrative Therapy is essentially a resource rehabilitation therapy with a particular focus on the rebuilding and rehabilitation of these socioemotional capacities and an integrated sense of self. Within this larger resource, loss frame is also a developmental theory framework that guides the identification and operationalization of the specific losses that occur and, to some extent, guides the nature of the interventions.

In this chapter, we first review the history of the development of STAIR Narrative Therapy and then expand upon the theoretical underpinnings through a resource loss and developmental (specifically attachment and narrative) frame. This theoretical frame is followed by a session-by-session overview of the treatment alongside a case example.

Treatment development and rationale

The treatment began as an intervention specifically for adults who had experienced childhood abuse with the goal of rehabilitating developmental injuries that had occurred as a result. Often, these injuries were evidenced by problems in the domains of emotional, social, and relational competencies. As the result of working with diverse trauma populations over several years, it became clear that emotional and social resource losses also occurred as a result of adulthood traumas. Often, individuals who develop PTSD as a result of adult trauma have a history of adverse childhood experiences or ACEs (e.g., parents with psychiatric disorders or with substance abuse disorders) that, like traumatic events, can undermine healthy development. Parents who are rageful, substance abusing, or neglectful provide poor role models for development. We have found these client populations have welcomed STAIR Narrative Therapy and have helped demonstrate its effectiveness (Levitt, Malta, Martin, Davis, & Cloitre, 2007; Weiss, Azevedo, Webb, Gimeno, & Cloitre, 2018). STAIR Narrative Therapy is distinct from other evidence-based therapies for PTSD in that it is a two-component treatment. In the first component, again unlike other therapies, there is an explicit and sustained focus on skills training for (1) improving emotional awareness and regulation skills and (2) increasing social awareness and interpersonal skills. This focus both narrows and broadens treatment in homing in on key complaints people have after a trauma—how they feel different, experience more sensitivity or numbness, and realize the negative impact on their relationships—while broadening the perspective of the therapist to address larger patterns of emotional and relationship problems that may have varying ties to any given Criterion A traumatic event.

The second component, Narrative Therapy, as with other evidence-based, trauma-focused cognitive behavioral therapies, involves an explicit verbal review of the trauma and exploration and reappraisal of its meaning. However, there are several differences in this approach compared with other established approaches. Instead of focusing primarily on habituation (like prolonged exposure; Foa, Hembree, & Rothbaum, 2007) or more conscious trauma-informed thought patterns (like cognitive processing therapy; Resick, Monson, & Chard, 2017), narrative therapy integrates these approaches systematically by reviewing a full range of traumatic memories organized by affective themes including those of fear, shame, and loss in the context of a safe, supportive therapeutic relationship. In addition, the creation of the narrative is supported and framed by the exploration of what have been called in attachment theory, *internal working models* of relating (as we will discuss in more detail later in this chapter). These working models (or what are called in the treatment more simply *relationship models*) are first identified during STAIR by discussing day-to-day interpersonal difficulties and systematically articulating the underlying interpersonal assumptions that are driving the difficulties. The link between STAIR and Narrative Therapy is the introduction of the identified relationship models during STAIR and evaluating their potential relevance to the trauma history. The relationship models driving current difficulties are often found to have their roots in traumatic events. This realization often provides validation to the client that their problems are not simply the result of a character flaw or personality disturbance but rather rooted in a traumatic event, whose analysis can help resolve the client's difficulties.

The narrative review process is iterative and incorporates role plays and STAIR skills review as needed. In this way, it is focused on not only the trauma memory itself but also the full context before, during, and after the trauma exposure. The overarching aim is to help the individual gain a sense of mastery while contextualizing the trauma and refocusing on creating one's present and future life narrative—one of recovery, hope, and agency. Thus STAIR's focus on emotions and relationships is not lost but instead weaved into the trauma-focused work of narrative therapy. Through this process the individual can regain emotional and social functioning alongside corrective emotional experiences with a supportive therapist. The interpersonal context is not a by-product of traditional therapeutic formats but a key ingredient in healing interpersonal trauma.

Why is STAIR Narrative Therapy an important treatment to consider? Emotion regulation and interpersonal problems have been shown to account for unique variance of adaptive functioning, even when including PTSD symptom severity as a predictor for people who have experienced childhood trauma (Cloitre, Miranda, Stovall-McClough, & Han, 2005). Clinicians and experts have also highlighted the need for more diverse treatment options, especially when multiple developmental traumas have occurred (e.g., Bradley, Greene, Russ, Dutra, & Westen, 2005; Cloitre, 2015). The benefit to the field in having diverse treatment options to offer is also important given the growing importance of client treatment choice and its impact on engagement (Adams & Drake, 2006; Mott, Stanley, Street Jr., Grady, & Teng, 2014).

Theoretical frame

Although sharing key features of modern cognitive behavioral and third-wave behavioral approaches, STAIR Narrative Therapy is generally informed by a resource loss model and developmental models of the formation of attachment and the self in a relational context. Many of these developmental models predate the official recognition in *DSM-III* (APA, 1980) of PTSD as a stand-alone diagnosis. Well before the field recognized PTSD, theorists and clinicians alike viewed trauma, especially early trauma, as a fundamental determinant of life-span development pathways. The mechanism of action, however, differed by model, from trauma causing a psychosexual fixation (e.g., Freud, 1920/1989) to it informing personality development and archetype-informed personal complexes (e.g., Jung, 1934/1981). The definition and nature of trauma has shifted over time, but what remains has always been its effect on the development of one's emotional landscape, sense of self, and sense of relating to the world. The connection between trauma exposure, PTSD, and emotion regulation difficulties is well established and indeed the focus of this entire book. Thus we do not independently review that literature here.

Conservation of resources and the resource loss model

A foundational influence on STAIR is the resource loss model described by Hobfoll's (2001) Conservation of Resources theory. In this framework, resources are conceptualized as intrapsychic or external sources of support and resilience. With stressful life events, including trauma, these resources are taxed. Subsequent stressors lead to an increasingly taxed system and eventually a breakdown of one's ability to cope effectively. Hobfoll, Mancini, Hall, Canetti, and Bonanno (2011) applied this model to trauma and found that repeated traumatic experiences led to a "spiral of loss" in resources and worsening of PTSD symptoms.

In the initial reaction following a trauma, a decrease in resources occurs, but community support appears to have a buffering effect on subsequent distress. For example, in a longitudinal sample of Vietnam veterans, community involvement upon homecoming was predictive of remission of PTSD symptoms at a later assessment, while perceived negative community attitudes predicted a more chronic course, suggesting that supportive social bonds play a role in resiliency (Koenen, Stellman, Stellman, & Sommer Jr., 2003). Complementary to these findings, as demonstrated in a very different sample of an inner city, primarily African American population with high trauma exposure rates, a lack of community support and cohesion has been shown to predict higher posttraumatic stress symptoms (Gapen et al., 2011).

Regardless of initial levels of community and social support, prolonged distress appears to lead to a loss in social status and dwindling social support over time. This phenomenon was illustrated by Kaniasty and Norris (2008), who conducted a longitudinal study of survivors of natural disaster. While initially the lack of social support was predictive of later degrees of distress, those survivors who experienced chronic distress experienced an erosion of their social support throughout the study. The authors suggested that this pattern may have been due to the suffering of chronic symptoms being seen by others as irritating or threatening to communal well-being. Another study of Gulf War military veterans also found that PTSD symptomology predicted a prospective decrease in social support over time, although social support was not a predictor of future posttraumatic stress severity (King, Taft, King, Hammond, & Stone, 2006).

The erosion of social support for those who experience posttraumatic symptoms appears to be at least partially related to the emotional dysregulation that occurs following a trauma. Some existing evidence supports that difficulties with emotional regulation, such as inappropriate or limited expression of positive affect, have a long-term negative impact on social adjustment and social competence over time. This pattern has been demonstrated in a sample of adult survivors of childhood sexual abuse (Bonanno et al., 2007) and in a New York City student sample following the 9/11 terrorist attacks (Papa & Bonanno, 2008).

The developmental literature has also demonstrated this phenomenon. For example, children who have experienced traumatic events have on average poorer emotion regulation and interpersonal skills (Shipman, Zeman, Penza, & Champion, 2000). Moreover, problematic affect regulation precedes and influences social difficulties. Prospective studies of children in middle school years have demonstrated that children assessed with poorer emotion regulation were, 2 years later, found to be less well liked by classmates and rated by teachers to have poorer social skills (Shipman & Zeman, 2001). Overall, these data indicate that trauma throughout the life span can lead to emotional dysregulation that in turn plays a key role in impairing social skills and eventually negatively impacting social bonds. The loss of social support and other relationship resources creates vulnerability to increased distress in the face of future traumas, leading to a chronic cycle of diminishing resources and a chronic PTSD course.

Attachment theory

The earlier resource loss model sets a frame for understanding the vicious cycle of trauma, its related relational and affective losses, and further risk for and poor response to future trauma. Preceding and supplementing the resource loss model, Bowlby's (1969, 1973, 1980) life-span developmental model of attachment theory outlines specific dynamics concerning how real experiences with significant others mold and shape one's emotional and social life. Bowlby described attachment as a behavioral system that starts in infancy to organize expectations and reactions to a caregiver's dependability and responses to calls for help, especially when distressed. By adapting to real experiences, an infant learns to initiate or shut off proximity-seeking behaviors and emotional expression at times of stress, separation, and reunion with a caregiver. These signal-response expectations are eventually internalized in the form of *internal working models* of both self (e.g., "Am I loveable?") and others (e.g., "If I cry out, will others respond to my needs?"). When early trauma is present, the attachment system internalizes lessons from these experiences and then starts to generalize about future interactions with close others. Individuals continue to adapt these internal working models throughout the life span and manifest their attachment expectations in thoughts and behaviors broadly organized in two dimensions: *attachment anxiety* and *attachment avoidance* (e.g., Bartholomew & Horowitz, 1991; Fraley, Waller, & Brennan, 2000).

From his original work, Bowlby emphasized the role of attachment across the entire life span. Although much early research focused on measuring attachment styles in early life (e.g., the strange-situation paradigm, Ainsworth & Bell, 1970; Ainsworth & Wittig, 1969), he identified direct connections to (1) emotional awareness and coping and (2) interpersonal expectations and outcomes into adolescence and adulthood. Perhaps the most impactful outcome of early attachment experiences is that of a *disorganized* attachment style, discovered many years after the initial Ainsworth categorization model (Main & Solomon, 1986). Bizarre behaviors suggesting a lack of an organized strategy to seek proximity and comfort from or, alternatively, avoid the caregiver were characteristic of infants with a disorganized style. Examples observed by Main and colleagues (Main & George, 1985; Main & Hesse, 1990) included becoming terrified by the caregiver, appearing disoriented or dissociated, or perhaps most striking walking backward toward the caregiver upon reunion—a very clear behavioral manifestation of psychological conflict. As such, the disorganized attachment style has been closely connected to severe childhood trauma and psychopathology (Baer & Martinez, 2006; Madigan et al., 2006; Main & Hesse, 1990) and poor outcomes later in life (e.g., Carlson, 1998; Hesse & Van IJzendoorn, 1998).

Even in less extreme cases of insecure attachment, however, the long-lasting effects of early negative childhood experiences and lessons learned from experiences with caregivers and other close attachment figures remain significant much later. After Bowlby, many theorists and researchers have explored how attachment in adulthood is related to object relations and social cognition (Calabrese, Farber, & Westen, 2003, 2005; Fonagy, Gergely, Jurist, & Target, 2002; Priel & Besser, 2001; Shaver & Mikulincer, 2005). Other models informed by attachment and the social context of development also support the idea that long-lasting effects of trauma are at least partly reliant on the impact on socioemotional skills or their development (see Charuvastra & Cloitre, 2008).

Consistent with these theories, in a study of a highly traumatized, low-socioeconomic status population, adult attachment variable's relationships with self-reported PTSD symptoms were partially mediated by object relation constructs of models of self and others (Ortigo, Westen, DeFife, & Bradley, 2013). This collective work has helped bridge the gap between early psychodynamic conceptualizations of trauma and development with that of modern attachment theory and adult development models of self and other.

Moreover, the tie between attachment and emotion regulation skills has become more evident (Mikulincer & Shaver, 2007; Mikulincer, Shaver, & Horesh, 2006). For example, in a longitudinal study from adolescence to adulthood, attachment styles predicted the use of emotion-oriented social support seeking wherein more anxiously attached individuals used more emotion-oriented coping strategies than their avoidantly attached peers (Pascuzzo, Cyr, & Moss, 2013). Attachment's influences on affect regulation can differ in specificity depending on the overarching attachment style or dimension. One study demonstrated that greater levels of avoidant attachment were associated with different patterns (suppression or dysregulation) specific to the type of emotion, whereas anxious attachment showed a more general pattern of emotional dysregulation (Brenning & Braet, 2012). These influences on emotion regulation are another route of impact on recovery efforts after trauma exposure, likely important when the "natural" posttraumatic recovery process is stalled, contributing to the development of PTSD.

Narrative Therapy

With its multidisciplinary roots in philosophy, psychology, and sociology, theories about the function of narratives, including their therapeutic counterparts, offer a way to tie together the influential frameworks of attachment and resource loss with pragmatic, grounded interventions such as found in STAIR Narrative Therapy. Narrative theory centralizes and examines the creation of a subjective understanding of personal experiences alongside the social aspect of sharing them with another person (for one theoretical overview, see Herman, Phelan, Rabinowitz, Richardson, & Warhol, 2012). Often heavily informed by existential philosophy and therapy (Frankl, 1959/1984; Yalom, 1980), clinical applications of narrative theory often offer ways to integrate (1) the acknowledgment of the potential "lack of ground" (nonessentialism) in one's default subjective beliefs, perspective, and memories with (2) the exploration of key themes of meaning, loss, loneliness, mortality, and freedom (Angus & McLeod, 2004; Richert, 2010).

Although these theories certainly are complex and involve esoteric concepts, narrative therapy in the context of trauma-informed treatment is fundamentally about approaching the experience and memory of trauma openly and courageously to create a narrative (or "story") that acknowledges the trauma while also finding meaning from what happened and/or how one recovered from it. The importance of narrative "story telling" has also been crucial in understanding attachment. In fact, one key approach for its assessment in adulthood, the Adult Attachment Interview (George, Kaplan, & Main, 1996; Roisman et al., 2007), elicits a person's memories and understanding of their caregivers, close others, and themselves to help an expert assessor categorize the individual's attachment style. The hallmark of the most impaired adult attachment style (labeled unresolved, similar to disorganized attachment) is a severely fragmented narrative and understanding of self and others. Furthermore, in a very different application, narrative therapy elements are critical to one of the more promising intervention methods for end-of-life distress and palliative care—therapeutic life review (Keall, Clayton, & Butow, 2015). If something is shared in the process of healing from trauma and the process of accepting death, it may very well be the meaning making that can occur through the creation of a personal life narrative that openly confronts deep existential themes and conflicts (mortality, loss, control, connection, etc.).

Consistent with this notion, narrative-informed approaches have been integrated with exposure models that highlight the central role of anxiety and avoidance in posttraumatic phenomena. Narrative exposure therapies were originally developed and applied to treat PTSD in refugees with multiple traumas (Neuner, Schauer, Klaschik, Karunakara, & Elbert, 2004). Thus they have been used most often with non-Western populations. In comparing narrative exposure therapy and prolonged exposure (Foa et al., 2007), Mørkved et al. (2014) identified similarities in the theoretical basis of the approaches with one highlighted distinction—narrative exposure approaches also emphasize the integration of "cold spot" (less emotionally triggering) and not just "hot spot" (more emotionally triggering) memories within the broader narrative. The goal is the creation of a coherent life narrative fully informed by the totality of traumas experienced across the life span, and not just habituation to the "hot spots" of the index traumatic memory or memories. For this reason, Mørkved et al.'s review of empirical studies of both treatments concluded that narrative exposure approaches may compliment prolonged exposure's long-established efficacy in treating PTSD by offering an approach more tailored to people with multiple traumatic experiences.

In STAIR Narrative Therapy, the narrative work is flexibly applied to individuals with varying levels of trauma exposure. For people with multiple traumas or a mix of childhood and adulthood traumas, the narratives incorporate all relevant experiences into the therapeutic process of creating a coherent life story. For people with a clearly defined index traumatic event, the approach and goal are similar. Because the creation of a life story inherently involves experiences throughout the life span, it is also by default a developmentally informed approach. The impact of the timing and intensity of any trauma is important, consistent with other theories described earlier, and is weaved into the narrative. Even without childhood trauma, adults experiencing severe trauma later in life can find themselves confronting deep existential themes surrounding death, loneliness, meaning, and agency and responsibility. No matter the details of the specific trauma history, creating a narrative also incorporates an understanding of the working models of relating with which the individual has lived. In STAIR Narrative Therapy, the client tells about a variety of traumatic experiences that, in total, comprise their autobiography. Each story represents a chapter in the autobiography, and each chapter typically contains an investigation of the traumatic relationship model embedded within it and the client's evaluation and potential reformulation of it. Each story becomes a part of the work and ideally results in a meaning-making process that empowers the survivor of trauma to forge their own path forward.

An overview of STAIR Narrative Therapy

To address such diverse and complex experiences of survivors of trauma, STAIR Narrative Therapy combines two modules typically delivered over 16–20 sessions, with each module consisting of 8–10 sessions. The first module, STAIR, focuses on skill acquisition, development of insight around relationship patterns, and application of emotional regulation skills to interpersonal problems. STAIR can be presented as a stand-alone treatment or can be followed by Narrative Therapy. The Narrative Therapy module is designed to directly follow the completion of STAIR. This module focuses on the narration and processing of traumatic events, with a continued emphasis on applying skills to address issues in emotional and interpersonal functioning. This narration "allows the client to visit the past with the tools of the present, in the safety of the present, and with the companionship of an ally from the present" (Cloitre et al., 2006, p. 243). Table 1 provides a general overview of STAIR Narrative Therapy by summarizing session content and focus within a 21-session delivery of the treatment.

Table 1 Session-by-session overview of STAIR Narrative Therapy.

Session number and name	Example new concepts and skills	Description of session
Module one: Skills Training in Affective and Interpersonal Regulation (STAIR)		
1 Introduction	• Treatment overview • Emotion regulation • Focused breathing	Initiate therapeutic alliance. Psychoeducation on the impact of trauma on emotions and emotional regulation introduced
2 Emotional awareness	• Three channels of emotion model • Emotional intensity scale • Feelings wheel and list • Emotion surfing • Feelings Monitoring Form	Psychoeducation on the function of emotions and importance of emotional awareness. Feeling Monitoring Form introduced for daily practice, alongside an example
3 Emotion regulation: focus on the body	• Self-care health plan • Sensory soothing • Progressive muscle relaxation	Focus on healthy and unhealthy efforts to cope along the body channel that are common in trauma survivors. Client makes a physical health self-care commitment for the week. Daily monitoring of feelings and skill practice continues
4 Emotion regulation: focus on thoughts and behavior	• Evidence technique • Positive imagery • Affirmations • Time out • Alternative action • Seeking support • Pleasurable activities	Focus on healthy and unhealthy efforts to cope along the thought and behavior channels. Skills for these channels introduced as relevant to client presentation, needs, and preferences. Daily monitoring of feelings and skill practice continues
5 Emotionally engaged living; distress tolerance	• Distress tolerance • Pros and cons	Distress tolerance introduced as an essential skill and option to encourage engaged and/or value-centered living. Client applies skills learned thus far toward one of the client's current goals

Continued

Table 1 Session-by-session overview of STAIR Narrative Therapy—cont'd

Session number and name	Example new concepts and skills	Description of session
6 Understanding relationship patterns	• Relationship models • Relationship patterns worksheet 1	Relationship models and patterns are introduced. Focus on healthy and unhealthy expectations in relationships. The client explores common expectations that survivors of trauma often share
7 Changing patterns: increasing assertiveness	• Basic personal rights • "I" messages • Relationship patterns worksheet 2	Psychoeducation on different communication styles and basic interpersonal rights. Practice assertiveness with explicit or covert roleplay exercise. Alternative relationship patterns introduced
8 Changing patterns: managing power	• Types of power balances • Flexibility • Respect bookends	Psychoeducation about power dynamics and impact of trauma on navigating power. Therapist explores current conflicts in client's life related to power, as relevant, and how to approach situations flexibly depending on the nature of the relationship (e.g., one's boss vs one's child)
9 Changing patterns: increasing closeness and intimacy	• Types of boundaries • Relationship guidelines	Focus on healthy and unhealthy boundaries in relationships. Client identifies a person with whom they would like to increase their sense of intimacy. Explore explicit strategies, such as repairing a relationship after an argument and beginning a new relationship
10 Summary of work and self-compassion	• Self-compassion • Summary of accomplishments • Preventing and responding to relapses • Planning next steps	Reflect on role of self-compassion in recovery. Review treatment progress, including therapist and client processing thoughts and feelings around transition from skills training. Discuss either termination (if not continuing to second module) or movement to narrative therapy module

Module two: narrative therapy

#	Topic	Content	Description
11	Introduction to narrative work	• Memory hierarchy • Skills practice (continued throughout narrative module) • Relationship patterns worksheet 2 (continued throughout narrative module)	Introduce rationale for narrative work, as a means of habituating fear response, emotionally processing memories, and making meaning. Up to five traumatic events selected for focus in second module
12	First memory narrative	• Subjective Units of Distress Scale (SUDS) for trauma narration • Narrative of a neutral memory (as a practice example) • Narrative processing of most distressing trauma memory	Introduce process for narrative work with a neutral example. Establish anchors for SUDS to track reaction to trauma memory narration. Client chooses a trauma memory to narrate. Explore meaning of narrative and relevant relationship models. Listening to audio recording assigned as daily practice
13–19	Continued narrative work	• Narrative processing of trauma memories • Selected relevant trauma themes: fear, shame, loss	Narrative work continues, moving to new memories as habituation is achieved. Depending on presentation, therapist may utilize probing questions or sharpen focus on "hot spots" in narratives. Continue exploring healthier relationship models
20	Closure	• Reflection and review of progress • Relapse prevention	Review of progress. Review relapse prevention strategies and identify next steps to maintain progress. Allow time for additional reflections and sense of closure

Module one: STAIR

The STAIR module follows a format that generally involves *gaining insight* around emotions and relationships; *acquiring skills* relevant for regulating these emotions, tolerating distress, and relationships; and then *applying skills* to reach valued goals. The first half of STAIR focuses on emotions, whereas the second focuses on relationships.

Emotion awareness and regulation skills (sessions 1–5)

The first two sessions introduce emotions as having a purpose and being functional in nature and being developed through social contexts. Emotional experiences have three components (what we call the three channels of emotion): *body*, *thought*, and *behavior* (see Fig. 1). The initial focus is on validating emotions and acknowledging how affective habits are formed through early-life experiences and impacted by traumatic events. The client and therapist then take a balanced approach to exploring the adaptive benefits and long-term consequences of one's emotional reactions and patterns.

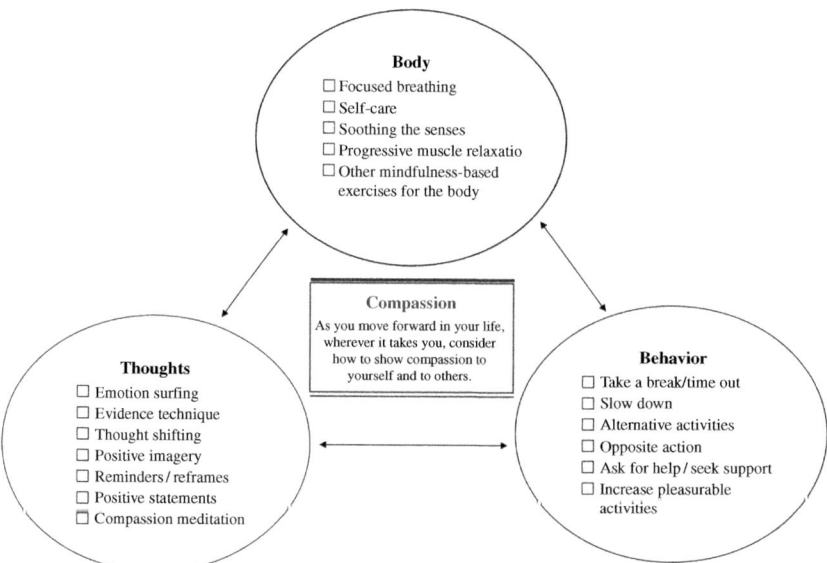

Fig. 1 The STAIR Three Channels of Emotion model and related skills. *Shared here with permission from Cloitre, M., Cohen, L. R., Ortigo, K. M., Koenen, K., & Jackson, C. (2020). Treating survivors of childhood abuse: Psychotherapy for the interrupted life (2nd ed.). New York, NY: Guilford Press. Permission to photocopy this box is granted only to purchasers of that book for personal use only (see The Guilford Press copyright page for details).*

The three channels of emotion model is presented and discussed in the introductory sessions as a foundational concept for the treatment. Fig. 1 summarizes the channels and their corresponding skills (which STAIR introduces systematically but not all at once). The model is a simple, easily understood formulation that has several functions, including the following: (1) to help organize emotional experiences, (2) to provide a rationale for solutions (like skills), and (3) to provide an easily remembered formulation and checklist for organizing solutions to emotional difficulties after treatment ends. The use of the three channels to organize experience is of substantial value to individuals who often feel overwhelmed by their emotions and by life. The three-channel formulation provides a way to organize what feels ungovernable to the client. The channels also create a direct link from problems to solutions and, as seen in Fig. 1, provide a summary of the interventions from which the client may choose. Lastly, toward the end of treatment, the client and therapist create a plan that identifies the skills the client found most helpful. A completed summary worksheet serves as a reminder to maintain practice and as a reference to which to refer for solutions in times of stress.

As part of the stepwise introduction of skills, sessions 3 and 4 involve the client learning a variety of skills for regulating emotions along each of the three channels. Introduced in session 2, the Feelings Monitoring Form provides a parallel tool to skills practice to further develop the client's insight around their emotional responses to present-day situations. This form, a modified and expanded trigger-response-consequence worksheet, helps the client uncover their triggers, evaluate the effectiveness of their current attempts to cope, and assist them in instilling new skills for the healthy management of their emotions. The essential components of the Feelings Monitoring Form include the triggering situation, the client's emotional response, intensity, and duration of feelings, any associated thoughts or beliefs, the chosen behavioral response (including whether skills were used), the resulting emotional intensity afterward, and the overall effectiveness of the strategy and response.

In session 5, the client explores the concept of *distress tolerance* and how to apply newly developed emotional regulation skills as a means to move toward their personal goals. This focus on goals works to empower the client and increase their sense of agency. The client's willingness to tolerate unavoidable, necessary distress when working toward their goals reinforces a key and possibly novel concept. That is, skills not only help a person to manage feelings about a painful past or difficult present for the sake of "just surviving" but also work in planning for a more positive future, a future created by setting meaningful goals, developing an effective strategy for anticipated distress, and executing the plan. As the client gains more experience with this process, they develop an increasing sense of self-efficacy and agency. At this stage in treatment, the client begins to look for more advanced ways to use emotional coping skills; rather than merely taking a reactionary stance to triggering events, they are now also encouraged to take preventative measures to address barriers in their life and relationships. Distress tolerance work prepares the client for the second half of the protocol, where the focus shifts to identifying relationship patterns and developing communication skills to increase interpersonal competence.

Relationship patterns and interpersonal skills (sessions 6–10)
The second half of STAIR shifts to applying the client's newfound emotion regulation skills to the work of understanding relationship patterns and making more conscious, skillful choices to improving them. In session 6, the client is invited to explore and identify habitual ways they interact with others and to distill the core underlying "working models of relationships" that are driving these interactions. The underlying patterns common to survivors of interpersonal trauma often relate to themes of safety, trust, power/control, and self-esteem. As with the identification of ineffective emotional coping efforts, the emphasis continues on both (1) validation of how these patterns developed and were likely adaptive in early developmental contexts and (2) acknowledgement of the importance of letting go of some old working models that may cause more harm than good in the client's current environment. At minimum the client may consider adding new working models to their repertoire.

Introduced in session 6, the Relationship Patterns Worksheet 1 lays the foundation for much of this interpersonal work. It helps the client log interpersonal difficulties and identifies the underlying relationship model. The structure of a worksheet builds on the format of the Feelings Monitoring Form. The client is now asked to describe not only their own thoughts and feelings but also their expectations or assumptions about the feelings and thoughts of the person with whom they are interacting. The Relationship Patterns Worksheet 1 later becomes the first half of the Relationship Patterns Worksheet 2, which is described later and presented in Fig. 2.

Sessions 7, 8, and 9 build on the client's growing insight and aims to increase the client's interpersonal repertoire by elucidating key interpersonal themes often relevant to survivors of trauma. Session 7 explores theme of

Relationship patterns worksheet 2

Interpersonal Situation	What did I feel and think about myself?		What were my expectations about the other person?		My resulting behavior
What happened?	**My Feelings**	**My Thoughts**	**Their Feelings**	**Their Thoughts**	**What did I do?**
My wife sat down next to me on the couch. I thought about telling her about my assault.	Anxiety Fear Shame	She'll never understand. If she knew what happened, she'd lose all respect for me.	Disgusted by what I shared Ashamed for me	How did she let that happen? Was she flirting with him? I can't think of Aisha like I did before.	Tensed up. Told her I wasn't in the mood to talk. Avoided her for the rest of the night.

Old model: WHEN I'm open with a loved one about something bad that happened to me THEN they'll be disgusted and ashamed of me and I'll get hurt.

Interpersonal goals for situation	Alternative beliefs and feelings about myself		Alternative beliefs and feelings about the other person?		Resulting actions
What are my goals in this situation?	*What else could I feel and believe about myself?*		*What else could I expect the other person to feel and think?*		
	My feelings:	*My thoughts:*	*Their feelings:*	*Their thoughts:*	*What else could I do?* *What else might they do?*
Tell my wife about the assault AND feel closer to her.	Love A little vulnerable Trust	My wife is a good person and will support me. She'll appreciate it if I share more with her.	Sad Angry at the guy who assaulted me Concerned Love Supportive	That was a horrible thing she had to go through. I want to know what's going on in her life. I will support her through this.	Tell her I love her. Tell her a little bit about what happened to me. Ask her to hug me. She could show she cares and understands.

Alternative model: WHEN I'm open with a loved one about something bad that happened to me THEN they'll appreciate my openness and feel closer to me.

Fig. 2 Example of completed Relationship Patterns Worksheet 2.

assertiveness, including effective communication strategies for making requests and sharing one's experience, and the concept of boundaries. It also introduces the Relationship Patterns Worksheet 2, an expanded version of the previous form (see Fig. 2), which allows the client to explore alternative, more helpful variations of their relationship patterns. Session 8 explores the role of power dynamics and how to balance elements of power in choosing strategies for communication. This theme also allows the opportunity to reflect on how different relationships require different approaches and styles of communication, that is, strategic flexibility in choosing how to respond. Session 9 explores the theme of emotional intimacy in one's relationships, whether increasing intimacy in existing, formerly closer relationships (e.g., repairing ruptures) or developing new relationships where intimacy needs to be fostered thoughtfully and at a reasonable pace.

Finally, session 10 emphasizes the importance of self-compassion in the recovery journey and how recovery continues through ongoing, more independent skills practice and refinement over time. The client learns that conscious efforts to show compassion to self and others support continued growth, even outside of the structure of therapy. Although first formally introduced in session 10, compassion is a concept experienced and modeled throughout the implementation of STAIR. Thus the client's experiences with their therapist prime them for the explicit introduction of self-compassion as an organizing principle moving forward.

As previously mentioned, STAIR can be used as a stand-alone skills treatment or as a precursor to Narrative Therapy. In the final session of STAIR, the client reviews their progress, skills learned, and steps forward. Regardless of whether the client completes treatment this session or moves onto the next module, the therapist emphasizes the need for continued practice of skills while having self-compassion in one's recovery journey. If the client will be continuing to the narrative module, this session is also when the therapist discusses the transition to narrative work.

Case example part 1: Response to the STAIR module

To illustrate the flow of treatment in this first module of STAIR, we introduce a case example to which we will return throughout the presentation of the treatment content.[a] Aisha is a 34-year-old married, lesbian-identified, African American woman who worked as a nurse for a large community

---

[a]Note that this case is a composite of real STAIR Narrative Therapy clinical cases, but details have been changed to protect the confidentiality of the individuals involved.

hospital and received a referral by her primary care physician after expressing concerns about ongoing relationship problems with her partner and at work. These problems began after she was sexually assaulted by a male coworker at her former place of work approximately 1 year prior. The assault occurred outside of work while having dinner with her perpetrator, who she had previously considered to be a close friend. She did not disclose this event to anyone, including her partner, and transferred to a new position in another hospital a couple months after the assault. She also reported a long-standing history of childhood physical abuse by her father, from whom she had been estranged since her late teens.

Aisha reported increasing depression and feelings of disconnection from her partner. She had also noticed an increase in her irritability and difficulty making friends at her current place of employment, where she had been working for almost a year. She recently stopped responding to her mother's phone calls, even though she used to speak to her mother on a weekly basis. She stated that speaking with her mother began bringing up memories of her physical abuse, which further led to strong and confusing emotional reactions, predominantly anger and sadness. She expressed anger with herself for creating distance with her mother. Aisha had no prior mental health treatment history. She was only open to meeting with a therapist because of her medical provider's recommendation.

The initial sessions of treatment allowed Aisha to notice more fully the range of emotions she was experiencing and to increase her vocabulary for affective experiencing. Labeling her feelings proved crucial to understanding the fear, anxiety, and shame that arose when she was with her wife, particularly during physical intimacy, as well as her irritability at work. When she entered therapy, she had only described these feelings as "being upset." Aisha then started using coping techniques across the three channels of emotion. Via the body channel, she practiced somatic approaches such as focused breathing to manage her irritability in the workplace. Via the behavior channel, she reengaged in hobbies such as riding her bicycle with her wife on the weekends, which served both to increase her physical exercise and reinforce an important source of social support and emotional connection. Via the thought channel, she began repeating positive affirmations regularly when she noticed herself becoming on edge at work, particularly around male colleagues. One of her most helpful affirmations was, "My feelings are just trying to keep me safe." This mantra helped her to accept and normalize her own reactions, without denying or harshly judging them, which in turn allowed her to refocus on any important tasks at hand.

In discussing relationship patterns, Aisha recognized that she had a history of distrusting people, which led to increasing emotional distance. Aisha and the therapist focused on the relationship with Aisha's wife. She realized that she was creating distance with her wife because she was fearful of what her wife's reaction would be if she really knew what had happened to her. The therapist guided Aisha through the top of the Relationships Patterns Worksheet 2 to identify Aisha's feelings and thoughts when she imagined telling her wife about what had happened, as well what she expected her wife's reaction (thoughts and feelings) to be (see Fig. 2). The completion of the form was extremely useful in helping identifying fearful feelings and thoughts Aisha was holding that she had not articulated. In addition, the work helped Aisha distill down the relationship model that seemed to be driving her feelings, beliefs, and actions: "If I'm open with a loved one about something bad that happened to me, then they'll be disgusted and ashamed of me and I'll get hurt." The therapist gently explored with Aisha whether and how much this relationship model was applicable to her wife. Aisha said she wasn't sure. The therapist then asked Aisha to identify an alternative relationship model she could build based on her knowledge of her wife and also on the kind of relationship she aspired to have with her wife. Aisha struggled with this and could not come up with an alternative. The therapist proposed that it was possible that she could both disclose the event *and* stay close with her wife.

Aisha liked this idea, and they worked together to align thoughts and feelings that Aisha could bring into the specific situation of disclosing her assault to maximize an outcome aligned with this aspiration. The alternative relationship model they constructed was "When I'm open with a loved one about something bad that happened to me, then they'll appreciate my openness and feel closer to me." They reviewed the kinds of thoughts and feelings Aisha could bring to the situation (e.g., love, vulnerability, and trust) and those she could imagine her wife might have (e.g., concern and support) that would allow the interaction to go well. After this session and with a few days of reflection, Aisha decided she wanted to share with her wife what had happened. In the next session, she practiced with her therapist exactly what she was going to say. She did not want to give details of the assault for fear that she would not get through the disclosure. The therapist helped her plan what she would say, and while practicing, Aisha's initial reluctance started to soften, and she gained confidence that she could act from a place that was consistent with the alternative model: that her wife would appreciate her openness and *not* be ashamed of her. While still difficult, Aisha disclosed the assault to her wife, who reacted empathically and then with anger toward the perpetrator rather than feeling disgusted or ashamed of Aisha. Her wife's expressed compassionate concern, warmth, and acceptance reinforced Aisha's choice to risk greater intimacy. The two hugged afterward, which was the first time in over a year that Aisha had felt safe in a physical embrace and allowed the compassion of her wife to be felt. The disclosure provided a corrective emotional experience supporting her new relationship model, as well as her broader commitment to healing.

Module two: Narrative Therapy

Narrative Therapy is a trauma-processing intervention that continues to build upon skills acquired in STAIR and to draw upon and expand the relationship patterns work to aid in reorganizing trauma-related memories and relationship models. The ultimate goal of this narrative approach is to help the client form an integrated life story that acknowledges the trauma(s) but serves to empower the client to write their own narrative for the future, one not controlled by the ills of the past.

In the first narrative session, the therapist connects the work completed in STAIR to the narrative intervention, emphasizing that processing traumatic memories allows the client to make contact with and appropriately work through emotions, leading to a decrease in the intensity of anxiety and fear. As the client learns to tolerate distress around discussing their trauma, the fear responses lessen through habituation. Another overarching goal is to explore the beliefs about self and others associated with the trauma itself, effectively expanding upon and deepening the work completed in the latter half of the skills training module. The therapist and client identify up to five traumas to review in the narrative work and collaboratively create a memory hierarchy. This hierarchy is based on expectations about the subjective distress each memory elicits at this point in treatment. It also establishes the order of processing, from most to least distressing traumatic memory.

In the subsequent sessions, the client begins to narrate memories aloud. The process is similar to prolonged exposure's imaginal exposure (Foa et al., 2007) in that the client completes the narrative work with eyes closed, focusing on reciting the memory in the present tense, in a vivid, detailed, moment-by-moment manner. Following completion of the narrative, the client and therapist work together on forming a narrative of how the traumatic experience connects to current relationship patterns, using the Relationship Patterns Worksheet 2 as introduced in STAIR sessions 7–10. This structured worksheet helps further clarify trauma-generated models and contrast them with current interpersonal goals. The therapist guides the client, in particular, to explore themes of fear, shame, and loss that may be contained in the narrative and then move toward healthier, more adaptive patterns of thought and behavior surrounding these themes. The approach is intentionally flexible and individualized to fit the needs, goals, and specific relationship patterns of the client.

Alongside the important and challenging trauma-processing work, the therapist continues to model and incorporate earlier relevant themes of compassion, flexibility, and agency, especially as they relate to the client's needs and personal goals. The client's process of internalizing these healthy attitudes parallels and complements their more explicitly manifested habituation to anxiety and fear as the narrative exposure work and meaning making deepens.

Case example part 2: Response to the Narrative Therapy module

In the case of Aisha, the narrative module came at a time when her skills were becoming more solidified and her experiences of compassion (for herself and from others) were blossoming, yet still new. The narrative work began by identifying several traumatic memories throughout her life span. In addition to the recent sexual assault by her former colleague, she also identified childhood memories in which she had been beaten by her father. As Aisha completed the narrative about her sexual assault, her feelings of fear reduced, but her feelings of shame remained and feelings of anger actually increased. The narrative about her physical abuse, however, led to much more insight about the source of her shame and anger.

As Aisha narrated the stories of her childhood abuse, she told how her mother had witnessed these events and had not intervened. During the appraisal of the meaning of the narrative, Aisha realized the reason she felt so much shame about herself was because her mother's failure to protect her made her feel that she was to blame. The recent assault had stirred up those feelings of shame and new feelings of anger toward her mother. She was able to complete the childhood abuse narrative by recalling that her mother did leave her father and took Aisha with her, which is how the beatings ended. As the reappraisal work progressed, Aisha was able to reconnect with her mother by accepting both her mother's part of the responsibility for the past and her still-present love for her mother despite her flaws.

The resolution of the shame she felt from her childhood trauma and the movement to more complex views of her mother reverberated in other areas, such as her growing compassion for herself and her own mistakes. The reevaluation she had made of herself during the narrative work allowed her to commit with more confidence to exploring the new relationship model "When I'm open with a loved one about something bad that happened to me, then they'll appreciate my openness and feel closer to me." She realized that sometimes when bad things happened, they were not her fault. She further realized that even when bad things happened that could be attributed to her, she did not really have to feel ashamed.

Aisha began sharing things with her wife that were embarrassing or did not put her in the best light (e.g., forgetting to pick up her wife's laundry on the way home and accidently oversleeping and being late to a meeting). She referenced the new relationship model as a guide for her interactions. She reported using STAIR skills, such as focused breathing and weighing pros and cons, to deal with the disclosures that she would have avoided in the past. For example, the positive side of telling her wife about forgetting the laundry was that doing so reflected a more authentic and open relationship with her wife; the negative was that her wife sometimes expressed irritation toward her. However, even here Aisha learned an important lesson: that her wife could be irritated at her and still love her and want to be in a relationship.

Evidence and applications of STAIR Narrative Therapy

The efficacy of STAIR Narrative Therapy and its variants has been assessed in three randomized controlled trials: two of STAIR Narrative Therapy in community samples and one of STAIR alone in a veteran sample. The first randomized controlled trial evaluated women with PTSD related to a history of childhood abuse and compared STAIR Narrative Therapy with a waitlist control (Cloitre, Koenen, Cohen, & Han, 2002). Results indicated that STAIR Narrative Therapy participants demonstrated significant improvements in PTSD symptoms, affect regulation, interpersonal problems, perceived social support, and overall functioning in family, work, and social domains; gains were maintained at 3- and 9-month follow-up periods.

In a second randomized controlled trial (RCT), Cloitre et al. (2010) conducted a component study and assessed STAIR Narrative Therapy as compared with two control conditions in which each active component was eliminated and replaced with an active nonspecific treatment. Results indicated that compared with the condition without STAIR (supportive counseling followed by Narrative Therapy), participants in STAIR

Narrative Therapy were more likely to achieve sustained and full remission of PTSD and demonstrated greater improvements in emotion regulation and interpersonal functioning, particularly at the 3- and 6-month follow-up assessments. This finding suggests that learning STAIR skills helped participants manage their lives and gain mastery over time. In a complementary finding, a functional magnetic resonance imaging (fMRI) evaluation of women with childhood trauma-related PTSD who completed STAIR Narrative Therapy found that treatment response was associated with reduced amygdala activation and increased activation of the ventral medial prefrontal cortex suggesting that posttreatment improvements were associated with greater emotion regulation with regard to fear responses (Brown et al., 2011).

The latest RCT evaluated the use of a five-session STAIR for Primary Care (STAIR-PC) delivered to veterans with PTSD symptoms and a range of traumas seeking care in a primary care setting. Given its brevity, the treatment goals focused on quickly identifying the most appropriate tools or skills for specific problems for which the veteran sought treatment. The therapist provided veterans an at-a-glance worksheet that identified problems (depression, anxiety, fear, insomnia, panic attacks, and social isolation) and matched them to at least two empirically based skill interventions from the STAIR emotion regulation sessions. The selection of skills was determined collaboratively by therapist and client. Compared with treatment as usual in primary care, STAIR-PC was found to reduce PTSD symptoms, emotion regulation problems, and social engagement difficulties (Cloitre, Gimeno, Ortigo, Weiss, & Jain, 2018).

Another study used a "benchmark design" to evaluate STAIR Narrative Therapy with survivors of the 9/11 World Trade Center terrorist attack (Levitt et al., 2007). In this design, STAIR Narrative Therapy was presented in a flexible format of 12–25 sessions, and study therapists were allowed to skip or repeat sessions depending on the needs and presentation of the client, which reflected a more individualized, flexible, and real-world application of evidence-based interventions. Results indicated significant improvements in PTSD, depression, and interpersonal problems, with equivalent effect sizes to those demonstrated in the first RCT presented earlier (Cloitre et al., 2002). This study also demonstrated the utility of STAIR Narrative Therapy for adult-onset trauma and not just childhood trauma as was the focus of the first RCTs for the intervention. In addition, it was the first study expanding the sample beyond women, and outcomes were equally good across genders.

Efforts to adapt the treatment have also expanded to adjusting the individual therapy format. Group STAIR (without the narrative component) has been investigated in three studies, one comparative study and two open trials. In a comparison study within an adult inpatient setting targeting comorbid schizoaffective disorders, Trappler and Newville (2007) found that clients who received the STAIR group treatment showed significant improvement on measures of PTSD and psychotic symptoms, as well as improvements in affect and emotional expression and management, in contrast to a group treatment as usual condition.

Two other open trial studies evaluated the effectiveness of Group STAIR in very different populations, one a community sample and the other a veteran sample, both with good outcomes. The community study evaluated the effectiveness of group STAIR with limited training (workshop and occasional supervision) among providers in an outpatient program for survivors of childhood trauma (MacIntosh, Cloitre, Kortis, Peck, & Weiss, 2016). Qualitative outcomes provided reports of high ease of delivery among members of the mental health team and high satisfaction among the clients. Significant reductions in PTSD, emotion dysregulation, dissociation, and interpersonal problems were observed. The second study evaluated Group STAIR in a VA PTSD clinic in which male and female veterans (including those with sexual trauma) participated together (Jackson, Weiss, & Cloitre, 2019). Both statistically and clinically significant reductions were obtained in PTSD and general distress symptoms, suggesting not only the benefits of STAIR but also the option of providing mixed-gender groups.

Lastly, two studies have evaluated Group STAIR adapted for trauma-exposed adolescents (STAIR-A), one a comparison study and the other an open trial. In the comparison study, STAIR-A was evaluated among girls in a low-socioeconomic status, public high school and compared with an assessment-only control group matched on age, trauma history, and baseline symptoms (Gudino, Leonard, Stiles, Havens, & Cloitre, 2017). Results indicated that STAIR-A provided significant improvements in depression and management of social stress, in comparison with the assessment-only group. There was little change in PTSD scores across conditions; however, baseline scores were relatively low, which may have impacted the ability to observe significant changes. In contrast, in an open trial of adolescents in a low-income inpatient setting, significant reductions in PTSD and depression symptoms were demonstrated (Gudino, Leonard, & Cloitre, 2015).

Future directions and adaptations of STAIR Narrative Therapy

Since the initial development of STAIR Narrative Therapy, there has been an increased and renewed focus on matching treatments to the individual needs of clients. While some individuals struggling with posttraumatic stress symptoms may benefit from in-depth processing of traumatic memories, others may be more concerned with the consequent disruption of socio-emotional engagement and prefer an approach that focuses on present-day rehabilitation of emotional and interpersonal skills. One direction for future research is to study the efficacy of STAIR as a stand-alone treatment, without the narrative therapy element, with a more active comparison intervention. It may also be beneficial to explore the benefits of matching treatments based on a client's current presenting concerns and individual preferences.

The skill-based approach within STAIR is highly flexible and adaptable to fit the needs of a variety of settings and clinical presentations. For example, an abbreviated format will likely better suit the needs of a primary care setting seeing more subthreshold symptoms, while a telehealth format may meet the needs of rural populations and others that have barriers to accessing in-person mental healthcare. In addition, self-directed formats through the use of web or mobile applications will increase the reach and accessibility of resources. Such technological adaptations of STAIR, which are currently live or in development, can also be bolstered by the use of trained coaches who can help reinforce engagement and application of skills to an individual's specific life circumstances and needs. STAIR also allows thoughtful adaptation to meet the unique presenting concerns of specific diverse groups, such as veterans, women, adolescents, parents, or LGBTQ people—just to highlight a few examples. Future research will continue to explore, expand, and evaluate the varied applications of STAIR and STAIR Narrative Therapy. All of this work, though, will continue to emphasize the importance of encouraging person-centered care and framing recovery from a developmentally informed, compassionate lens.

Acknowledgments

Writing of this chapter was supported by the VA Office of Rural Health Enterprise-Wide Initiative funding for webSTAIR implementation and Connecting Women to Care Health Services Research and Development (HSR&D) grant (#IIR 16-070).

Disclaimer: The views and opinions of authors expressed herein do not necessarily reflect those of the Department of Veterans Affairs.

References

Adams, J. R., & Drake, R. E. (2006). Shared decision-making and evidence-based practice. *Community Mental Health Journal, 42*(1), 87–105. https://doi.org/10.1007/s10597-005-9005-8.

Ainsworth, M. D. S., & Bell, S. M. (1970). Attachment, separation, and exploration: Illustrated by the behavior of one-year-olds in a strange situation. *Child Development, 41*, 49–67.

Ainsworth, M. D. S., & Wittig, B. A. (1969). Attachment and exploratory behavior of one-year-old children in a strange situation. *Determinants of Infant Behavior, 4*, 111–136.

American Psychiatric Association. (1980). *Diagnostic and statistical manual of mental disorders* (3rd ed.). Washington, DC: American Psychiatric Association.

Angus, L. E. & McLeod, J. (Eds.), (2004). *The handbook of narrative and psychotherapy: Practice, theory and research.* Thousand Oaks, CA: Sage.

Baer, J. C., & Martinez, C. D. (2006). Child maltreatment and insecure attachment: A meta-analysis. *Journal of Reproductive and Infant Psychology, 24*(3), 187–197.

Bartholomew, K., & Horowitz, L. M. (1991). Attachment styles among young adults: A test of a four-category model. *Journal of Personality and Social Psychology, 61*, 226.

Bonanno, G., Çolak, D., Keltner, D., Shiota, M. N., Papa, A., Noll, J. G., … Trickett, P. K. (2007). Context matters: The benefits and costs of expressing positive emotion among survivors of childhood sexual abuse. *Emotion,* 7(4), 824–837.

Bowlby, J. (1969). *Attachment. Vol. 1.* New York: Hogarth.

Bowlby, J. (1973). *Separation: Anxiety and anger. Vol. 2.* London: Hogarth Press.

Bowlby, J. (1980). *Loss: Sadness and depression. Vol. 3.* New York: Basic Books.

Bradley, R., Greene, J., Russ, E., Dutra, L., & Westen, D. (2005). A multidimensional meta-analysis of psychotherapy for PTSD. *American Journal of Psychiatry, 162*, 214–227.

Brenning, K. M., & Braet, C. (2012). The emotion regulation model of attachment: An emotion-specific approach. *Personal Relationships, 20*(1), 107–123. https://doi.org/10.1111/j.1475-6811.2012.01399.x.

Brown, A. D., Root, J. C., Cloitre, M., Perez, D., Teuscher, O., Pan, H., & Stern, E. (2011). *Changes in fear reactivity in response to STAIR/NST: A preliminary analysis.* In: Paper presented at the symposium at the annual conference International Society for Traumatic Stress Society, Baltimore.

Calabrese, M. L., Farber, B., & Westen, D. (2003). *Implicit versus explicit internal working models: Integrating attachment and object relations to representations of self, others, and relationships.* Columbia University. Unpublished Manuscript.

Calabrese, M. L., Farber, B. A., & Westen, D. (2005). The relationship of adult attachment constructs to object relational patterns of representing self and others. *The Journal of the American Academy of Psychoanalysis and Dynamic Psychiatry, 33*, 513–530. https://doi.org/10.1521/jaap.2005.33.3.513.

Carlson, E. A. (1998). A prospective longitudinal study of attachment disorganization/disorientation. *Child Development, 69*, 1107–1128.

Charuvastra, A., & Cloitre, M. (2008). Social bonds and posttraumatic stress disorder. *Annual Review of Psychology, 59*, 301–328. https://doi.org/10.1146/annurev.psych.58.110405.085650.

Cloitre, M. (2015). The "one size fits all" approach to trauma treatment: Should we be satisfied? *European Journal of Psychotraumatology, 6*, 27344. https://doi.org/10.3402/ejpt.v6.27344.

Cloitre, M., Cohen, L. R., & Koenen, K. C. (2006). *Treating survivors of childhood abuse: Psychotherapy for the interrupted life.* New York, NY: Guilford Press.

Cloitre, M., Cohen, L. R., Ortigo, K. M., Koenen, K., & Jackson, C. (2020). *Treating survivors of childhood abuse: Psychotherapy for the interrupted life* (2nd ed.). New York, NY: Guilford Press.

Cloitre, M., Gimeno, J., Ortigo, K. M., Weiss, B., & Jain, S. (2018). *STAIR as a stand-alone treatment: Results from a randomized controlled trial.* In: *Presented in Symposium (Chair: C. Rosen), What if we don't talk about trauma? Evidence-based alternatives to trauma-focused psychotherapy, at the annual conference of the International Society of Traumatic Stress Studies, Washington, DC.*

Cloitre, M., Koenen, K. C., Cohen, L. R., & Han, H. (2002). Skills Training In Affective and Interpersonal Regulation followed by exposure: A module-based treatment for PTSD related to child abuse. *Journal of Consulting and Clinical Psychology, 70*(5), 1067–1074.

Cloitre, M., Miranda, R., Stovall-McClough, K. C., & Han, H. (2005). Beyond PTSD: Emotion regulation and interpersonal problems as predictors of functional impairment in survivors of childhood abuse. *Behavior Therapy, 36*, 119–124.

Cloitre, M., Stovall-McClough, K. C., Nooner, K., Zorbas, P., Cherry, S., Jackson, C. L., … Petkova, E. (2010). Treatment for PTSD related to childhood abuse: A randomized controlled trial. *American Journal of Psychiatry, 167*(8), 915–924.

Cook, A., Spinazzola, J., Ford, J., Lanktree, C., Blaustein, M., Cloitre, M., … van der Kolk, B. (2005). Complex trauma in children and adolescents. *Psychiatric Annals, 35*(5), 390–398. https://doi.org/10.3928/00485713-20050501-05.

Foa, E. B., Hembree, E. A., & Rothbaum, B. O. (2007). *Prolonged exposure therapy for PTSD: Emotional processing of traumatic experiences, Therapist guide*. New York, NY: Oxford University Press.

Fonagy, P., Gergely, G., Jurist, E. L., & Target, M. (2002). *Affect regulation, mentalization, and the development of the self*. New York, NY: Other Press.

Fraley, R. C., Waller, N. G., & Brennan, K. A. (2000). An item response theory analysis of self-report measures of adult attachment. *Journal of Personality and Social Psychology, 78*, 350.

Frankl, V. E. (1959/1984). *Man's search for meaning*. New York, NY: Pocket Books (Simon & Schuster).

Freud, S. (1920/1989). Fixation to traumas—The unconscious. [J. Strachey, Trans.]. In Strachey, J. (Ed.), *Introductory lectures on psycho-analysis*. (Standard edition, pp. 338–353). New York, NY: W. W. Norton & Company.

Gapen, M., Cross, D., Ortigo, K. M., Graham, A., Johnson, E., Evces, M., … Bradley, B. (2011). Perceived neighborhood disorder, community cohesion, and PTSD symptoms among low-income African Americans in an urban health setting. *American Journal of Orthopsychiatry, 81*(1), 31–37. https://doi.org/10.1111/j.1939-0025.2010.01069.x.

George, C., Kaplan, N., & Main, M. (1996). *Adult attachment interview* (3rd ed.). Berkeley: University of California. [Unpublished Manuscript].

Gudino, O. G., Leonard, S., & Cloitre, M. (2015). STAIR-A for girls: A pilot study of a skills-based group for traumatized youth in an urban school setting. *Journal of Child & Adolescent Trauma, 9*(1), 67–79.

Gudino, O. G., Leonard, S., Stiles, A. A., Havens, J. F., & Cloitre, M. (2017). Skills Training in Affective and Interpersonal Regulation for Adolescents (STAIR-A). In M. Landolt, U. Schnyder, & M. Cloitre (Eds.), *Evidence-based treatments for trauma-related disorders in children and adolescents*. New York: Springer.

Herman, D., Phelan, J., Rabinowitz, P., Richardson, B., & Warhol, R. (2012). *Narrative theory: Core concepts and critical debates*. Columbus, OH: Ohio State University Press

Hesse, E., & Van IJzendoorn, M. H. (1998). Parental loss of close family members and propensities towards absorption in offspring. *Developmental Science, 1*(2), 299–305.

Hobfoll, S. E. (2001). The influence of culture, community, and the nested-self in the stress process: Advancing Conservation of Resources theory. *Applied Psychology, 50*(3), 337–421. https://doi.org/10.1111/1464-0597.00062.

Hobfoll, S. E., Mancini, A. D., Hall, B. J., Canetti, D., & Bonanno, G. A. (2011). The limits of resilience: Distress following chronic political violence among Palestinians. *Social Science & Medicine, 72*(8), 1400–1408. https://doi.org/10.1016/j.socscimed.2011.02.022.

International Society for Traumatic Stress Studies. (2018). *ISTSS PTSD prevention and treatment guidelines methodology and recommendations*. Oakbrook Terrace, IL: International Society for Traumatic Stress Studies.

Jackson, C., Weiss, B. J., & Cloitre, M. (2019). STAIR Group treatment for veterans with PTSD: Efficacy and impact of gender on outcome. *Military Medicine, 184*(1–2), e143–e147. https://doi.org/10.1093/milmed/usy164.

Jung, C. G. (1934/1981). The development of personality. [R. F. C. Hull, Trans.] In Read, H., Fordham, M., Adler, G., & McGuire, W. (Eds.), *The development of personality: Papers on child psychology, education, and related subjects, Volume 17 of the collected works of C. G. Jung* (pp. 165–186). Princeton, NJ: Princeton University Press (Bollingen Series).

Kaniasty, K., & Norris, F. H. (2008). Longitudinal linkages between perceived social support and posttraumatic stress symptoms: Sequential roles of social causation and social selection. *Journal of Traumatic Stress, 21*(3), 274–281.

Keall, R. M., Clayton, J. M., & Butow, P. N. (2015). Therapeutic life review in palliative care: A systematic review of quantitative evaluations. *Journal of Pain and Symptom Management, 49*(4), 747–761.

King, D. W., Taft, C., King, L. A., Hammond, C., & Stone, E. R. (2006). Directionality of the association between social support and Posttraumatic Stress Disorder: A longitudinal investigation. *Journal of Applied Social Psychology, 36*(12), 2980–2992.

Koenen, K. C., Stellman, J. M., Stellman, S. D., Sommer, J. F., Jr. (2003). Risk factors for course of Posttraumatic Stress Disorder among Vietnam veterans: A 14-year follow-up of American Legionnaires. *Journal of Consulting and Clinical Psychology, 71*(6), 980–986.

Levitt, J. T., Malta, L. S., Martin, A., Davis, L., & Cloitre, M. (2007). The flexible application of a manualized treatment for PTSD symptoms and functional impairment related to the 9/11 World Trade Center attack. *Behaviour Research and Therapy, 45*(7), 1419–1433.

MacIntosh, H. B., Cloitre, M., Kortis, K., Peck, A., & Weiss, B. J. (2016). Implementation and evaluation of the Skills Training in Affective and Interpersonal Regulation (STAIR) in a community setting in the context of childhood sexual abuse. *Research on Social Work Practice, 28*(5), 595–602.

Madigan, S., Bakermans-Kranenburg, M. J., Van IJzendoorn, M. H., Moran, G., Pederson, D. R., & Benoit, D. (2006). Unresolved states of mind, anomalous parental behavior, and disorganized attachment: A review and meta-analysis of a transmission gap. *Attachment & Human Development, 8*(2), 89–111.

Main, M., & George, C. (1985). Responses of abused and disadvantaged toddlers to distress in agemates: A study in the day care setting. *Developmental Psychology, 21*(3), 407–412.

Main, M., & Hesse, E. (1990). Parents' unresolved traumatic experiences are related to infant disorganized attachment status: Is frightened and/or frightening parental behavior the linking mechanism? In M. Greenberg, D. Cicchetti, & E. M. Cummings (Eds.), *Attachment in the preschool years: Theory, research, and intervention* (pp. 161–182). Chicago, IL: The University of Chicago Press.

Main, M., & Solomon, J. (1986). Discovery of an insecure-disorganized/disoriented attachment pattern. In T. B. Brazelton & M. W. Yogman (Eds.), *Affective development in infancy* (pp. 95–124). Norwood, NJ: Ablex.

Mikulincer, M., & Shaver, P. R. (2007). *Attachment processes and emotion regulation*. In *Attachment in adulthood: Structure, dynamics, and change* (pp. 188–218). New York, NY: Guilford Press.

Mikulincer, M., Shaver, P. R., & Horesh, N. (2006). Attachment bases of emotion regulation and posttraumatic adjustment. In D. K. Snyder, J. A. Simpson, & J. N. Hughes (Eds.), *Emotion regulation in families: Pathways to dysfunction and health* (pp. 77–99). Washington, DC: American Psychological Association.

Mørkved, N., Hartmann, K., Aarsheim, L. M., Holen, D., Milde, A. M., Bomyea, J., & Thorp, S. R. (2014). A comparison of narrative exposure therapy and prolonged exposure therapy for PTSD. *Clinical Psychology Review, 34*(6), 453–467. https://doi.org/10.1016/j.cpr.2014.06.005.

Mott, J. M., Stanley, M. A., Street, R. L., Jr., Grady, R. H., & Teng, E. J. (2014). Increasing engagement in evidence-based PTSD treatment through shared decision-making: A pilot study. *Military Medicine, 179*(2), 143–149. https://doi.org/10.7205/MILMED-D-13-00363.

Neuner, F., Schauer, M., Klaschik, C., Karunakara, U., & Elbert, T. (2004). A comparison of narrative exposure therapy, supportive counseling, and psychoeducation for treating posttraumatic stress disorder in an african refugee settlement. *Journal of Consulting and Clinical Psychology, 72*(4), 579–587. https://doi.org/10.1037/0022-006X.72.4.579.

Ortigo, K. M., Westen, D., DeFife, J. A., & Bradley, B. (2013). Attachment, social cognition, and posttraumatic stress symptoms in a traumatized, urban population: Evidence for the mediating role of object relations. *Journal of Traumatic Stress, 26*(3), 361–368. https://doi.org/10.1002/jts.21815.

Papa, A., & Bonanno, G. A. (2008). Smiling in the face of adversity: The interpersonal and intrapersonal functions of smiling. *Emotion, 8*(1), 1–12.

Pascuzzo, K., Cyr, C., & Moss, E. (2013). Longitudinal association between adolescent attachment, adult romantic attachment, and emotion regulation strategies. *Attachment & Human Development, 15*(1), 83–103. https://doi.org/10.1080/14616734.2013.745713.

Priel, B., & Besser, A. (2001). Bridging the gap between attachment and object relations theories: A study of the transition to motherhood. *British Journal of Medical Psychology, 74*, 85–100. https://doi.org/10.1348/000711201160821.

Resick, P. A., Monson, C. M., & Chard, K. M. (2017). *Cognitive Processing Therapy for PTSD: A comprehensive manual*. New York, NY: Guilford Press.

Richert, A. J. (2010). *Integrating existential and narrative therapy: A theoretical base for eclectic practice*. Pittsburgh, PA: Duquesne University Press.

Roisman, G. I., Holland, A., Fortuna, K., Fraley, R. C., Clausell, E., & Clarke, A. (2007). The Adult Attachment Interview and self-reports of attachment style: An empirical rapprochement. *Journal of Personality and Social Psychology, 92*(4), 678–697.

Shaver, P. R., & Mikulincer, M. (2005). Attachment theory and research: Resurrection of the psychodynamic approach to personality. *Journal of Research in Personality, 39*, 22–45. https://doi.org/10.1016/j.jrp.2004.09.002.

Shay, J. (1994/2010). *Achilles in vietnam: Combat trauma and the undoing of character*. New York, NY: Simon & Schuster.

Shipman, K. L., & Zeman, J. (2001). Socialization of children's emotion regulation in mother–child dyads. A developmental psychopathology perspective. *Development and Psychopathology, 13*, 317–336.

Shipman, K., Zeman, J., Penza, S., & Champion, K. (2000). Emotion management skills in sexually maltreated and nonmaltreated girls: A developmental psychopathology perspective. *Development and Psychopathology, 12*, 47–62.

Trappler, B., & Newville, H. (2007). Trauma healing via cognitive behavior therapy in chronically hospitalized patients. *Psychiatric Quarterly, 78*(4), 317–325.

Weiss, B. J., Azevedo, K., Webb, K., Gimeno, J., & Cloitre, M. (2018). Telemental health delivery of Skills Training in Affective and Interpersonal Regulation (STAIR) for rural women veterans who have experienced military sexual trauma. *Journal of Traumatic Stress, 31,* 620–625.

Yalom, I. D. (1980). *Existential psychotherapy.* New York, NY: Basic Books.

CHAPTER 19

Acceptance-based behavioral therapy for PTSD

Elizabeth Coe[a], Sonja V. Batten[b], Eric C. Meyer[a,c,d,e]

[a]Warriors Research Institute, Baylor Scott & White Health, Waco, TX, United States
[b]Booz Allen Hamilton, Washington, DC, United States
[c]U.S. Department of Veterans Affairs, VISN 17 Center of Excellence for Research on Returning War Veterans, Waco, TX, United States
[d]Department of Psychiatry and Behavioral Science, Texas A&M University Health Science Center, College of Medicine, College Station, TX, United States
[e]Department of Psychology and Neuroscience, Baylor University, Waco, TX, United States

> *Healing is a coming to terms with things as they are, rather than struggling to force them to be as they once were, or as we would like them to be to feel secure, or to have what we sometimes think of as our own way.*
>
> Jon Kabat-Zinn (2005)

Acceptance- and mindfulness-based treatments for posttraumatic stress seek to promote an adaptive response style by bolstering a set of interrelated psychological constructs including mindfulness and psychological flexibility (Meyer, Frankfurt, et al., 2018; Yadavaia, Hayes, & Vilardaga, 2014). Recent research indicates that these constructs form a higher-order factor that, while strongly related to PTSD symptom severity, also uniquely predicts long-term functioning and quality of life over time after accounting for PTSD symptom severity (Meyer, Frankfurt, et al., 2018) and can be addressed as change mechanisms in acceptance- and mindfulness-based interventions (e.g., Meyer, Walser, et al., 2018; Yadavaia et al., 2014). The practice of mindfulness forms the foundation of many acceptance-based behavioral therapies. These interventions build on this foundation by scaffolding these interrelated constructs onto one another to build a more adaptive, workable stance in relation to trauma-related internal experiences. This chapter provides an overview of several acceptance-based behavioral approaches as they have been applied to PTSD, the treatment outcome literature on each treatment in this area, the benefits of these approaches, and suggestions for future research within this growing field.

Overview of acceptance-based behavioral therapies
Acceptance and commitment therapy
Overview of the treatment model
Acceptance and commitment therapy (ACT) is a behavioral therapy based on functional contextualism. For a full description of the conceptual model, readers are referred to the second edition of the ACT text (Hayes, Strosahl, & Wilson, 2012). ACT holds that psychological inflexibility, including cognitive fusion (i.e., the inability to separate one's thoughts from one's self or reality) and avoidance of one's negatively evaluated internal experiences, underlies many forms of psychopathology. Psychological flexibility, on the other hand, is defined as "the ability to fully contact the present moment and the thoughts and feeling it contains without needless defense, and, depending on what the situation affords, persisting in or changing behavior in the pursuit of goals and values" (Hayes, Luoma, Bond, Masuda, & Lillis, 2006). This flexibility is viewed as a pivotal component of healthy human functioning, behavioral change, and resilience in response to stressful or traumatic life events. Psychological flexibility involves an open, willing, nonjudgmental stance in relation to the full range of one's internal experiences, including— but, importantly, not limited to—trauma-related internal experiences such as intrusive memories, emotional distress, and physiological hyperarousal. This openness allows one the flexibility to make behavioral choices based on personally defined values, goals, and other contingencies, as opposed to being locked in a rigid struggle to try to control or avoid unwanted internal experiences. ACT and other acceptance-based behavioral therapies for PTSD seek to bolster psychological flexibility through experiential activities, fostering willingness to approach (rather than avoid) one's negatively evaluated internal experiences to increase valued, meaningful living.

A growing body of research supports the key influence of psychological inflexibility on the development and course of PTSD symptoms over time. In a study of trauma-exposed war veterans, Meyer and colleagues (Meyer, Morissette, Kimbrel, Kruse, & Gulliver, 2013) found that psychological inflexibility, measured using the Acceptance and Action Questionnaire-II (AAQ-II, Bond et al., 2011), was associated with PTSD symptom severity, even after accounting for personality factors and the strongest pretrauma, peritrauma, and posttrauma risk factors identified in metaanalyses of PTSD risk factors (e.g., trauma severity, perceived threat, peritraumatic dissociation, level of social support, and recent life stress). This finding was subsequently replicated in a larger, longitudinal study that also accounted for additional, military-specific PTSD risk factors (Meyer, La Bash, et al., 2018).

However, these studies are limited by assessing psychological inflexibility following exposure to trauma. One prospective study found that psychological inflexibility measured prior to a mass shooting predicted long-term trajectories of posttraumatic stress (Orcutt, Bonanno, Hannan, & Miron, 2014). These findings underscore the key role of psychological inflexibility in the development and maintenance of PTSD over time, particularly compared with more static or nonmodifiable factors. Numerous studies have demonstrated that psychological flexibility is modifiable through ACT and other types of interventions (e.g., Arch et al., 2012; Gloster, Meyer, & Lieb, 2017; Hayes et al., 2006; Meyer, Walser, et al., 2018).

The ACT model includes six interconnected core processes that collectively buttress psychological flexibility. These include acceptance of or willingness to experience negatively evaluated emotional experiences; present-moment awareness (i.e., mindfulness); self-as-context (i.e., experiential contact with the self as the continuous experiencer of an ongoing flow of experiences, including both wanted and unwanted experiences, as distinguished from the content of these experiences); thought defusion (i.e., experiential contact with thoughts as mental events, as distinguished from the self); clarification of personal values as aspirational guides for behavioral choices; and engaging in committed action in the service of pursuing one's personal values (Hayes et al., 2012). ACT sessions typically include brief formal mindfulness practices, as well as experiential activities, discussions, and use of metaphors aimed at teaching and practicing the ACT core processes. Rather than focusing on symptom reduction as the primary goal, ACT focuses on increasing clients' psychological flexibility, or their acceptance of and willingness to have the full range of internal experiences, including those that clients evaluate negatively (i.e., "unwanted" or "bad" thoughts, emotions, and physiological sensations) while making and maintaining commitments to behavioral change in the service of enacting their personal values (Orsillo & Batten, 2005). Clients are guided through structured exercises aimed at identifying, clarifying, and behaviorally operationalizing their personal values. Thus ACT involves a process of shifting energy previously spent on unsuccessful or otherwise unworkable (from a functional perspective) attempts to avoid, control, or escape from negatively evaluated internal experiences toward intentional behavior change in the service of personal values. Through an ACT lens, improved functioning and life satisfaction are more important outcomes than symptom reduction. That said, symptom reduction is often observed as clients become less entangled with, judgmental toward, and distressed by trauma-related and other negatively evaluated internal experiences.

As with many other treatments for PTSD, exposure to negatively evaluated memories and emotions plays an important role in ACT. However, rather than aiming toward habituation of fear responses, the purpose of exposure-based and other experiential exercises used in ACT is to help the client develop a stance of willingness toward their own internal experiences (e.g., thoughts, feelings, and memories) to live in accord with their values. Depending on the case, these experiential activities may be more or less focused on specific trauma-related material (Orsillo & Batten, 2005). Importantly, exposure exercises within ACT are conducted to teach that willingness and valued actions may be practiced regardless of one's internal experiences; that is, regardless of whether distress decreases, increases, or stays the same. It is one's willingness to experience distress, rather than the distress itself, that is the target of exposure exercises in ACT. This stands in contrast to the emphasis on habituation typical of traditional exposure-based treatments, such as prolonged exposure (PE). Although the effectiveness of such treatments is well established, some individuals do not experience full or lasting benefits, and it is thought that these individuals may experience deficits in the inhibitory learning processes thought to underlie the effectiveness of exposures (Craske, Treanor, Conway, Zbozinek, & Vervliet, 2014). Acceptance-based approaches, including ACT, may serve to enhance inhibitory learning in a way that is distinct from the habituation model (Craske et al., 2014; Treanor, 2011). The ACT concepts of mindfulness, acceptance, and willingness mesh well with newer models of exposure that emphasize acceptance or tolerance of distress over habituation (Craske et al., 2014; Thompson, Luoma, & LeJeune, 2013). Thus the underlying mechanisms of acceptance-based approaches can be thought of as complementary, rather than contradictory, to the mechanisms of traditional exposure-based approaches. Thompson et al. (2013) highlight ways in which ACT may provide a more flexible foundation from which to conceptualize and conduct exposure exercises along with a broader range of other value-consistent behavioral exercises.

Review of treatment outcomes

There is a strong evidence base for the use of ACT to treat problems that frequently co-occur with PTSD. In a growing body of published trials, ACT has been shown to be superior to waitlist control, placebo, and treatment as usual (TAU) and equivalent to well-established treatments (e.g., cognitive behavioral therapy) for depression, anxiety, stress, general distress, substance use, and chronic pain (A-Tjak et al., 2015; Forman, Herbert, Moitra,

Yeomans, & Geller, 2007; Hayes et al., 1999; Hayes et al., 2006; Öst, 2008; Öst, 2014; Powers, Zum Vörde Sive Vörding, & Emmelkamp, 2009). The evidence base for ACT for treating co-occurring conditions such as substance use and depression is of import to the treatment of PTSD, since such comorbidities are quite common among individuals with PTSD (Brady, Killeen, Brewerton, & Lucerini, 2000; Kimbrel et al., 2015). The use of ACT to treat people living with PTSD is an emerging area. A small randomized controlled trial (RCT; Boals & Murrell, 2016) examined a modified, brief version of ACT as an adjunct to TAU as a treatment for people exposed to traumatic events or other major life stressors (e.g., divorce or loss of child custody) who screened positive for PTSD. The modified version of ACT placed increased emphasis on the core process of self-as-context. TAU consisted of various cognitive behavioral and psychoeducational interventions. TAU plus ACT led to greater reductions in symptoms of PTSD and depression at posttreatment compared with TAU, though at follow-up this group difference remained significant for depression only. A recent RCT with veterans and military personnel with a range of diagnoses, many of whom screened positive for a probable PTSD diagnosis, found that ACT led to improvements in PTSD symptoms, functioning, and quality of life that were similar in magnitude to those who received present-centered therapy (PCT; Lang et al., 2017).

Recent studies have also included an uncontrolled pilot study of ACT with veterans diagnosed with co-occurring PTSD and alcohol use disorder (AUD; Meyer, Walser, et al., 2018). Co-occurring PTSD-AUD is associated with more severe symptoms, worse functioning, increased suicide risk, greater treatment utilization, and greater treatment dropout than either disorder alone (Roberts, Roberts, Jones, & Bisson, 2015; Rojas, Bujarski, Babson, Dutton, & Feldner, 2014; Sells et al., 2016), underscoring the pervasive negative impact of maladaptive emotional control attempts in people with PTSD. This study achieved a treatment completion rate (67%) for a 12-session outpatient individual therapy protocol that was higher than is typical in this population (Roberts et al., 2015), accompanied by high treatment credibility and satisfaction ratings. Treatment completers reported sustained improvements in functional disability and quality of life (medium effects), as well as in PTSD symptoms, AUD symptoms, and suicidal ideation (large effects). Additional randomized trials are warranted to examine the utility of ACT for promoting functional recovery in people living with PTSD, among whom a broad range of co-occurring mental and physical health challenges are typically present. Meyer and colleagues are currently building on this study by further adapting and pilot-testing ACT to address severe functional impairment among war veterans exposed to any combination of the most common mental and physical wounds of war (i.e., PTSD, depression, AUD, chronic pain, and traumatic brain injury; NCT03615222).

Consistent with the theory underlying ACT, numerous studies indicate that mindfulness and psychological flexibility—major therapeutic targets in ACT—mediate response to treatment for individuals with PTSD, anxiety, and depression (Berking, Neacsiu, Comtois, & Linehan, 2009; Flaxman & Bond, 2010; Hayes et al., 2006; Hayes, Orsillo, & Roemer, 2010; Meyer, Walser, et al., 2018). For example, a study of ACT with a large sample of veterans with depression (Walser et al., 2015) showed that improvements in depression symptoms, quality of life, and suicidal ideation were linked with improvements in self-reported psychological flexibility measured using the AAQ-II and mindfulness using the Five Facet Mindfulness Questionnaire (FFMQ; Baer, Smith, Hopkins, Krietemeyer, & Toney, 2006). In our study of ACT for PTSD-AUD (Meyer, Walser, et al., 2018), improvements in psychological flexibility using the AAQ-II were associated with greater improvements in symptoms of PTSD, AUD, and depression, as well as improvements in functional disability and quality of life. In addition, more between-session mindfulness practice was associated with greater reductions in drinking and AUD symptoms.

Mindfulness-based stress reduction
Overview of the treatment model
Mindfulness-based stress reduction (MBSR) is an intervention that teaches individuals to attend to the present moment (i.e., immediate emotional and physical states, including discomfort) in a nonjudgmental, accepting manner (Kabat-Zinn, 1990). Like other acceptance-based interventions, MBSR targets avoidance of negatively evaluated internal experiences, a key factor in the development and maintenance of PTSD, by encouraging acceptance of thoughts, feelings, and experiences without avoidance (Polusny et al., 2015) MBSR sessions are often delivered in group format and typically include didactic content and formal meditation practice, including body scans (systematically directing attention through various areas of the body), sitting meditation (sustained self-observation through directing attention to specific experiences such as one's breathing or thoughts), and mindful yoga (gentle movement and stretching while maintaining present-moment attention) (Polusny et al., 2015). MBSR

programs may also incorporate informal mindfulness practices aimed at cultivating present-moment awareness during tasks of daily living, such as eating and walking (Polusny et al., 2015).

Review of treatment outcomes

Mindfulness meditation programs have demonstrated small to moderately sized improvements in depression, anxiety, pain, stress/distress, and quality of life (Goyal et al., 2014). In a RCT of 116 veterans with PTSD, Polusny et al. (2015) found that group MBSR, compared with present-centered group therapy (PCGT), resulted in a greater decrease in self-reported PTSD symptom severity, and these changes were associated with increases in self-reported mindfulness using the FFMQ (Baer et al., 2006). Moreover, the MBSR group reported additional symptom decreases through 2 months posttreatment, whereas the PCGT condition did not. The MBSR group also reported significantly greater improvements in quality of life. Clinical response rates for the MBSR group were comparable with those found in studies of first-line PTSD treatments (i.e., CPT and PE). Although treatment dropout was higher among the MBSR group than the PCGT group (22.4% vs. 6.9%), dropout rates were lower than those typically reported for PE (28.1%–44%) and CPT (26.8%–35%) (Polusny et al., 2015). Similarly, a recent pilot study of MBSR for veterans with PTSD incorporating brain imaging techniques found that, compared with PCGT, MBSR treatment was associated with improvements in PTSD symptoms up to 6 months after treatment and increased self-reported mindfulness (using the FFMQ; Baer et al., 2006), as well as theoretically expected increased anterior cingulate and inferior parietal lobule functioning and decreased insula and precuneus functioning in response to trauma reminders (Bremner et al., 2017).

Mantram repetition program
Overview of the treatment model

Mantram repetition program (MRP) is another mindfulness-based intervention that has been applied to populations experiencing PTSD. The primary aim of MRP is to teach individuals to silently repeat a *mantram*, or sacred word or phrase (Easwaran, 2008) to cultivate mindful awareness (Bormann, Oman, Walter, & Johnson, 2014). MRP typically includes modules addressing psychoeducation about PTSD symptoms, what is a mantram and how to choose one, using and tracking mantram repetition, the stress response as related to PTSD and mantram repetition, using mantram repetition to slow down cognitive, emotional, and behavioral reactivity, and practicing single-pointed attention (Bormann et al., 2014). Clients are encouraged to repeat the mantram in a range of settings and contexts. This includes routine practice during times when they are not experiencing a strong response (e.g., before bedtime) so that mantram repetition is a more readily available response when they experience a strong stress response (e.g., when they awake from a nightmare; Bormann et al., 2014).

Review of treatment outcomes
Two RCTs have demonstrated the efficacy of MRP for the treatment of PTSD. MRP was initially proposed and evaluated as a complementary intervention. In a RCT comparing medication and case management alone (TAU) with MRP+TAU for the treatment of PTSD in a sample of 146 veterans, significantly greater improvements in self-reported and clinician-rated PTSD symptoms, depression, and psychological well-being were reported for the MRP+TAU group compared with TAU alone (Bormann, Thorp, Wetherell, Golshan, & Lang, 2013). Both treatment conditions evidenced a 7% dropout rate. These improvements were mediated by increases in mindful attention (Bormann et al., 2014). Building on this work, MRP has recently been evaluated as a stand-alone intervention for PTSD. A RCT conducted with a sample of 173 veterans across two sites compared MRP with PCT, each of which were delivered as individual treatments in 8 weekly sessions (Bormann et al., 2018). Superior improvements in clinician- and self-reported PTSD symptoms, loss of PTSD diagnosis, and insomnia were observed in the MRP group (Bormann et al., 2018). Thus MRP appears to be a highly promising intervention.

Behavioral activation
Overview of the treatment model
Behavioral activation is based on behavioral reinforcement theory. In the context of PTSD, it emphasizes overcoming patterns of withdrawal and avoidance that may follow a precipitating event (e.g., trauma and loss of a loved one) by coaching clients to reengage in rewarding and meaningful activities (Martell, Addis, & Jacobson, 2001). This application of behavioral activation is based on the theory that increases in these types of activities may break the avoidance patterns that maintain PTSD symptoms (Wagner, Zatzick, Ghesquiere, & Jurkovich, 2007). Behavioral activation differs from other treatments that focus on reducing avoidance in its "outside in" approach to change (Martell et al., 2001). That is, instead of exploring cognitive schemas, emotions, or physical responses to trauma cues, behavioral

activation uses active problem-solving to increase engagement in meaningful activities (Jakupcak, Wagner, Paulson, Varra, & McFall, 2010), even in the presence of trauma cues. Behavioral activation differs from exposure-based PTSD treatments (e.g., PE) in that activities are chosen based on their likelihood of producing pleasure or mastery, rather than their likelihood of producing distress (Jakupcak et al., 2010). For example, a veteran with PTSD who avoids attending church due to distress around crowds but reengages in spiritual activities through individual prayer and reading and small-group bible studies would be considered to have successfully completed the behavioral assignment because he employed problem-solving to successfully reengage in a meaningful life domain, even though exposure to the feared situation did not occur. Behavioral activation can also be distinguished from ACT in that, while valued living is seen as an end in itself from an ACT perspective, behavioral activation practitioners are concerned with valued living insofar as it increases a client's sense of pleasure and/or mastery (Jakupcak et al., 2010). Nevertheless, behavioral activation can still be viewed as an acceptance-based treatment because it emphasizes acceptance of short-term distress while engaging in activities that foster longer-term pleasure and/or mastery. Although most commonly used to treat depression, behavioral activation has also been applied to anxiety disorders (Hopko, Robertson, & Lejuez, 2006) and PTSD (Jakupcak et al., 2006; Wagner et al., 2007), as avoidance and withdrawal are common maintaining features across all of these disorders.

Review of treatment outcomes
Behavioral activation (BA) has successfully been used to treat disorders that commonly co-occur with PTSD, such as depression (e.g., Dimidjian et al., 2006; Ekers et al., 2014) and anxiety (e.g., Hopko et al., 2006). BA has shown promise as a treatment for PTSD, although most published studies to date have been small and/or nonrandomized (Jakupcak et al., 2006; Jakupcak et al., 2010; Wagner et al., 2007). Initial evidence suggests that BA is typically well tolerated (Jakupcak et al., 2006), and because of the relatively straightforward, intuitive nature of its intervention principles, it may be particularly well-suited for delivery in nonspecialty mental health settings, such as primary care clinics (Hopko, Bell, Armento, Hunt, & Lejuez, 2005; Jakupcak et al., 2010). For example, in a small, preliminary (noncontrolled) study of eight veterans receiving BA delivered in a postdeployment primary care clinic, Jakupcak et al. (2010) found significant and meaningful reductions in PTSD symptoms on structured clinical assessments and

self-report measures at posttreatment and 3-month follow-up. The majority of veterans also demonstrated meaningful improvements in depression and quality of life and reported high satisfaction with treatment (Jakupcak et al., 2010). In a pilot study of BA for early intervention for PTSD and depression, Wagner et al. (2007) randomized a small group of physically injured survivors of traumatic injury ($N=8$) to BA or TAU. Compared with TAU, the BA group showed greater improvement in PTSD symptom severity. A larger RCT comparing BA with usual care for the treatment of PTSD among veterans is currently underway (NCT03615222).

Dialectical behavior therapy
Overview of the treatment model
Dialectical behavior therapy (DBT; Linehan, 1993a, 1993b), the most widely studied evidence-based approach for the treatment of borderline personality disorder (BPD), is also the only approach to have been studied in terms of its impact on co-occurring BPD-PTSD (Harned, Korslund, & Linehan, 2014). Like other treatments for BPD, DBT emphasizes the "here and now" rather than the past, including past trauma (Harned et al., 2014). DBT addresses, in the following order in terms of priorities, (1) life threatening behaviors such as suicide attempts and nonsuicidal self-injury (NSSI); (2) treatment interfering behaviors such as treatment noncompliance; and (3) severe quality-of-life interfering behaviors, including behaviors associated with other disorders such as PTSD. Considered a form of cognitive behavioral treatment, DBT includes acceptance strategies such as mindful awareness of the present moment and tolerating, rather than changing or avoiding, distress. Standard DBT consists of 1 hour of weekly individual psychotherapy, 2.5 hours of weekly group skills training, phone consultations as needed, and a weekly therapist consultation team meeting (Linehan, 1993a, 1993b). DBT may have meaningful applications for PTSD treatment, especially considering that BPD, suicidality/NSSI, and PTSD commonly co-occur (Harned, Rizvi, & Linehan, 2010; Kimbrel et al., 2015; Pagura et al., 2010). Compared with individuals with BPD or PTSD alone, individuals with comorbid BPD and PTSD are two and five times more likely to make a suicide attempt, respectively (Pagura et al., 2010). However, first-line PTSD treatment guidelines explicitly list active suicidality as a contraindication (e.g., Foa, Keane, Friedman, & Cohen, 2009; National Institute for Clinical Excellence, 2005), and studies of these treatments have historically excluded patients with serious suicidality or NSSI (Bradley, Greene, Russ, Dutra, & Westen, 2005). Patients with these comorbidities are

often referred out for stabilization before entering PTSD-specific treatment (Harned et al., 2014). Considering these standard referral procedures and the high rates of comorbidity, DBT providers are likely to treat a sizable trauma-affected population. The standard practice of DBT involves several PTSD-relevant techniques, including using self-soothing skills to manage anxiety, challenging dysfunctional beliefs by "checking the facts," reducing avoidance by engaging in actions that are opposite to emotion-driven behaviors, and acceptance-based skills such as mindfulness and distress tolerance (Harned et al., 2014).

Review of treatment outcomes

DBT PE (a treatment protocol that incorporates DBT strategies into PE therapy for PTSD) has demonstrated positive outcomes in case studies (Harned & Linehan, 2008), an open trial (Harned, Korslund, Foa, & Linehan, 2012), and a pilot RCT (Harned et al., 2014). In the pilot RCT, which included a sample of clients ($N=26$) with comorbid BPD, PTSD, and NSSI, DBT PE produced greater improvement than DBT alone, with DBT PE producing moderate to large effect sizes for dissociation, trauma-related guilt cognitions, shame, anxiety, depression, and global functioning. The researchers concluded that DBT PE is a feasible approach for treating PTSD that co-occurs with BPD and NSSI and may produce larger improvements than DBT alone (Harned et al., 2014). However, it should be noted that both groups (DBT PE and DBT alone) demonstrated improvements in PTSD severity, suggesting that standard DBT skills can be successfully applied to trauma-related symptoms.

Benefits of acceptance-based behavioral therapies

Acceptance-based treatments address the interconnected processes and mechanisms associated with the development and maintenance of PTSD symptoms. As detailed earlier, PTSD is associated with high levels of psychological inflexibility, as well as distress intolerance, experiential avoidance, and deficits in emotion regulation skills. Experiential avoidance, which is viewed as a core facet of the broader construct of psychological inflexibility, is often considered a core feature of the disorder in that PTSD symptoms are developed and maintained by efforts to avoid unwanted trauma memories and related thoughts and feelings (Orsillo & Batten, 2005). Active efforts to avoid or escape trauma-related thoughts, feelings, or reminders are included in the diagnostic criteria for PTSD (Criterion C, *DSM-5*, APA, 2013).

Consistent with that, but even more broadly speaking, experiential avoidance refers to efforts to avoid, control, or suppress any unwanted internal experience including, but not limited to, trauma-related internal experiences. At least two studies have demonstrated that psychological inflexibility, which encompasses the narrower construct of experiential avoidance, predicts total PTSD symptom severity even after accounting for the avoidance symptoms of PTSD (Meyer et al., 2013; Meyer, La Bash, et al., 2018). Thought suppression, which is one type of experiential avoidance, has been shown to paradoxically increase unwanted thoughts and related distress, particularly when used as a strategy to cope with a traumatic memory (Aikins et al., 2009; Beck, Gudmundsdottir, Palyo, Miller, & Grant, 2006; Garland & Roberts-Lewis, 2013; Shipherd & Beck, 2005). In fact, use of thought suppression as a coping mechanism has been found to be significantly more characteristic of individuals with PTSD than those with other types of anxiety disorders (Amir et al., 1997).

Emotion regulation deficits, or the inability to flexibly apply effective strategies to regulate emotions, are also theoretically and empirically associated with PTSD (Boden et al., 2013; Eftekhari, Zoellner, & Vigil, 2009; Shepherd & Wild, 2014) and are considered another core feature of the disorder (Frewen & Lanius, 2006). Such deficits, including irritability, hypervigilance, and decreased positive affect are categorized in PTSD Criteria D and E of the *DSM-5* (APA, 2013). Whereas thought suppression and other forms of experiential avoidance are viewed as active, effortful strategies to avoid or control negatively evaluated internal experiences, emotion regulation deficits are seen as automatic processes in response to emotional distress (Feeny, Zoellner, Fitzgibbons, & Foa, 2000). Distress intolerance, or the perceived inability to withstand distressing emotional states, is another characteristic feature of PTSD that has been associated with overall symptom severity and severity of specific PTSD symptom clusters (e.g., avoidance and hyperarousal) (Vujanovic et al., 2013; Vujanovic, Bonn-Miller, Potter, Marshal, & Zvolensky, 2011). Distress intolerance, emotion regulation deficits, and experiential avoidance may hold particular clinical relevance since they not only contribute to the maintenance of PTSD symptoms but also may prevent individuals with PTSD from seeking or completing treatment that involves confronting distressing memories, thoughts, and feelings.

Acceptance-based treatments address these factors associated with PTSD symptom development and maintenance in several ways that differ from current first-line PTSD treatments. First, mindfulness practice is a key component of these interventions and is a foundation for all other skill building.

Emerging evidence supports the notion that practicing mindfulness supports effective emotion regulation (Grecucci, Pappaianni, Siugzdaite, Theuninck, & Job, 2015; Lutz et al., 2014). Avoidance of negatively evaluated internal experiences is addressed not from a habituation standpoint, but rather by emphasizing mindful awareness of these experiences. Mindful, nonjudgmental awareness fosters increased tolerance, willingness, and self-compassion in relation to having these experiences. Next, particularly relevant to ACT, values clarification work serves as a guide and motivating factor for subsequent behavioral exercises. ACT emphasizes the necessity of allowing negatively evaluated internal experiences to engage in value-consistent actions. Experiential exercises, including those involving intentional exposure to trauma-related distress, are used to help the client develop a stance of willingness toward their internal experiences. Over the course of treatment and across different clients, exercises vary in terms of being more or less "trauma focused" (Orsillo & Batten, 2005). Acceptance-based approaches include several elements that represent strong matches for the preferences of many clients and may be useful for those who decline to engage in current first-line exposure-based treatments for PTSD (Orsillo & Batten, 2005).

Several other potential benefits of acceptance-based treatments arise from their transdiagnostic approach. There are compelling theoretical and pragmatic reasons why transdiagnostic treatments may offer benefits for individuals struggling with PTSD (Gutner, Galovski, Bovin, & Schnurr, 2016). In fact, this may be especially relevant to the treatment of PTSD, which is associated with high rates of comorbidity (Brady et al., 2000; Kessler, Sonnega, Bromet, Hughes, & Nelson, 1995; Orsillo et al., 1996). In one study of Iraq and Afghanistan war veterans who completed a psychiatric screening questionnaire, about half (51.2%) screened positive on at least three clinical subscales, while more than one-third (34.9%) screened positive on five or more clinical subscales (Kimbrel et al., 2015), illustrating the importance of addressing comorbid conditions in this population. Acceptance-based approaches can be applied to several presenting problems simultaneously, without needing to first determine and target the "primary" disorder (Lang et al., 2012). Moreover, addressing the full range of a client's presenting concerns using a transdiagnostic approach may promote treatment satisfaction and engagement compared with diagnosis-specific approaches (Barlow et al., 2017). While the rationale for taking a transdiagnostic approach in treating PTSD is strong, controlled research is warranted to determine whether it indeed confers these benefits compared with current first-line PTSD treatments.

Acceptance-based approaches may offer benefits to clinicians and agencies with diverse clientele in terms of facilitating training and dissemination. Because these approaches may be used to improve functioning-based outcomes across a broad range of presenting problems, including complex combinations of symptoms, it may be efficient for clinicians to develop expertise in these approaches as opposed to a greater number of disorder-specific protocols. Efficient allocation of training resources increases the likelihood that clinicians will provide clients with evidence-based treatment and provides potential cost savings for agencies (Lang et al., 2012). Finally, acceptance-based approaches may offer benefits in terms of both client and clinician preferences. In addition to the client-based reasons listed previously, there are also clinician-based reasons (e.g., idiosyncratically preferring not to use a first-line trauma-specific approach) that alternatives to more well-established, trauma-specific treatments are needed (Lang et al., 2012; Steenkamp & Litz, 2013, 2014). Dissemination and acceptability are important factors to consider when evaluating the effectiveness of treatment (Chambless & Hollon, 1998), and acceptance-based approaches seem to offer benefits on both fronts.

Summary of the literature, limitations, and areas for future research

The theoretical rationale for applying acceptance-based approaches to promoting recovery in people living with PTSD and commonly co-occurring mental and physical health challenges is strong, and preliminary empirical evidence suggests these approaches may foster clinically meaningful change. In two RCTs that examined ACT in the treatment of people with PTSD symptoms, ACT plus TAU was superior to TAU (Boals & Murrell, 2016) and equivalent to present-centered therapy (Lang et al., 2017). While both of these studies included participants with PTSD, neither was a rigorous trial of ACT for PTSD per se. Two RCTs comparing MBSR with present-centered therapy found that MBSR led to greater improvements in PTSD symptoms and mindfulness (Bremner et al., 2017; Polusny et al., 2015). One of these studies also found support for hypothesized brain changes in the MBSR group (Bremner et al., 2017). Two RCTs have now demonstrated the efficacy of MRP as an adjunctive and as a stand-alone treatment for PTSD (Bormann et al., 2013, 2018). In a pilot RCT, Wagner et al. (2007) found that BA was associated with greater reductions in PTSD symptoms compared with TAU in physically injured survivors of traumatic injury.

On balance, acceptance-based approaches are less well established than first-line, exposure-based PTSD treatments in terms of the number of RCTs showing superiority to active comparison treatments. However, they appear highly promising with results that are theoretically consistent with the patterns of exacerbation of and recovery from PTSD. Because of the strong theoretical rationale for applying these approaches, promising preliminary evidence from clinical trials, and limitations on the acceptability, reach, and effectiveness of current first-line treatments for PTSD, further research in this area is warranted. Suggested directions for continuing the examination of acceptance-based interventions for addressing posttrauma recovery include additional RCTs, comparative effectiveness, and noninferiority studies. In addition, given the unique treatment components included in acceptance-based interventions, it has frequently been posited that they may be beneficial for people who have not responded sufficiently to first-line PTSD treatments; however, this hypothesis has yet to be tested empirically.

We view the continued integration of acceptance-based and compassion-focused interventions as a fruitful area for further treatment development (Tirch, Schoendorff, & Silberstein, 2014). One prominent conceptualization characterizes self-compassion as including three main components: (1) mindful, nonjudgmental awareness of emotional distress, as contrasted with fusion or overidentification with distress; (2) intentionally turning kindness inward when experiencing emotional distress, as opposed to being self-critical; and (3) a sense of common humanity, or seeing one's emotional distress as part of the universal human experience, as opposed to feeling isolated as a result of these experiences (Neff, 2003a, 2003b). The conceptual overlap among mindfulness, acceptance, psychological flexibility, and self-compassion is apparent and has received recent theoretical and empirical attention (Meyer, Frankfurt, et al., 2018; Neff & Tirch, 2013; Yadavaia et al., 2014). For example, as mentioned previously, in a sample of combat veterans, measures of mindfulness (Mindfulness Attention Awareness Scale (MAAS); Brown & Ryan, 2003), psychological flexibility (AAQ-II; Bond et al., 2011), and self-compassion (Self-Compassion Scale (SCS); Neff, 2003b) formed a single latent factor that predicted better functioning and quality of life over time even after accounting for the influence of PTSD symptom severity (Meyer, Frankfurt, et al., 2018). Another study demonstrated that lower self-compassion predicted increases in PTSD symptoms over time after accounting for level of combat exposure (Hiraoka et al., 2015). These findings suggest that targeting self-compassion may lead to meaningful improvements in mental health and functional outcomes in trauma-exposed populations. A recent RCT tested an adaptation of ACT that aimed to bolster self-compassion (Yadavaia et al., 2014). In the Yadavaia et al. (2014) trial, compared with a wait list control, those who received ACT reported significant improvements in self-compassion, general psychological distress, anxiety, and depression. Moreover, increased psychological flexibility mediated these changes. Additional research examining the role of self-compassion in the development, maintenance, and treatment of PTSD is warranted (see Chapter 21 for an extended discussion regarding the role of self-compassion in PTSD).

In closing

Acceptance-based interventions comprise a set of evidence-based approaches that likely represent a strong fit for many people living with the full spectrum of trauma-related health conditions, including PTSD. It is probable that these interventions may benefit some people who opt not to access first-line treatments or who do not respond particularly well to such treatments. Acceptance-based interventions may offer unique benefits beyond those of traditional PTSD treatments, such as improving broadly applicable mindfulness skills and increasing engagement in valued life domains. Additional research in this area can address several key questions including whether these interventions offer efficacy beyond those conferred by existing trauma treatments, whether the effectiveness of acceptance-based interventions may be equivalent to first-line treatments when factors such as client choice and treatment completion rate are considered, and the role of mechanisms such as mindfulness, psychological flexibility, and self-compassion in PTSD treatment. Finally, emerging evidence indicates that acceptance-based interventions are amenable to deliver via telehealth and other newer delivery mechanisms for a range of problems that frequently co-occur with PTSD. Our preliminary experience delivering ACT for people with PTSD via telehealth indicates that this is an acceptable delivery method. Given the complexity of the mental and physical health challenges associated with trauma exposure, combined with limitations associated with even the most well-established, efficacious treatments for PTSD, having a broad range of treatment options appears to be important. Acceptance-based interventions, by virtue of their strong acceptability for many people and their transdiagnostic approach, represent an important set of options for promoting recovery in trauma survivors.

References

Aikins, D. E., Johnson, D. C., Borelli, J. L., Klemanski, D. H., Morrissey, P. M., Benham, T. L., ... Tolin, D. F. (2009). Thought suppression failures in combat PTSD: A cognitive load hypothesis. *Behaviour Research and Therapy, 47,* 744–751. https://doi.org/10.1016/j.brat.2009.06.006.

American Psychiatric Association. (2013). *Diagnostic and statistical manual of mental disorders* (5th ed.). Arlington, VA: American Psychiatric Publishing.

Amir, M., Kaplan, Z., Efroni, R., Levine, Y., Benjamin, J., & Kotler, M. (1997). Coping styles in post-traumatic stress disorder (PTSD) patients. *Personality and Individual Differences, 23,* 399–405. https://doi.org/10.1016/S0191-8869(97)80005-0.

Arch, J. J., Eifert, G. H., Davies, C., Plumb Vilardaga, J. C., Rose, R. D., & Craske, M. G. (2012). Randomized clinical trial of cognitive behavioral therapy (CBT) versus acceptance and commitment therapy (ACT) for mixed anxiety disorders. *Journal of Consulting and Clinical Psychology, 80,* 750–765.

A-Tjak, J. G., Davis, M. L., Morina, N., Powers, M. B., Smits, J. A., & Emmelkamp, P. M. (2015). A meta-analysis of the efficacy of acceptance and commitment therapy for clinically relevant mental and physical health problems. *Psychotherapy and Psychosomatics, 84,* 30–36. https://doi.org/10.1159/000365764.

Baer, R. A., Smith, G. T., Hopkins, J., Krietemeyer, J., & Toney, L. (2006). Using self-report assessment methods to explore facets of mindfulness. *Assessment, 13,* 27–45.

Barlow, D. H., Farchione, T. J., Bullis, J. R., Gallagher, M. W., Murray-Latin, H., Sauer-Zavala, S., ... Cassiello-Robbins, C. (2017). The unified protocol for transdiagnostic treatment of emotional disorders compared with diagnosis-specific protocols for anxiety disorders: A randomized clinical trial. *JAMA Psychiatry, 74,* 875–884. https://doi.org/10.1001/jamapsychiatry.2017.2164.

Beck, J. G., Gudmundsdottir, B., Palyo, S. A., Miller, L. M., & Grant, D. M. (2006). Rebound effects following deliberate thought suppression: Does PTSD make a difference? *Behavior Therapy, 37,* 170–180. https://doi.org/10.1016/j.beth.2005.11.002.

Berking, M., Neacsiu, A., Comtois, K., & Linehan, M. (2009). The impact of experiential avoidance on the reduction of depression in treatment for borderline personality disorder. *Behaviour Research and Therapy, 47,* 663–670. https://doi.org/10.1016/j.brat.2009.04.011.

Boals, A., & Murrell, A. R. (2016). I am > trauma: Experimentally reducing event centrality and PTSD symptoms in a clinical trial. *Journal of Loss and Trauma, 21,* 471–483. https://doi.org/10.1080/15325024.2015.1117930.

Boden, M. T., Westermann, S., McRae, K., Kuo, J., Alvarez, J., Kulkarni, M. R., ... Bonn-Miller, M. O. (2013). Emotion regulation and posttraumatic stress disorder: A prospective investigation. *Journal of Social and Clinical Psychology, 32,* 296–314. https://doi.org/10.1521/jscp.2013.32.3.296.

Bond, F. W., Hayes, S. C., Baer, R. A., Carpenter, K. M., Guenole, N., Orcutt, H. K., ... Zettle, R. D. (2011). Preliminary psychometric properties of the acceptance and action questionniare – II: A revised measure of psychological flexibility and experiential avoidance. *Behavior Therapy, 42,* 676–688.

Bormann, J. E., Oman, D., Walter, K. H., & Johnson, B. D. (2014). Mindful attention increases and mediates psychological outcomes following mantram repetition practice in veterans with posttraumatic stress disorder. *Medical Care, 52,* S13–S18. https://doi.org/10.1097/MLR.0000000000000200.

Bormann, J. E., Thorp, S. R., Wetherell, J. L., Golshan, S., & Lang, A. J. (2013). Meditation-based mantram intervention for veterans with posttraumatic stress disorder: A randomized trial. *Psychological Trauma Theory Research Practice and Policy, 5,* 259–267. https://doi.org/10.1037/a0027522.

Bormann, J. E., Thorp, S. R., Smith, E., Glickman, M., Beck, D., Plumb, D., ... Elwy, A. R. (2018). Individual treatment of posttraumatic stress disorder using mantram repetition: A randomized clinical trial. *American Journal of Psychiatry, 175,* 979–988. https://doi.org/10.1176/appi.ajp.2018.17060611. Epub 2018 Jun 20.

Bradley, R., Greene, J., Russ, E., Dutra, L., & Westen, D. (2005). A multidimensional meta-analysis of psychotherapy for PTSD. *American Journal of Psychiatry, 162,* 214–227.

Brady, K. T., Killeen, T. K., Brewerton, T., & Lucerini, S. (2000). Comorbidity of psychiatric disorders and posttraumatic stress disorder. *Journal of Clinical Psychiatry, 61,* 22–32.

Bremner, J. D., Mishra, S., Campanella, C., Shah, M., Kasher, N., Evans, S., ... Carmody, J. (2017). A pilot study of the effects of mindfulness-based stress reduction on post-traumatic stress disorder symptoms and brain response to traumatic reminders of combat in operation enduring freedom/operation Iraqi freedom combat veterans with post-traumatic stress disorder. *Frontiers in Psychiatry, 8.* https://doi.org/10.3389/fpsyt.2017.00157.

Brown, K. W., & Ryan, R. M. (2003). The benefits of being present: Mindfulness and its role in psychological well-being. *Journal of Personality and Social Psychology, 84,* 822–848.

Chambless, D. L., & Hollon, S. D. (1998). Defining empirically supported therapies. *Journal of Consulting and Clinical Psychology, 66,* 7–18.

Craske, M. G., Treanor, M., Conway, C. C., Zbozinek, T., & Vervliet, B. (2014). Maximizing exposure therapy: An inhibitory learning approach. *Behaviour Research and Therapy, 58,* 10–23. https://doi.org/10.1016/j.brat.2014.04.006.

Dimidjian, S., Hollon, S. D., Dobson, K. S., Schmaling, K. B., Kohlenberg, R. J., Addis, M. E., ... Jacobson, N. S. (2006). Randomized trial of behavioral activation, cognitive therapy, and antidepressant medication in the acute treatment of adults with major depression. *Journal of Consulting and Clinical Psychology, 74,* 658–670. https://doi.org/10.1037/0022-006X.74.4.658.

Easwaran, E. (2008). *The mantram handbook: A practical guide to choosing your mantram and calming your mind.* 5th ed) Tomales: Nilgiri Press.

Eftekhari, A., Zoellner, L. A., & Vigil, S. A. (2009). Patterns of emotion regulation and psychopathology. *Anxiety, Stress, and Coping, 22,* 571–586. https://doi.org/10.1080/10615800802179860.

Ekers, D., Webster, L., Van Straten, A., Cuijpers, P., Richards, D., & Gilbody, S. (2014). Behavioural activation for depression: An update of meta-analysis of effectiveness and sub group analysis. *PLoS One, 9.* https://doi.org/10.1371/journal.pone.0100100.

Feeny, N. C., Zoellner, L. A., Fitzgibbons, L. A., & Foa, E. B. (2000). Exploring the roles of emotional numbing, depression, and dissociation in PTSD. *Journal of Traumatic Stress, 13,* 489–498. https://doi.org/10.1023/A:1007789409330.

Flaxman, P. E., & Bond, F. W. (2010). A randomised worksite comparison of acceptance and commitment therapy and stress inoculation training. *Behavior Research and Therapy, 48,* 816–820.

Foa, E. B., Keane, T. M., Friedman, M. J., & Cohen, J. A. (2009). *Effective treatments for PTSD: Practice guidelines from the International Society for Traumatic Stress Studies* (2nd ed.). New York, NY: Guilford Press.

Forman, E. M., Herbert, J. D., Moitra, E., Yeomans, P. D., & Geller, P. A. (2007). A randomized controlled trial of acceptance and commitment therapy and cognitive therapy for anxiety and depression. *Behavior Modification, 31,* 772–799.

Frewen, P. A., & Lanius, R. A. (2006). Toward a psychobiology of posttraumatic self-dysregulation: Reexperiencing, hyperarousal, dissociation, and emotional numbing. *Annals of the New York Academy of Sciences, 1071,* 110–124. https://doi.org/10.1196/annals.1364.010.

Garland, E., & Roberts-Lewis, A. (2013). Differential roles of thought suppression and dispositional mindfulness in posttraumatic stress symptoms and craving. *Addictive Behaviors, 38*(2), 1555–1562. https://doi.org/10.1016/j.addbeh.2012.02.004.

Gloster, A. T., Meyer, A. H., & Lieb, R. (2017). Psychological flexibility as a malleable public health target: Evidence from a representative sample. *Journal of Contextual Behavioral Science, 6*, 166–171. https://doi.org/10.1016/j.jcbs.2017.02.003.

Goyal, M., Singh, S., Sibinga, E. M., Gould, N. F., Rowland-Seymour, A., Sharma, R., … Haythornthwaite, J. A. (2014). Meditation programs for psychological stress and well-being: A systematic review and meta-analysis. *JAMA Internal Medicine, 174*, 357–368.

Grecucci, A., Pappaianni, E., Siugzdaite, R., Theuninck, A., & Job, R. (2015). Mindful emotion regulation: Exploring the neurocognitive mechanisms behind mindfulness. *BioMed Research International, 2015*, 670724https://doi.org/10.1155/2015/670724.

Gutner, C. A., Galovski, T., Bovin, M. J., & Schnurr, P. P. (2016). Emergence of transdiagnostic treatments for PTSD and posttraumatic distress. *Current Psychiatry Reports, 18*, https://doi.org/10.1007/s11920-016-0734-x.

Harned, M. S., Korslund, K. E., & Linehan, M. M. (2014). A pilot randomized controlled trial of dialectical behavior therapy with and without the dialectical behavior therapy prolonged exposure protocol for suicidal and self-injuring women with borderline personality disorder and PTSD. *Behaviour Research and Therapy, 55*, 7–17. https://doi.org/10.1016/j.brat.2014.01.008.

Harned, M. S., Korslund, K. E., Foa, E. B., & Linehan, M. M. (2012). Treating PTSD in suicidal and self-injuring women with borderline personality disorder: Development and preliminary evaluation of a dialectical behavior therapy prolonged exposure protocol. *Behaviour Research and Therapy, 50*, 381–386.

Harned, M., & Linehan, M. (2008). Integrating dialectical behavior therapy and prolonged exposure to treat co-occurring borderline personality disorder and PTSD: Two case studies. *Cognitive and Behavioral Practice, 15*, 263–276.

Harned, M. S., Rizvi, S. L., & Linehan, M. M. (2010). Impact of co-occurring posttraumatic stress disorder on suicidal women with borderline personality disorder. *American Journal of Psychiatry, 167*, 1210–1217. https://doi.org/10.1176/appi.ajp.2010.09081213.

Hayes, S. A., Orsillo, S. M., & Roemer, L. (2010). Changes in proposed mechanisms of action in an acceptance-based behavior therapy for generalized anxiety disorder. *Behaviour Research and Therapy, 48*, 238–245.

Hayes, S. C., Bissett, R., Korn, Z., Zettle, R. D., Rosenfarb, I., Cooper, L., & Grundt, A. (1999). The impact of acceptance versus control rationales on pain tolerance. *The Psychological Record, 49*(1), 33–47.

Hayes, S. C., Luoma, J. B., Bond, F. W., Masuda, A., & Lillis, J. (2006). Acceptance and commitment therapy: Model, processes, and outcomes. *Behaviour Research and Therapy, 44*, 1–25.

Hayes, S. C., Strosahl, K. D., & Wilson, K. G. (2012). *Acceptance and commitment therapy: The process and practice of mindful change*. New York: The Guilford Press.

Hiraoka, R., Meyer, E. C., Kimbrel, N. A., DeBeer, B. B., Gulliver, S. B., & Morissette, S. B. (2015). Self-compassion as a prospective predictor of PTSD symptom severity among trauma-exposed U.S. Iraq and Afghanistan war veterans. *Journal of Traumatic Stress, 28*, 127–133.

Hopko, D. R., Bell, J. L., Armento, M. E. A., Hunt, M. K., & Lejuez, C. W. (2005). Behavior therapy for depressed cancer patients in primary care. *Psychotherapy: Theory, Research, Practice, Training, 42*, 236–243.

Hopko, D. R., Robertson, S. M. C., & Lejuez, C. W. (2006). Behavioral activation for anxiety disorders. *The Behavior Analyst Today, 7*, 212–224.

Jakupcak, M., Roberts, L. J., Martell, C., Mulick, P., Michael, S., Reed, R., … McFall, M. (2006). A pilot study of behavioral activation for veterans with posttraumatic stress disorder. *Journal of Traumatic Stress, 19*, 387–391. https://doi.org/10.1002/jts.20125.

Jakupcak, M., Wagner, A., Paulson, A., Varra, A., & McFall, M. (2010). Behavioral activation as a primary care-based treatment for PTSD and depression among returning veterans. *Journal of Traumatic Stress, 23*, 491–495.

Kabat-Zinn, J. (1990). *Full catastrophe living: Using the wisdom of your body and mind to face stress, pain, and illness.* New York, NY: Random House Publishing Group.

Kabat-Zinn, J. (2005). *Coming to our senses: Healing ourselves and the world through mindfulness.* New York, NY: Hyperion.

Kessler, R. C., Sonnega, A., Bromet, E., Hughes, M., & Nelson, C. B. (1995). Posttraumatic stress disorder in the National Comorbidity Study. *Archives of General Psychiatry, 52*, 1048–1060.

Kimbrel, N. A., Gratz, K. L., Tull, M. T., Morissette, S. B., Meyer, E. C., DeBeer, B. B., ... Beckham, J. C. (2015). Non-suicidal self-injury as a predictor of active and passive suicidal ideation among Iraq/Afghanistan veterans. *Psychiatry Research, 227*, 360–362. https://doi.org/10.1016/j.psychres.2015.03.026.

Lang, A. J., Strauss, J. L., Bomyea, J., Bormann, J. E., Hickman, S. D., Good, R. C., & Essex, M. (2012). The theoretical and empirical basis for meditation as an intervention for PTSD. *Behavior Modification, 36*, 759–786.

Lang, A. J., Schnurr, P. P., Jain, S., He, F., Walser, R. D., Bolton, E., ... Chard, K. M. (2017). Randomized controlled trial of acceptance and commitment therapy for distress and impairment in OEF/OIF/OND veterans. *Psychological Trauma Theory Research Practice and Policy, 9*, 74–84. https://doi.org/10.1037/tra0000127.

Linehan, M. M. (1993a). *Cognitive-behavioral treatment of borderline personality disorder.* New York, NY: Guilford Press.

Linehan, M. M. (1993b). *Skills training manual for treating borderline personality disorder.* New York, NY: Guilford Press.

Lutz, J., Herwig, U., Opialla, S., Hittmeyer, A., Jäncke, L., Rufer, M., ... Brühl, A. B. (2014). Mindfulness and emotion regulation—an fMRI study. *Social Cognitive and Affective Neuroscience, 9*, 776–785. https://doi.org/10.1093/scan/nst043.

Martell, C. R., Addis, M. E., & Jacobson, N. S. (2001). *Depression in context: Strategies for guided action.* New York: Norton.

Meyer, E. C., Morissette, S. B., Kimbrel, N. A., Kruse, M. I., & Gulliver, S. B. (2013). Acceptance and action questionnaire – II scores as a predictor of posttraumatic stress disorder symptoms among war veterans. *Psychological Trauma Theory Research Practice and Policy, 5*, 521–528. https://doi.org/10.1037/a0030178.

Meyer, E. C., Frankfurt, S. B., Kimbrel, N. A., DeBeer, B. B., Gulliver, S. B., & Morissette, S. B. (2018). The influence of mindfulness, self-compassion, psychological flexibility, and posttraumatic stress disorder on disability and quality of life over time in war veterans. *Journal of Clinical Psychology, 74*, 1272–1280. https://doi.org/10.1002/jclp.22596.

Meyer, E. C., La Bash, H., DeBeer, B. B., Kimbrel, N. A., Gulliver, S. B., & Morissette, S. B. (2018). Psychological inflexibility predicts PTSD symptom severity in war veterans after accounting for established PTSD risk factors and personality. *Psychological Trauma Theory Research Practice and Policy, 11*(4), 383–390. https://doi.org/10.1037/tra0000358.

Meyer, E. C., Walser, R. D., Hermann, B., La Bash, H., DeBeer, B. B., Morissette, S. B., ... Schnurr, P. P. (2018). Acceptance and commitment therapy for co-occurring PTSD and alcohol use disorders in veterans: Pilot treatment outcomes. *Journal of Traumatic Stress, 31*(5), 781–789. https://doi.org/10.1002/jts.22322.

National Institute for Clinical Excellence. (2005). *Post-traumatic stress disorder (PTSD): The management of PTSD in adults and children in primary and secondary care.* London: National Institute for Clinical Excellence.

Neff, K. (2003a). Self-compassion: An alternative conceptualization of a healthy attitude toward oneself. *Self and Identity, 2*, 85–101. https://doi.org/10.1080/15298860390129863.

Neff, K. D. (2003b). The development and validation of a scale to measure self-compassion. *Self and Identity, 2*, 223–250. https://doi.org/10.1080/15298860309027.

Neff, K., & Tirch, D. (2013). Self-compassion and ACT. In T. B. Kashdan & J. Ciarrochi (Eds.), *Mindfulness, acceptance, and positive psychology: The seven foundations of well-being* (pp. 78–106). Oakland, CA: Context Press/New Harbinger Publications.

Orcutt, H. K., Bonanno, G. A., Hannan, S. M., & Miron, L. R. (2014). Prospective trajectories of posttraumatic stress in college women following a campus mass shooting. *Journal of Traumatic Stress, 27*, 249–256. https://doi.org/10.1002/jts.21914.

Orsillo, S. M., Weathers, F. W., Litz, B. T., Steinberg, H. R., Huska, J. A., & Keane, T. M. (1996). Current and lifetime psychiatric disorders among veterans with war zone related posttraumatic stress disorder. *Journal of Nervous and Mental Disease, 184*, 307–313.

Orsillo, S. M., & Batten, S. V. (2005). Acceptance and commitment therapy in the treatment of posttraumatic stress disorder. *Behavior Modification, 29*, 95–129.

Öst, L.-G. (2008). Efficacy of the third wave of behavioral therapies: A systematic review and meta-analysis. *Behaviour Research and Therapy, 46*, 296–321.

Öst, L.-G. (2014). The efficacy of acceptance and commitment therapy: An updated systematic review and meta-analysis. *Behaviour Research and Therapy, 61*, 105–121. https://doi.org/10.1016/j.brat.2014.07.018.

Pagura, J., Stein, M. B., Bolton, J. M., Cox, B. J., Grant, B., & Sareen, J. (2010). Comorbidity of borderline personality disorder and posttraumatic stress disorder in the U.S. population. *Journal of Psychiatry Research, 44*, 1190–1198.

Polusny, M. A., Erbes, C. R., Thuras, P., Moran, A., Lamberty, G. J., Collins, R. C., ... Lim, K. O. (2015). Mindfulness-based stress reduction for posttraumatic stress disorder among veterans: A randomized clinical trial. *Journal of the American Medical Association, 314*, 456–465. https://doi.org/10.1001/jama.2015.8361.

Powers, M. B., Zum Vörde Sive Vörding, M. B., & Emmelkamp, P. M. G. (2009). Acceptance and commitment therapy: A meta-analytic review. *Psychotherapy and Psychosomatics, 78*, 73–80.

Roberts, N. P., Roberts, P. A., Jones, N., & Bisson, J. I. (2015). Psychological interventions for PTSD and comorbid substance use disorder: A systematic review and meta-analysis. *Clinical Psychology Review, 38*, 25–38. https://doi.org/10.1016/j.cpr.2015.02.007.

Rojas, S. M., Bujarski, S., Babson, K. A., Dutton, C. E., & Feldner, M. T. (2014). Understanding PTSD comorbidity and suicidal behavior: Associations among histories of alcohol dependence, major depressive disorder, and suicidal ideation and attempts. *Journal of Anxiety Disorders, 28*, 318–325.

Sells, J. R., Waters, A. J., Schwandt, M. L., Kwako, L. E., Heilig, M., George, D. T., & Ramchandani, V. A. (2016). Characterization of comorbid PTSD in treatment-seeking alcohol dependent inpatients: Severity and personality trait differences. *Drug and Alcohol Dependence, 163*, 242–246. https://doi.org/10.1016/j.drugalcdep.2016.03.016.

Shepherd, L., & Wild, J. (2014). Emotion regulation, physiological arousal and PTSD symptoms in trauma-exposed individuals. *Journal of Behavior Therapy and Experimental Psychiatry, 45*, 360–367. https://doi.org/10.1016/j.jbtep.2014.03.002.

Shipherd, J. C., & Beck, J. G. (2005). The role of thought suppression in posttraumatic stress disorder. *Behavior Therapy, 36*, 277–287.

Steenkamp, M. M., & Litz, B. T. (2013). Psychotherapy for military-related posttraumatic stress disorder: Review of the evidence. *Clinical Psychology Review, 33*, 45–53. https://doi.org/10.1016/j.cpr.2012.10.002.

Steenkamp, M. M., & Litz, B. T. (2014). One-size-fits-all approach to PTSD in the VA not supported by the evidence. *American Psychologist, 69*, 706–707. https://doi.org/10.1037/a0037360.

Thompson, B. L., Luoma, J. B., & LeJeune, J. T. (2013). Using acceptance and commitment therapy to guide exposure-based interventions for posttraumatic stress disorder. *Journal of Contemporary Psychotherapy, 43*, 133–140.

Tirch, D., Schoendorff, B., & Silberstein, L. R. (2014). *The ACT practitioner's guide to the science of compassion*. Oakland, CA: New Harbinger.

Treanor, M. (2011). The potential impact of mindfulness on exposure and extinction learning in anxiety disorders. *Clinical Psychology Review, 31*, 617–625. https://doi.org/10.1016/j.cpr.2011.02.003.

Vujanovic, A. A., Bonn-Miller, M. O., Potter, C. M., Marshal, E. C., & Zvolensky, M. J. (2011). An evaluation of the relation between distress tolerance and posttraumatic stress within a trauma-exposed sample. *Journal of Psychopathology and Behavioral Assessment, 33*, 129–135. https://doi.org/10.1007/s10862-010-9209-2.

Vujanovic, A. A., Hart, A. S., Potter, C. M., Berenz, E. C., Niles, B., & Bernstein, A. (2013). Main and interactive effects of distress tolerance and negative affect intensity in relation to PTSD symptoms among trauma-exposed adults. *Journal of Psychopathology and Behavioral Assessment, 53*, 235–243. https://doi.org/10.1007/s10862-012-9325-2.

Wagner, A., Zatzick, D., Ghesquiere, A., & Jurkovich, G. (2007). Behavioral activation as an early intervention for posttraumatic stress disorder and depression among physically injured trauma survivors. *Cognitive and Behavioral Practice, 14*, 341–349.

Walser, R. D., Garvert, D. W., Karlin, B. E., Trockel, M., Ryu, D. M., & Taylor, C. B. (2015). Effectiveness of acceptance and commitment therapy in treating depression and suicidal ideation in veterans. *Behaviour Research and Therapy, 74*, 25–31.

Yadavaia, J. E., Hayes, S. C., & Vilardaga, R. (2014). Using acceptance and commitment therapy to increase self-compassion: A randomized controlled trial. *Journal of Contextual Behavioral Science, 3*, 248–257. https://doi.org/10.1016/j.jcbs.2014.09.002.

CHAPTER 20

Self-compassion in PTSD[*]

Christine Braehler[a], Kristin Neff[b]
[a]University of Glasgow, United Kingdom
[b]University of Texas at Austin, United States

> *Compassion is the antitoxin of the soul:*
> *where there is compassion even the most poisonous impulses remain*
> *relatively harmless.*
>
> Eric Hoffer (1955)

Defining self-compassion

Self-compassion is formally defined and measured by Neff (Neff, 2003; Neff et al., 2018) as representing the balance between increased positive and decreased negative self-responding in times of personal struggle. Self-compassion entails being kinder and more supportive toward oneself and less harshly judgmental. It involves greater recognition of the shared human experience, understanding that all humans are imperfect and lead imperfect lives, and fewer feelings of being isolated by one's imperfection. It entails mindful awareness of personal suffering and ruminating less about negative aspects of oneself or one's life experience. The six components of self-compassion are conceptually distinct yet operate as a system, tapping into different ways that individuals emotionally respond to suffering (with kindness or judgment), cognitively understand their predicament (as part of the human experience or as isolating), and pay attention to pain (in a mindful or overidentified manner).

Self-kindness versus self-judgment

Most of us try to be kind and considerate toward our friends and loved ones when they make a mistake, feel inadequate, or suffer some misfortune. We may offer words of support and understanding to let them know we care—perhaps even a physical gesture of affection such as a hug. We might

---

[*]Note: Client's stories have been altered and blended to protect clients' identities. All clients were treated by Christine Braehler, Clinical Psychologist, DClinPsy, PhD.

ask them "what do you need right now?" and consider what we can do to help. Curiously, we often treat ourselves very differently. We say harsh and cruel things to ourselves that we would never say to a friend. In fact, we're often tougher on ourselves than we are with people that we don't like very much. The kindness inherent to self-compassion, however, puts an end to the constant self-judgment and disparaging internal commentary that most of us have come to see as normal. Our internal dialogues become benevolent and encouraging rather than punishing or belittling, reflecting a friendlier and more supportive attitude toward ourselves. We begin to understand our weaknesses and failures instead of condemning them. We acknowledge our shortcomings while accepting ourselves unconditionally as flawed, imperfect human beings. Most importantly, we recognize the extent to which we harm ourselves through relentless self-criticism and choose another way.

Self-kindness involves more than merely ending self-criticism, however. It involves actively opening up our hearts to ourselves and responding to our suffering as we would to a dear friend in need. Beyond accepting ourselves without judgment, we may also soothe and comfort ourselves in the midst of emotional turmoil. We are motivated to try to help ourselves, to ease our own pain if we can. Normally, even when we experience unavoidable problems like having an unforeseeable accident, we focus more on fixing the problem than caring for ourselves. We treat ourselves with cold stoicism rather than warmth or tender concern and move straight into problem-solving mode. With self-kindness, however, we learn to nurture ourselves when life is difficult, offering ourselves support and encouragement. We allow ourselves to be emotionally moved by our own pain, stopping to say, "This is really difficult right now. How can I care for myself in this moment?" If we are being threatened in some way, we actively try to protect ourselves from harm. We can't be perfect, and our lives will always involve struggle. When we deny or resist our imperfections, we exacerbate our suffering in the form of stress, frustration, and self-criticism; however, when we respond to ourselves with benevolence and goodwill, we generate positive emotions of love and care that help.

Common humanity versus isolation

Self-compassion is embedded within a sense of interconnection rather than separation. One of the biggest problems with harsh self-judgment is that it tends to make us feel isolated and cut off from others. When we fail or feel inadequate in some way, we irrationally feel like everyone else is just fine and it's only me who is such a hopeless loser. This creates a frightening sense of disconnection and loneliness that greatly exacerbates our suffering.

With self-compassion, however, we recognize that life challenges and personal failures are part of being human, an experience we all share. In fact, our flaws and weaknesses are what make us card-carrying members of the human race. The element of common humanity also helps to distinguish self-compassion from mere self-acceptance or self-love. While self-acceptance and self-love are important, they are incomplete by themselves. They leave out an essential factor—other people. Compassion is, by definition, relational. It implies a basic mutuality in the experience of suffering and springs from the acknowledgement that the human experience is imperfect. Why else would we say "it's only human" to console someone who has made a mistake? Self-compassion honors the fact that all human beings are fallible and that taking wrong turns is an inevitable part of living. When we're in touch with our common humanity, we remember that everyone has feelings of inadequacy and disappointment. The pain I feel in difficult times is the same pain that you feel in difficult times. The triggers are different, the circumstances are different, the degree of pain is different, but the process is the same. With self-compassion, every moment of suffering is an opportunity to feel closer and more connected to others. It reminds us that we are not alone.

Mindfulness versus overidentification

To have compassion for ourselves, we need to be willing to turn toward our own pain and to acknowledge it with mindfulness. Mindfulness is a type of balanced awareness that neither resists, avoids, nor exaggerates our moment-to-moment experience. In this receptive mind state, we become aware of our negative thoughts and feelings and are able to just be with them as they are, without fighting or denying them. We recognize when we're suffering, without immediately trying to fix our feelings and make them go away.

We might think that we don't need to become mindfully aware of our suffering. Suffering is blindingly obvious, isn't it? Not really. We certainly feel the pain of falling short of our ideals, but our mind tends to focus on the failure itself, rather than the pain caused by failure. This is a crucial difference. When our attention becomes completely absorbed by our perceived inadequacies, we can't step outside ourselves. We become overly identified with our negative thoughts or feelings and are swept away by our aversive reactions. This type of rumination narrows our focus and exaggerates implications for self-worth (Nolen-Hoeksema, 1991). Not only did I fail, but also I am a failure. Not only am I disappointed, but also my life is disappointing. Overidentification means that we reify our moment-to-moment experience, perceiving transitory events as definitive and permanent.

With mindfulness, however, everything changes. Rather than confusing our negative self-concepts with our actual selves, we can recognize that our thoughts and feelings are just that—thoughts and feelings—which help us to drop our absorption in the storyline of our inadequate, worthless selves. Like a clear, still pool without ripples, mindfulness mirrors what's occurring without distortion so that we can take a more objective perspective on ourselves and our lives.

Mindfulness also provides the mental spaciousness and equanimity needed to see and do things differently. When we are mindful, we can also wisely determine the best course of action to help ourselves when in need, even if that means simply holding our experience in gentle, loving awareness. It takes courage to turn toward our pain and acknowledge it, but this act of courage is essential if our hearts are to open in response to suffering. We can't heal what we can't feel. For this reason, mindfulness is the pillar on which self-compassion rests.

Higher levels of self-compassion have repeatedly been associated with psychological well-being (MacBeth & Gumley, 2012; Zessin, Dickhäuser, & Garbade, 2015), while lower levels of self-compassion have been associated with increased symptoms of depression (Ehret, Joormann, & Berking, 2015; Krieger, Altenstein, Baettig, Doerig, & Holtforth, 2013), generalized anxiety disorder (Hoge et al., 2013), psychosis (Eicher, Davis, & Lysaker, 2013), eating disorders (Kelly & Tasca, 2016), bipolar disorder (Døssing et al., 2015), and PTSD (Hiraoka et al., 2015). Self-compassion therefore appears to increase emotional (Dahm et al., 2015; Zeller, Yuval, Nitzan-Assayag, & Bernstein, 2015) and physical resilience (Breines et al., 2014) by buffering against the pain we all inevitably experience. It would seem logical to conclude that individuals with psychological disorders—especially those affected by PTSD—may benefit from (1) becoming aware of their pain instead of avoiding it or being overidentified with it, (2) feeling more connected instead of isolated, and (3) caring for themselves instead of judging themselves.

Germer and Neff (Germer & Neff, 2019; Neff, 2018; Neff & Germer, 2018) note that there is both a yin and a yang aspect to self-compassion. Yin self-compassion involves "being with" pain by validating, soothing, and comforting the self. The three components of kindness, common humanity, and mindfulness manifest in yin self-compassion as loving, connected presence. When individuals hold their pain in loving, connected presence, they begin to heal. Yang self-compassion involves acting in the world by protecting oneself, providing what is needed in the moment, and motivating change. The components of yang self-compassion can manifest as fierce, empowered truth (Neff, 2018). Self-kindness means we fiercely protect ourselves to prevent harm. Common humanity helps us to recognize that we are not alone and that we don't need to hang our heads in shame. We can stand together with others in the experience of being harmed and become empowered as a result. And mindfulness manifests as clearly seeing and speaking the truth.

Research indicates that both aspects of self-compassion lead to well-being. For instance, yin self-compassion reduces depression, anxiety, and shame (Johnson & O'Brien, 2013; MacBeth & Gumley, 2012) by replacing self-judgment with self-acceptance. When individuals take refuge in the safety of their own warmth and care, they become happier and more satisfied with their lives as a result (Neff, Kirkpatrick, & Rude, 2007). Yang self-compassion allows individuals to actively cope with life challenges. Whether it's combat (Hiraoka et al., 2015), divorce (Sbarra, Smith, & Mehl, 2012), cancer (Pinto-Gouveia, Duarte, Matos, & Fráguas, 2014), or parenting a special-needs child (Neff & Faso, 2015), self-compassion provides people with the resilience needed to stand strong without becoming overwhelmed. Yang self-compassion motivates individuals to keep going even after failure and setbacks, providing grit and perseverance in the face of adversity.

Self-compassion and PTSD

What do we currently know about the relationship between self-compassion and PTSD? Although research on this topic is still new, the body of literature is growing. Thompson and Waltz found self-compassion to be negatively associated with the avoidance cluster of PTSD but unrelated to the hyperarousal and reexperiencing clusters in a student sample (Thompson & Waltz, 2008). Maheux and Price (2015) found self-compassion to be negatively associated with PTSD symptoms in two nonclinical samples when using both DSM-IV and DSM-5 definitions. Scoglio et al. (2018) found self-compassion to be negatively related to PTSD symptoms among women who were seeking treatment after experiencing interpersonal violence. Karatzias et al. (2017) examined the link between self-compassion and complex PTSD. According to the forthcoming 11th version of the

International Classification of Diseases (ICD-11), complex PTSD is defined as consisting of the existing PTSD criteria plus a set of symptoms referred to as "disturbances in self-organization," which summarize the pervasive dysregulating effects of chronic victimization on affect, the sense of self, and relational functioning (Maercker et al., 2013) (Herman, 1992). Interestingly, self-compassion was found to be negatively associated with the "disturbance in self-organization" factor, but not with a general PTSD factor in this largely female sample referred to trauma therapy services (Karatzias et al., 2017). More specifically, low self-compassion was linked to negative self-concept, relationship difficulties, and affect dysregulation—particularly hypoactivation strategies. Hypoactivation of distress involves attempts to downregulate, numb, and turn the distress inward, which are in keeping with shame, defeat, dissociation, and feelings of depression, whereas hyperactivating strategies involve increasing arousal, expressing distress, and potentially becoming aggressive. Such an internalization of the abusive treatment by others is a common consequence of chronic interpersonal trauma, especially when it occurs early in life and in caregiving contexts (van der Kolk, Roth, Pelcovitz, Sunday, & Spinazzola, 2005). In contrast, increasing self-compassion among individuals with PTSD is hypothesized to help them to improve their affect regulation, self-concept, and relational functioning by reducing their feelings of shame, guilt, failure, and defeat (Lee, Scragg, & Turner, 2001).

Self-compassion, shame, and PTSD

Self-compassion may be an effective antidote to the shame experienced by those with PTSD, where feelings of kindness, common humanity, and mindfulness can come to replace feelings of self-blame, isolation, and emotional avoidance. Although shame is known to exacerbate many types of psychological disorders, including PTSD, it is still frequently overlooked in treatment, which has historically prioritized anxiety, fear, and anger (Taylor, 2015).

What is shame? Shame is a universal emotion (Sznycer et al., 2018) characterized by a state of hypoarousal and submission that evolved to avert attack by members from one's own group (Keltner & Harker, 1998). To escape the attack, shame activates submissive or aggressive defenses. If those are ineffective and the individual cannot find safety or support with another person or group, then primary consciousness shuts down leading to dissociation (Schore, 2015). Shame can have a paralyzing effect, and thus being shamed can be considered traumatic if it happens early in development and if done by a caregiver (Matos & Pinto-Gouveia, 2010). Submissive strategies involve pleasing powerful others by adapting and giving up one's own will, internalizing their opinions about oneself, correcting or punishing oneself, suppressing anger and self-protective impulses in the body, losing any boundaries with powerful others, and becoming complacent. Aggressive strategies involve some form of counterattack such as blaming others overtly or covertly by talking negatively about people behind their backs or by using physical aggression.

Shame may help to maintain and exacerbate PTSD by increasing the severity and duration of illness (Brewin & Holmes, 2003) over and above the impact of exposure to trauma (DePrince, Chu, & Pineda, 2011). For example, intrusive memories of traumatic experiences were found to be accompanied more often by feelings of shame than by feelings of fear, horror, or helplessness (Holmes, Grey, & Young, 2005). Other recent research indicates that shame may mediate the relationship between PTSD symptom severity and suicidal ideation among veterans with PTSD (Cunningham, Davis, Wilson, & Resick, 2018). In contrast, self-compassion might help veterans to better adjust once back home by reducing their shame, as higher self-compassion has been linked to lower PTSD symptoms, general psychopathology, and better functioning among returning veterans, irrespective of trauma exposure (Dahm et al., 2015; Hiraoka et al., 2015).

Shame is often a result of childhood trauma, which is a risk factor for the later development of PTSD and other pathologies. Childhood maltreatment has been linked to higher shame and lower self-compassion and worse PTSD symptoms (Andrews, Brewin, Rose, & Kirk, 2000) via emotion dysregulation (Barlow, Goldsmith Turow, & Gerhart, 2017; Scoglio et al., 2018; Vettese, Dyer, Li, & Wekerle, 2011). Children may internalize their caregivers' contemptuous and hostile intentions by developing deeply shaming core beliefs such as "I am bad/evil/disgusting/unworthy," which resolves some of the cognitive dissonance (Briere, 1992). Not surprisingly, higher doses of early trauma are associated with lower levels of self-compassion (Játiva & Cerezo, 2014; Tanaka, Wekerle, Schmuck, & Paglia-Boak, 2011). For example, highly critical or otherwise dysfunctional family environments and emotional abuse have been linked to low self-compassion and insecure attachment (Neff & McGehee, 2010; Tanaka et al., 2011). In turn, low levels of self-compassion have been associated with depression and anxiety (Joeng et al., 2017), self-harming behavior (Jiang et al., 2016), anxiety (Berryhill, Hayes, & Lloyd, 2018), and PTSD symptoms (Bistricky et al., 2017).

Self-compassion has the potential to buffer against the impact of traumatic stress through the process of cognitive appraisal and emotion regulation (Barlow et al., 2017; Játiva & Cerezo, 2014; Zeller et al., 2015) and by helping individuals to make better use of social support (Maheux & Price, 2015). Much like attachment security, self-compassion offers an inner safe haven, where individuals can seek refuge and recover when distressed, and a secure base, from which they can explore the world and connect to others to feel energized again (Bowlby, 1988). Capacities for emotion regulation and mentalizing/psychological flexibility develop in early attachment contexts and are critical for later emotional and interpersonal functioning (Fonagy, Gergely, Jurist, & Target, 2002). Self-compassion involves both emotional regulation (self-kindness) and mentalizing (mindfulness and common humanity). Research demonstrates that attachment styles are, however, fluid across the lifespan and that a secure attachment style can be developed later in life through corrective experiences with another attachment figure, such as a teacher, a romantic partner, or a spiritual being—an "earned secure" style (Roisman, Padrón, Sroufe, & Egeland, 2002). It could be argued that self-compassion is the result of such a corrective experiences, which have been shown to buffer against the impact of dysfunctional family experiences (Berryhill et al., 2018; Homan, 2016; Jiang, You, Zheng, & Lin, 2017).

Self-compassion and the treatment of PTSD

A number of psychotherapies have been found to be efficacious treatments for PTSD (Cusack et al., 2016). What is the role of self-compassion, if any, in standard treatments for PTSD? Two individual trauma-focused approaches recommended for PTSD sufferers with childhood trauma are DBT-PTSD (Bohus et al., 2013) and STAIR/MPI Training (Cloitre, Koenen, Cohen, & Han, 2002), which both include phases of alliance building and skills training, followed by exposure. STAIR/MPI invites patients to create narratives of shame but does not directly train in self-compassion. DBT-PTSD has more recently begun to include self-compassion modules to reduce shame.

Hoffart, Øktedalen, and Langkaas (2015) randomized 65 PTSD patients to either standard prolonged exposure, which includes imaginal exposure to the traumatic memory, or modified prolonged exposure, with imagery rescripting of the memory instead of imaginal exposure in a 10-week residential program. Improvements in self-kindness, self-judgment, isolation, and overidentification each had significant effects on risk for subsequent PTSD symptoms across therapies and most strongly in patients with higher initial self-judgment; however, what remains unclear is what helped the patients to develop self-compassion? Was it a function of the patients' or the therapists' characteristics, the therapeutic setting, the therapeutic relationship, or other events in the patients' lives? Kearney et al. (2013) offered war veterans with PTSD group-based loving-kindness meditation (LKM) training. By the end of the training, self-compassion had increased, with a large effect size. At the 3-month follow-up, increases in self-compassion were associated with reduced PTSD and depressive symptoms. However, this study was limited by the fact that veterans received concurrent treatment making it difficult to assess any specific effect of LKM.

Compassion-focused therapy was developed to reduce shame and increase self-compassion in psychiatric groups as an adjunct to cognitive behavior therapy (CBT) (Gilbert, 2010). Beaumont and colleagues (Beaumont, Galpin, & Jenkins, 2012) compared individual CBT versus CBT plus 12 sessions of individual compassionate mind training (CMT) in 32 individuals referred following a traumatic incident. Both treatment groups experienced a comparable reduction in anxiety, depression, hyperarousal, reexperiencing, and avoidance symptoms. Both treatment groups also demonstrated an increase in self-compassion, which was statistically greater in the CBT plus CMT group but unrelated to symptoms.

The strongest evidence to date of the potential of CFT to reduce shame and to increase compassion comes from two pilot RCTs focusing on psychosis (Braehler, Gumley, et al., 2013) and eating disorders (Kelly, Wisniewski, Martin-Wagar, & Hoffman, 2017). Kelly had previously shown shame to trigger episodes of disordered eating and demonstrated that TAU (individual CBT with DBT elements) plus 12 sessions of group-based compassionate mind training adapted for eating disorders resulted in significantly greater improvements in self-compassion, shame, and pathology than TAU alone. Braehler et al. compared 16-session CFT group therapy adapted for psychosis (Braehler, Harper, & Gilbert, 2013) with TAU involving only nursing support but not therapy. Individuals in the CFT group showed significantly greater clinical improvements than TAU and greater increases in self-compassion and decreases in perceived sense of marginalization (external shame) and depressive symptoms (Braehler, Gumley, et al., 2013). Although these results do not directly translate to the treatment of PTSD, these studies suggest that clinical adaptations of group-based compassion-based trainings and group therapies are safe, acceptable, and beneficial if (1) delivered by clinicians who can determine the timing of the intervention to suit the individual's needs and capacities and (2) offered as part of routine mental health care with options to offer individual therapy (Kirby, Tellegen, & Steindl, 2017).

Although the research to date clearly suggests that self-compassion may protect against the toxic effects of shame in PTSD, the preliminary intervention data are limited. Nonetheless, Judith Herman, a pioneer of trauma research and therapy, summarizes the work of trauma recovery as involving "overcoming barriers to shame and secrecy, making intolerable feelings bearable through connection with others, grieving the past, and coming to a new perspective with a more compassionate view of oneself in the present" (Herman, 2015, p. 276), thus placing self-compassion at the heart of trauma recovery.

Understanding fears and barriers to self-compassion

Asking a person who is more familiar with being in abusive than nonabusive relationships to "treat herself like a dear friend," as is done in some self-compassion practices, may simply be impossible for the person to do, as she has no template to draw upon. The repair of the attachment system of this person must start by slowly developing a secure attachment to the therapist. How should a patient hold herself in a kind embrace when she has never been held unconditionally? Learning to trust and to receive kindness and care from another seems the logical first step in treatment. Unfortunately, individuals with attachment trauma are often deeply mistrustful and afraid of compassion and kindness from others, which is the real starting point of therapy.

Why are some people afraid of kindness and care? Such difficulties are often rooted in early experiences of receiving care and likewise may make it difficult to extend compassion. Our attachment system is shaped by our experience of how our caregivers responded to us when we were in distress. Consequently, our reactions to compassion and our obstacles to developing self-compassion will vary as a function of our early attachment experiences. If we experienced neglect or some form of emotional or physical abuse at times of distress, we will have formed emotional memories linking the experience of needing and/or receiving care with negative emotions of shame, anger, loneliness, fear, or vulnerability. Our ability to feel affiliative emotions such as love, compassion, longing, and grief toward others and ourselves can be severely compromised. Germer and Neff (Germer & Neff, 2019; Neff & Germer, 2018) summarize this paradoxical experience by saying, "When we give ourselves unconditional love, we uncover the conditions under which we have not been loved."

Receiving kindness from a therapist may be a first step in helping patients to desensitize to the care that most longed for, but many clients experience receiving compassion from another as aversive. Why would clients be rejecting of the very quality they need to alleviate their suffering? Whereas many people experience ease when imagining a compassionate friend, individuals with more insecure attachments experience it as subjectively and objectively stressful (Rockliff, Gilbert, McEwan, Lightman, & Glover, 2008). Common metacognitive beliefs people hold are as follows: "I will become dependent on it." "I do not deserve it." "It will make me weak/lazy/selfish." "I will be overwhelmed by distress." "I will let myself off the hook." "Others will take advantage of me" (Gilbert, McEwan, Matos, & Rivis, 2011). Such fears and resistance to receiving and giving compassion have been associated with greater anxiety and depression in the general population (Gilbert et al., 2012) and in clinical cohorts (Gilbert, McEwan, Catarino, Baião, & Palmeira, 2014).

Early memories of shame and fewer memories of warmth and safeness have been linked to being more afraid of receiving compassion from others and receiving compassion from oneself and to increased anxiety, depression, and paranoia (Matos, Duarte, & Pinto-Gouveia, 2017). In both nondepressed and clinically depressed samples, fear of receiving compassion from others has been strongly correlated with fear of receiving compassion from oneself. A combined fear of compassion factor predicted 53% of variance in depressive symptoms and correlated strongly with self-criticism, a known predictor of depression (Gilbert, McEwan, Catarino, & Baião, 2014a). Fear of compassion has also been associated with greater attachment insecurity in clinically depressed individuals (Gilbert, McEwan, Catarino, & Baião, 2014a; Gilbert, McEwan, Catarino, Baião, & Palmeira, 2014).

Although research refers to "fears of compassion," the concept really taps into people's fears about what reactions they may have to compassion. More in-depth research shows that fear of sadness, in particular, is associated with depression and that fear and avoidance of sadness and anger correlate with fears of compassion (Gilbert, McEwan, Catarino, & Baião, 2014b). This makes intuitive sense: if one has no early memories of safeness and warmth when being distressed to draw upon, then experiencing compassion might undermine one's emotion regulation capacities. Thus the expression of difficult emotions can be threatening if we cannot build on representations of others lovingly validating and holding our emotions.

Another obstacle to receiving compassion from others can be related to mentalizing difficulties, such as a compromised ability to infer and reflect on the mental states of oneself and others and to take different perspectives (Fonagy et al., 2002), which is similar to psychological inflexibility (Miron, Seligowski, Boykin, & Orcutt, 2016). Gilbert found that individuals with greater fears of compassion not only were more self-critical but also struggled to label their emotions and to talk about their feelings making it more difficult to notice and feel their pain (Gilbert et al., 2012). Fear of self-compassion and high psychological inflexibility have been shown to interact to predict PTSD symptom severity in students with trauma exposure (Boykin et al., 2018; Hiraoka et al., 2015). Clients with the highest levels of shame and self-criticism experience the lowest level of self-compassion and the greatest fears of receiving compassion from others including from oneself, which correlates with insecure attachment (Gilbert, McEwan, Catarino, Baião, & Palmeira, 2014), low mentalizing ability (Boykin et al., 2018), and worse outcomes (Kelly & Carter, 2013; Miron et al., 2016; Vettese et al., 2011). Therefore adding self-compassion to the clinical tool box requires, first and foremost, awareness and skills for overcoming these fears and barriers while taking into account their clients' mentalizing capacity. Compassion-based work is attachment-based work and taps directly into early experiences of attention, care, love, and appreciation or lack thereof. As with all trauma-based work, it requires clinicians to be sensitive, flexible, diligent, and willing to become a safe haven and secure base for their clients until they learn to become a safe haven and secure base for themselves.

Safely navigating unchartered attachment trauma territory

Research shows that isolated self-compassion exercises can activate threat states in individuals with early attachment trauma instead of creating safeness (Rockliff et al., 2008). Research also suggests that receiving compassion from others—including a therapist—may evoke fears in clients with childhood trauma and high shame. If clients fear the care they long for, then how can we safely integrate self-compassion into therapy? The main question to ask is as follows: "How can I, as a therapist, safely access the care system in this person without unnecessarily activating the threat system?"

If a person presents with complex PTSD, shame, and self-blame, it is important to gauge if they have any fears of compassion from others, for others, or for self. Self-report questionnaires might help to assess conscious fears (Gilbert et al., 2011); however, these fears of receiving kindness often show up relationally between client and therapist. The following variables are important in understanding how these fears might manifest between a therapist and a client.
- Attachment
- Emotion regulation
- Mentalization or psychological flexibility

Attachment

If securely attached, clients are less likely to show such fears as they can draw on an inner working model that considers others and oneself as reliable sources of support at times of distress. The person considers herself worthy of care and is therefore open to accepting help from the therapist. The person also considers herself to be capable of supporting herself and therefore is likely to find it easier to offer herself compassion when in distress.

Anxious-preoccupied attachment tendencies might manifest as a sense of needing the therapist or other people in her life for support, guidance, reassurance, and comfort as the person does not trust her own ability of supporting herself. The idea of giving herself what she needs may evoke barriers such as not knowing what she needs and wants, not trusting her own feelings and needs, and fearing that if she did care for herself that she would lose the attachment bonds with significant others. The person may require the supportive encouragement from the therapist to gradually develop self-knowledge, trust in her own experiences and capacities to care, and self-efficacy. Therapists at first require compassion for the restlessness and clinging that stems from the lack of an inner safe haven to avoid being dismissive and to trust in the person's potential to develop such an inner safe haven to avoid staying in a parental role that would foster the sense of dependence on others. Adopting the role of a supportive, encouraging sports coach who often says "I trust that you can do it. Give it a try. I am here to catch you" is likely to support clients.

Dismissive-avoidant attachment tendencies are the opposite of the anxious-preoccupied ones. The person has come to exclusively rely on herself as other human beings could not be relied upon at times of need. Seeking help from others will evoke fears of being abandoned, let down, or disappointed and will likely evoke great shame because the person must acknowledge that her assumption that she can take care of all her needs has been undermined. Expressing respect for and genuinely honoring the strength that the person has shown to survive without others' help are important in beginning to engage clients with dismissive-avoidant attachment strategies. Treating the person with respect—including respecting their need for autonomy—will be critical throughout. Therapists can engage the person by giving sensitive and attuned care from a place of being an authentic human being who sometimes struggles, instead of a parent who is helping a child. Using focused self-disclosure can help to reduce shame in the client. For example, if a therapist feels the client hesitating to open up, the therapists may ask how they are feeling about coming to see them and seeking help. The therapist may normalize that experience in different ways including sharing their own shame about seeking help or fears of being vulnerable at different times in their lives.

Fearful-avoidant attachment tendencies in adults result from a sense of being unable to trust neither themselves nor others to care for them at times of distress, creating confusion, dissociation, and intense despair. Such expectations are strongly associated with sexual, physical, or emotional abuse and neglect by caregivers during childhood (Van Ijzendoorn, Schuengel, & Bakermans-Kranenburg, 1999). The attachment figure was a source of fear accounting for the intensely ambivalent and confusing behavior in relationships. Seeking help is likely to be a struggle for clients with this attachment conditioning as they touch on the fear of being abused again. Significant courage and motivation to help oneself are needed to overcome this fear. Expressions of care from the therapist are likely to be met with intense fear and reactivation of early relational trauma, such as flashbacks, dissociation, or numbing. Rage or anger might be another response to create distance again from the therapist who is seeking to come closer. Clients with such distressing early experiences require the utmost stability, attunement, flexibility, patience, and good will from the therapist. Any rigid expectations with regard to how the client should respond and behave or how therapy should unfold may be met with increased dysregulation and could potentially result in the client dropping out of therapy prematurely.

Emotion regulation

Insecure and disorganized attachment styles are linked to emotion regulation difficulties. If caregivers did not sufficiently soothe the distress of the child, the child needed to develop alternative ways to regulate the distress, either by making it disappear or by increasing its display to get help. Hypoactivating emotion regulation strategies involve the downregulation of affect to hide it not only from others but also from oneself. These can include withdrawal from others, not speaking about feelings to others, experiential avoidance, or numbing feelings through substances or food. Hyperactivating emotion regulation strategies involve the upregulation of affect to elicit a care response from others. These can include worry, impulsiveness, high emotional expression, or frequent seeking of closeness including sexual intimacy.

Mentalization

Attachment style partly predicts the ability to infer one's own and other mental states and to shift perspectives and hold different perspectives in mind (i.e. mentalizing), which is an emotion regulation capacity as it helps to make sense of our relationships (Fonagy et al., 2002). Good mentalizing involves being able to make sense of one's own and others' behavior in terms of mental states, such as intentions, feelings, and thoughts. Low mentalizing ability shows up in different ways. When clients are stuck in self-blame, they tend to get overidentified with stories of suffering, which serves to maintain their distress instead of putting their heart to rest. Exaggerated self-blame distorts the actual attribution of responsibility in a given situation. Alternatively, low mentalizing may manifest as clients who struggle to shift perspectives and who might project their perception onto others or lack the ability to see the world in terms of mental states but rather in concrete behaviors.

Principles for selecting a starting point in treatment

The following descriptions apply to complex PTSD clients with shame who have experienced childhood abuse from a caregiver and have not had any corrective interpersonal experiences. Hence it makes sense to enhance evidence-based treatments for complex PTSD by including self-compassion when dealing with shame (Lee et al., 2001). Following an assessment and formulation, including experiences of safeness, warmth, and care, the therapist might want to formulate some hypotheses about where roadblocks may occur. Despite the best analysis and preparation, many attachment traumas remain hidden until they are activated during therapy. The following three principles developed by me (CB) may help to prevent unnecessary activation of the threat system and help overcome fears of compassion safely.

Desensitization: Warming up slowly

As the client with attachment trauma may experience a threat state at any time when opening the attachment system, the therapist needs to "warm up slowly." First, this means that the therapist and client look for an easy and safe entry point to bring the care system "online" without evoking any distress. Therapists might want to normalize the paradoxical effect that the client may at first feel worse as the care system goes online before they start to feel better: "Imagine you have been out in the snow and cold without gloves on for a while. Your hands will start to feel ice cold. As you enter the warm house to warm your hands over the fire, you might feel some pain as your hands receive the warmth they so urgently needed. To titrate the pain, you approach the fire slowly so your hands can slowly absorb the warmth.

It is like a temporary pain of healing, that is, in and of itself, is not harmful. The warmth is what you need to survive in the long term. Starting with lukewarm is perfectly okay and wise."

The care system is "online" when the person experiences any, some, or all of the following sensations and emotions: safeness, trust, ease, warmth, connection, calm, peace, openness, acceptance, love, joy, or compassion. By bringing to bear the principle of desensitization to the fear of care itself, we can help to avoid overwhelming the care system of the traumatized person and stay within the window of tolerance. To familiarize the person to the felt sense of care, therapists can ask clients to describe what sensations they notice in the body all the while ensuring that no distress arises. Therapist and client can then shift to a metalevel to address any fears of compassion that are present, by asking directly what it felt like and how that relate to their worry. For instance, if the fear is that receiving kindness or compassion makes one weak and vulnerable, the client might discover the quality of calm strength in the care state during the session. Clients can be invited to engage in behavioral experiments in between sessions to further test out the beliefs that compassion might have negative consequences. Once the care state can be brought online by the client without any distress, the person has discovered a wonderful resource that feels empowering.

Direction: Who is caring for whom?

The principle of "warming up slowly" can further be implemented by considering in which direction the client can feel one or several qualities of care most easily. Care can only be experienced in relationship with another: other to other, self to other, other to self, and self to self. The client can choose how far away or how close to them they want to experience the care. Clients with intense mistrust or paranoia benefit from observing caring interactions between others from a distance as an initial exposure exercise to get themselves used to the feeling without their fears of being used or bullied or being triggered. For instance, a client might observe a friendly interaction between a shop assistant and a customer at a shop or between a mother and a child in a park (Braehler, Gumley, et al., 2013; Braehler, Harper, & Gilbert, 2013). She is likely to feel free from threat and can let herself feel some of the qualities of good will or kindness vicariously as she is not involved in this interaction. Imagining and feeling what it is like to give care to others (self to other) such as to a child, a pet, or a good friend (self to other) moves the compassion closer in. Clients usually find this direction easy as memories typically evoke their strengths and less their vulnerability since the focus is on the pain of the other. Imagining or recalling receiving compassion from another (other to self) directly activates early memories of receiving care, including any moments of abuse or neglect or other traumatic events that may have occurred in the context of receiving care. Since this is likely to be the most challenging direction to train and to "detoxify," we require the third principle of differentiation (see in the succeeding text). Giving oneself care (self to self) is effectively self-compassion. For clients with avoidant attachment tendencies, it can be a safe starting point to view themselves as the source of the care for oneself instead of risking disappointment or hurt from letting in a real or imaginary other. Giving yourself sensitive and attuned care that you would have hoped to have gotten from another is a goal that many therapies pursue. Ask yourself: "In which direction can the client feel trust, kindness, compassion, or happiness most easily while experiencing the least threat?"

Differentiation: Who needs what from whom?

Over the course of therapy, clients become more aware of their emotions, sensations, thoughts, and links to earlier experiences. Some clients achieve a degree of differentiation that allows them to identify and distinguish an adult self with the capacity for compassion or extra helper selves with the compassion still "outsourced," protector parts, and vulnerable exiled parts. The protector and exiled parts are connected to specific autobiographical memories. Clients with a good level of mentalizing, reasonable stability (no dissociation, no self-harm, or suicidality), and a reasonably compassionate adult self may therefore be able to work with parts-based psychotherapies to help integrate traumatic memories. These approaches allow the client to care for specific younger selves in a very individual and developmentally sensitive way; however, since the differentiation of parts usually involves contacting specific pain-laden autobiographical memories, individuals with lower mentalizing may not have the ability to switch perspectives and disidentify from the old pain so easily. Working with differentiated self-compassion practices such as the compassionate friend might evoke more distress. Thus, when working with individual with low mentalizing abilities, it is recommended that you make any intervention concrete, visible (interpersonal), and focused on the here and now instead of making it abstract and invisible (intrapersonal) and focusing it on distress in the there and then (Bateman, 2006).

The principle of blended versus differentiated can also be applied to the resource of self-compassion instead of the degree of differentiated awareness of the suffering. In interpersonal trauma, victims usually become mistrustful of others. Building up trust in other human beings is a critical part of thriving and building support network beyond therapy. The degree of human mindedness can slowly be increased with imagery (see "titrating the receiving of compassion" in the succeeding text) or in reality by simply giving the client time and space to return to the same practice and to build on it. Over time the mindedness of sources of care—real or imaginary—may increase, thus having a corrective effect on the attachment system.

Differentiation may also be applied to the quality of care. What is needed? The following are some qualities of care that are either more archetypically feminine (Yin) or masculine (Yang).
- Insecurity/encouragement (Yang)
- Abuse/protection (Yang)
- Neglect/providing (Yang)
- Grief/comfort (Yin)
- Fear/calming (Yin)
- Shame/validating and belonging (Yin)

Safeness and trust are the foundation of care. Physical affection reflects the earliest form of care when a parent soothes the undifferentiated distress of a preverbal infant who cannot yet express distress in a differentiated way. Soothing or supportive touch or a compassionate body scan may therefore be good blended practices for clients with low mentalizing capacities.

Integrating self-compassion into treatment of complex PTSD

We now highlight how self-compassion can enhance working with shame during any phase-based treatment of complex PTSD (Judith Lewis Herman, 1992). Self-compassion for the therapist strengthens compassionate presence, which, in turn, strengthens the therapeutic relationship and lastly allows for sensitive direct interventions with the client.

Attuned presence helps to establish safety for shame and perpetrator parts

Nora presented with outbursts of rage that she turned against herself and others that were interfering with her ability to relate to others.

Engagement. At our first meeting, it became apparent that having eye contact with me (CB) was difficult for her as she gazed down or to the side beginning to shuffle restlessly in her chair and eventually having to walk around the room to relieve the obvious tension. I sensed the intense discomfort she was experiencing. Without yet knowing any of her history, I had to make sense of her behavior through what I saw and felt in that moment "neck down" in my body (resonance) and my previous clinical experience (perspective). Clearly, she felt unsafe, and it was my job to invite her to make herself feel as safe as possible in my office. Having worked with people with complex mental health issues, I was familiar with people struggling to engage with the therapist in this classic therapy setting and knew that having more casual arrangements like going for walk or sitting facing in the same direction and chatting helped as the person could avoid direct eye contact. I therefore suggested to her to change the seating arrangement so that she would feel as comfortable as possible involving us sitting more side by side, thus avoiding direct eye contact and her being free to get up, stand, and walk around whenever she wanted to. What was key here was to let her control the degree of contact. Nora's window of tolerance was standing or sitting side by side and looking out of the window together and ending the session when she felt it was enough. Being attuned to her and my own body allowed me to tune into her and respond sensitively to her needs in the situation, which she did not dare to voice yet due to being in a submissive state. I then realized that the simple setup of being seen and having all the attention focused on her triggered shame.

Despite giving her control over the setting, her fearful–avoidant attachment tendencies and difficulties mentalizing led her to sometimes see me as the perpetrator leading to fear, anger, and then dissociation. My personal practice of mindfulness and compassion helped me to notice my slight anxiety at her getting worse, and I would breathe in for myself and out for her to calm my anxiety and to stay connected with Nora in good will and trust instead of worry and dread as if saying to her in my mind "I do not yet understand why this is happening but that's okay. I trust and I will stay here and am willing to figure out together with you what you need now." I was then able to ground her by inviting her to stand up and walk around, which she did, and by guiding her to talk about daily events at work that helped her to connect with her adult self that was less burdened.

What helped me not to get panicky was not just clinical experience but more importantly an informal practice called Compassion with Equanimity that I had learned in the mindful self-compassion (MSC) training program developed by Germer and Neff (Germer & Neff, 2019; Neff & Germer, 2018).

Self-reflection: Strengthening compassionate presence with high shame clients

- Letting go of the script of how therapy should unfold that resides "neck up" and dropping "neck down" to feeling one's body and emotions to resonate with the client. Resonance gives us cues for what to say and do next, and the feedback from our client gives us yet another cue, therefore allowing an interactive unfolding dance moment to moment. The guiding intention from the therapist is to support the person in an attuned way.
- Expecting self-consciousness or shame to arise in clients during engagement instead of being taken aback by it. Respecting shame without questioning it and flexibly adapting situation to ensure client feels safe.
- Taking responsibility for when one's own difficult emotions or vulnerabilities get triggered. For instance, we may feel shamed by a client not engaging with ourselves. In return, we may shame the client by saying things such as "it does not seem like we will be able to work with each other." Or "You do not seem to be ready yet to engage in therapy." To name just one of many responses. Learning about shame in therapy starts with learning about our shame conditioning. Explore the following questions in peer supervision where you feel safe:
  - How do I know that I feel shame?
  - What triggers feelings of shame in me?
  - How do I compensate feelings of shame? Submissive or aggressive or both?
- Staying with not knowing.
- Connecting to good will.
- Trusting emergence.
- Committing to working together and being flexible.

Titrating the receiving of compassion from another

Warmth, kindness, and wisdom are qualities we would hope to find in all therapists. What is important is that any warmth does not emerge from a script of how one should be as a therapist but emerges from resonance with the client and good will, which helps therapists to provide sensitive, attuned, and thus flexible care. Flexible means also toning down one's expression of kindness if it is "too much of a good thing" for the client, as I learned with Nora.

Nora had started to feel safer in our sessions but still experienced direct eye contact as triggering an aversive body memory. She wanted to be able to look at me more often as she found it helped her to stay in her adult self instead of dissociating. To help tolerate looking at me more often, she asked me "to not look in such a kind way, for instance, by not smiling so strongly." That made a lot of sense. She wanted to connect with me as the source of care from other to self as a starting point of desensitizing to these qualities that were conditioned with fear and traumas. However, she needed to titrate the dose and asked me to tone down the dose of warmth and kindness to a level that she could tolerate without her threat system being activated.

Nora feared that kindness would make her vulnerable. She felt safe to begin with compassion for others. She could recall times when she effectively comforted a distressed toddler at the nursery she worked at. This memory provided her with a visceral experience of warmth, calm, and being soothed and tapped into her motivation to help others. Over time, she was able to let in more care from myself, other therapists, and her friends. Imagery exercises were too difficult as mentalizing broke down when the affect was too high. Here-and-now relationships were easier to work with as they were more concrete.

Another client, Lucy, had avoidant attachment tendencies and good enough mentalizing. Due to her high level of autonomy, her safe entry point into her care system was imagery (self to self), which became increasingly more differentiated and human. We worked with the compassionate friend practice (Germer & Neff, 2019; Neff & Germer, 2018) or ideal compassionate other (Gilbert, 2010) in stages and over several months during weekly sessions to allow her to gradually let in the care and love from another without getting distressed. As is often the case with interpersonal trauma, images were at first impersonal (a golden white light) and later became human (Phoebe from TV series "Friends") and greater in number (different kind, wise, "clumsy," and funny TV characters). These images elicited a strong visceral experience of relaxation, warmth around the heart, feeling safe, connected, sharing joy, and accepted as she is. She concluded that since she could trust these characters she may as well try to trust others in real life. The growth of her inner affiliative system seems to have acted as imaginary exposure to external affiliation. Drawing on these embodied experiences of feeling supported, accepted, joyous, and strong helped her to feel more self-confident and content and to start to take care of her body, of which she had previously been deeply ashamed.

Empowerment through fierce self-compassion

Survivors of high betrayal traumas need to activate healthy anger (fierce self-compassion) to move from dissociation, shame, and self-attacking into righteous outrage about the injustices suffered and eventually into self-protection and self-empowerment. If a therapist cannot tolerate a patient's anger well, then the therapist is likely to shame the client for being angry or to react with anger. As a consequence the patient is likely to feel invalidated or shamed, thus dropping out of therapy or returning to submissive relational patterns. A common misunderstanding among therapists is that compassion means only to be nurturing, calming, and soothing instead of also including assertiveness, protection, and encouragement. Compassion means being sensitive to suffering and alleviating it with a quality of care that is appropriate to the suffering. What is needed in the case of anger in high betrayal traumas is to validate the outrage, the injustice, the betrayal, and all the pain it caused and still causes fully and as long as the person needs to have this validated, not as long as I as the therapist think this should last. In fact, my role as a therapist is often to mirror and model healthy self-protective anger on the patient's behalf to let them realize that being raped, locked up, beaten, or having food or medical help withheld from them is not normal and not a sign of good will. Eventually, once clients believe themselves and have realized that what was done to them was wrong and in most cases a criminal offense, they grow in their ability to validate their own anger. This is the first component of yang self-compassion as described in the yang self-compassion break (Neff, 2019). Whenever the shame creeps back in, the person returns to feeling guilty for having been angry with the perpetrator and a submissive defeat state reactivates depressive mood (Catarino, Gilbert, McEwan, & Baião, 2014). Therapists require patience to explore the fear of anger that many such clients carry and help them slowly to teach clients to protect themselves to avoid retraumatization including body-based work and self-defense techniques.

Informal practice for trauma survivors: Yang self-compassion break for protection

Once clients are no longer in the grip of shame and guilt and have learned to seize the power of their anger to protect and to assert themselves, they might feel confident enough to practice fierce self-compassion when they feel like somebody is acting in unjust ways toward them or threatening to hurt or harm them in some way. The following self-compassion break was developed by Neff (2019) to help validate one's truth, to evoke courage, to empower oneself to connect with others and feel solidarity, and to connect with one's own wisdom and vow to protect oneself. Even though the fierceness in this practice may seem strong or partly angry, it is never aggressive and never disconnecting. It is a stable, solid, calm, and strong stance of self-assertion.

1. Mindfulness of suffering
   a. Validating the pain or hurt or injustice you are experiencing
      "This is my truth. I believe what I experienced. I trust myself despite what others might be saying to invalidate me!"
   b. Courage
      "I dare speak my truth starting with myself."
2. Common humanity
   a. Empowering yourself to reach out to and to trust in others to share your truth.
   b. Feeling connected in suffering to others who experience similar suffering and to feel solidarity in protecting yourself against future hurts.
3. Fierce self-kindness
   a. What do I truly need to protect myself or to support myself or to stand up for myself? To say "no!" to draw my boundaries?
   b. What is the wisest thing for me to do in this situation in the short term and in the longer term?

Mourning the life lost

Once shame and guilt have abated and righteous anger has been transformed into assertiveness and self-protection, many trauma survivors experience grief about the horrors they had to endure and the negative impacts these experiences have had on their quality of life. Mourning the life lost is an important part of trauma recovery, and the quality of care needed is comfort. Therapists' compassionate presence will allow clients to gradually grieve while feeling held by the therapist's witnessing and being with the pain. Over time, when the fears of receiving compassion have reduced, clients might feel able to comfort themselves.

Informal practice for trauma survivors: Yin self-compassion break for grief

The self-compassion break developed by Neff and Germer (Germer & Neff, 2019; Neff & Germer, 2018) for the MSC program might help a client to comfort themselves when grief about life lost emerges. The client requires an ability to feel their body without dissociating and to be able to name their emotions to themselves and to have found safe access to their

care system in therapy. This informal practice is well suited for when clients end therapy and need tools to help them maintain what they have learned.
1. Mindfulness: To notice, name, and validate the distress
   "It is tragic that this happened to me. It is so painful to have missed out on happiness, joy, and ease. It is so understandable that I would feel like this at this point."
2. Common humanity: To help you feel connected instead of isolated
   "Even though it feels like I am all alone, I also feel some connection to my two friends from group therapy. They are people who really understand what it is like to have experienced these unspeakable things."
3. Soothing touch: To bring the care physiology online
   Client may place the hand on a part of the body where touch feels comforting or touches a warm blanket or warm mug of tea or strokes a pet.
4. Self-kindness: To offer yourself inner guidance and comfort
   "I am here for you, my dear. Luckily the hurt is over. May you be extra gentle and kind with yourself as you go through this grief remembering that it will pass, too. I will take good care of you and allow you to feel whatever you are feeling."

Allowing in happiness and play

Just when the final phase of reconnecting is reached in trauma therapy, the final phase focuses on learning to thrive instead of merely surviving. Most survivors of complex PTSD have spent most of their lives in a threat state because their survival was threatened when they relaxed and were playful and carefree. Chronic tension and hyperarousal, sleep and digestive problems, restlessness, and hypervigilance constitute a familiar way of being. A set of specific fears may arise about what might happen when the client feels joy and happiness depending on early conditioning (Şar, Türk, & Öztürk, 2019). In addition to the general unease about relaxing and needing to slowly habituate to these new states, specific conditioned memories may manifest. For instance, if a client was humiliated through words and physically punished after happily singing along to a cheerful song and laughing when a child, the client quickly learns to suppress any expression of happiness and instead habituates to a hypoaroused defeat states. Fear of happiness shows similar relationships to attachment insecurity, alexithymia in clinically depressed groups as do fears of compassion (Joshanloo, 2018). Letting joy in when we are afraid of it or simply unfamiliar with it requires modeling from the therapist and reinforcement in other relationships, just as with compassion. Rejoicing with our clients and acknowledging their strengths and resources as well as inviting humor and lightness are part of such modeling and counteract shame.

Natalie had been a naturally cheerful and happy child who had developed complex PTSD as an adult when early trauma was triggered by a sexual assault. The final phase of therapy focused on gradually building up more periods of rest and spontaneous play time with friends and her partner. The moment she sat down to rest instead of work, a defeat state set in accompanied by an urge to binge-eat. After spending several sessions integrating the old memories of being humiliated and punished for simply being a happy cheerful child playing, she was gradually able to better relax and to enjoy "doing nothing" for short periods of time on her own. As she had been rather isolated as a child during those states, we worked on her seeking out contact with others including from a distance such as lying in a park in the sun among other people and later on to spend time with friends to "play with." Despite her happy child having been silenced in such cruel ways, she was able to reclaim her own voice by singing and literally playing music with her friends.

Conclusion

We cannot change our clients' painful past experiences, nor is it entirely within our power—no matter how effective the treatments are—to fully alleviate their suffering. Self-compassion can provide a resource for therapists to protect themselves against compassion fatigue, to increase therapists' holding capacity to avoid shaming or retraumatizing their clients, and to offer a safe haven until client have developed safe havens within themselves. The suggested principles for integrating compassion into treatment for PTSD can help to strengthen clients' attachment systems, improve their capacity for emotion regulation, and enhance their interpersonal functioning in a safe way and ultimately to improve the qualities of lives of people with complex PTSD.

References

Andrews, B., Brewin, C. R., Rose, S., & Kirk, M. (2000). Predicting PTSD symptoms in victims of violent crime: The role of shame, anger, and childhood abuse. *Journal of Abnormal Psychology, 109*(1), 69.

Barlow, M. R., Goldsmith Turow, R. E., & Gerhart, J. (2017). Trauma appraisals, emotion regulation difficulties, and self-compassion predict posttraumatic stress symptoms following childhood abuse. *Child Abuse & Neglect, 65*, 37–47.

Bateman, A. (2006). Mentalization-based treatment for borderline personality disorder: A practical guide. OUP Oxford.

Beaumont, E., Galpin, A., & Jenkins, P. (2012). 'Being kinder to myself': A prospective comparative study, exploring post-trauma therapy outcome measures, for two groups of clients, receiving either cognitive behaviour therapy or cognitive behaviour therapy and compassionate mind training. *Counselling Psychology Review*, 27(1), 31–43.

Berryhill, M. B., Hayes, A., & Lloyd, K. (2018). Chaotic-enmeshment and anxiety: The mediating role of psychological flexibility and self-compassion. *Contemporary Family Therapy*, 1–12.

Bistricky, S. L., Gallagher, M. W., Roberts, C. M., Ferris, L., Gonzalez, A. J., & Wetterneck, C. T. (2017). Frequency of interpersonal trauma types, avoidant attachment, self-compassion, and interpersonal competence: A model of persisting posttraumatic symptoms. *Journal of Aggression, Maltreatment & Trauma*, 26(6), 608–625.

Bohus, M., Dyer, A. S., Priebe, K., Krüger, A., Kleindienst, N., Schmahl, C., ... Steil, R. (2013). Dialectical behaviour therapy for post-traumatic stress disorder after childhood sexual abuse in patients with and without borderline personality disorder: A randomised controlled trial. *Psychotherapy and Psychosomatics*, 82(4), 221–233.

Bowlby, J. (1988). *A secure base: Clinical applications of attachment theory (collected papers)*. London: Tavistock.

Boykin, D. M., Himmerich, S. J., Pinciotti, C. M., Miller, L. M., Miron, L. R., & Orcutt, H. K. (2018). Barriers to self-compassion for female survivors of childhood maltreatment: The roles of fear of self-compassion and psychological inflexibility. *Child Abuse & Neglect*, 76, Supplement C, 216–224.

Braehler, C., Gumley, A., Harper, J., Wallace, S., Norrie, J., & Gilbert, P. (2013). Exploring change processes in compassion focused therapy in psychosis: Results of a feasibility randomized controlled trial. *British Journal of Clinical Psychology*, 52(2), 199–214.

Braehler, C., Harper, J., & Gilbert, P. (2013). Compassion focused group therapy for recovery after psychosis. In *Cognitive behaviour therapy for schizophrenia: Evidence based interventions and future directions* (pp. 236–266). West Sussex, UK: Wiley.

Breines, J. G., Thoma, M. V., Gianferante, D., Hanlin, L., Chen, X., & Rohleder, N. (2014). Self-compassion as a predictor of interleukin-6 response to acute psychosocial stress. *Brain, Behavior, and Immunity*, 37, 109–114.

Brewin, C. R., & Holmes, E. A. (2003). Psychological theories of posttraumatic stress disorder. *Clinical Psychology Review*, 23(3), 339–376.

Briere, J. N. (1992). *Child abuse trauma: Theory and treatment of the lasting effects*. Sage.

Catarino, F., Gilbert, P., McEwan, K., & Baião, R. (2014). Compassion motivations: Distinguishing submissive compassion from genuine compassion and its association with shame, submissive behavior, depression, anxiety and stress. *Journal of Social and Clinical Psychology*, 33(5), 399–412.

Cloitre, M., Koenen, K. C., Cohen, L. R., & Han, H. (2002). Skills training in affective and interpersonal regulation followed by exposure: A phase-based treatment for PTSD related to childhood abuse. *Journal of Consulting and Clinical Psychology*, 70(5), 1067.

Cunningham, K. C., Davis, J. L., Wilson, S. M., & Resick, P. A. (2018). A relative weights comparison of trauma-related shame and guilt as predictors of DSM-5 posttraumatic stress disorder symptom severity among US veterans and military members. *British Journal of Clinical Psychology*, 57(2), 163–176.

Cusack, K., Jonas, D. E., Forneris, C. A., Wines, C., Sonis, J., Middleton, J. C., ... Gaynes, B. N. (2016). Psychological treatments for adults with posttraumatic stress disorder: A systematic review and meta-analysis. *Clinical Psychology Review*, 43, 128–141.

Dahm, K. A., Meyer, E. C., Neff, K. D., Kimbrel, N. A., Gulliver, S. B., & Morissette, S. B. (2015). Mindfulness, self-compassion, posttraumatic stress disorder symptoms, and functional disability in US Iraq and Afghanistan war veterans. *Journal of Traumatic Stress*, 28(5), 460–464.

DePrince, A. P., Chu, A. T., & Pineda, A. S. (2011). Links between specific posttrauma appraisals and three forms of trauma-related distress. *Psychological Trauma Theory Research Practice and Policy*, 3(4), 430.

Døssing, M., Nilsson, K. K., Svejstrup, S. R., Sørensen, V. V., Straarup, K. N., & Hansen, T. B. (2015). Low self-compassion in patients with bipolar disorder. *Comprehensive Psychiatry*, *60*, 53–58.

Ehret, A. M., Joormann, J., & Berking, M. (2015). Examining risk and resilience factors for depression: The role of self-criticism and self-compassion. *Cognition and Emotion*, *29*(8), 1496–1504.

Eicher, A. C., Davis, L. W., & Lysaker, P. H. (2013). Self-compassion: A novel link with symptoms in schizophrenia? *The Journal of Nervous and Mental Disease*, *201*(5), 389–393.

Fonagy, P., Gergely, G., Jurist, E. L., & Target, M. (2002). *Affect regulation, mentalization, and the development of the self*.

Germer, C. K., & Neff, K. D. (2019). *Teaching the mindful self-compassion program: A guide for professionals*. New York: Guilford Press.

Gilbert, P. (2010). *Compassion focused therapy: Distinctive features*. Routledge.

Gilbert, P., McEwan, K., Catarino, F., & Baião, R. (2014a). Fears of compassion in a depressed population implication for psychotherapy. *Journal of Depression and Anxiety*, *2014*. https://doi.org/10.4172/2167-1044.S2-003.

Gilbert, P., McEwan, K., Catarino, F., & Baião, R. (2014b). Fears of negative emotions in relation to fears of happiness, compassion, alexithymia and psychopathology in a depressed population: A preliminary study. *Journal of Depression and Anxiety*, *2014*. https://doi.org/10.4172/2167-1044.S2-004.

Gilbert, P., McEwan, K., Catarino, F., Baião, R., & Palmeira, L. (2014). Fears of happiness and compassion in relationship with depression, alexithymia, and attachment security in a depressed sample. *British Journal of Clinical Psychology*, *53*(2), 228–244.

Gilbert, P., McEwan, K., Gibbons, L., Chotai, S., Duarte, J., & Matos, M. (2012). Fears of compassion and happiness in relation to alexithymia, mindfulness, and self-criticism. *Psychology and Psychotherapy: Theory, Research and Practice*, *85*(4), 374–390.

Gilbert, P., McEwan, K., Matos, M., & Rivis, A. (2011). Fears of compassion: Development of three self-report measures. *Psychology and Psychotherapy: Theory, Research and Practice*, *84*(3), 239–255.

Herman, J. L. (1992). Complex PTSD: A syndrome in survivors of prolonged and repeated trauma. *Journal of Traumatic Stress*, *5*(3), 377–391.

Herman, J. L. (2015). *Trauma and recovery: The aftermath of violence—From domestic abuse to political terror*. UK: Hachette.

Hiraoka, R., Meyer, E. C., Kimbrel, N. A., DeBeer, B. B., Gulliver, S. B., & Morissette, S. B. (2015). Self-compassion as a prospective predictor of PTSD symptom severity among trauma-exposed U.S. Iraq and Afghanistan war veterans. *Journal of Traumatic Stress*, *28*(2), 127–133.

Hoffart, A., Øktedalen, T., & Langkaas, T. F. (2015). Self-compassion influences PTSD symptoms in the process of change in trauma-focused cognitive-behavioral therapies: a study of within-person processes. *Frontiers in Psychology*, *6*, 1273.

Hoffer, E. (1955). *The passionate state of mind*. Section 139.

Hoge, E. A., Hölzel, B. K., Marques, L., Metcalf, C. A., Brach, N., Lazar, S. W., & Simon, N. M. (2013). Mindfulness and self-compassion in generalized anxiety disorder: Examining predictors of disability. *Evidence-based Complementary and Alternative Medicine*, *2013*.

Holmes, E. A., Grey, N., & Young, K. A. D. (2005). Intrusive images and "hotspots" of trauma memories in posttraumatic stress disorder: An exploratory investigation of emotions and cognitive themes. *Journal of Behavior Therapy and Experimental Psychiatry*, *36*(1), 3–17.

Homan, K. J. (2016). Self-compassion and psychological well-being in older adults. *Journal of Adult Development*, *23*(2), 111–119.

Játiva, R., & Cerezo, M. A. (2014). The mediating role of self-compassion in the relationship between victimization and psychological maladjustment in a sample of adolescents. *Child Abuse & Neglect*, *38*(7), 1180–1190.

Jiang, Y., You, J., Hou, Y., Du, C., Lin, M.-P., Zheng, X., & Ma, C. (2016). Buffering the effects of peer victimization on adolescent non-suicidal self-injury: The role of self-compassion and family cohesion. *Journal of Adolescence, 53*, 107–115.

Jiang, Y., You, J., Zheng, X., & Lin, M.-P. (2017). The qualities of attachment with significant others and self-compassion protect adolescents from non suicidal self-injury. *School Psychology Quarterly, 32*(2), 143.

Joeng, J. R., Turner, S. L., Kim, E. Y., Choi, S. A., Lee, Y. J., & Kim, J. K. (2017). Insecure attachment and emotional distress: Fear of self-compassion and self-compassion as mediators. *Personality and Individual Differences, 112*, 6–11.

Johnson, E. A., & O'Brien, K. A. (2013). Self-compassion soothes the savage ego-threat system: Effects on negative affect, shame, rumination, and depressive symptoms. *Journal of Social and Clinical Psychology, 32*(9), 939.

Joshanloo, M. (2018). Fear and fragility of happiness as mediators of the relationship between insecure attachment and subjective well-being. *Personality and Individual Differences, 123*, 115–118.

Karatzias, T., Shevlin, M., Fyvie, C., Hyland, P., Efthymiadou, E., Wilson, D., ... Cloitre, M. (2017). Evidence of distinct profiles of posttraumatic stress disorder (PTSD) and complex posttraumatic stress disorder (CPTSD) based on the new ICD-11 trauma questionnaire (ICD-TQ). *Journal of Affective Disorders, 207*, 181–187.

Kearney, D. J., Malte, C. A., McManus, C., Martinez, M. E., Felleman, B., & Simpson, T. L. (2013). Loving-kindness meditation for posttraumatic stress disorder: A pilot study. *Journal of Traumatic Stress, 26*(4), 426–434.

Kelly, A. C., & Carter, J. C. (2013). Why self-critical patients present with more severe eating disorder pathology: The mediating role of shame. *British Journal of Clinical Psychology, 52*(2), 148–161.

Kelly, A. C., & Tasca, G. A. (2016). Within-persons predictors of change during eating disorders treatment: An examination of self-compassion, self-criticism, shame, and eating disorder symptoms. *International Journal of Eating Disorders, 49*(7), 716–722.

Kelly, A. C., Wisniewski, L., Martin-Wagar, C., & Hoffman, E. (2017). Group-based compassion-focused therapy as an adjunct to outpatient treatment for eating disorders: A pilot randomized controlled trial. *Clinical Psychology & Psychotherapy, 24*(2), 475–487.

Keltner, D., & Harker, L. (1998). The forms and functions of the nonverbal signal of shame. In P. Gilbert & B. Andrews (Eds.), *Series in affective science. Shame: Interpersonal behavior, psychopathology, and culture* (pp. 78–98). New York, NY, US: Oxford University Press.

Kirby, J. N., Tellegen, C. L., & Steindl, S. R. (2017). A meta-analysis of compassion-based interventions: Current state of knowledge and future directions. *Behavior Therapy, 48*(6), 778–792.

Krieger, T., Altenstein, D., Baettig, I., Doerig, N., & Holtforth, M. G. (2013). Self-compassion in depression: Associations with depressive symptoms, rumination, and avoidance in depressed outpatients. *Behavior Therapy, 44*(3), 501–513.

Lee, D. A., Scragg, P., & Turner, S. (2001). The role of shame and guilt in traumatic events: A clinical model of shame-based and guilt-based PTSD. *British Journal of Medical Psychology, 74*(4), 451–466.

MacBeth, A., & Gumley, A. (2012). Exploring compassion: A meta-analysis of the association between self-compassion and psychopathology. *Clinical Psychology Review, 32*(6), 545–552.

Maercker, A., Brewin, C. R., Bryant, R. A., Cloitre, M., van Ommeren, M., Jones, L. M., ... Rousseau, C. (2013). Diagnosis and classification of disorders specifically associated with stress: Proposals for ICD-11. *World Psychiatry, 12*(3), 198–206.

Maheux, A., & Price, M. (2015). Investigation of the relation between PTSD symptoms and self-compassion: Comparison across DSM IV and DSM 5 PTSD symptom clusters. *Self and Identity, 14*(6), 627–637.

Matos, M., Duarte, J., & Pinto-Gouveia, J. (2017). The origins of fears of compassion: Shame and lack of safeness memories, fears of compassion and psychopathology. *The Journal of Psychology, 151*(8), 804–819.

Matos, M., & Pinto-Gouveia, J. (2010). Shame as a traumatic memory. *Clinical Psychology & Psychotherapy, 17*(4), 299–312.

Miron, L. R., Seligowski, A. V., Boykin, D. M., & Orcutt, H. K. (2016). The potential indirect effect of childhood abuse on posttrauma pathology through self-compassion and fear of self-compassion. *Mindfulness, 7*(3), 596–605.

Neff, K. (2003). Self-compassion: An alternative conceptualization of a healthy attitude toward oneself. *Self and Identity, 2*(2), 85–101.

Neff, K. (2018). Why women need fierce self-compassion. *Greater Good Magazine*, (October). Greater Good Science Center University of California at Berkeley.

Neff, K. (2019). *The yin and yang of self-compassion (audio recording)*. Boulder, CO: Sounds True.

Neff, K., & Germer, C. (2018). *The mindful self-compassion workbook: A proven way to accept yourself, build inner strength, and thrive*. Guilford Press.

Neff, K. D., & Faso, D. J. (2015). Self-compassion and well-being in parents of children with autism. *Mindfulness, 6*(4), 938–947.

Neff, K. D., Kirkpatrick, K. L., & Rude, S. S. (2007). Self-compassion and adaptive psychological functioning. *Journal of Research in Personality, 41*(1), 139–154.

Neff, K. D., & McGehee, P. (2010). Self-compassion and psychological resilience among adolescents and young adults. *Self and Identity, 9*(3), 225–240.

Neff, K. D., Tóth-Király, I., Yarnell, L., Arimitsu, K., Castilho, P., Ghorbani, N., ... Mantios, M. (2018). Examining the factor structure of the self-compassion scale using exploratory SEM bifactor analysis in 20 diverse samples: Support for use of a total score and six subscale scores. *Psychological Assessment, 31*(1), 27–45.

Nolen-Hoeksema, S. (1991). Responses to depression and their effects on the duration of depressive episodes. *Journal of Abnormal Psychology, 100*(4), 569.

Pinto-Gouveia, J., Duarte, C., Matos, M., & Fráguas, S. (2014). The protective role of self-compassion in relation to psychopathology symptoms and quality of life in chronic and in cancer patients. *Clinical Psychology & Psychotherapy, 21*(4), 311–323.

Rockliff, H., Gilbert, P., McEwan, K., Lightman, S., & Glover, D. (2008). A pilot exploration of heart rate variability and salivary cortisol responses to compassion-focused imagery. *Clinical Neuropsychiatry, 5*(3), 132–139.

Roisman, G. I., Padrón, E., Sroufe, L. A., & Egeland, B. (2002). Earned–secure attachment status in retrospect and Prospect. *Child Development, 73*(4), 1204–1219.

Şar, V., Türk, T., & Öztürk, E. (2019). Fear of happiness among college students: The role of gender, childhood psychological trauma, and dissociation. *Indian Journal of Psychiatry, 61*(4), 389–394.

Sbarra, D. A., Smith, H. L., & Mehl, M. R. (2012). When leaving your ex, love yourself observational ratings of self-compassion predict the course of emotional recovery following marital separation. *Psychological Science*. 0956797611429466.

Schore, A. N. (2015). *Affect regulation and the origin of the self: The neurobiology of emotional development*. Routledge.

Scoglio, A. A. J., Rudat, D. A., Garvert, D., Jarmolowski, M., Jackson, C., & Herman, J. L. (2018). Self-compassion and responses to trauma: The role of emotion regulation. *Journal of Interpersonal Violence, 33*(13), 2016–2036.

Sznycer, D., Xygalatas, D., Agey, E., Alami, S., An, X.-F., Ananyeva, K. I., ... Tooby, J. (2018). Cross-cultural invariances in the architecture of shame. *Proceedings of the National Academy of Sciences, 115*(39), 9702–9707.

Tanaka, M., Wekerle, C., Schmuck, M. L., & Paglia-Boak, A. (2011). The linkages among childhood maltreatment, adolescent mental health, and self-compassion in child welfare adolescents. *Child Abuse & Neglect, 35*(10), 887–898.

Taylor, T. F. (2015). The influence of shame on posttrauma disorders: Have we failed to see the obvious? *European Journal of Psychotraumatology, 6*, 10.

Thompson, B. L., & Waltz, J. (2008). Self-compassion and PTSD symptom severity. *Journal of Traumatic Stress, 21*(6), 556–558.

van der Kolk, B. A., Roth, S., Pelcovitz, D., Sunday, S., & Spinazzola, J. (2005). Disorders of extreme stress: The empirical foundation of a complex adaptation to trauma. *Journal of Traumatic Stress, 18*(5), 389–399.

Van Ijzendoorn, M. H., Schuengel, C., & Bakermans-Kranenburg, M. J. (1999). Disorganized attachment in early childhood: Meta-analysis of precursors, concomitants, and sequelae. *Development and Psychopathology, 11*(2), 225–250.

Vettese, L. C., Dyer, C. E., Li, W. L., & Wekerle, C. (2011). Does self-compassion mitigate the association between childhood maltreatment and later emotion regulation difficulties? A preliminary investigation. *International Journal of Mental Health and Addiction, 9*(5), 480.

Zeller, M., Yuval, K., Nitzan-Assayag, Y., & Bernstein, A. (2015). Self-compassion in recovery following potentially traumatic stress: Longitudinal study of at-risk youth. *Journal of Abnormal Child Psychology, 43*(4), 645–653.

Zessin, U., Dickhäuser, O., & Garbade, S. (2015). The relationship between self-compassion and well-being: A meta-analysis. *Applied Psychology. Health and Well-Being, 7*(3), 340–364.

CHAPTER 21

Culture, PTSD, and emotion regulation: An anthropological perspective

Andrea Chiovenda[a], Devon E. Hinton[b], Byron J. Good[a]
[a]Department of Global Health and Social Medicine, Harvard Medical School, Boston, MA, United States
[b]Center for Anxiety and Traumatic Stress Disorders, Massachusetts General Hospital, Harvard Medical School, Boston, MA, United States

Introduction

This chapter looks at culture, posttraumatic stress disorder (PTSD), and emotion regulation from the perspective of medical and psychological anthropology. As such, it does not provide a traditional review of the extant scientific literature but rather investigates culture, PTSD, and emotion regulation from an ethnographic perspective. Though ethnographic research produces results that can be difficult to quantify, it has the advantage of providing information from a naturalistic environment in which the human subjects are able to freely behave with authenticity and consistency. The ethnographic perspective also enables researchers to discover relevant aspects of informants' lives and experiences that would be difficult, if not impossible, to identify with standardized instruments based on preconceived hypotheses. We believe that the ethnographic approach is particularly relevant to this topic, given how little empirical research has been conducted to date concerning the specific intersection of culture, PTSD, and emotion regulation. While some prior work has examined the relation between PTSD and emotion regulation, PTSD and cultural cues (i.e., cultural prompts), and the influence of cultural backgrounds on the deployment of emotion regulation strategies, these three units of analysis have seldom been considered in conjunction with each other in prior studies. Accordingly, the present work will focus on ethnographic accounts that have the potential to shed light on how these elements come together in the lived experiences of individuals across diverse sociocultural settings.

From the perspective of cultural engagement, there now exists a considerable literature that shows how cultural background and social expectations

(1) impact individuals' capability to perceive certain emotional expressions in others, (2) impact individuals' ability to give meaning to these expression, (3) inform the response that the individual should give to certain emotional inputs according to what is accepted and expected by a specific cultural environment, and (4) shape the regulatory techniques that one is supposed to engage in to address the effects of such emotions. For example, the work by De Leersnyder, Boiger, and Mesquita (2013) has demonstrated the degree to which emotions are culturally shaped, or at least displayed publicly. Specifically, they have demonstrated that cultural cues and constraints are so strong that emotion regulation often starts even before there is an emotional experience at all by demonstrating that certain cultural cues may serve the purpose of avoiding escalation to the stage of emotional expression. Similarly, Boiger, Mesquita, Uchida, and Feldman Barrett (2013) found that cultural affordances[a] for emotions of anger and shame in Japan and the United States predicted emotional reactions in individuals from the respective cultural milieus (see also Mesquita, Boiger, & De Leersnyder, 2016), whereas Barrett, Lindquist, and Gendron (2007) and Barrett and Kensinger (2010) have worked intensely on the idea of proximate context as a crucial variable in the perception of emotions in others and in oneself, even beyond cultural specificities (see also Masuda et al., 2008).

In this chapter, we present three ethnographic studies concerning emotion regulation in the settings of violence and conflict. We will suggest two major points. First, ethnographic evidence can and should be integrated in experimental studies such as the ones we have briefly summarized, particularly because it may open avenues of investigation that preconceived experimental study designs do not allow access to and also because it may contribute constructively to rethinking certain foundational concepts that are seldom questioned in the field. Second, a more circumscribed point we will make concerns the specific emotion of anger, which the individuals we will present, all from very different cultural and social contexts, had to deal with and regulate to function effectively. We will see that anger seems to be a constitutive emotional aspect of a syndrome such as PTSD, wherein the role and import of anger should be interpreted and evaluated on a case-by-case basis, according to the specific cultural, social, and even political realities of the context under observation. In the case of the Cambodian refugees we discussed in the succeeding text, anger is one of the end products of past experiences of profound

---

[a] By "cultural affordances," we mean the advantages and disadvantages that certain behaviors and emotional expressions bring to the actor in his/her own social and cultural context.

physical and psychological trauma and feeds into a self-perpetuating cycle of maintenance and even deterioration of symptoms consistent with PTSD. On the other hand, the example from Afghanistan's rural context may point to an adaptive function that anger might fulfill, in a milieu of widespread and entrenched collective trauma within entire communities. Ethnographic investigation can elucidate how such dynamics unfold and can provide answers for the questions that arise from them, particularly when in the presence of data and information that is hidden from sight, kept private by the direct actors, until a deeper relationship is established with the researcher (see Good (2012) for a case study of this phenomenon).

In the succeeding text, we present three ethnographic studies that speak to the importance of cultural context in understanding the role of emotion and emotion regulation among individuals with PTSD. Please note that the original material provided here is taken from fieldwork research conducted by each of three authors that has been previously published in different contexts (e.g., Chiovenda, 2020, B. Good, 2012, 2015; Hinton, Rasmussen, Nou, Pollack, & Good, 2009).

Emotion regulation in context: Part 1

Andrea Chiovenda carried out extensive fieldwork research in Afghanistan from 2009 to 2013, for a total of over 18 months. He worked in a Pashtun-majority environment, in both rural and urban areas of a very volatile south-eastern province of the country, Nangarhar, on the border with Pakistan. Pashtun society is strongly androcentric, where norms dictating a strong and somewhat belligerent masculinity are constitutive of the ideal image of culturally appropriate "manly man." Values that revolve around honor and shame are still very deeply engrained in such context, though, in practice, no two individuals will operationalize and interpret them in the same way.

Pashtun in Afghanistan (and the country as a whole) is a deeply scarred sociopolitical and cultural environment. Forty years of continuous war have shaped two generations of Afghans. Violence over the years has visited one way or another the great majority of households in the country. This is especially true in areas of the Pashtun belt at the border with Pakistan. Traumatic stress and PTSD symptomatology are widespread (Eggerman & Panter-Brick, 2010, Panter-Brick, Goodman, Tol, & Eggerman, 2011, Panter-Brick, Grimon, Kalin, & Eggerman, 2015). In fact, since 2006, when the insurgency led by the remnants of the Taliban movement and their new acolytes gained ground again, the level of violence in the province where

Chiovenda worked increased steeply (Giustozzi, 2007). Taliban followers and government forces fought fiercely in the rural areas of the province and threatened the capital, Jalalabad.

One of Chiovenda's main informants was a Pashtun man in his early thirties, Rohullah (a pseudonym), who hails from a small town in Gardez, a Pashtun-majority province bordering Pakistan that has undergone similar conflict vicissitudes as Nangarhar. He was mostly raised in Kabul, in a more cosmopolitan and culturally permissive environment than the rural districts, by his father, a university professor, and his mother, an illiterate Pashtun woman, alongside three sisters and three brothers. His Pashtun cultural upbringing was characterized by a degree of sophistication and education that was not common in most rural areas of the Pashtun belt. Due to the civil war that was raging in Kabul (1992–96), the family had to relocate to their ancestral village in Gardez province when he was about 6 years old. This is when the problems started for Rohullah. The harsher sociocultural context of the village, in which the violence of the two previous decades of conflict had taken a much higher toll than in Kabul, deeply affected Rohullah. Untrained in the aggressive and competitive nature of relationships that he found in the village, Rohullah was constantly and violently bullied by his classmates during his first few years in the village. They, in Rohullah's opinion, resented his more refined demeanor and educational skills and took advantage of his unwillingness (or inability) to respond with violence to their persecutions. After 4 or 5 years of this treatment, Rohullah decided that he would no longer be the subject of his peers' persecution. Instead, he started to play by the rules of the village and resolved to transform himself into one of the very people that had been harassing him until that point. By then, he had become "culturally fluent" in the local modus operandi, and he had learned how to assert himself accordingly. Presented in the succeeding text are verbatim excerpts of the numerous interviews that Chiovenda had with Rohullah, over a long period of time and in a private setting, where Rohullah felt secure in disclosing personal information (Chiovenda, 2020).

> Rohullah: I did not want to be considered weak any longer. The fact that I was quiet, shy, and did not want to end up in fights was mistaken for weakness. I wanted that to stop. My father had raised me stressing values of peacefulness and calmness, unlike the fathers of the other kids did, and these values were what I considered right. But I was having a hard life outside the house, it was a nightmare… and I wanted it to stop. I chose to learn how to be more aggressive and assertive. I started following more closely what my elders and the stronger men in the village were doing, how they were behaving. I started attending the marakas and the jirgas in the village [councils of elders to discuss and solve intracommunity conflicts], so that I could see how they were asserting themselves and fighting for their rights. I also started to respond to the other kids' provocations, and to fight back. I fought more often over time.

Andrea: How did this new behavior of yours make you feel? How did you feel about fighting?

Rohullah: I did not like it. I felt there was something wrong about it, and I remember that when I got home after a brawl [lanja] I felt unhappy [khapa]. But I had to do it. It was about survival. I had to survive in that world, I could not just let things go like they had gone until that time. I had to survive a bad situation. I used to tell myself that it was not my fault, I had to do it.

Andrea: Was it like this all the times you came back home after a fight?

Rohullah: Yes, but after some time something changed as well. You see, a Pashtun man has to be ghairati, he has to have ghairat [the courage to fight for one's own rights]. It is something very important for a Pashtun man. This is important for me as well. If need be [ka cherta da pa kaar wi], you have to show that you are a ghairati man. You have to protect your rights. When you do so, people around you admire you, and give you respect. The more you show ghairat, the more people talks about you in a respectful way. You start to have a reputation [nuum], you start to have power [qowat]. Until I kept quiet and did not fight, I had very few friends. After I started fighting back and standing up to the bullies who bothered me, a lot of other kids started hanging out with me, looked for my company. I felt I had gained power. It made me feel good. After some time, I think I became like addicted [amali] to the power that my aggressive behavior was giving me. I enjoyed a lot of popularity among my schoolmates. They wanted to play with me.

Andrea: You did not feel bad any longer for fighting with other kids?

Rohullah: Well, yes, I still felt bad about it, I still felt I should not have behaved like that. My father would not have liked it. But it was a strange feeling, it gave me pleasure to have all those kids following me as if I was their leader. I would also feel pleasure when I told my friends the stories of my brawls and fights. They were all excited and in admiration of me. So I think I really got out of control, and started behaving more and more violently and aggressively. Too much. You know, there are two kinds of ghairat. There is good ghairat, and bad ghairat. For example, imagine you are walking in the street with your sister, and a strange man passes by and looks at your sister intensely, as if he knew her. If you start speculating that they are having an illicit relationship, and kill them both without even inquiring with any of them, you are not a ghairati man, you are ignorant and stupid [besauada aw kamaqal]. On the contrary, if you discover that your wife has been the object of bad verbal or physical harassment against her will, and you kill the culprit, this is good ghairat, something obligatory [majbur]. However, this being ghairati, it also brings lust for power [ghoror], as it has happened for me. It makes you feel good, important. And also, you have to continuously demonstrate you are ghairati to all the other people, so you try to always top the action that you have undertaken the previous times, to impress others, and fulfill expectations. Violence escalates in this way. But I also think that greed for power is something inherent in human nature. Everybody is somehow affected by it. If a structure like the state does not exist, or is weak, like here [in Afghanistan], there will be no restraint in people. There should be a power like the state to punish people and prevent them from becoming so violent.

Rohullah was probably referring to himself in the last sentence, given that he admitted to retaining some of the self that he construed for himself in the village after his family relocated to Kabul. The training he gave himself had marked him and had become "authentic" enough to resurface occasionally. He remained somewhat of a troublemaker, he confessed, susceptible to display violent behavior. He gave an example of it describing something that had happened when he was finishing high school in Kabul (he has since graduated from college and now works in a private company).

> *One time in Kabul, I was in my last year of high school, when Karzai was president for the second year [2003]. I always sat in the front seats, because I was smart, I knew everything the teacher was talking about ... The other students were lazy ... One morning I found one of my classmates, a Hazara guy [one of the ethnic minorities of Afghanistan], sitting in my seat. I told him to move, but he refused to move. I asked him again, and he did not move. So I punched him in the face, and, without saying anything, he got up and left my seat. For a couple of months afterwards I was afraid that he would gather his friends and relatives and come to school with them to take revenge for what had happened. But nothing happened. Kabul is different from my village in Paktia.*[b]

From the narrative that Rohullah gives us, however brief and incomplete these excerpts might be, about his awakening from a condition of punishing social subordination to his own peers, we can already recognize some stages of emotional awareness and modification that scientific literature has expounded over the past two decades. Particularly the well-established model presented by James Gross seems to be relevant to us here (Gross, 1998; Gross & Thompson, 2007; John & Gross, 2007), not because we can isolate the single instances in which emotional modification processes occurred in Rohullah, but rather because the whole personal trajectory that he describes in the excerpt seems to represent a sublimation of Gross's model, spread over a protracted period of time—the model writ large, so to speak. Rohullah certainly did not have the luxury to go through an exercise in situation selection, given his full immersion in the village context. He did go through a period

---

[b]In analyzing this incident, it is useful to consider also the ethnic component. Pashtuns have historically been the dominant ethnicity in Afghanistan, both from the political and social point of view. Conversely, Hazaras have been the downtrodden, subject to decades of discrimination and persecution because of their Shi'a faith and supposed Mongol ancestry. Only after the demise of the Taliban, under the protection of the international military forces, Hazaras have managed to win for themselves new political and social opportunities. To this day, however, many Pashtuns express (either publicly or privately) contempt, acrimony, and resentment toward Hazaras (all the more because of their newly acquired visibility). Such attitude might have been present also in Rohullah the moment of his act of bullying against his Hazara classmate.

of attempts at situation modification that, according to what he recounted to Chiovenda, only offered counterproductive outcomes. Attentional deployment, either in the form of distraction or concentration, did not work in his case for the same reason that situation selection failed. At that point, he was left with the stage of reappraisal/cognitive change, which is when we start receiving the most details about his emotional trajectory. Rohullah did reappraise the situation and did so in a momentous way. He finally looked for long-term solutions to his plight and saw no other way, to "socially" and physically survive in the village, than to "become" one of his peers, aligning himself with the local modus operandi. He had by then, whether consciously or unconsciously (and going through a process of veritable renewed enculturation), gained knowledge and awareness of the villagers' expectations and requirements of an appropriate "manly man." Such knowledge he used to shape for himself an alternative, or rather complementary, self, more in line with the expectations of his surroundings: an emotionally filled subjectivity that contained elements of contradiction with his "original" self (with which he arrived to the village in the first place), but at the same time also very authentic elements, that embodied his new life reality replete with traumatic experiences and suffering. Thus he entered the stage of response modulation and started playing the game by the local rules. He excelled at it to the point of becoming a ringleader among the local youth.

Despite such a situation, if we analyze in more intimate details the dynamics that Rohullah describes about his emotional trajectory, things become less clear-cut. He feels he is doing something wrong when he starts participating in brawls and beating up his peers. Faced with what his father had taught him as a child, he feels he is disappointing him. At the same time though, he needs this new, if abusive, self to survive, which is somewhat culturally accepted and socially functional. Additionally, after a while, he starts finding personal pleasure in such displays of aggression, which gain him so much popularity among his peers, respect, and power. Thus he finds himself straddling two clashing cultural worlds. Moreover, his emotional reactions vary accordingly—apparently inconsistent, in reality each legitimized by their own cultural context.

Emotion regulation in context: Part 2

The second case that we present is based on the work conducted by anthropologists Byron Good and Mary-Jo DelVecchio Good in the Indonesian province of Aceh, on the island of Sumatra. The Goods began intensive work in Indonesia in 1996, settling in Yogyakarta, a center of Javanese culture, and began collaborations with colleagues in Gadjah Mada University. Byron Good in particular developed a research team focused on early psychosis and conducted combined ethnographic, clinical, and survey research, with a particular interest in very rapid onset psychoses. In 2005 the Goods shifted their focus from Yogyakarta to Aceh following an intense earthquake on December 26, 2004, off Aceh's west coast that produced a massive tsunami. Three waves, nearly 30 m each, devastated the coast, leaving mass destruction and approximately 180,000 people missing and dead. The scope of the ensuing humanitarian response was unprecedented. By the end of February 2005, just 2 months after the tsunami, the UN's Humanitarian Information Center in Banda Aceh listed 320 organizations working in Aceh (Hedman, 2008).

The influx of international organizations brought to greater awareness a military conflict that had been largely hidden from view. Acehnese had been living under martial law since May 2003, while Indonesian security forces carried out a massive counterinsurgency campaign against Aceh's separatist rebels in the Free Aceh Movement (GAM, *Gerakan Aceh Merdeka*)—a protracted conflict that was launched in 1976 and intensified in the late 1980s. The massive humanitarian influx placed pressure on the Government of Indonesia and GAM to reach a peace accord. With the signing of a peace agreement between representatives of the two parties in Helsinki on August 15, 2005, less than 8 months after the tsunami, the role of national and international agencies and humanitarian groups was dramatically increased in scope. Whereas the tsunami relief focused on the coastal communities, the peace process focused on communities up in the hills of Aceh, where the violence had been most intense.

Given their near decade of work on mental illness and mental health services in Indonesia, the Goods were invited by the International Organization for Migration, IOM, to serve as consultants as they developed psychosocial responses to communities affected by the tsunami and then to those affected by the conflict. In November 2005 they were invited to form a team to conduct a Psychosocial Needs Assessment (PNA) in the three districts most affected by the violence. This survey (of 596 individuals) found extraordinarily high rates of exposure to traumatic violence, depression, anxiety, and PTSD (Good, Good, Grayman, & Lakoma, 2006).

Perhaps equally important for the members of the research team were the experiences of hearing stories of what these villagers had lived through during the years prior to the signing of the MOU.[c] As the Indonesian military became frustrated at its inability to find and defeat the shadowy military forces of GAM, soldiers increasingly turned their violence onto civilian communities believed to be the "bases" for GAM—the source of soldiers, food, and money. All men in a village would be rounded up and tortured, in an effort to gather information about GAM. The whole villages would be forced to flee their homes, returning to find their houses empty, schools burned, and all food and animals confiscated. Women learned to survive, defending their homes and when possible their men and protecting each other from sexual violence (M. Good, 2015). Men, women, and young people, even children, were threatened, beaten, and terrified, and many were forced to watch family and friends tortured and killed. Families were increasingly prevented from maintaining their fields, rice paddies, and gardens.

This initial work was followed by an extension of the PNA research to 75 more villages in the remaining 11 districts of Aceh (Good, Good, Grayman, & Lakoma, 2007). This was followed (2006–07) by the development of a pilot program of mental health outreach teams, using general practitioners and nurses, trained and supervised by an Acehnese psychiatrist, to provide general mental health services for 25 villages in areas our research had shown to have suffered the highest levels of violence. When this proved feasible, the team extended this work to 50 more villages in 2008–09, this time with a carefully designed evaluation documenting levels of violence experienced and symptoms and social functioning before and after the intervention program. During both phases of the intervention program, the Goods carried out intensive interviews with individual patients being treated by mental health outreach teams. It is this work, linking anthropologists with years of ethnographic experience with local, Acehnese public mental health practitioners that the Goods have analyzed, focusing on the cultural phenomenology of trauma-related mental illnesses (including PTSD), coping strategies, and the effectiveness of medical interventions. Here, we link these to investigations of PTSD, cultural shaping of emotions, and emotion regulation, drawing in particular on the ethnographic studies conducted with persons in a treatment program.

The local phenomenology of trauma-related disorders in these communities emerged in the context of providing care (see Good, Good, &

[c]Discussion of such stories is found in Good, Good, and Grayman (2010), B. Good (2012, 2015), M. Good (2010, 2015), and Good and Good (2017).

Grayman, 2015; cf. Good & Good, 2017). An extremely common initial presentation of distress, reflected in ethnographic interviews, in clinical interactions, and in the medical records, would begin with a simple statement—*jantung berdebar debar*—my heart pounds.[d] Those diagnosed as suffering a trauma-related disorder or PTSD would often go on: *Saya sering takut*, I am often afraid. *Teringat*, I have memories that come unbidden to me. *Tidak bisa tidur dengan enak*, I can't sleep well at night; *ada mimpi buruk*, I have nightmares or bad dreams, wake up feeling frightened, and cannot sleep again. *Gelisah*, I often feel restless, anxious, and worried. My body feels weak, *lemah*; I lack spirit or energy, *semangat*, so that I am unable to go off to work in the rice paddies or the gardens. In some cases these symptoms were presented as such—as symptoms—to the physicians or a member of the medical team, with narrative content emerging after several meetings with the clinicians, when a close enough relationship was established to recount horrifying memories, such as those described in the beginning of this chapter. In other cases the narratives came first, with symptoms essentially describing the embodied response to the events that had occurred. Physicians would then inquire further to determine more specific diagnoses.

Two characteristics of those whose condition most represented that described clinically as PTSD are especially noteworthy. First, the symptoms of intense, intrusive memories were clearly marked by the Indonesian verb "*teringat*," "to remember" in the sense of intrusive memories, coming unbidden, in contrast with the verb "*mengingat*," to remember in the sense of an active remembering process. In some cases individuals described intrusive memories as producing acute episodes of extreme fear or anxiety, with symptoms meeting criteria for panic attacks and nightmares in which these events were vividly reexperienced. The boundary between acute remembering, often with intense anxiety, and reexperiencing of the kind popularly described as "flashbacks" is often unclear in Aceh. Patients being treated would describe acute, intrusive remembering of terrible events they had witnessed directly, things that had been done to them, or in some cases events they had only heard about when a family member was tortured or killed. Some would describe seeing such events being played out as though on a video—in some cases even if they had not seen the events directly. Many would describe becoming anxious in specific places in their villages

---

[d]This analysis reports symptoms in Bahasa Indonesia, or Indonesian language. Local villages in our region spoke primarily Acehnese. The Indonesian terms, here, are translations of Acehnese and the terms used when Acehnese spoke Indonesian.

or their homes where terrible events had happened or in some cases having extremely acute memories be triggered when they were in such settings. And many described avoiding going out in crowds or trying to avoid the specific places where these events had occurred.

Second, equally striking was the fact that many of those persons the clinicians labeled as suffering PTSD would, at some point early in their care, tell stories of what they had witnessed or experienced as though they had occurred very recently. Good and Good (2017) describe cases in which stories were told to them or to clinicians (who retold the stories to them) with great vividness and a sense that they had occurred in the past days or in recent months, even though the war had ended at least 18 months earlier. The ethnographers and the clinicians would later learn that the events had happened years before, in some especially memorable cases up to 16 years earlier. Although Indonesian language does not neatly distinguish present and past tense, these stories were told as though in the vivid present, as recent occurrences that were cause for current, ongoing anxiety. After the treatment the same patients would tell or refer to these stories as clearly being a part of their past.

The cultural phenomenology of PTSD experiences emerged during the second phase of the intervention program, which included the effectiveness evaluation. Several things should be noted. First, this intervention began 18 months after the peace agreement, which had led to a complete cessation of the fighting and, after some months, withdrawal of the Indonesian special forces from the region. Mental health workers in the IOM project diagnosed and treated approximately 11% of the total adult population of the villages in these high-conflict areas—in areas in which over 80% of adults reported having experienced combat (Good, Good, & Grayman, 2015). Of all patients treated for mental health problems, 33% (40% of males and 29% of females) were given a diagnosis of PTSD by the clinicians—less than 4% of the total adult population. The others were given diagnoses of depression, anxiety disorders, or mixed depression and anxiety. Only 3% of the patients treated were diagnosed with psychotic or organic disorders.

It is important to note that—whereas rates of traumatic violence were extremely high— during the initial PNA research, 86% of those in the two districts in which the intervention was carried out reported experiencing combat or gun fights, 51% reported being beaten by the military, 25% reported being tortured, 5% reported having a spouse killed, and 6% reported having a child killed (cf. Good, Grayman, & Good, 2015). In this context, we need to ask how it is that so *few* persons were actually identified as having mental health problems 18 months after such violence, raising questions about the differences between those whose symptoms remitted versus those who developed more persistent PTSD or other mental health problems.

The intervention consisted of teams of trained general practitioners and nurses holding general medical clinics in the 50 villages, identifying those with mental health problems for follow-up, then making diagnoses, and providing care for those identified and others who came forward for treatment when it became known that these teams were providing care for *trauma* or *stress*, terms that came into general use in Aceh during the conflict. The teams provided medical care, with regular follow-up visits to the villages (initially twice per month and later once per month) and supportive counseling. Data indicated that 47% of the patients were given an antidepressant medication, 44% an antianxiety medication, and 3% an antipsychotic medication. Seventy-four percent of the sample reported having used a medication for 6 months or less, with antianxiety drugs usually being provided for quite short periods. At the end of the intervention program, 1% of patients indicated their symptoms were worse than when they entered treatment; 16% were the same; and 83% reported their symptoms were better (37%), much better (38%), or extremely improved (9%). Whereas the patients estimated that at the time they were ill their ability to work had declined to a mean of 10h per week, at the end of the program, that mean increased to 41 h per week.

In this research and in research conducted by a Dutch anthropologist (Annemarie Samuels) with families who suffered severe losses during the tsunami, indigenous forms of emotion regulation were local, often religious psychocultural processes. Although the words *trauma* and *stress* are used commonly in Aceh—though with somewhat different meanings than in North America—the term "recovery" is seldom used to describe coming to terms with a profound loss or trauma. Instead, Acehnese regularly use the religious terms *pasrah* (surrender to God and attain acceptance) and *ikhlas* (sincere—in one's moral commitment to accepting what God has given), as well as *sabar* (patience) (see Samuels (2019, Chapter 3) for a full discussion), to describe the process of working through trauma or loss. One experiences the immediate shock of *trauma*, but must come to accept this by sincerely surrendering to God. The alternative is to risk becoming *stress*, permanently crazy, as a result of these events. Attending religious services and engaging in Islamic meditation (*zikhir*) and personal contemplation and conversations with friends and family members were all seen as important for achieving acceptance (*pasrah*).

Acehnese never talk of "forgetting"—who could forget the effects of the tsunami or the conflict—but of accepting, surrendering oneself to what was given by God, and thus working through the experience. It was widely agreed by Acehnese that despite the terrible magnitude of the tsunami—a nurse told us of losing 27 members of her family, while one of the physicians who became a psychiatrist told of losing 300 members of her extended kin when the city of Calang was wiped out—that it was easier to accept the tsunami than the trauma associated with the conflict. Many of those interviewed expressed continued anger at the Indonesian military for what they had done, and many Acehnese feel that only a genuine acknowledgement of the human rights violations would allow them to come fully to terms with the violence. "If you know someone has done something to you, and they come into your house and say, 'I did such and such, I'm sorry'," a former GAM leader told us, "the matter would be easily finished. But if that person comes into your house and sits with you, but does not acknowledge what he has done, how can the matter be finished?" Trauma and trauma treatment, or "trauma healing," as friends from government ministries would say wistfully of this work, focuses attention on individual suffering, often allowing the historical events and the structures that produce and reproduce violence to remain unchallenged.

It is unclear whether we should hypothesize that failures of emotion regulation place individuals at greater risk for developing persistent PTSD or whether the symptoms of PTSD—intrusive memories provoking renewed fear akin to panic attacks and long periods with sleep loss—made persons unable to engage in local forms of emotion regulation. The Goods have hypothesized that the medical treatments reduced panic attacks and led to increased ability to sleep and that, when this occurred, individuals were again able to engage in religious activities and local forms of emotion regulation that led to recovery (Good, Good, & Grayman, 2015).

Our conclusion from this work is that ethnographic research is complementary with the research on emotion regulation, that trauma-related disorders and emotion regulation are deeply cultural, and that both cultural meanings and personal interpretations and appropriations of these cultural forms need to be explored if we are to understand the processes of emotion regulation as they bear on PTSD.

Emotion regulation in context: Part 3

The last case discussed here is derived from Devon Hinton's work with Cambodian immigrants and/or refugees in the United States. In particular, we will take into consideration one specific study related to the perception and expression of anger (Hinton et al., 2009). Hinton is a psychiatrist and medical anthropologist who has extensive ethnographic experience working in Cambodia with Khmer populations and is fluent in Khmer language. His research emphasizes mixed methods; both quantitative and qualitative methodologies are utilized for collecting data, and we will see how this can indeed make a significant difference in the study of emotions as well. Hinton and his colleagues (2009) worked with refugees who relocated to the United States after escaping very traumatic experiences of physical, sexual, and/or psychological abuse and death of family members and personal near-death experiences. The study addressed a specific and underanalyzed (to this day) aspect of the plight of people who find themselves in these circumstances: that of the expression of anger, including explosive anger, and the triggering by members of their own nuclear family. An important aspect of the study is that Hinton and colleagues looked into the links between anger triggers, anger severity, the presence of anger-associated symptoms and behaviors, and PTSD severity. All of the previously mentioned, in turn, had to be appropriately viewed through the lens of the cultural background of the main actors involved in the study, who often had very limited knowledge of the English language and had retained a fairly strong identification with their Khmer self. Hinton and colleagues' article may well be (to the best of our knowledge) the only one extant in the experimental literature that specifically brings together emotion regulation, PTSD, and cultural constraints.

The authors of the study chose 143 patients from a psychiatric clinic in Lowell, a town in northern Massachusetts with a large minority of Khmer population. The individuals who agreed to be part of the study (73% women) were administered 12 scales. Eight of the scales profiled the anger episodes; three assessed the degree to which a language gap existed between the individual subject to the study and his/her (younger) family members; the last scale assessed PTSD severity. A further dimension to the study was given by qualitative interviews in which the subject reported the methods and techniques he/she used to death with an episode of overt anger and regain composure and the perceived reasons why he/she erupted with anger against a nuclear family member.

Of the 143 individuals surveyed, 68 (48%) had experienced at least one episode of anger against a nuclear family member in the previous 4 weeks. Of these 68, 45 (66%) displayed symptoms consistent with a PTSD diagnosis. Anger itself was found to predict 24% of the variance in PTSD severity, but anger-induced trauma recall and catastrophic cognitions accounted for an additional 30% of the variance. Techniques to control the anger episode varied from coining techniques (a local healing practice to open up channels in the body for, among other things, the flow of *khyal*, a wind-like substance) to Buddhist meditation, cold showers, self-imposed change of attention, and many others. The reasons given as provoking the anger episode (mostly against children) were often mundane and rather commonplace—from acting disrespectfully to making noise and from not doing assigned chores to fighting among siblings.

The study showed how anger came to represent a central aspect of PTSD symptomatology in the population under investigation. However, our purpose here is not to go further into the details of the study by Hinton and colleagues; instead, we wish to emphasize the ethnographic soundness of the premises on which the measures themselves and the overall design of the study rested. Hinton and his colleagues based the choice of items to put in their scales and the conceptual formulation of them, based upon years of ethnographic experience with traumatized Cambodian refugees. In fact, all the refugees surveyed in the study had experienced personally the horror of the Khmer Rouge dictatorship in the country (1975–79), when one-quarter of the 7.8 million people living then in Cambodia died by either execution, torture, slave labor, disease, or starvation. In gathering data to understand the degree to which anger episodes triggered flashbacks of past traumatic experiences and, in turn, how these affected PTSD severity, the authors made use of their knowledge of the context and cultural–historical background. For example, there was the link between personal disrespect from a naughty child in the United States and the personal disrespect suffered in Cambodia (e.g., being forced to work under conditions of quasi slavery and performing tasks traditionally reserved for work animals), associations that triggered trauma recall and contributed to worsening PTSD symptoms. Thus, while it is true that this particular study by Hinton and colleagues was not carried out through a strictly ethnographic methodology, it is evident how the previous ethnographic research carried out by the authors and the expertise derived from it were able to inform the development and execution of a culturally sensitive study of emotion regulation processes among a highly traumatized population of Cambodian immigrants/refugees.

Conclusion

In this chapter, we have attempted to highlight how insights gained from ethnographically grounded research on PTSD and emotion regulation can and should be used to inform future culturally sensitive quantitative work on these constructs in other cultures. We believe that the types of qualitative and ethnographic research described in the present chapter are complementary to the traditional quantitative approaches to emotion regulation research employed in the majority of other chapters in this book and can only lead to greater insight in these other important areas of study. We also strongly advocate that the term "culture" should never be reduced to stereotypical traits that are so wide ranging that they encompass populations inhabiting entire continents (e.g., "individualists" vs. "collectivists" or "East" vs. "West"). Instead, we believe that culture refers to very specific, local contexts of meaning and meaning-making. Finally, we believe that ethnographic research provides clear evidence that there is always much variation not only *between* human groups but also *within* human groups.

References

Barrett, L. F., & Kensinger, E. (2010). Context is routinely encoded during emotion perception. *Psychological Science*, *21*, 595–599.
Barrett, L. F., Lindquist, K., & Gendron, M. (2007). Language as a context for emotion perception. *Trends in Cognitive Sciences*, *11*, 327–332.
Boiger, M., Mesquita, B., Uchida, Y., & Feldman Barrett, L. (2013). Condoned or condemned: The situational affordance of anger and shame in the United States and Japan. *Personality and Social Psychology Bulletin*, *39*, 540–553.
Chiovenda, A. (2020). *Crafting masculine selves: Culture, war and psychodynamics in Afghanistan*. New York: Oxford University Press.
De Leersnyder, J., Boiger, M., & Mesquita, B. (2013). *Cultural regulation of emotion: Individual, relational and structural sources*. [Unpublished manuscript].
Eggerman, M., & Panter-Brick, C. (2010). Suffering, Hope and entrapment: Resilience and cultural values in Afghanistan. *Social Science and Medicine*, *71*(1), 71–83.
Giustozzi, A. (2007). *Koran, kalashnikov and laptop*. London: Hurst.
Good, B. (2012). Phenomenology, psychoanalysis and subjectivity in Java. *Ethos*, *40*, 24–36.
Good, B. (2015a). Haunted by Aceh: Specters of violence in post-Suharto Indonesia. In H. Devon & A. Hinton (Eds.), *Genocide and mass violence: Memory, symptom, and recovery*. Cambridge: Cambridge University Press.
Good, B. J., Good, M.-J.D.V., Grayman, J. H., & Lakoma, M. (2006). *Psychosocial needs assessment of communities affected by the conflict in the districts of Pidie, Bireuen, and Aceh Utara*. Jakarta: International Organization for Migration. Available at http://ghsm.hms.harvard.edu/uploads/pdf/good_m_pna1_iom.pdf.
Good, B. J., Good, M.-J.D.V., & Grayman, J. (2015). Is PTSD a "Good enough" concept for Postconflict mental health care? Reflections on Work in Aceh, Indonesia" In Hinton, D. & Good, B. (Eds.), *Culture and PTSD: Trauma in global and historical perspective*. Philadelphia: University of Pennsylvania Press.
Good, B. J., Grayman, J., & Good, M.-J.D.V. (2015). Humanitarianism and 'Mobile sovereignty. In S. Abramowitz & C. Panter-Brick (Eds.), *Strong state settings: Reflections on medical humanitarianism in Aceh, Indonesia*. Philadelphia: University of Pennsylvania Press. Medical Humanitarianism: Ethnographies of Practice.
Good, D., & Good, M.-J.D.V. (2017). Toward a cultural psychology of trauma and trauma-related disorders. In J. Cassaniti & U. Menon (Eds.), *Universalism without uniformity: Exploration in mind and culture*. Chicago: University of Chicago Press.

Good, M.-J.D.V. (2010). Trauma in post-conflict Aceh and Psychopharmaceuticals as a medium of exchange. In J. Jenkins (Ed.), *Pharmaceutical self: The global shaping of experience in an age of psychopharmacology*. Santa Fe, N.M: SAR Press.

Good, M.-J.D.V. (2015b). Acehenese Women's narratives of traumatic experience, resilience and recovery', in Devon E. Hinton and Alexander L. Hinton, eds. In *Genocide and mass violence: Memory, symptom, recovery*. Cambridge: Cambridge University Press.

Good, M.-J.D., Good, B. J., Grayman, J. H., & Lakoma, M. (2007). *A psychosocial needs assessment of communities in 14 conflict-affected districts in Aceh*. Jakarta: International Organization for Migration. Available at http://ghsm.hms.harvard.edu/uploads/pdf/good_m_pna2_iom.pdf.

Good, M.-J.D.V., Good, B. J., & Grayman, J. (2010). Complex engagements: Responding to violence in Postconflict Aceh. In D. Fassin & M. Pandolfi (Eds.), *Contemporary states of emergency: The politics of military and humanitarian interventions*. New York: Zone.

Gross, J. (1998). The emerging field of emotion regulation: An integrative review. *Review of General Psychology, 2*, 271–299.

Gross, J., & Thompson, R. (2007). Emotion regulation: Conceptual foundations. In J. Gross (Ed.), *Handbook of emotion regulation*. New York: The Guilford Press.

Hedman, E.-L. (2008). Back to the barracks: Relokasi Pengungsi in post-tsunami Aceh. In E. Hedman (Ed.), *Conflict, violence, and displacement in Indonesia*. Ithaca, NY: Cornell Southeast Asia Program Publications.

Hinton, D., Rasmussen, A., Nou, L., Pollack, M., & Good, M.-J.D.V. (2009). Anger, PTSD and the nuclear family: A study of Cambodian refugees. *Social Science and Medicine, 69*, 1387–1394.

John, O., & Gross, J. (2007). Individual differences in emotion regulation. In J. Gross (Ed.), *Handbook of emotion regulation*. New York: The Guilford Press.

Masuda, T., Ellsworth, P., Mesquita, B., Leu, J., Tanida, S., & Van De Veerdonk, E. (2008). Placing the face in context: Cultural differences in the perception of facial emotion. *Journal of Personality and Social Psychology, 94*, 365–381.

Mesquita, B., Boiger, M., & De Leersnyder, J. (2016). The cultural construction of emotion. *Current Opinion in Psychology, 8*, 31–36.

Panter-Brick, C., Goodman, A., Tol, W., & Eggerman, M. (2011). Mental health and childhood adversities: A longitudinal study in Kabul, Afghanistan. *Journal of the American Academy of Child and Adolescent Psychiatry, 50*(4), 349–363.

Panter-Brick, C., Grimon, M.-P., Kalin, M., & Eggerman, M. (2015). Trauma memories, mental health and resilience: A prospective study of Afghan youth. *The Journal of Child Psychology and Psychiatry, 56*(7), 814–825.

Samuels, A. (2019). *After the tsunami: The remaking of everyday life in Banda Aceh. Indonesia*. Honolulu: University of Hawa'i Press.

Index

Note: Page numbers followed by *f* indicate figures and *t* indicate tables.

A

Acceptance and Action Questionnaire, revised (AAQ-II), 19–22*t*, 29–30, 546–547
Acceptance-based behavioral therapy, 546–555
    acceptance and commitment therapy (ACT), 546–550
    behavioral activation (BA), 552–554
    benefits of, 555–558
    dialectical behavior therapy (DBT), 554–555
    future research, 558–560
    mantram repetition program (MRP), 551–552
    mindfulness-based stress reduction (MBSR), 550–551
Acceptance-based interventions, 559–560
Adaptive disclosure (AD), 134–135
ADCYAP1R1 gene, 212–213, 237
Adrenocorticotrophin-releasing hormone (ACTH), 276–277
Adult Attachment Interview, 520–521
Adverse childhood experiences (ACEs), 514–515
Affiliative emotions, 576
Aggression, 65, 68–69, 72–78
Ainsworth categorization model, 519
Alcohol use disorder (AUD), 549–550
Alexithymia, 18–23, 381–383
α-amino-3-hydroxy-5-methyl-4-isoxazolepropionic acid receptors (AMPA), 279–280
Amygdala, 44–46, 175–179, 502
Amygdala–insula functional connectivity, 504
Anger, 4–9, 65–66, 498–499, 609–611
    associations with PTSD, 66–68, 70–73
    cognitive appraisal theories, 72–73
    fear avoidance theory, 70–71
    social information processing theories, 72
    survival mode theory, 71
    as cause/consequence of PTSD, 68–70
    defining, 66
    future directions, 80–81
    impact of PTSD treatment on, 73–75
    intimate partner violence (IPV), 77–78
    measures of, 4–8
    novel approaches to treating, 79–80
    role of gender in treatment for, 78–79
    Strength at Home Program (SAH), 77–78
    treatments, 75–77
Anger Expression Index of the STAXI-2, 76
Angry faces, 392
Anhedonia/emotional numbing, 23–25, 94–99
    costs of, 98–99
    measures of, 23–25
    theoretical underpinnings of, 96–97
Animal-reminder disgust, 118–119, 121
Anterior cingulate cortex (ACC), 178, 191–192
Anxiety and fear, 9–10, 43–57, 125, 255–256, 500
    avoidance, excessive, 52–53
    fear overgeneralization, 50–51
    heightened sensitivity to threat, 47–50
        anxiety sensitivity, 47–48
        attentional bias to threat, 48–50
    impaired fear extinction, 53–56
    implications of, 56–57
    measures of, 9–10
    and mood disorders, 395–396
    neurobiology of, 44–47
Anxiety sensitivity (AS), 25–26, 47–48, 353, 390–391
Anxiety Sensitivity Index (ASI), 19–22*t*, 25–26

Anxious-preoccupied attachment
    tendencies, 579
Apolipoprotein E (APOE), 212–213
Approach-avoid task (AAT), 53
Arousal focus, 379, 389–391
Assessment of emotion-related processes,
    3–30, 5–7t, 19–22t
  alexithymia, 18–23
  anger, 4–9
  anhedonia/numbing, 23–25
  anxiety, 9–10, 25–26
  distress tolerance, 26–28
  emotion regulation, 28–29
  experiential avoidance, 29–30
  fear, 10–12
  guilt and shame, 12–14
  negative and positive affect, 16–17
  sadness, 15
Attachment, 579–580
  anxiety, 385–386, 518–519
  avoidance, 518–519
  and emotion regulation skills, 520
Attachment theory, 515, 518–520
Attentional bias, 48–50, 392–393
Attentional bias to threat (ABT), 315
  dual-process models of, 317–320
  measuring ABT in PTSD, 320–326
Attentional control theory, 311–312,
    317–318, 326–333
Attentional deployment, 311–312
Attentional matrix, 394–395
Attention bias modification (ABM),
    333–335
Attention maintenance, 49–50
  vigilance-avoidance versus, 316
Avoidance, excessive, 52–53
Avoidance/numbing symptoms, 23, 382
Avoidant behaviors, 103–104
AX+/BX− model, 272–273

B

BEAM approach, 184
Behavioral activation (BA), 552–554
Behavioral contributions, 278–279
  sleep disturbances, 278–279
Behavioral genetics, 211–212
Behavioral Indicator of Resiliency to
    Distress (BIRD) task, 358

Behavioral inhibition system (BIS), 424
Behavioral reinforcement theory, 552–553
Between-session reduction, 468–469
Blended versus differentiated, principle of, 584
Blood-injection-injury phobia, 119–120
Body-based work, 587–588
Body listening, 398–399
Body, thought, and behavior, 526
Bordering constructs of EA, 431
Bordering psychological constructs,
    413–414
Borderline pathology, Kernberg's
    description of, 384–385
Borderline personality disorder (BPD), 474,
    554–555
Brain pathways, 444–445
Brain regions and networks implication,
    175–183
  amygdala, 176–178
  fear extinction model, 179–180
  hippocampus, 175–176
  network connectivity, role of, 181–183
    resting-state functional connectivity,
      182–183
    task-related functional connectivity,
      181–182
  prefrontal regions' role in emotion
    regulation, 180–181
  ventromedial prefrontal cortex (vmPFC),
    178–179
Breath-holding task, 355–356
Brief eclectic psychotherapy, 107
Brief Experiential Avoidance Questionnaire
    (BEAQ), 430
Brooding, 102
Buss-Durkee Hostility Inventory (BDHI),
    8–9
Buss-Perry Aggression Questionnaire
    (BPAQ), 8–9

C

Candidate gene by environment (cGxE)
    studies, 213–214, 234
Candidate-gene methylation studies,
    212, 213, 231–232
Cardiovascular indices, 252–253
Catechol-O-methyltransferase (COMT) gene,
    232–233

Causal attributions, 146
Central executive network, 181
Childhood maltreatment, 573
Childhood sexual abuse (CSA), 129, 154–155, 366, 500–501
Childhood trauma, 449–450, 514, 516, 535–537
Classical conditioning, 255
Clinician-Administered PTSD Scale (CAPS), 178
Clinician-Administered PTSD Scale for DSM-5 (CAPS-5), 3
Close/attachment relationships, 384–386
Cognitive appraisals, 72–73, 492
Cognitive avoidance, 418
Cognitive behavioral therapy (CBT), 76, 264–266, 504, 548–549, 575
Cognitive bias modification (CBM) techniques, 79–80
Cognitive fusion, 546
Cognitive paired stimulation, 191
Cognitive processing therapy (CPT), 73–74, 157–159, 491–492
   emotion in, 491–493
   limitations and future directions, 505–506
   mechanisms of action in, 502–505
      biological mechanisms, 504–505
      research on natural and manufactured emotions, 492–493
   specific emotional changes in, 494–502
      emotion regulation, 502
      empirical support, 495–502
      theoretical rationale, 494–495
Cognitive restructuring and imagery modification (CRIM) approach, 135
Cognitive restructuring (CR) in prolonged exposure, 473
Cognitive therapy for battered women (CTT-BW), 504–505
Coheritability, 215
Cold spot, 521
Common cause model, 414–415
Common humanity, 589–590
   vs. isolation, 568–569
Community support, 517
Comorbid depression, 89–90

Comorbidity of PTSD and depression, 89–90, 103–104, 106
Compassion, 530, 586–587
Compassionate mind training (CMT), 575
Compassion-based work, 577–578
Compassion-focused interventions, 559–560
Compassion-focused therapy (CFT), 160, 575
Compassion with Equanimity, 585
Conditioned response (CR), 122–123
Conditioned stimulus (CS), 122–123, 255
Conditioning model, 44, 45f, 53–54
Confirmatory factor analysis studies, 395–396
Connectivity-based rTMS, 203
Conscious fears, 578
Conservation of resources and resource loss model, 517–518
Conservation of Resources theory, 517
Constructs, 380–383
Contamination-based disgust reactions, 123–124
Contamination-based obsessive compulsive disorder (OCD), 119–120, 132–133
Co-occurring PTSD, 549–550
Coping skills, 527–528
Coping techniques, 531
Core affect, 395–396
Core disgust, 118
Corticotropin-releasing hormone (CRH), 276–277
Counterfactual thinking (CFT), 357–358
C-phosphate-G (CpG) dinucleotides, 231
CRHR1, 231–232
Criterion A traumatic exposure, 446–447
Cultural influences on emotional regulation, 597–599

D

D-cycloserine (DCS), 55–56, 280–282
Deactivating, 385
Deep brain stimulation (DBS), 197–199
Default mode network, 181
Defensive survival circuit, 46
Delayed stopping, 332
Depersonalization symptoms, 100–101

Depression, 89–90, 104–106, 500–502
  PTSD and depression, comorbidity of, 89–90, 103–104
  and suicide, 105
  symptoms, 471, 500
  treatment implications, 106–108
Derealization symptoms, 100–101
Developmental trauma, 516
Dexamethasone, 281–282
Diagnostic and Statistical Manual (DSM) of Mental Disorder version IV, 237
Diagnostic and Statistical Manual of Mental Disorders, Fifth Edition (DSM-5), 259
Dialectical behavior therapy (DBT), 367–368, 397–398, 554–555, 574
  treatment model, 554–555
  treatment outcomes, 555
Difficulties in Emotion Regulation Scale (DERS), 19–22t, 28–29, 302–303, 380–381, 438
  DERS-Positive, 304–305
Dimensions of Anger Reaction Scale, 75
Direct eye contact, 584–587
Directionality, 449–450
Disaster recovery workers in the World Trade Center attacks
  numbing/avoidance symptoms in, 99
Discomfort Intolerance Scale (DIS), 356
Disgust, 117–120
  clinical assessment and treatment implications, 131–135
  cultural considerations, 135–136
  disgust proneness, 125–127
  models, 122–125
  peritraumatic disgust, 127–128
  posttraumatic disgust, 128–131
  relevance to trauma and PTSD, 120–122
sensitivity, 119, 126–127
Dismissive-avoidant attachment tendencies, 579–580
Dissociation, 99–102
Distress
  guilt-related, 497
  intolerance, 556
Distress Intolerance Index (DII), 369
Distress Overtolerance Scale (DOS), 345, 369
Distress tolerance (DT), 26–28, 343–344, 527–528, 533

  adolescents, 358–359
  community adults, 352–354
    interactive effects, 353–354
    longitudinal studies, 354
  defining and measuring, 346
  limitations and future directions, 366–369
  measures of, 26–27
  outpatient treatment-seeking adults, 357–358
  psychiatric inpatient adults, 356–357
  substance use, applications to, 362–366
    community adults, 362–363
    military veterans, 365–366
    treatment-seeking adults, 364–365
    undergraduate students, 363–364
  suicidal ideation and behavior, applications to, 359–362
  theoretical framework, 344–345
  undergraduate students, 355–356
Distress Tolerance Scale (DTS), 19–22t, 27
Disturbances in self-organization (DSO), 367
DNA methylation (DNAm), 230–233
  candidate-gene methylation studies, 231–232
  considerations and limitations in, 233
  epigenome-wide methylation studies (EWAS), 232
  methylation age, 232
  translating epigenetic findings, 232–233
Dopamine beta-hydroxylase (DBH), 229
Dopamine receptor genes, 445
Dopamine systems, 444
Dorsolateral prefrontal cortex (DLPFC), 183–184
Dot-probe task, 322
DRD2, 445
DRD3, 445
DRD4, 445
Dysfunctional arousal, 391
Dysphoria, 95, 98, 395–396

E

Ecological momentary assessment (EMA), 377, 430
Electroconvulsive therapy (ECT)
  efficacy, to treat PTSD, 196–197
  and magnetic seizure therapy, 195–197

Electrodermal indices, 253–254
Electromyogram, 10
Electromyographic indices, 254–255
Electromyography (EMG), 251–252
Emotional activation, 474
Emotional avoidance, 572
Emotional awareness, 398–399
Emotional engagement, 468
Emotional granularity, 377–379, 396–397
  alexithymia, 381–383
  arousal focus, PTSD and, 389–391
  close/attachment relationships, 384–386
  and coping with traumatic stressors, 386–387
  emotional clarity, 380–381
  flexibility, 386–387
  as moderator between trauma exposure and PTSD, 387–388
  psychological constructionism, PTSD, and, 393–396
  PTSD symptoms and, 389
  trauma exposure, 384–386
  valence focus, PTSD and, 392–393
Emotional numbing, 94–99, 105–106, 108–109
  costs of, 98–99
  theoretical underpinnings of, 96–97
Emotional processing theory (EPT), 24–25, 30–31, 465–469, 499
Emotion awareness and regulation skills, 526–528
Emotion differentiation, 377–379
Emotion-driven impulsivity, 437, 439–440, 445–446
  biological mechanisms, 443–445
    brain pathways, 444–445
    gene polymorphisms, 445
    neurotransmitters, 443–444
  clinical implications, 450–451
  conceptualizing, 437–439
  empirical evidence, 445–448
  future research, important avenues for, 448–450
    biological mechanisms, 449
    directionality, 449–450
    positive emotions, 448–449
  psychological mechanisms, 440–443

Emotion dysfunction, 304–305
Emotion in cognitive processing therapy, 491–493
Emotion network, 467–468
  activation of, 468
Emotion-oriented social support seeking, 520
Emotion reduction, 468–469
Emotion regulation, 28–29, 311–312, 332, 387–388, 410–411, 437, 474, 502, 514, 516, 518, 520, 535–536, 580, 597–611
  abilities, in PTSD, 302–304
  deficits, 556
  definitions of, 296–298
    emotion regulation abilities model, 297–298
    emotion regulation strategy model, 296–297
  evaluating emotion regulation strategy and abilities models, 299–300
  implications of research, 305–306
  integrating models of, 298–299
  measures of, 28–29
  positive emotion regulation, 304–305
  strategies, and PTSD, 300–302
Emotion Regulation Interview, 29
Emotion regulation therapy (ERT), 161–162
Empirical evidence, 445–448
Empirical support, 495–502
  anger, 498–499
  depression, 500–502
  fear, 499–500
  guilt, 495–497
  shame, 497–498
Engagement, 584–585
Enhanced treatment as usual (ETAU), 77–78
Epigenome-wide association studies (EWAS), 230
Epigenome-wide methylation studies, 232
Erroneous cognitions, modification of, 468
Ethnographic research, 597, 609, 611–612
Evaluative conditioning, 123–125
Evidence-based psychotherapy, 491–492
Evidence-based therapies, 513

Experiential avoidance (EA), 29–30, 409–413
  clinical implications, 429–430
  distinguishing from emotion regulation, 409–411
  EA PTSD relationship, possible mechanisms of, 421–426
  inflexible use of, 422–423
  language, role of, 411–413
  limitations and future directions, 430–431
  malleability of, 426
  measures of, 29–30
  models of, and PTSD, 414–415
  posttrauma, 417–418
  pretrauma, 415–417
  and rumination, 424–425
  and subjective arousal, 423–424
  and treatment response, 418–421
Exposure-based PTSD treatment, 152
Exposure therapies, 513
The Expression of the Emotions in Man and Animals (1872/2003), 117
Extinction learning, 261
  individual differences in, 263–276
    extinction recall, 266
    fear inhibition, 272–274
    fear load, 266–267
    generalization of fear, 274–275
    reinstatement, 270–271
    renewal, 267–269
    return of fear, 267
    reversal learning, 276
    spontaneous recovery, 269–270
    within-session, 261–263
Extinction of fear, 44, 45f, 53–56
Extinction recall, 266
Eyeblink, measures of, 10
Eye movement, desensitization, and reprocessing therapy (EMDR), 74, 157–158, 418–419, 470
Eye movement desensitization, 107
Eye-tracking technology, 325, 329–330

F

Fear, 10–12, 499–500, 576–578
  acquisition, 256–260
  and anxiety (*see* Anxiety and fear)
  of compassion, 577–578
  conditioning, 255–256
  generalization of, 274–275
  of happiness, 590–591
  inhibition, 272–274
  load, 266–267
  measures of, 10, 12
  overgeneralization, 50–51
  return of, 267
  structures, 466–467
Fear avoidance theory, 70–71
Fear extinction, 175, 179–180, 261–263
  extinction learning, 261
  impaired, 53–56
  within-session extinction learning, 261–263
Fearful-avoidant attachment tendencies, 580
Fear inhibition theory of anger, 73–74
Fear network, 313–315
Fear processing, psychobiological factors that moderate, 276–282
  behavioral contributions, 278–279
  sleep disturbances, 278–279
  D-cycloserine, 280–282
  dexamethasone, 281–282
  frontiers in pharmacotherapy, 279–280
  hormones, 276–278
    sex hormones, 277–278
    stress hormones, 276–277
Feelings Monitoring Form, 527–528
Fierce self-kindness, 589
First-line PTSD treatments, 551
Five Facet Mindfulness Questionnaire (FFMQ), 550–551
FKBP5, 213–214, 231–233, 238–239
Flexible care, 586
Fluid vulnerability theory (FVT), 156–157
Fluoxetine, 107
Food rejection model, 117–118
Four-factor dysphoria model, 98
Four-factor emotional numbing model, 98–99
Frontiers in pharmacotherapy, 279–280
Functional magnetic resonance imagery (fMRI), 11–12, 177

G

Gene expression, 233–236
  candidate gene expression studies, 234
  considerations and limitations in expression studies, 235–236
  genome-wide gene expression studies, 234–235
  network and pathway analyses, 235
Gene polymorphisms, 445
Generalization stimuli (GSs), 274–275
Generalized anxiety disorder (GAD), 10, 161–162
Genetic correlations, 215
Genetic influences
  behavioral genetics of PTSD, 211–212
  DNA methylation, 231–233
    candidate-gene methylation studies, 231–232
    considerations and limitations in DNAm, 233
    epigenome-wide methylation studies (EWAS), 232
    methylation age, 232
    translating epigenetic findings, 232–233
  future directions, 237–239
  gene expression, 233–236
    candidate gene expression studies, 234
    considerations and limitations in expression studies, 235–236
    genome-wide gene expression studies, 234–235
    network and pathway analyses, 235
  genomic platforms, 230
  molecular genetic studies, 212–215
    candidate gene by environment (cGxE) studies, 213–214
    candidate gene studies, 212–213
    genome-wide association studies (GWAS), 214–215, 230f
  novel statistical genetic procedures using GWAS data, 215–229
    genomic relatedness matrix restricted maximum likelihood (GREML), 227
    Linkage Disequilibrium Score Regression (LDSC/LDSR), 227–228
  Mendelian randomization (MR), 229
  polygenic risk scores (PRSs), 215–226
  unique considerations for PTSD genomics, 236–237
Genetic instrumental variables (GIV), 229
Genetic overlap, 215
Genome-wide association studies (GWAS), 214–215, 230f
  novel statistical genetic procedures using, 215–229
    genomic relatedness matrix restricted maximum likelihood (GREML), 227
    Linkage Disequilibrium Score Regression (LDSC/LDSR), 227–228
    Mendelian randomization (MR), 229
    polygenic risk scores (PRSs), 215–226
Genome-wide gene expression studies, 234–235
Genomic relatedness matrix restricted maximum likelihood (GREML), 227
Glucocorticoids, 175–176, 281
Glucocorticoid signaling (GR), 234
Goal-directed behavior, 474
Group STAIR, 537
  adapted for trauma-exposed adolescents, 537
Guilt, 12–14, 150–153, 495–497
  measures of, 13–14
  theoretical models, 148

H

Habituation, 468
Hebbian-like plasticity, 190
Hebb's rule, 279–280
Hedonic Deficit and Interference Scale (HDIS), 24–25
Hippocampus, 46–47, 175–176, 179
Hopelessness, 503
Hormones, 276–278
  sex hormones, 277–278
  stress hormones, 276–277
Hostile attribution bias, 72
Hostile interpretation bias, 72
Hostility, 65–67, 69
Hot spot, 521

Hyperactivating emotion regulation strategies, 580
Hyperarousal, 97
Hypoactivating emotion regulation strategies, 580
Hypothalamic-pituitary-adrenal (HPA) axis, 44–46, 175–176, 212–213, 276–277

I

Idiographic trauma imagery, 130
IGF2, 231–232
Illness anxiety, 119–120
Imaginal exposure, 465, 473, 476–481
  goal of, 464–465
Infant stress-regulatory responses, 385–386
Information processing models of emotion regulation, 313–315
Inhibition of fear, 44–46, 53–56
Inhibitory learning processes, 468–469, 483, 548
Insecure and disorganized attachment styles, 580
Internalized Shame Scale (ISS), 13–14, 149–150, 157
Internal working models, 515, 518–519
International Affective Picture System (IAPS), 16–17
International Classification of Diseases (ICD-11), 571–572
Interoception, 398–399
Interpersonal disgust, 118–119
Interpersonal relationships, effects of anger on, 4
Interpersonal theory of suicide (IPTS), 156–157
Interpersonal trauma, 516, 528
Intimate partner violence (IPV), 77–78
Intimate relationships, social impairment in, 104–105
Intraclass correlations (ICCs), 379
Invasive brain stimulation techniques, 197–201
  deep brain stimulation (DBS), 197–199
    to treat PTSD, 198–199
  vagus nerve stimulation (VNS), 199–201
    to treat PTSD, 200–201
Inventory of depression and anxiety symptoms (IDAS), 355–356

In vivo exposure, 463–465, 475–476, 483
Isolation, 572

K

Kernberg's object-relation theory, 384–385

L

Latent-growth mixture modeling (LGMM), 263–264
Learning model, 44, 54–57
LeDoux's proposed fear system, 46
Life threatening behaviors, 554–555
Likert scale, 8–9, 11, 13
Linkage disequilibrium (LD), 215–226
Linkage Disequilibrium Score Regression (LDSC/LDSR), 227–228
Long-term depression (LTD), 279–280
Long-term potentiation (LTP), 279–280
Loving-kindness meditation (LKM) training, 574–575

M

Magnetic resonance imaging (MRI), 178
Magnetic seizure therapy, 195–197
Major depressive disorder (MDD), 10, 15, 17, 54, 89, 93–94, 161–162, 273–274
Maladaptive cognitive appraisals, 442
Mantram repetition program (MRP), 551–552
  review of treatment outcomes, 552
  treatment model, 551–552
Manufactured emotions, 492
  research on, 492–493
Measures
  of anger, 4–8
  of anhedonia, 23–25
  of anxiety, 9–10
  of distress tolerance, 26–27
  of emotion regulation, 28–29
  of experiential avoidance, 29–30
  of fear, 10, 12
  of sadness, 15
  of shame and guilt, 13–14, 149–150
Medial prefrontal cortex (mPFC), 46, 175, 179
Memory encoding, 466–467
Memory hierarchy, 533

Mendelian randomization (MR), 229
Mental contamination, 123–124, 136
Mentalization, 580–581
Mental states, 393
Men who have sex with men (MSM), 366
Methylation age, 232
Methylation quantitative trait loci
    (mQTLs), 238–239
Methylene blue, 55–56
Military-related PTSD, 481–482
Military sexual trauma, 501
Mindfulness, 547, 550, 560, 569–570
Mindfulness Attention Awareness Scale
    (MAAS), 559–560
Mindfulness-based stress reduction
    (MBSR), 550–551
  treatment model, 550–551
  treatment outcomes, 551
Mindfulness meditation programs, 551
Mindfulness of suffering, 589
Mindfulness training, 394–395
Mindful self-compassion (MSC) training
    program, 585
Minimal attention condition (MA), 496
Mirror-Tracing Persistence Task-
    Computerized Version (MTPT-C),
    355
Mirror Tracing Task, 19–22t, 27
Mobile Intervention for Reducing Anger
    (MIRA), 79–80
Molecular genetic studies, 212–215
  candidate gene by environment (cGxE)
    studies, 213–214
  candidate gene studies, 212–213
  genome-wide association studies
    (GWAS), 214–215, 230f
Moral disgust, 118–119, 121
Moral injury, 124–125, 151–152
Multidimensional Assessment of
    Interoceptive Awareness (MAIA),
    398–399
Multidimensional Experiential Avoidance
    Questionnaire (MEAQ), 19–22t,
    30, 430

N

Narrative exposure therapy, 107, 521
Narrative therapy, 515, 520–522, 533–535

Natural emotions, 492
  research on, 492
Negative affect, reduction of, 467–469, 473
Negative affective interference, 24–25
Negative affectivity, 395–396
Negative alterations in cognition in mood
    (NACM), 95–96
Negative and positive affect, 16–17
Negative and positive urgency, 438
Negative cognitions, 125
Negative emotion differentiation, 379, 396
Negative self-cognitions, 156–157
Negative thought, 102
Network and pathway analyses, 235
Network connectivity, role of, 181–183
  resting-state functional connectivity,
    182–183
  task-related functional connectivity,
    181–182
Neurobiology and neuromodulation of
    emotion
  brain regions and networks implication,
    175–183
    amygdala, 176–178
    fear extinction model, 179–180
    hippocampus, 175–176
    network connectivity, role of, 181–183
    prefrontal regions, role of, 180–181
    ventromedial prefrontal cortex
      (vmPFC), 178–179
  future directions, 201–204
  invasive brain stimulation techniques,
    197–201
    deep brain stimulation (DBS),
      197–199
    vagus nerve stimulation (VNS),
      199–201
  noninvasive and convulsive brain
    stimulation techniques, 195–197
    electroconvulsive therapy and
      magnetic seizure therapy, 195–197
  noninvasive and nonconvulsive brain
    stimulation techniques, 183–195
    repetitive transcranial magnetic
      stimulation, 183–192
    transcranial electrical stimulation,
      192–195
Neuroticism, 126

Neurotransmitters, 443–444
N-methyl-D-aspartate (NMDA), 279–280
Nonfear-based emotions, 471–472
Noninvasive and convulsive brain stimulation techniques, 195–197
  electroconvulsive therapy
    efficacy, to treat PTSD, 196–197
    and magnetic seizure therapy, 195–197
Noninvasive and nonconvulsive brain stimulation techniques, 183–195
  repetitive transcranial magnetic stimulation (rTMS), 183–192
    acute and maintenance rTMS treatment, 187, 188*t*
    neural mechanisms underlying response to rTMS, 191–192
    optimizing rTMS efficacy, 190–191
    rTMS for the treatment, 187–189
    spatial parameters, 183–185
    stimulation waveform parameters, 185–186
  transcranial electrical stimulation, 192–195
Nonspecific distress, 396
Nonspecific emotional distress, symptoms of, 395–396
Nonsuicidal self-injury (NSSI), 345, 554–555
Northern Illinois University (NIU) Trauma Study, 415
Nosology, 94–95
Novaco Anger Scale, 76
NPU-threat test, 10–11
NR3C1 gene, 231–232

O

Operation Iraqi Freedom (OIF)/Operation Enduring Freedom (OEF), 497–498
Orbicularis oculi muscle, 254–255
Orbitofrontal cortex (OFC), 444–445
Overidentification, 569–570
"Overmodulation" pattern of reactivity, 390

P

Paced Auditory Serial Addition Task (PASAT), 19–22*t*, 27–28, 356
Paroxetine, 107
Pathoplasticity model, 414–415

Pavlovian threat conditioning, 255
Peritraumatic disgust, 127–128
Peritraumatic dissociation (PD), 101
Peritraumatic sadness, 92
Polygenic risk scores (PRSs), 215–226
Positive affect, 96–97
Positive and Negative Affect Schedule (PANAS), 11–12, 15
Positive emotion differentiation, 379
Positive emotion regulation, 304–305
Positive emotions, 3, 98, 448–449
Positron emission tomography (PET), 11–12, 176–177
Posttrauma experiential avoidance, 417–418
Posttrauma phenomena, 513
Posttraumatic alexithymia, 18
Posttraumatic cognitions, 491–492
Posttraumatic Cognitions Inventory (PTCI), 72–73
Posttraumatic disgust, 125–126, 128–131
Posttraumatic Experience of Mental Contamination Scale (PEMC), 132
Posttraumatic stress, 546–547
Potentially traumatic events (PTE), 343–344, 360–361
Predisposition model, 414–415
Prefrontal regions' role in emotion regulation, 180–181
Present-centered group therapy (PCGT), 551
Present-centered therapy (PCT), 79, 501–502, 548–549, 558–559
Pretrauma experiential avoidance, 415–417
Primary emotions, 492
Probabilistic approach, 184–185
Prolonged exposure therapy (PE), 73, 154–155, 157–158, 418–419, 463–466, 548
  case study, 474–481
  emotional processing theory (EPT), 466–469
  impact of culture on PE on emotion, 481–482
  influence of PE on specific emotions, 469–474
  limitations and future research, 482–485
  treatment components, 464*t*
Psychiatric Genomic Consortium (PGC), 214–215

Psychological constructionist model of
    emotion, 393
Psychological flexibility, 546–547
Psychological inflexibility, 546–547,
    555–556, 577–578
Psychophysiological assessment, 11–12
Psychophysiology of emotional responding
    in PTSD, 251–252
  cardiovascular indices, 252–253
  electrodermal indices, 253–254
  electromyographic indices, 254–255
  extinction learning, 261, 263–276
    extinction recall, 266
    fear inhibition, 272–274
    fear load, 266–267
    generalization of fear, 274–275
    reinstatement, 270–271
    renewal, 267–269
    return of fear, 267
    reversal learning, 276
    spontaneous recovery, 269–270
    within-session, 261–263
  fear extinction, 261–263
  fear processing, 276–282
    behavioral contributions, 278–279
    D-cycloserine, 280–282
    dexamethasone, 281–282
    frontiers in pharmacotherapy,
      279–280
    hormones, 276–278
    translational fear conditioning models,
      256–260
    translational psychophysiological
      research, 255–256
Psychosis, 575
Psychosocial Needs Assessment
    (PNA), 604
PTSD Checklist (PCL-5), 3, 365–366
PTSD symptoms (PTSS), 413–414,
    417–418, 421

R

Randomized control trials (RCTs), 229,
    333–334, 548–549, 552
Rapid eye movement (REM) phase of
    sleep, 278
Reinstatement, 270–271
Relational frame theory (RFT), 411

Relationship models, 515
Relationship patterns and interpersonal
    skills, 528–530
Religiosity, 136
Repetitive transcranial magnetic stimulation
    (rTMS), 183–192
  acute and maintenance rTMS treatment,
    187, 188t
  neural mechanisms underlying response
    to, 191–192
  optimizing rTMS efficacy, 190–191
  spatial parameters, 183–185
  stimulation waveform parameters,
    185–186
  treatment, 187–189
Reprocessing therapy (EMDR), 107
Rescripting-based imagery, 470
Resonance, 586
Responses to Script-Driven Imagery Scale
    (RSDI), 12
Response styles theory, 102
Resting motor threshold (rMT), 186
Resting-state functional connectivity,
    182–183
Retrospective self-report, 128
Reversal learning, 276
Rumination, 102–103
  experiential avoidance and, 424–425

S

Sad mood, prolonged, 94
Sadness and depression, 15, 89–94
  anhedonia/emotional numbing, 94–99
    costs of emotional numbing in, 98–99
    theoretical underpinnings of, 96–97
  comorbidity of PTSD and depression,
    89–90
  dissociation, 100–101
  measures of, 15
  rumination, 102–103
  social functioning deficits, 103–105
  structures, 467
  suicide, 105–106
  treatment implications, 106–108
Safeness and trust, 584
Secondary emotions, 492–493
Seeking safety (SS), 427–428
Self-acceptance, 571

Self-Assessment Manikin (SAM), 16
Self-attributions, 147
Self-blame, 572
Self-compassion, 160–161, 530, 559–560, 567, 571–572, 583
  break, 589–590
  common humanity versus isolation, 568–569
  defined, 567–571
  exercises, 578
  integrating into treatment of complex PTSD, 584–591
    allowing in happiness and play, 590–591
    attuned presence, 584–585
    empowerment through fierce self-compassion, 587–589
    mourning the life lost, 589–590
    self-reflection, 586
    titrating the receiving of compassion from another, 586–587
  mindfulness vs. overidentification, 569–571
  principles for selecting a starting point in treatment, 581–584
    desensitization, 581–582
    differentiation, 583–584
    direction, 582–583
  safely navigating unchartered attachment trauma territory, 578–581
    attachment, 579–580
    emotion regulation, 580
    mentalization, 580–581
  self-kindness versus self-judgment, 567–568
  shame, and PTSD, 572–574
  understanding fears and barriers to, 576–578
  and treatment, 574–576
Self-Compassion Scale (SCS), 559–560
Self-conscious emotions, 146
Self criticism, 568, 577
Self-defense techniques, 587–588
Self-disgust, 123
Self-evaluative social emotions, shame and guilt as, 146–147
Self-judgment, 567–568, 571
Self-kindness, 568, 570–571, 589–590

Self-reflection, 586
Self-regulation, 398–399
Self-report measures, 505
Semistructured Emotion Regulation Interview, 29
Serotonin system, 443–444
Serotonin transporter gene (5HTTPLR), 445
Sertraline, 107
Sex hormones, 277–278
Sexual assault histories, 498
Sexual assault victims, 123–124, 126–127
Sexual revictimization in adulthood, 382
Sexual trauma, women with a history of, 133–134
Sexual violence, 132–133
Shame and guilt, 12–14, 145–146, 153–155, 497–498, 572–574
  complimentary treatment approaches for, 159–162
  differentiating, 147–148
  future directions, 162–163
  measurement challenges, 149–150
  measures of, 13–14
  in PTSD treatment, 157–159
  self-evaluative social emotions, 146–147
  sex differences, 155–156
  and suicide risk, 156–157
  theoretical models, 149
Shame and Guilt After Trauma Scale (SGATS), 150
Shame-free guilt, 151
Signal-response expectations, 518–519
Situationally accessible memory (SAM) system, 313–314
Skills training in affective and interpersonal regulation (STAIR) Narrative Therapy, 522–535, 523–525t
  emotion awareness and regulation skills, 526–528
  evidence and applications of, 535–537
  future directions and adaptations of, 538
  Narrative Therapy module, response to, 534–535
  relationship patterns and interpersonal skills, 528–530
  STAIR module, response to, 530–533
  theoretical frame, 516–522

attachment theory, 518–520
conservation of resources and the resource loss model, 517–518
narrative therapy, 520–522
treatment development and rationale, 514–516
Skin conductance response, 253–254
SLC6A3, 231–232
SLC6A4, 231–232
Sleep disturbances, 278–279, 431
Snaith-Hamilton Pleasure Scale (SHAPS), 23–24
Social functioning deficits, 103–105
Social information processing theories, 72
Social support, 517–518
Socratic questioning and challenging, 492
Soothing touch, 590
STAIR for Primary Care (STAIR-PC), 536
STAIR/MPI Training, 574
State anger, 66
State-Trait Anger Expression Inventory (STAXI), 8, 75
  STAXI-2, 76
State-Trait Anxiety Inventory (STAI), 9–10
Stimulation intensity, 186
Stimulus functions, 411–412
Strength at Home Program (SAH), 77–78
Stress hormones, 276–277
Stress inoculation therapy (SIT), 74
Stroop task, 49, 320–321
Stuck points, 491–492
Subjective arousal, experiential avoidance and, 423–424
Subjective Units of Distress Scale (SUDS), 463–464
Substance use disorders (SUD), 159–160, 343–344, 362, 427–428
Suffering, 569–570
Suicide, 105–106, 345
  attempts, 554–555
  risk, 156–157, 162–163
Survival mode theory, 71

T

Task-related functional connectivity, 181–182
Test of Self-Conscious Affect (TOSCA), 14, 149–150

Threat cues, 392
Three-channel formulation, 527
Toronto Alexithymia Scale, 381
Toronto Alexithymia Scale, 20-item (TAS-20), 18–23, 19–22*t*
Transcranial direct current stimulation (tDCS), 193–195
Transcranial electrical stimulation, 183, 192–195
Translating epigenetic findings, 232–233
Translational fear conditioning models, 256–260
  fear acquisition, 256–260
Translational psychophysiological research, 255–256
Trauma, 66–73, 77–79, 491–492, 608
  childhood, 449–450
  disgust in relation to, 121–122
Trauma exposure, 384–388
Trauma-focused CBT, 504
Traumagenerated models, 533–534
Trauma-informed thought patterns, 515
Trauma management therapy (TMT), 158
Trauma-related cognitions, 491–492, 503–504
Trauma-related disorders, 605–606
Trauma-related guilt, 151–152
  reductions in, 470
  sex differences in, 155–156
Trauma-Related Guilt Inventory (TRGI), 13, 148, 150
Trauma-related regulatory mechanism, attentional control as, 326–333
Trauma-related shame, 149
Trauma-Related Shame Inventory (TRSI), 14, 150, 505–506
Trauma-related triggers, 11
Trauma script-driven imagery, 11–12
Trauma scripts, 92
Trauma survivors, informal practice for, 588–591
Traumatic relationship model, 522
Traumatic stressors, emotional granularity and coping with, 386–387
Treatment as usual (TAU), 548–549, 552
Treatment response, experiential avoidance (EA) and, 418–421

Triple network model, 175, 181–183
  resting-state functional connectivity, 182–183
  task-related functional connectivity, 181–182

U

Ultrabrief exposure therapy, 190–191
Unconditioned response (UCR), 122–123, 259–260
Unconditioned stimulus (US), 43–44, 122–123, 255
"Undermodulation" pattern of reactivity, 390
Undifferentiated negative affect, 395–396
Unique considerations for PTSD genomics, 236–237
UPPS-P Impulsive Behavior Scale, 438
Urgency traits, 438–441, 444–445

V

Vagus nerve stimulation (VNS), 199–201
Valence dimension, 377–378
Valence focus, 379, 392–393
Venlafaxine, 107
Ventromedial prefrontal cortex (vmPFC), 177–179, 444–445
Verbally accessible memory (VAM) system, 313–314
Veterans affairs healthcare system, 491–492
Vietnam veterans, 54
Vigilance-avoidance vs. attention maintenance, 316
Violations of conscience, 124–125

W

Weighted gene coexpression network analysis (WGCNA), 235
Within-session extinction learning, 261–263
Working memory, 394–395
Working models of relationships, 528

Y

Yang self-compassion, 567–568, 570–571
Yin self-compassion, 570–571, 589–590